P9-AQV-457

Fundamental constants

Quantity	Symbol	Value	Unit
Speed of light	c	$2.997\,925 \times 10^{8}$	$\mathrm{m\,s^{-1}}$
Charge of proton[1]	e	$1.602\,19 \times 10^{-19}$	C
Planck's constant	h	$6.626\,18 \times 10^{-34}$	J s
	$\hbar = h/2\pi$	$1.054\,59 \times 10^{-34}$	J s
Boltzmann's constant	k	$1.380\,66 \times 10^{-23}$	$\mathrm{J\,K^{-1}}$
Avogadro's constant	L	$6.022\,05 \times 10^{23}$	$\mathrm{mol^{-1}}$
Mass of electron	m_e	$9.109\,53 \times 10^{-31}$	kg
proton	m_p	$1.672\,65 \times 10^{-27}$	kg
neutron	m_n	$1.674\,95 \times 10^{-27}$	kg
nuclide[2]	$M_r u$; $u =$	$1.660\,56 \times 10^{-27}$	kg
Vacuum permittivity	ε_0	$8.854\,188 \times 10^{-12}$	$\mathrm{J^{-1}\,C^2\,m^{-1}}$
	$4\pi\varepsilon_0$	$1.112\,650 \times 10^{-10}$	$\mathrm{J^{-1}\,C^2\,m^{-1}}$
Vacuum permeability[3]	μ_0	$4\pi \times 10^{-7}$	$\mathrm{J\,s^2\,C^{-2}\,m^{-1}}$
Bohr magneton	$\mu_B = e\hbar/2m_e$	$9.274\,08 \times 10^{-24}$	$\mathrm{J\,T^{-1}}$
Nuclear magneton	$\mu_N = e\hbar/2m_p$	$5.050\,82 \times 10^{-27}$	$\mathrm{J\,T^{-1}}$
Bohr radius	$a_0 = 4\pi\varepsilon_0\hbar^2/m_e e^2$	$5.291\,77 \times 10^{-11}$	m
Rydberg constant	$hcR_\infty = m_e e^4/8h^2\varepsilon_0^2$	$2.179\,908 \times 10^{-18}$	J
	R_∞	$1.097\,373 \times 10^{5}$	$\mathrm{cm^{-1}}$
Fine structure constant	$\alpha = e^2/4\pi\hbar c\varepsilon_0$	$7.297\,351 \times 10^{-3}$	
	α^{-1}	137.0360	
Compton wavelength of electron	$\lambda_C = h/m_e c$	$2.426\,309 \times 10^{-12}$	m
Stefan–Boltzmann constant	$\sigma = 2\pi^5 k^4/15h^3c^2$	$5.670\,32 \times 10^{-8}$	$\mathrm{W\,m^{-2}\,K^{-4}}$
Second radiation constant[4]	$c_2 = hc/k$	$1.438\,786 \times 10^{-2}$	m K

[1] The charge on the electron is $-e$.
[2] M_r is the relative molecular mass.
[3] ε_0 and μ_0 are related by $1/\varepsilon_0\mu_0 = c^2$.
[4] $c_2/5$ is the constant in the Wien displacement law; the first radiation constant is $c_1 = 2\pi hc^2$.

Molecular Quantum Mechanics

Molecular Quantum Mechanics

SECOND EDITION

P. W. ATKINS

Oxford New York
OXFORD UNIVERSITY PRESS

Oxford University Press, Walton Street, Oxford OX2 6DP
London Glasgow New York Toronto
Delhi Bombay Calcutta Madras Karachi
Kuala Lumpur Singapore Hong Kong Tokyo
Nairobi Dar es Salaam Cape Town
Melbourne Auckland
and associate companies in
Beirut Berlin Ibadan Mexico City

First edition 1970
Second edition 1983
Reprinted 1983

ISBN 0-19-855171-1
ISBN 0-19-855170-3 Pbk

British Library Cataloguing in Publication Data

Atkins, P. W.
Molecular quantum mechanics.—2nd ed.
1. Quantum chemistry
I. Title
541.2′8 QD462

ISBN 0-19-855171-1
ISBN 0-19-855170-3 (pbk.)

Library of Congress Cataloging in Publication Data

Atkins, P. W. (Peter William), 1940–
Molecular quantum mechanics.
Includes bibliographies and index.
1. Quantum chemistry. I. Title.
QD462.A84 1983 541.2′8 82-18998
ISBN 0-19-855171-1
ISBN 0-19-855170-3 (pbk.)

Set by The Universities Press, Belfast
Printed in Great Britain by Butler and Tanner Ltd, Frome

Preface to the second edition

THIS is more a total rewriting than a second edition. I have taken the text of the first edition as my theme, but have worked extensive variations on it. In so doing I have tried to preserve and enhance what reviewers remarked was one of the principal virtues of the first edition—the way it imparted insight and not merely algebra. I have also aimed to make the book more accessible as a text for introductory courses in quantum mechanics and quantum chemistry.

Much of the material has been reorganized, but the main structure is largely unchanged (except for some obvious differences of format and layout). *Foundations 1* introduces elementary quantum theory from an historical point of view, sets up the Schrödinger equation, and presents some of its consequences. It concludes with a straightforward account of the quantum mechanics of the three basic types of motion: translation, vibration, and rotation. The division of the old Chapter 3 into two should make the material more digestible, and has allowed me to include more material. *Foundations 2* presents the more formal aspects of quantum theory. I have introduced angular momentum before group theory in this edition, because it can then be seen to be an extended example of the manipulation of operators introduced in the first chapter of Part 2. In the chapter on group theory, I have switched from the passive to the active convention, and so have brought it into line with the way it is normally taught. The chapter on techniques of approximation now deals much more fully with time-dependent processes; I have used the device of introducing both time-independent and time-dependent perturbation theory with a discussion of a two-level, exactly solvable system, and then generalizing and approximating. Part 3 still deals with *Applications*, but I have distinguished the elementary applications, the ones most likely to be encountered in undergraduate courses, from the rest, and have put the latter into Part 4, of which more shortly. The principal change in what remains of Part 3 is the division of the molecular spectroscopy chapter into two, one dealing with molecular rotations and vibrations, and the other with electronic processes. The sharp eye will also notice the changes of title: those two chapters are now called '... transitions', because I want to emphasize that this is a book dealing with the background quantum mechanics of chemistry and not a textbook of spectroscopy. I have also totally reorganized the chapter on molecular structure, bringing molecular orbitals to the front of the stage and allowing valence bond theory to recede. As in the case of spectroscopy, this is not intended to be a book dealing with the highly developed and specialized techniques for calculating molecular structures: my object is to describe the basis of valence theory, not its practical implementation. Part 4, a new part, is on *Advanced applications*, and is the home of what used to be the single chapter on electric and magnetic properties. The name of this part is a signal that the material goes beyond what undergraduate courses normally cover. It is included for several reasons. One is that I think readers ought to have the opportunity to see how the techniques they have acquired can be adapted to discuss these more involved (and more interesting) properties even though they might lie outside the course. Another is to help the reader bridge the gap between undergraduate and graduate work, and to do so with the minimum of difficulty: with Part 4 the book becomes a text that can accompany the reader from the first encounter with quantum theory to the commencement of research. I adopt a different style in Part 4: there, much more than elsewhere, I allow the algebra to speak for itself.

Several other changes should be noticed. I have made quite heavy use of appendices, which are now collected together at the end of the book. They contain a number of detailed points, and should extend the book's usefulness without breaking the flow of the text. In the second place, I have inserted a fair number (54) of worked *Examples*. These do disrupt the flow, but I think that a small price to pay for the opportunities they provide for developing points which students often find difficult, and they should make the subject much easier to learn. Many of the *Examples* are designed to extend the reader's range of information, and sometimes touch on new material and techniques. The *Further reading* sections at the end of each chapter have been brought up to date (and restructured). Virtually the entire set of *Problems* (numbering 267) is new; there are hints pegging out the ground in the more involved problems; a *Solutions Manual* is available which provides the detailed solutions to the problems. All the illustrations have been redrawn.

I have many debts it is my pleasure to acknowledge. First, there are all those who took the trouble to write to me with comments on the first edition: such information is crucial in the preparation of a new edition, and although I cannot name them all here, I hope they and their successors will continue to write, knowing that their comments are appreciated. A number of people read the draft of this edition and commented in detail. I wish to thank Professor D. L. Beveridge (Hunter College, City University of New York), Dr J.-L. Calais (University of Uppsala), Dr D. B. Cook (University of Sheffield), Professor D. P. Craig (Australian National University, Canberra), Professor S. F. A. Kettle (University of East Anglia, Norwich), Dr A. J. MacDermott (University of Oxford), Professor I. M. Mills (University of Reading), Professor J. P. Simons (University of Utah), and Professor J. S. Winn (University of California, Berkeley), all of whom have made a significant contribution for which I am deeply grateful. Dr M. P. Allen kindly prepared the computer graphics and Caron Crisp typed with her usual meticulous care. I also thank my publishers, yet again, for their continuing helpfulness and responsiveness.

Oxford 1982 P.W.A.

From the preface to the first edition

THE aim of this book is to present an outline of the quantum-mechanical principles that are fundamental to an understanding of the properties of molecules. The text is designed to lead an undergraduate reader from the point where the failures of classical physics may be recognized to the construction and application of a theory that seems to be adequate to account for the spectra and the electric and magnetic properties of molecules. The emphasis of the later sections of the book is on properties rather than structure, for many excellent books describing the latter already exist; also, no attention is paid to the theory of chemical reactions, which, however fascinating, is still an infant subject. The theory of chemical bonding cannot sensibly be omitted, but the chapter that is included is designed to provide an account of the principles of bonding rather than its details.

. . . the material of Part III is such that much of its early parts can be read and understood without the material of Part II, but as the reader goes further into it he should realize that the material of Part II is relevant and important and be stimulated to master its contents. . . .

One principle omitted from the first two Parts is the Pauli principle. It is to be found, however, in the Applications section, Part III. This rather odd arrangement is not accidental, but arguments in support of such an apparently whimsical position must clearly be put forward. The basic argument is that if it is to be presented as a consequence of observation, then the account of the spectrum of helium provides the first opportunity to discuss it: throughout the book I have tried to show how experiments have been the guide in the construction of quantum mechanics, and the Pauli principle is an excellent and rather exciting example of this because it is so unexpected. It appears in the chapter on atomic spectra and structure, where I have sought to show that even though in Part I we saw that we could solve the Schrödinger equation for the hydrogen atom, we had insufficient information to account for the details of its spectrum: in a similar vein the spectrum of helium can be explained only by invoking the Pauli principle. . . .

Having learned about atoms, and from their spectra obtained the final key to their structure and the explanation of the periodic table, the book passes on to molecules, their structure, spectra, and electric and magnetic properties. Each chapter turns one's attention back to Part II, but most, possibly all, can be read first. Thus, in the chapter on molecular structure I have drawn on the notation and occasionally the conclusions of group theory, but the essential points can be obtained without an appreciation of the latter. . . .

Finally I must acknowledge the help I have received in writing this book. First I must thank Professor M. C. R. Symons who guided my thoughts on chemistry initially, and who is a man of considerable chemical insight. My thanks must also go to Professor C. A. Coulson who, together with his colleagues at two Oxford Summer Schools in Theoretical Chemistry and in later conversations, in his inimitable way gave impetus and direction to my thinking about theoretical chemistry. I must thank my pupils in Lincoln College who have read much of the manuscript and helped to remove infelicities and ambiguities (by which I mean, I suppose, errors) and who have during tutorials forced me to explain what previously I had thought I understood. . . .

Oxford 1969 P.W.A.

Contents

'Experiments are the only means of knowledge at our disposal. The rest is poetry, imagination.'

MAX PLANCK

Part 1 Foundations 1

IN THIS Part we see how the need for quantum mechanics emerges from the experiments which overthrew classical mechanics. We shall see that a very small modification of classical theory—the replacement of a zero by a new, small, fundamental constant—had immense implications. In place of Newton's equations, the basic equation of mechanics becomes Schrödinger's equation. We shall see how it emerges from straightforward ideas about the propagation of particles, and how it leads to the quantization of energy. There are three basic types of motion—translation, rotation, and vibration—and many of the features of real systems can be discussed once the quantum mechanics of these modes have been determined. We shall see that they can be solved exactly. Furthermore, in each case we shall see how the quantum mechanics of motion can have, under some circumstances, properties almost exactly the same as classical mechanics predicts. We shall therefore see the quantum basis of the everyday world as well as the quantum mechanics of the world of the atom.

1 Historical introduction

THERE are two equally beautiful approaches to quantum mechanics. One is to follow the historical development of the theory from the first indication that the whole fabric of classical mechanics and electrodynamics should be held in doubt to the resolution of the problem in the work of Planck, Einstein, Heisenberg, Schrödinger, and Dirac. The other is to stand back at a point late in the development of the theory and to seek its underlying theoretical structure. The first is exciting because the theory is seen gradually emerging from confusion and dilemma. One sees experiment and intuition determining the form of the theory and, above all, one appreciates the necessity of replacing the classical theory. The second, more formal approach is exciting and beautiful in a different sense: there is logic and elegance in a scheme that starts from only a few postulates, yet reveals, as their implications are unfolded, a rich, experimentally verifiable structure. The links of the new theory with classical physics are often clear, and the difference between the old and new is seen to be one more of development than of discontinuity.

Our initial approach to quantum mechanics will be historical: this will let us appreciate the physical content of the equations we later derive and use. By Chapter 5 we shall be in a much better position to comprehend the more formal approach considered there.

1.1 Black-body radiation

The expression 'black body' signifies something that absorbs and emits radiation without favouring particular frequencies. A pin-hole in an otherwise sealed container, Fig. 1.1, is a good approximation. The pin-hole acts as a source of radiation (and as a sink) and the intensity of radiation emitted is a measure of the intensity of radiation inside the container. Black-body radiation had intrigued a number of workers since the middle of the nineteenth century when Kirchhoff had shown (in 1859) that for radiation of the same wavelength, and matter at the same temperature, the energy emitted and the fraction of the incident radiation absorbed were in the same ratio whatever material the black body was constructed from (this is *Kirchhoff's law*). This, at least, is a simplification, for if we can discover information about such radiation then we have something of fundamental importance about the radiation field, and not merely information about one particular experimental material.

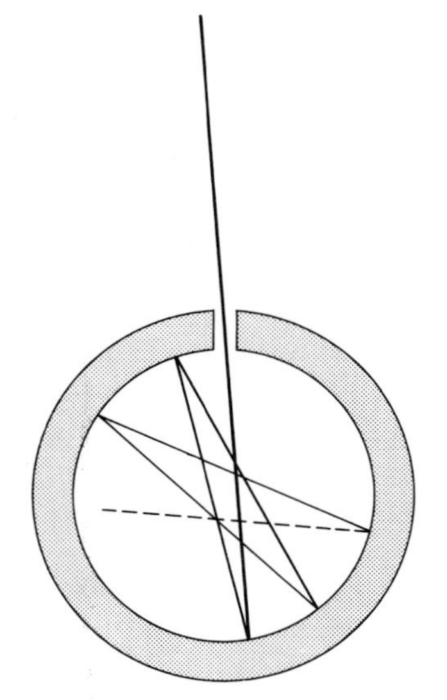

Fig. 1.1. A hole simulating a black body.

Stefan (in 1879) was able to formulate a quantitative law concerning the *total energy density*, the radiation energy per unit volume $\mathcal{U}$, in equilibrium with a black body at a temperature T:

Stefan's law: $\mathcal{U} \propto T^4$.

A more convenient formulation of the law refers to the *excitance* M, the power emitted per unit area. Since the emitted power is proportional to the energy density, we also have $M \propto T^4$. This is normally written

$$M = \sigma T^4. \tag{1.1.1}$$

The constant σ, the *Stefan–Boltzmann constant*, is independent of the material, and its modern value is $5.67 \times 10^{-8}\,\mathrm{W\,m^{-2}\,K^{-4}}$. This means that each $1\,\mathrm{cm}^2$ of the surface of a black body heated to 1000 K radiates about 5.7 W (and one hundred times less at room temperature). Stefan based his conclusion on the observation that the total energy emitted by a platinum wire at 1473 K was 11.7 times that emitted at 798 K and $11.7 \approx (1473/798)^4$, whereas a modern experiment would yield a ratio of about 18.6 because the wire is not a true black body! Boltzmann was able to derive Stefan's law on the basis of purely thermodynamic reasoning, and we shall soon see that it can be obtained in another manner and that it is now securely based, both experimentally and theoretically.

Wien (in 1894) provided another useful piece of quantitative information in the form of his *displacement law*, a simple form of which is that the wavelength corresponding to the maximum of the energy distribution of black-body radiation, $\lambda_{\max}$, obeys the relation:

Wien's law: $\lambda_{\max} T = \text{constant}.$ (1.1.2)

An alternative expression of this law is that the energy density in the range λ to $\lambda + \mathrm{d}\lambda$, $\mathrm{d}\mathcal{U}$, has the form

$$\mathrm{d}\mathcal{U} = (\phi/\lambda^5)\,\mathrm{d}\lambda. \tag{1.1.3}$$

where ϕ is a function of the product λT. Wien's displacement law was confirmed experimentally by Lummer and Pringsheim, although it was noted that there were marked deviations at long wavelengths (low frequencies). Furthermore, the function $\phi(\lambda T)$ could not be found on the basis of a classical mechanical calculation, yet it was realized that because of its generality it must have a fundamental significance.

The climax of this discontent came in 1900, when Rayleigh made an attempt to calculate the form of the energy distribution. This calculation consists of two parts and is based on the view that the electromagnetic field can be thought of as a collection of oscillators, each one corresponding to a particular wavelength (or frequency) of radiation. In the first step, one calculates the number of oscillators in an enclosure that correspond to a wavelength λ. In the second step, in accord with the classical equipartition principle, an average energy kT is ascribed to each oscillator. Rayleigh had considerable experience with calculations that involved waves within containers, but even he made a mistake in the calculation: 'It seems to me' said Jeans, 'that Lord Rayleigh has introduced an unnecessary factor 8 by counting negative as well as positive values of his integers.' With this comment the Rayleigh law became the Rayleigh–Jeans law (which is wrong anyway!). Their combined effort yielded the number of modes of radiation per unit volume with wavelength in the range λ to $\lambda + d\lambda$ as

$$d\mathcal{N} = (8\pi/\lambda^4)\, d\lambda. \tag{1.1.4}$$

(This result is obtained in Appendix 1.) Consequently, as the energy per mode is kT, the energy density arising from radiation in the range λ to $\lambda + d\lambda$ is

$$\textit{Rayleigh–Jeans law:}\ d\mathcal{U} = (8\pi kT/\lambda^4)\, d\lambda \tag{1.1.5a}$$

or, in terms of the frequency $\nu = c/\lambda$, since then $d\mathcal{N} = (8\pi\nu^2/c^3)\, d\nu$,

$$d\mathcal{U} = (8\pi\nu^2 kT/c^3)\, d\nu. \tag{1.1.5b}$$

In each of these expressions the factor multiplying $d\lambda$ or $d\nu$ is called the *density of states*, and written either $\rho(\lambda)$ or $\rho(\nu)$. Then we can write

$$d\mathcal{U} = \rho(\lambda)\, d\lambda, \qquad \rho(\lambda) = 8\pi kT/\lambda^4 \tag{1.1.6a}$$

$$d\mathcal{U} = \rho(\nu)\, d\nu, \qquad \rho(\nu) = 8\pi\nu^2 kT/c^3. \tag{1.1.6b}$$

The energy distribution this law predicts is obviously absurd, Fig. 1.2, where the experimentally determined curve is also shown. Note, however, that the absurdity is apparent only at high frequencies, for at low frequencies the agreement between the curves is striking. The principal absurdity is that the law predicts a virtually infinite energy density at very high frequencies. This implies that the total energy emitted, which is the integral of eqns (1.1.5a) or (1.1.5b) over the entire range of wavelength or frequency, is infinite. The accumulation of the energy in the high frequency region was termed by Ehrenfest the *ultraviolet catastrophe*, and arises, we may presume at this stage,

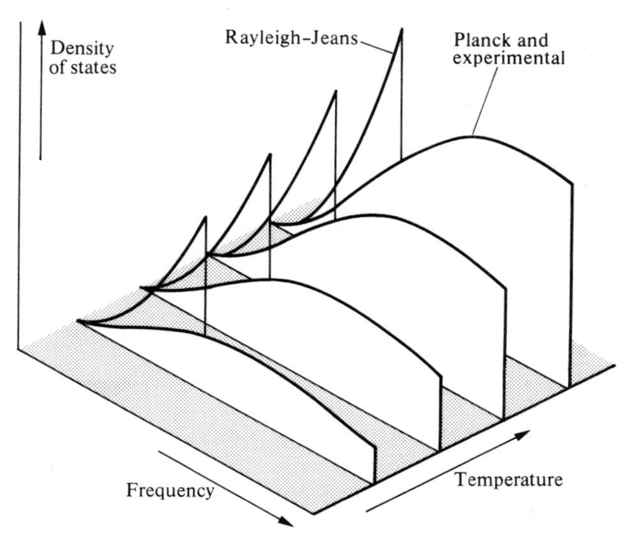

Fig. 1.2. Densities of states at various temperatures.

because in some way the contribution of the high-frequency oscillators has been over-emphasized.

At this point Planck made his historic contribution. His intervention took the form of an interpolation between two formulae, of which one led to the Rayleigh–Jeans law and the other to the Wien displacement law. (The first, remember, is correct at low frequencies and the second is correct at high frequencies.) The interpolation is equivalent to the assumption that *the energy of an oscillator of frequency ν is not continuously variable, but is restricted to integral multiples of the quantity $h\nu$* where h is a universal constant. $h\nu$ is referred to as a *quantum* of energy.

Common experience is quite contrary to this assumption. Thus experience suggests that the energy of a radio wave of some frequency may be increased continuously, that a vibrating string may be excited more vigorously (at the same frequency) in a continuous manner, and that a pendulum may swing at its characteristic frequency with any reasonable degree of violence. But it is naïve to suppose that the unknown can be explained in terms of the known, and to suppose that the behaviour of systems of an unfamiliar scale mirror the behaviour of familiar systems. We have seen that classical (that is 'familiar' physics) fails in its attempt to explain the experimental results; therefore we must construct a new theory.

The number of radiation oscillators is left unchanged by Planck's hypothesis, only the extent of their excitation is affected, and through that their mean energy. The calculation therefore involves finding the average energy of an oscillator of frequency ν when it can possess only the discrete energies $0, h\nu, 2h\nu, \ldots$ or $vh\nu$ in general ($v = 0, 1, 2, \ldots$). If the energy of an oscillator in the *quantum state* labelled v is ε_v, and the probability that it is in that state is p_v, then its average energy, ε is

$$\varepsilon = \sum_v p_v \varepsilon_v. \tag{1.1.7}$$

At thermal equilibrium at a temperature T the probability that a system is in a state with energy ε_v is given by the

Boltzmann distribution: $$p_v = \frac{e^{-\varepsilon_v/kT}}{q}, \qquad q = \sum_v e^{-\varepsilon_v/kT} \tag{1.1.8}$$

where k is Boltzmann's constant ($k = 1.380\,66 \times 10^{-23}\ \mathrm{J\,K^{-1}}$). The use of Boltzmann's distribution is justified retrospectively in a weak sense because the formula we shall derive will be seen to agree with experiment, and justified in a much stronger sense in that it may be shown that only this choice leads to thermodynamic equilibrium.

As an oscillator may possess only the energies $\varepsilon_v = vh\nu$, $v = 0, 1, 2, \ldots$, by hypothesis, where h is an as yet undetermined fundamental constant, the expression for the mean energy takes the following form:

$$\varepsilon=\left\{\sum_{v}\varepsilon_v \mathrm{e}^{-\varepsilon_v/kT}\right\}\Big/\left\{\sum_{v}\mathrm{e}^{-\varepsilon_v/kT}\right\}$$
$$=\left\{\sum_{v} vh\nu \mathrm{e}^{-vh\nu/kT}\right\}\Big/\left\{\sum_{v}\mathrm{e}^{-vh\nu/kT}\right\}=h\nu\left\{\sum_{v} vx^{v}\right\}\Big/\left\{\sum_{v} x^{v}\right\}.$$

where $x=\mathrm{e}^{-h\nu/kT}$. If we recall that for $x<1$ the expansion of $(1-x)^{-1}$ is

$$(1-x)^{-1}=1+x+x^2+\ldots=\sum_{v} x^{v},$$

and note that $\sum_v vx^v = x(\mathrm{d}/\mathrm{d}x)\sum_v x^v$, we obtain

$$\varepsilon = h\nu\{x(\mathrm{d}/\mathrm{d}x)(1-x)^{-1}\}/(1-x)^{-1}$$
$$= h\nu x/(1-x) = h\nu \mathrm{e}^{-h\nu/kT}/\{1-\mathrm{e}^{-h\nu/kT}\}. \quad (1.1.9)$$

This is the mean energy of an oscillator of frequency ν. Since we know $\mathrm{d}\mathcal{N}$, the number density of oscillators in the range ν to $\nu+\mathrm{d}\nu$, the energy density in this range is $\varepsilon\ \mathrm{d}\mathcal{N}$, or

Planck distribution: $$\mathrm{d}\mathcal{U}=\varepsilon\ \mathrm{d}\mathcal{N}=\frac{8\pi h\nu^3}{c^3}\left(\frac{\mathrm{e}^{-h\nu/kT}}{1-\mathrm{e}^{-h\nu/kT}}\right)\mathrm{d}\nu. \quad (1.1.10)$$

In future applications we shall need the density of states: clearly

$$\rho(\nu)=(8\pi h\nu^3/c^3)\left(\frac{\mathrm{e}^{-h\nu/kT}}{1-\mathrm{e}^{-h\nu/kT}}\right). \quad (1.1.11)$$

For wavelengths the energy density takes the form

$$\mathrm{d}\mathcal{U}=\rho(\lambda)\,\mathrm{d}\lambda,\qquad \rho(\lambda)=(8\pi hc/\lambda^5)\left(\frac{\mathrm{e}^{-hc/\lambda kT}}{1-\mathrm{e}^{-hc/\lambda kT}}\right). \quad (1.1.12)$$

Example. The background radiation in the universe is equivalent to a black-body radiator at a temperature of 2.7 K. Calculate (a) the energy density of the cosmic background; (b) the number of photons it contributes to a cube of side 1 cm^3.

• *Method.* (a) The total energy density of black-body radiation is obtained by integrating eqn (1.1.10) over all frequencies. The integral is standard (i.e., listed in tables) and has the form

$$\int_0^\infty \left\{\frac{x^{2n-1}}{\mathrm{e}^x-1}\right\}\mathrm{d}x=(2\pi)^{2n}B_n/4n;\qquad B_1=\tfrac{1}{6},\quad B_2=\tfrac{1}{30},\quad B_3=\tfrac{1}{42},\ldots.$$

(b) Each photon carries an energy $h\nu$, and so the number density is $\mathrm{d}\mathcal{N}=\rho(\nu)\,\mathrm{d}\nu/h\nu$. Integrate this over all frequencies for the total number density; multiply by the volume of the cube for the number itself. The integral required is standard, and has the form

$$\int_0^\infty \left\{\frac{x^{2n}}{e^x-1}\right\} dx = (2n)!\,\zeta(2n+1); \qquad \zeta(3)=1.202\,06, \quad \zeta(5)=1.036\,93, \ldots.$$

• *Answer.*

(a) $$\mathscr{U} = \int_0^\infty (8\pi h/c^3)\left\{\frac{\nu^3\, d\nu}{e^{h\nu/kT}-1}\right\} = (\pi^2/15)(k^4/c^3\hbar^3)T^4$$

$$= 7.566\times 10^{-16}(T/\mathrm{K})^4\ \mathrm{J\,m^{-3}} = 4.02\times 10^{-14}\ \mathrm{J\,m^{-3}}\ \text{at}\ T=2.7\ \mathrm{K}.$$

(b) $$\mathscr{N} = \int_0^\infty (8\pi/c^3)\left\{\frac{\nu^2\, d\nu}{e^{h\nu/kT}-1}\right\} = (2\zeta(3)/\pi^2)(k^3/c^3\hbar^3)T^3$$

$$= 2.029\times 10^7(T/\mathrm{K})^3\ \mathrm{m^{-3}} = 4.0\times 10^8\ \mathrm{m^{-3}}\ \text{at}\ T=2.7\ \mathrm{K}.$$

Hence in a volume 1 cm^3 the total number of black-body photons is 400.

• *Comment.* The B_n are *Bernouilli numbers* and the $\zeta(n)$ are *Riemann zeta-functions*; their values are listed in M. Abramowitz and I. A. Stegun, *Handbook of mathematical functions*, Dover (1965). The collection of fundamental constants $2\pi^5k^4/15c^2h^3=\sigma$, the *Stefan–Boltzmann constant.*

Two questions now arise: do these expressions agree with experiment, and what is the value of h? It is interesting to explore the effect of letting $h\nu/kT$ approach zero, corresponding to low frequencies. An exponential e^x may be expanded as follows:

$$e^x = 1 + x + x^2/2! + \ldots$$

and so, in the limit of $\nu\to 0$, eqn (1.1.10) becomes

$$\lim_{\nu\to 0} d\mathscr{U} = \frac{8\pi h\nu^3}{c^3}\left(\frac{1-h\nu/kT+\ldots}{1-1+h\nu/kT-h^2\nu^2/2k^2T^2\ldots}\right)d\nu$$
$$= (8\pi h\nu^2kT/c^3h)\, d\nu = (8\pi\nu^2kT/c^3)\, d\nu,$$

which is exactly the classical result. We know already that the Rayleigh–Jeans distribution is correct at low frequencies (Fig. 1.2), and so we seem to be on the right track. Furthermore, when the frequency of the radiation is high, the exponential term in the numerator of eqn (1.1.10) is very small and smothers the tendency of ν^3 to increase. This eliminates the ultraviolet catastrophe. Finally, in the high frequency (short wavelength) limit we should expect that the Planck distribution leads to the Wien law. In this limit we may approximate eqn (1.1.12) by

$$\rho(\lambda) \approx (8\pi hc/\lambda^5)e^{-hc/\lambda kT}$$

because the exponential in the denominator is then much less than unity. The condition for this expression to have its maximum value is

$$(d\rho/d\lambda)_{\lambda=\lambda_{max}} = 0, \text{ or } -5 + hc/\lambda_{max}kT = 0.$$

Therefore,

$$\lambda_{max}T = hc/5k = \text{constant}, \qquad (1.1.13)$$

which is the Wien displacement law, eqn (1.1.2).

Equation (1.1.13) lets us estimate the size of h, for the constant in the Wien law is found experimentally to have the value 2.9 mm K; as both Boltzmann's constant and the speed of light are known, we deduce that $h \approx 6.7 \times 10^{-34}$ J s. It is now clear why common experience with pendulums, etc. does not conflict with Planck's hypothesis: h is so small that for an oscillator of frequency 1 Hz neighbouring energy levels are separated by only about 7×10^{-34} J and the discontinuity in the permissible energies is unnoticeably small. For oscillators with optical frequencies, when $\nu \approx 10^{15}$ Hz, the separation of the energy levels is of the order of 10^{-19} J (about 60 kJ mol^{-1}), and is much more significant.

The ultimate test of Planck's hypothesis is to see whether it reproduces the observed energy distribution of radiation from a black body. This is does triumphantly,† and the best modern value of h is $h = 6.626\,18 \times 10^{-34}$ J s. Its dimensions are those of *action* (energy × time, ML^2T^{-1}).

The physical basis for the success of the quantum hypothesis is that at a given temperature there may be insufficient energy available to excite the higher frequency oscillators, for on the quantum hypothesis they can be excited only by absorbing not less than one quantum of energy, $h\nu$. On the classical theory the oscillators could be excited in a continuous manner and even the highest frequency oscillators could be excited and possess the same energy, kT, as the low frequency oscillators, and contribute to the radiation field. Planck's quantum hypothesis therefore has the effect of damping out the high-frequency oscillations, just as we asserted was necessary.

It has already been emphasized that black-body radiation is a fundamental problem. Planck arrived at a solution by making a radical alteration to classical theory, and we should therefore expect to discover ramifications of the hypothesis in other parts of physics and chemistry, and therefore throughout science. We shall now see how in its early days the quantum hypothesis gained in stature and structure by solving several outstanding problems.

1.2 Heat capacities

In 1819 science had a deceptive simplicity. Dulong and Petit, for example, were able to propose the law that 'the atoms of all simple bodies have exactly the same heat capacity'. This statement is easily given a theoretical foundation if we suppose that the N atoms that constitute the 'simple body' behave as a collection of $3N-6$ oscillators. (Each atom has three degrees of freedom and so the body as a whole has $3N$, but three of these are translations of the body rigidly and three are rigid rotations, none of which contributes to the heat

† The distribution that has been derived here is not quite correct: we shall see in Chapter 3 that an oscillator can never lose all its energy and this 'zero point energy' (p. 53) should be included. The final result is changed only a little, and this extra complexity need not distract us.

capacity if it is at rest in the laboratory.) As N is so large we may write $3N-6\approx 3N$. By the classical equipartition theorem each oscillator has an energy kT, and so the internal vibrational energy of the body is $3NkT$. The heat capacity is therefore $(\mathrm{d}/\mathrm{d}T)3NkT$, or $3Nk$. For unit amount of substance N is replaced by Avogadro's constant L, and so the molar heat capacity is $3Lk$, or $3R$. (R is the molar gas constant, $R=kL$.)

Dulong and Petit's law was based on observations at room temperature, and it was unfortunate for classical physics that when more sophisticated experiments were performed at lower temperatures, deviations from the law, and therefore also from the predictions of classical physics, were observed. It is easy to appreciate the similarity of this problem to the one that attracted Planck's attention: whereas his problem was the distribution of energy among the oscillators that comprise the radiation field (although that was not his original approach), the heat capacity problem deals with the energy distribution among material oscillators. Einstein recognized the analogy, and in 1906 was able to derive a result which, although it was not in exact agreement with experiment, suggested that the quantum approach was valid.

The essential point of Einstein's theory was to suppose that all $3N$ atomic oscillators of the crystal have the same frequency ν. The quantum hypothesis was invoked to restrict the energy of the oscillators to integral multiples of $h\nu$. Then the vibrational energy of the crystal is $3N\varepsilon$, where ε is the mean vibrational energy of an oscillator of frequency ν, and is given by eqn (1.1.9). The (constant volume) heat capacity of the crystal is therefore

$$C_V=\left(\frac{\partial U}{\partial T}\right)_V=3N\frac{\mathrm{d}\varepsilon}{\mathrm{d}T}=\frac{3Nk(h\nu/kT)^2\mathrm{e}^{h\nu/kT}}{(1-\mathrm{e}^{h\nu/kT})^2}. \tag{1.2.1}$$

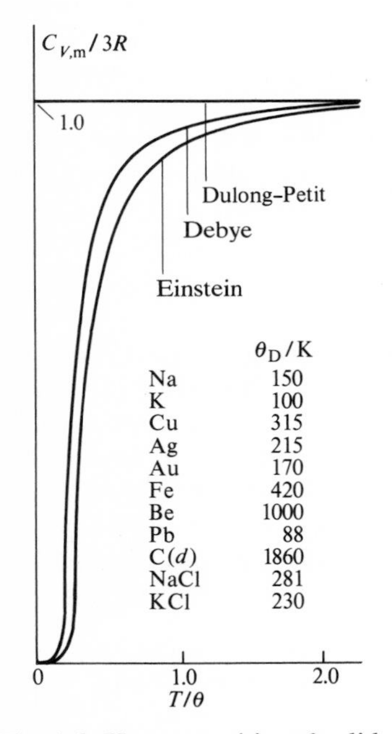

Fig. 1.3. Heat capacities of solids.

It is convenient to introduce both the *Einstein temperature* θ_E, $\theta_E=h\nu/k$, and the

Einstein heat capacity function: $c^{E}(T)=(\theta_E/T)^2\left\{\dfrac{\mathrm{e}^{\theta_E/T}}{(1-\mathrm{e}^{\theta_E/T})^2}\right\}$. (1.2.2)

Then the molar heat capacity is

$$C_{V,m}=3Rc^{E}(T). \tag{1.2.3}$$

This function, which is plotted in Fig. 1.3, closely resembles the experimental curve except at low temperatures, where it falls off too rapidly.

The reasons for both the success and the failure of this model are not hard to find. It is successful because, whereas classical theory presupposes the full excitation of each oscillator however tiny the amount of heat supplied (so that they act as a heat sink at even the lowest temperatures and always contribute to the heat capacity) quantum theory permits excitation only when at least an energy $h\nu$ has been supplied. In this problem, as in the radiation problem, the susceptibility of the oscillators to the stimulation is reduced. The partial failure stems from the simplicity of Einstein's model: in a real crystal the atoms are coupled by the interatomic forces and do not oscillate independently. Debye allowed for these interactions and showed that the atoms oscillate collectively, and a crystal oscillates with all frequencies up to a maximum, $\nu_{\max}$.† The derivation of Debye's formula would take us too far astray and we shall merely quote the results in terms of the *Debye temperature,* $\theta_{\mathrm{D}} = h\nu_{\max}/k$, which is related to the rigidity of the crystal, see Fig. 1.3, and the *Debye heat capacity function,* $c^{\mathrm{D}}(T)$:

$$C_{V,m} = 3Rc^{\mathrm{D}}(T); \qquad c^{\mathrm{D}}(T) = 3(T/\theta_{\mathrm{D}})^3 \int_0^{\theta_{\mathrm{D}}/T} \frac{x^4 \mathrm{e}^x \, \mathrm{d}x}{(1-\mathrm{e}^x)^2}, \tag{1.2.4}$$

which is also plotted in Fig. 1.3 (x in the integral is an integration variable and comes from $x = h\nu/kT$ and the integration over the frequencies ν of the atomic vibrations of the lattice). At high temperatures ($T \gg \theta_{\mathrm{D}}$) x is small throughout the range of integration and the integrand may be approximated by x^2. The integral then has the value $\frac{1}{3}(\theta_{\mathrm{D}}/T)^3$, and so $C_{V,m}/3R = 1$, in accord with the classical result. At low temperatures ($T \ll \theta_{\mathrm{D}}$) the upper limit of the integral may be replaced by infinity; the integral is then standard, and has the value $4\pi^4/5$. Then we arrive at the

Debye T^3-law: $C_{V,m}/3R = (12\pi^4/5)(T/\theta_{\mathrm{D}})^3$. (1.2.5)

This relation is used in thermodynamics to obtain entropies at low temperatures. The fact that the Debye function is larger than the Einstein function at a given temperature is a consequence of the model: the Debye model permits lower frequency vibrations than the single-frequency Einstein model, and so more oscillators may be stimulated at low temperatures.

1.3 The photoelectric effect

In those enormously productive months of 1905–6, when Einstein produced not only his heat capacity theory but also the special theory of relativity, he found time to make another fundamental contribution to modern physics. His achievement was to relate Planck's quantum hypothesis to the phenomenon of the *photoelectric effect,* the emission

† To see that there is a maximum frequency in the crystal it is probably easier to see that there is a minimum wavelength. Consider a linear array of atoms, and vibrations that consist of transverse displacements. The minimum wavelength of the displacements is that in which alternate atoms are displaced above and below the line, for then the displacements may be denoted $+-+-+-\ldots$ and the wavelength is twice the lattice spacing. No more rapidly varying phase can be constructed.

of electrons from metals when they are irradiated with ultraviolet light. The puzzling features of the effect were that the emission was instantaneous when the light was applied, however low its intensity, but there was no emission, whatever the intensity of the radiation, unless the frequency of the light was greater than a threshold value typical of each element.

Einstein pointed out that if the radiation field is quantized, so that it can accept or supply energy only in discrete amounts, then exactly this behaviour should be expected. In order for an electron to be ejected from a metal it must receive an energy at least equivalent to its binding energy, the *work function* for the metal involved. If the frequency of the light is so low that the energy $h\nu$ is less than the work function, then the electron cannot be ejected, hence the threshold frequency is accounted for. Emission occurs when the energy $h\nu$ exceeds the work function, but if the entire quantum of energy is transferred from the radiation any excess of $h\nu$ over the work function, Φ, must appear as the kinetic energy, $\frac{1}{2}m_e v^2$, of the emitted electron. We should therefore expect the following equation to hold:

$$\tfrac{1}{2}m_e v^2 = h\nu - \Phi. \tag{1.3.1}$$

This behaviour is observed, Fig. 1.4.

The quantum hypothesis accounts satisfactorily for the photoelectric effect and goes further. The implication of the theory that a quantum of energy must be transferred *in its entirety* to a *single* electron, is that the light quantum is confined to a small region of space. The energy cannot be spread evenly across an entire wave-front because then the energy to eject an electron would accumulate only slowly and the instantaneous emission could not be explained. This implies that the energy is localized in packets. In other words, the light beam should be considered to be a stream of corpuscles. These packets of energy $h\nu$ were called *photons* by Lewis (in 1926).

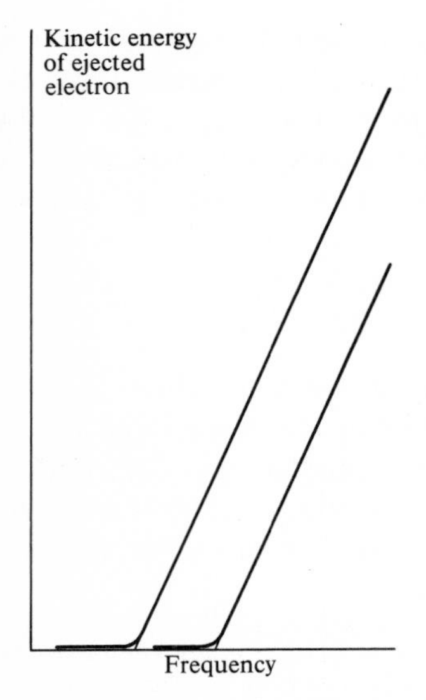

Fig. 1.4. Experimental data on the photoelectric effect for two metals.

Such an interpretation seems to be a denial of the extensive accumulation of data that apparently provided unequivocal support for a wave theory of light. By following the implications of Planck's quantum hypothesis we seem to have uncovered another deficiency of classical physics, but at the same time to have accounted quantitatively for a phenomenon for which classical theories could not supply even a qualitative explanation.

1.4 The Compton effect

If photons are truly particles they should possess a linear momentum, p. The relativistic expression relating a particle's energy to its mass and momentum is $E=(m^2c^4+p^2c^2)^{\frac{1}{2}}$. In the case of a photon $E=h\nu$ and $m=0$; hence $p=h\nu/c$. This momentum should be detectable if a beam of light falls on a stream of electrons: a transfer of momentum during the photon–electron collisions should appear as a scattering of the

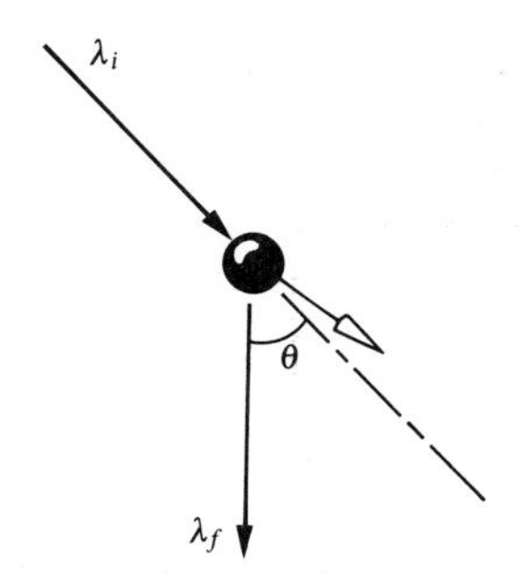

Fig. 1.5. The Compton effect.

photons by the electrons. In 1922 Compton performed this experiment using X-rays, and the results he obtained were in complete agreement with predictions. On the basis of the conservation of momentum and energy the change in the wavelength of the incident light scattered through an angle θ, Fig. 1.5, should be given by

$$\lambda_{\text{final}} - \lambda_{\text{initial}} = (2h/m_e c)\sin^2 \tfrac{1}{2}\theta, \tag{1.4.1}$$

where m_e is the mass of the electron. This expression was verified. (Notice that there would be no wavelength shift in a 'classical' universe in which $h=0$.) Only one value of the wavelength shift is predicted and observed at a given scattering angle. This is consistent with the model of the process, in which transfer takes place in a single blow during the collision, and rules out a gradual transfer as would be implied by a classical wave theory.

1.5 Atomic spectra

There was another body of data that physics could not elucidate before the introduction of quantum theory. This was the observation that the spectra of atoms appear to be a collection of discrete lines and not a continuous emission or absorption. The spectrum of atomic hydrogen had the simplest appearance, and Balmer had already noticed (in 1885) that the wavelengths of the lines in the spectrum could be related to a simple formula based on integers. The empirical expression he proposed, in modern notation, was

$$\frac{1}{\lambda} = R_H\left(\frac{1}{2^2} - \frac{1}{n^2}\right) \tag{1.5.1}$$

where R_H is the *Rydberg constant* for hydrogen ($R_H = 1.096\,78 \times 10^5\,\text{cm}^{-1}$) and $n = 3, 4, 5, \ldots$. Rydberg generalized this observation when in 1890 he proposed the rule that 'in the spectra of the elements there are series of rays whose wavelengths are functions of consecutive integral numbers'. The culmination of these manipulations was the *combination principle* proposed by Ritz in 1908, which stated that the wavelength of any spectral line could be expressed as the difference between two quantities or *terms*.

$$\textit{Ritz combination principle:}\ \frac{1}{\lambda} = T_1 - T_2. \tag{1.5.2}$$

As any one term could contribute to a number of lines, this reduced considerably the complexity of the description of a spectrum.

The explanation of the spectra of the elements is bound up with the structure of their atoms. Rutherford's nuclear model of the atom, that it consisted of a small positively charged nucleus surrounded by a swarm of electrons, was a major step forward, for it had been thought that the spectral lines must be associated with harmonic oscillations of the atomic electrons, and Rutherford's model could, perhaps, provide a basis for a calculation of their frequencies. But the Rutherford atom itself appeared to conflict with classical physics. There is a theorem in classical electrodynamics, Earnshaw's theorem, which states that no static system of charges can be in equilibrium; therefore the electrons in an atom must be moving, and so they must also be accelerating. (If they are to remain in the vicinity of the nucleus they cannot travel indefinitely in a straight line, and as a change of direction requires a force there must be an acceleration.) But if an electron is accelerated, classical physics demands that it radiate electromagnetic energy. Therefore the Rutherford atom is inherently an unstable system, and the electrons should collapse on to the nucleus.

It was at this point (in 1913) that Bohr made his contribution by arbitrarily welding Planck's quantum hypothesis to classical mechanics. His calculation, more a kind of shotgun marriage, was not what today we call quantum-mechanical, but it was the first synthesis of quantum theory and mechanics. Bohr had realized that a calculation on the Rutherford atom should yield a quantity of the dimensions of length to represent the distance of the electron from the nucleus. (He considered the hydrogen atom, and so too do we.) The only quantities that occur in the classical calculation are the mass m_e and charge $-e$ of the electron and the permittivity of free space ε_0 (which enters through the Coulomb potential): a quantity with dimensions of length cannot be constructed from these alone. A length can be constructed from m_e, e, and ε_0 if one also includes a quantity with the dimensions of action. This points clearly to the involvement of the fundamental unit of action, h. The combination $\varepsilon_0 h^2/m_e e^2$ has dimensions of length, and its magnitude is about 0.16 nm, which is similar to atomic dimensions (*c.* 0.1 nm).

Bohr then made a set of assumptions that enabled him to render quantitative his basic idea that there exists in the atom a discrete set of stable, *stationary orbits.*

(1) Energy is emitted only when an electron makes a transition between two stationary orbits, and not continuously.
(2) The frequency of the radiation emitted when a transition occurs between two orbits that differ in energy by ΔE is $\Delta E/h$.
(3) The dynamical equilibrium that governs the stationary orbits is determined by classical mechanics, by balancing the electrostatic attraction against the centrifugal effect of the orbital motion.
(4) The stationary states are determined by the condition that the ratio of the total energy of the electron to its orbital frequency shall be an integral multiple of $\frac{1}{2}h$. For circular orbits this is equivalent to the restriction of the angular momentum of the electron to integral multiples of $h/2\pi$.

The actual calculation on the basis of these postulates leads to the result that the energy of the electron in a hydrogen atom is restricted to the values

$$E_n = -\mu e^4/8n^2h^2\varepsilon_0^2, \tag{1.5.3}$$

where μ is the *reduced mass* $m_e m_p/(m_e + m_p)$, m_p being the mass of the proton, and where the *quantum number* n is restricted to the values $1, 2, 3, \ldots$. If an electron makes a transition from an orbit of quantum number n_1 to one of quantum number n_2, the atom's energy changes by

$$\Delta E = \left(\frac{\mu e^4}{8h^2\varepsilon_0^2}\right)\left(\frac{1}{n_2^2} - \frac{1}{n_1^2}\right). \tag{1.5.4}$$

Then, with $h\nu = \Delta E$, and $\nu = c/\lambda$, we predict a spectral line of wavelength

$$\frac{1}{\lambda} = \left(\frac{\mu e^4}{8h^3 c\varepsilon_0^2}\right)\left(\frac{1}{n_2^2} - \frac{1}{n_1^2}\right) \tag{1.5.5}$$

just as the Balmer formula requires. Furthermore, R_H is predicted to have the value $\mu e^4/8h^3c\varepsilon_0^2$, or $1.096\,78 \times 10^5\ \text{cm}^{-1}$, exactly as found experimentally. This formula gives very good agreement with experiment in so far as it predicts the wavelengths of the spectral lines with high precision, but it is clearly only a first step along the way to understanding both structure and spectra, because the fact that it predicts lines is no more than one of the postulates of the model. The most that we can conclude is that the theory is on the right track. We shall see soon, however, that it contains a fundamental error.

Example. Find an expression for the energy levels of a hydrogen-like atom of atomic number Z based on Bohr's model.

- *Method.* Find the radius r_* at which the centrifugal force $J^2/\mu r^3$ balances the Coulomb force $-Ze^2/4\pi\varepsilon_0 r^2$. Set angular momentum $J = nh/2\pi$, then find the corresponding total energy from $E = T + V$, where T is the kinetic energy $J^2/2\mu r_*^2$, and V is the Coulomb potential energy $-Ze^2/4\pi\varepsilon_0 r_*$. Use $\mu = m_e m_N/(m_e + m_N)$, the reduced mass, m_N being the mass of the nucleus.
- *Answer.* $J^2/\mu r^3 = Ze^2/4\pi\varepsilon_0 r^2$ when $r = 4\pi\varepsilon_0 J^2/Ze^2\mu$.

  Therefore $r_* = 4\pi\varepsilon_0 n^2h^2/4\pi^2 Ze^2\mu = n^2(\varepsilon_0 h^2/\pi\mu e^2 Z)$.

  Consequently $E = J^2/2\mu r_*^2 - Ze^2/4\pi\varepsilon_0 r_*$

  $= (\mu e^4 Z^2/8\varepsilon_0^2 h^2)(1/n^2) - (\mu e^4 Z^2/4\varepsilon_0^2 h^2)(1/n^2)$

  $= -(\mu e^4 Z^2/8\varepsilon_0^2 h^2)(1/n^2)$.
- *Comment.* When $Z = 1$ the answer is the same as that quoted in eqn (1.5.3). Notice that $T = -\frac{1}{2}V$: this is an aspect of the *virial theorem*, which is discussed later and derived in Appendix 5.

There were a number of attempts to account for the finer details of the spectrum of atomic hydrogen, and also the structures and gross details of the spectra of atoms containing more than one electron. These were unsuccessful and we shall not pursue them here since, as we have suggested, Bohr's synthesis of classical and quantum physics was fundamentally false. The important feature of his work is that in it we see a theory developed to account for properties of radiation applied to the restriction of the mechanical properties of a system.

1.6 The wave nature of matter

The grand synthesis of these theories and the demonstration of the links that exist between radiation propagation and the dynamics of particles began with de Broglie, when he recognized the similarity that existed between *Fermat's principle* of least time, which governed the propagation of light, and Maupertuis's principle (less metaphysically, *Hamilton's principle*) of least action, which governed the propagation of particles. In 1924 he proposed that with any moving body there is 'associated a wave', and that the momentum of the particle and the wavelength are related by the

de Broglie relation: $p = h/\lambda$. (1.6.1)

Here we see a fusion of opposites, the momentum is a property of particles, the wavelength is a property of waves, and this *duality*, the possession of both wave- and particle-like properties, is a persistent theme in the remainder of the book.

The elucidation of the significance of the de Broglie waves is the subject of the next chapter, but we can already begin to see that their existence goes some way towards explaining Bohr's fourth postulate. If we consider a particle in an orbit, then for most radii and momenta the waves associated with the particle interfere destructively as they wrap round each orbit, Fig. 1.6; only for some radii and momenta will the ends of the waves on each orbital circuit join so that non-destructive interference occurs. Of course, this is nothing more than a plausibility argument because so far we have no reason to suppose that waves which interfere destructively should be omitted.

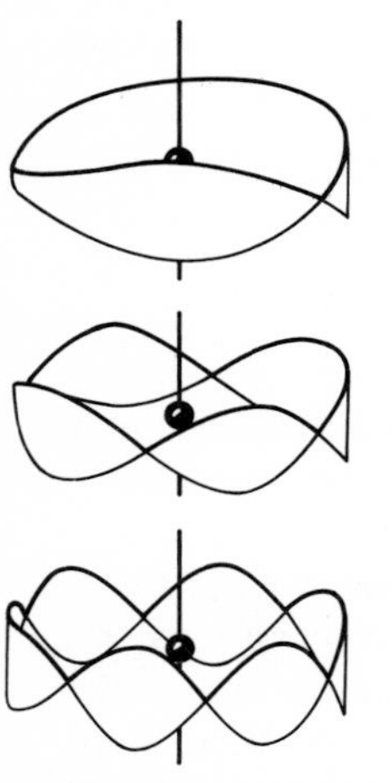

Fig. 1.6. Standing waves and permitted orbits.

The de Broglie waves seem to point to a solution of the quantization problem in mechanics, but we need evidence that they exist. We have already seen evidence for the association of particles with wave motion: we saw the photoelectric effect demands the existence of light particles, and light is known to be wave-like because it can be diffracted. This in turn points to a possible test of de Broglie's conjecture: can a beam of particles be diffracted? In fact a diffraction experiment using electrons had already been performed in 1921, although the phenomenon had not been recognized. Franck realised that an anomalous scattering peak in Davisson's experiment was probably due to diffraction. In 1925 Davisson and Germer were able to perform an experiment which demonstrated the diffraction of electrons in quantitative agreement with a wavelength calculated from de Broglie's relation. (Their experiment owed its success to an accident which caused their polycrystalline sample to fuse into a single crystal.) Another confirmation of de

Broglie's hypothesis was obtained in 1927 by G. P. Thomson who observed the diffraction of electrons by a thin film of celluloid.†

While these results confirm de Broglie's conjecture, we have not yet discovered how the waves govern the propagation of the particles or how quantization is a consequence of their existence. We shall be able to do both once we know their law of propagation. The first successful proposal about how to do quantum mechanical calculations was due to Heisenberg (1926), who put forward the scheme known as 'matrix mechanics'. In the same year Schrödinger put forward his rival theory of 'wave mechanics'. It appeared to be quite different from matrix mechanics. Shortly, however, the two approaches were shown to be mathematically equivalent by Schrödinger himself. Each has its advantages for different types of problem, but Schrödinger's formulation is the simpler conceptually and we shall concentrate on it, and develop it in Chapter 2, but not hesitate to use Heisenberg's approach when it seems appropriate.

1.7 The uncertainty principle

Before this chapter closes there is one more deeply significant topic to consider. It will complete our presentation of the historical development of quantum mechanics, the central theme of which has been the abnegation of the validity of the assumption that casually extrapolated common sense is a guide in the discussion of the unfamiliar. The topic is Heisenberg's famous *uncertainty principle*, or the *principle of indeterminacy*. The former name is more common although the latter is better because the principle is not concerned with whether we know something, but whether there is anything to know.

The uncertainty principle denies us the ability to determine simultaneously and with arbitrarily high precision the values of particular pairs of observables. Observables restricted in this way are called *complementary*. We shall see that position and linear momentum form a pair of complementary observables, and hence that the position and linear momentum of a particle cannot be determined simultaneously to arbitrarily high precision. This restriction is not merely a recognition of our technological incompetence, but arises because some observations are mutually incompatible. Some pairs of properties are not just simultaneously unknown, they are unknowable. Heisenberg formulated his principle in 1927 and we shall explore its content in a later chapter (Chapter 5); but already we should be sufficiently aware of the direction that quantum theory is taking to be able to establish a qualitative understanding of the basis of the principle.

We investigate the validity of the following question: 'What is the linear momentum and the position of a particle at some instant?' In order for there to be a precise answer to the second part of the

† It has been pointed out by M. Jammer that J. J. Thomson was awarded the Nobel Prize for showing that the electron is a particle, and G. P. Thomson, his son, was awarded the Prize for showing that the electron is a wave.

question the particle must be localized at a definite point, an infinitesimal region, in space. It is a plausible assumption that the de Broglie wave associated with such a particle will also be localized in the same region. (This assumption will be confirmed in the next chapter where we learn the significance of the waves.) But what is meant by a wave that is localized in an infinitesimal region? In acoustics it corresponds to a short, sharp sound—we should hear a bang as a sharply defined pressure wave passed. Such a wave can be regarded as a collection of a large number of waves, each with a different wavelength, superimposed so that constructive interference creates the pressure zone but destructive interference occurs everywhere else, Fig. 1.7. By analogy, the de Broglie waves associated with a localized particle should therefore be considered as a superposition of waves of a wide range of wavelengths. But if there is a range of wavelengths then there is also a corresponding range of momenta (through the de Broglie relation $p = h/\lambda$). Consequently, we cannot associate a definite momentum with that localized particle.

If, on the other hand, the particle had a definite momentum, its de Broglie wave would be of uniform wavelength and spread throughout space. But because it spreads everywhere, the location of the particle cannot be specified. Hence preparing the particle in a given momentum state entails being unable to specify its location.

This qualitative discussion shows that it makes no sense to speak in terms of the location *and* the linear momentum. The observables are *complementary*: *specification* (meaning measurement or prediction of its value) *of one observable rules out the specification of the other.* We shall later prove that the fundamental constraint on the simultaneous specification of location and momentum is

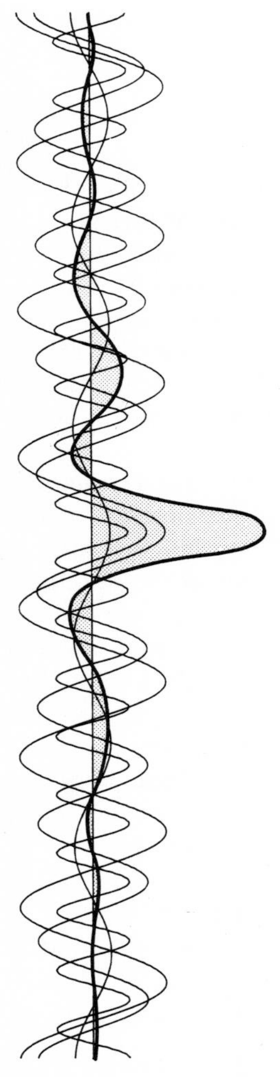

Fig. 1.7. The formation of a localized wavepacket.

$$\delta p\,\delta x \gtrsim h/4\pi \qquad (1.7.1)$$

where δp and δx are the uncertainties (specifically the root mean square deviations from the mean, $\delta A = \sqrt{(\langle A^2\rangle - \langle A\rangle^2)}$) of the momentum and position. For this to be satisfied when either observable is precisely specified (e.g. $\delta x = 0$), the other must be completely unspecified (e.g. $\delta p \sim \infty$). An electron confined to an atom ($\delta x \approx 0.1$ nm) cannot have a linear momentum specified more precisely than $\delta p \approx h/4\pi(0.1\text{ nm}) \approx 0.5\times10^{-24}$ J s, corresponding to a speed which cannot be specified more precisely than about 10^6 m s^{-1}.

Example. Show that the uncertainty principle is consistent with an attempt to measure the position and momentum of a particle simultaneously using a microscope.

- *Method.* Consider how the resolving power depends on the wavelength of the light used in the microscope and investigate how the momentum of the particle is affected by the position measurement.
- *Answer.* The resolving power of a lens of angular aperture α (Fig. 1.8) is $(\sin\alpha)/\lambda$. This means that after interaction with the particle the photon must

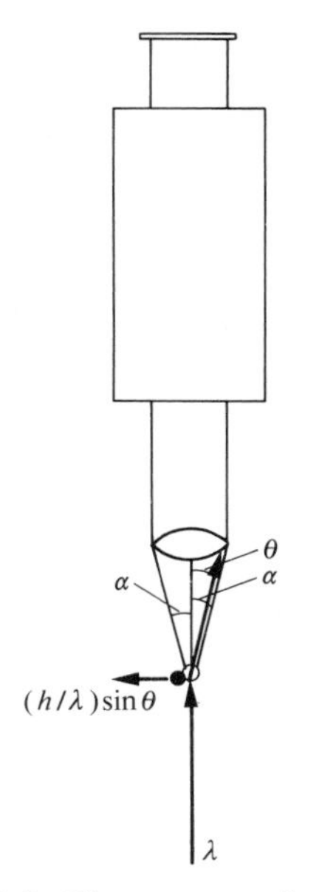

Fig. 1.8. The γ-ray-microscope thought experiment.

travel at an angle θ, with θ anywhere within the range 2α, to appear in the microscope. When a photon of wavelength λ (and momentum h/λ) is scattered through an angle θ it imparts a momentum $\Delta p = (h/\lambda)\sin\theta$ to the particle with θ within the same range (2α). Therefore $\Delta p\,\Delta x \approx h$, in accord with the uncertainty relation.

• *Comment.* This style of 'thought experiment' shows that the complementarity of momentum and position is in accord with any attempt to measure them simultaneously. Numerous analogous experiments have been proposed. See D. Bohm, *Quantum theory*, Prentice-Hall (1951).

The position–momentum uncertainty relation is only one of a number of such relations that apparently restrict our knowledge of the world, and later chapters illustrate the principle more fully. We can, however, see one of its far-reaching implications: *the classical idea of the trajectory of a particle is incompatible with quantum mechanics.* For, in order to define the trajectory of a particle, it is necessary to specify at each instant along its path, and therefore simultaneously, both the position and the momentum of the particle. The ability to do this is denied by the uncertainty principle. One consequence is the recognition of the fundamental internal inconsistency of Bohr's model of the hydrogen atom which required the concept of 'orbit', a periodic trajectory.

The overall implication of the uncertainty princple was summarized by Heisenberg by saying that in the view of classical physics given by the statement 'if the present is known exactly then the future may be predicted' it was the premise that was false. I think this is a false summary. The uncertainty principle puts no restraint on our ability to know the present exactly—it simply modifies our conception of what is 'complete' knowledge. Classical mechanics attempts to force on nature an overcomplete description.

Further reading

Historical background

The conceptual development of quantum mechanics. M. Jammer; McGraw Hill, New York, 1966.
Black-body theory and the quantum discontinuity, 1894–1912. T. G. Kuhn; Oxford University Press, New York, 1978.
The history of quantum theory. F. Hund, Harrap, London, 1974.

Further details of material

Physical chemistry. P. W. Atkins; Oxford University Press and W. H. Freeman and Company, San Francisco, 2nd edition 1982.
Quantization (lecture cassette and workbook). P. W. Atkins; Royal Society of Chemistry, London, 1981.
Quanta: a handbook of concepts. P. W. Atkins, Oxford University Press, 1974.
Modern atomic physics. B. Cagnac and J. C. Pebay-Peyroula, Macmillan, London, 1975.

Problems

1.1. Calculate the size of the quanta involved in the excitation of an electronic motion of period 10^{-15} s, a molecular vibration of period 10^{-14} s, and a pendulum of period 1 s.

1.2. Calculate the mean energy of an oscillator of frequency (a) 1 Hz, (b) 10^{14} Hz at temperatures of (i) 100 K, (ii) 1000 K. Compare the result with the value predicted by the equipartition principle.

1.3. In the argument that leads to eqn (1.1.10) assume (as will turn out to be the case) that the energy of a harmonic oscillator of frequency ν is limited to the values $\varepsilon_v = (v + \frac{1}{2})h\nu$. Find the correct form of the Planck distribution.

1.4. The peak in the sun's emitted energy occurs at about 480 nm. Estimate the temperature of its surface on the basis of it being regarded as a black-body emitter.

1.5. Find the (a) low temperature, (b) high temperature forms of the Einstein heat capacity function.

1.6. Estimate the molar heat capacities of metallic sodium ($\theta_D = 150$ K) and diamond ($\theta_D = 1860$ K) at room temperature (300 K).

1.7. Calculate the molar entropy of an Einstein solid at $T = \theta_E$. *Hint.* The entropy is $S = \int_0^T (C_V/T)\,dT$. Evaluate the integral numerically (i.e. either graphically, or by using Simpson's rule on a programmable calculator).

1.8. How many photons would be emitted each second by a 100 W sodium lamp which radiated all its energy with 100 per cent efficiency as yellow 589 nm light?

1.9. Calculate the speed of an electron emitted from a clean potassium surface ($\Phi \triangleq 2.3$ eV; $\triangleq$ means 'corresponds to') by light of wavelength (a) 300 nm, (b) 600 nm.

1.10. Deduce eqn (1.4.1) for the Compton effect on the basis of the conservation of energy and linear momentum. *Hint.* use the relativistic expressions. Initially the electron is at rest with energy $m_e c^2$. When it is travelling with momentum p its energy is $(p^2c^2 + m_e^2c^4)^{\frac{1}{2}}$. The photon, initial momentum h/λ_i and energy $h\nu_i$ strikes the stationary electron, is deflected through an angle θ, and emerges with momentum h/λ_f and energy $h\nu_f$. The electron is initially stationary ($p = 0$) but moves off with an angle θ' to the incident photon. Conserve energy and both components of linear momentum. Eliminate θ', then p, and so arive at an expression for $\delta\lambda$.

1.11. The first few lines of the visible (Balmer) series in the spectrum of atomic hydrogen lie at λ/nm = 656.46, 486.27, 434.17, 410.29, ... Find a value of R_H, the Rydberg constant for hydrogen. The ionization energy, I, is the minimum energy required to remove the electron. Find it from the data and express its value in electronvolts. How is I related to R_H? *Hint.* The ionization limit corresponds to $n \to \infty$ for the final state of the electron.

1.12. Calculate the de Broglie wavelength of (a) a mass of 1 g travelling at 1 cm s^{-1}, (b) the same at 95 per cent of the speed of light, (c) a hydrogen atom at room temperature (300 K; estimate the mean speed from the equipartition principle), (d) an electron accelerated from rest through a potential difference of a (i), 1V, (ii) 10 kV. *Hint.* For the momentum in (b) use $p = m_e v/(1 - v^2/c^2)^{\frac{1}{2}}$ and for the speed in (d) use $\frac{1}{2}m_e v^2 = e\Delta\phi$, $\Delta\phi$ being the potential difference.

1.13. An electron is confined to a linear box of dimension 0.10 nm. What is the minimum uncertainty in its velocity and its kinetic energy?

1.14. Use the uncertainty principle to estimate the order of magnitude of the diameter of an atom. Compare the result with the radius of the first Bohr orbit of hydrogen (which can be calculated from the information in the *Example* on p. 15). *Hint.* Suppose the electron to be confined to a region of extent δx, this implies a non-zero kinetic energy. There is also a potential energy of order of magnitude $-e^2/4\pi\varepsilon_0\,\delta x$. Find δx such that the total energy is a minimum, and evaluate the expression.

2 The Schrödinger equation

WHERE do we find the Schrödinger equation? One approach is to pluck it out of the air (which is a euphemism for another book) and to present it as a fundamental postulate of quantum mechanics. Such a procedure is perfectly valid, but the equation is so elaborate that it is hard to believe there is nothing behind it. This chapter puts the equation into the context of classical physics, and shows how it emerges from familiar equations when small adjustments are made. Such a procedure also has the advantage of emphasizing that classical physics is an approximation to quantum mechanics: we shall see that it was a very near miss, and fails largely because it approximates some very small quantities by zero.

2.1 Heisenberg's suggestion

In classical physics we deal with position, linear momentum, and other observables by regarding them as functions of time. The time evolution of a system is then calculated by solving various equations and reporting the results as the functions $x(t)$ and $p(t)$. For example, in the case of a particle of mass m undergoing simple harmonic motion, the basic equation is Newton's second law

$$m\,\mathrm{d}^2x/\mathrm{d}t^2 = F \tag{2.1.1}$$

with the force F proportional to the displacement, $F = -kx(t)$. A solution is

$$x(t) = A\cos\omega t, \qquad \omega = (k/m)^{\frac{1}{2}}, \tag{2.1.2}$$

which is the displacement expressed as a function of time. The linear momentum of the particle is $p = m\dot{x}(t)$, or

$$p(t) = -mA\omega\sin\omega t. \tag{2.1.3}$$

With these two functions we can set about exploring the properties of the oscillator. In particular we might find that in the discussion of some property we have to deal with the product xp, which in the present case is

$$xp = -mA^2\omega\cos\omega t\sin\omega t = -\tfrac{1}{2}mA^2\omega\sin(2\omega t).$$

It is obviously trivial to point out that the quantity px has exactly the same value:

$$px = -mA^2\omega\sin\omega t\cos\omega t = -\tfrac{1}{2}mA^2\omega\sin(2\omega t),$$

and therefore that

$$xp - px = 0. \tag{2.1.4}$$

Trivial it may be, but true it is not. Heisenberg's contribution to quantum mechanics was equivalent to saying that the right-hand side of this equation is not zero but is actually a small, non-zero quantity. Classical mechanics makes the approximation of neglecting this quantity; quantum mechanics takes it into account. Of course, since the last equation follows from Newton's equation, Heisenberg's suggestion implies that Newton's equation must be wrong: it must be only an approximation to the true equation of motion. Finding the true equation turns out to be the same as finding the Schrödinger equation.

The dimensions of the product xp are $\mathrm{L(MLT^{-1}) = ML^2T^{-1}}$, which are the same as the dimensions of action. The zero on the right of $xp - px = 0$ is therefore to be replaced, according to Heisenberg, by a small quantity with the dimensions of action. The natural choice is Planck's constant h. In fact it turns out that in order to get agreement with experiment the right-hand side should be replaced by $\mathrm{i}h/2\pi$. Since $h/2\pi$ occurs throughout quantum mechanics, it is convenient to give it a special symbol, and to write $\hbar = h/2\pi$. Then Heisenberg's suggestion (as formulated by Born and Jordan) amounts to demanding that x and p should satisfy the

Commutation relation: $xp - px = \mathrm{i}\hbar.$ (2.1.5)

This is the single most important expression in quantum mechanics, and the entire theory flows from it.

An immediate major implication of the commutation relation is that x and p cannot any longer be regarded as functions of time, because for functions it is always true that the order of multiplication is unimportant. Hence, the change from eqn (2.1.4) to (2.1.5) represents far more than a minor modification of the form of the observables. There are two possible interpretations. One is that x and p should be regarded as *matrices*, because it is a property of matrices that the order in which they are multiplied affects the outcome, and in general AB is not equal to BA. This identification of x and p as matrices is the source of the name *matrix mechanics*. The other interpretation, the one we shall use, is to regard x and p as *operators*. An operator is simply an instruction to do something (e.g., multiply, differentiate, etc.), and in general the order in which the instructions are performed, as in a computer program, affects the outcome. Hence, if xp is interpreted as operation p followed by operation x, it will have a different outcome from px, operation x followed by operation p.

One choice of the explicit form of the operators x and p is as follows. The operator corresponding to position, x, is *multiplication by the coordinate* x. In other words x is to be interpreted as the operator 'multiplication by x'. Linear momentum, on the other hand, is to be interpreted as the operation *differentiation with respect to* x, or, to be precise,

$$p = (\hbar/\mathrm{i})(\mathrm{d}/\mathrm{d}x). \tag{2.1.6}$$

That this pair of choices satisfies eqn (2.1.5) can be checked as follows.

First, note that if x and p are operators, they need something, a function, to operate on. We denote this function Ψ. Then eqn (2.1.5) becomes

$$(xp - px)\Psi = \mathrm{i}\hbar\Psi$$

when we allow each side to operate on Ψ. Now test the choice. We have

$$xp\Psi = x(\hbar/\mathrm{i})(\mathrm{d}/\mathrm{d}x)\Psi = (\hbar/\mathrm{i})x\,\mathrm{d}\Psi/\mathrm{d}x;$$

$$px\Psi = (\hbar/\mathrm{i})(\mathrm{d}/\mathrm{d}x)x\Psi = (\hbar/\mathrm{i})\{\Psi + x(\mathrm{d}\Psi/\mathrm{d}x)\}.$$

Therefore,

$$(xp - px)\Psi = -(\hbar/\mathrm{i})\Psi = \mathrm{i}\hbar\Psi,$$

as we required.

Quantum mechanics demands that we do our calculations with the operators corresponding to the observables, not, as in classical mechanics, with the observables themselves. We can glance at what this means by considering the simple harmonic oscillator again. The total energy of the oscillator is the sum of its kinetic and potential energies. The kinetic energy is $\frac{1}{2}mv^2$, or, in terms of the momentum $p = mv$, $p^2/2m$. The potential energy is the parabolic function $\frac{1}{2}kx^2$ (this comes from the classical result that $F = -\mathrm{d}V/\mathrm{d}x$, F being the force $-kx$). The total energy is therefore

$$E = p^2/2m + \tfrac{1}{2}kx^2. \tag{2.1.7}$$

When p and x are interpreted as operators the right-hand side becomes the operator

$$(1/2m)\{(\hbar/\mathrm{i})(\mathrm{d}/\mathrm{d}x)\}^2 + \tfrac{1}{2}kx^2 = -(\hbar^2/2m)(\mathrm{d}^2/\mathrm{d}x^2) + \tfrac{1}{2}kx^2$$

(the term $\frac{1}{2}kx^2$ being interpreted as the operator 'multiply by $\frac{1}{2}kx^2$'). If we write the function on which this operator operates as Ψ, eqn (2.1.7) turns into

$$E\Psi = -(\hbar^2/2m)(\mathrm{d}^2\Psi/\mathrm{d}x^2) + \tfrac{1}{2}kx^2\Psi. \tag{2.1.8}$$

This is nothing other than the time-independent Schrödinger equation for the harmonic oscillator.

If we were dealing with a system other than a simple harmonic oscillator, the potential energy would have a general form $V(x)$, and its corresponding operator would be multiplication by the function $V(x)$. We shall see plenty of examples of this in due course. The general form of the last equation is therefore

$$E\Psi = -(\hbar^2/2m)(\mathrm{d}^2\Psi/\mathrm{d}x^2) + V(x)\Psi. \tag{2.1.9}$$

The right-hand side is normally simplified by introducing the

Hamiltonian operator: $H = -(\hbar^2/2m)(\mathrm{d}^2/\mathrm{d}x^2) + V(x)$, (2.1.10)

when the equation becomes the

Time-independent Schrödinger equation: $H\Psi = E\Psi$. (2.1.11)

Most of quantum mechanics is concerned with finding solutions of this deceptively simple equation.

What we have achieved so far is the simplification of the form of the basic postulate of quantum mechanics. Instead of having to pull the Schrödinger equation out of the air complete in all its complexity, eqn (2.1.9), we have seen that it is necessary only to propose the commutation relation, eqn (2.1.5). Nevertheless, we are not yet at the complete form of the Schrödinger equation because eqn (2.1.11) makes no reference to the time evolution of the system. We now turn to this aspect.

2.2 Hamilton's contribution

The Astronomer Royal of Ireland, W. R. Hamilton, came remarkably close to discovering quantum mechanics long before the experimental data required it. He expressed classical mechanics in a form that makes the transition from classical to quantum mechanics as easy as Heisenberg's little but implication-packed modification of replacing 0 by $i\hbar$. Furthermore, his formulation leads directly to the time-dependent form of the Schrödinger equation. We concentrate on the qualitative aspects in the rest of this section; the calculations themselves can be found in Appendix 2.

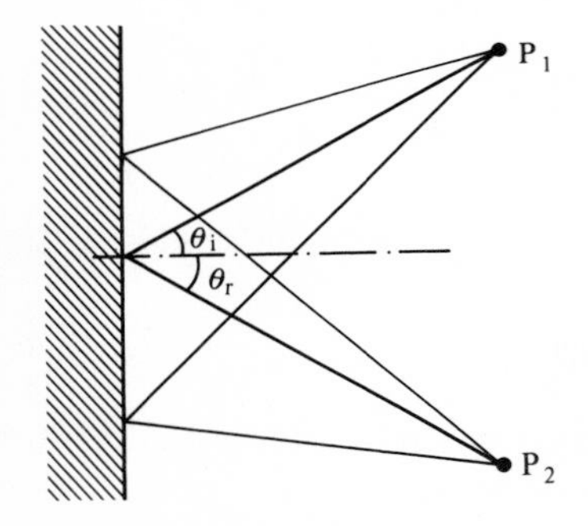

Fig. 2.1. Paths for the reflection of light.

In geometrical optics light travels in straight lines in a uniform medium, and we know that the physical nature of light is a wave motion. In classical mechanics particles travel in straight lines unless a force is present, and the evidence in the last chapter pointed towards an underlying wave nature of matter. There are clearly deep analogies here. We shall first establish how, in optics, wave motion can lead to straight line motion, and then argue by analogy about the wave nature of particles.

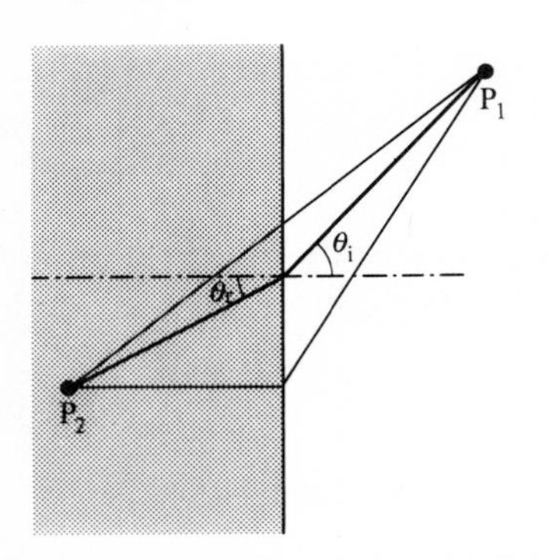

Fig. 2.2. Paths for the refraction of light.

The basic rule governing light propagation in geometrical optics is *Fermat's principle of least time.* A simple form of the principle is that the path taken by a light ray through a medium is such that its time of passage is a minimum. As an example, consider the relation between the angles of incidence and reflection, Fig. 2.1. The briefest path between source, mirror, and observer is clearly the one corresponding to equal angles of incidence and reflection. In the case of refraction it is necessary to take into account the different speeds of propagation in the two media. In Fig. 2.2 the geometrically straight path between source and observer is not necessarily the briefest, because the light travels relatively slowly through the dense medium. The briefest path is easily shown to be the one where the angles of incidence (θ_i) and refraction (θ_r) are related by Snell's law, that $\sin\theta_r/\sin\theta_i = n_1/n_2$; n_1 and n_2 are the refractive indices of the less and more dense media respectively (they come in because the speed of light in a medium is c/n).

How does the wave nature of light account for this behaviour? Consider the case illustrated in Fig. 2.3 where we are interested in the propagation of light between two fixed points, P_1 and P_2. A wave of electromagnetic radiation travelling along some general path A arrives at P_2 with some phase. A wave travelling along a neighbouring path A′ arrives with a different phase; there is destructive interference if the amplitudes (in this case, of electric field) cancel, and constructive interference if they add. Path A has many neighbouring paths, and the overall effect of the superposition of all the waves is cancellation. Hence an observer is led to the view that the light does not travel along a path like A. The same argument applies to every path between the two points, with one exception, the straight line path B. The neighbours of B do not interfere destructively with B itself, and it survives. The reason for this exceptional behaviour can be seen as follows.

The amplitude of a wave at some point x can be written $a\mathrm{e}^{2\pi\mathrm{i}x/\lambda}$ where λ is its wavelength. It follows that the amplitude at P_1 is $a\mathrm{e}^{2\pi\mathrm{i}x_1/\lambda}$ and the amplitude at P_2 is $a\mathrm{e}^{2\pi\mathrm{i}x_2/\lambda}$. The two amplitudes are related as follows:

$$\Psi(\mathrm{P}_2) = a\mathrm{e}^{2\pi\mathrm{i}x_2/\lambda} = a\mathrm{e}^{2\pi\mathrm{i}(x_2-x_1)/\lambda}\mathrm{e}^{2\pi\mathrm{i}x_1/\lambda} = \mathrm{e}^{2\pi\mathrm{i}(x_2-x_1)/\lambda}\Psi(\mathrm{P}_1),$$

which can be written more simply as

$$\Psi(\mathrm{P}_2) = \mathrm{e}^{\mathrm{i}\phi}\Psi(\mathrm{P}_1), \qquad \phi = 2\pi(x_2 - x_1)/\lambda. \tag{2.2.1}$$

The function ϕ is called the *phase length* of the straight line path. The relative phases at P_2 and P_1 for waves that travel by curved paths are also related by an expression of the kind $\Psi(\mathrm{P}_2) = \mathrm{e}^{\mathrm{i}\phi}\Psi(\mathrm{P}_1)$, but the phase length is now determined by the length L of the curving paths:

$$\phi = 2\pi L/\lambda. \tag{2.2.2}$$

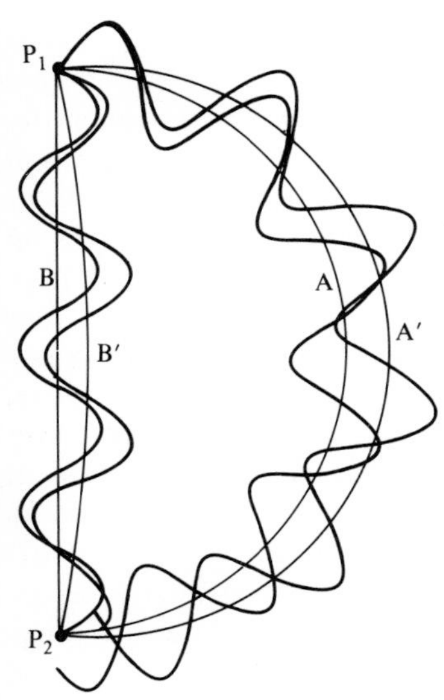

Fig. 2.3. Interfering paths and the propagation of light.

Now consider how ϕ depends on the distortion of the path from a straight line. If we distort the path from B to A or to C in Fig. 2.4(a), ϕ changes as depicted in Fig. 2.4(b). Quite clearly ϕ goes through a minimum at B. Now we arrive at the crux of the argument. Consider the relative phase lengths of paths in the vicinity of A. The phase length of path A′ can be related to that of path A by the following Taylor series

$$\phi(\mathrm{A}') = \phi(\mathrm{A}) + \phi'(\mathrm{A})\,\delta s + \tfrac{1}{2}\phi''(\mathrm{A})\,\delta s^2 + \ldots$$

where δs is a measure of the distortion of the path (Fig. 2.4a). Compare this to the relative phase lengths of path B and its neighbours:

$$\phi(\mathrm{B}') = \phi(\mathrm{B}) + \phi'(\mathrm{B})\,\delta s + \tfrac{1}{2}\phi''(\mathrm{B})\,\delta s^2 + \ldots = \phi(\mathrm{B}) + \tfrac{1}{2}\phi''(\mathrm{B})\,\delta s^2 + \ldots$$

because the first derivative is zero at the minimum of a curve. In other words, to first order in the displacement, *straight line paths have*

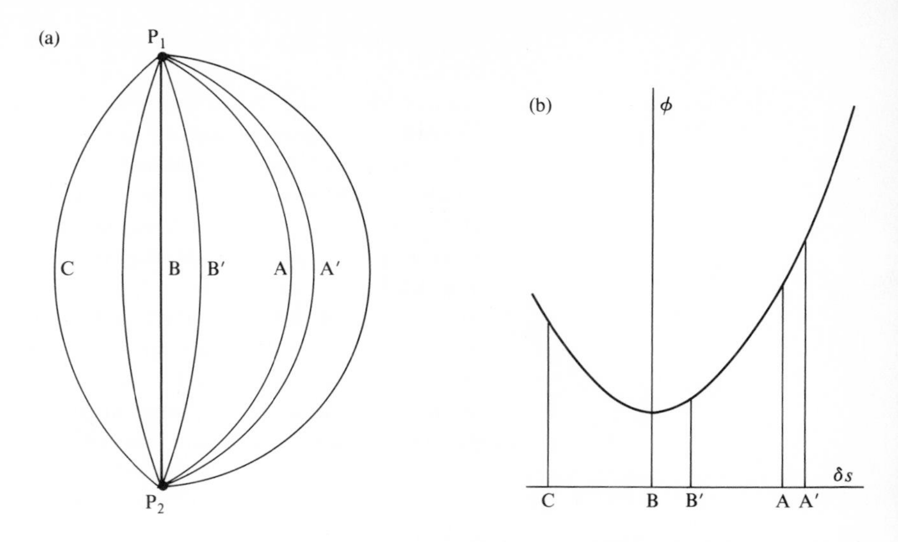

Fig. 2.4. (a) Various distorted paths, and (b) the corresponding phase lengths.

neighbours with the same phase length, but *curved line paths have neighbours with different phase lengths*. This is the reason why straight line propagation survives while curved paths do not: the latter have annihilating neighbours.

Two further points may now be made. When the medium is not uniform the wavelength of a wave depends on position. Since $\lambda = v/\nu$ and v, the speed of propagation, is equal to c/n, and n is a function of position, a more general form of the phase length is

$$\phi = 2\pi \int_{P_1}^{P_2} \frac{dx}{\lambda(x)} = \left(\frac{2\pi\nu}{c}\right) \int_{P_1}^{P_2} n(x)\, dx. \tag{2.2.3}$$

The same argument applies, but because of the dependence of the refractive index on position, *a curved path may turn out to correspond to a minimum phase length*, and therefore have, to first order, no destructive neighbours. Hence the path apparently taken by the light will be curved. Focusing by a lens is a manifestation of this effect.

The second point concerns the stringency of the conclusion that minimum-phase-length paths have non-destructive neighbours. Since the wavelength appears in the denominator of the expression for ϕ, waves of short wavelength have greater phase lengths for a given path than waves of long wavelength. The dependence of phase length on wavelength is indicated in Fig. 2.5. It should be clear that neighbours annihilate each other much more effectively in the case of short wavelengths than in the case of long. Therefore the rule that light (or any other wave motion) propagates itself in straight lines becomes increasingly valid as its wavelength shortens. Sound waves travel in more or less straight lines; their wavelengths are long. Light waves travel in almost exactly straight lines, but small objects can interfere

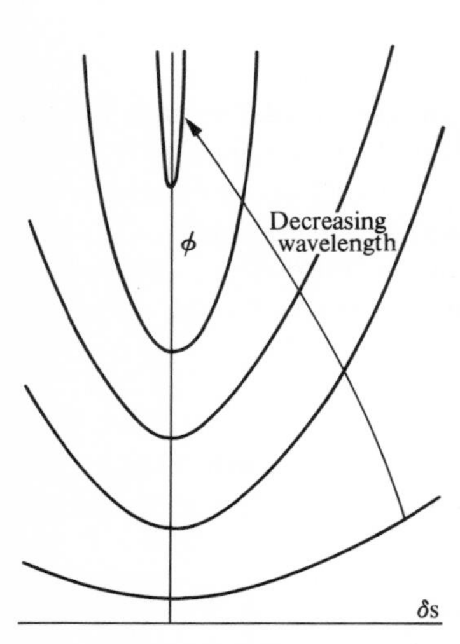

Fig. 2.5. The dependence of the phase length on wavelength.

and show up as diffraction effects. Geometrical optics is the limit of infinitely short wavelength, where the annihilation by neighbours is so effective that the light appears to travel in perfectly straight lines and diffraction effects do not occur.

We can summarize this account as follows. Light consists of waves which can be expressed in the form $\Psi(x) = a e^{2\pi i x/\lambda}$, and the amplitude at one point can be related to that at another by a relation of the form $\Psi(P_2) = e^{i\phi}\Psi(P_1)$. Light takes paths such that the phase length ϕ is a minimum (this is the more precise form of Fermat's principle). In the geometrical optics limit, where diffraction effects are absent, the phase length is infinitely large.

Now we turn to the propagation of particles. The path taken by particles in classical mechanics is determined by Newton's laws, but it turns out that these are equivalent to *Hamilton's principle*, that particles select paths between two points such that the *action* associated with the path is a minimum. The equivalence of these two approaches is demonstrated in Appendix 2: here we are interested in the striking analogy between Fermat's principle of least time and Hamilton's principle of least action.

The formal definition of action is given in Appendix 2, where it is seen to be an integral taken along the path of the particle, just like the phase length in optics. When we turn to the question as to why particles adopt the path involving least action, we can hardly avoid the conclusion that the reason must be the same as why light adopts path of least time, or, as we refined it, of least phase length. But in order to apply that argument to particles we have to suppose that particles are associated with a wave motion. You can see that this attempt to 'explain' classical mechanics leads almost unavoidably to the heart of quantum mechanics: we have experimental evidence to encourage us; Hamilton did not.

The hypothesis we now make is that a particle is described by some kind of amplitude Ψ, and amplitudes at different points are related by an expression of the form $\Psi(P_2) = e^{i\phi}\Psi(P_1)$. By analogy with the optical case we say that the wave is propagated along the path that makes ϕ a minimum. But we also know that in the classical limit the particle propagates along the path that makes the action, S, a minimum. We therefore suppose that ϕ is proportional to S. As ϕ is dimensionless (it has to be, if it is to appear as an exponent, because $e^{i\phi} = 1 + i\phi + \ldots$) the constant of proportionality must have the dimensions of 1/(action). Furthermore, we have seen that geometrical optics, diffraction-free optics, corresponds to the limit of short wavelengths, or large phase lengths. In classical mechanics, which is diffraction-free mechanics, particles travel along 'geometrical' trajectories. Therefore ϕ must be large; hence the quantity with the dimensions of action must be small. The natural quantity to introduce is Planck's constant, or some small multiple of it. It turns out that agreement with experiment (i.e., the correct form of the Schrödinger equation) is obtained if we use $\hbar$; we therefore conclude that we should write $\phi = S/\hbar$.

You should notice the relation between this approach and Heisenberg's. In his, a 0 was replaced by $\hbar$, and classical mechanics turned into quantum mechanics. In this a 0 has also been replaced by $\hbar$, for had we wanted *precise* geometrical trajectories (as in classical mechanics) we would require infinitely large phase lengths ϕ, which could be achieved by dividing S by zero. Quantum mechanics emerges when we replace that zero by a small quantity.

We have arrived at the stage where the amplitude describing a particle is described by relations of the form

$$\Psi(\mathrm{P}_2) = \mathrm{e}^{\mathrm{i}S/\hbar}\Psi(\mathrm{P}_1), \tag{2.2.4}$$

where S is the action associated with the path from P_1 (a point at x_1, t_1) to P_2 (at x_2, t_2). This form lets us develop an equation of motion, because we can differentiate with respect to the time t_2:

$$(\partial/\partial t_2)\Psi(x_2, t_2) = (\mathrm{i}/\hbar)(\partial S/\partial t_2)\mathrm{e}^{\mathrm{i}S/\hbar}\Psi(x_1, t_1) = (\mathrm{i}/\hbar)(\partial S/\partial t_2)\Psi(x_2, t_2).$$

One of the results derived in Appendix 2 is that the rate of change of the action is equal to $-E$, E being the total energy $T+V$:

$$(\partial S/\partial t) = -E.$$

Therefore the equation of motion (dropping the subscript 2 from x and t) is

$$(\partial/\partial t)\Psi(x, t) = -(\mathrm{i}/\hbar)E\Psi(x, t).$$

The final step involves combining this expression for E with the one derived in the preceding section, which instructed us to interpret E as the hamiltonian operator H, eqn (2.1.10). When $E\Psi$ is replaced by $H\Psi$ we arrive at the

Time-dependent Schrödinger equation: $H\Psi = \mathrm{i}\hbar(\partial\Psi/\partial t)$. (2.2.5)

2.3 Schrödinger's equation

There are several points worth making about the Schrödinger equation before we start looking for its explicit solutions.

For the one-dimensional systems we have been considering the hamiltonian takes the form given in eqn (2.1.11), but it is quite easy to appreciate that in three dimensions it has the form

$$H = -(\hbar^2/2m)\left\{\frac{\partial^2}{\partial x^2} + \frac{\partial^2}{\partial y^2} + \frac{\partial^2}{\partial z^2}\right\} + V(x, y, z). \tag{2.3.1}$$

The appearance of this expression is normally simplified by introducing the

Laplacian: $\nabla^2 = (\partial^2/\partial x^2) + (\partial^2/\partial y^2) + (\partial^2/\partial z^2)$. (2.3.2)

(∇^2 is read 'del-squared', and named after the French mathematician and physicist Laplace.) With this symbol the hamiltonian is written

$$H = -(\hbar^2/2m)\nabla^2 + V. \tag{2.3.3}$$

Example. Write the hamiltonians for (a) the hydrogen atom, (b) the hydrogen molecule, assuming the nuclei to be stationary.

- *Method.* Express the kinetic energy of each electron as $-(\hbar^2/2m_e)\nabla^2$ and the various contributions to the potential energy as $q_1q_2/4\pi\varepsilon_0 r_{12}$, where q_1, q_2 are the charges (e.g. $-e$ for an electron, e for a proton) and r_{12} their separation.

- *Answer*

(a) $H(\text{hydrogen atom}) = -(\hbar^2/2m_e)\nabla^2 - e^2/4\pi\varepsilon_0 r$

(b) $H(\text{H}_2\ \text{molecule}) = -(\hbar^2/2m_e)\nabla_1^2 - (\hbar^2/2m_e)\nabla_2^2$

$$-(e^2/4\pi\varepsilon_0)\left\{\frac{1}{r_{1a}} + \frac{1}{r_{1b}} + \frac{1}{r_{2a}} + \frac{1}{r_{2b}} - \frac{1}{r_{12}} - \frac{1}{R}\right\}$$

where $r_{i\nu}$ is the separation of electron i from nucleus ν, r_{12} is the electron–electron distance, R is the proton–proton distance, and ∇_i^2 is the laplacian for electron i.

- *Comment.* The hydrogen-molecule hamiltonian (a fearsome operator even for this simple system) gives rise to a second-order differential equation in six variables $(x_1, y_1, \ldots, z_2)$ and one parameter (R): quantum chemistry is largely the search for methods of finding approximate solutions, because it is hopeless to expect to be able to find analytical solutions.

We have attempted to put the Schrödinger equation into a physically plausible setting by showing how a modification of classical mechanics (the replacement of a 0 by a small quantity which went unnoticed until experimental techniques were refined during the nineteenth century) lead to an interpretation of $T+V$ as an operator on a function, and how the time dependence of the function can be related to $T+V$ by exploring the analogies between classical optics and classical mechanics. We should therefore not be surprised if the derivation has not captured some purely quantum mechanical properties having no classical analogue. We shall see later that the property of electron 'spin' has been missed because, even though it has an evocative name, it has no classical counterpart.

A further point is that the derivation of the equation has been entirely non-relativistic. A glance at eqn (2.2.5) is enough to show that it does not conform to the requirements of relativity, the symmetrical treatment of space and time: whereas second derivatives with respect to space variables appear, it has first derivatives with respect to time. This dissymmetry has to be removed if quantum theory is to be compatible with relativity, and that was the contribution of Dirac. The *Dirac equation* is essentially a symmetrized Schrödinger equation, which treats space and time even-handedly.

Finally, the Schrödinger equation is not a wave equation. A wave equation has a second derivative with respect to time (see the first equation in Appendix 1, for example); the Schrödinger equation has a first derivative, and is therefore a kind of diffusion equation. There is perhaps some intuitive satisfaction in the notion that the solutions of the basic equation of mechanics evolve by diffusing. Of course, the Schrödinger equation is often referred to as a wave equation, and we have been using the term 'wavefunction' for its solutions. There is a sense in which this is valid, as may be seen by considering the form of the equation when the potential energy is independent of time. The one-dimensional form of the equation then has the form

$$-(\hbar/2m)(\partial^2\Psi/\partial x^2)+V(x)\Psi=\mathrm{i}\hbar(\partial\Psi/\partial t). \tag{2.3.4}$$

Equations of this kind can be solved by the technique called *separation of variables* in which a trial solution takes the form

$$\Psi(x,t)=\psi(x)\theta(t). \tag{2.3.5}$$

When this substitution is made, and the resulting equation is divided through by $\psi\theta$ one obtains

$$-(\hbar^2/2m)\left(\frac{1}{\psi}\frac{\mathrm{d}^2\psi}{\mathrm{d}x^2}\right)+V(x)=\mathrm{i}\hbar\frac{1}{\theta}\left(\frac{\mathrm{d}\theta}{\mathrm{d}t}\right).$$

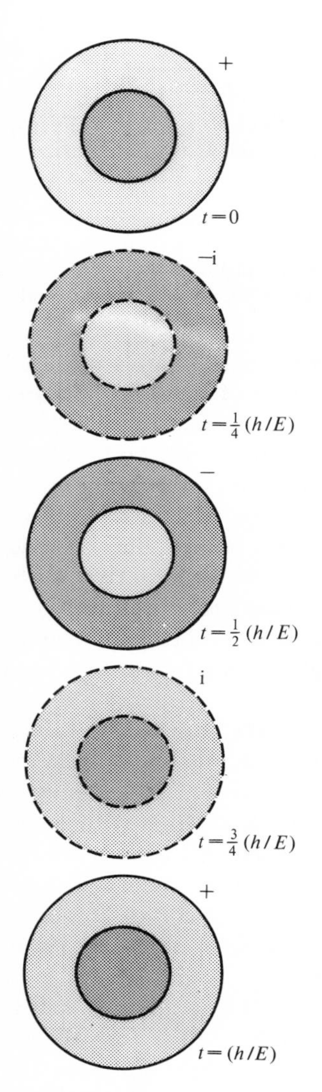

Fig. 2.6. The modulation of the phase of a wavefunction.

The left-hand side of a function of x alone, and so when x changes only the left can change. But the left is equal to the right, which does not change when x changes. Therefore it must itself be equal to a constant. The right-hand side is equal to the left, and so it is equal to the same constant. The dimensions of the constant are those of energy (V on the left is an energy), and so it is natural to denote it E. In other words, the time-dependent equation separates into two equations:

$$-(\hbar^2/2m)(\mathrm{d}^2\psi/\mathrm{d}x^2)+V(x)\psi=E\psi, \tag{2.3.6a}$$

$$\mathrm{i}\hbar\,\mathrm{d}\theta/\mathrm{d}t=E\theta. \tag{2.3.6b}$$

The second of these equations has the solution

$$\theta(t)=C\mathrm{e}^{-\mathrm{i}Et/\hbar},\qquad C\text{ a constant,} \tag{2.3.7}$$

and so the complete wavefunction, $\psi\theta$, has the form

$$\Psi(x,t)=\psi(x)\mathrm{e}^{-\mathrm{i}Et/\hbar} \tag{2.3.8}$$

where the constant C has been absorbed into ψ.

This calculation stimulates several remarks. Equation (2.3.6a) has the form of a *standing-wave equation:* therefore, so long as we are interested only in the spatial dependence of the wavefunction it is perfectly proper to regard the Schrödinger equation as a wave equation. Secondly, when the potential energy of the system does not change with time (so that the separation of variables procedure works) and the system is in a state of definite energy E, it is a trivial matter to construct the time-dependent form of the wavefunction: merely form the product $\psi(x)\mathrm{e}^{-\mathrm{i}Et/\hbar}$. The time-dependence of such systems is simply a modulation of the phase of the wavefunction: $\mathrm{e}^{-\mathrm{i}Et/\hbar}=\cos(Et/\hbar)-\mathrm{i}\sin(Et/\hbar)$ oscillates periodically from 1 to $-\mathrm{i}$ to -1 to i and back to 1 with a frequency $E/\hbar$ and period h/E. This is depicted in Fig. 2.6. Therefore, if you want to imagine the full time dependence of a wavefunction, think of it as flickering from positive to negative amplitude with a frequency determined by its energy. If a 1s-orbital of a hydrogen atom is thought of as lying at zero energy, its phase remains constant; the 2s-orbital lies 1.6×10^{-18} J higher in energy (a result from spectroscopy or eqn (1.5.3)), and so its phase oscillates at 1.6×10^{16} rad s^{-1}, or 2.5×10^{15} Hz.

Although the phase of wavefunctions Ψ of definite energy E oscillates, the product $\Psi^*\Psi=|\Psi|^2$ remains constant:

$$\Psi^*\Psi=(\psi^*e^{iEt/\hbar})(\psi e^{-iEt/\hbar})=\psi^*\psi. \tag{2.3.9}$$

Since, as we shall see, the product $\Psi^*\Psi$ plays a crucial role in the interpretation of quantum mechanics, the time-independence of $\Psi^*\Psi$ is an important feature. States having wavefunctions for which $\Psi^*\Psi$ is independent of time are called *stationary states*. From what we have seen so far, it follows that systems with definite energies and in which the potential energy does not depend on the time are in stationary states: their wavefunctions flicker from one phase to another but $\Psi^*\Psi$ is stationary. This constancy will turn out to be the true significance of Bohr's 'stationary orbits' of the hydrogen atom.

From now until Chapter 8 we shall concentrate on time-independent potentials. Solving the Schrödinger equation therefore amounts to solving eqn (2.3.6a), because we can always graft on the phase factor corresponding to the time-dependent component of the function.

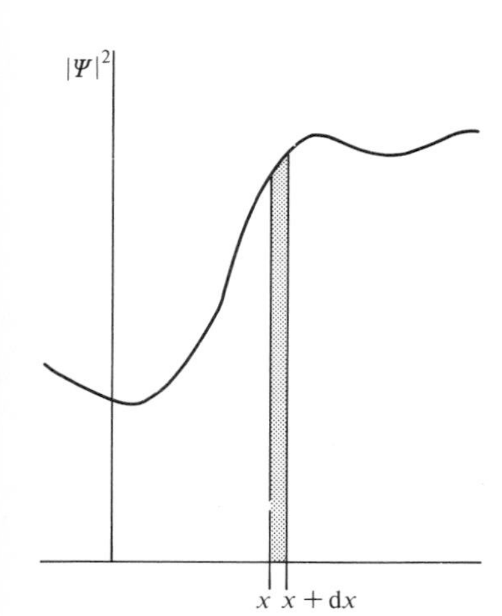

Fig. 2.7. The basis of the Born interpretation.

2.4 Born's interpretation

The basis of the interpretation of the significance of Ψ is a suggestion made by Born in 1926 that $\Psi^*\Psi$ should be regarded as a *probability density*. This means the following. In the case of a one-dimensional system the particle is free to be found anywhere along a line. Divide this line into infinitesimal segments of length dx, Fig. 2.7. Then *the probability that at a time t the particle is in the segment between x and* $x+dx$ *is* $\Psi^*(x,t)\Psi(x,t)\,dx$. Note that $\Psi^*\Psi$ is not itself a probability: it is a probability *density* in the sense that it gives a probability when it is multiplied by the volume of the region of interest (just as ordinary mass density gives a mass when it is multiplied by a volume). In three dimensions the wavefunction depends on the coordinates x, y, and z, or collectively $\mathbf{r}$, and Born's interpretation of $\Psi(\mathbf{r},t)$ is that $\Psi^*(\mathbf{r},t)\Psi(\mathbf{r},t)\,d\tau$ *is the probability of the particle being in the infinitesimal volume* $d\tau$ *at the point* $\mathbf{r}$ *at time* t, Fig. 2.8.†

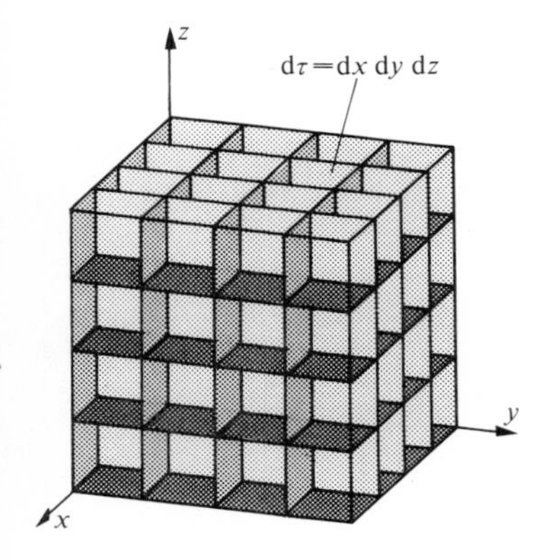

Fig. 2.8. The Born interpretation in three dimensions; $d\tau$ is the volume element.

Example. The wavefunction for the ground state of the hydrogen atom is $\psi(r)=(1/\pi a_0^3)^{\frac{1}{2}}e^{-r/a_0}$ where $a_0=53$ pm. (a) Where is the most probable location

† A more precise statement, which gets away from worries about the nature of differentials and their relation to volume elements, is that the probability of the particle being in the volume δV is the integral $\int_{\delta V}\Psi^*\Psi\,d\tau$, where the integration is over δV (and which may itself be virtually infinitesimal). We shall always use the more succinct $\Psi^*\Psi\,d\tau$ form, but it is only shorthand for this more precise form, and the latter should be returned to when worries arise.

of the electron? (b) What is the probability of finding the electron in a volume $1\ \mathrm{pm}^3$ at (i) $r=0$, (ii) $r=a_0$? (c) What is the probability of finding the electron anywhere in a sphere of radius a_0 centred on the nucleus?

- *Method.* (a) The most probable point at which the electron will be found is at the maximum of the probability density $\psi^2(r)$. (b) Evaluate $\psi^2(0)\,\delta V$ with $\delta V = 1\ \mathrm{pm}^3$, since δV is almost infinitesimal (on the scale of the atom) and ψ^2 is virtually constant over its extent. (c) Evaluate the integral $\int \psi^2(r)\,\mathrm{d}\tau$ from $r=0$ to $r=a_0$ and over all angles. The volume element in polar coordinates is $\mathrm{d}\tau = r^2 \sin\theta\,\mathrm{d}\theta\,\mathrm{d}\phi$ with $0 \le \theta \le \pi$, $0 \le \phi \le 2\pi$.

- *Answer.*

(a) $\psi^2(r) = (1/\pi a_0^3)\mathrm{e}^{-2r/a_0}$ is a maximum at $r=0$.

(b) $\psi^2(0)\,\delta V = (1/\pi a_0^3)\,\delta V = 2.14 \times 10^{-6}$;
$\psi^2(a_0)\,\delta V = (1/\pi a_0^3\mathrm{e}^2)\,\delta V = 2.89 \times 10^{-7}$.

(c)
$$\int_0^{a_0} \mathrm{d}r \int_0^{2\pi} \mathrm{d}\phi \int_0^{\pi} \mathrm{d}\theta r^2 (1/\pi a_0^3)\mathrm{e}^{-2r/a_0} \sin\theta = (4/a_0^3)\int_0^{a_0} r^2 \mathrm{e}^{-2r/a_0}\,\mathrm{d}r$$
$$= 1 - 5/\mathrm{e}^2 = 0.323.$$

- *Comment.* Go on to show that the probability of finding the electron within a sphere of radius R is $P(R) = 1 - \{1 + (2R/a_0) + 2(R/a_0)^2\}\mathrm{e}^{-2R/a_0}$. This has the value unity only at infinity, but has reached 0.90 at $R = 2.665a_0 = 141.0\ \mathrm{pm}$ and 0.999 99 at $R = 450\ \mathrm{pm}$. Therefore atoms, while formally infinite in size, are in practice minute.

Born was led to his interpretation by Einstein's correlation of the number of photons in a light beam with its intensity, and the fact that the intensity is proportional to the *squares* of the amplitudes of the electric and magnetic fields. It is an interpretation based on analogy, and is acceptable because it leads to experimentally verifiable conclusions, as we shall now see.

If we accept Born's interpretation as a working hypothesis we can make some powerful remarks about wavefunctions which lead, in the end, to the explanation of quantization of energy. For the interpretation to be tenable it must be the case that when the probability of finding the particle in each segment of the universe is summed over all the segments, the result is unity: the particle has to be somewhere, the probability of finding it somewhere is unity. The mathematical expression of this requirement is the *normalization condition*, that the integral $\int_{-\infty}^{\infty} \Psi^*(x,t)\Psi(x,t)\,\mathrm{d}x$ must exist, and be equal to unity. In three dimensions we write

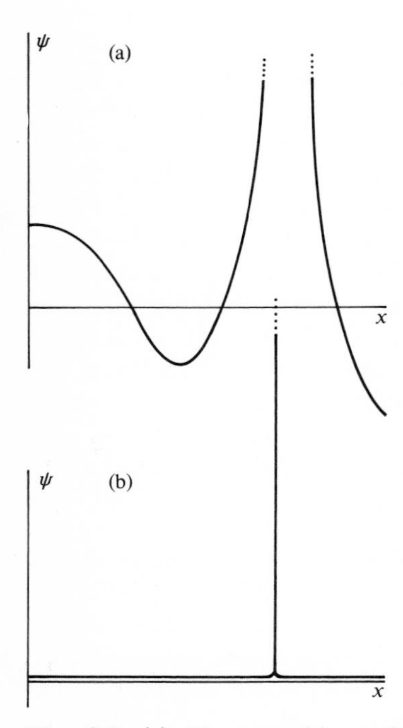

Fig. 2.9. (a) Unacceptable and (b) acceptable wavefunctions.

$$\textit{Normalization condition:}\ \int \Psi^*\Psi\,\mathrm{d}\tau = 1, \tag{2.4.1}$$

the integration being over all space.

Example. The wavefunction $\psi = Nx\mathrm{e}^{-\alpha r}$ describes the distribution of an electron in an excited state of the hydrogen atom. Find N such that it is normalized to unity.

• *Method.* Evaluate the integral in eqn (2.4.1) using spherical polar coordinates in which $x = r \sin\theta \cos\phi$ and $d\tau = r^2\, dr \sin\theta\, d\theta\, d\phi$, and choose N such that its value is unity. In spherical coordinates the entire space is spanned by $0 \leq r \leq \infty$, $0 \leq \theta \leq \pi$, and $0 \leq \phi \leq 2\pi$. Useful integrals, which appear frequently in calculations of this kind, are

$$\int_0^\infty x^n e^{-ax}\, dx = n!/a^{n+1}; \qquad \int_0^\pi \cos^n\theta \sin\theta\, d\theta = \{1 + (-1)^n\}/(n+1).$$

• *Answer.*

$$\begin{aligned}\int \psi^2\, d\tau &= N^2 \int x^2 e^{-2\alpha r}\, d\tau = N^2 \int_0^\infty dr \int_0^\pi d\theta \int_0^{2\pi} d\phi r^4 \sin^3\theta \cos^2\phi e^{-2\alpha r}\\ &= N^2 \int_0^\infty r^4 e^{-2\alpha r}\, dr \int_0^\pi (1 - \cos^2\theta)\sin\theta\, d\theta \int_0^{2\pi} \cos^2\phi\, d\phi\\ &= N^2\{4!/2^5\alpha^5\}\{2 - \tfrac{2}{3}\}\{\pi\} = N^2\pi/\alpha^5.\end{aligned}$$

Therefore $N = (\alpha^5/\pi)^{\frac{1}{2}}$.

• *Comment.* N is called a *normalization constant* for the wavefunction. A good source of tabulated integrals is I. S. Gradshteyn and I. W. Ryzhik, *Tables of integrals and products*, Academic Press (1965), as well as in M. Abramowitz and I. A. Stegun, *Handbook of mathematical functions*, Dover (1965).

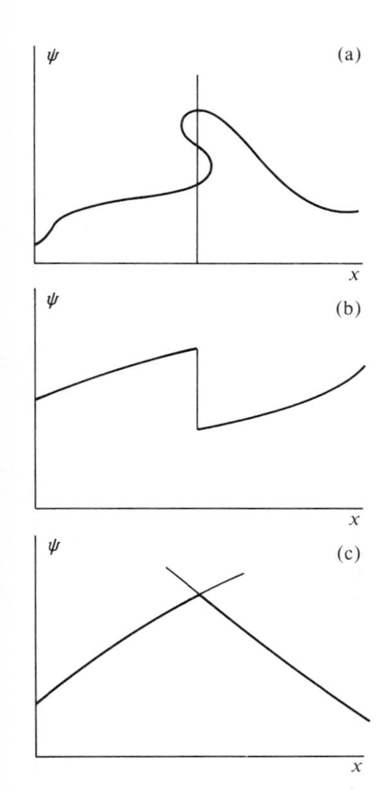

Fig. 2.10. Wavefunctions which are (a) multiple-valued, (b) discontinuous, (c) of discontinuous slope.

The implication of the normalization condition is that the wavefunction cannot become infinite over a finite region of space, as in Fig. 2.9(a), for if it did the integral over $\Psi^*\Psi$ would be infinite and the Born interpretation would be untenable. This does not rule out the possibility that the wavefunction could be infinite over an *infinitesimal* region of space, Fig. 2.9(b), because then the integral over $\Psi^*\Psi$ can be finite even though $\Psi^*\Psi$ is infinite (infinitely high × infinitely narrow = finite area in some cases). Such a wavefunction corresponds to a particle being localized at a single, precise point, like a speck of dust on a table at absolute zero.

Another implication of the Born interpretation is that if $\Psi^*\Psi$ is to tell us the probability of finding the particle, then it must be *single-valued.* It would be absurd if $\Psi^*\Psi$ could take more than one value at each point of space: the Born interpretation would be untenable. In simple applications the single-valuedness of $\Psi^*\Psi$ implies the single-valuedness of Ψ itself (complications arise when electron spin is incorporated into the wavefunction), and we shall normally impose the single-valuedness of Ψ as the appropriate condition.

There are two other constraints on the form that the wavefunction may take: these stem from the requirement that it is a solution of a second-order differential equation, and therefore that its second derivative should exist. In the first place, in order to define a second derivative it is necessary that the function itself should be continuous. Therefore Ψ *must be continuous,* Fig. 2.10. A weaker condition is that *its first derivative should also be continuous,* Fig. 2.10. This is weaker because there are systems, those with ill-behaved potentials, where the restriction is too severe. For example, when we deal with a particle in a box we encounter a potential that is excessively ill-behaved because it jumps from zero to infinity in an infinitesimal distance (when the particle touches the wall of the box). In such a case the wavefunction need not have a continuous first derivative. There are not many such pathological cases, but they are just common (and elementary) enough for them to be held in mind as a possibility.

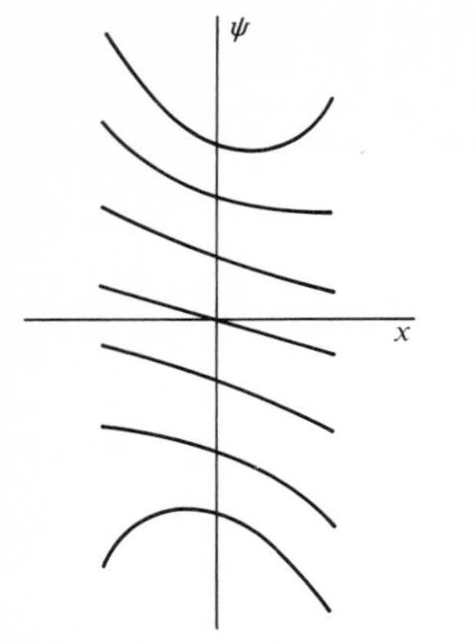

Fig. 2.11. The relation between curvature and amplitude.

2.5 Quantization

The time-independent Schrödinger equation, eqn (2.3.6a), is an equation for the curvature of the wavefunction. With this idea established it is possible to guess the form of its solutions even when the potential energy is complicated and, in particular, to see the reason why energy is quantized.

The curvature of a function is its second derivative.† A function with positive curvature looks like ⌣ and one with negative curvature looks like ⌢. The one-dimensional Schrödinger equation expresses the curvature as

$$d^2\psi/dx^2=(2m/\hbar^2)(V-E)\psi, \qquad (2.5.1)$$

and so if we know the values of $V-E$ and ψ at some point we can state the curvature there. In this section we concentrate on the qualitative features of the equation, because they show us how to unfold the form of the wavefunction without the clutter of detailed solutions.

First, note that *the curvature of ψ is proportional to its amplitude.* Therefore, for a given value of $|V-E|$, when ψ is large the curvature is large. Where ψ falls towards zero its curvature falls, Fig. 2.11. Where ψ is zero it is uncurved. Next, note that when E is greater than V the factor $V-E$ is negative, and so the sign of the curvature of ψ is opposite to the sign of ψ itself. That is, *if $E>V$, and ψ is positive, it has negative curvature*, and looks like ⌢. On the other hand, if E happens to be less than V, $V-E$ is positive, and the curvature has the same sign as the amplitude. A wavefunction with positive amplitude would then have positive curvature and look like ⌣. Finally, *the curvature is proportional to the difference* $|V-E|$: if the total energy is greatly in excess of the potential energy, the curvature is large. These points are collected in Fig. 2.12, which contains all the information we need to solve the Schrödinger equation qualitatively.

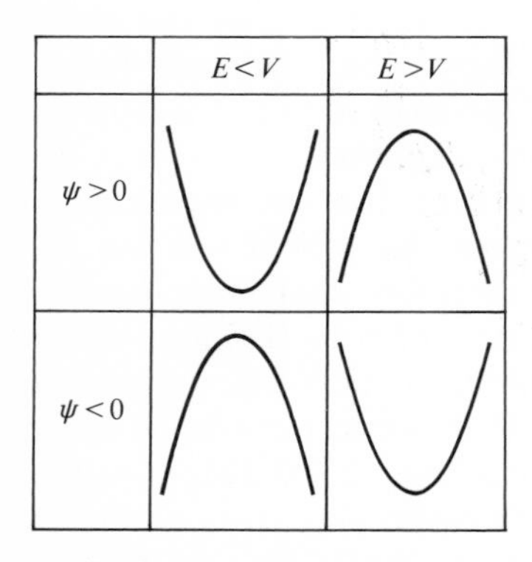

Fig. 2.12. The relation between curvature and $E-V$.

Consider a system in which the potential energy depends on position as depicted in Fig. 2.13. Suppose that at x'' the wavefunction has the amplitude and slope shown as A, and that the energy of the particle is E. Note that E exceeds V for positions to the left of x', but is less than V to the right of x': $E-V$ therefore changes sign at x'. ψ_A is positive at x'' and $V<E$; therefore its curvature is negative. ψ_A remains

† This use of 'curvature' is colloquial. In fact *curvature* is a precisely defined concept in the theory of surfaces; in one dimension the curvature of a function f is equal to $(d^2f/dx^2)/\{1+(df/dx)^2\}^{\frac{3}{2}}$. We shall invariably employ the colloquial meaning.

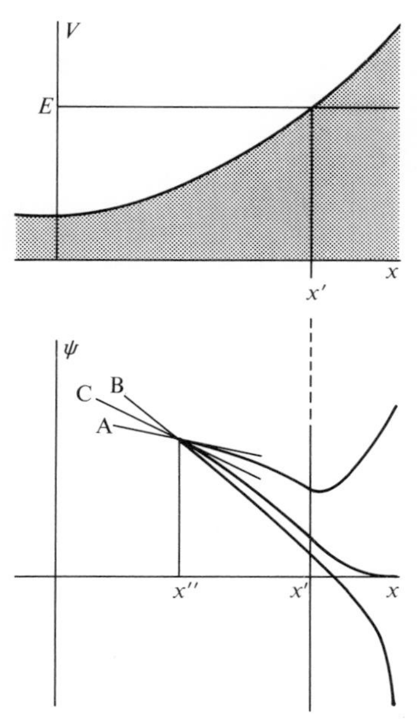

Fig. 2.13. Wavefunctions in a potential.

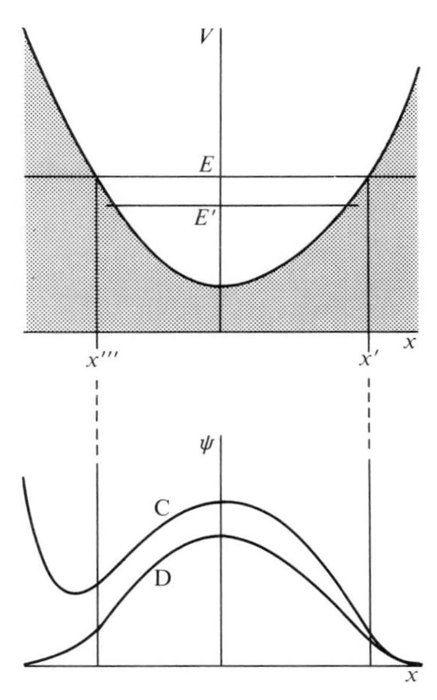

Fig. 2.14. Quantization in a bounded system.

positive at x', but after that point $V > E$. Its curvature therefore becomes positive, and it bends away from the x-axis and goes off to infinity as x tends to infinity. Therefore, according to the Born interpretation, ψ_A is an inadmissible function. With this failure in mind we select a function that has a different slope at x'' but the same magnitude. ψ_B starts with a negative curvature ($E > V$). It becomes positive at x' because ψ_B is still positive but now $V > E$. The change in curvature is insufficient to prevent ψ_B falling through zero to a negative value, and as it does so its curvature changes sign. This negative curvature forces ψ_B to a negatively infinite value as x tends to infinity, and is therefore inadmissible. Learning from our mistakes we now select a function ψ_C that has a slope intermediate between those of ψ_A and ψ_B. Its curvature changes sign at x', but does so in such a way that ψ_C approaches the x-axis asymptotically. As it does so its curvature declines (because ψ approaches zero) and it curls off to neither positive nor negative infinity. Such a function is acceptable. Note that, for a potential of the form shown in Fig. 2.13, a well-behaved function can be found for *any* value of E simply by modifying the slope or amplitude of the function at x''. Therefore the energy of such systems is not quantized.

Having seen the sensitivity of the wavefunction to a potential that rises to a large value on only one side, it is easy to appreciate the difficulty of fitting a function to a system in which the potential confines the particle at both sides. Such a potential is shown in Fig. 2.14. The function ψ_C has been traced to x''' where V begins to exceed E again. We see that its behaviour at the new boundary renders it inadmissible. In fact it may be impossible to find a well-behaved function for an arbitrary value of E. *Only for some values of the energy E is it possible to construct a well-behaved function.* One such function is ψ_D in Fig. 2.14 corresponding to the energy E'. In other words, *in a system with boundaries the energy is quantized.*

The importance of this conclusion cannot be overemphasized. The Schrödinger equation, being a differential equation, has an infinite number of solutions, and in particular has a solution for any value of E. The Born interpretation has the effect of imposing a restriction on the solutions. When the system has boundaries so that the particle is confined to some finite region, almost all the solutions are unacceptable—acceptable solutions occur only for special values of E. Hence, *energy quantization is a consequence of boundary conditions.*

In Fig. 2.15 we depict the effect of the boundaries on the quantization of the energy of a particle. Quantization occurs only when the particle is confined to a finite region of space. When its energy exceeds E^* it can escape to all positive values of x, and when it exceeds E^{**} it can travel indefinitely to positive and negative values. Furthermore, as the potential gets less confining, the separation between the quantized levels is reduced because it gets progressively easier to find energies which give well-behaved functions.

There is one final point before we proceed to the business of the

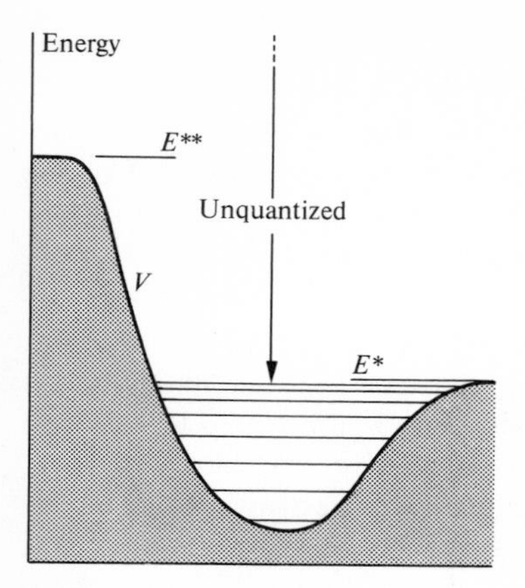

Fig. 2.15. Quantization and confinement.

book. This is to remark that a glance at Fig. 2.14 shows that the wavefunction may be non-zero even where $E < V$; that is, where classically the kinetic energy is negative. The non-vanishing probability that a particle may be found in classically forbidden regions is called *tunnelling*, and we shall encounter several examples in due course. The fact that the kinetic energy is negative in such regions is no particular cause for alarm. We shall see that (a) *observable* energies are *average values*, and the average kinetic energy, its average over the regions where it is positive and where it is negative, is always positive; and (b) any attempt to confine the particle into the non-classical region, and then to measure its kinetic energy would run into the uncertainty principle. That is, the confinement would have to be to such a small region that the corresponding uncertainty in the momentum (and therefore the kinetic energy) would be so great that we would be unable to conclude that it was negative.

Further reading

Classical mechanics background

Classical mechanics. H. Goldstein; McGraw-Hill, New York, 1950.
Quantum chemistry. W. Kauzmann; Academic Press, New York, 1957.

Matrix mechanics

Matrix mechanics. H. S. Green; Noordhoff, Groningen, 1965.

Minimum principles

Classical mechanics. H. Goldstein; McGraw-Hill, New York, 1950.
The Feynman lectures in physics. R. P. Feynman, R. B. Leighton, and M. Sands; Addison-Wesley, Reading, Mass., 1964.
Quantum mechanics and path integrals. R. P. Feynman and A. R. Hibbs; McGraw-Hill, New York, 1965.
Variation principles in dynamics and quantum theory. W. Yourgrau and S. Mandelstam; Pitman, London, 1955.
Classical mechanics, quantum mechanics, field theory. A. Katz; Academic Press, New York, 1965.

Properties and interpretation of the Schrödinger equation

Introduction to quantum mechanics. L. Pauling and E. B. Wilson; McGraw-Hill, New York, 1935.
Quantum theory. D. Bohm; Prentice-Hall, Englewood Cliffs, 1951.
Quantum mechanics. A. I. M. Rae; McGraw-Hill, London, 1981.
The principles of quantum mechanics. P. A. M. Dirac; Oxford University Press, 4th edition, 1958; reprinted 1982.

Problems

2.1. Find the operator for position x if the operator for momentum p is taken to be multiplication by p. *Hint.* Find an operator x such that eqn (2.1.5) is satisfied when momentum is multiplication by p.

2.2. Write the time-independent Schrödinger equations for (a) the hydrogen atom, (b) the helium atom, (c) the hydrogen molecule, (d) a free particle, (e) a particle subjected to a constant, uniform force.

2.3. Explore the concept of phase length as follows. First, consider two points p_1, p_2 separated by a distance l and let the paths taken by waves of wavelength λ be a straight line from p_1 to a point a distance d above the midpoint of the line p_1p_2, and then on to p_2. Find an expression for the phase length and sketch it as a function of d for various values of λ. Confirm explicitly that $\phi' = 0$ at $d = 0$.

2.4. Confirm that the path of minimum phase length for light passing from one medium to another is refracted at their interface in accord with Snell's law.

2.5. We have seen that the time-dependent Schrödinger equation is separable when V is independent of time. Show that it is also separable when V is a function only of time (and is uniform in space; $V = V(t)$) and solve the pair of equations. Let $V(t) = V \cos \omega t$; find an expression for $\Psi(x, t)$ in terms of $\Psi(x, 0)$. Is $\Psi(x, t)$ stationary in the sense of eqn (2.3.9)?

2.6. Show that if the Schrödinger equation had the form of a true wave equation, the integrated probability would be time-dependent. *Hint.* A wave equation has $\kappa\, \partial^2/\partial t^2$ in place of $\partial/\partial t$, where κ is some constant with the appropriate dimensions (what are they?). Solve the time component of the separable equation and investigate the behaviour of $\int \Psi^* \Psi \, d\tau$.

2.7. The ground-state wavefunction of the hydrogen atom has the form $\psi(r) = N e^{-ar}$, $1/a$ being a collection of fundamental constants with the magnitude 53 pm. Normalize this spherically symmetrical function. *Hint.* The volume element is $d\tau = \sin\theta \, d\theta \, d\phi \, r^2 dr$, with $0 \leq \theta \leq \pi$, $0 \leq \phi \leq 2\pi$, $0 \leq r \leq \infty$. 'Normalize' always means 'normalize to unity'.

2.8. A particle in an infinite one-dimensional system was described by the wavefunction $\psi(x) = N e^{-x^2/2\Gamma^2}$. Normalize this function. Calculate the probability of finding the particle in the range $-\Gamma \leq x \leq \Gamma$. *Hint.* The integral encountered in the second part is the error function. It is defined and tabulated in M. Abramowitz and I. A. Stegun, *Handbook of mathematical functions*, Dover (1965).

2.9. An excited state of the same system is described by the wavefunction $\psi(x) = N x e^{-x^2/2\Gamma^2}$. What is the most probable location of the particle?

2.10. On the basis of the information in Problem 2.7, calculate the probability density of finding the electron (a) at the nucleus, (b) at a point in space 53 pm from the nucleus. Calculate the probabilities of finding the electron inside a 1 pm^3 volume element located at these points. Calculate the probability of the electron being found anywhere within a sphere of radius 53 pm. If the radius of the atom is defined as the radius of the sphere inside which there is a 90 per cent probability of finding the electron, what is the atom's radius?

3 Exact solutions: linear motion

IN THIS chapter we consider the quantum mechanics of translation and vibration. Both types of motion can be solved exactly, and both are important not only in their own right but because together with the problems treated in the next chapter they form a basis for the description of the more complicated types of motion encountered in quantum chemistry, such as the motion of electrons near nuclei and the rotations and vibrations of molecules. Translational motion also has the advantage of introducing in a simple way many of the striking features of quantum mechanics, especially its differences from classical mechanics and the way it blends into and accounts for classical behaviour when macroscopic objects are involved.

3.1 Translational motion

The easiest type of motion to consider is that of a completely free particle travelling in one dimension. Since the potential energy is constant and may be chosen to be zero, the hamiltonian operator for the system is

$$H = -(\hbar^2/2m)(\mathrm{d}^2/\mathrm{d}x^2). \tag{3.1.1}$$

The time-independent Schrödinger equation, $H\psi = E\psi$, therefore has the explicit form

$$-(\hbar^2/2m)(\mathrm{d}^2\psi/\mathrm{d}x^2) = E\psi,$$

and its solutions are

$$\psi(x) = A\mathrm{e}^{\mathrm{i}kx} + B\mathrm{e}^{-\mathrm{i}kx}, \qquad k = (2mE/\hbar^2)^{\frac{1}{2}}, \tag{3.1.2}$$

as may readily be checked by substitution. Since $\mathrm{e}^{\mathrm{i}x} = \cos x + \mathrm{i}\sin x$, an alternative solution is

$$\psi(x) = C\cos kx + D\sin kx. \tag{3.1.3}$$

In both forms of the solutions the coefficients A, B, C, D are to be determined by the boundary conditions which we turn to shortly.

The first point about the solutions is that, as the motion is completely unconfined, the energy is unquantized. A solution can be found for any value of E: we simply use the appropriate value of k in eqn (3.1.2) or (3.1.3).

The classical connection between the energy of a free particle and its linear momentum p is $E = p^2/2m$. In the present case E is related to the parameter k by $E = k^2\hbar^2/2m$, and so the magnitude of the linear momentum of a particle described by the wavefunction in eqn (3.1.2) or (3.1.3) is

$$p = k\hbar. \tag{3.1.4}$$

This can be developed in a number of ways. For instance, we can turn it round, and say that the form of the wavefunction of a particle with linear momentum of magnitude p is given by eqn (3.1.2) with $k = p/\hbar$. A second point is that the wavefunctions in eqns (3.1.2) and (3.1.3) have a definite wavelength. This may be easier to see in the case of eqn (3.1.3) because a wave of wavelength λ is normally written as $\cos(2\pi x/\lambda)$ or as $\sin(2\pi x/\lambda)$, although $e^{2\pi ix/\lambda}$ and $e^{-2\pi ix/\lambda}$ would do just as well. The wavelength of the wavefunction in eqn (3.1.3) is clearly $\lambda = 2\pi/k$. In other words, the wavefunction of a particle with momentum $p = k\hbar$ has a wavelength $\lambda = 2\pi/k$, and so wavelength and linear momentum are related by

$$p = (2\pi/\lambda)\hbar = h/\lambda, \tag{3.1.5}$$

which is just the de Broglie relation.

A third point lets us understand the role of the coefficients A, B, etc. The Schrödinger equation has the form $H\psi = E\psi$. That is, when the hamiltonian operator operates it gives the same function back but multiplied by the energy:

(energy operator) × (function) = (energy value) × (same function).

Equations of this form, called *eigenvalue equations*, crop up throughout quantum mechanics, and we study them in Chapter 5. For the present we note that exactly the same kind of equation arises when we use the form of the operator for the linear momentum introduced in the last chapter, eqn (2.1.6). For example, set $B = 0$ in eqn (3.1.2), then

$$p_{\text{operator}}\psi = (\hbar/\mathrm{i})(\mathrm{d}/\mathrm{d}x)A\,\mathrm{e}^{\mathrm{i}kx} = k\hbar A\,\mathrm{e}^{\mathrm{i}kx} = k\hbar\psi.$$

We have already established that $k\hbar$ is the magnitude of the linear momentum, hence the last expression has the form of an eigenvalue equation. But suppose we had set $A = 0$: what then? We find

$$p_{\text{operator}}\psi = (\hbar/\mathrm{i})(\mathrm{d}/\mathrm{d}x)B\mathrm{e}^{-\mathrm{i}kx} = -k\hbar B\mathrm{e}^{-\mathrm{i}kx} = -k\hbar\psi.$$

The distinction between the two results is the sign. Since linear momentum is a vector quantity we are immediately led to the conclusion that *the two wavefunctions correspond to states of the particle with the same magnitude of linear momentum but opposite directions*. This is a very important point, for it lets us write down the wavefunctions for particles that not only have a definite kinetic energy and therefore magnitude of momentum, but for which we can also specify direction of travel, Fig. 3.1.

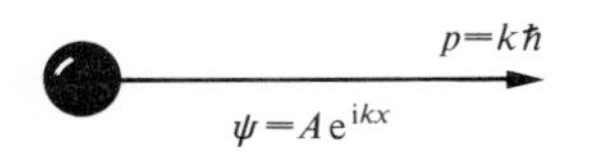

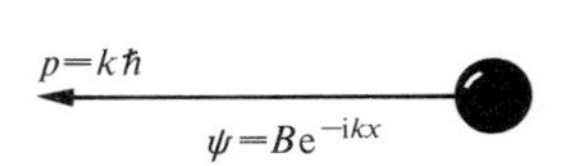

Fig. 3.1. Wavefunctions for linear motion.

The significance of the coefficients A and B should now be clearer: they depend on how the state of the particle was prepared. If it was shot from a gun in the direction of positive x, then $B = 0$. If it had been shot in the opposite direction (by the duelling partner) its state would be described by eqn (3.1.2) with $A=0$. (The value of the non-zero coefficient, A and B respectively, is determined by the normalization condition: we shall put that question off until later.†)

What, though, is the significance of the coefficients C and D in the alternative form of the solution? Suppose $D=0$ so that the particle is described by the wavefunction $C\cos kx$. When we examine the effect of the momentum operator we find

$$p_{\text{operator}}\psi=(\hbar/\mathrm{i})(\mathrm{d}/\mathrm{d}x)C\cos kx=-(k\hbar/\mathrm{i})C\sin kx.$$

This does not have the form of an eigenvalue equation (the function on the right is not the same as the one on the left), and so we cannot conclude anything about the linear momentum. We do know from the earlier remarks that the magnitude of the momentum is $k\hbar$. Therefore the only thing we are ignorant about is the direction of the motion. In other words, a wavefunction $C\cos kx$, although it corresponds to a definite magnitude of momentum, does not correspond to a definite direction of motion. Another way of seeing this is to expand the cosine function as follows:

$$\psi=C\cos kx=\tfrac{1}{2}C\mathrm{e}^{\mathrm{i}kx}+\tfrac{1}{2}C\mathrm{e}^{-\mathrm{i}kx}.$$

This shows that it is a *superposition* of two states, one corresponding to motion to the right, the other to motion to the left. In other words $C\cos kx$ is a *standing wave*. Similar remarks apply to $\sin kx$.

There are several important points attached to this result. One general feature that can be distilled (and proved in Chapter 5) is that *complex wavefunctions* (such as $\mathrm{e}^{\mathrm{i}kx}$) correspond to a definite state of linear motion, while *real wavefunctions* (such as $\cos kx$) do not. A second point is that suppose we attempted to measure the direction of motion of a particle in a state with wavefunction $\psi=C\cos kx$, what would be the outcome? Since $C\cos kx$ is a superposition of e^{ikx} and e^{-ikx} with equal weight, we should obtain 'motion to the right' as the outcome in half the observations, and 'motion to the left' in the other half. We are totally incapable, however, of predicting which outcome we should expect in a given experiment: we can only predict average outcomes over many trials. This is the basic *indeterminance* of quantum mechanics, an aspect that we shall see in more detail later.

† The normalization condition is a little tricky. This is because $\psi^*\psi=A^2$, and so the normalization integral, the integral of the constant A^2 over all space, diverges. Disaster is avoided by the device of pretending that space extends over some finite region from $x=-\tfrac{1}{2}L$ to $x=\tfrac{1}{2}L$. Then $\int\psi^*\psi\,\mathrm{d}\tau=\int_{-\frac{1}{2}L}^{\frac{1}{2}L}A^2\,\mathrm{d}x=A^2L$; which implies that $A=1/L^{\frac{1}{2}}$. It is always found that when L is allowed to approach infinity at the end of the calculation of some observable quantity, that the result becomes independent of its value. For example, suppose we have a beam of N particles, each one travelling to the right with momentum $k\hbar$. The number of particles in the region between x and $x+\mathrm{d}x$ is $N\times$(probability density for one particle)$\times\mathrm{d}x=N\psi^*\psi\,\mathrm{d}x=NA^2\,\mathrm{d}x$. But since $A=1/L^{\frac{1}{2}}$, the number is $(N/L)\,\mathrm{d}x$. The ratio N/L is the number density of particles (ρ, the number per unit length), a quantity that remains constant as L increases (because lengthening the line captures more particles, and so N increases as L increases). Therefore the number of particles in the range x to $x+\mathrm{d}x$ is $\rho\,\mathrm{d}x$, a result independent of L (and of the value of x), and corresponding to a uniform distribution of particles in the beam.

So far we have considered the case when the energy of the particle is specified exactly. But suppose the particle had been prepared, as in real life, with an imprecisely specified energy. What is the form of its wavefunction? Since the energy is imprecise, the wavefunction has to be expressed as a superposition of wavefunctions, each one corresponding to a different energy. That is, we form a *wavepacket*. For example, suppose the particle corresponds to a bullet fired towards positive x, then we know that its wavefunction must be a superposition of functions of the form e^{ikx} with a range of values of k corresponding to the range of momenta (or kinetic energies) over which the preparation of the state is uncertain (e.g. uncertainties in the explosion characteristics of the gunpowder). We know that bullets appear to move, and this is a signal that we are dealing with time-dependent behaviour. Therefore we need to express the superposition in terms of time-dependent wavefunctions.

We have already seen how to add the time-dependent phase factor to wavefunctions of definite energy (p. 30). In the present case we know that a wavefunction e^{ikx} corresponds to an energy $E = k^2\hbar^2/2m$, and so its full, time-dependent form is

$$\Psi_k(x, t) = Ae^{ikx}e^{-iEt/\hbar} = Ae^{ikx - ik^2\hbar t/2m}. \qquad (3.1.6)$$

The superposition corresponding to the wavepacket prepared when the trigger was pulled is a sum of such functions, each one weighted with some coefficient, which we write $g(k)$, called the *shape function* of the packet. Since k is a continuously variable parameter the sum is really an integral over k, and so the wavepacket has the mathematical form

$$\Psi = \int g(k)\Psi_k(x, t)\,dk. \qquad (3.1.7)$$

The pictorial form of such a packet is shown in Fig. 3.2(a). You can see that on account of the interference between the waves, at one instant it has a strong amplitude at one point of space (inside the barrel), but because the time-dependent factor affects the phases of the waves forming the superposition, the region of constructive interference changes with time, Fig. 3.2(b). It is not hard to believe that the centre of the packet moves to the right (towards the target), and this is confirmed by the mathematical analysis of its motion given in Appendix 3 (the packet also spreads). The classical motion of a particle (such as a bullet) is captured by the motion of the wavepacket, and once again we see how classical mechanics emerges as an imprecise form of quantum mechanics. In the present case, the imprecision is in the specification of the initial conditions, the state of preparation, of the projectile. This aspect will recur throughout this chapter.

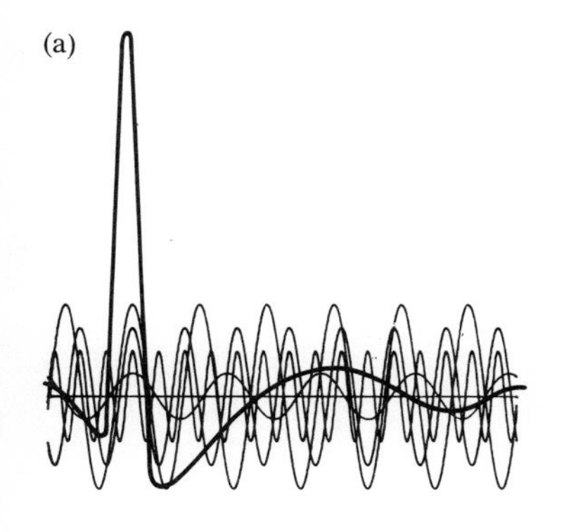

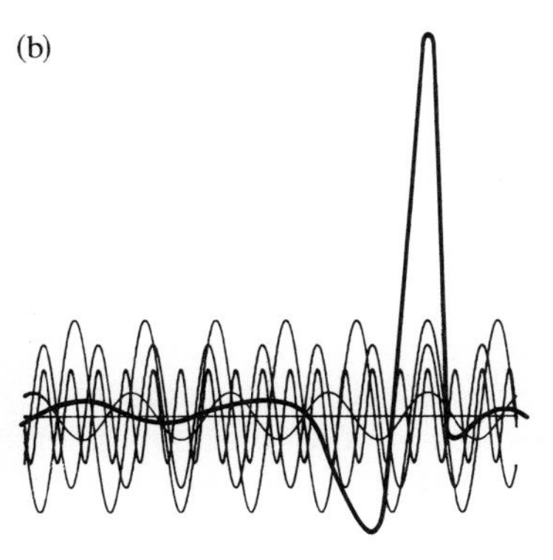

Fig. 3.2. A wavepacket (a) initially and (b) later.

3.2 Potential barriers and tunnelling

A highly instructive extension of the results for free translational motion is to the case where the potential energy varies as depicted in

Fig. 3.3(a). This can be regarded as an idealization of a particle penetrating a thin foil. Classically we know what happens: if the particle is shot from the left, it will penetrate only if its initial energy is greater than the potential energy it possesses when embedded in the foil, otherwise it will be reflected. But what does quantum mechanics predict?

The Schrödinger equation for the problem falls apart into three equations, one for each of the zones. The hamiltonians for each zone are

$$\text{Zone A:}\quad H=-(\hbar^2/2m)(\mathrm{d}^2/\mathrm{d}x^2),\qquad x<0 \tag{3.2.1a}$$

$$\text{Zone B:}\quad H=-(\hbar^2/2m)(\mathrm{d}^2/\mathrm{d}x^2)+V,\qquad 0\leq x\leq l \tag{3.2.1b}$$

$$\text{Zone C:}\quad H=-(\hbar^2/2m)(\mathrm{d}^2/\mathrm{d}x^2)\qquad x>l. \tag{3.2.1c}$$

The corresponding equations each have the form for free-particle motion (but with $E-V$ in place of E for Zone B). Therefore we can write down the solutions by referring to eqn (3.1.2):

$$\text{Zone A:}\quad \psi=A\mathrm{e}^{\mathrm{i}kx}+B\mathrm{e}^{-\mathrm{i}kx},\qquad k=(2mE/\hbar^2)^{\frac{1}{2}}, \tag{3.2.2a}$$

$$\text{Zone B:}\quad \psi=A'\mathrm{e}^{\mathrm{i}k'x}+B'\mathrm{e}^{-\mathrm{i}k'x},\qquad k'=\{2m(E-V)/\hbar^2\}^{\frac{1}{2}}, \tag{3.2.2b}$$

$$\text{Zone C:}\quad \psi=A''\mathrm{e}^{\mathrm{i}kx}+B''\mathrm{e}^{-\mathrm{i}kx},\qquad k=(2mE/\hbar^2)^{\frac{1}{2}}. \tag{3.2.2c}$$

We shall concentrate on the case when E is less than V, so that classically the particle cannot penetrate the barier. This has an immediate and important effect on the form of the wavefunction for Zone B, because when $E<V$, k' is imaginary. Therefore, if we write $k'=\mathrm{i}\kappa$ where κ is the real quantity $\{2m(V-E)/\hbar^2\}^{\frac{1}{2}}$, we find

$$\text{Zone B:}\quad \psi=A'\mathrm{e}^{-\kappa x}+B'\mathrm{e}^{\kappa x}, \tag{3.2.3}$$

which is a mixture of decaying and increasing exponential functions: *a wavefunction does not oscillate when* $E<V$.

If the barrier is infinitely wide (when the foil is armour plate) the increasing exponential must be ruled out because it leads to an infinite amplitude, which conflicts with the Born interpretation. Therefore inside a barrier like that shown in Fig. 3.3(b) the wavefunction must be simply a *decaying* exponential function $\mathrm{e}^{-\kappa x}$. The important point is that the amplitude is not zero inside the barrier, and so *the particle may be found inside a classically forbidden region.* This effect is called *penetration* of a barrier. The rapidity with which the wavefunction decays to zero depends sharply on the mass of the particle and the value of $V-E$, and macroscopic particles have wavefunctions that decay so quickly with distance that for all practical purposes they do not penetrate the classically forbidden zone. An electron or a proton, on the other hand, may penetrate a forbidden zone to an appreciable extent. For example, a 1 eV electron (an electron accelerated through a potential difference of 1 V and therefore having an energy $E=e\times(1\text{ V})=1.6\times10^{-19}\text{ C V}=1.6\times10^{-19}\text{ J}$) incident on a potential barrier equivalent to 2 V (and therefore to a potential energy 3.2×10^{-19} J

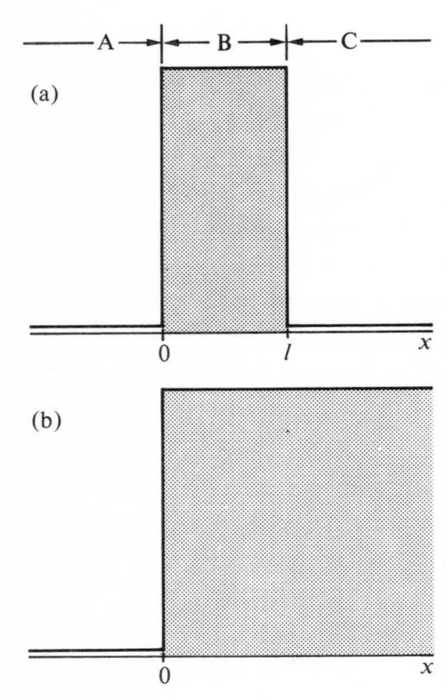

Fig. 3.3. (a) A potential barrier; (b) an infinitely thick wall.

when the electron is inside it) has a wavefunction that decays with distance as $e^{-5.12(x/\mathrm{nm})}$, and so it has decayed to $1/e$ of its initial value only after 0.20 nm, or about the diameter of an atom. Hence penetration can have very important effects on processes at surfaces (e.g. electrodes) and for all events on an atomic scale.

When the barrier is not of infinite width, the increasing exponential component of the Zone B wavefunction is not ruled out because it does not necessarily rise to an infinite amplitude before the potential has dropped back to zero again. Therefore the coefficient B' is not necessarily zero. The values of the coefficients are established by using the other criteria for acceptable wavefunctions set out in Chapter 2, and in particular the requirement that they and their slopes be continuous. This continuity condition lets us match the wavefunction at the points where the zones meet, and therefore to find values for the coefficients. For example, the continuity of the amplitudes at $x=0$ and at $x=1$ leads to the two conditions

$$A+B=A'+B'; \qquad A'e^{-\kappa l}+B'e^{\kappa l}=A''e^{ikl}+B''e^{-ikl}, \tag{3.2.4a}$$

and the continuity of the slopes at the same two points leads to

$$ikA-ikB=-\kappa A'+\kappa B'; \qquad -\kappa A'e^{\kappa l}+\kappa B'e^{\kappa l}=ikA''e^{ikl}-ikB''e^{ikl} \tag{3.2.4b}$$

which is almost enough to determine the six coefficients.

Consider the case where particles are prepared with a momentum carrying them towards the right. Then we can immediately infer that the coefficient B'' must be zero, because the wavefunction $B''e^{-ikx}$ denotes the presence of particles travelling to the left on the right-hand side of the barrier, and there can be no such particles. There may be particles on the left of the barrier travelling to the left, because the barrier reflects the incoming particles. We can therefore identify the coefficient B as determining (via $|B|^2$) the probability that the prepared state particles are *reflected* from the barrier, and the coefficient A'' as the probability (via $|A''|^2$) that the particle *penetrates* the barrier and emerges on the other side to travel there with momentum $k\hbar$ to the right. We shall use the relations between the coefficients established above to compute the ratio $|A''|^2/|A|^2=P$, the *transmission probability*, the probability that a particle incident on the left of the barrier emerges on the right.

The calculation involves elementary manipulations of the relations in eqn (3.2.4); the result is that

$$P=1/(1+G); \qquad G=\frac{(e^{\kappa l}-e^{-\kappa l})^2}{4(E/V)(1-E/V)}, \qquad \kappa=\{2m(V-E)/\hbar^2\}^{\frac{1}{2}}. \tag{3.2.5}$$

The important feature of this result is that P may be non-zero; that is, the particle may tunnel through the barrier, even though E is less than V. This phenomenon is an entirely quantum mechanical property, and depends on the fact that a wavefunction does not fall abruptly to zero inside a region where its potential energy exceeds its total energy. Its amplitude may be non-zero on the other side of the barrier, where it may burst into undamped oscillation again.

Tunnelling is sharply dependent on the mass of the particle and the height of the barrier relative to the incident energy. The behaviour of P for a proton and a deuteron is shown in Fig. 3.4. The diagram also shows how P depends on the energy when $E > V$ so that the particle would, according to classical mechanics, certainly be able to surmount the barrier and therefore have a transmission probability of unity. (The calculation carries through in the same way as before, but k' is no longer imaginary.) The interesting feature is that $P < 1$ in the quantum calculation, and even though the energy is sufficient to carry the particle over the barrier, there is a high probability that the particle will be reflected. Therefore, although quantum mechanics predicts an enhanced tunnelling when $E < V$, it predicts an enhanced reflection, a kind of *antitunnelling*, when $E > V$. This is reminiscent of the way that light reflects from an abrupt change of refractive index at the interface of two transparent media: the wave nature of the incident state is manifesting itself in the same way in the two cases.

Tunnelling is an important property of electrons, protons, and deuterons, and has a number of effects in chemistry. Although the potential energies then have a more complex position dependence than the one we have treated, the broad conclusions are the same: only light particles tunnel significantly, and then only if the barrier is neither too high nor too wide.

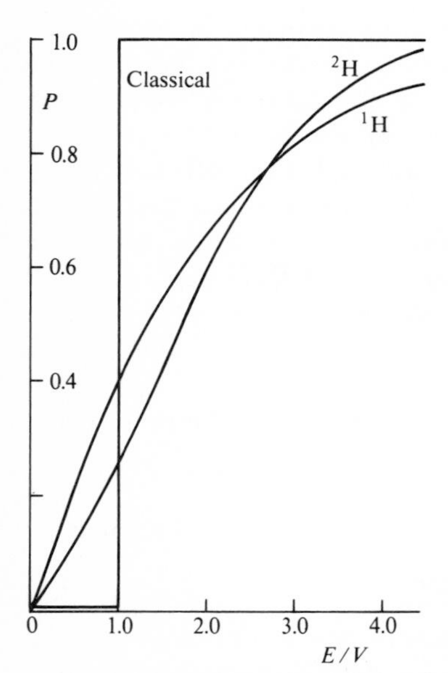

Fig. 3.4. Tunnelling probabilities for protons and deuterons ($V = 5$ eV, $l = 0.1$ nm).

3.3 Particle in a box

Now we turn to the case where the particle is confined by walls to a region of space of length L. The walls can be represented by a potential energy that is zero inside the region, but which rises sharply to infinity at the edges, Fig. 3.5. This is the *one-dimensional square well*, or the *particle in a box*, the squareness referring to the steepness with which the potential energy of the particle goes off to infinity at the edges of the box. Since the particle is confined, its energy is quantized. We shall see how the boundary conditions select the permissible energies.

The hamiltonian operator for the infinitely deep square well is

$$H = -(\hbar^2/2m)(d^2/dx^2) + V(x); \qquad (3.3.1)$$

$$V(x) = \begin{cases} 0 & \text{for} \quad 0 \leq x \leq L \\ \infty & \text{for} \quad x < 0 \text{ and } x > L. \end{cases}$$

Because the potential energy of a particle that touches the walls is infinitely great, it cannot penetrate them. This remark is justified by the behaviour of the wavefunction in the previous section where we saw that there is no penetration of an infinitely high barrier. It follows that the hamiltonian for the region where the potential energy is not infinite, and therefore the only region where the wavefunction is

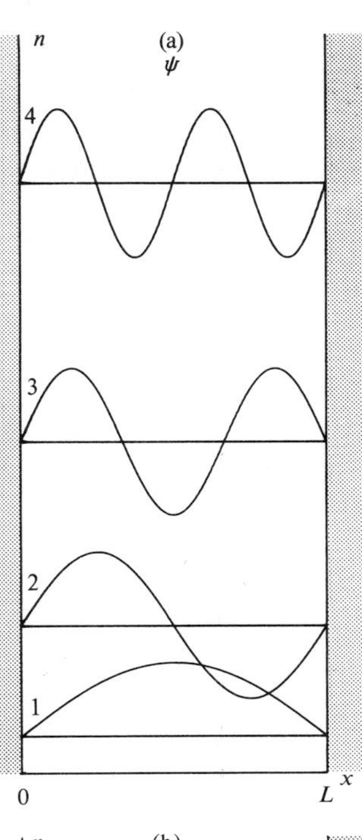

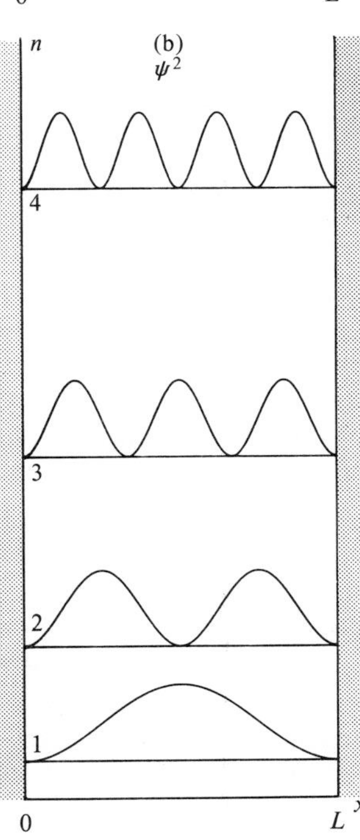

Fig. 3.5. (a) Wavefunctions and (b) probability densities for a particle in a box.

non-zero, is

$$H = -(\hbar^2/2m)(d^2/dx^2). \tag{3.3.2}$$

This is the same as the hamiltonian for free translational motion, and therefore we know at once that the solutions are of the form given in eqns (3.1.2) or (3.1.3). However, in the present case there are boundary conditions to satisfy, and these will have the effect of discarding most of the possible solutions.

The wavefunctions are zero within the walls. Wavefunctions are everywhere continuous. Therefore the wavefunctions are zero at the walls. The walls are located at $x=0$ and at $x=L$; therefore the boundary conditions to be satisfied are $\psi(0)=0$ and $\psi(L)=0$. We apply each in turn to the form of the solution in eqn (3.1.3):

$$\psi(x) = C\cos kx + D\sin kx, \qquad k = (2mE/\hbar^2)^{\frac{1}{2}}.$$

First, at $x=0$, since $\cos 0 = 1$ and $\sin 0 = 0$, $\psi(0) = C$. Therefore $C=0$ in order for $\psi(0)=0$. The wavefunction must also vanish at $x=L$, and so it must satisfy the condition

$$\psi(L) = D\sin kL = 0.$$

One way of achieving this is to set $D=0$ too, but that would imply that the wavefunction was zero everywhere, and so the particle would be found nowhere. The only alternative is to require that the sine function itself vanishes. This is the case if kL is an integral multiple of π. That is, we must permit k to take the values

$$k = n\pi/L, \qquad n = 1, 2, 3, \ldots \tag{3.3.3}$$

and no others. Since $E = k^2\hbar^2/2m$, this implies that the energy is confined to the values

$$E_n = n^2(\hbar^2\pi^2/2mL^2) = n^2(h^2/8mL^2), \qquad n = 1, 2, 3, \ldots. \tag{3.3.4}$$

There remains only the constant D to determine before the solution is complete. The probability of finding the particle somewhere within the box must be unity. Therefore the integral of $\psi_n^*\psi_n$ over the region $0 \le x \le L$ must be equal to 1:

$$1 = \int_0^L \psi_n^*(x)\psi_n(x)\,dx = \int_0^L D^2\sin^2 n\pi x/L\,dx = \tfrac{1}{2}LD^2.$$

Therefore $D = (2/L)^{\frac{1}{2}}$. The complete solution is therefore

$$\psi_n = (2/L)^{\frac{1}{2}}\sin(n\pi x/L),$$
$$E_n = n^2(h^2/8mL^2), \qquad n = 1, 2, \ldots. \tag{3.3.5}$$

The number n, which labels the wavefunction and the energy, is called a *quantum number.* When the quantum number is specified we can quote the corresponding energy and wavefunction using the formulas in eqn (3.3.5). Some of the wavefunctions and their energies are depicted in Fig. 3.5. The squares of the wavefunctions are also shown: these denote the probability densities for the location of the particle when it occupies the specified state. Note how it apparently tends to avoid the walls in the lower energy states, but is distributed increasingly uniformly as n increases. The general result, that classical behaviour emerges as n approaches infinity, is called the *correspondence principle.*

Example. Evaluate the probability of finding a particle in (a) the central third of a box and (b) the right-hand third. Show that these probabilities become equal as n increases.

- *Method.* The probability of finding the particle in the region $a \leq x \leq b$ is the integral of $\psi_n^2(x)$ between these limits. Use the integrals

$$\int \sin^2 nz \; dz = \tfrac{1}{2}z - (1/4n)\sin(2nz)$$

with (a) $a = \frac{1}{3}L$, $b = \frac{2}{3}L$, (b) $a = \frac{2}{3}L$, $b = L$ for the integration limits.

- *Answer.*

$$P_n = \int_a^b \psi_n^2(x)\,dx = (2/L)\int_a^b \sin^2(n\pi x/L)\,dx$$

$$= (2/\pi)\int_{\pi a/L}^{\pi b/L} \sin^2 nz\; dz = (b-a)/L - (1/2\pi n)\{\sin(2n\pi b/L) - \sin(2n\pi a/L)\}.$$

Therefore

(a) $P_n = \frac{1}{3} - (1/2\pi n)\{\sin(4n\pi/3) - \sin(2n\pi/3)\} = \frac{1}{3} - (-1)^n\left\{\dfrac{\sin(n\pi/3)}{\pi n}\right\},$

(b) $P_n = \frac{1}{3} - (1/2\pi n)\{\sin(2n\pi) - \sin(4n\pi/3)\} = \frac{1}{3} + (-1)^n\left\{\dfrac{\sin(n\pi/3)}{2\pi n}\right\}.$

The second term vanishes in each case as $n \to \infty$, and so $P_n(\text{a}) \sim P_n(\text{b})$.

- *Comment.* In the case $n = 1$, $P_1(\text{a}) = 0.609$ while $P_1(\text{b}) = 0.196$, reflecting a marked accumulation of the particle in the central zone. Note that $P_n(\text{a}) + 2P_n(\text{b}) = 1$: always check calculations for consistency.

The most striking property of the solutions is that, as a result of the boundary conditions, the energy is restricted to discrete values and cannot be varied continuously. That is, *the energy is quantized.* Note too that the minimum energy a particle can have when it is confined to a region of length L is $E_1 = h^2/8mL^2$. This minimum, irremoveable energy is called the *zero-point energy*. It is a purely quantum mechanical property, and in an imaginary universe in which $h = 0$ all the energy could be removed ($E_1 = 0$). The uncertainty principle gives some understanding of the reason for the existence of the zero-point energy, because we have seen that the range of uncertainty in the values of the linear momentum and the position of a particle are connected by an expression of the form $\delta p\, \delta x \geq \hbar/2$. Since the particle is within the box, the range of uncertainty in its position is of the order of L, the length of the box. The uncertainty in the momentum is therefore not less than about $\delta p \approx \hbar/L$. But saying that the momentum is exactly zero, as would be the case if the zero-point energy were zero, would conflict. Therefore, the energy cannot be zero. A better way of understanding its source is to note that the wavefunction is zero at the walls, cannot be zero everywhere, and has to be continuous; therefore it must be curved. A wavefunction's curvature is proportional to the kinetic energy (the kinetic energy term in the hamiltonian is proportional to d^2/dx^2); hence every wavefunction in the box corresponds to non-zero energy.

Another point is that the energy separations between neighbouring quantized levels, $E_{n+1}-E_n$, increases as the walls confine the particle to a smaller region. This is because, in order to fulfil the boundary conditions as L decreases, it is necessary for the wavefunction to have shorter wavelengths; by the de Broglie relation the particle's momentum increases, and hence the kinetic energy increases. Conversely, as the walls allow greater freedom to the particle, the energy separations decrease, and in the limit of infinite L, all values of the energy are allowed because the energy separations are infinitesimal. We regain the case of a free particle.

When the box is so long that the energy separations are small, the state of a macroscopic particle, which has only a fairly precisely defined energy, is described not by a single wavefunction corresponding to a precise energy, but by a superposition of many. This leads to the formation of a wavepacket, the centre of which behaves like the centre of mass of a classical particle.

The final point concerning the solutions may at first appear to be merely a formal property, but later we shall see that it is of great consequence. This property is the *orthogonality* of the wavefunctions corresponding to different energies. Two functions ψ_m and ψ_n are *orthogonal* if the integral of the product $\psi_m^*\psi_n$ over all space is zero:

Orthogonality: $$\int \psi_m^*\psi_n \, d\tau = 0. \qquad (3.3.6)$$

We can readily confirm the orthogonality of the wavefunctions of a particle in a box

$$\int_0^L \psi_m^*(x)\psi_n(x)\, dx = (2/L)\int_0^L \sin(n\pi x/L)\sin(m\pi x/L)\, dx = 0$$

where $n \neq m$. The symbol δ_{nm} is called the

Kronecker delta: $$\delta_{nm} = \begin{cases} 0 & \text{if} \quad n \neq m \\ 1 & \text{if} \quad n = m. \end{cases} \qquad (3.3.7)$$

It is a very useful notational device because it lets us combine the orthogonality and normalization conditions into the *orthonormality* condition:

Orthonormality: $$\int \psi_m^*\psi_n \, d\tau = \delta_{mn}. \qquad (3.3.8)$$

3.4 The two-dimensional square well

Interesting new features arise when we turn from the one-dimensional square well to the two-dimensional square well, in which the particle is confined to a rectangular surface with the linear dimensions L_1 and L_2. The potential energy is illustrated in Fig. 3.6. Just as in a one-dimensional system the wavefunctions look like a vibrating string, in two-dimensions they can be expected to correspond to a vibrating rectangular plate whose edges are rigidly clamped. Rectangular drums are not common, but it is fairly easy to visualize the vibrations they would perform.

The hamiltonian for the two-dimensional, infinitely deep square well is

$$H=-(\hbar^2/2m)(\partial^2/\partial x^2+\partial^2/\partial y^2)+V(x,y), \tag{3.4.1}$$

$$V(x,y)=\infty, \qquad x<0, x>L_1 \text{ or } y<0, y>L_2$$

$$V(x,y)=0, \qquad 0\leq x\leq L_1 \quad\text{and}\quad 0\leq y\leq L_2.$$

The Schrödinger equation is therefore

$$(\partial^2\psi/\partial x^2)+(\partial^2\psi/\partial y^2)=(-2mE/\hbar^2)\psi. \tag{3.4.2}$$

The boundary conditions are that the wavefunction must disappear at all the walls.

In order to solve this equation in two variables, we try the separation of variables technique described in Chapter 2 and attempt a solution in the form $\psi(x, y)=X(x)Y(y)$, where X is a function only of x, and Y a function only of y. On inserting this trial solution into eqn (3.4.2) and dividing through by XY we obtain

$$Y''/Y+X''/X=-2mE/\hbar^2,$$

where $Y''=\mathrm{d}^2Y/\mathrm{d}y^2$ and $X''=\mathrm{d}^2X/\mathrm{d}x^2$. Now use the same argument as in Chapter 2. That is, say that as x appears only in X''/X, when x varies only X''/X can change. But the sum of X''/X and Y''/Y is a constant; therefore X''/X must itself be a constant. We write it $-2mE^X/\hbar^2$. Likewise, Y''/Y must also be a constant, and we call it $-2mE^Y/\hbar^2$. The sum of the E^X and E^Y is equal to E. Therefore the equation may be separated into two parts:

$$\left.\begin{aligned} X''(x)&=-(2mE^X/\hbar^2)X(x),\\ Y''(y)&=-(2mE^Y/\hbar^2)Y(y),\end{aligned}\right\} \tag{3.4.3}$$

with $E^X+E^Y=E$. Each of these equations has the same form as the one-dimensional box equation; since the boundary conditions are the same the permissible solutions are the same. Therefore the complete solutions, $\psi=XY$ and $E=E^X+E^Y$, are

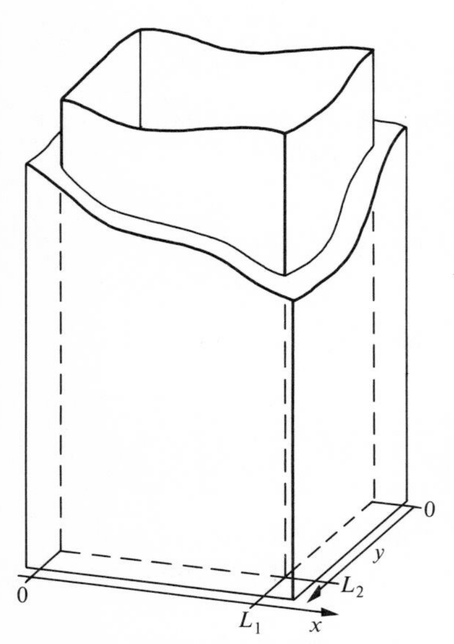

Fig. 3.6. The two-dimensional square well.

$$\psi_{n_1n_2}(x,y)=X_{n_1}(x)Y_{n_2}(y)=(4/L_1L_2)^{\frac{1}{2}}\sin(n_1\pi x/L_1)\sin(n_2\pi y/L_2), \tag{3.4.4}$$

$$E_{n_1n_2}=E^X_{n_1}+E^Y_{n_2}=(h^2/8)(n_1^2/L_1^2+n_2^2/L_2^2),$$

$$n_1=1,2,\ldots;\ n_2=1,2,\ldots.$$

Many of the features of the one-dimensional problem are reproduced in the two-dimensional. There is a zero-point energy (E_{11}), and the energy separations increase as the walls become more confining. The energy is quantized as a consequence of the boundary conditions. The shapes of a few of the low-energy wavefunctions are illustrated in Fig. 3.7: the upper diagram in each case is a drawing of the function and the lower diagram is a contour map. As in the one-dimensional

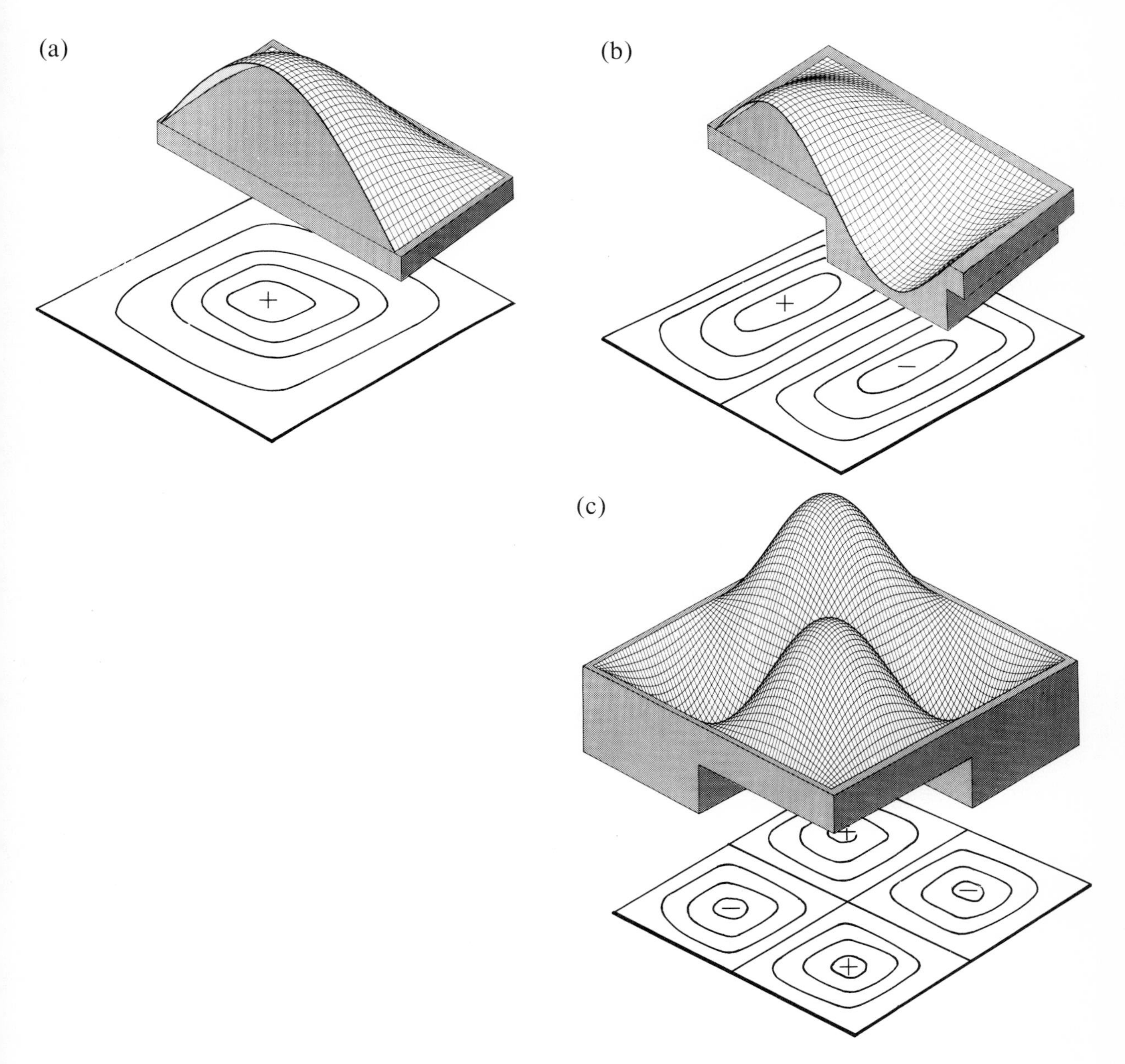

Fig. 3.7. (a) ψ_{11}, (b) ψ_{21}, (c) ψ_{22} for a two-dimensional square well.

case, the particle is distributed more uniformly at high energies than at low.

One feature found in two dimensions and not in one becomes apparent when the box is geometrically square (a *square square well*). Then $L_1 = L_2 = L$, and the energies are given by

$$E_{n_1 n_2} = (h^2/8mL^2)(n_1^2 + n_2^2). \tag{3.4.5}$$

This implies that a state with the quantum numbers $n_1 = a$ and $n_2 = b$ has exactly the same energy as one with $n_1 = b$ and $n_2 = a$. When $a = b$ that is a trivial remark, but it is important when a and b are unequal. It means that *different states may have the same energy*. This property is called *degeneracy*, and different states corresponding to the same energy are said to be *degenerate*.

As an example, consider the two states with $n_1 = 1$, $n_2 = 2$ and $n_1 = 2$, $n_2 = 1$. Each corresponds to the energy $5h^2/8mL^2$, but their two wavefunctions are different:

$$\psi_{12} = (2/L)\sin(\pi x/L)\sin(2\pi y/L), \qquad \psi_{21} = (2/L)\sin(2\pi x/L)\sin(\pi y/L).$$

Inspection of Fig. 3.8 shows the basis of this degeneracy: one wavefunction may be changed into the other by a *symmetry transformation* of the system, the rotation of the box by 90°. Degeneracy can always be traced to an aspect of the system's symmetry, as we shall see in more detail in Chapter 7. In the case of a rectangular but not square box, this ability to transform one function into the other is lost, and the degeneracy disappears too. Sometimes we encounter degeneracy when it might not be expected, and call it *accidental*. Sometimes it is unexpected because the full symmetry of the system has not been recognized. When one explores more deeply one finds a symmetry transformation that does interrelate the functions of the degenerate set. This occurs in the hydrogen atom, and we shall continue this discussion there.

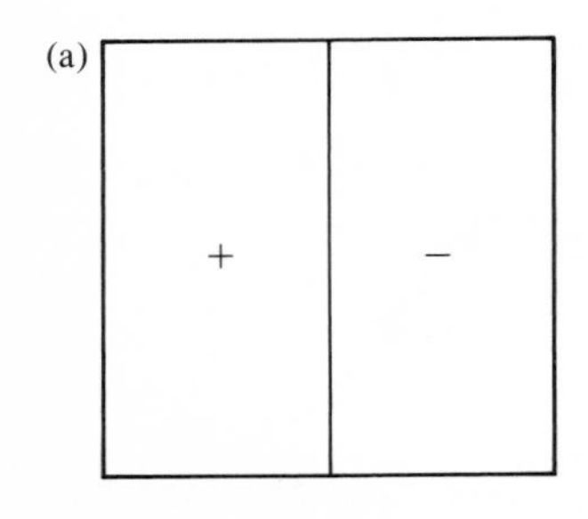

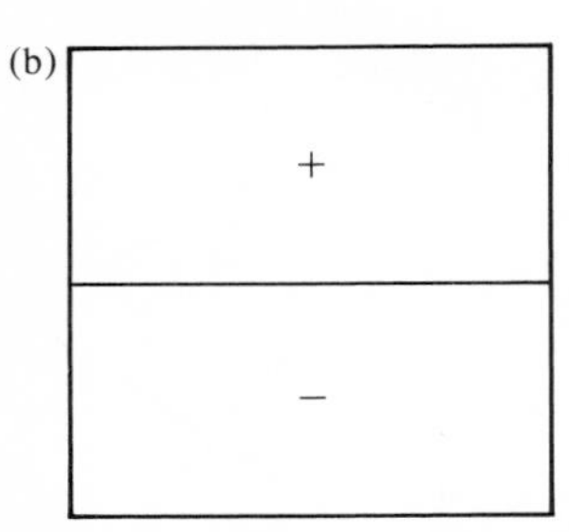

Fig. 3.8. (a) ψ_{21}, (b) ψ_{12} for a square two-dimensional square well.

3.5 The harmonic oscillator

We turn now to one of the most important topics of quantum mechanics, the harmonic oscillator. Harmonic oscillations occur when a system contains a part that experiences a restoring force proportional to the displacement from equilibrium. Pendulums and vibrating strings are familiar macroscopic examples. An example of obvious chemical importance is the vibration of the two atoms in a chemical bond within a molecule. A less obvious example is the electromagnetic field, which can be regarded as a collection of harmonic oscillators. The importance of the harmonic oscillator also lies in the way that the same algebra occurs in a variety of different problems; for example, we shall recognize it when we treat rotational motion.

The restoring force in the harmonic oscillator is $-kx$, where the constant of proportionality k is called the *force constant*, this implies that the potential energy of the particle depends on the displacement as

$$V(x)=\tfrac{1}{2}kx^2, \tag{3.5.1}$$

because the force acting on a particle is the negative of the gradient of the potential ($F=-\mathrm{d}V/\mathrm{d}x=-kx$). This *parabolic potential* is illustrated in Fig. 3.9. The difference between this and the square well potential is the rapidity with which it rises to infinity: the 'walls' of the oscillator are much softer and so we should expect the wavefunctions of the oscillating particle to penetrate them slightly. In other respects the two potentials are similar, and we may imagine the slow deformation of the square well into the smooth parabola of the oscillator. The wavefunctions of one system should change slowly into those of the other; they will have the same general form, but penetrate to classically disallowed displacements. Steal a glance at the diagram showing the final form of the solutions (Fig. 3.10 on p. 52) and compare it with Fig. 3.5 (p. 45).

Another point to make about the harmonic oscillator is that it is really much too simple. Its simplicity arises from the symmetrical occurrence of momentum and displacement in the expression for the total energy. Classically the energy is $E=p^2/2m+\tfrac{1}{2}kx^2$, and both p and x occur quadratically. This hidden symmetry has important implications, one being that if one has a new theory which can be applied to the harmonic oscillator and solved, it may still be unsolvable for other systems. Another implication involves the uncertainty principle, for in the ground state of the harmonic oscillator the product of the uncertainties δp and δx is *equal* to $\hbar/2$.

Since the potential energy has the form $V=\tfrac{1}{2}kx^2$, the hamiltonian operator for the harmonic oscillator is

$$H=-(\hbar^2/2m)(\mathrm{d}^2/\mathrm{d}x^2)+\tfrac{1}{2}kx^2 \tag{3.5.2}$$

and so the Schrödinger equation is

$$-(\hbar^2/2m)\,\mathrm{d}^2\psi/\mathrm{d}x^2+\tfrac{1}{2}kx^2\psi=E\psi. \tag{3.5.3}$$

The best method of solving this equation—a method that works for rotational motion (as we shall see in detail in Chapter 6) and for the hydrogen atom—is set out in Appendix 4; it depends on looking for a way of factorizing the hamiltonian (which is the sum of two squares) and introduces the concept of *creation and annihilation operators*. That algebra need not divert us from the main thread of this chapter, the discussion of the solutions themselves. It turns out that these are remarkably simple.

The *energy* of an harmonic oscillator is quantized (as we expected) and confined to the values

$$E_v=(v+\tfrac{1}{2})\hbar\omega, \qquad v=0,1,2,\ldots \tag{3.5.4}$$

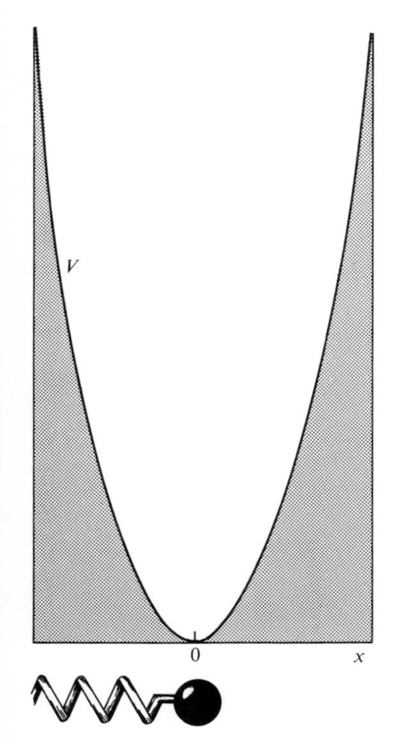

Fig. 3.9. The parabolic potential energy of a harmonic oscillator.

with $\omega=(k/m)^{\frac{1}{2}}$. The *wavefunctions* are no longer the simple sine functions of the square well, but resemble them fairly closely, and can be thought of as sine curves that collapse towards zero at large displacements, Fig. 3.10. Their precise form is that of a bell-shaped *gaussian function* multiplied by a polynomial in the displacement:

$$\psi_v(x) = N_v H_v(y) e^{-y^2/2}, \qquad y = (m\omega/\hbar)^{\frac{1}{2}} x. \tag{3.5.5}$$

N_v is a normalization constant; $N_v = (1/2^v v!\, \pi^{\frac{1}{2}})^{\frac{1}{2}}$. The $H_v(y)$ are called *Hermite polynomials*. When $v = 0$, $H_0(y) = 1$, and so the wavefunction is proportional to the gaussian function $e^{-y^2/2}$. When $v = 1$, $H_1(y) = 2y$, and so the wavefunction is the same gaussian multiplied by $2y$: it is plotted in Fig. 3.10 and the point where it passes through zero (at $x = 0$) is called a ***node***. (Notice that the wavefunction is not curved at the node, in accord with the general discussion in the last chapter.) The Hermite polynomials get progressively more complicated as v increases, and the first few are set out in Table 3.1. The wavefunctions they give rise to are drawn in Fig. 3.10.

We can now discuss the properties of the harmonic oscillator in terms of the solutions outlined above. As to the quantized energy levels, the most significant point is that each level is equidistant from its neighbours, the separation being $E_{v+1} - E_v = \hbar\omega$ whatever the value of v. This means that the energy levels form a uniform ladder, Fig. 3.10. (This uniformity is another aspect of the hidden symmetry of the oscillator hamiltonian.) As the force constant increases so the frequency ω increases ($\omega \propto \sqrt{k}$); therefore the separation between neighbours also increases. As k decreases (or as the mass increases) ω becomes smaller, and so the separation of neighbours decreases. In the limit of zero force constant, the parabolic potential fails to confine the particle (it corresponds to an infinitely weak spring) and the energy can

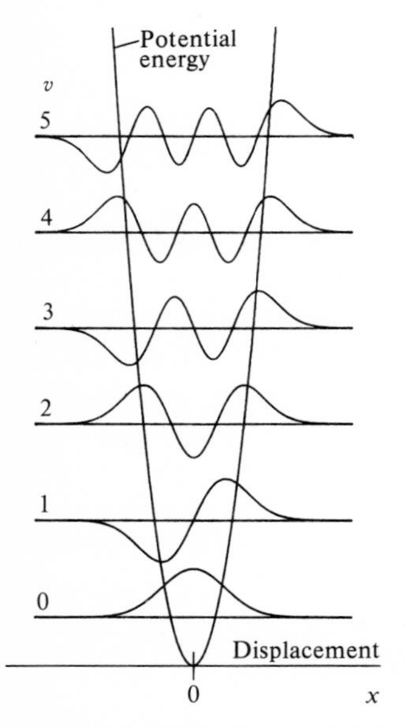

Fig. 3.10. Harmonic oscillator wavefunctions and energies.

Table 3.1 Hermite polynomials

v	$H_v(y)$
0	1
1	$2y$
2	$4y^2 - 2$
3	$8y^3 - 12y$
4	$16y^4 - 48y^2 + 12$
5	$32y^5 - 160y^3 + 120y$
6	$64y^6 - 480y^4 + 720y^2 - 120$
7	$128y^7 - 1344y^5 + 3360y^3 - 1680y$
8	$256y^8 - 3584y^4 + 13440y^4 - 13440y^2 + 1680$

Differential equation: $H_v'' - 2yH_v' + 2vH_v = 0$
Recursion relation: $H_{v+1} = 2yH_v - 2vH_{v-1}$
Orthogonality relation: $\int_{-\infty}^{\infty} H_v(y) H_{v'}(y) e^{-y^2}\, dy = 0$ for $v \neq v'$
Further information: M. Abramowitz and I. A. Stegun, *Handbook of mathematical functions*, Dover (1965), Chapter 22.

vary continuously: there is no quantization in this limit; the particle is free.

In thinking about the contributions to the total energy of a harmonic oscillator we have to take account of both the kinetic energy (which depends on the curvature of the wavefunction) and the potential energy (which depends on the probability of the particle being found at large displacements from equilibrium). The greater the force constant, the more confining the potential, and therefore the more sharply curved the wavefunctions. But whereas in the square well the walls were a stringent limitation on the distribution of the particle, in the harmonic oscillator they can be penetrated. Therefore, as the force constant increases, the curvature need not follow the confinement so obediently, and some of the sharpness of the curvature can be lost by the wavefunction spreading further out into greater displacements, but at the cost of acquiring a greater average potential energy. This discussion is simplified by the *virial theorem* which although originally a result from classical mechanics, has a quantum mechanical counterpart, Appendix 5. The virial theorem implies that if the potential energy can be expressed in the form $V=x^s$, then the mean potential and kinetic energies are related by $2\bar{T}=s\bar{V}$. It follows that the total mean energy is $\bar{E}=\bar{T}+\bar{V}=(1+2/s)\bar{T}$. In the case of the harmonic oscillator, where $s=2$, we have $\bar{T}=\bar{V}$ (hidden symmetry again) and $\bar{E}=2\bar{T}$. Therefore, as the *total* energy increases (as it does as k increases for a given quantum state), both the kinetic energy and the potential energy increase. That is, not only does the curvature of the wavefunction increase, but it also spreads into regions of higher potential energy. In classical terms, this corresponds to a pendulum swinging more rapidly *and* with greater amplitude (but at the same frequency) as its energy is increased.

A harmonic oscillator has a zero-point energy of magnitude $\frac{1}{2}\hbar\omega$, (corresponding to $v=0$ in the energy expression). Classically one would say that the particle can never be made to stop fluctuating about its equilibrium position. The reason for its existence is the same as for a particle in a square well. The wavefunction must be zero at large displacements in either direction (because the potential energy rises towards infinity), non-zero in between (because the particle has to be somewhere), and continuous (a general requirement about wavefunctions). This can be achieved only if the wavefunction has curvature; hence the oscillator must possess kinetic energy. By the virial theorem it must possess the same potential energy; hence it must possess at least some energy, its zero-point energy.

The shapes of the wavefunctions have already been drawn in Fig. 3.10. Their similarities to the square well wavefunctions should be noted. The major difference between the two kinds of wavefunctions is the penetration of the harmonic oscillator's into the classically forbidden regions (the regions where E is less then V and the kinetic energy is negative). In the same way as in the square well, the particle clusters away from the walls of the potential in the lowest energy states. This is the behaviour one might expect for a truly stationary oscillator, for it is most likely to be found at its equilibrium displacement. When the oscillator is moving, the classical prediction would be that it should most probably be found at its maximum displacement, the turning-points of the classical trajectory. The behaviour of the oscillator for small quantum numbers is, therefore, quite different from that expected classically. This peculiar behaviour disappears as it is excited into higher quantum levels. Figure 3.10 shows that the wavefunction for high quantum numbers has maxima near the classical turning points and resembles the classical distribution much more closely than the curves for the low quantum numbers. The quantum distribution therefore increasingly resembles the average classical distribution as the quantum number increases. This was also true of the particle in a box.

When the energy levels of the oscillator are close in comparison with the precision with which its state can be prepared, for example when the parabola is so broad or the mass so great that the levels lie close together, the state of the oscillating particle must be expressed as a superposition of the functions considered so far. For example, we saw on p. 9 that the energy levels are only about 10^{-34} J apart for a pendulum of period 1 s, and we cannot hope to make it swing with such precision. As usual, the components in the superposition interfere in such a way as to create a wavepacket. The time-dependence of the centre of the packet causes it to move from one side of the potential to the other with a frequency ω. That is, for coarse preparations of initial states, there is a sharply defined wavepacket which oscillates in the potential with the frequency $\omega = (k/m)^{\frac{1}{2}}$. This is precisely the classical behaviour, for the wavepacket corresponds to the position of the classical particle. Not only that, but the frequency, $(k/m)^{\frac{1}{2}}$, is exactly what classical calculations predict for an oscillator of force constant k and mass m. In other words, when we watch a pendulum swing, we are seeing the display of its quantum frequency.

Example. Show that whatever superposition of harmonic oscillator states is used to construct a wavepacket, it is localized at the same place at the times $0, T, 2T, \ldots$, where T is the classical period of the oscillator.

- *Method.* The classical period is $T = 2\pi/\omega$. Form a time-dependent wavepacket using $\Psi_v(t)$, and evaluate it at $t = nT$, $n = 0, 1, \ldots$.
- *Answer*

$$\Psi(x, t) = \sum_v c_v \Psi_v(x, t) = \sum_v c_v \Psi_v(x) e^{-iE_v t/\hbar}$$
$$= \sum_v c_v \Psi_v(x) e^{-i\omega(v+\frac{1}{2})t}.$$

Therefore $\Psi(x, nT) = \sum_v c_v \Psi_v(x) e^{-n2\pi i(v+\frac{1}{2})} = \sum_v c_v \Psi_v(x)(-1)^n = (-1)^n \Psi(x, 0)$, because $e^{2\pi i} = 1$ and $e^{i\pi} = -1$.

- *Comment.* The wavefunction changes sign after each period T, but is otherwise unchanged. Since the location is proportional to $|\Psi|^2$, it follows that the initial distribution is recovered at the end of each period. This is exactly the behaviour of a swinging pendulum.

The harmonic oscillator is so important that we summarize its properties:

1. The energy is quantized with $E_v = (v + \frac{1}{2})\hbar\omega$, where v is an integer from 0 upwards and $\omega = (k/m)^{\frac{1}{2}}$.
2. There is a zero-point energy $E_0 = \frac{1}{2}\hbar\omega$ which is a quantum phenomenon, for E_0 vanishes in a world in which h is zero.
3. The oscillating particle penetrates into classically forbidden regions.
4. The distribution of the particle approaches the classically predicted average distribution as the quantum number v becomes large.
5. The classical picture of an oscillating particle is obtained for coarse initial state preparations. The correct quantum description of the distribution of the particle is then in terms of a wavepacket oscillating with a frequency ω (which is the frequency corresponding to the separation of the quantum levels).
6. As the rigidity of the restraining force increases (as the force constant increases) or the mass decreases, the zero-point energy and the level separations increase. In the opposite limit where the force constant decreases and the potential becomes shallower, or the mass increases, the levels converge and in the limit of zero k or infinite mass, merge to a continuum.

Further reading

General information

Introduction to quantum mechanics. L. Pauling and E. B. Wilson; McGraw-Hill, New York, 1935.
Quantum mechanics. L. I. Schiff; McGraw-Hill, New York, 1968.
Quantum mechanics. A. Messiah; Wiley, New York, 1961.
Quantum mechanics. A. S. Davydov; Pergamon, Oxford, 2nd edition, 1976.
Quantum mechanics. L. D. Landau and E. M. Lifshitz; Pergamon, Oxford, 1958.

Potential barriers and tunnelling

Quantum theory. D. Bohm; Prentice-Hall, Englewood Cliffs, 1951.
Problems in quantum mechanics. I. I. Gol'dman and V. D. Krivchenkov; Pergamon, Oxford, 1961.
The tunnel effect in chemistry. R. P. Bell; Chapman and Hall, London, 1980.

Virial theorems

Classical mechanics. H. Goldstein; McGraw-Hill, New York, 1950.
Classical and quantum mechanical virial theorems. J. O. Hirschfelder; *J. chem. Phys.*, **33,** 1462 (1960).

Problems

3.1. Write the wavefunctions for (a) an electron travelling to the right ($x > 0$) after being accelerated from rest through a potential difference of (i) 1 V, (ii) 10 kV, (b) a particle of mass 1 g travelling to the right at $10\ \mathrm{m\ s^{-1}}$.

3.2. Find expressions for the probability densities of the particles in the preceding Problem.

3.3. A particle was prepared travelling to the right with all momenta between $(k-\frac{1}{2}\delta k)\hbar$ and $(k+\frac{1}{2}\delta k)\hbar$ contributing equally to the wavepacket. Find the explicit form of the wavepacket at $t=0$, normalize it, and estimate the range of positions, δx, within which the particle is likely to be found. Compare the last conclusion with a prediction based on the uncertainty principle. *Hint.* Use eqn (3.1.7) with $g=B$, a constant, inside the range $k-\frac{1}{2}\delta k$ to $k+\frac{1}{2}\delta k$ and zero elsewhere, and eqn (3.1.6) with $t=0$ for Ψ_k. In order to evaluate $\int|\Psi|^2\,\mathrm{d}\tau$ (for the normalization step) use the integral $\int_{-\infty}^{\infty}(\sin x/x)^2\,\mathrm{d}x=\sqrt{\pi}$. Take δx to be determined by the locations where $|\Psi|^2$ falls to $\frac{1}{2}$ its value at $x=0$. For the last part use $\delta p\approx\hbar\,\delta k$.

3.4. Sketch the form of the wavepacket constructed in Problem 3. Sketch its form a short time after, when t is non-zero but still small. *Hint.* For the second part use eqn (3.1.6) but with $e^{-ik^2t/2m}\approx 1-ik^2t/2m$. If you have access to a computer, program it to draw the wavepacket at longer times, evaluating the appropriate integrals numerically.

3.5. Repeat the evaluation leading to eqn (3.2.5) for the case $E>V$ and calculate the tunnelling probability.

3.6. A particle of mass m is incident from the left on a wall of infinite thickness and which may be represented by a potential energy V. Calculate the reflection coefficient for (a) $E\le V$, (b) $E>V$. For electrons incident on a metal surface $V\approx 10\,\mathrm{eV}$. Evaluate and plot the reflection coefficient. *Hint.* Proceed as in the last Problem but consider only two domains, inside the barrier and outside it. The reflection coefficient is the ratio $|B|^2/|A|^2$ in the notation of eqn (3.2.2).

3.7. A particle of mass m is confined to a one-dimensional box of length L. Calculate the probability of finding it in the following regions: (a) $0\le x\le\frac{1}{2}L$, (b) $0\le x\le\frac{1}{4}L$, (c) $\frac{1}{2}L-\delta\le x\le\frac{1}{2}L+\delta$. Evaluate the expressions for general values of n, and then specialize to $n=1$.

3.8. An electron is confined to a one-dimensional box of length L. What should be the length of the box in order for its zero-point energy to be equal to its rest mass energy (m_ec^2)? Express the result in terms of the *Compton wavelength*, $\lambda_C=h/m_ec$.

3.9. Energy is required to compress the box when a particle is inside: this suggests that the particle exerts a force on the walls. On the basis that when the length of the box changes by $\mathrm{d}L$ the energy changes by $\mathrm{d}E=-F\,\mathrm{d}L$, find an expression for the force. At what length does $F=1\,\mathrm{N}$ when an electron is in the state $n=1$?

3.10. The mean position $\langle x\rangle$ of a particle in a one-dimensional well can be calculated by weighting its position x by the probability that it will be found in the region $\mathrm{d}x$ at x, which is $\psi^2(x)\,\mathrm{d}x$, and then summing (i.e. integrating) these values. Show that $\langle x\rangle=\frac{1}{2}L$ for all values of n. *Hint.* Evaluate $\int_0^L x\psi_n^2(x)\,\mathrm{d}x$.

3.11. The root mean square deviation of the particle from its mean position is $\delta x=\sqrt{\{\langle x^2\rangle-\langle x\rangle^2\}}$. Evaluate this quantity for a particle in a well and show that it approaches its classical value as $n\to\infty$. *Hint.* Evaluate $\langle x^2\rangle=\int_0^L x^2\psi^2(x)\,\mathrm{d}x$. In the classical case the distribution is uniform across the box, and so in effect $\psi^2(x)=1/L$.

3.12. The mean value and mean square value of the linear momentum are given by $\int_0^L\psi^*p\psi\,\mathrm{d}x$ and $\int_0^L\psi^*p^2\psi\,\mathrm{d}x$ respectively. Evaluate these quantities, form the r.m.s. deviation $\delta p=\sqrt{\{\langle p^2\rangle-\langle p\rangle^2\}}$, and investigate the consistency of the outcome with the uncertainty principle. *Hint.* Use $p=(\hbar/i)(\mathrm{d}/\mathrm{d}x)$. For $\langle p^2\rangle$ notice that $E=p^2/2m$ and we already know E for each n. For the last part, form $\delta x\,\delta p$ and show that $\delta x\,\delta p\ge\frac{1}{2}\hbar$, the precise form of the principle, for all n; evaluate $\delta x\,\delta p$ for $n=1$.

3.13. Calculate the energies and wavefunctions for a particle in a one-dimensional square well in which the potential energy rises to a finite value V at each end, and is zero inside the well. Show that for any V and L there is always at least one bound level, and that as $V \to \infty$ the solutions coincide with those in eqn (3.3.5). *Hint.* This is a difficult Problem. Divide space into three zones, solve the Schrödinger equations, and impose the boundary conditions (finiteness of ψ and its continuity and the continuity of $d\psi/dx$ across the zone boundaries: combine the latter requirements into the continuity of the logarithmic derivatives $(1/\psi)(d\psi/dx)$). After some algebra arrive at

$$kL + 2\arcsin\{k\hbar/(2mV)^{\frac{1}{2}}\} = n\pi, \qquad \hbar k = (2mE)^{\frac{1}{2}}.$$

Solve this graphically for k and hence find the energies for each value of the integer n.

3.14. A very simple model of polyenes is the *free electron molecular orbital model.* Regard a chain of N conjugated carbon atoms, bond length R_{CC}, as forming a box of length $L = (N-1)R_{CC}$. Find the wavefunctions and their energies. Suppose that the electrons enter the states in pairs so that the lowest $\frac{1}{2}N$ states are occupied. Estimate the wavelength of the lowest energy transition. What colour are carrots? *Hint.* Carrots owe their colour to carotene,

β-carotene:

Sometimes the length of the chain is taken to be $(N+1)R_{CC}$, allowing for electrons to spill over the ends slightly.

3.15. Show that the variables in the Schrödinger equation for a cubic box may be separated and the overall wavefunctions expressed as $X(x)Y(y)Z(z)$. Deduce the energy levels and wavefunctions. Show that the functions are orthonormal. What is the degeneracy of the level with $E = 14(h^2/8mL^2)$?

3.16. The Hermite polynomials $H_v(y)$ satisfy the differential equation

$$H''_v(y) - 2yH'_v(y) + 2H_v(y) = 0.$$

Confirm that the wavefunctions in eqn (3.7.5) are solutions of the harmonic oscillator Schrödinger equation.

3.17. The oscillation of the atoms around their equilibrium positions in the molecule HI can be modelled as a harmonic oscillator of mass $m \approx m_H$ (the iodine atom is almost stationary) and force constant $k \approx 313.8\ \mathrm{N\,m^{-1}}$. Evaluate the separation of the energy levels and predict the wavelength of the light needed to induce a transition between neighbouring levels. What the populations of the $v = 0$ and $v = 1$ states at 300 K? *Hint.* For the last part, calculate the Boltzmann distribution, eqn (1.1.8).

3.18. What is the relative probability of finding the HI molecule with its bond length 10 per cent greater than its equilibrium value and its equilibrium bond length (161 pm) when it is in (a) the $v = 0$ state, (b) the $v = 4$ state?

3.19. Calculate the values of $\langle x\rangle$, $\langle x^2\rangle$, $\langle p\rangle$, $\langle p^2\rangle$ for an harmonic oscillator in its ground state by evaluation of the appropriate integrals (as in Problems 3.11, 3.12) and examine the value of $\delta x\,\delta p$ in the light of the uncertainty principle. *Hint.* Use the integrals

$$\int_{-\infty}^{\infty} e^{-\alpha x^2}\,dx = (\pi/\alpha)^{\frac{1}{2}}; \qquad \int_{0}^{\infty} x e^{-\alpha x^2}\,dx = \tfrac{1}{2}\alpha; \qquad \int_{-\infty}^{\infty} x^2 e^{-\alpha x^2}\,dx = \tfrac{1}{2}(\pi/\alpha^3)^{\frac{1}{2}}.$$

4 Exact solutions: rotational motion

THE second class of motion, which completes the treatment of the basic types, is rotational motion, the motion of an object around a fixed point. With this problem we encounter *angular momentum*, which is one of the most important topics in quantum mechanics. In this chapter we approach it by solving the Schrödinger equation. Then in Chapter 6 we return to it and see how many of its features emerge from the properties of operators. This is a chapter for pictures: later we see the algebra beneath the pictures.

The material we describe here recurs throughout quantum mechanics. In particular it crops up whenever we are interested in the motion of a particle under the influence of a *central field* of force. The basic example of this is the motion of a charged particle in the presence of an electrostatic Coulomb field, such as the electron in a hydrogen atom. This problem is also exactly solvable, and we consider it here.

4.1 Particle on a ring

As a first step we consider the quantum mechanics of a particle constrained to move on a circular ring. This is more general than it might seem, for although it applies to a bead on a circle of wire, it also applies to any body rotating in a plane (e.g. a gramophone record), Fig. 4.1. This is because any such body can be represented by a mass point moving in a circle of radius r, its *radius of gyration*, about its centre of mass. We shall see, in fact, that the quantity that determines the energy levels of the system is the *moment of inertia* $I=mr^2$, and it is not necessary to enquire into whether the value of I comes from a true mass point on a circle of radius r or from a body of mass m and radius of gyration r rotating about its own centre of mass. We shall refer to both interpretations as a particle on a ring.

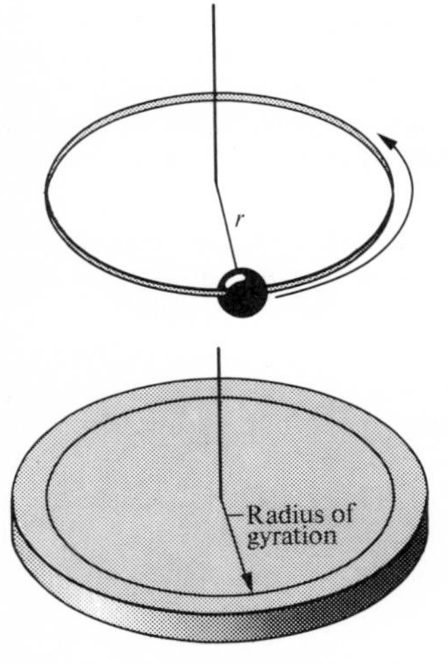

Fig. 4.1. A particle on a ring and the equivalent rotating disc.

The particle of mass m moves on a circle of radius r lying in the x, y-plane. Its potential energy is constant, and may be taken to be zero. The hamiltonian is therefore

$$H=-(\hbar^2/2m)\{(\partial^2/\partial x^2)+(\partial^2/\partial y^2)\}. \tag{4.1.1}$$

Since the motion is confined to a circle, a simpler description can be expected if we adopt cylindrical coordinates and write $x=r\cos\phi$ and $y=r\sin\phi$. Then the second derivatives are expressed in terms of r and ϕ. As r is constant, derivatives with respect to r play no role in the problem, and the hamiltonian turns into

$$H=-(\hbar^2/2mr^2)(\mathrm{d}^2/\mathrm{d}\phi^2)=-(\hbar^2/2I)(\mathrm{d}^2/\mathrm{d}\phi^2). \tag{4.1.2}$$

The wavefunction depends only on the angle ϕ, and so we denote it

$\Phi(\phi)$. The Schrödinger equation is therefore

$$d^2\Phi/d\phi^2 = -(2IE/\hbar^2)\Phi. \tag{4.1.3}$$

The solutions are

$$\Phi(\phi) = Ae^{im_l\phi} + Be^{-im_l\phi}, \qquad m_l = (2IE/\hbar^2)^{\frac{1}{2}}. \tag{4.1.4}$$

The quantity m_l is dimensionless, but at this stage it is completely unrestricted in value.

Example. The wavefunctions for a particle on a ring also arise in connection with a particle confined to a circular region (a circular square well). Show that the Schrödinger equation is separable, and find equations for the radial and angular components.

- *Method.* Express the wavefunction in the form $\psi(r, \phi) = R(r)\Phi(\phi)$ and show that the Schrödinger equation separates. The hamiltonian for the problem is $H = T$ (because $V = 0$ where the particle may be found), and the kinetic energy operator is proportional to $\partial^2/\partial x^2 + \partial^2/\partial y^2$, which should be expressed in terms of r and ϕ using $x = r\cos\phi$, $y = r\sin\phi$.
- *Answer.*

$$(\partial^2/\partial x^2) + (\partial^2/\partial y^2) = (\partial^2/\partial r^2) + (1/r)(\partial/\partial r) + (1/r^2)(\partial^2/\partial\phi^2).$$

$$H = -(\hbar^2/2m)\{(\partial^2/\partial r^2) + (1/r)(\partial/\partial r) + (1/r^2)(\partial^2/\partial\phi^2)\}; \qquad H\psi(r, \phi) = E\psi(r, \phi).$$

Try $\psi(r, \phi) = R(r)\Phi(\phi)$ on both sides, and divide by $R\Phi$:

$$-(\hbar^2/2m)(1/R)\{R'' + (1/r)R'\} - (\hbar^2/2mr^2)\Phi''/\Phi = E.$$

$1/r^2$ occurs in the second term; therefore multiply through by r^2 and rearrange:

$$-(\hbar^2/2m)(1/R)\{r^2R'' + rR'\} - Er^2 = (\hbar^2/2m)\Phi''/\Phi.$$

This is separable. Write $\Phi'' = -m_l^2\Phi$ and find

$$r^2R'' + rR' + (2mE/\hbar^2)r^2R = m_l^2R$$

as the radial equation.

- *Comment.* Go on to solve the radial equation. Identify the form of the equation by reference to M. Abramowitz and I. A. Stegun, *Handbook of mathematical functions*, Dover (1965) (try their Chapter 9). Devise and impose the approximate boundary conditions and use their table of zeros to find the zero-point energy.

Now we introduce the boundary conditions. There are no barriers to the particle's motion so long as it remains on the ring, and so there is no requirement for the wavefunction to vanish at some point. One of the restrictions on acceptable wavefunctions introduced in Chapter 2 is that they should give rise to a single-valued probability density. In the present case this corresponds to the requirement that the wavefunction must be single-valued, or $\Phi(\phi + 2\pi) = \Phi(\phi)$. This can be expressed as

$$\left.\begin{aligned}\Phi(\phi + 2\pi) &= Ae^{im_l\phi}e^{2\pi im_l} + Be^{-im_l\phi}e^{-2\pi im_l}\\ \Phi(\phi) &= Ae^{im_l\phi} + Be^{-im_l\phi}\end{aligned}\right\}\Phi(\phi + 2\pi) = \Phi(\phi)$$

which is satisfied only if m_l is an integer (since then $e^{2\pi im_l} = 1$). That is, the *cyclic boundary conditions* impose the restriction that

$$m_l = 0, \pm 1, \pm 2, \ldots,$$

and therefore the energies permitted to the system are

$$E = m_l^2(\hbar^2/2I), \qquad m_l = 0, \pm 1, \pm 2, \ldots. \tag{4.1.5}$$

Hence the energy is quantized.

The significance of the two components with different signs of m_l in the exponent can be discovered in exactly the same way as in the case of linear motion. There e^{ikx} and e^{-ikx} corresponded to the same magnitude of *linear* momentum but in opposite directions. $e^{im_l\phi}$ and $e^{-im_l\phi}$ correspond to the same magnitude of *angular* momentum, but in opposite directions. This can be confirmed in the same way as in the linear case. First, the classical expression for the rotational energy of an object confined to motion in a plane is $J^2/2I$ (the analogue of $p^2/2m$), J being the angular momentum. It follows by comparison with eqn (4.1.5) that the magnitude of the angular momentum is related to the value of m_l by $J^2 = m_l^2\hbar^2$. Opposite signs of m_l therefore correspond to the same *magnitude* of angular momentum. In the linear case we found an eigenvalue equation which let us come to a conclusion about the meaning of the sign of k. We can do the same thing for m_l by setting up the operator for the angular momentum. The motion is confined to the x, y-plane, and the classical expression for the only non-zero component of angular momentum, the z-component J_z, is

$$J_z = xp_y - yp_x,$$

p_x and p_y being the x- and y-components of the linear momentum. (We formalize this definition in Chapter 6.) The corresponding operator is therefore

$$J_{z,\,\text{operator}} = x(\hbar/\mathrm{i})(\partial/\partial y) - y(\hbar/\mathrm{i})(\partial/\partial x) = (\hbar/\mathrm{i})(\partial/\partial\phi), \tag{4.1.6}$$

where the second equality follows from the use of cylindrical coordinates. Now consider the case where the wavefunction has $B = 0$. We find

$$J_{z,\,\text{operator}}\Phi = (\hbar/\mathrm{i})(\partial/\partial\phi)A\mathrm{e}^{\mathrm{i}m_l\phi} = m_l\hbar A\mathrm{e}^{\mathrm{i}m_l\phi} = m_l\hbar\Phi,$$

which is an eigenvalue equation, and we see that the wavefunction $A\mathrm{e}^{\mathrm{i}m_l\phi}$ corresponds to a state with angular momentum $m_l\hbar$. In exactly the same way we deduce that $B\mathrm{e}^{-\mathrm{i}m_l\phi}$ corresponds to a state with angular momentum $-m_l\hbar$. Hence the two functions correspond to the same magnitude of angular momentum, but to different *senses* of rotation, Fig. 4.2.

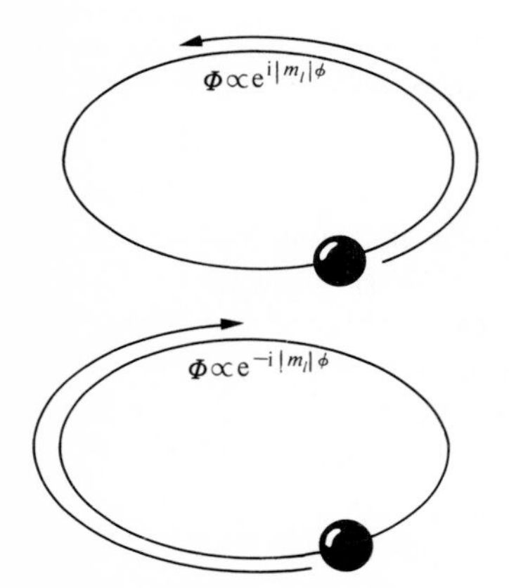

Fig. 4.2. Wavefunctions for a particle on a ring.

The only remaining task is to establish the normalization constants. Suppose the wavefunction is $A\mathrm{e}^{\mathrm{i}m_l\phi}$, then the normalization condition is

$$\int_0^{2\pi} A^2\mathrm{e}^{-\mathrm{i}m_l\phi}\mathrm{e}^{\mathrm{i}m_l\phi}\,\mathrm{d}\phi = A^2\int_0^{2\pi}\mathrm{d}\phi = 2\pi A^2 = 1; \quad \text{hence } A = (1/2\pi)^{\frac{1}{2}}.$$

It is an easy task to show that the wavefunctions are also mutually orthogonal.

The physical basis of the quantization of rotational energy becomes clear once we investigate the shapes of the wavefunctions. The wavefunction corresponding to a definite state of angular momentum $m_l\hbar$ is

$$\Phi_{m_l}(\phi)=(1/2\pi)^{\frac{1}{2}}e^{im_l\phi}=(1/2\pi)^{\frac{1}{2}}(\cos m_l\phi+i\sin m_l\phi). \quad (4.1.7)$$

We shall think about the cosine part of this function, but similar remarks apply to the sine component. (In passing, note again that a state representing motion is described by a complex wavefunction, just as in the linear case.) When m_l is an integer, the cosine functions define a wave that has an *integral* number of wavelengths between the points $\phi=0$ and $\phi=2\pi$: the 'ends' join, and the function reproduces itself on its next circuit, Fig. 4.3(b). When m_l is not an integer the wavefunction has an incomplete number of wavelengths between 0 and 2π, and does not reproduce itself on the next circuit of the ring, Fig. 4.3(a). At the single point denoted $\phi=0$ or $\phi=2\pi$ the wavefunction is double valued; hence the Born interpretation is untenable. This is avoided only when m_l is an integer.

A glance at the expression for the energy shows that all the levels except the lowest ($m_l=0$) are doubly degenerate. (E is proportional to the square of m_l, m_l may be positive or negative, and the functions $e^{im_l\phi}$ and $e^{-im_l\phi}$ are different so long as $m_l\neq 0$.) This degeneracy results from the fact that the particle may circulate in either sense and therefore be in states of different angular momentum, but travel with the same speed, and therefore have the same kinetic energy. The absence of degeneracy in the ground state is because there is then no angular momentum, and so the question of its direction does not arise.

There are several ways of depicting the shapes of the wavefunctions. The simplest is to plot the amplitude on the perimeter of the ring. One problem is that the wavefunctions are complex (unless $m_l=0$). But since $e^{im_l\phi}=\cos m_l\phi+i\sin m_l\phi$ we can avoid the difficulty by plotting only the cosine component and remembering that there is also a sine component of exactly the same shape but displaced by a quarter of a wavelength.† One hazard of this procedure is that the cosine part of the wavefunction gives the impression that the particle is distributed non-uniformly around the ring ($\cos^2 m_l\phi$ varies with ϕ). The square modulus itself, $\Phi^*\Phi=(1/2\pi)^{\frac{1}{2}}e^{-im_l\phi}(1/2\pi)^{\frac{1}{2}}e^{im_l\phi}=1/2\pi$, is independent of the value of ϕ. Hence, *when the particle is in a state of definite angular momentum its distribution is completely uniform.* Always bear

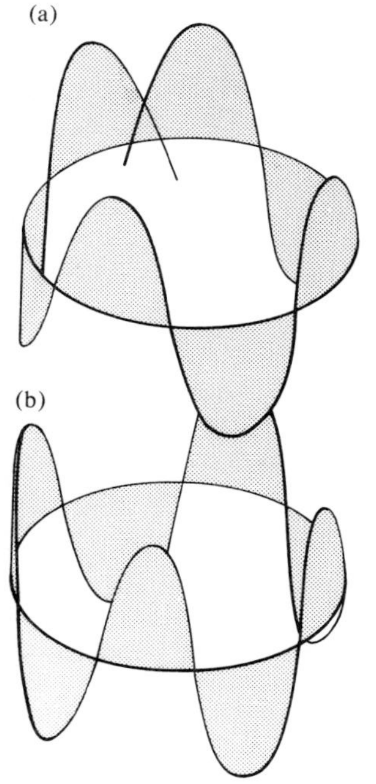

Fig. 4.3. (a) Unacceptable and (b) acceptable wavefunctions.

† Note that if $m_l=1$, the wavefunction is $\cos\phi+i\sin\phi$, and if $m_l=-1$ it is $\cos\phi-i\sin\phi$. That is, for $m_l=1$ (or any positive value), corresponding to clockwise motion seen from below, the imaginary component *leads* the real by 90°, but if $m_l=-1$ (or any negative value), corresponding to anticlockwise motion, the imaginary *lags* by 90°. The *relative* phase of the real and the imaginary components indicates the direction of circulation. The same is true for linear motion.

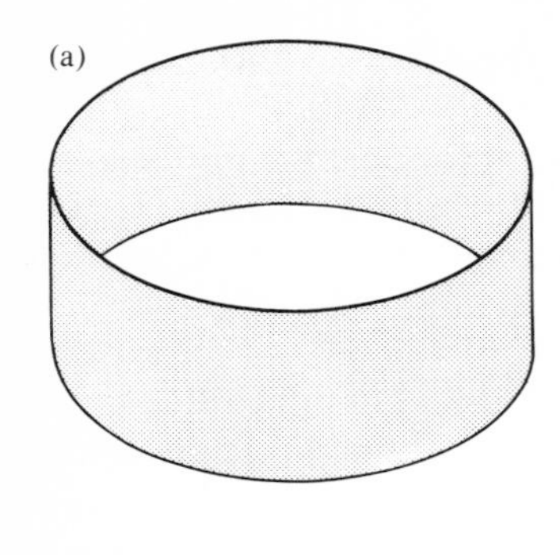

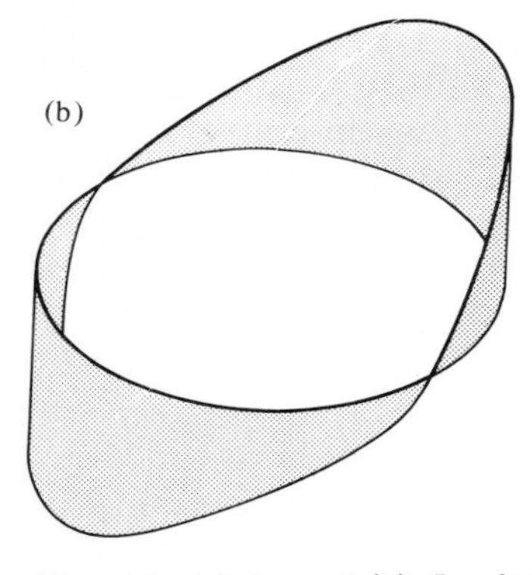

Fig. 4.4. (a) Φ_0 and (b) $\Phi_{\pm 1}$ for a particle on a ring.

this in mind when thinking about the wavefunctions for particles in states of rotational motion.

The simple representation of the real part of the wavefunction is depicted in Fig. 4.4 for $m_l = 0$ and $m_l = \pm 1$. A neater way of representing them is to connect each point of the perimeter to the centre, and to mark on this radius a length proportional to the amplitude at that point. If all such points are joined one obtains a curve of the shape shown in Fig. 4.5 for $\Phi_{\pm 1}$. The corresponding curve for Φ_0 is a circle. It is possible to represent the functions by the *auxiliary functions* alone, and these are shown in the figure. Notice that the two nodal points of the original wavefunction become a single *nodal line* in the auxiliary function. The signs marked on the lobes of the functions denote the *phases* of the wavefunction: a positive sign indicates that the wavefunction has a positive amplitude for that range of angles. The complete set of wavefunction and auxiliary function for $\Phi_{\pm 2}$ is shown in Fig. 4.6. Note that the four nodal points of the wavefunction become the two nodal lines of the auxiliary function.

There are several significant features of the wavefunctions. The first is that the number of nodes increases as the energy increases. This is an example of the behaviour we have already discussed: the kinetic energy is proportional to the curvature of the wavefunction, and its curvature increases if it is buckled so that it bends backwards and forwards more often across zero. Altenatively, because the kinetic energy is due entirely to the angular momentum, we can say that *the angular momentum increases as the number of nodes increases.* The number of nodal lines in the auxiliary functions is equal to the magnitude of m_l, and hence to the magnitude of the angular momentum. Finally, notice that there is symmetry in the wavefunctions, for on replacing ϕ by $\phi + \pi$ and using $e^{i\pi} = -1$,

$$\Phi_{m_l}(\phi + \pi) = (1/2\pi)^{\frac{1}{2}} e^{im_l(\phi+\pi)} = (1/2\pi)^{\frac{1}{2}} e^{im_l\phi} e^{i\pi m_l} = (-1)^{m_l} \Phi_{m_l}(\phi).$$

That is, points separated by 180° (across any diameter of the circle) have the same magnitude of amplitude but opposite signs when m_l is odd and the same signs when m_l is even. Inspection of the pictures confirms this. The importance of this observation will become clearer as we explore the role of symmetry in Chapters 6 and 7.

In the system we are considering there is no zero-point energy, because m_l and hence E is free to equal zero. This is because, unlike in the square well, the wavefunction need not be curved to fit between its 'endpoints' at $\phi = 0$ and 2π.†

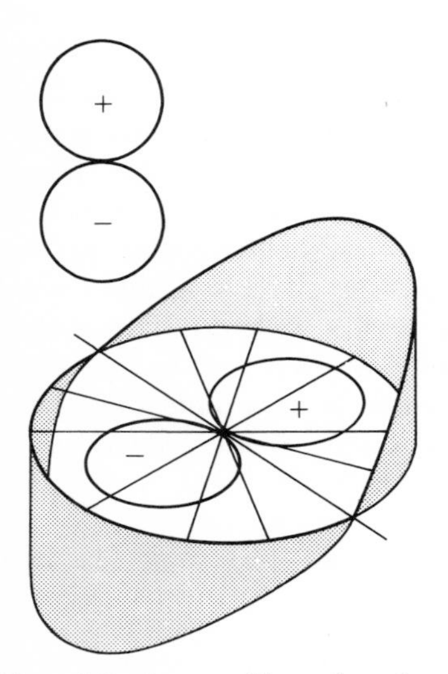

Fig. 4.5. the auxiliary function representation of $\Phi_{\pm 1}$.

† Sometimes the non-appearance of the zero-point energy is based on the uncertainty principle, in the form that since the particle may be anywhere in the infinite range $\phi = -\infty$ to $+\infty$, the angular momentum may be precisely zero. This is specious. Great care must be taken with the uncertainty principle when the observables are periodic, as location is in this case. By using more complex arguments using the variables $\sin\phi$ and $\cos\phi$ as position observables, and denoting the z-component of the angular momentum as l_z, one can deduce that the correct form of the uncertainty relation for a particle on a ring is $(\delta l_z)^2(\delta \sin\phi)^2 \geq \frac{1}{4}\hbar^2 \langle\cos\phi\rangle^2$; see P. Carruthers and M. M. Nieto, *Rev. Mod. Phys.*, **40,** 411 (1968).

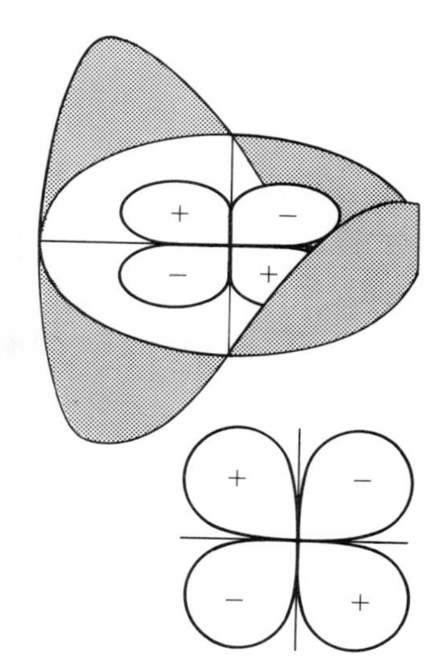

Fig. 4.6. The auxiliary function for $\Phi_{\pm 2}$.

The final point concerns the classical limit of the quantum mechanical description of rotational motion. The tale should be a familiar one by now. When the particle is prepared with an energy that is imprecise in comparison with the energy-level separations (as when a gramophone turntable is switched on) the correct description of the system is as a superposition of wavefunctions which interfere to produce a wavepacket. This packet is localized in space (it may represent the mass point modelling the turntable) and because of the time-dependence of the component wavefunctions (each one has a time-dependence of the form $e^{-iEt/\hbar} = e^{-im_l^2\hbar t/2I}$) its centre, the point of maximum constructive interference, travels in a circle according to the predictions of classical physics.

Rotational motion in classical physics is normally denoted by a vector representing the state of angular momentum. In the case of motion confined to a plane, the vector is perpendicular to the plane of rotation. Hence, in the present case, it lies parallel to the z-axis. The *length* of the vector represents that *magnitude* of the momentum, and its direction represents the sense of the rotation. On the basis of the right-hand screw rule, a vector pointing in the direction of *positive z* indicates *clockwise motion seen from below*, Fig. 4.7(a). A vector pointing towards *negative z* indicates *anticlockwise motion seen from below*, Fig. 4.7(b). This gives a simple way of representing the permitted states of rotational motion for a particle in two dimensions. The difference between the classical and quantum mechanical pictures is that, whereas in the former the length of the vector is continuously variable, in quantum mechanics it is confined to the discrete values corresponding to the permitted values of $m_l\hbar$.

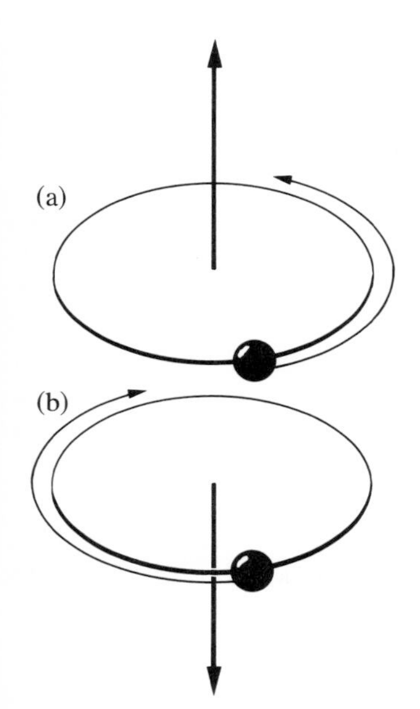

Fig. 4.7. The vector representation of angular momentum.

4.2 Particle on a sphere

Now we consider the case of a particle free to move in three dimensions on the surface of a sphere. Just as the particle on a ring is a model of an actual solid body represented by a mass point moving at the radius of gyration about the centre of mass, so the particle on a sphere models a sphere rotating about its own centre. The 'particle' is then a particle of mass m at the radius of gyration, the motion of the sphere being represented by this point moving over the corresponding spherical surface.† As we shall see later in the chapter, this calculation has a wider scope than is suggested by this model, and the calculation we are about to review is of considerable importance and wide application.

Since the particle is free to move in three dimensions, but does so at constant (zero) potential energy (the potential energy of a ball is

† From elementary classical mechanics we know that the radius of gyration of a solid sphere of mass m and uniform density is $r = (\frac{2}{5})^{\frac{1}{2}}R$, where R is its actual radius, and so we can represent its rotational motion by a particle of mass m moving over a sphere of radius $r = (\frac{2}{5})^{\frac{1}{2}}R$.

independent of its orientation in space), the hamiltonian operator is

$$H = -(\hbar^2/2m)\{\partial^2/\partial x^2 + \partial^2/\partial y^2 + \partial^2/\partial z^2\} = -(\hbar^2/2m)\nabla^2. \tag{4.2.1}$$

We have to remember that the distance of the particle from the origin, r, is constant, and this suggests that we should express the laplacian in terms of *spherical polar coordinates* in which r and derivatives with respect to r occur explicitly. The spherical polar coordinates are expressed in terms of the radius r, the colatitude θ, and the azimuth ϕ as shown in Fig. 4.8. Explicitly,

$$x = r \sin\theta \cos\phi, \qquad y = r \sin\theta \sin\phi, \qquad z = r \cos\theta. \tag{4.2.2}$$

Standard manipulation of the differentials leads to the following expression for the laplacian in terms of these coordinates:

$$\nabla^2 = (1/r)(\partial^2/\partial r^2)r + (1/r^2)\Lambda^2 \tag{4.2.3}$$

where Λ^2 is the

legendrian: $$\Lambda^2 = (1/\sin^2\theta)(\partial^2/\partial\phi^2) + (1/\sin\theta)(\partial/\partial\theta)\sin\theta(\partial/\partial\theta), \tag{4.2.4}$$

the angular part of the laplacian. The condition that the particle is confined to the surface of fixed radius is equivalent to ignoring the radial derivatives, and hence we retain only the legendrian part of the laplacian, and treat r as a constant. The hamiltonian is therefore

$$H = -(\hbar^2/2mr^2)\Lambda^2, \tag{4.2.5}$$

and since $mr^2 = I$, the moment of inertia, the Schrödinger equation is

$$\Lambda^2\psi = -(2IE/\hbar^2)\psi. \tag{4.2.6}$$

The wavefunction is a function of the two angles θ and ϕ: $\psi = \psi(\theta, \phi)$.

There are several ways of solving this second-order partial differential equation. One is to realize that the functions should resemble the solutions we have already found for the particle on a ring, for from one point of view (from any point of view) a sphere can be regarded as a collection of rings. The difference betwen the two cases is that now the particle can travel from ring to ring. This approach resembles the two-dimensional square well, where the solutions can be considered to be a collection of one-dimensional-well wavefunctions stacked together. That problem was solved by the separation of variables technique, and we can expect the same approach to work here. We therefore take a trial solution of the form $\psi(\theta, \phi) = \Theta(\theta)\Phi(\phi)$, hoping that the Schrödinger equation will separate into two, one for Θ and the other for Φ. The equation for Φ is:

$$d^2\Phi/d\phi^2 = (\text{const.})\Phi \tag{4.2.7}$$

and its solutions are $\Phi = e^{im_l\phi}$, just as for the particle on a ring. The

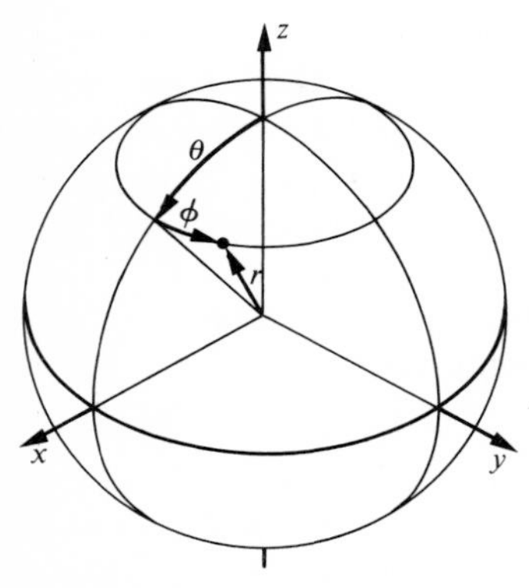

Fig. 4.8. Spherical polar coordinates.

equation for Θ is much more involved, and elementary techniques require a long and involved calculation to find its solutions.

The second method of solution of eqn (4.2.6) is to look for factors of the hamiltonian operator which turn the Schrödinger equation into a first-order equation, just as we did in the case of the harmonic oscillator. That the same technique works here should also be expected: in classical physics simple harmonic motion can be thought of as the projection of uniform circular motion on to a diameter, and so it should not be surprising that harmonic motion and angular momentum have deep similarities in their mathematical descriptions. This is the technique we use in the chapter devoted to angular momentum, Chapter 6.

The third method is to make the straightforward claim that we recognize eqn (4.2.6) as a well-known equation in mathematics, so that we can simply look up its solutions in tables.† This is the technique we use here, even though it may not seem to be the most convincing. There seems to be little point in solving the equation by elementary but tedious techniques if in a couple of chapters we are to deal with it in a quick, elegant fashion; furthermore, solution by recognition is in fact the way that many differential equations are solved by professional theoreticians, and is a method not to be scorned!

The equation to recall, identify, or now learn for the first time, is the one satisfied by the *spherical harmonics*, $Y_{lm_l}(\theta, \phi)$:

$$\Lambda^2 Y_{lm_l}(\theta, \phi) = -l(l+1) Y_{lm_l}(\theta, \phi), \tag{4.2.8}$$

where the labels l and m_l are integers and limited to the values $l = 0, 1, 2, \ldots$ and $m_l = l, l-1, \ldots, -l$ (giving $2l+1$ values for a given value of l). It is obvious that this equation has exactly the same form as eqn (4.2.6), and so the wavefunctions ψ are proportional to the spherical harmonics. The harmonics themselves consist of two factors, one a function of ϕ alone, and the other a function of θ. That is,

$$Y_{lm_l}(\theta, \phi) = \Theta_{lm_l}(\theta)\Phi_{m_l}(\phi), \tag{4.2.9}$$

just as the separation of variables technique indicated. The Φ functions are the ones we have already found for the particle on a ring, eqn (4.1.8), and the Θ functions are standard functions known as the *associated Legendre functions*. The first few are listed in Table 4.1.

We can check that individual spherical harmonics are solutions of eqn (4.2.8). For example, in the case of Y_{10}:

$$\begin{aligned}\Lambda^2 Y_{10} &= (1/\sin\theta)(\partial/\partial\theta)(3^{\frac{1}{2}}/2\pi^{\frac{1}{2}})\cos\theta + (1/\sin^2\theta)(\partial^2/\partial\phi^2)(3^{\frac{1}{2}}/2\pi^{\frac{1}{2}})\cos\theta\\ &= -(3^{\frac{1}{2}}/2\pi^{\frac{1}{2}})(1/\sin\theta)(\mathrm{d}/\mathrm{d}\theta)\sin^2\theta\\ &= -2(3^{\frac{1}{2}}/2\pi^{\frac{1}{2}})(1/\sin\theta)\sin\theta\cos\theta = -2Y_{10},\end{aligned}$$

which is consistent with eqn (4.2.8) with $l = 1$.

† This is a very common way of solving differential equations, and the *Handbook of Mathematical Functions*, M. Abramowitz and I. A. Stegun, Dover, is one of the principal sources for identifying solutions. It is an ideal desert-island book for shipwrecked quantum chemists.

Table 4.1 Spherical harmonics

l	m_l	$Y_{lm_l}(\theta, \phi)$
0	0	$1/2\pi^{\frac{1}{2}}$
1	0	$\frac{1}{2}(3/\pi)^{\frac{1}{2}}\cos\theta$
	±1	$\mp\frac{1}{2}(3/2\pi)^{\frac{1}{2}}\sin\theta e^{\pm i\phi}$
2	0	$\frac{1}{4}(5/\pi)^{\frac{1}{2}}(3\cos^2\theta-1)$
	±1	$\mp\frac{1}{2}(15/2\pi)^{\frac{1}{2}}\cos\theta\sin\theta e^{\pm i\phi}$
	±2	$\frac{1}{4}(15/2\pi)^{\frac{1}{2}}\sin^2\theta e^{\pm 2i\phi}$
3	0	$\frac{1}{4}(7/\pi)^{\frac{1}{2}}(2-5\sin^2\theta)\cos\theta$
	±1	$\mp\frac{1}{8}(21/\pi)^{\frac{1}{2}}(5\cos^2\theta-1)\sin\theta e^{\pm i\phi}$
	±2	$\frac{1}{4}(105/2\pi)^{\frac{1}{2}}\cos\theta\sin^2\theta e^{\pm 2i\phi}$
	±3	$\mp\frac{1}{3}(35/\pi)^{\frac{1}{2}}\sin^3\theta e^{\pm 3i\phi}$

Structure: $Y_{lm_l}(\theta, \phi)=\Theta_{lm_l}(\theta)\Phi_{m_l}(\phi)$

$$\Phi_{m_l}(\phi)=(1/2\pi)^{\frac{1}{2}}e^{im_l\phi}, \qquad \Theta_{l,m_l}(\theta)=\left\{\frac{(2l+1)(l-|m_l|)!}{2(l+|m_l|)!}\right\}^{\frac{1}{2}}P_l^{|m_l|}(\cos\theta)$$

$P_l^{|m_l|}(\cos\theta)$ are the associated Legendre functions.

Orthogonality relation:
$$\int_{-1}^{1} P_{l'}^{|m_l'|}(z)P_l^{|m_l|}(z)\,dz = \begin{cases} 0 & \text{for } l\neq l'. \\ \left(\dfrac{2}{2l+1}\right)\dfrac{(l+|m_l|)!}{(l-|m_l|)!} & \text{for } l=l' \end{cases}$$

Recursion relation:
$$zP_l^{|m_l|}(z)=\left(\frac{l+|m_l|}{2l+1}\right)P_{l-1}^{|m_l|}(z)+\left(\frac{l-|m_l|+1}{2l+1}\right)P_{l+1}^{|m_l|}.$$

Further information: M. Abramowitz and I. A. Stegun, *Handbook of mathematical functions*, Dover (1965), Chapter 8.

Comparison of eqn (4.2.8) and eqn (4.2.6) shows that the energies are confined to the values

$$E=(\hbar^2/2I)l(l+1), \qquad l=0, 1, 2, \ldots. \tag{4.2.10}$$

l is therefore a quantum number for the energy. Notice that E is independent of the value of m_l. Since for each value of l there are $2l+1$ permitted values of m_l, and therefore $2l+1$ different wavefunctions, we conclude that *each energy level is* $(2l+1)$*-fold degenerate.*

The labels l and m_l have a further significance. The rotational energy of a spherical body of moment of inertia I is given by classical physics as $\frac{1}{2}I\omega^2$, where ω is its angular velocity. The magnitude of the angular momentum is $J=I\omega$, and so the energy is $J^2/2I$. If this is compared with eqn (4.2.10) we see that in quantum mechanics

$$\textit{magnitude of the angular momentum}=\hbar\surd\{l(l+1)\}. \tag{4.2.11}$$

It too is quantized. The label l is therefore also a quantum number for the magnitude of the angular momentum, and hence is called the *angular momentum quantum number* (its old name is *azimuthal quantum number*). Note that angular momentum has the same dimensions

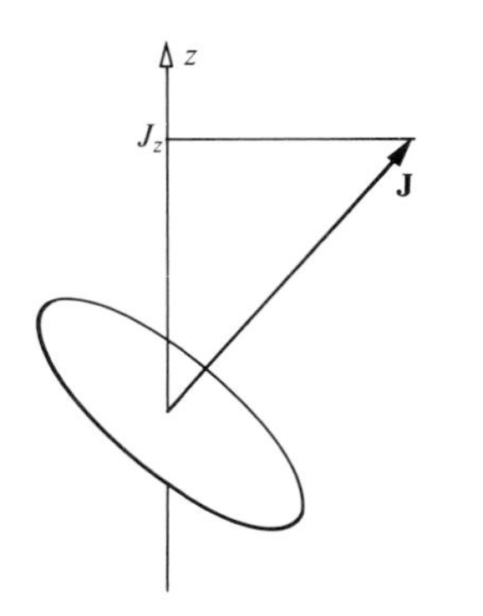

Fig. 4.9. The angular momentum vector and its z-component.

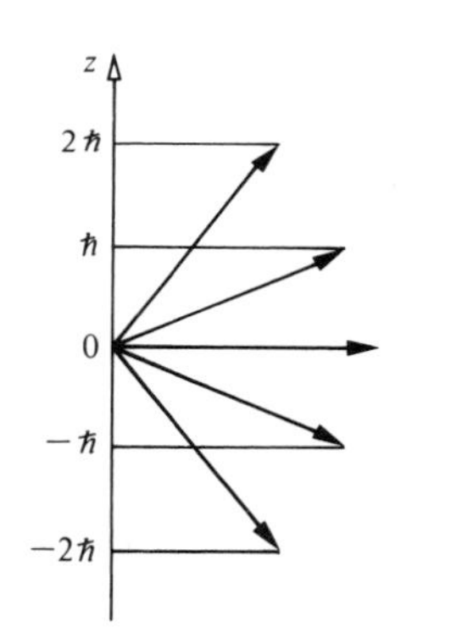

Fig. 4.10. Space quantization.

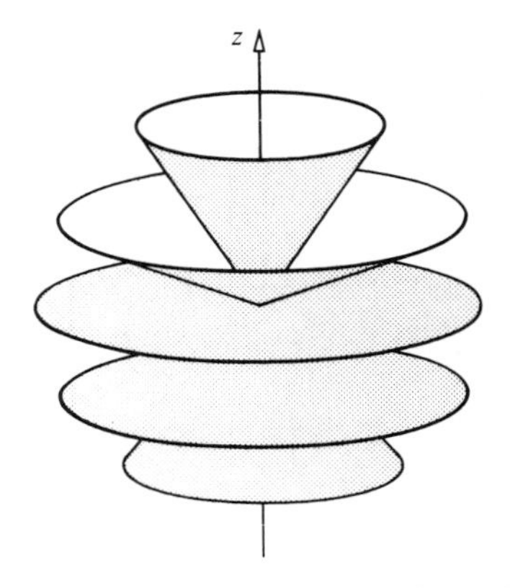

Fig. 4.11. Cones of location of the angular momentum vector.

as action (ML^2T^{-1}) and so is normally expressed as so many joule-seconds, J s.

When the particle is confined to a ring (i.e. when the body can rotate in only two dimensions) the angular momentum is fully determined by the value of m_l, which also carries information about the direction of the angular momentum as well as its magnitude. In three dimensions the vector representing the angular momentum need not be parallel to the z-axis. Nevertheless, we can still speak of the *component* of angular momentum with respect to the z-axis, Fig. 4.9, and m_l specifies its value. A very important point is that since m_l is allowed only discrete, integral values, the z-component of angular momentum is also quantized, and can take only discrete values. This is called *space quantization*, since in classical terms it appears to signify that a rotating body can adopt only a discrete number of orientations in space. For example, when $l=2$, so that m_l may take the five values $2, 1, 0, -1, -2$, the magnitude of the angular momentum is $\hbar\sqrt{6} = 2.45\hbar$ and the z-component may be one of the five values $2\hbar$, $\hbar$, 0, $-\hbar$, $-2\hbar$. This is illustrated in Fig. 4.10. You should note, however, that in the absence of externally applied fields, the z-direction is entirely arbitrary, and space itself does not have any discrete structure.

So far we have spoken only of the z-component of angular momentum. This is the only component quantum mechanics allows us to specify (although, as mentioned above, which direction we select as z is entirely arbitrary, and different choices can be made for every object in a collection). The quantum numbers l and m_l give us no information about the values of the x- and y-components of angular momentum. Therefore, if we want to depict the angular momentum of an object, we have to leave unspecified the x- and y-components. This is not just a deficiency of the calculation: later we shall see that it is an intrinsic feature of the quantum mechanics of angular momentum. Therefore, a *complete* depiction of the angular momentum is as one of the cones drawn in Fig. 4.11 where each surface indicates the possible but unspecifiable location of the vector of length $\hbar\sqrt{6}$ and z-component $m_l\hbar$. (At this stage you should not think of the angular momentum vector as sweeping out one of the cones, but simply as lying at some orientation. We shall add the idea of this motion, which is called *precession*, when we study the effects of externally applied fields.) Notice that the angular momentum vector can never lie exactly parallel to the z-axis: its maximum projection is $l\hbar$ (when $m_l = l$), but its length is $\hbar\sqrt{\{l(l+1)\}}$, which is always greater. Only for very large values of l, such as those for macroscopic objects like balls, is $\sqrt{\{l(l+1)\}}$ so close in value to l that to all intents the object is able to rotate solely around the z-axis.

Example. A solid ball of mass 250 g, radius 4.0 cm is spinning at 5.0 revolutions a second. Estimate the value of l and the minimum angle its angular momentum vector can make with respect to some axis.

• *Method.* The moment of inertia of a solid sphere of radius R is mr^2, $r = (\frac{2}{5})^{\frac{1}{2}}R$. Its angular momentum is $I\omega$, ω being the angular velocity (in radians per second). Estimate l from eqn (4.2.11). The maximum z-component is $m_l\hbar$ with $m_l = l$. The corresponding angle is given by $\cos\theta = m_l/\{l(l+1)\}^{\frac{1}{2}}$.

- *Answer.* For a sphere of radius $R = 4.0$ cm, $r = 2.5$ cm and so $I = 1.6\times10^{-4}$ kg m^2. 5.0 revolutions/second corresponds to 31 rad s^{-1}. Hence $J = 5.0\times10^{-3}$ kg m^2 s^{-1}. Therefore, by eqn (4.2.11), $l \approx 6.9\times10^{15}$. When $l \gg 1$, as in this case, and $m_l = l$,

$$\cos\theta = l/\{l^2(1+1/l)\}^{\frac{1}{2}} = 1/(1+1/l)^{\frac{1}{2}} \approx 1/(1+1/2l) \approx 1-1/2l,$$

$$\cos\theta = 1-\tfrac{1}{2}\theta^2 + \ldots \approx 1-\tfrac{1}{2}\theta^2.$$

Hence $\theta \approx 1/\sqrt{l} \approx 1.2\times10^{-8}$ rad.

- *Comment.* 1.2×10^{-8} rad corresponds to 0.000 000 69°, which for all practical purposes is zero. Hence a macroscopic object can rotate solely around a specified axis, with negligible deviation.

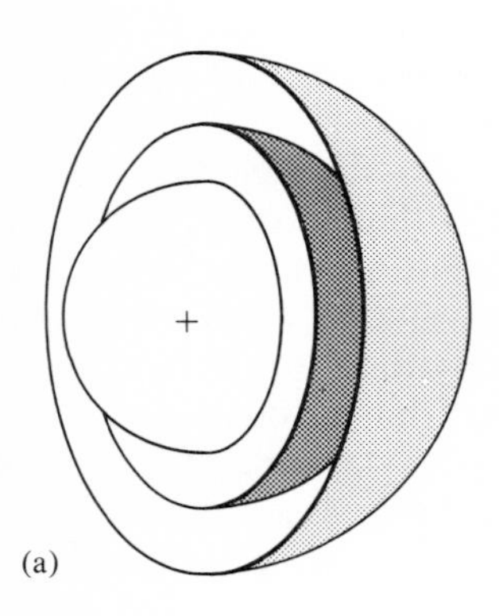

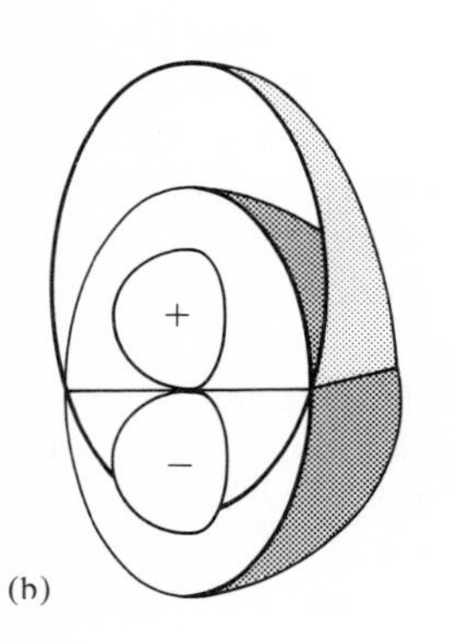

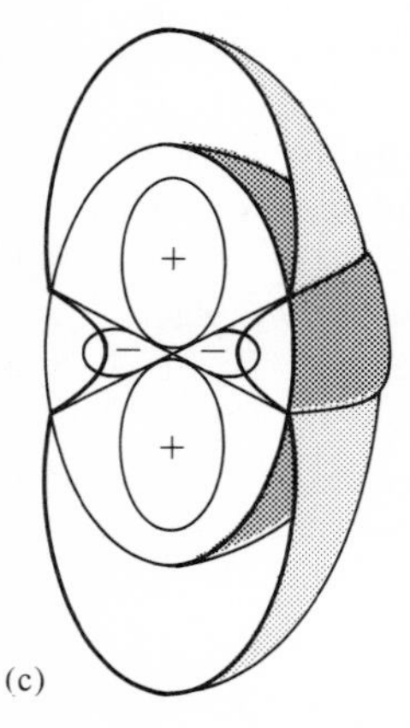

Fig. 4.12. Wavefunctions and auxiliary functions for (a) Y_{00}, (b) Y_{10}, (c) Y_{20}.

The vector model of angular momentum gives rise to pictures that mirror to some extent the distributions predicted on the basis of the wavefunctions. The spherical harmonics can be drawn in the same ways as we used for the particle on the ring, either by plotting their amplitudes over the surface of a sphere or by use of an auxiliary function. In each case we plot the real part of the function (the $\cos m_l\phi$ coming from the $\Phi = (1/2\pi)^{\frac{1}{2}}e^{im_l\phi}$ factor), and the imaginary component $i\sin m_l\phi$, must be held in mind. The amplitudes of Y_{00}, Y_{10}, and Y_{20}, corresponding to $l = 0, 1, 2$ respectively, and $m_l = 0$, are drawn in Fig. 4.12. The number of nodal lines increases with l, the greater buckling of the functions that results accounting for the increasing magnitude of the angular momentum. The auxiliary functions are constructed in the same way as before, but now the amplitude is marked off along a radius extending from the centre of the sphere, and gives rise to a three-dimensional figure. Note that the nodal lines of the actual functions become *nodal planes* of the auxiliaries, and *the number of nodal planes is equal to the value of l.*

The probability distributions of the particle in its various angular momentum states are obtained by forming the square modulus of the wavefunctions, $Y^*_{lm_l}Y_{lm_l}$. The resulting distributions for $l = 0, 1, 2$ and $m_l = 0$ are shown in Fig. 4.13. Notice that the distribution is completely uniform when $l = 0$, and moves progressively towards the poles as l increases. The distribution for other values of m_l are also shown. Notice how the distribution shifts towards the equator as m_l deviates from zero. This corresponds to the reduced tilt of the plane of rotation in the classical case, and the distributions (c)–(e) in Fig. 4.13 for $l = 2$ should be compared with the five paths in Fig. 4.11.

There are two further points. One is that there is no zero-point energy in the system, because the wavefunction need not be curved (recall the argument on p. 62). The second is that the classical limit is reached when the particle is set rotating with a poorly specified energy, when its state is described by a time-dependent wavepacket, the centre of which moves in accord with the predictions of classical physics. Since macroscopic objects are excited to large values of l for even

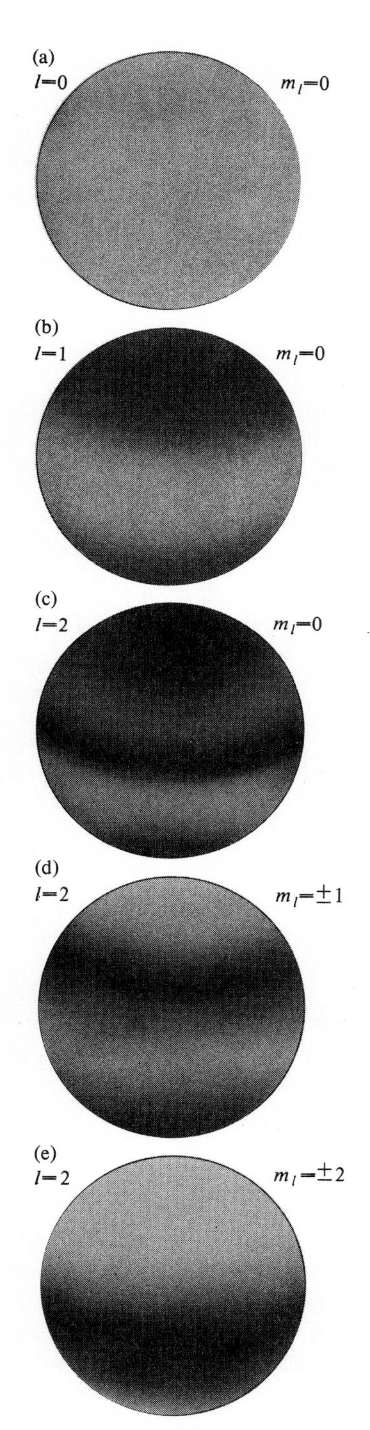

Fig. 4.13. Probability distributions of a particle with $m_l = 0$ and $l =$ (a), 0, (b) 1, (c) 2, and (d) $l = 2$, $m_l = \pm 1$, (e) $l = 2$, $m_l = \pm 2$.

quite modest energies, the wavepackets show no signs of space quantization ($2l+1$ orientations are indistinguishable from a continuum of orientations when l is large), and they appear to rotate smoothly through all angles.

A general comment may be made to conclude this section. It is always pleasing to solve a number of problems simultaneously. We should therefore be delighted that in this section we have made great progress not only with angular momentum but also in the mathematical description of vibrating rings, bouncing balls, tidal motion, and nuclear fission! This is because the spherical harmonics occur in the description of all these phenomena. This is most easily seen if one looks at the drawing of Y_{10} in Fig. 4.12, for the shape there resembles the distortion of a ball that has been dropped on to a surface. The drawing of Y_{00} also resembles the distortion that a symmetrically struck ball would suffer (for example, if a balloon were suddenly subjected to a change in pressure). Figure 4.12 shows another symmetrical distortion of a sphere: if one imagines this distortion being continued, then eventually the sphere breaks into two equal parts. Such a description is the basis of the *liquid drop model* of the nucleus and nuclear fission. The shape in Fig. 4.12 is also what the oceans became under the influence of the gravitational attraction of the sun and moon. As this distortion rotates the tides rise and fall.

4.3 Motion in a Coulomb field: the hydrogen atom

The motion of an electron in a Coulomb field, one where the force varies as $1/r^2$ and hence where the potential energy varies as $1/r$ (because $F = -\mathrm{d}V/\mathrm{d}r$), is of central importance in quantum chemistry. Most of the work involved in solving the Schrödinger equation for this problem has been done in the preceding sections because the electron in the hydrogen atom (the basic example of this type of motion) behaves like a particle on a collection of concentric spheres. Therefore the wavefunctions can be expected to contain as a factor the solutions for a particle on a sphere. The only new work is to account for the *radial* dependence of the wavefunction, the extra degree of freedom which enables the electron to travel between the infinitesimally close spherical surfaces.

The hamiltonian for the hydrogen atom regarded as stationary in space (so that we do not have to worry about the irrelevant kinetic energy of the atom as a whole) is the sum of the kinetic energy and the potential energy. The latter is equal to $-(e^2/4\pi\varepsilon_0)(1/r)$, where ε_0 is the permittivity of free space. Therefore

$$H = -(\hbar^2/2\mu)\nabla^2 - (e^2/4\pi\varepsilon_0)(1/r). \tag{4.3.1}$$

μ is the *reduced mass*, $\mu = m_e m_p/(m_e + m_p)$, which takes account of the fact that the electron (of mass m_e) and the proton (of mass m_p) both rotate about a common centre, and that the motion may be modelled by a single particle of mass μ rotating about a single point (like the sun in the solar system, the nucleus is almost but not quite stationary). The derivation of the form of the reduced mass is given in Appendix 6: we shall encounter it often. The spherical symmetry of the system (that is, the spherical symmetry of the Coulomb field) dictates that we choose the spherical polar form of the laplacian, eqn (4.2.3). Therefore the Schrödinger equation for the hydrogen atom becomes

$$-(\hbar^2/2\mu)\nabla^2\psi-(e^2/4\pi\varepsilon_0 r)\psi=E\psi, \tag{4.3.2}$$

or

$$(1/r)(\partial^2/\partial r^2)r\psi+(1/r^2)\Lambda^2\psi+(\mu e^2/2\pi\varepsilon_0\hbar^2 r)\psi=-(2\mu E/\hbar^2)\psi.$$

We expect the angular part of the solution to be the spherical harmonics, $Y(\theta, \phi)$, corresponding to the motion of a particle on a sphere. Therefore we see whether the solution $\psi(r, \theta, \phi)$ can be expressed in the form

$$\psi(r, \theta, \phi)=R(r)Y(\theta, \phi). \tag{4.3.3}$$

When this expression is substituted in the equation, and we use $\Lambda^2 Y=-l(l+1)Y$ (which is eqn (4.2.8)) it turns into

$$(1/r)(\partial^2/\partial r^2)rRY-(1/r^2)l(l+1)RY+(\mu e^2/2\pi\varepsilon_0\hbar^2 r)RY \\ =-(2\mu E/\hbar^2)RY.$$

Y may be cancelled throughout, leaving an equation for the function R:

$$(1/r)(\mathrm{d}^2/\mathrm{d}r^2)rR+\{(\mu e^2/2\pi\varepsilon_0\hbar^2 r)-l(l+1)/r^2\}R=-(2\mu E/\hbar^2)R.$$

If the product $rR(r)$ is written as the function $P(r)$ the equation becomes

$$\mathrm{d}^2P/\mathrm{d}r^2+(2\mu/\hbar^2)\{(e^2/4\pi\varepsilon_0 r)-l(l+1)\hbar^2/2\mu r^2\}P=-(2\mu E/\hbar^2)P, \tag{4.3.4}$$

which is a one-dimensional Schrödinger equation that would be obtained if, instead of choosing the pure Coulombic potential in the hamiltonian, we had chosen an *effective potential energy*, V_{eff}, which depended on the angular momentum of the electron:

$$V_{\text{eff}}=-e^2/4\pi\varepsilon_0 r+l(l+1)\hbar^2/2\mu r^2. \tag{4.3.5}$$

The effective potential energy may be given a simple physical interpretation. The first part is the attractive Coulomb potential energy and the second angular-momentum-dependent part, is repulsive and corresponds to a centrifugal force. (In classical mechanics the centrifugal force acting on a particle of mass μ is (angular momentum)$^2/\mu r^3$; if this is integrated to find the corresponding potential and the quantum mechanical expression for the angular momentum inserted, we obtain the repulsive part of the potential.)

When $l=0$ the electron has no angular momentum and the force on it is everywhere attractive because there is no motion to fling it away from the nucleus. The potential energy for this special case is everywhere negative, and is shown in Fig. 4.14(a). When l is not zero the electron possesses an angular momentum which tends to fling it from the vicinity of the nucleus, and there is a competition between this effect and the attractive part of the potential. At very short distances from the nucleus the repulsive component of the effective potential tends more strongly to infinity than the attractive part (its radial dependence is r^{-2} whereas the latter's is only r^{-1}). At large distances the repulsive part decays to zero more rapidly than the attractive part and so the latter dominates. The shape of the effective potential energy for non-zero angular momenta is shown in Fig. 4.14(b). The two potential energies (for $l=0$ and $l\neq 0$) are qualitatively quite different and we investigate them separately.

When $l=0$ the repulsive part of the effective potential energy is absent. When r is close to zero, the potential energy $(e^2/4\pi\varepsilon_0 r)$ so dominates E that the latter may be neglected in eqn (4.3.4). The equation therefore becomes

$$d^2P/dr^2+(\mu e^2/2\pi\varepsilon_0\hbar^2)P/r\sim 0 \quad \text{for} \quad l=0,\ r\sim 0. \tag{4.3.6}$$

A solution of this equation is $P\sim Ar+\frac{1}{2}Ar^2$ (check by substitution followed by taking the limit $r\rightarrow 0$). Therefore the wavefunction itself has the form $R=P/r\sim A$, a non-zero constant. In other words, *when* $l=0$ *there is a non-zero probability of finding the electron at the nucleus.* This is an echo of the shape of the orbit in the Bohr theory in which the electron swings in a straight line through the nucleus, and which he dismissed as unacceptable. (The spherical symmetry reflects the uncertainty of the orientation of the straight line.)

When $l\neq 0$ the large repulsive component of the effective potential energy of the electron in the vicinity of the nucleus (see Fig. 4.14(b)) has the effect of forbidding it to be found there. In classical terms, the centrifugal force of an electron with non-zero angular momentum is strong enough at short distances to overcome the attractive Coulombic force. When $l\neq 0$ but r is close to zero, eqn (4.3.4) becomes

$$d^2P/dr^2-\{l(l+1)\hbar^2/r^2\}P\sim 0$$

because $1/r^2$ is the dominant term. The solution has the form

$$P\sim Ar^{l+1}+Br^{-l} \quad \text{for} \quad l\neq 0 \quad \text{and} \quad r\sim 0.$$

Since $P=rR$, $P=0$ at $r=0$, and so $B=0$ necessarily. Therefore the radial function has the form

$$R=P/r\sim Ar^l \quad \text{for} \quad r\sim 0.$$

Since $l\neq 0$ it follows that $R=0$ at $r=0$. This implies that there is a node at $r=0$ in all wavefunctions with $l\neq 0$, and the electron will never be found at the nucleus.

Explicit solutions of the radial wave equation can be found in the same ways as for the harmonic oscillator and the particle on a sphere. One technique involves finding factors of the hamiltonian, and turning the second-order differential equation into a much simpler first-order

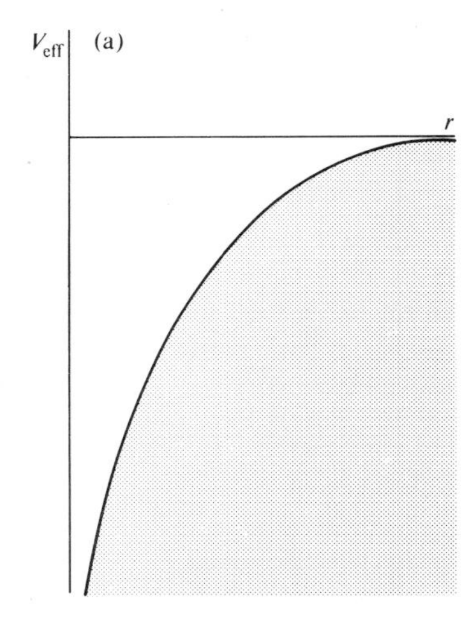

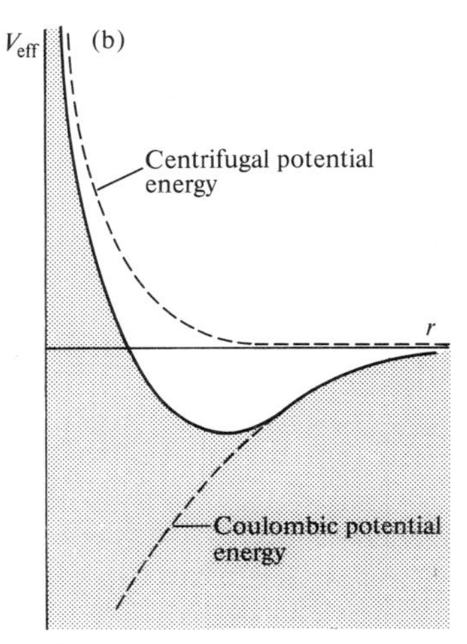

Fig. 4.14. The effective potential energy for (a) $l=0$, (b) $l\neq 0$.

equation. Another is to recognize that the equation has a standard mathematical form, and that its solutions are 'well known'. We adopt the latter course here, but give the explicit factorization solution in Appendix 7.

Example. Show that at large distances from the nucleus, bound state atomic wavefunctions decay exponentially towards zero.

- *Method.* Identify the terms that survive in eqn (4.3.4) as $r \to \infty$, and solve the resulting equation.
- *Answer.* When $r \to \infty$ eqn (4.3.4) reduces to $d^2P/dr^2 \sim -(2\mu E/\hbar^2)P$. But $P = rR$, and so in the same limit, since $d^2P/dr^2 = 2R' + rR'' \sim rR''$ it is also equal to

$$d^2R/dr^2 \sim -(2\mu E/\hbar^2)R = +(2\mu\,|E|/\hbar^2)R$$

because E is negative (for bound states). This is satisfied by $R \sim e^{-(2\mu E/\hbar^2)^{\frac{1}{2}}r}$; hence the wavefunction decays exponentially at large distances.

- *Comment.* The unbound states correspond to $E > 0$, and the same analysis leads to the conclusion that $R \sim e^{\pm i(2\mu E/\hbar^2)^{\frac{1}{2}}r}$, which are oscillating functions corresponding to travelling waves. If (for $E < 0$) we write $R \sim f(r)e^{-(2\mu|E|/\hbar^2)^{\frac{1}{2}}r}$, the same analysis also identifies $f(r) \sim r^l$; hence the detailed asymptotic form of hydrogen-like atomic orbitals is $R(r) \sim r^l e^{-\kappa r}$, $\kappa = (2\mu\,|E|/\hbar^2)^{\frac{1}{2}}$; the decaying exponential factor dominates r^l for all l.

The radial wave equation is the same as the differential equation in mathematics which has the *associated Laguerre functions* as acceptable solutions.† The first few of the functions are given in Table 4.2, they

Table 4.2 Hydrogen-like radial wavefunctions

n	l	$R_{nl}(r)$
1	0(1s)	$(Z/a)^{\frac{3}{2}}2e^{-\rho/2}$
2	0(2s)	$(Z/a)^{\frac{3}{2}}(1/2\sqrt{2})(2-\rho)e^{-\rho/2}$
	1(2p)	$(Z/a)^{\frac{3}{2}}(1/2\sqrt{6})\rho e^{-\rho/2}$
3	0(3s)	$(Z/a)^{\frac{3}{2}}(1/9\sqrt{3})(6-6\rho+\rho^2)e^{-\rho/2}$
	1(3p)	$(Z/a)^{\frac{3}{2}}(1/9\sqrt{6})(4-\rho)\rho e^{-\rho/2}$
	2(3d)	$(Z/a)^{\frac{3}{2}}(1/9\sqrt{30})\rho^2 e^{-\rho/2}$

$\rho = (2Z/na)r$; $a = 4\pi\varepsilon_0\hbar^2/\mu e^2$.
For an infinitely heavy nucleus $\mu = m_e$ and $a = a_0$, the Bohr radius.
Relation to associated Laguerre functions:

$$R_{nl}(r) = -\left(\frac{2Z}{na}\right)\left\{\frac{(n-l-1)!}{2n[(n+l)!]^3}\right\}\rho^l L_{n+l}^{2l+1}(\rho)e^{-\rho/2}.$$

Further information: M. Abramowitz and I. A. Stegun, *Handbook of mathematical functions*, Dover (1965), Chapter 22.

consist of an exponential function multiplied by a simple polynomial. Each one is specified by two labels, n and l. n is restricted to the values

† See M. Abramowitz and I. A. Stegun, *Handbook of mathematical functions*, Dover, (1965), Chapter 22. By 'acceptable' we mean (a) conform to the Born interpretation and (b) correspond to states of negative energy; that is, to bound states of the atom.

1, 2, 3, . . . , and for any given n the label l can take the values $l=0, 1, 2, \ldots, n-1$. Some of the functions are plotted in Fig. 4.15. Notice that those with $l=0$ are non-zero (and finite) at $r=0$ while all those with $l \neq 0$ are zero at $r=0$. This accords with the discussion above. Some of the functions are zero at other radii: these *radial nodes* occur at values of r where the polynomial component of the associated

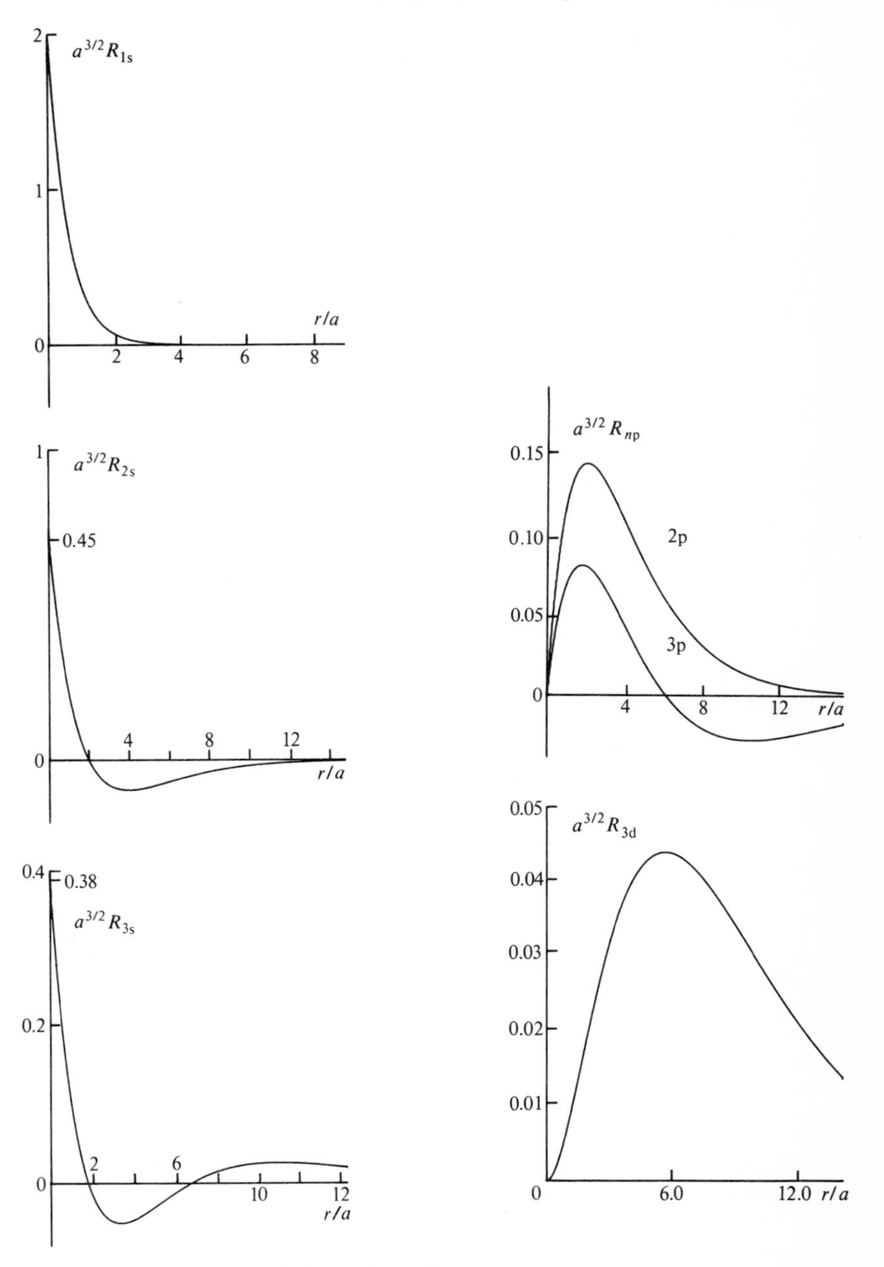

Fig. 4.15. Hydrogen radial wavefunctions.

Laguerre functions vanish. Notice that the functions with the highest permitted values of l (those with $l = n - 1$) have no radial node except the one at the nucleus.

Insertion of the radial wavefunctions into eqn (4.3.4) lets us arrive at an expression for the energy. We find

$$E = -(\mu e^4/32\pi^2\varepsilon_0^2\hbar^2)(1/n^2) \qquad n = 1, 2, \ldots \tag{4.3.7}$$

irrespective of the value of l. Hence n plays the role of a quantum number for the energy of the electron in the hydrogen atom, and is called the *principal quantum number.* It is most important to note that the energy depends only on n and on neither l nor m_l. This result is peculiar to the Coulomb potential and we return to it shortly.

4.4 Atomic orbitals

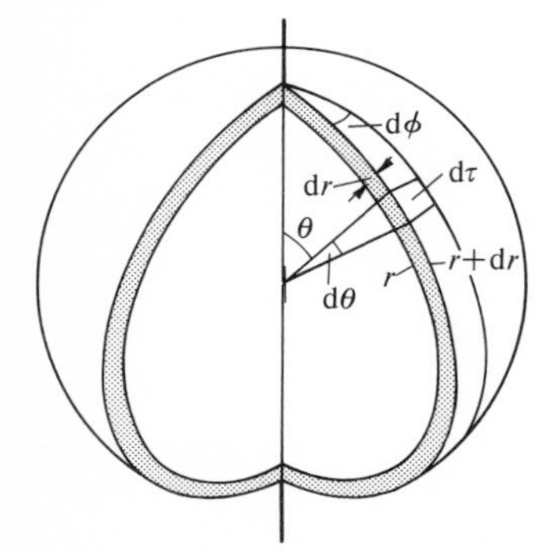

Fig. 4.16. The volume element in spherical polar coordinates.

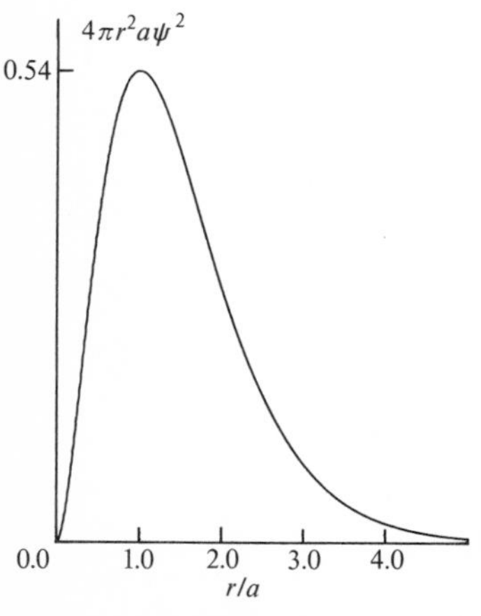

Fig. 4.17. The radial distribution function for the hydrogen 1s-orbital.

We are now in a position to discuss the complete solutions of the states of the hydrogen atom. The complete wavefunctions are the functions $\psi_{nlm_l}(r, \theta, \phi) = R_{nl}(r)Y_{lm_l}(\theta, \phi)$. In atoms wavefunctions are called *orbitals,* a name selected to convey a less precise meaning than the classical term 'orbit'. For historical reasons orbitals with $l = 0$ are called *s-orbitals,* those with $l = 1$, *p-orbitals,* those with $l = 2$, *d-orbitals,* and those with $l = 3$, *f-orbitals* (after that we have g-, h-, . . . , orbitals). The square modulus $|\psi|^2 = \psi^*\psi$ is the probability density for the electron in the sense that the probability of the electron being in an infinitesimal volume element $d\tau$ located at the point r, θ, ϕ is $|\psi_{nlm_l}(r, \theta, \phi)|^2\, d\tau$ if the electron is in a state with quantum numbers n, l, m_l.

In a number of applications it is interesting to know not the most probable *point* of finding the electron but the most probable *distance* from the nucleus irrespective of direction. We shall confine our attention to s-orbitals, which have spherical symmetry. The probability of finding the particle in a volume element $d\tau$ at some single point r, θ, ϕ is $|\psi(r, \theta, \phi)|^2\, d\tau$. Since ψ is real for an s-orbital we can drop the modulus signs. In spherical polar coordinates the volume element is $d\tau = r^2\, dr \sin\theta\, d\theta\, d\phi$, Fig. 4.16. The probability of finding the particle in the spherical shell of thickness dr and radius r is the sum of all these probabilities as θ and ϕ move over all their values ($0 \le \theta \le \pi, 0 \le \phi \le 2\pi$). This is the integral $\int_0^\pi d\theta \sin\theta \int_0^{2\pi} d\phi\, |\psi(r, \theta, \phi)|^2 r^2\, dr$. In the case of s-orbitals the wavefunction is independent of θ and ϕ, and so they may be integrated immediately to give 4π. The probability of finding the particle at any location on the spherical shell is therefore $4\pi r^2\psi^2(r)\, dr$, and the quantity $4\pi r^2\psi^2(r)$ is called the *radial distribution function.* It is plotted for the 1s-orbital in Fig. 4.17: whereas the probability density itself is a maximum at the nucleus (that being the most probable *point* at which the electron will be found) the radial distribution function is zero there (because of the r^2 factor) and goes through a maximum at $r = 8\pi\varepsilon_0\hbar^2/\mu e^2$. By coincidence (not the only one where the Coulomb problem is concerned) this is also the radius of the lowest energy orbit in the Bohr theory of the atom.

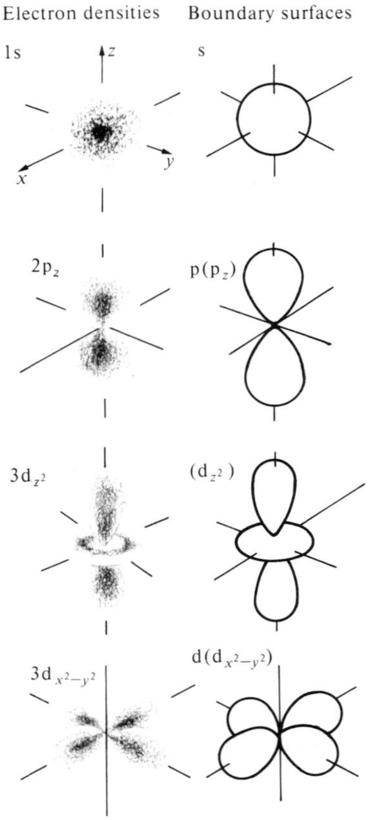

The shapes of the atomic orbitals may be expressed in a variety of ways. One way is to denote the probability amplitude by the density of shading, and the resulting diagrams for some of the orbitals are shown in Fig. 4.18. Such diagrams convey too much information for most qualitative applications, and so orbitals are often represented by their *boundary surfaces.* These are the surfaces within which there is a high probability (often 90 per cent is the value taken) of finding the electron. The diagrams that result are shown in Fig. 4.19. Often the phase of the wavefunction itself is denoted by a + or − sign.

Fig. 4.18. (*Left*) Electron density depiction of orbitals.
Fig. 4.19. (*Right*) Boundary surfaces corresponding to the distributions in Fig. 4.18.

An s-orbital with principal quantum number n is called the *n*s-*orbital.* An electron that occupies it (i.e. an electron in the hydrogen atom in the state $n, l=0$) is called an *n*s-*electron.* Since when $l=0$ the only value allowed to m_l is zero, there is only one s-orbital for each value of n.

What accounts for the stability of s-orbitals? We find the answer in the competition between the electron's kinetic and potential energies. In order that the particle may cluster more closely to the nucleus, and so reach a state of lower potential energy, its wavefunction must be confined to the vicinity of the nucleus, Fig. 4.20. As the wavefunction's confinement increases its curvature must increase; therefore the kinetic energy associated with it increases too, just as happens for a particle in a box as a wall is pushed in. Therefore, although the potential energy decreases as the electron clusters more closely to the nucleus, the kinetic energy increases and the state of minimum *total* energy is the one with the electron clustering in the general vicinity of the nucleus, but not confined exactly to it. The stability of the ground state of the hydrogen atom (and of all the s-orbitals, and in part of the other orbitals too) stems from the rise in kinetic energy that accompanies *radial confinement* of a particle, and to that extent it is a purely quantum mechanical phenomenon. That atoms, and therefore things, are not points but have bulk is entirely due to the fact that h is not zero.

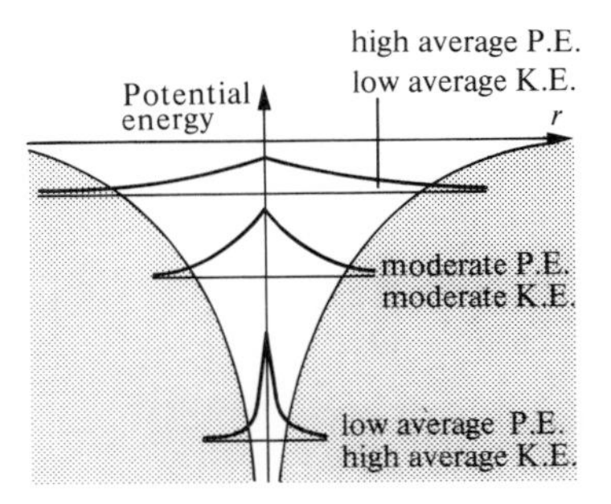

Fig. 4.20. Competition between kinetic energy and potential energy.

A p-orbital with principal quantum number n is called an *n*p-*orbital* and an electron that occupies it an *n*p-*electron.* Since when $l=1$ m_l may take the values $-1, 0, 1$, there are three p-orbitals for each value of n (for $n>1$). The p-orbital with $m_l=0$, which has its principal amplitude along the z-axis is called the p_z-*orbital.* The orbitals with $m_l=\pm1$, p_+ and p_-, are complex, and have their maximum amplitude in the x, y-plane, just as in Fig. 4.13. The linear combinations $p_+ + p_-$ and $i(p_+ - p_-)$ are real (they correspond to the $\cos\phi$ and $\sin\phi$ components of the spherical harmonics) and have their maxima along the x- and y-directions respectively. They are therefore called the p_x- and p_y-*orbitals* respectively, Fig. 4.21. In free atoms the original p_+ and p_- orbitals, with their well-defined angular momenta around the z-axis, are the most natural ones to use; we shall see that in molecules, where neighbouring atoms help to define specific directions in space, it is

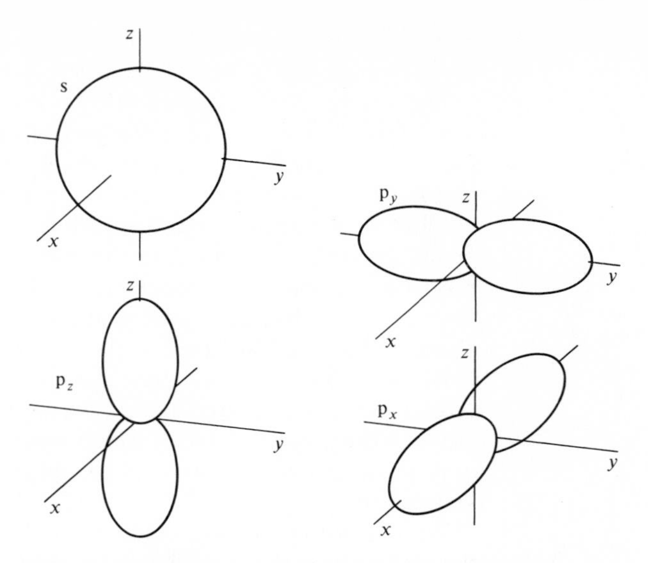

Fig. 4.21. Boundary surfaces of s- and p-orbitals.

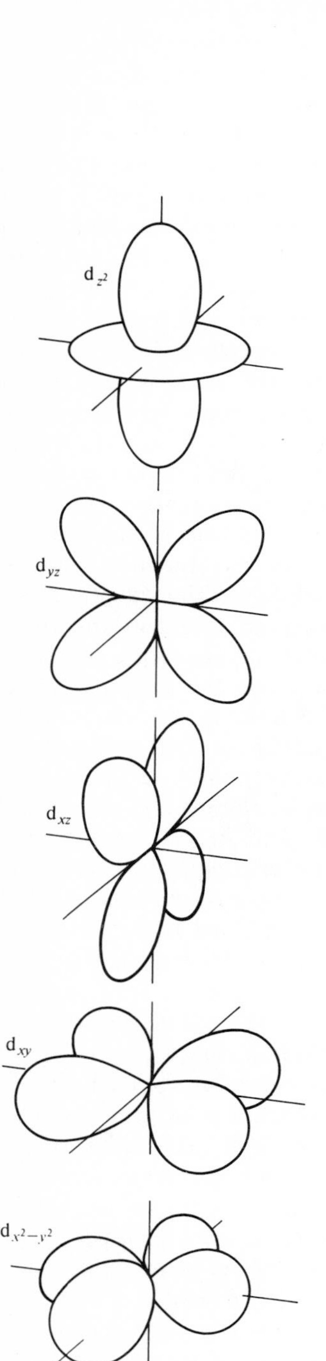

Fig. 4.22. Boundary surfaces for d-orbitals.

often more appropriate to use the linear combinations p_x and p_y. That is why in most accounts of quantum chemistry only the p_x and p_y forms are drawn and discussed.

Example. What is the most probable point at which an electron in a $2p_z$-orbital will be found? What is the probability of finding an electron within a sphere of radius R centred on the nucleus?

- *Method.* Locate the maximum of $\psi^2_{2p_z}(\mathbf{r})$. Construct $\psi_{2p_z}(\mathbf{r})$ from the information in Tables 4.1 and 4.2 noting that $2p_z$ corresponds to $n=2$, $l=1$, $m_l=0$; hence $\psi_{2p_z}=R_{21}Y_{10}$. For the second part, integrate ψ^2 over a sphere of radius R.
- *Answer.* $\psi_{2p_z}(\mathbf{r})=\frac{1}{2}(1/6a_0^3)^{\frac{1}{2}}\rho e^{-\rho/2}\frac{1}{2}(6/2\pi)^{\frac{1}{2}}\cos\theta,\ \rho=r/a_0$

  $=\frac{1}{4}(1/2\pi a_0^3)^{\frac{1}{2}}\rho\cos\theta e^{-\rho/2}.$

  Hence $\psi^2_{2p_z}(\mathbf{r})=(1/32\pi a_0^3)\rho^2\cos^2\theta e^{-\rho}$.

This has a maximum when $\cos^2\theta=1$. Then its maximum is where

$$(d/d\rho)\rho^2e^{-\rho}=0;\quad \text{or } 2\rho e^{-\rho}-\rho^2e^{-\rho}=0.$$

Therefore the probability is greatest when $\theta=0$ or π and $\rho=2$; that is, at a distance $2a_0$ from the nucleus along the z-axis (in either direction).

$$P(R)=\int_0^R dr r^2\int_0^\pi d\theta\sin\theta\int_0^{2\pi}d\phi(1/32\pi a_0^3)\rho^2\cos^2\theta e^{-\rho}$$

$$=(1/32\pi a_0^3)(1/a_0^2)\left\{\int_0^R r^4e^{-r/a_0}dr\right\}\left\{\int_0^\pi\cos^2\theta\sin\theta\,d\theta\right\}\left\{\int_0^{2\pi}d\phi\right\}$$

$$=(1/32\pi a_0^5)(24a_0^5)\{1-[1+(R/a_0)+\tfrac{1}{2}(R/a_0)^2+\tfrac{1}{6}(R/a_0)^3+\tfrac{1}{24}(R/a_0)^4]e^{-R/a_0}\}\{\tfrac{2}{3}\}\{2\pi\}$$

$$=1-\{1+(R/a_0)+\tfrac{1}{2}(R/a_0)^2+\tfrac{1}{6}(R/a_0)^3+\tfrac{1}{24}(R/a_0)^4\}e^{-R/a_0}.$$

When $R=2a_0$, $P=1-7/e^2=0.053$.

• *Comment.* Note that there is little significance in the radial distribution function when the distribution is not isotropic. It can still be defined (as $4\pi r^2 R^2(r)$), but not interpreted as the probability of occurrence at a radius. Note that $P=0.90$ when $R=8a_0=423$ pm, reflecting the diffuseness of the 2p-orbitals (compare with the result in the *Example* on p. 32).

A d-orbital with principal quantum number n is called an *nd-orbital*, and an electron occupying it is an *nd-electron*. Since when $l=2$, m_l may take one of the values $-2, -1, 0, 1, 2$, there are five d-orbitals for each value of n (so long as n is greater than 2). The orbitals with $m_l \neq 0$ are complex, and it is normal to represent them by the real combinations d_2+d_{-2}, d_1+d_{-1}, etc. The boundary surfaces for these five functions are shown in Fig. 4.22. Note that d-orbitals have two *angular nodes* (arising from $l=2$ in the spherical harmonics).

The energies of the orbitals depend only on their principal quantum number (E in eqn (4.3.7) depends only on n, not on l or m_l). Therefore, although the lowest energy level ($n=1$) is non-degenerate (there is only one orbital corresponding to $n=1$, the 1s-orbital) the next higher level ($n=2$) corresponds to any one of *four* states (the 2s-orbital and the three 2p-orbitals). It is therefore four-fold degenerate. The next higher level (with $n=3$) corresponds to any one of *nine* states (the 3s-orbital, the three 3p-orbitals, and the five 3d-orbitals). It is therefore nine-fold degenerate. In general, the level corresponding to the principal quantum number n is *n^2-fold degenerate.* (When electron spin is taken into account the degeneracies double to $2n^2$.) These degeneracies can be seen very easily in the energy-level diagram, Fig. 4.23.

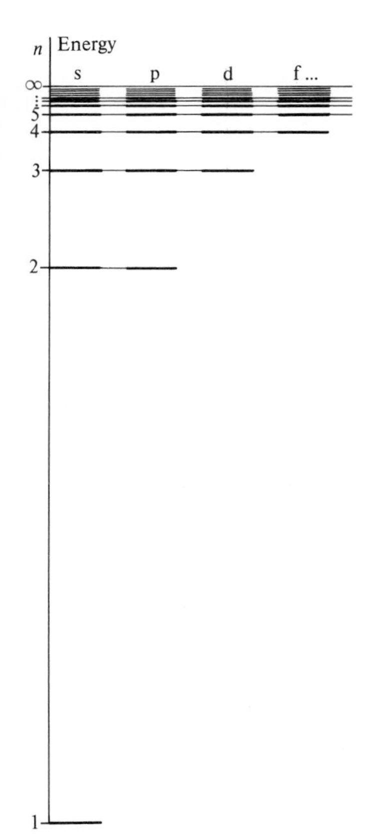

Fig. 4.23. The energy levels of atomic hydrogen.

We have already emphasized that the extremely high degeneracy of the hydrogen atom is a special property of the Coulomb potential, and would not be present if the potential energy varied as $1/r^s$, $s \neq 1$. It may be viewed as an example of *accidental degeneracy* because there appears to be no way of relating s, p, d, . . . orbitals to each other by symmetry transformations (e.g. rotations) of the atom. (Recall the discussion of the degeneracy of the 'square square' well.) This, however, is too superficial a view, for the Coulomb potential is quite remarkable in having very special symmetries that show up only when expressed in higher than three dimensions. What this means is that a four-dimensional being would be able to see at a glance that a 2p-orbital could be rotated into a 2s-orbital, and therefore would be no more surprised at their degeneracy than we are at the degeneracies of the three 2p-orbitals themselves (p_x, p_y, and p_z can obviously be rotated into each other). A glimpse of how this comes about is shown in Fig. 4.24 where we have imagined how a two-dimensional being might experience the projection of a patterned sphere. It is quite easy for us to see that one of the rotations of the sphere leads to a change in the projection of the diagram which would lead the Flatlander to think that a p-orbital has been transformed into an s-orbital. We, in our three dimensions, can easily see that they are related by symmetry; he, in two, might not. The hydrogen atom has exactly this kind of higher-dimensional symmetry.

Finally, we have to emphasize that we have restricted attention to states of negative energy (see footnote p. 72). These are the *bound states* of the hydrogen atom. The difference in energy between the lowest ($n=1$) and highest ($n=\infty$) levels is $\mu e^4/32\pi^2\varepsilon_0^2\hbar^2 = 2.2\times10^{-18}$ J, a quantity called the *ionization energy* of the atom. (This energy corresponds to a potential difference of 13.6 V for a bound electron being removed to infinity, and so the *ionization potential* is 13.6 V.) But suppose the electron is given more than that energy (for instance, by collision with a cosmic ray particle): what is its state? The *unbound states* of the hydrogen atom, those corresponding to positive energy (or an energy greater than 2.2×10^{-18} J above the lowest bound state) are unquantized, and are referred to as the *continuum states*. They are of great importance in scattering theory, but for many problems in chemistry it is safe (or, if not safe, simple) to neglect them. For the present we shall pretend that they are too high in energy to be important, but we should remember their existence and note that they play a central role in problems involving the collisions of electrons with atoms.

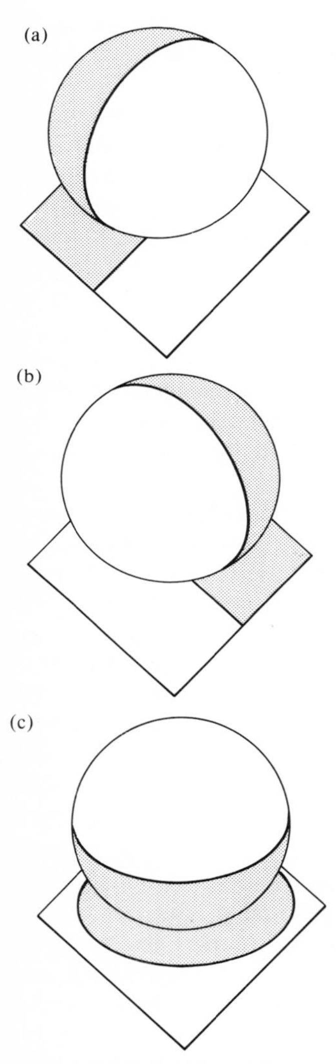

Fig. 4.24. The rotation from (a) to (b) models $2p_x \rightarrow 2p_y$; that from (b) to (c) models $2p \rightarrow 2s$.

Further reading

Angular momentum (*see also Chapter 6*)

Introduction to quantum mechanics. L. Pauling and E. B. Wilson; McGraw-Hill, New York, 1935.
Quantum chemistry. W. Kauzmann; Academic Press, New York, 1957.
Quantum chemistry. J. P. Lowe; Academic Press, New York, 1978.
Quantum chemistry. I. N. Levine; Allyn and Bacon, Boston, 2nd edition, 1974.

The hydrogen atom

The books referred to above, together with:
Atomic structure. E. U. Condon and H. Odabaşi; Cambridge University Press, 1980.
Atomic spectra and atomic structure. G. Herzberg; Dover, New York, 1944.
Symmetry and degeneracy. H. V. MacIntosh; *Group theory and its applications* (ed. E. M. Loebl), II, Academic Press, New York, 1971.
Group theory and the hydrogen atom. M. Bander and C. Itzykson; *Rev. Mod. Phys.*, **38,** 330 and 346 (1966).

Problems

4.1. The rotation of the HI molecule can be pictured as an orbiting of the hydrogen atom at a radius of 160 pm about a virtually stationary I atom. If the rotation is thought of as taking place in a plane (a restriction removed later), what are the rotational energy levels? What wavelength radiation is emitted in the transition $m_l = 1 \rightarrow m_l = 0$?

4.2. Show that the wavefunctions in eqn (4.1.7) are mutually orthogonal.

4.3 Calculate the rotational energy levels of a gramophone record of radius 10 cm, mass 150 g free to rotate in a plane. To what value of m_l does a rotation rate of 33 r.p.m. correspond?

4.4. Construct the analogues of Figs. 4.4–4.6 for the states of a rotor with $m_l = 3$ and 4.

4.5. Construct a wavepacket $\Psi = N\sum_{m_l}(1/m_l!)e^{im_l\phi}$ and normalize it to unity. Sketch the form of $|\Psi|^2$ for $0 \leq \phi \leq 2\pi$. Calculate $\langle\phi\rangle$, $\langle\sin\phi\rangle$, and $\langle l_z\rangle$. Why is $\langle l_z\rangle < \hbar$? ***Hint.*** Draw on a variety of pieces of information, including $\sum_{n=0}^{\infty}(x^n/n!) = e^x$, and the following integrals:

$$\int_0^{2\pi} e^{z\cos\phi}\,d\phi = 2\pi I_0(z), \qquad \int_0^{2\pi}\cos\phi\, e^{z\cos\phi}\,d\phi = 2\pi I_1(z)$$

with $I_0(2) = 2.280$, $I_1(2) = 1.591$ (M. Abramowitz and I. A. Stegun, *Handbook of mathematical functions*, Dover (1965), Table 9.8; the $I(z)$ are *modified Bessel functions*).

4.6. Investigate the properties of the more general wavepacket $\sum_{m_l=0}^{\infty}(\alpha^{m_l}/m_l!)e^{im_l\phi}$, and show that when α is large $\langle l_z\rangle \sim \alpha\hbar$. ***Hint.*** Proceed as in the last *Example.* The large value expansions of $I_0(z)$ and $I_1(z)$ are $I_0(z) \sim I_1(z) \sim e^z/\sqrt{(2\pi z)}$. (Abramowitz and Stegun (see Problem 4.6) (9.7.1).)

4.7. Confirm that the wavefunctions for a particle on a sphere may be written $\psi(\theta, \phi) = \Theta(\theta)\Phi(\phi)$ by the method of separation of variables, and find the equation for Θ.

4.8. Confirm that Y_{11} and Y_{20} as listed in Table 4.1 are solutions of the Schrödinger equation for a particle on a sphere. Confirm by explicit integration that Y_{11} and Y_{20} are normalized and mutually orthogonal. *Hint.* The volume element for the integration is $\sin\theta\, d\theta\, d\phi$, with $0 \leq \phi \leq 2\pi$ and $0 \leq \theta \leq \pi$.

4.9. Modify Problem 4.1 so that the molecule is free to rotate in three dimensions. Calculate the energies and degeneracies of the lowest three rotational levels, and predict the wavelength of radiation involved in the $l = 1 \rightarrow 0$ transition.

4.10. Calculate the angle that the angular momentum vector makes with the z-axis when the system is described by the wavefunction ψ_{l,m_l}. Show that the minimum angle approaches zero as l approaches infinity. Calculate the allowed angles for the cases $l = 1, 2$, and 3.

4.11. Draw the analogues of Fig. 4.13 for $l = 3$. Notice how the maxima (of $|Y|^2$) move into the equatorial plane as $|m_l|$ increases, and relate the diagrams to the conclusions drawn in Problem 4.10.

4.12. Obtain an expression for the energy levels of a hydrogen-like one-electron atom with atomic number Z. *Hint.* Replace $e^2/4\pi\varepsilon_0 r$ in eqn (4.3.1) by $Ze^2/4\pi\varepsilon_0 r$, and decide how Z will then appear in eqn (4.3.7).

4.13. Calculate the mean kinetic and potential energies of an electron in the ground state of the hydrogen atom, and confirm that the virial theorem is satisfied. *Hint.* Evaluate $\langle T\rangle = \int\psi_{1s}^*(-\hbar^2/2\mu)\nabla^2\psi_{1s}\,d\tau$ and $\langle V\rangle = \int\psi_{1s}^2(-e^2/4\pi\varepsilon_0 r)\,d\tau$; the laplacian is given in eqn (4.2.3). The virial theorem is dealt with in Appendix 5.

4.14. Confirm that the radial wavefunctions R_{10}, R_{20}, and R_{31} do satisfy the radial wave equation, eqn (4.3.4). Use Table 4.2.

4.15. Locate the radial nodes of the (a) 2s-orbital, (b) 3s-orbital.

4.16. Calculate (a) the mean radius, (b) the mean square radius, and (c) the most probable radius of the 1s-, 2s-, and 3s-orbitals of a hydrogen-like atom of atomic number Z. *Hint.* For the most probable radius look for the principal maximum of the radial distribution function.

4.17. A quantity important in some branches of spectroscopy (Chapter 14) is the probability of an electron being found at the same location as the nucleus. Evaluate this probability density for an electron in the 1s-, 2s-, and 3s-orbitals of a hydrogen-like atom.

4.18. Confirm that ψ_{1s} and ψ_{2s} are mutually orthogonal.

4.19. Another quantity of interest in spectroscopy is the average value of $1/r^3$ (for example, the average magnetic dipole interaction between the electron and nuclear magnetic moments depends on it). Evaluate $\langle 1/r^3 \rangle$ for an electron in a 2p-orbital of a hydrogen-like atom.

4.20. Calculate in joules and cm^{-1} the energies of an electron in the 1s-, 2s-, 2p-, 3s-, 3p-, and 3d-orbitals of the hydrogen atom, and the wavenumbers and wavelengths of the radiation emitted during the transitions 3d $\rightarrow$ 2p, 3p $\rightarrow$ 2s, 2p $\rightarrow$ 1s.

4.21. Calculate the difference in ionization energies of ^{1}H and ^{2}H.

4.22. For a given principal quantum number n, l takes the values $0, 1, \ldots, n-1$, and for each l, m_l takes the values $m_l = l, l-1, \ldots, -l$. Confirm that the degeneracy of the term with principal quantum number n is equal to n^2.

4.23. Confirm, by drawing pictures like those in Fig. 4.24, that a Flatlander might be shown that 3s-, 3p-, and 3d-orbitals are degenerate.

Part 2 Foundations 2

THE elementary aspects of quantum mechanics have now been established. In this part we refine them, and in the process introduce some extremely powerful techniques. These techniques—operator theory, angular momentum, and group theory, will be encountered wherever quantum theory is applied. There comes a point, however, where we have to accept that it is a fruitless task looking for exact solutions to the Schrödinger equation for chemically interesting problems, and so we have to see how to construct approximate solutions. Formulating and assessing approximations is probably the most important aspect of quantum chemistry, and at the end of this part we shall see something of the techniques involved. That chapter is a springboard for the discussion of applications in Part 3.

5 The properties of operators

Now we collect the material of the preceding chapters into a more systematic form. We do so by introducing quantum mechanics as a set of postulates which, when developed, imply all the material we have covered so far. Apart from being a tidying-up operation, this serves to introduce the language we encounter throughout the remainder of the book. We shall see that a great deal of quantum mechanics can be deduced from a knowledge of the properties of the operators that correspond to observables, and that it is often unnecessary to go through the stage of solving the Schrödinger equation. General principles are introduced in this chapter, and a particularly important illustration is developed in Chapter 6.

5.1 The postulates of quantum mechanics

Postulates in science may be of two types. There is the obvious kind, in which the content is self-evident. An example is the Second Law of thermodynamics in the form 'heat flows spontaneously from a hot to a cold body'. Then there is the subtle kind, from which the content has to be unfolded by a chain of argument. An example is the Second Law in the form 'entropy increases during a spontaneous process'.

The postulates we use as a basis for quantum mechanics are of the second, subtle, sort. They are by no means the most subtle that have been devised, but they are strong enough for what we have to do.

Postulate 1. The state of a system is fully described by the wavefunction $\Psi(\mathbf{r}_1, \mathbf{r}_2, \ldots; t)$.

$\mathbf{r}_1, \mathbf{r}_2 \ldots$ are the coordinates of particles $1, 2, \ldots$ that constitute the system; t is the time. The time-independent component is denoted $\psi(\mathbf{r}_1, \mathbf{r}_2, \ldots)$. The state may also depend on some internal variables of the particles (their spin): we ignore that for now but return to it later. By 'describe' we mean that the function contains all the information open to determination by experimental observations. The wavefunction will turn out to be specified by a set of labels called *quantum numbers* from which the values of observables may be calculated, and so the wavefunction corresponding to the state of the system with quantum numbers $l, m, n, \ldots$ can be denoted either as $\Psi_{l,m,n,\ldots}$ or as $\Psi_{a,b,c,\ldots}$ where $a, b, c, \ldots$ are the values of the observables themselves. Sometimes we shall refer to the *state* of the system without referring to the corresponding wavefunction, and for this we use the notation $|l, m, n, \ldots\rangle$ or $|a, b, c, \ldots\rangle$. That is, $|l, m, n, \ldots\rangle$ specifies the state of the system, and $\Psi_{l,m,n,\ldots}$ denotes the corresponding wavefunction. This shift of emphasis away from the wavefunction is the underlying theme of this chapter, and is the first step in the direction of avoiding having to solve the Schrödinger equation. It is also the first step in the introduction of a new notation, Dirac's bracket notation, which we shall meet in more detail shortly.

Postulate 2. Observables are represented by operators chosen to satisfy the commutation relation $xp - px = i\hbar$.

Similar expressions apply to the y- and z-components; we shall see soon how to express a general observable in terms of x and p, and so this rule has wide applicability. We saw in Chapter 2 how Heisenberg modified classical mechanics by demanding that x and p should be regarded in this way. Here we are regarding the commutation relation as a basic, unprovable, and underivable postulate. In order to develop the content of this postulate we need to introduce some concepts.

A function f is an *eigenfunction* of some operator Ω if it satisfies an

Eigenvalue equation: $\Omega f = \omega f$, (5.1.1)

where ω is a number. As an example, e^{ax} is an eigenfunction of the operator d/dx because $(d/dx)e^{ax} = ae^{ax}$, a number multiplying the original function. e^{ax^2} is not an eigenfunction of d/dx because $(d/dx)e^{ax^2} = 2axe^{ax^2} = 2a(xe^{ax^2})$, a number times a different function of x. The number ω in the eigenvalue equation is called the *eigenvalue* of the operator corresponding to the eigenfunction f.

A further point is that a function can be expanded in terms of the eigenfunctions of an operator. That is, if f_n are eigenfunctions of an operator Ω and satisfy the eigenvalue equations $\Omega f_n = \omega_n f_n$, then (subject to some fairly straightforward restrictions) some function g can be written as the sum (or *linear combination*)

$$g = \sum_n c_n f_n, \tag{5.1.2}$$

where the c_n are the combination coefficients. In other words, the eigenfunctions of an operator form a *complete set* in the sense that they can be used to build up another function. For example, the straight line $g = ax$ can be constructed by superimposing an infinite number of sine functions, which are eigenfunctions of d^2/dx^2, or an infinite number of e^{ikx} functions, which are eigenfunctions of d/dx. The advantage of this procedure is that it lets us deduce the effect of an operator on a function that is not one of its eigenfunctions. The effect of Ω on g in eqn (5.1.2) is simply

$$\Omega g = \Omega \sum_n c_n f_n = \sum_n c_n \Omega f_n = \sum_n c_n \omega_n f_n. \tag{5.1.3}$$

One way of interpreting this result is that if we write $c'_n = c_n \omega_n$, then

$$\Omega g = \sum_n c'_n f_n = g', \tag{5.1.4}$$

and so the effect of Ω on g is to generate some new function g′. We arrive at a more powerful interpretation in a moment.

A special case of these linear combinations is when we have a set of *degenerate eigenfunctions.* Suppose $f_1, f_2, \ldots, f_k$ are all eigenfunctions of the operator Ω and correspond to the same eigenvalue ω:

$$\Omega f_n = \omega f_n, \qquad n = 1, 2, \ldots, k.$$

Then the states are said to be *k-fold degenerate.* (We saw several examples of this in Chapters 3 and 4; the orbitals of the hydrogen atom, for example, are n^2-fold degenerate eigenfunctions of the hamiltonian.) Now consider linear combinations of these functions: we can show that *any linear combination of a set of degenerate eigenfunctions is also an eigenfunction corresponding to the same eigenvalue.* The proof is trivial. Consider the linear combination $g = \sum_{n=1}^{k} c_n f_n$; then

$$\begin{aligned}\Omega g &= \Omega \sum_{n=1}^{k} c_n f_n = \sum_{n=1}^{k} c_n \Omega f_n \\ &= \sum_{n=1}^{k} c_n \omega f_n = \omega \sum_{n=1}^{k} c_n f_n = \omega g,\end{aligned}$$

which has the form of an eigenvalue equation. Note that a linear combination of a set of *basis functions* (the f_n in this example) are *linearly independent* if none of them can be expressed as a linear combination of the others. From a set of k functions it is possible to choose an infinite number of sets of linearly independent combinations, but each set can have no more than k members. Thus, from three 2p-orbitals we can form any number of sets of linearly independent functions, but each set has no more than three members.

The next postulate brings together the wavefunction and the operators and makes the link between formal calculation and experimental observation.

Postulate 3. When a system is described by a wavefunction ψ, the mean value of the observable Ω is equal to the expectation value of the corresponding operator.

By *expectation value* of an operator Ω in a state ψ we mean the integral

Expectation value: $$\langle\Omega\rangle = \int \psi^* \Omega \psi \, d\tau \Big/ \int \psi^* \psi \, d\tau. \qquad (5.1.5)$$

If the wavefunction has been normalized to unity (p. 32), the denominator is unity and we have the simpler form:

$$\langle\Omega\rangle = \int \psi^* \Omega \psi \, d\tau, \quad \psi \text{ normalized to unity.} \qquad (5.1.6)$$

We shall assume that this is the case from now on.

The meaning of this postulate can be unravelled as follows. First, suppose that ψ is an eigenfunction of Ω, then

$$\langle\Omega\rangle=\int\psi^*\Omega\psi\,\mathrm{d}\tau=\int\psi^*\omega\psi\,\mathrm{d}\tau=\omega\int\psi^*\psi\,\mathrm{d}\tau=\omega. \tag{5.1.7}$$

That is, a series of experiments to determine Ω will have the average value ω. Now suppose that the system is not in an eigenstate of Ω but that its wavefunction can be expressed as a linear combination of eigenfunctions:

$$\psi=\sum_n c_n\psi_n \quad \text{where} \quad \Omega\psi_n=\omega_n\psi_n. \tag{5.1.8}$$

Then the expectation value is

$$\langle\Omega\rangle=\int\Big(\sum_m c_m\psi_m\Big)^*\Omega\Big(\sum_n c_n\psi_n\Big)\,\mathrm{d}\tau=\sum_n\sum_m c_m^*c_n\int\psi_m^*\Omega\psi_n\,\mathrm{d}\tau$$

$$=\sum_n\sum_m c_m^*c_n\omega_n\int\psi_m^*\psi_n\,\mathrm{d}\tau.$$

Since the eigenfunctions are orthogonal (p. 47; we shall prove later that this is generally so), the integral vanishes if $m\neq n$, and so the double sum reduces to a single sum:

$$\langle\Omega\rangle=\sum_n c_n^*c_n\omega_n\int\psi_n^*\psi_n\,\mathrm{d}\tau.$$

Since the eigenfunctions are normalized to unity, each integral is equal to unity. Therefore

$$\langle\Omega\rangle=\sum_n c_n^*c_n\omega_n=\sum_n|c_n|^2\,\omega_n. \tag{5.1.9}$$

That is, the expectation value is a *weighted sum* of the eigenvalues of Ω, the contribution to the sum being determined by the value of the square modulus of the coefficient c_n in the wavefunction expansion.

We can now interpret the difference between eqn (5.1.7) and eqn (5.1.9). This is really a subsidiary postulate:

> *Postulate 3′*. When ψ is an eigenfunction of the operator Ω corresponding to the observable of interest, the determination of Ω always yields one result, the corresponding eigenvalue of Ω. When ψ is not an eigenfunction of Ω, a single measurement of Ω yields a single result which is one of the eigenvalues of Ω, and the probability that a particular eigenvalue ω_n is measured is equal to $|c_n|^2$, where c_n is the coefficient of the eigenfunction ψ_n in the expansion of the wavefunction.

One experiment can give only one result: a pointer can indicate only one value on a dial at any one time. A series of experiments can lead to a series of results with some mean value. The subsidiary postulate asserts that a measurement of the observable Ω always results in the pointer indicating one of the eigenvalues of the corresponding operator

Ω. If the function that describes the state of the system is an eigenvalue of Ω, then every pointer reading is exactly ω, and the mean value is also ω. If the system has been prepared in a state which is not an eigenfunction of the operator of interest, then different measurements give different values, but every value is an eigenvalue of Ω, and the probability that a particular outcome ω_n is obtained is determined by the expansion coefficient c_n as $|c_n|^2$. In both cases the mean value of the observation is given by the expectation value.

The only remaining work of this section is to put forward some of the explicit forms of the operators we shall meet. Postulate 2 requires them to satisfy a commutation rule. It can be satisfied in a variety of ways. In Chapter 2 we saw that $xp - px = i\hbar$ could be satisfied by choosing the operator for position to be multiplication by the value of the coordinate and the operator for momentum to be differentiation with respect to that coordinate. This choice is the *position representation* of the operators:

Observable:	position, x	momentum, p	(5.1.10)
Operator:	$x\cdot$	$(\hbar/\mathrm{i})(\partial/\partial x)$	

(and likewise for the y- and z-components). An alternative choice is to represent the momentum by multiplication. Then, in order to satisfy the commutation relation, we need to choose the position operator to be differentiation with respect to the momentum: this is the *momentum representation:*

Observable:	position, x	momentum, p	(5.1.11)
Operator:	$-(\hbar/\mathrm{i})(\partial/\partial p)$	$p\cdot$	

where p is the x-component of the linear momentum. There are other representations. We shall normally use the position representation, but we shall also see that many of the calculations in quantum mechanics can be done without deciding on a representation.

Other observables of interest can be calculated on the basis of the operators for position and momentum. For instance, the *kinetic energy operator* can be formed by noting that kinetic energy is related to linear momentum by $T = p^2/2m$, and so in one dimension the kinetic energy operator in the position representation is

$$T = p^2/2m = (1/2m)\{(\hbar/\mathrm{i})(\mathrm{d}/\mathrm{d}x)\}^2 = -(\hbar^2/2m)(\mathrm{d}^2/\mathrm{d}x^2), \tag{5.1.12}$$

and in three dimensions

$$\begin{aligned} T &= (1/2m)(p_x^2 + p_y^2 + p_z^2) = -(\hbar^2/2m)\{(\partial/\partial x)^2 + (\partial/\partial y)^2 + (\partial/\partial z)^2\} \\ &= -(\hbar^2/2m)\nabla^2, \end{aligned} \tag{5.1.13}$$

as we used throughout Chapters 3 and 4. The operator for *potential energy*, $V(x)$, in the position representation is multiplication by the function $V(x)$ in one dimension and by $V(\mathbf{r})$ in three dimensions. The operator for the *dipole moment* of a collection of charges e_j can be constructed similarly. Since the classical expression for the dipole moment is

$$\boldsymbol{\mu}=\sum_j e_j\mathbf{r}_j, \tag{5.1.14}$$

where the $\mathbf{r}_j$ are the locations of the particles, the operator for the dipole moment is multiplication by $\boldsymbol{\mu}$.

5.2 Hermitian operators

The class of operators called *hermitian operators* plays a special role in quantum mechanics. This is because they have real eigenvalues, and hence correspond to observables (the outcome of an experimental determination of a property must be real quantity). We shall sometimes meet non-hermitian operators, but as their eigenvalues are not real they do not correspond to physical observables. Furthermore, although all observables correspond to hermitian operators, not all hermitian operators correspond to observables.

First, the *definition* of a hermitian operator. An operator Ω is hermitian if the following equation is satisfied:

$$\textit{Hermiticity}: \int \psi_m^*\Omega\psi_n\,\mathrm{d}\tau=\left\{\int \psi_n^*\Omega\psi_m\,\mathrm{d}\tau\right\}^*. \tag{5.2.1}$$

An alternative version of the definition is

$$\int \psi_m^*\Omega\psi_n\,\mathrm{d}\tau=\int (\Omega\psi_m)^*\psi_n\,\mathrm{d}\tau, \tag{5.2.2}$$

which is obtained by taking the complex conjugate of each term in the integral on the right of eqn (5.2.1).

We are on the edge of getting lost in a complicated notation. The appearance of many quantum mechanical expressions and derivations is greatly simplified by a change of notation, and henceforth we shall frequently use the *Dirac bracket notation* in which integrals such as the ones on the left of the last two equations are written as follows:

$$\textit{Bracket notation}: \langle m|\,\Omega\,|n\rangle\equiv\int \psi_m^*\Omega\psi_n\,\mathrm{d}\tau. \tag{5.2.3}$$

The symbol $|n\rangle$ is called a *ket*, and denotes the state with wavefunction ψ_n. The conjugate of the wavefunction, ψ_n^* is denoted by the *bra* $\langle n|$. When bras and kets are strung together with an operator between them, such as in the *bracket* $\langle m|\,\Omega\,|n\rangle$, the integral in eqn (5.2.3) is to be understood. When the operator is simply multiplication by 1, as in the orthonormality integral $\int \psi_m^*\psi_n\,\mathrm{d}\tau$, it is omitted from the bracket.

That is, we use the convention

$$\langle m \mid n\rangle \equiv \int \psi_m^* \psi_n \, \mathrm{d}\tau. \tag{5.2.4}$$

This notation is very elegant. For example, the orthonormality condition, eqn (3.3.8), becomes

Orthonormality: $\langle m \mid n\rangle = \delta_{mn}$. (5.2.5)

Note too that $\langle n \mid m\rangle = \langle m \mid n\rangle^*$. (Relations such as these can all be established by expressing the bracket in terms of the defining integral.)

In terms of the bracket notation, the definition of hermiticity is

Hermiticity: $\langle m|\, \Omega \,|n\rangle = \langle n|\, \Omega \,|m\rangle^*$. (5.2.6)

This property has far-reaching implications.

As a first step we check whether the position and momentum operators in the position representation are hermitian. That the position operator is hermitian is obvious from inspection:

$$\langle m|\, x \,|n\rangle = \int \psi_m^*(x) x \psi_n(x) \, \mathrm{d}x = \int \psi_n(x) x \psi_m^*(x) \, \mathrm{d}x$$

$$= \left\{ \int \psi_n^*(x) x \psi_m(x) \, \mathrm{d}x \right\}^* = \langle n|\, x \,|m\rangle^*.$$

because $(\psi^*)^* = \psi$ and x is real. The demonstration of the hermiticity of p involves an integration by parts ($\int u \, \mathrm{d}v = uv - \int v \, \mathrm{d}u$):

$$\langle m|\, p \,|n\rangle = \int \psi_m^*(x)(\hbar/\mathrm{i})(\mathrm{d}/\mathrm{d}x)\psi_n(x) \, \mathrm{d}x$$

$$= (\hbar/\mathrm{i}) \int \psi_m^* \, \mathrm{d}\psi_n = (\hbar/\mathrm{i}) \left\{ \psi_m^* \psi_n - \int \psi_n \, \mathrm{d}\psi_m^* \right\} \Bigg|_{x=-\infty}^{x=+\infty}$$

$$= (\hbar/\mathrm{i}) \left\{ \psi_m^*(x)\psi_n(x) \Big|_{x=-\infty}^{x=\infty} - \int \psi_n(x)(\mathrm{d}/\mathrm{d}x)\psi_m^*(x) \, \mathrm{d}x \right\}.$$

The first term on the right is zero (when x is infinite the wavefunction is vanishingly small). Therefore

$$\langle m|\, p \,|n\rangle = -(\hbar/\mathrm{i}) \int \psi_n(x)(\mathrm{d}/\mathrm{d}x)\psi_m^*(x) \, \mathrm{d}x$$

$$= \left\{ \int \psi_n^*(x)(\hbar/\mathrm{i})(\mathrm{d}/\mathrm{d}x)\psi_m(x) \, \mathrm{d}x \right\}^* = \langle n|\, p \,|m\rangle^*,$$

as required.

Now we move on to discovering why hermitian operators are so important. First we establish two of their important properties.

Property 1. The eigenvalues of hermitian operators are real.

In order to prove this result consider the eigenvalue equation

$$\Omega\,|\omega\rangle=\omega\,|\omega\rangle$$

where the ket $|\omega\rangle$ denotes an eigenfunction of the operator Ω corresponding to the eigenvalue ω (in other words, we are labelling the states with their eigenvalues). This corresponds to the equation

$$\Omega\psi_\omega=\omega\psi_\omega$$

where ψ_ω is the wavefunction corresponding to the state $|\omega\rangle$. Since ψ_ω is an eigen*function*, the state $|\omega\rangle$ to which it corresponds is naturally called an *eigenstate*. Multiply from the left by ψ_ω^* and integrate over all space:

$$\int\psi_\omega^*\Omega\psi_\omega\,\mathrm{d}\tau=\omega\int\psi_\omega^*\psi_\omega\,\mathrm{d}\tau,\quad\text{or}\quad\langle\omega|\,\Omega\,|\omega\rangle=\omega\langle\omega\mid\omega\rangle.$$

In the first place $\langle\omega\mid\omega\rangle=1$ (normalization). In the second, if Ω is hermitian, then $\langle\omega'|\,\Omega\,|\omega\rangle=\langle\omega|\,\Omega\,|\omega'\rangle^*$ in general and therefore $\langle\omega|\,\Omega\,|\omega\rangle=\langle\omega|\,\Omega\,|\omega\rangle^*$ in the special case of $\omega'=\omega$. Therefore we have

$$\omega=\langle\omega|\,\Omega\,|\omega\rangle=\langle\omega|\,\Omega\,|\omega\rangle^*=\omega^*.$$

Since $\omega=\omega^*$ it follows that ω is real, as we were to prove.

Property 2. The eigenstates corresponding to different eigenvalues of hermitian operators are orthogonal.

Suppose we have two eigenstates $|\omega_1\rangle$ and $|\omega_2\rangle$ of the hermitian operator Ω, and $\omega_1\neq\omega_2$. They satisfy $\Omega\,|\omega_1\rangle=\omega_1\,|\omega_1\rangle$ and $\Omega\,|\omega_2\rangle=\omega_2\,|\omega_2\rangle$. Multiply the first of these from the left by $\langle\omega_2|$ and the second by $\langle\omega_1|$. This gives

$$\langle\omega_2|\,\Omega\,|\omega_1\rangle=\omega_1\langle\omega_2\mid\omega_1\rangle\quad\text{and}\quad\langle\omega_1|\,\Omega\,|\omega_2\rangle=\omega_2\langle\omega_1\mid\omega_2\rangle.$$

Take the complex conjugate of the second, $\langle\omega_1|\,\Omega\,|\omega_2\rangle^*=\omega_2^*\langle\omega_1\mid\omega_2\rangle^*=\omega_2^*\langle\omega_2\mid\omega_1\rangle$, and subtract it from the first:

$$\langle\omega_2|\,\Omega\,|\omega_1\rangle-\langle\omega_1|\,\Omega\,|\omega_2\rangle^*=(\omega_1-\omega_2^*)\langle\omega_2\mid\omega_1\rangle.$$

Now use hermiticity. By hermiticity ω_2 is real, hence $\omega_2^*=\omega_2$. By hermiticity $\langle\omega_2|\,\Omega\,|\omega_1\rangle=\langle\omega_1|\,\Omega\,|\omega_2\rangle^*$, hence the left-hand side is zero. That is,

$$0=(\omega_1-\omega_2)\langle\omega_2\mid\omega_1\rangle.$$

But $\omega_1\neq\omega_2$, by hypothesis; therefore $\langle\omega_2\mid\omega_1\rangle=0$, and the eigenstates are orthogonal, as was to be proved.

5.3 The specification of states: complementarity

The question that now arises is the following. The state of a system can be specified as $|a, b, \ldots\rangle$ where each of the eigenvalues $a, b, \ldots$ corresponds to the operators representing the different observations $A, B, \ldots$

that may be made on the system. If the system is in a state $|a, b, \ldots\rangle$ then when we measure A we shall get precisely a as the outcome, when we measure property B we shall get b, and so on. But can a state be specified *arbitrarily* completely? Can it be *simultaneously* an eigenstate of all possible properties $A, B, \ldots$ or is there some restriction? We are moving into the domain of the uncertainty principle.

As a first step we establish the conditions under which two observables may be specified simultaneously to arbitrary precision. That is, we establish the conditions for a function ψ to be simultaneously an eigenfunction of two operators A and B. We shall prove the following:

Property 3. If two observables are to have simultaneously precisely defined values, their corresponding operators must commute; that is, AB must equal BA.

First assume that ψ *is* an eigenfunction of both A and B so that we know that $A\psi = a\psi$ and $B\psi = b\psi$. Then we may write the following chain:

$$AB\psi = Ab\psi = bA\psi = ba\psi = ab\psi = aB\psi = Ba\psi = BA\psi.$$

Therefore if ψ is an eigenfunction of A and B it certainly follows that $AB = BA$. That is, $AB = BA$ is a *necessary* condition.† But is it *sufficient*? That is, does the condition $AB = BA$ guarantee that A and B have simultaneous eigenfunctions? In other words, if $A\psi = a\psi$, is ψ also an eigenfunction of B given that $AB = BA$? We confirm this as follows. Since $A\psi = a\psi$ we can write

$$BA\psi = Ba\psi = aB\psi.$$

Since, by hypothesis, $AB = BA$, it follows that

$$AB\psi = aB\psi; \quad \text{that is} \quad A(B\psi) = a(B\psi).$$

But since we know that $A\psi = a\psi$ it follows that the function $B\psi$ must be proportional to ψ itself; hence $B\psi \propto \psi$, or $B\psi = b\psi$ where b is some coefficient of proportionality. That is, ψ is also an eigenfunction of B, as was to be proved.‡ Notice that in neither stage of the proof is any reference made to hermiticity: the conclusion therefore applies to all operators, and to hermitian operators in particular.

We are now in a position to determine which observables may be specified simultaneously: it is merely necessary to inspect the commutator $AB - BA$. If it is zero, then A and B may both be specified.

As an example, consider the specification of the position of a

† The conclusion is valid only if $AB\psi_i = BA\psi_i$ for *every* member ψ_i of a complete set of functions. We meet several counter-examples below. For instance, it turns out that $l_z\psi_{1s} = 0$ and $l_x\psi_{1s} = 0$ even though $l_xl_z - l_zl_x \neq 0$.

‡ The conclusion is valid only if ψ is a non-degenerate eigenfunction of A: if ψ is degenerate $B\psi$ may be any linear combination of the degenerate set. For instance, take $A = l^2$, $B = l_z$, and $\psi = \psi_{2p_x}$; then $l^2\{l_z\psi_{2p_x}\} = \hbar^2 l(l+1)\{l_z\psi_{2p_x}\}$, but ψ_{2p_x} is not an eigenfunction of l_z.

particle. To be complete we have to specify the x-, y-, and z-coordinates simultaneously. Do the corresponding operators all commute? The basic commutation relation puts no constraint on $xy - yx$ etc., and so x, y, and z may have simultaneously precisely specified values. The same is true of the three components of linear momentum, because the operators p_x, p_y, and p_z all commute among themselves. We may also specify simultaneously the location of the particle along the x-axis and its momentum parallel to the y-axis, because $xp_y - p_yx = 0$. But we cannot specify x-location and x-momentum simultaneously because $xp_x - p_xx \neq 0$. Now we are at the heart of the uncertainty principle and the complementarity of position and momentum. *Complementary observables,* we see, *are observables represented by non-commuting operators.*

5.4 The uncertainty principle

Now we turn to the case of two observables represented by non-commuting operators. Although we cannot specify both precisely, is there anything that can be said about the *relative* precision with which the specifications can be made? This is motivated by the thought that if we were willing to give up all the precision in the specification of property A, then we might be able to specify B completely, and vice versa. We might also be willing to accept some blurring of A, in which case a blurred value of B might be open to specification. What we are about to do is to derive the precise form of the uncertainty principle.

Consider two observables represented by the hermitian operators A and B. Since the commutator is so central to the argument (and to the whole of quantum mechanics) we introduce the notation:

$$\textit{Commutator of A and B}: \; [A, B] = AB - BA. \qquad (5.4.1)$$

It is important to master manipulations of commutator expressions: these can always be established by expanding them in terms of the definition. For example,

$$[A, B] = AB - BA = -(BA - AB) = -[B, A]. \qquad (5.4.2a)$$

If b is a number, then

$$[A, bB] = AbB - bBA = b(AB - BA) = b[A, B]. \qquad (5.4.2b)$$

If A, B, and C are all operators, then

$$\begin{aligned}[A, BC] &= ABC - BCA = ABC - BAC + BAC - BCA \\ &= [A, B]C + B[A, C] \qquad (5.4.2c)\end{aligned}$$

$$\begin{aligned}[A, B + C] &= A(B + C) - (B + C)A = AB + AC - BA - CA \\ &= [A, B] + [A, C] \qquad (5.4.2d)\end{aligned}$$

$$\begin{aligned}[A, [B, C]] &= [A, BC - CB] = [A, BC] - [A, CB] \\ &= ABC - BCA - ACB + CBA. \qquad (5.4.2e)\end{aligned}$$

If A, B, C, and D are all operators, then

$$[A+B, C+D]=[A, C]+[A, D]+[B, C]+[B, D]. \qquad (5.4.2\text{f})$$

We shall use these relations frequently.

Example. Evaluate the commutators $[p, x^2]$ and $[p^2, x]$.

- *Method.* Use eqn (5.4.2c) to expand the commutators, and then use the basic commutator $[x, p]=\mathrm{i}\hbar$.
- *Answer.*

$$\begin{aligned}[p, x^2]&=[p, xx]=[p, x]x+x[p, x]\\&=-[x, p]x-x[x, p]=-2\mathrm{i}\hbar x.\\ [p^2, x]&=[pp, x]=-[x, pp]\\&=-[x, p]p-p[x, p]=-2\mathrm{i}\hbar p.\end{aligned}$$

- *Comment.* The commutators have been evaluated without resorting to any representation. An alternative approach would be to use $p=(\hbar/\mathrm{i})\,\mathrm{d}/\mathrm{d}x$, and to evaluate the derivatives. Never forget the unwritten function on which the operators ultimately operate. For instance,

$$\begin{aligned}[p^2, x]\psi&=-\hbar^2[\mathrm{d}^2/\mathrm{d}x^2, x]\psi\\&=-\hbar^2\{(\mathrm{d}^2/\mathrm{d}x^2)x\psi-x(\mathrm{d}^2/\mathrm{d}x^2)\psi\}\\&=-\hbar^2\{(\mathrm{d}/\mathrm{d}x)[\psi+x(\mathrm{d}/\mathrm{d}x)\psi]-x(\mathrm{d}^2/\mathrm{d}x^2)\psi\}\\&=-\hbar^2\{(\mathrm{d}/\mathrm{d}x)\psi+(\mathrm{d}/\mathrm{d}x)\psi+x(\mathrm{d}^2/\mathrm{d}x^2)\psi-x(\mathrm{d}^2/\mathrm{d}x^2)\psi\}\\&=-\hbar^2 2(\mathrm{d}/\mathrm{d}x)\psi=-2\hbar\mathrm{i}p\psi,\end{aligned}$$

in accord with the representation-free result.

In the present case we are dealing with operators that have a non-vanishing commutator. We shall suppose that $AB-BA=\mathrm{i}C$, where C is some other operator (the i is there for future convenience: it mirrors $xp-px=\mathrm{i}\hbar$, in which case $C=\hbar$, the operator 'multiply by $\hbar$'). That is, we suppose that

$$[A, B]=\mathrm{i}C. \qquad (5.4.3)$$

Let the system be prepared in some arbitrary state described by a wavefunction ψ which is not necessarily an eigenfunction of either A or B. The results of measuring the properties *separately* are expressed by the expectation values

$$\langle A\rangle=\int\psi^*A\psi\,\mathrm{d}\tau \quad\text{and}\quad \langle B\rangle=\int\psi^*B\psi\,\mathrm{d}\tau. \qquad (5.4.4)$$

We shall be interested in a measure of the spread of values around each of these mean values, and therefore introduce the operators for the deviation of the observables from their mean values:

$$\Delta A=A-\langle A\rangle, \qquad \Delta B=B-\langle B\rangle. \qquad (5.4.5)$$

It is easy to verify that because expectation values are just numbers and therefore commute with everything, then ΔA and ΔB have the same commutation relation as A and B themselves:

$$\begin{aligned}[\Delta A, \Delta B] &= [A-\langle A\rangle, B-\langle B\rangle] \\ &= [A, B]-[\langle A\rangle, B]-[A, \langle B\rangle]+[\langle A\rangle, \langle B\rangle] \\ &= [A, B] = \mathrm{i}C. \end{aligned} \tag{5.4.6}$$

Now consider the properties of the following integral:

$$I = \int |(\alpha\, \Delta A - \mathrm{i}\, \Delta B)\psi|^2 \, \mathrm{d}\tau \tag{5.4.7}$$

where α is a real but otherwise arbitrary parameter. The first point about I is that it is obviously not negative: $I \geq 0$. The next point is that the square modulus in the integrand can be expanded into

$$|(\alpha\, \Delta A - \mathrm{i}\, \Delta B)\psi|^2 = \{(\alpha\, \Delta A^* + \mathrm{i}\, \Delta B^*)\psi^*\}\{(\alpha\, \Delta A - \mathrm{i}\, \Delta B)\psi\}.$$

Since the operators ΔA and ΔB are hermitian (they correspond to observables) it follows that the integral may be expressed as

$$\begin{aligned} I &= \alpha \int (\Delta A^*\psi^*)(\alpha\, \Delta A - \mathrm{i}\, \Delta B)\psi \, \mathrm{d}\tau + \mathrm{i} \int (\Delta B^*\psi^*)(\alpha\, \Delta A - \mathrm{i}\, \Delta B)\psi \, \mathrm{d}\tau \\ &= \alpha \int \psi^*\, \Delta A(\alpha\, \Delta A - \mathrm{i}\, \Delta B)\psi \, \mathrm{d}x + \mathrm{i} \int \psi^*\, \Delta B(\alpha\, \Delta A - \mathrm{i}\, \Delta B)\psi \, \mathrm{d}\tau \\ &= \int \psi^*(\alpha\, \Delta A + \mathrm{i}\, \Delta B)(\alpha\, \Delta A - \mathrm{i}\, \Delta B)\psi \, \mathrm{d}\tau, \end{aligned}$$

where the definition of hermiticity in eqn (5.2.2) has been used for the individual operators ΔA and ΔB. Still $I \geq 0$. Now expand the integrand as follows:

$$\begin{aligned} I &= \int \psi^*\{\alpha^2(\Delta A)^2 + (\Delta B)^2 - \mathrm{i}\alpha(\Delta A\, \Delta B) + \mathrm{i}\alpha(\Delta B\, \Delta A)\}\psi \, \mathrm{d}\tau \\ &= \alpha^2\langle(\Delta A)^2\rangle + \langle(\Delta B)^2\rangle - \mathrm{i}\alpha\langle[\Delta A, \Delta B]\rangle \\ &= \alpha^2\langle(\Delta A)^2\rangle + \langle(\Delta B)^2\rangle + \alpha\langle C\rangle. \end{aligned}$$

Still $I \geq 0$. The equation can be arranged into the following form:

$$I = \langle(\Delta A)^2\rangle\{\alpha + \langle C\rangle/2\langle(\Delta A)^2\rangle\}^2 + \langle(\Delta B)^2\rangle - \langle C\rangle^2/4\langle(\Delta A)^2\rangle \geq 0.$$

(We have 'completed the square' for the first term.) The inequality is true for all real α, and so we may choose a value that makes the left-hand side a minimum. This is the value that causes the first term to vanish, because it adds a positive quantity to the left-hand side. Making this choice of α turns the inequality into

$$\langle(\Delta A)^2\rangle\langle(\Delta B)^2\rangle \geq \tfrac{1}{4}\langle C\rangle^2,$$

and a familiar expression is beginning to emerge.

The expectation values on the left can be put into a simpler form by writing them as follows:

$$\begin{aligned} \langle(\Delta A)^2\rangle &= \langle(A - \langle A\rangle)^2\rangle = \langle(A^2 - 2A\langle A\rangle + \langle A\rangle^2)\rangle \\ &= \langle A^2\rangle - 2\langle A\rangle^2 + \langle A\rangle^2 = \langle A^2\rangle - \langle A\rangle^2, \end{aligned}$$

the *mean square deviation* of A from its mean value. We shall define the *root mean square deviation* of A as the square root of this quantity:

$$\delta A = \sqrt{\langle(\Delta A)^2\rangle} = \sqrt{\{\langle A^2\rangle - \langle A\rangle^2\}} \tag{5.4.8}$$

and likewise for B. Then the inequality becomes

The uncertainty principle: $\delta A\,\delta B \geq \frac{1}{2}|\langle C\rangle|$ (5.4.9)

with $C=[A, B]/\mathrm{i}$. This is an exact and precise form of the uncertainty principle: note that the terms δA and δB are precisely defined (as root mean square deviations), and the right-hand side is a rigorous lower bound on the product $\delta A\,\delta B$.

The first point to note is that the uncertainty principle is consistent with Property 3, because if A and B commute, $C=0$, and so there is no lower bound (except 0) to the product of δA and δB. This means that there is no inconsistency in having both $\delta A=0$ and $\delta B=0$ when $[A, B]=0$. On the other hand, when A and B do not commute the values of δA and δB are related. For instance, while it may be possible to prepare a system in a state in which $\delta A=0$, the uncertainty principle implies that the value of δB must then be infinite so that the product $\delta A\,\delta B$ is not less than $\frac{1}{2}\langle C\rangle$. For example, suppose A corresponds to position along x and B corresponds to momentum parallel to x, then since $[x, p_x]=\mathrm{i}\hbar$ it follows that $C=\hbar$, and therefore that

$$\delta x\,\delta p_x \geq \tfrac{1}{2}\hbar, \tag{5.4.10}$$

the precise form of the position–momentum uncertainty relation.

Example. A particle was prepared in a state with wavefunction $\psi(x)=N\mathrm{e}^{-x^2/2\Gamma}$; $N=(1/\pi\Gamma)^{\frac{1}{4}}$. Evaluate δx and δp and confirm that the uncertainty principle is satisfied.

- *Method.* Evaluate $\langle x\rangle$, $\langle x^2\rangle$, $\langle p\rangle$, $\langle p^2\rangle$ by integration. There are two short-cuts. For $\langle x\rangle$, note that $\psi(x)$ is symmetrical around $x=0$, and so $\langle x\rangle=0$. For $\langle p\rangle$, note that ψ is real; hence $\langle p\rangle=0$. For the remaining integrals use

$$\int_{-\infty}^{\infty}\mathrm{e}^{-ax^2}\,\mathrm{d}x=(\pi/a)^{\frac{1}{2}};\qquad \int_{-\infty}^{\infty}x^2\mathrm{e}^{-ax^2}\,\mathrm{d}x=\tfrac{1}{2}(\pi/a^3)^{\frac{1}{2}}.$$

- *Answer.*

$$\langle x^2\rangle = N^2\int_{-\infty}^{\infty}x^2\mathrm{e}^{-x^2/\Gamma}\,\mathrm{d}x=\tfrac{1}{2}\Gamma;\qquad \langle x\rangle=0.$$

$$\langle p^2\rangle = N^2\int_{-\infty}^{\infty}\mathrm{e}^{-x^2/2\Gamma}(-\hbar^2\,\mathrm{d}^2/\mathrm{d}x^2)\mathrm{e}^{-x^2/2\Gamma}\,\mathrm{d}x$$

$$= N^2\hbar^2\left\{\int_{-\infty}^{\infty}\mathrm{e}^{-x^2/\Gamma}\,\mathrm{d}x/\Gamma-\int_{-\infty}^{\infty}x^2\mathrm{e}^{-x^2/\Gamma}\,\mathrm{d}x/\Gamma^2\right\}=\hbar^2/2\Gamma.$$

Hence $\delta x\,\delta p=\langle x^2\rangle^{\frac{1}{2}}\langle p^2\rangle^{\frac{1}{2}}=\hbar/2$.

- *Comment.* In this example $\delta x\,\delta p$ has its minimum permitted value. This is a special feature of gaussian wavefunctions. A gaussian wavefunction is encountered in the ground state of the harmonic oscillator.

The uncertainty principle in the form eqn (5.4.9) can be applied to all pairs of observables: all it is necessary to do is to evaluate the relevant commutator.† We shall see further examples when we turn to angular momentum in the next chapter.

Finally, a word about the 'energy–time uncertainty principle' is appropriate here. It does not exist. An expression of the form $\delta E\,\delta t \geq \hbar$ if often referred to as an uncertainty principle, and interpreted as signifying a complementarity between the lifetime and energy of a system. This is not so in a technical sense, but it is so in a loose, practical sense. In order for $\delta E\,\delta t \geq \hbar$ to be an uncertainty relation it would be necessary for the operators for energy and time to have non-zero commutator. Whereas the energy operator is the hamiltonian, there is no operator for time in quantum mechanics and hence the question of its commutator does not arise. (This dismissal of the principle also avoids the obvious conceptual difficulty of knowing what is meant by the 'simultaneous' specification of a lifetime and an energy.) In fact the so-called uncertainty relation between energy and time is an aspect of the time-dependent Schrödinger equation, and is therefore a consequence of the $[x, p]$ commutation relation. We shall come to grips with it and see its true significance and interpretation when we turn to time-dependent processes in Chapter 8.

5.5 Time-evolution and conservation laws

The commutator of two operators also plays a role in determining the time-evolution of systems, and in particular the time evolution of the expectation values of observables. The question we examine here is the rate of change of the expectation value of some observable Ω. In order to calculate $(\mathrm{d}/\mathrm{d}t)\langle\Omega\rangle$ we use the definition of the expectation value, eqn (5.1.6) adapted to the case of a time-dependent wavefunction Ψ, and combine it with the time-dependent Schrödinger equation $H\Psi = \mathrm{i}\hbar(\partial\Psi/\partial t)$:

$$\begin{aligned}(\mathrm{d}/\mathrm{d}t)\langle\Omega\rangle &= (\mathrm{d}/\mathrm{d}t)\int \Psi^*(t)\Omega\Psi(t)\,\mathrm{d}\tau \\ &= \int (\partial\Psi^*/\partial t)\Omega\Psi\,\mathrm{d}\tau + \int \Psi^*\Omega(\partial\Psi/\partial t)\,\mathrm{d}\tau \\ &= (1/\mathrm{i}\hbar)\left\{-\int (H\Psi)^*\Omega\Psi\,\mathrm{d}\tau + \int \Psi^*\Omega H\Psi\,\mathrm{d}\tau\right\} \\ &= (1/\mathrm{i}\hbar)\left\{-\int \Psi^* H\Omega\Psi\,\mathrm{d}\tau + \int \Psi^*\Omega H\Psi\,\mathrm{d}\tau\right\} \\ &= (\mathrm{i}/\hbar)\int \Psi^*(H\Omega - \Omega H)\Psi\,\mathrm{d}\tau = (\mathrm{i}/\hbar)\langle[H, \Omega]\rangle. \qquad (5.5.1)\end{aligned}$$

† A special word of warning is appropriate here. It may be the case that even though the commutator $[A, B]$ is non-zero, when it acts on a function it gives rise to zero; recall the footnote on p. 91. That is, even though $[A, B] \neq 0$, $[A, B]\psi = 0$. Then $\langle[A, B]\rangle = 0$ also, and $\delta A\,\delta B \geq 0$ even though $[A, B] \neq 0$. In this special case it is possible to specify A and B precisely. This might seem a niggling point, but we shall see an important example of it when we turn to angular momentum (see p. 119).

In the fourth equality we have used the hermiticity of the hamiltonian. Note that the operator Ω itself is independent of time (e.g. it is the derivative operator $\mathrm{d}/\mathrm{d}x$ or x itself) and its expectation value acquires its time-dependence from the time-evolution of the wavefunction.

Equation (5.5.1) shows that the expectation value is constant in time if Ω commutes with the hamiltonian. Observables with operators which commute with H are called *constants of the motion* and their expectation values are *conserved*:

Constant of the motion: $(\mathrm{d}/\mathrm{d}t)\langle\Omega\rangle = 0$ if $[H, \Omega] = 0$. (5.5.2)

As an important example, consider the case of the linear momentum in a one-dimensional system. Is linear momentum a constant of the motion? The commutator takes the form

$$[H, p] = [-(\hbar^2/2m)(\mathrm{d}^2/\mathrm{d}x^2) + V(x), (\hbar/\mathrm{i})(\mathrm{d}/\mathrm{d}x)] = (\hbar/\mathrm{i})[V(x), \mathrm{d}/\mathrm{d}x]$$

because the derivatives commute. The commutator involving the potential energy can be evaluated by remembering that there is an unwritten function on the right on which the operators operate:

$$\begin{aligned}[H, p] &= (\hbar/\mathrm{i})\{V(\mathrm{d}/\mathrm{d}x) - (\mathrm{d}/\mathrm{d}x)V\} = (\hbar/\mathrm{i})\{V(\mathrm{d}/\mathrm{d}x) - (\mathrm{d}V/\mathrm{d}x) - V(\mathrm{d}/\mathrm{d}x)\} \\ &= -(\hbar/\mathrm{i})(\mathrm{d}V/\mathrm{d}x). \end{aligned} \quad (5.5.3)$$

Therefore the linear momentum is a constant of the motion only if the potential energy is uniform and $\mathrm{d}V/\mathrm{d}x = 0$. Hence p is a constant of the motion in free space. Furthermore, since in classical physics we call the quantity $-\mathrm{d}V/\mathrm{d}x$ the *force*, we can immediately conclude that *the linear momentum is a constant of the motion in the absence of a force.* This is Newton's first law! We can go on: the rate of change of the expectation value of p in the presence of a force is given by eqn (5.5.1) as

$$(\mathrm{d}/\mathrm{d}t)\langle p\rangle = (\mathrm{i}/\hbar)\langle[H, p]\rangle = \langle-(\mathrm{d}V/\mathrm{d}x)\rangle = \langle F\rangle. \quad (5.5.4)$$

That is, *the rate of change of the expectation value of the momentum is equal to the expectation value of the force:* this is Newton's second law! Now the relation of quantum to classical mechanics should be clear. *Classical mechanics deals with average values; quantum mechanics deals with the details.* Furthermore, when we are thinking about (and picturing) expectation values, there is no harm in thinking classically; but we cannot use classical pictures when thinking about the details of the motion.

5.6 Matrices in quantum mechanics

The basic equation of quantum mechanics, $xp - px = \mathrm{i}\hbar$, can be satisfied by representing x and p as operators. In Chapter 2 we pointed out that an alternative method is to represent them as *matrices.* In this section we give a brief introduction to this approach because it introduces some more language.

A *matrix* is an array of numbers. It is normally written as a two-dimensional square array of *elements*, and each element can be specified by quoting the row and column it occupies. Thus the matrix **M** is the array of its elements M_{rc}, where r labels the row and c the column. Matrices are a kind of generalized numbers and may be added and multiplied according to the rules set out in Appendix 8. For our present purposes it is sufficient to emphasize the multiplication rule: the *product* of two matrices **M** and **N** is another matrix **P** with elements given by the rule

$$P_{rc}=\sum_s M_{rs}N_{sc} \quad \text{defines } \mathbf{P}=\mathbf{MN}. \tag{5.6.1}$$

Matrix multiplication is *non-commutative*. This means that the order in which the matrices **M** and **N** are multiplied affects the result. Thus in general the matrix $\mathbf{R}=\mathbf{NM}$, which is given by

$$R_{rc}=\sum_s N_{rs}M_{sc} \quad \text{defines } \mathbf{R}=\mathbf{NM} \tag{5.6.2}$$

is not the same as the matrix $\mathbf{P}=\mathbf{MN}$. Hence $\mathbf{MN}-\mathbf{NM}\neq 0$ in general, and finding matrices for x and p such that $\mathbf{xp}-\mathbf{px}=i\hbar$ was the basis of Heisenberg's approach.

Throughout the first parts of this chapter we have met quantities like $\langle m|\,\Omega\,|n\rangle$. These are sometimes written Ω_{mn}, which immediately suggests that they are elements of a matrix. For this reason the brackets $\langle m|\,\Omega\,|n\rangle$ are called *matrix elements* of the operator Ω. We shall often encounter products of matrix elements, especially in sums of products such as

$$\sum_s \langle r|\,A\,|s\rangle\langle s|\,B\,|c\rangle.$$

If these are regarded as matrix elements and written in terms of A_{rs} and B_{sc}, it follows that we can use the matrix multiplication rule as follows:

$$\sum_s \langle r|\,A\,|s\rangle\langle s|\,B\,|c\rangle=\sum_s A_{rs}B_{sc}=(AB)_{rc}=\langle r|\,AB\,|c\rangle. \tag{5.6.3}$$

That is, the sum is equal to the single matrix element (bracket) of the product of operators AB. Comparing the two ends of the last line of equations leads us to the

Completeness relation: $\sum_s |s\rangle\langle s|=1.$ (5.6.4)

This relation is exceptionally useful for developing quantum mechanical equations. It is often used in reverse: the bracket $\langle r|\,AB\,|c\rangle$ can always be split into factors by regarding it as $\langle r|\,A1B\,|c\rangle$ and then replacing 1 by the sum over a complete set of states.

We can use this result to show that *the eigenvalues of squares of hermitian operators are positive;* that is,

$$\text{for } \Omega^2 |a\rangle = a\,|a\rangle, \quad a \geq 0 \text{ if } \Omega \text{ is hermitian.} \tag{5.6.5}$$

This is equivalent to saying that the *diagonal matrix elements of the square of a hermitian operator are positive.* For on multiplying both sides by $\langle a|$,

$$\langle a|\,\Omega^2\,|a\rangle = a, \quad a \geq 0 \text{ if } \Omega \text{ is hermitian.} \tag{5.6.6}$$

The proof runs as follows:

$$\begin{aligned}\langle a|\,\Omega^2\,|a\rangle &= \langle a|\,\Omega\Omega\,|a\rangle = \sum_n \langle a|\,\Omega\,|n\rangle\langle n|\,\Omega\,|a\rangle \\ &= \sum_n \langle a|\,\Omega\,|n\rangle\langle a|\,\Omega\,|n\rangle^* = \sum_n |\langle a|\,\Omega\,|n\rangle|^2 \geq 0,\end{aligned}$$

as was to be proved. We shall use these results in the discussion of angular momentum in Section 6.3.

The time-independent form of the Schrödinger equation can be given a matrix interpretation. The equation we have to solve has the form $H\psi = E\psi$. Suppose we express ψ as a linear combination (sum) of states $|n\rangle$, then

$$H\psi = \sum_n c_n H\,|n\rangle = E\sum_n c_n\,|n\rangle.$$

Now multiply from the left by some bra $\langle m|$:

$$\sum_n c_n \langle m|\,H\,|n\rangle = E\sum_n c_n \langle m \mid n\rangle = Ec_m.$$

In matrix notation this is

$$\sum_n H_{mn}c_n = Ec_m. \tag{5.6.7}$$

Now, suppose we can find the set of states such that the hamiltonian has a *diagonal matrix;* that is, $H_{mn} = 0$ unless $m = n$. Then the left-hand side takes the form

$$H_{mm}c_m = Ec_m, \tag{5.6.8}$$

and the eigenvalue E is simply the diagonal matrix element of the hamiltonian. In other words, *solving the Schrödinger equation is the same problem as diagonalizing the hamiltonian matrix* (see Appendix 8). This is yet another link between the Schrödinger and Heisenberg forms of quantum mechanics. Indeed, it is reported that when Heisenberg was looking for ways of diagonalizing his matrices, Hilbert suggested to him that he should look for the corresponding differential equation instead. Had he done so, wave mechanics would have been Heisenberg's too.

Example. In a system consisting of only two states (such as an electron spin) the hamiltonian had the following matrix elements: $H_{11}=a$, $H_{22}=b$, $H_{12}=H_{21}=d$. Find the energy levels and eigenstates.

• *Method.* The energy levels are the eigenvalues of the matrix. Use Appendix 8 to find them (by solving the secular determinant $\det|H-ES|=0$). Find the eigenstates in the form $c_1|1\rangle+c_2|2\rangle$ by solving the secular equations for each eigenvalue. The best procedure is to express them as $|a\rangle=|1\rangle\cos\beta+|2\rangle\sin\beta$ and $|b\rangle=-|1\rangle\sin\beta+|2\rangle\cos\beta$ because this ensures that they are orthonormal, β being a parameter we have to determine. These forms of the eigenstates enable us to establish the transformation matrix **s** such that $\mathbf{s}^{-1}\mathbf{Hs}$ is diagonal, its diagonal elements being the eigenvalues. Hence form $\mathbf{s}^{-1}\mathbf{Hs}$, equate it to the matrix $\begin{pmatrix} E_+ & 0 \\ 0 & E_- \end{pmatrix}$, and solve for β.

• *Answer.* Since $|1\rangle$ and $|2\rangle$ are orthonormal, $S_{11}=S_{22}=1$, $S_{12}=S_{21}=0$; hence

$$\det|H-ES|=\begin{vmatrix} a-E & d \\ d & b-E \end{vmatrix}=(a-E)(b-E)-d^2=0.$$

This quadratic equation for E has the roots

$$E_\pm=\tfrac{1}{2}(a+b)\pm\tfrac{1}{2}\surd\{(a-b)^2+4d^2\}=\tfrac{1}{2}(a+b)\pm\tfrac{1}{2}\Delta$$

where $\Delta^2=(a-b)^2+4d^2$. These are the eigenvalues and hence the energy levels. According to Appendix 8, the matrix that diagonalizes H is formed from the coefficients of $|1\rangle$ and $|2\rangle$ as follows:

$$\mathbf{s}=\begin{pmatrix} \cos\beta & -\sin\beta \\ \sin\beta & \cos\beta \end{pmatrix}, \qquad \mathbf{s}^{-1}=\begin{pmatrix} \cos\beta & \sin\beta \\ -\sin\beta & \cos\beta \end{pmatrix}.$$

$$\begin{pmatrix} E_+ & 0 \\ 0 & E_- \end{pmatrix}=\mathbf{s}^{-1}\mathbf{Hs}=\begin{pmatrix} \cos\beta & \sin\beta \\ -\sin\beta & \cos\beta \end{pmatrix}\begin{pmatrix} a & d \\ d & b \end{pmatrix}\begin{pmatrix} \cos\beta & -\sin\beta \\ \sin\beta & \cos\beta \end{pmatrix}$$

$$=\begin{pmatrix} a\cos^2\beta+b\sin^2\beta+2d\cos\beta\sin\beta & d(\cos^2\beta-\sin^2\beta)+(b-a)\cos\beta\sin\beta \\ d(\cos^2\beta-\sin^2\beta)+(b-a)\cos\beta\sin\beta & a\sin^2\beta+b\cos^2\beta-2d\cos\beta\sin\beta \end{pmatrix}.$$

Consequently, by equating corresponding off-diagonal elements,

$$(b-a)\cos\beta\sin\beta=-d(\cos^2\beta-\sin^2\beta), \quad \text{or} \quad \tan 2\beta=-2d/(b-a).$$

• *Comment.* The two-level system occurs widely in quantum mechanics, and we study it in detail in Chapter 8. The *parametrization* of the states in terms of the angle $\beta=\tfrac{1}{2}\arctan\{2d/(a-b)\}$ is a very useful device. Once systems have more than two levels it becomes almost essential to deal with them numerically using library programs.

Further reading

Postulates

The conceptual development of quantum mechanics. M. Jammer; McGraw-Hill, New York, 1966.

Aspects of quantum theory. A. Salam and E. P. Wigner (eds.); Cambridge University Press, 1972.

Philosophy of quantum mechanics. M. Jammer; Wiley, New York, 1975.
Foundations of quantum mechanics. J. M. Jauch; Addison-Wesley, Reading, Mass., 1968.
Conceptual foundation of quantum mechanics. B. D'Espagnat; Benjamin, Mass., 1976.

Operators

Techniques of applied quantum mechanics. J. P. Killingbeck; Butterworths, 1975.
Mathematical foundations of quantum mechanics. J. von Neumann; Princeton University Press, 1955.
Quantum mechanics. A. S. Davydov; Pergamon, Oxford, 2nd edition, 1976.
Matrix mechanics. H. S. Green; Noordhoff, Groningen, 1965.

Uncertainty principle

Physical principles of the quantum theory. W. Heisenberg; Dover, New York, 1949.
Quantum theory. D. Bohm; Prentice-Hall, Englewood Cliffs, 1951.
The Feynman lectures in physics. R. P. Feynman, R. B. Leighton, and M. Sands; Addison-Wesley, Reading, Mass., 1964.
The uncertainty principle and foundations of quantum mechanics. W. C. Price and S. S. Chissick; Wiley, London, 1977.

Problems

5.1. Which of the following functions are eigenfunctions of (a) $\mathrm{d}/\mathrm{d}x$, (b) $\mathrm{d}^2/\mathrm{d}x^2$: e^{ax}, e^{ax^2}, x, x^2, $ax+b$, $\sin x$?

5.2. Construct quantum mechanical operators in the position representation for the following observables: (a) kinetic energy in one and in three dimensions, (b) the inverse separation, $1/x$, (c) electric dipole moment, (d) z-component of angular momentum, (e) the mean square deviations of the position and momentum of a particle from the mean value.

5.3. Repeat Problem 2 but find operators in the momentum representation. *Hint.* The observable $1/x$ should be regarded as x^{-1}; hence the operator is the inverse of the operator for x.

5.4. In relativistic mechanics, energy and momentum are related by the expression $E^2=p^2c^2+m^2c^4$. Show that when $p^2c^2 \ll m^2c^4$ this reduces to $E=p^2/2m+mc^2$. Construct the relativistic analogue of the Schrödinger equation from the relativistic expression. What can be said about the conservation of probability?

5.5. Confirm that the operators (a) $T=-(\hbar^2/2m)(\mathrm{d}^2/\mathrm{d}x^2)$ and (b) $l_z=(\hbar/\mathrm{i})(\partial/\partial\phi)$ are hermitian. *Hint.* Consider the integrals $\int_0^L \psi^* T\psi\,\mathrm{d}x$ (or $\int_{-\infty}^{\infty} \psi^* T\psi\,\mathrm{d}x$) and $\int_0^{2\pi} \psi^* l_z\psi\,\mathrm{d}\phi$, and integrate by parts.

5.6. Demonstrate that the linear combinations $A+\mathrm{i}B$ and $A-\mathrm{i}B$ are not hermitian if A and B are hermitian operators.

5.7. Evaluate the commutators (a) $[x, y]$, (b) $[p_x, p_y]$, (c) $[x, p]$, (d) $[x^2, p]$, (e) $[x^n, p]$, (f) $[(1/x), p]$, (g) $[(1/x), p^2]$, (h) $[xp_y-yp_x, yp_z-zp_y]$, (i) $[x^2(\partial^2/\partial y^2), y(\partial/\partial x)]$. ($p$ denotes p_x, unless otherwise specified.)

5.8. Show that (a) $[A, B]=-[B, A]$, (b) $[A^m, A^n]=0$ for all m, n, (c) $[A^2, B]=A[A, B]+[A, B]A$, (d) $[A,[B, C]]+[B,[C, A]]+[C,[A, B]]=0$. Evaluate $[l_y,[l_y, l_z]]$ given that $[l_x, l_y]=\mathrm{i}\hbar l_z$, $[l_y, l_z]=\mathrm{i}\hbar l_x$, and $[l_z, l_x]=\mathrm{i}\hbar l_y$.

5.9. The operator e^A has a meaning if it is expanded as a power series: $\mathrm{e}^A=\sum_n (1/n!)A^n$. Show that if $|a\rangle$ is an eigenstate of A with eigenvalue a, then it is also an eigenstate of e^A. Find the latter's eigenvalue.

5.10. Show that $e^A e^B = e^{A+B}$ only if $[A, B] = 0$. If $[A, B] \neq 0$, but $[A, [A, B]] = [B, [A, B]] = 0$, show that $e^A e^B = e^{A+B} e^f$, where f is a simple function of $[A, B]$. *Hint.* This is another example of the differences between operators (*q-numbers*) and ordinary numbers (*c-numbers*). The simplest approach is to expand the exponentials and to collect and compare terms on both sides of the equality. Note that $e^A e^B$ will give terms like $2AB$ while e^{A+B} will give $AB + BA$. Be careful with order.

5.11. Evaluate the commutators $[H, p_x]$ and $[H, x]$, where $H = p^2/2m + V(x)$. Choose (a) $V(x) = V$, (b) $V(x) = \frac{1}{2}kx^2$, (c) $V(x) \rightarrow V(r) = e^2/4\pi\varepsilon_0 r$.

5.12. Evaluate (via eqn (5.4.9)) the limitation on the simultaneous specification of the following observables: (a) the position and momentum of a particle, (b) the three components of linear momentum of a particle, (c) the kinetic energy and potential energy of a particle, (d) the electric dipole moment and the total energy of a one-dimensional system, (e) the kinetic energy and the position of a particle in one dimension.

5.13. Use eqn (5.5.1) to find expressions for the rate of change of the expectation values of position and momentum of a harmonic oscillator; solve the pair of differential equations, and show that the expectation values change in time in the same way as for a classical oscillator.

5.14. Confirm that the z-component of angular momentum is a constant of the motion for a particle on a ring.

5.15. The only non-zero matrix elements of x and p for a harmonic oscillator are $\langle v+1|\, x\, |v\rangle = (\hbar/2m\omega)^{\frac{1}{2}}(v+1)^{\frac{1}{2}}$, $\langle v-1|\, x\, |v\rangle = (\hbar/2m\omega)^{\frac{1}{2}}v^{\frac{1}{2}}$, $\langle v+1|\, p\, |v\rangle = i(\hbar m\omega/2)^{\frac{1}{2}}(v+1)^{\frac{1}{2}}$, $\langle v-1|\, p\, |v\rangle = -i(\hbar m\omega/2)^{\frac{1}{2}}v^{\frac{1}{2}}$ (and their hermitian conjugates); see p. 299. Write out the matrices of x and p explicitly (label the rows and columns $v = 0, 1, 2, \ldots$) up to about $v = 4$, and confirm by matrix multiplication that they satisfy the commutation rule. Construct the hamiltonian matrix by forming $p^2/2m + \frac{1}{2}kx^2$ by matrix multiplication and addition, and infer the eigenvalues.

5.16. Use the completeness relation, eqn (5.6.4), and the information in Problem 5.15 to deduce the value of the matrix element $\langle v|\, xp^2x\, |v\rangle$.

6 Angular momentum

IN THIS chapter we illustrate the material described in the last by developing the quantum mechanics of angular momentum. We saw in Chapter 4 that angular momentum is quantized, and that its magnitude and z-component are restricted to values determined by the quantum numbers l and m_l. Those results were obtained by solving the Schrödinger equation. In this chapter we see that the operators for angular momentum carry this information, and that it is not necessary even to set up, let alone solve, the Schrödinger equation in order to discover the properties of angular momentum. A further point is that the development will be based on the commutation properties of the operators for angular momentum, and on nothing else. Therefore, whenever we meet a set of operators that have the same commutation rule, even though they seem not to refer to 'angular momentum' we immediately know *all* the properties of the observables the operators represent. This is one of the reasons why angular momentum is of such central importance in quantum mechanics.

Angular momentum theory also has many more mundane applications. It is central to the discussion of the structure of atoms (we caught a glimpse of that with the hydrogen atom), to the discussion of the rotation of molecules, as well as to virtually all forms of spectroscopy. We shall draw heavily on this material when we turn to the applications of quantum mechnics in Part 3.

6.1 The angular momentum operators

In accord with the discussion in the last chapter, we can find an expression for the angular momentum operators if the classical angular momentum is expressed in terms of the position and linear momentum but ensure that x and p satisfy $[x, p] = i\hbar$ wherever they occur. In classical mechanics the angular momentum $\mathbf{l}$ is defined as the vector product $\mathbf{r} \wedge \mathbf{p}$. Figure 6.1 shows how this definition captures the sense of motion of a particle. Note that $\mathbf{l}$ shows the sense of rotation according to the right-hand screw rule: $\mathbf{l}$ points in the direction a right-hand (ordinary) screw travels when it is turned in the same sense as the rotation.

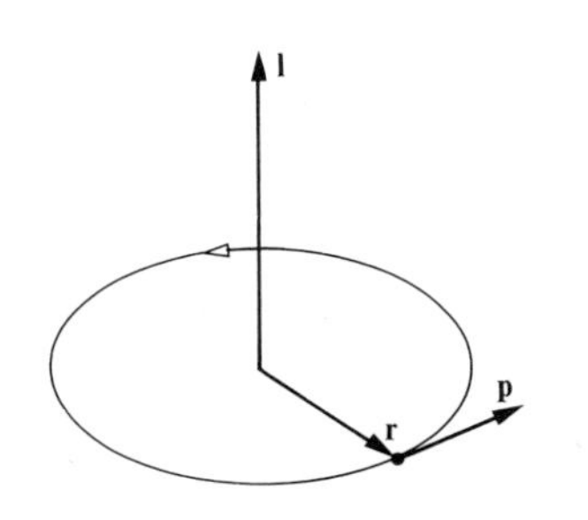

Fig. 6.1. The classical definition of angular momentum, $\mathbf{l} = \mathbf{r} \wedge \mathbf{p}$.

If the position of the particle is expressed in terms of its components x, y, and z as the vector $\mathbf{r} = x\hat{\mathbf{i}} + y\hat{\mathbf{j}} + z\hat{\mathbf{k}}$ and the linear momentum is expressed in terms of its components p_x, p_y, and p_z as the vector $\mathbf{p} = p_x\hat{\mathbf{i}} + p_y\hat{\mathbf{j}} + p_z\hat{\mathbf{k}}$, the angular momentum vector can be expressed in terms of its components $\mathbf{l} = l_x\hat{\mathbf{i}} + l_y\hat{\mathbf{j}} + l_z\hat{\mathbf{k}}$ using the standard definition of the vector product as a determinant:

$$\mathbf{l}=\mathbf{r}\wedge\mathbf{p}=\begin{vmatrix}\hat{\mathbf{i}} & \hat{\mathbf{j}} & \hat{\mathbf{k}}\\ x & y & z\\ p_x & p_y & p_z\end{vmatrix}$$

$$=(yp_z-zp_y)\hat{\mathbf{i}}+(zp_x-xp_z)\hat{\mathbf{j}}+(xp_y-yp_x)\hat{\mathbf{k}}. \qquad (6.1.1)$$

This identifies the three components as

$$l_x=yp_z-zp_y, \qquad l_y=zp_x-xp_z, \qquad l_z=xp_y-yp_x. \qquad (6.1.2)$$

(Note how to go from one to the other by cyclic permutation of x, y, and z: noticing relations like that saves a lot of time with calculations.) These expressions for the angular momentum components show that they are the *moments of linear momentum* about some point, Fig. 6.2. The *magnitude* of the angular momentum in classical physics is the length of the vector $\mathbf{l}$, which in terms of its components is $l=(l_x^2+l_y^2+l_z^2)^{\frac{1}{2}}$. The *angular velocity* ω of a sphere of moment of inertia I is then $\omega=l/I$ and is normally expressed as so many rad s^{-1}. Classical mechanics puts no restrictions on the magnitude of the angular momentum, the angular velocity, or the three components of the angular momentum vector: objects can be excited to any rate of rotation, at any orientation, and with all three components of angular momentum precisely specified.

The definitions remain when we turn to quantum mechanics, but the properties of the observables are extensively modified. We shall see how all these changes stem from the imposition of the basic commutation relations

$$[x,p_x]=i\hbar, \qquad [y,p_y]=i\hbar, \qquad [z,p_z]=i\hbar. \qquad (6.1.3)$$

We could proceed by choosing a specific representation of the operators; for instance the position representation would lead to the replacement of p_x by $(\hbar/i)(\partial/\partial x)$, etc., and hence to the operators

$$l_x=(\hbar/i)\{y(\partial/\partial z)-z(\partial/\partial y)\} \qquad (6.1.4a)$$

$$l_y=(\hbar/i)\{z(\partial/\partial x)-x(\partial/\partial z)\} \qquad (6.1.4b)$$

$$l_z=(\hbar/i)\{x(\partial/\partial y)-y(\partial/\partial x)\}, \qquad (6.1.4c)$$

but it is interesting, instructive, and time-saving to see that the angular momentum properties can be derived without making any choice of representation.

The material of Chapter 5 showed that a central aspect of operators are their commutation properties: they determine whether observables can be specified simultaneously, and they control the time evolution of the system. We shall also see in this chapter that they govern its permitted states. Therefore we begin the study of angular momentum by evaluating the commutation relations of the angular momentum operators.

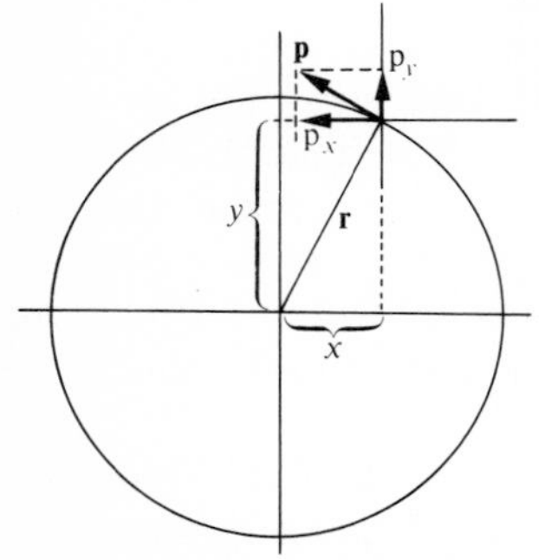

Fig. 6.2. The moment of momentum about the origin, $l_z=xp_y-yp_x$.

Consider first the commutator of the operators l_x and l_y:

$$\begin{aligned}[l_x, l_y] &= [(yp_z - zp_y), (zp_x - xp_z)] \\ &= [yp_z, zp_x] - [yp_z, xp_z] - [zp_y, zp_x] + [zp_y, xp_z] \\ &= y[p_z, z]p_x - 0 - 0 + xp_y[z, p_z] \\ &= i\hbar(-yp_x + xp_y) = i\hbar l_z. \end{aligned} \tag{6.1.5}$$

In line 1 we have inserted the definitions; in line 2 we used eqn (5.4.2f) to expand the commutator; in line 3 we have used the fact that y and p_x commute with each other and with p_z and z, likewise for x and p_y.

The calculation shows that the commutator $[l_x, l_y]$ is equal to another operator, which we recognize as $i\hbar l_z$. The other commutators can be derived in the same way, but it is quicker to realize that because the operators are defined by cylically permuting the subscripts, the commutators can be obtained in the same way. Both approaches lead to the conclusion that

$$[l_x, l_y] = i\hbar l_z, \qquad [l_y, l_z] = i\hbar l_x, \qquad [l_z, l_x] = i\hbar l_y. \tag{6.1.6}$$

These three relations are the foundations for the entire theory of angular momentum. Whenever we encounter three operators having these three commutation relations we know that the properties they represent are identical to the properties we are about to derive. Therefore we shall say that an observable *is* an angular momentum if its three components satisfy these commutation relations.†

The immediate conclusion from the commutation relations is that the three components of angular momentum cannot be specified simultaneously. We may choose to specify one component; then having done so, the other two components must remain unspecified. *A rotating body cannot have more than one of its components of angular momentum specified precisely*. The z-component is the one normally taken to be specified (although in the absence of externally applied fields, what we call the z-axis is entirely arbitrary). With l_z specified, the components l_x and l_y cannot have sharply defined values. This was the basis of the vector model of angular momentum introduced in Chapter 4, where the angular momentum was represented by a vector lying at an unspecified orientation on a cone.

But, if we know l_z, can we simultaneously know the magnitude of the angular momentum? To answer this we need the operator for the magnitude. We use the classical definition $l = (l_x^2 + l_y^2 + l_z^2)^{\frac{1}{2}}$, but concentrate on the operator for the square of the magnitude, $l^2 = l_x^2 + l_y^2 + l_x^2$,

† Since all the properties of the observables are identical, there seems no other course but to give the observables the same name. Nevertheless, when we do so we capture some odd bed-fellows. Electric charge, for example, is described by the same set of commutation rules and is therefore indistinguishable from angular momentum—but should we regard it, or imagine it, as an angular momentum? Electron 'spin' likewise is described by the same set of commutation relations—but should we regard it, or imagine it, as an angular momentum?

in order to avoid having to worry about the meaning of the square root of an operator. The first step is to evaluate the commutator of l_z and l^2, using the basic commutation relations in eqn (6.1.6). First, expand the commutator as

$$[l^2, l_z]=[l_x^2+l_y^2+l_z^2, l_z]=[l_x^2, l_z]+[l_y^2, l_z]+[l_z^2, l_z].$$

Concentrate on the first term. The definition of the commutator enables one to write

$$\begin{aligned}[l_x^2, l_z] &= l_xl_xl_z - l_zl_xl_x = l_xl_xl_z - l_xl_zl_x + l_xl_zl_x - l_zl_xl_x \\ &= l_xl_xl_z - l_xl_zl_x + [l_x, l_z]l_x \\ &= l_xl_xl_z - l_xl_xl_z + l_xl_xl_z - l_xl_zl_x + [l_x, l_z]l_x \\ &= l_xl_xl_z - l_xl_xl_z + l_x[l_x, l_z]+[l_x, l_z]l_x = l_x[l_x, l_z]+[l_x, l_z]l_x \\ &= -\mathrm{i}\hbar(l_xl_y + l_yl_x).\end{aligned}$$

and similarly,

$$[l_y^2, l_z]=\mathrm{i}\hbar(l_xl_y + l_yl_x); \qquad [l_z^2, l_z]= l_zl_zl_z - l_zl_zl_z = 0.$$

Consequently,

$$[l^2, l_z]=0 \tag{6.1.7}$$

As x, y, and z occur symmetrically in l^2 we can conclude that it also commutes with l_x and l_y.

The answer to the question is now clear: the magnitude of the angular momentum can be specified together with any of its components. The component commutation relations rule that only one component may be specified. This is further justification of the vector model of angular momentum, for it lets us represent angular momentum by a vector of precise length and with one component specified.

At this point, though, we can begin to see that the vector model must be regarded with caution. The commutation rules in eqn (6.1.6) may be written as a single equation if we use the definition of the vector product. They become

$$\mathbf{l}\wedge\mathbf{l}=\mathrm{i}\hbar\mathbf{l}. \tag{6.1.8}$$

(Check this by expanding the left-hand side and comparing x-, y-, z-components of both sides of the equation.) But the vector product of a vector with itself is zero, as may be seen by looking at the defining determinant with two rows equal, or knowing that the magnitude of the vector $\mathbf{a}\wedge\mathbf{b}$ is $ab \sin\theta$, where θ is the angle between the vectors $\mathbf{a}$ and $\mathbf{b}$. But the vector product of $\mathbf{l}$ with itself is not zero, and so we must conclude that $\mathbf{l}$ is not really a vector. The vector model is useful only if we realize that it is not the whole truth, and the fact that $\mathbf{l}$ is a *vector operator* and not a classical vector is kept at the back of our minds, and sometimes at the front. We shall have more to say on this when we discuss the vector model of the coupling of angular momenta.

The next task is to see how the commutation relations govern the forms of the eigenfunctions and the eigenvalues of the operators l^2 and l_z: these two operators commute, and so the eigenfunctions of l_z are eigenfunctions of l^2 (p. 91). We shall discover that the solutions found in Chapter 4 were incomplete in a very important respect. In the process we shall also discover an elegant way of finding the spherical harmonics and the values of some useful matrix elements.

6.2 The shift operators

It will prove expedient to introduce two new operators into the discussion. These are the *shift operators*, or the *raising and lowering operators*, l^+ and l^-:

$$l^+ = l_x + il_y; \qquad l^- = l_x - il_y. \tag{6.2.1}$$

The inverse relations are useful:

$$l_x = \tfrac{1}{2}(l^+ + l^-), \qquad l_y = (1/2i)(l^+ - l^-). \tag{6.2.2}$$

We shall require their commutation properties, which they are easily deduced from the fundamental relations in eqn (6.1.6):

$$[l^+, l_z] = [l_x, l_z] + i[l_y, l_z] = -i\hbar l_y - \hbar l_x = -\hbar(l_x + il_y) = \hbar l^+.$$

The others are obtained similarly, and the three are

$$[l^+, l_z] = -\hbar l^+, \qquad [l^-, l_z] = \hbar l^-, \qquad [l^+, l^-] = 2\hbar l_z. \tag{6.2.3}$$

Furthermore, because l^2 commutes with all components of $\mathbf{l}$, it also commutes with l^+ and l^-. This point is important and will be used later.

6.3 The eigenvalues of the angular momentum

We suppose that the simultaneous eigenstates of l^2 and l_z are distinguished by two quantum numbers which for the time being we denote n and m. The state of the system is therefore denoted $|n, m\rangle$. (We could always include more labels if needed.) We shall define m through the statement that†

$$l_z\,|n, m\rangle = m\hbar\,|n, m\rangle. \tag{6.3.1}$$

That is, $m\hbar$ is the eigenvalue of l_z for the state $|n, m\rangle$. This is entirely valid, because l_z has the same dimensions as $\hbar$, and so its eigenvalues

† Although we are writing operators and states in an abstract way, you can always think of equations like this, and all those that follow, as definite mathematical operations on functions. Thus, in the position representation l_z is represented by $(\hbar/i)(\partial/\partial\phi)$ and $|n, m\rangle$ corresponds to some function of ϕ (and θ). It will turn out to be a spherical harmonic $Y(\theta, \phi)$. In the position representation the equation $l_z\,|n, m\rangle = m\hbar\,|n, m\rangle$ becomes $(\hbar/i)(\partial\psi_{n,m}/\partial\phi) = m\hbar\psi_{n,m}$; likewise for the rest of the abstract equations in this chapter.

must be some multiple of $\hbar$: we are saying nothing at this stage about whether m can vary continuously or not, nor are we restricting its range in any way. The only thing we know is that m is a real number (because l_z is hermitian). Since l^2 and l_z commute, $|n, m\rangle$ is also an eigenstate of l^2. For the time being we allow for the possibility that the eigenvalues of l^2 for the state $|n, m\rangle$ might depend on the values of both n and m, and write

$$l^2 |n, m\rangle = \hbar^2 f(n, m) |n, m\rangle \tag{6.3.2}$$

where $f(n, m)$ is some dimensionless function of n and m (we know from Chapter 4 that it will turn out to be $l(l+1)$ and be independent of m, but that has yet to be derived). Since l^2 is hermitian, f is real. Because l^2 is also the square of a hermitian operator, f is also positive (eqn (5.6.5)).

Since $l^2 = l_x^2 + l_y^2 + l_z^2$ there is a restriction on the relative magnitudes of f and m. First, note that

$$\langle n, m| l^2 - l_z^2 |n, m\rangle = \langle n, m| f(n, m)\hbar^2 - m^2\hbar^2 |n, m\rangle = \{f(n, m) - m^2\}\hbar^2$$

and also

$$\begin{aligned}\langle n, m| l^2 - l_z^2 |n, m\rangle &= \langle n, m| l_x^2 + l_y^2 |n, m\rangle \\ &= \langle n, m| l_x^2 |n, m\rangle + \langle n, m| l_y^2 |n, m\rangle \geq 0,\end{aligned}$$

because we know (eqn (5.6.6)) that the diagonal elements of the square of any hermitian operator are positive. Comparing the two lines shows that for any value of m,

$$f(n, m) \geq m^2. \tag{6.3.3}$$

Now bring in the commutation relations. To do so we introduce the shift operators (and see why they are so called). Consider the effect of l^+ on the state $|n, m\rangle$. Because l^+ commutes with l^2 it cannot affect the magnitude of the angular momentum of the state. The formal proof of this runs as follows. Apply l^+ to both sides of eqn (6.3.2) and use the fact that $l^+l^2 = l^2l^+$:

$$l^+l^2 |n, m\rangle = \hbar^2 f(n, m) l^+ |n, m\rangle \quad \text{and} \quad l^+l^2 |n, m\rangle = l^2 l^+ |n, m\rangle;$$

comparing the two equations shows that

$$l^2\{l^+ |n, m\rangle\} = \hbar^2 f(n, m)\{l^+ |n, m\rangle\}, \tag{6.3.4}$$

and so the state $\{l^+ |n, m\rangle\}$ is an eigenstate of l^2 with the same eigenvalue as the state $|n, m\rangle$ itself. Hence $l^+ |n, m\rangle$ has the same magnitude of angular momentum as $|n, m\rangle$, as was to be proved. The argument fails for l_z because l^+ and l_z do not commute, and so the states $l^+ |n, m\rangle$ and $|n, m\rangle$ correspond to different eigenvalues of l_z. That is, l^+ (and by a similar argument l^-) *shift* the value of m but leave the magnitude of the momentum unchanged.

To what eigenvalue of l_z does the state $l^+ |n, m\rangle$ correspond? We use eqn (6.3.1) to find out, and form $l_z l^+ |n, m\rangle$. The l_z operator can be brought through l^+ to operate on $|n, m\rangle$ if we use the commutation relations:

$$l_z l^+ |n, m\rangle = \{l^+ l_z + [l_z, l^+]\} |n, m\rangle = \{l^+ l_z + \hbar l^+\} |n, m\rangle$$
$$= \{l^+ m\hbar + \hbar l^+\} |n, m\rangle = (m+1)\hbar l^+ |n, m\rangle.$$

But eqn (6.3.1) implies that the state $|n, m+1\rangle$ has the same eigenvalue, $(m+1)\hbar$, as the one just found. Therefore $l^+|n, m\rangle$ is proportional to $|n, m+1\rangle$. We write

$$l^+ |n, m\rangle = \hbar c^+_{n,m} |n, m+1\rangle, \quad (6.3.5)$$

where $c^+_{n,m}$ is some dimensionless numerical coefficient. This explains why l^+ is called a *raising operator*: when it operates on a state with z-component $m\hbar$ it generates from it the state with the same magnitude but with z-component one unit greater, $(m+1)\hbar$, Fig. 6.3. In exactly the same way the operator l^- may be shown to generate the state with z-component one unit lower:

$$l^- |n, m\rangle = \hbar c^-_{n,m} |n, m-1\rangle. \quad (6.3.6)$$

Hence l^- is called a *lowering operator.*

The shift operators step the value of m up and down by unity each time they operate. It would appear that any value of m could be reached by applying them a sufficient number of times. But we have already established that m is bounded, and that m^2 cannot exceed $f(n, m)$. Suppose we try to raise the value of m from its maximum value $m_{\max}$ (whatever that happens to be). There is no state with a greater value of m, and so we must have

$$l^+ |n, m_{\max}\rangle = 0. \quad (6.3.7)$$

We can establish the value of $m_{\max}$ as follows. Operate on both sides of the last equation with l^-; we have

$$l^- l^+ |n, m_{\max}\rangle = 0.$$

The operator product $l^- l^+$ can be expanded as follows:

$$l^- l^+ = (l_x - \mathrm{i}l_y)(l_x + \mathrm{i}l_y) = l_x^2 + l_y^2 + \mathrm{i}l_x l_y - \mathrm{i}l_y l_x$$
$$= l^2 - l_z^2 + \mathrm{i}[l_x, l_y] = l^2 - l_z^2 - \hbar l_z. \quad (6.3.8)$$

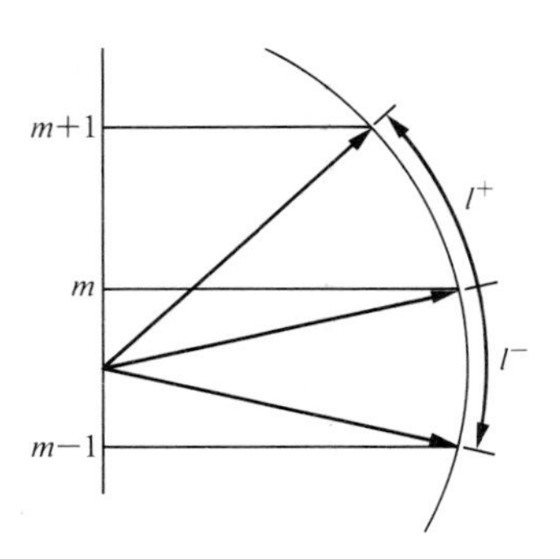

Fig. 6.3. The effects of the shift operators.

Therefore the last equation is the same as

$$(l^2 - l_z^2 - \hbar l_z) |n, m_{\max}\rangle = 0,$$

or

$$l^2 |n, m_{\max}\rangle = (l_z^2 + \hbar l_z) |n, m_{\max}\rangle = (m_{\max}^2 \hbar^2 + m_{\max}\hbar^2) |n, m_{\max}\rangle$$
$$= m_{\max}(m_{\max}+1)\hbar^2 |n, m_{\max}\rangle.$$

It follows by comparing this result with eqn (6.3.2) that

$$f(n, m_{\max}) = m_{\max}(m_{\max}+1).$$

Now, we have already established that when l^- operates it leaves unchanged the eigenvalue of l^2. But when l^- operates on $|l, m_{\max}\rangle$ it generates in turn $|n, m_{\max}-1\rangle$, $|n, m_{\max}-2\rangle$, and so on. All these states correspond to the same value of f, namely $m_{\max}(m_{\max}+1)$. Therefore, for a given value of $m_{\max}$ we have

$$l^2\,|n, m\rangle = \hbar^2 m_{\max}(m_{\max}+1)\,|n, m\rangle, \qquad m = m_{\max}, m_{\max}-1, \ldots$$

and the magnitude of the angular momentum is determined by the value of $m_{\max}$ (whatever that is). Now we put the quantum number n to work. We set $n = m_{\max}$, in which case we have

$$l^2\,|m_{\max}, m\rangle = \hbar^2 m_{\max}(m_{\max}+1)\,|m_{\max}, m\rangle, \quad m = m_{\max}, m_{\max}-1, \ldots.$$

But it would be much simpler to express $m_{\max}$ by a single letter, and we choose l. Then the last expression becomes

$$l^2\,|l, m\rangle = \hbar^2 l(l+1)\,|l, m\rangle, \qquad m = l, l-1, \ldots. \tag{6.3.9}$$

This is exactly the form of the expression we used in Chapter 4, but notice how the emphasis has been shifted: there l was the dominant quantity, and m (there called m_l) was allowed to range through $l, l-1, \ldots$. Here m is the dominant quantity: l is simply its maximum value.

The calculation is not yet complete because we have not yet established that m and l are integers; nor have we found the lower bound of m (we know it exists because of the condition $m^2 \leq f$). The two points are related, and they can be tackled by considering the values of the coefficients c^+ and c^- in eqns (6.3.5) and (6.3.6). We shall now show that not only does the ladder of m values have a top rung at $m_{\max} = l$, but it also has a bottom rung at $m_{\min} = -l$: the ladder is symmetrical about $m = 0$.

The first step involves finding two expressions for the matrix elements of l^-l^+. We can use eqn (6.3.8) to write

$$\begin{aligned} l^-l^+\,|l, m\rangle &= (l^2 - l_z^2 - \hbar l_z)\,|l, m\rangle = \{\hbar^2 l(l+1) - m^2\hbar^2 - m\hbar^2\}\,|l, m\rangle \\ &= \hbar^2\{l(l+1) - m(m+1)\}\,|l, m\rangle, \end{aligned}$$

so that

$$\langle l, m|\, l^-l^+\,|l, m\rangle = \hbar^2\{l(l+1) - m(m+1)\}. \tag{6.3.10}$$

We can then use eqns (6.3.5) and (6.3.6) to write

$$\begin{aligned} \langle l, m|\, l^-l^+\,|l, m\rangle &= \langle l, m|\, l^-\hbar c_{l,m}^+\,|l, m+1\rangle = \hbar c_{l,m}^+\langle l, m|\, l^-\,|l, m+1\rangle \\ &= \hbar^2 c_{l,m}^+ c_{l,m+1}^-\langle l, m \mid l, m\rangle = c_{l,m}^+ c_{l,m+1}^-\hbar^2. \end{aligned}$$

It follows that

$$c_{l,m}^+ c_{l,m+1}^- = l(l+1) - m(m+1). \tag{6.3.11}$$

The next step involves finding a connection between c^+ and c^-. First note that

$$\langle l, m|\, l^- \,|l, m+1\rangle = \hbar c^-_{l,m+1}\langle l, m \mid l, m\rangle = \hbar c^-_{l,m+1}. \tag{6.3.12}$$

Then make use of the hermiticity of l_x and l_y (*not* the hermiticity of l^- and l^+: they are not hermitian, and do not correspond to observables; they are each others *hermitian conjugate* in the sense that we are about to show). Consider the following:

$$\begin{aligned}\langle l, m|\, l^- \,|l, m+1\rangle &= \langle l, m|\, l_x - \mathrm{i}l_y \,|l, m+1\rangle \\ &= \langle l, m|\, l_x \,|l, m+1\rangle - \mathrm{i}\langle l, m|\, l_y \,|l, m+1\rangle \\ &= \langle l, m+1|\, l_x \,|l, m\rangle^* \\ &\qquad - \mathrm{i}\langle l, m+1|\, l_y \,|l, m\rangle^* \quad \text{(hermiticity)} \\ &= \{\langle l, m+1|\, l_x \,|l, m\rangle + \mathrm{i}\langle l, m+1|\, l_y \,|l, m\rangle\}^* \\ &= \langle l, m+1|\, (l_x + \mathrm{i}l_y) \,|lm\rangle^* = \langle l, m+1|\, l^+ \,|l, m\rangle^*.\end{aligned} \tag{6.3.13}$$

(Note that the matrix elements of the two operators are related by an expression of the form $\langle a|\, l^+ \,|b\rangle = \langle b|\, l^- \,|a\rangle^*$, and that $\langle a|\, l^+ \,|b\rangle$ is *not* equal to $\langle b|\, l^+ \,|a\rangle^*$; this is what we mean by two operators being hermitian conjugates as distinct from hermitian.) Since eqn (6.3.5) implies that

$$\langle l, m+1|\, l^+ \,|l, m\rangle = \hbar c^+_{l,m}\langle l, m+1 \mid l, m+1\rangle = \hbar c^+_{l,m} \tag{6.3.14}$$

we have arrived at the conclusion that

$$\hbar c^{+*}_{l,m} = \langle l, m|\, l^- \,|l, m+1\rangle = \hbar c^-_{l,m+1}, \quad \text{or} \quad c^-_{l,m+1} = c^{+*}_{l,m}. \tag{6.3.15}$$

When the last result is combined with eqn (6.3.11) we obtain

$$c^+_{l,m}c^{+*}_{l,m} = |c^+_{l,m}|^2 = l(l+1) - m(m+1)$$

and hence on making an arbitary but convenient choice of phase for c^+ (choosing it to be real and positive)

$$c^+_{l,m} = \{l(l+1) - m(m+1)\}^{\frac{1}{2}} \tag{6.3.16a}$$

$$c^-_{l,m} = c^{+*}_{l,m-1} = \{l(l+1) - m(m-1)\}^{\frac{1}{2}}. \tag{6.3.16b}$$

Example. Evaluate the matrix elements (a) $\langle l, m+1|\, l_x \,|l, m\rangle$, (b) $\langle l, m+2|\, l_x \,|l, m\rangle$, and (c) $\langle l, m+2|\, l_x^2 \,|l, m\rangle$. Evaluate the results for $l=1$, $m=-1$.

- *Method.* Express l_x in terms of l^+ and l^- using eqn (6.2.2) and construct the matrix elements from eqns (6.3.5) and (6.3.6). Note that $l_x^2 = l_x l_x$ and $\langle l, m \mid l, m'\rangle = 0$ unless $m = m'$.

- *Answer.*

(a)
$$\begin{aligned}\langle l, m+1|\, l_x \,|l, m\rangle &= \tfrac{1}{2}\langle l, m+1|\, l^+ + l^- \,|l, m\rangle \\ &= \tfrac{1}{2}\langle l, m+1|\, l^+ \,|l, m\rangle + \tfrac{1}{2}\langle l, m+1|\, l^- \,|l, m\rangle \\ &= \tfrac{1}{2}c^+_{l,m}\hbar\langle l, m+1 \mid l, m+1\rangle + \tfrac{1}{2}c^-_{l,m}\hbar\langle l, m+1 \mid l, m-1\rangle \\ &= \tfrac{1}{2}c^+_{l,m}\hbar = \tfrac{1}{2}\{l(l+1) - m(m+1)\}^{\frac{1}{2}}\hbar.\end{aligned}$$

The matrix element has the value $\hbar/\sqrt{2}$ for $l=1$, $m=-1$.

(b) $\langle l, m+2|\, l_x\, |l, m\rangle = \frac{1}{2}c^+_{l,m}\hbar\langle l, m+2 \mid l, m+1\rangle + \frac{1}{2}c^-_{l,m}\hbar\langle l, m+2 \mid l, m-1\rangle = 0.$

(c) $$\begin{aligned}\langle l, m+2|\, l_x^2\, |l, m\rangle &= \tfrac{1}{4}\langle l, m+2|\, l^+l^+ + l^-l^- + l^+l^- + l^-l^+\, |l, m\rangle \\ &= \tfrac{1}{4}\langle l, m+2|\, l^+l^+\, |l, m\rangle \\ &= \tfrac{1}{4}c^+_{l,m}\hbar\langle l, m+2|\, l^+\, |l, m+1\rangle \\ &= \tfrac{1}{4}c^+_{l,m}c^+_{l,m+1}\hbar^2\langle l, m+2 \mid l, m+2\rangle \\ &= \tfrac{1}{4}c^+_{l,m}c^+_{l,m+1}\hbar^2 \\ &= \tfrac{1}{4}\{[l(l+1)-m(m+1)][l(l+1)-(m+1)(m+2)]\}^{\frac{1}{2}}\hbar^2.\end{aligned}$$

The matrix element has the value $\frac{1}{2}\hbar^2$ for $l=1$, $m=-1$.

- *Comment.* Note that it is easy to spot short-cuts, as in the third example, where it is obvious that only l_+^2 contributes. Once we have the matrix elements of l_+ and l_- the matrix elements of all angular momentum operators and their combinations can be calculated.

Now that we have the coefficients we can consider the effect of l^- on the state with the lowest possible value of m, $m = m_{\min}$ on the ladder of states with a top rung corresponding to $m = m_{\max} = l$. When l^- operates on $|l, m_{\min}\rangle$ it generates 0, because a lower state does not exist. That is,

$$l^-\,|l, m_{\min}\rangle = 0.$$

This means that the matrix element $\langle l, m_{\min}-1|\, l^-\, |l, m_{\min}\rangle$ is also equal to zero. But according to eqn (6.3.6) it is also equal to $\hbar c^-_{l,m_{\min}}$. Therefore

$$0 = c^-_{l,m_{\min}} = \{l(l+1) - m_{\min}(m_{\min}-1)\}^{\frac{1}{2}}, \quad \text{or} \quad m_{\min} = -l.$$

We have therefore shown that *the ladder of m values runs in unit steps from the top rung at m = l down to the bottom rung at m = −l.*

The symmetrical shape of the ladder has an important implication. In order to form such a ladder there are only two choices for l: it must be an *integer* (so that for $l=2$, for example, m ranges over $2, 1, 0, -1, -2$) or a *half-integer* (so that for $l=\frac{3}{2}$, for example, m ranges over $\frac{3}{2}, \frac{1}{2}, -\frac{1}{2}, -\frac{3}{2}$). There are no other possibilities ($l=\frac{3}{4}$, for example, would give rise to the unsymmetrical ladder $l=\frac{3}{4}, -\frac{1}{4}$).

We can now summarize the results. On the basis of the commutation relations and the hermicity of the angular momentum operators we have come to the conclusion that the magnitude and the z-component of angular momentum are quantized, that the magnitude of the angular momentum is given by the expression $\{l(l+1)\}^{\frac{1}{2}}\hbar$ with l an integer or half-integer $(0, \frac{1}{2}, 1, \ldots)$ and the z-component given by $m\hbar$ with m allowed the values $l, l-1, l-2, \ldots, -l$ for each value of l.

These conclusions differ in only one detail from those we drew in Chapter 4. There l was confined to *integral* values $l = 0, 1, 2, \ldots$. There, however, we arrived at the values of l on the basis of the cyclic boundary conditions on the wavefunctions. So far we have not imposed any boundary conditions: when we do so we shall see that only the integral values of l survive. Nevertheless, the point should not be missed that the commutation relations do allow l to take half-integral values, and if the boundary conditions for some reason do not apply, then half-integral angular momenta should not be excluded. Half-integral angular momenta do occur: electron spin, an 'internal' motion of the electron for which the cyclic spatial boundary conditions are irrelevant, turns out to be described by a half-integral angular momentum quantum number. It is quite remarkable how often it turns out that what mathematics allows, Nature shows!

In order to emphasize that there is a distinction between angular momenta according to whether or not particular boundary conditions have to be satisfied, we shall use the following notation. Whenever we want to emphasize that the angular momentum is confined to integral values, as in orbital angular momenta of electrons around nuclei, we shall denote the quantum numbers as l and m_l (hence the notation in Chapter 4). When internal ('spin') angular momentum is being discussed we shall use s and m_s. When the results are generally applicable and we do not want to specify what kind of motion is under consideration we shall write j and m_j. *In general*, therefore,

$$j^2\,|j, m_j\rangle = \hbar^2 j(j+1)\,|j, m_j\rangle, \qquad j = 0, \tfrac{1}{2}, 1, \ldots \tag{6.3.17}$$

$$j_z\,|j, m_j\rangle = m_j\hbar\,|j, m_j\rangle, \qquad m_j = j, j-1, \ldots, -j \text{ (i.e. } 2j+1 \text{ values)} \tag{6.3.18}$$

$$\langle j, m_j+1|\,j^+\,|j, m_j\rangle = \hbar\{j(j+1) - m_j(m_j+1)\}^{\frac{1}{2}} \tag{6.3.19}$$

$$\langle j, m_j-1|\,j^-\,|j, m_j\rangle = \hbar\{j(j+1) - m_j(m_j-1)\}^{\frac{1}{2}}. \tag{6.3.20}$$

The last four lines, solely consequences of the hermiticity of the operators j_q and of the relations $[j_x, j_y] = i\hbar j_z$ and its cyclic permutations, summarize the principal content of angular momentum theory, the only things lacking being the forms of the eigenfunctions themselves. To this we now turn.

6.4 The eigenfunctions of the angular momentum

We consider the case of *orbital angular momentum*; that is, of angular momentum involving motion through space, and therefore one described by integer values of the angular momentum quantum numbers. Whereas in Chapter 4 we found the wavefunctions as solutions of the second-order partial differential equation, here we shall discover them by solving a first-order equation. This is very much simpler, and corresponds to the technique used on the harmonic oscillator in Appendix 4.

It is sufficient to find the wavefunction for the state $|l, l\rangle$, the one with maximum z-component, because all the other states of different m_l but the same l can be generated by repeated application of l^- and the use of eqn (6.3.6). We can find $|l, l\rangle$ by solving the equation $l^+\,|l, l\rangle = 0$.

Everything up to now has been abstract in the sense that we have not needed to choose a representation of the angular momentum operators. Now we have to, because we want to find the solutions as functions of position in space (see footnote on p. 107). Therefore we adopt the position representation of the operators set out in eqn (6.1.4). There they are expressed in terms of the cartesian coordinates; for the present purpose, because we want to find the wavefunctions as functions of the angles θ and ϕ, we convert the derivatives to polar coordinate form. This leads to

$$l_x = -(\hbar/\mathrm{i})\{\sin\phi(\partial/\partial\theta) + \cot\theta\cos\phi(\partial/\partial\phi)\} \quad (6.4.1a)$$

$$l_y = (\hbar/\mathrm{i})\{\cos\phi(\partial/\partial\theta) - \cot\theta\sin\phi(\partial/\partial\phi)\} \quad (6.4.1b)$$

$$l_z = (\hbar/\mathrm{i})(\partial/\partial\phi). \quad (6.4.1c)$$

(The l_z operator was encountered in Chapter 4.) The explicit forms of the shift operators can be deduced by forming the appropriate combinations $l_x + \mathrm{i}l_y$ and using $\cos\phi \pm \mathrm{i}\sin\phi = \mathrm{e}^{\pm\mathrm{i}\phi}$:

$$l^+ = \hbar\mathrm{e}^{\mathrm{i}\phi}\{(\partial/\partial\theta) + \mathrm{i}\cot\theta(\partial/\partial\phi)\} \quad (6.4.2a)$$

$$l^- = -\hbar\mathrm{e}^{-\mathrm{i}\phi}\{(\partial/\partial\theta) - \mathrm{i}\cot\theta(\partial/\partial\phi)\}. \quad (6.4.2b)$$

The state $|l, l\rangle$ satisfies the relation $l^+|l, l\rangle = 0$ because there is no higher state. In terms of the wavefunction $\psi_{l,l}(\theta, \phi)$, the equation reads

$$\hbar\mathrm{e}^{\mathrm{i}\phi}\{(\partial/\partial\theta) + \mathrm{i}\cot\theta(\partial/\partial\phi)\}\psi_{l,l}(\theta, \phi) = 0.$$

We try the separation of variables technique on this first-order partial differential equation, and write $\psi(\theta, \phi) = \Theta(\theta)\Phi(\phi)$. A little manipulation then yields

$$(\tan\theta/\Theta)(\mathrm{d}\Theta/\mathrm{d}\theta) = -\mathrm{i}(1/\Phi)(\mathrm{d}\Phi/\mathrm{d}\phi).$$

According to the usual separation of variables argument, both sides are equal to the same constant (c), and so we have

$$\tan\theta(\mathrm{d}\Theta/\mathrm{d}\theta) = c\Theta, \qquad \mathrm{d}\Phi/\mathrm{d}\phi = \mathrm{i}c\Phi.$$

The two equations integrate immediately to

$$\Theta \propto \sin^c\theta, \qquad \Phi \propto \mathrm{e}^{\mathrm{i}c\phi},$$

and so the complete function is

$$\psi_{l,l}(\theta, \phi) = N\sin^c\theta\,\mathrm{e}^{\mathrm{i}c\phi}, \quad (6.4.3)$$

where N is a normalization constant. The value of c is obtained by choosing it so that $l_z\psi_{l,l} = l\hbar\psi_{l,l}$. With l_z in eqn (6.4.1c) it follows at once that $c = l$, and so we have the complete form of $\psi_{l,l}$. The function corresponding to $m_l = l - 1$ can now be generated by acting on $\psi_{l,l}$ with l^- in the explicit form given in eqn (6.4.2b). In this way all the wavefunctions can be generated for any value of l and m_l, and they turn out to be the spherical harmonics described in Chapter 4.

Example. Construct the wavefunction for the state with $m_l = l - 1$.

• *Method.* Act on eqn (6.4.3) with l^- in the form of eqn (6.4.2b) and use $l^-|l, l\rangle = c_{ll}^-\hbar|l, l-1\rangle$.

• *Answer.*

$$l^-\psi_{l,l} = -\hbar e^{-i\phi}\{(\partial/\partial\theta) - i\cot\theta(\partial/\partial\phi)\}N\sin^l\theta e^{il\phi}$$
$$= -N\hbar e^{-i\phi}\{l\sin^{l-1}\theta\cos\theta - i(il)\cot\theta\sin^l\theta\}e^{il\phi}$$
$$= -2N\hbar l\sin^{l-1}\theta\cos\theta e^{i(l-1)\phi}.$$

$c_{ll}^- = \{l(l+1) - l(l-1)\}^{\frac{1}{2}} = \surd(2l)$; therefore $|l, l-1\rangle = (1/\hbar\surd(2l))l^-|l, l\rangle$, and so

$$\psi_{l,l-1} = -(2l)^{\frac{1}{2}}N\sin^{l-1}\cos\theta e^{i(l-1)\phi}.$$

• *Comment.* If ψ_{ll} is normalized to unity (with $N = (1/2^l l!)\{(2l+1)!/4\pi\}^{\frac{1}{2}}$) then so too is $\psi_{l,l-1}$ (with the same value of N). Compare the outcome of this calculation with the expressions for the spherical harmonics in Table 4.1.

6.5 Spin

Uhlenbeck and Goudsmit realized in 1925 that a great simplification of the description of atomic spectra could be obtained if it was assumed that an electron had an *intrinsic angular momentum* with the quantum number $j = \frac{1}{2}$, and that it could be oriented in one of two directions with projections $+\frac{1}{2}\hbar$ and $-\frac{1}{2}\hbar$. They also needed to propose that there was a magnetic moment associated with this motion, but that seemed reasonable since the electron is charged: a magnetic field arises from an electric current, and therefore from moving charges. Normally the letter s is used for the intrinsic angular momentum quantum number, which is called the *spin quantum number*. This is an evocative name, but the footnote on p. 105 should be recalled. Furthermore, as $s = \frac{1}{2}$ for an electron, in an imaginary classical universe in which $\hbar = 0$ the spin angular momentum would be zero,† and so it is a phenomenon without a classical counterpart. Uhlenbeck and Goudsmit's proposal was no more than an hypothesis, but when Dirac welded quantum mechanics to relativity the existence of particles with half-integral quantum numbers for the angular momentum appeared automatically.‡ We should not be too surprised that in our derivation of the Schrödinger equation we missed the spin, for we warned that we were deriving it by referring to a classical limit, and things that disappeared in that limit might be missed.

† Bodies may rotate in a classical universe for their angular momentum is given by $\surd\{l(l+1)\}\hbar$ and as $\hbar \to 0$ we could take $l \to \infty$ and obtain a non-zero answer. The spin quantum number s is a fundamental property of the electron, like its charge, and cannot be increased to stop the disappearance of the product $\surd\{s(s+1)\}\hbar$ as $\hbar \to 0$.

‡ Dirac's equation is not difficult to construct, nor is it difficult to extract from it the main features of its solutions. In Chapter 2 we saw that the Schrödinger equation did not satisfy the requirement of relativity, that the space and time coordinates must occur symmetrically. In the absence of a potential energy we could have derived the equation from the relation $E = p^2/2m$ by replacing E by the corresponding operator, and p likewise. In relativity theory the energy and momentum are related by the equation $E^2 = p^2c^2 + m^2c^4$, where c is the speed of light: this reduces to the former equation when p is small compared with mc. The insertion of the operators into this equation

Continued on p. 116.

The angular momentum operator equations apply to spin, and when $s=\frac{1}{2}$ they do so in a very simple way. As the case $s=\frac{1}{2}$ is so important, it has proved useful to introduce the symbol α to denote the state with $s=\frac{1}{2}$ and $m_s=\frac{1}{2}$, $\alpha\equiv|\frac{1}{2},\frac{1}{2}\rangle$, and the symbol β to denote the state $|\frac{1}{2},-\frac{1}{2}\rangle$. These states obey the following eigenvalue equations:

$$s_z\alpha=\tfrac{1}{2}\hbar\alpha, \qquad s_z\beta=-\tfrac{1}{2}\hbar\beta, \qquad s^2\alpha=\tfrac{3}{4}\hbar^2\alpha, \qquad s^2\beta=\tfrac{3}{4}\hbar^2\beta, \tag{6.5.1}$$

and the effect of the shift operators is as follows (use eqns (6.3.19) and (6.3.20)):

$$s^+\alpha=0, \qquad s^+\beta=\hbar\alpha, \qquad s^-\alpha=\hbar\beta, \qquad s^-\beta=0 \tag{6.5.2}$$

so that the only non-zero matrix elements of s^+ and s^- are

$$\langle\alpha|\, s^+\,|\beta\rangle=\hbar, \qquad \langle\beta|\, s^-\,|\alpha\rangle=\hbar. \tag{6.5.3}$$

We shall use these results in Part 3.

6.6 The angular momenta of composite systems

We now consider a system in which there are two sources of angular momentum, $\mathbf{j}_1$ and $\mathbf{j}_2$. The system might be a single particle that possesses both spin and orbital angular momenta, or it might consist of two particles each with spin or each with orbital momentum, or a particle with spin carried on a rotating molecule. What do the angular momentum commutation relations imply about the *total* angular momentum of the system?

If the angular momentum state of one part of the system is fully specified, can the state of the other part be specified simultaneously? The state of particle 1 is fully specified by reporting its quantum numbers j_1

Continued from p. 115.

leads to the *Klein–Gordon equation.* This leads to a difficulty that was discussed on p. 37. Dirac said that one should try to take an equation of the form $\partial\psi/\partial t=\alpha_x\partial\psi/\partial x+\alpha_y\partial\psi/\partial y+\alpha_z\partial\psi/\partial z+\beta mc^2$, where α_x, α_y, α_z, β are constants. This equation is relativistically acceptable and leads to a conserved probability, but as the Klein–Gordon equation comes straight from the energy expression the solutions of the Dirac equation must also satisfy the Klein–Gordon equation. Using this idea one can find the values of the four constants in the Dirac equation. It turns out that they cannot be simple numbers, but must be at least 4×4 matrices. Therefore the Dirac equation is really four equations in one, and so there are four solutions. In the absence of a magnetic field the solutions separate at low kinetic energies into two doubly degenerate sets, one of which can be associated with negative kinetic energies and the other with positive kinetic energies. Leaving aside the former and its suggestions of anti-matter we consider the latter. This double degeneracy can be removed by the application of a magnetic field, and so one may associate with the particle a magnetic moment that may assume two orientations; that is, the angular momentum quantum number must be $s=\frac{1}{2}$. The magnetic moment required by Uhlenbeck and Goudsmit was predicted correctly, and will be mentioned again in Chapter 9.

and m_{j1}; likewise for particle 2. In order to be able to specify the entire state as $|j_1 m_{j1}; j_2 m_{j2}\rangle$ we have to investigate the commutation relations. Do the operators j_1^2 and j_{1z} commute with j_2^2 and j_{2z}? If they do, then the specification is permissible. Operators for independent sources of angular momentum do commute. One way of seeing that this is so for two orbital angular momenta is to note that the operators can be expressed in terms of the coordinates of the two particles and derivatives with respect to these coordinates, and the x_1, etc. terms commute with derivatives with respect to x_2, etc. Likewise the spin operators act on the 'internal' coordinates of a particle while the orbital operators act on its coordinates in space, and so the two are independent and therefore commute. *Operators referring to independent components of a system always commute.* We can summarize this by saying that each component j_{1q}, $q = x, y, z$ commutes with each component j_{2q}:

$$[j_{1q}, j_{2q'}] = 0 \quad \text{for all } q, q' = x, y, z. \tag{6.6.1}$$

Since j_1^2 and j_2^2 are both defined in terms of their respective components they also commute. Hence all four operators j_1^2, j_{1z}, j_2^2, and j_{2z} commute among themselves, and so the state may be specified as $|j_1 m_{j1}; j_2 m_{j2}\rangle$.

Can the total angular momentum be specified? We have to decide what we mean by total angular momentum. The obvious definition is $\mathbf{j} = \mathbf{j}_1 + \mathbf{j}_2$ (in the sense $j_x = j_{1x} + j_{2x}$, etc). But is this operator an angular momentum? This can be answered by calculating the commutation relations for the components of j in terms of the commutation relations for $\mathbf{j}_1$ and $\mathbf{j}_2$. For example,

$$\begin{aligned}[j_x, j_y] &= [j_{1x} + j_{2x}, j_{1y} + j_{2y}] = [j_{1x}, j_{1y}] + [j_{1x}, j_{2y}] + [j_{2x}, j_{1y}] + [j_{2x}, j_{2y}] \\ &= i\hbar j_{1z} + 0 + 0 + i\hbar j_{2z} = i\hbar(j_{1z} + j_{2z}) = i\hbar j_z. \end{aligned} \tag{6.6.2}$$

The commutation relation is characteristic of an angular momentum, and so $\mathbf{j} = \mathbf{j}_1 + \mathbf{j}_2$ is an angular momentum ($\mathbf{j}_1 - \mathbf{j}_2$, on the other hand, is not). Since $\mathbf{j}$ *is* an angular momentum we may conclude without further work that the magnitude of the total angular momentum is $\{j(j+1)\}^{\frac{1}{2}}\hbar$, where j is the *total angular momentum quantum number* and may take the values $j = 0, \frac{1}{2}, 1, \frac{3}{2}, \ldots$, and that the z-component of the total angular momentum is restricted to the values $m_j\hbar$, where $m_j = j, j-1, \ldots, -j$.

Can we specify the value of j if we also specify the values of j_1 and j_2? That is, can we speak of a well-defined total angular momentum if we specify the magnitudes of the contributing angular momenta? Since j_1^2 commutes with all its components, and j_2^2 likewise, and since j^2 can be expressed in terms of the same components, it is easy to see that

$$[j^2, j_1^2] = [j^2, j_2^2] = 0. \tag{6.6.3}$$

Therefore the eigenvalues of j^2, j_1^2, and j_2^2 can be specified simultaneously. In other words, a p-electron ($s = \frac{1}{2}$, $l = 1$) can be regarded as having a well-defined total angular momentum with a magnitude given by some value of j (the actual permitted value we have yet to find).

Can we specify the z-component of the total angular momentum if we also specify the magnitudes of the total and component momenta? Since $j_z = j_{1z} + j_{2z}$, it commutes with both j_1^2 and j_2^2 (as well as with j^2), and so we can.

At this stage we can conclude that the state of angular momentum can be specified as $|j_1, j_2; j, m_j\rangle$. *Can we simultaneously specify the z-components of the individual momenta*, and specify the state as $|j_1 m_{j1}; j_2 m_{j2}; jm_j\rangle$? We examine this question in the usual way by calculating the commutators of j_{1z} and j_{2z} with j^2. As the indices 1 and 2 occur symmetrically in j^2 it is sufficient to determine one of the commutators. We find:

$$\begin{aligned}
[j_{1z}, j^2] &= [j_{1z}, j_x^2] + [j_{1z}, j_y^2] + [j_{1z}, j_z^2] \\
&= [j_{1z}, (j_{1x} + j_{2x})^2] + [j_{1z}, (j_{1y} + j_{2y})^2] + [j_{1z}, (j_{1z} + j_{2z})^2] \\
&= [j_{1z}, j_{1x}^2 + 2j_{1x}j_{2x}] + [j_{1z}, j_{1y}^2 + 2j_{1y}j_{2y}] + 0 \text{ terms} \\
&= [j_{1z}, j_{1x}^2 + j_{1y}^2] + 2[j_{1z}, j_{1x}]j_{2x} + 2[j_{1z}, j_{1y}]j_{2y} \\
&= [j_{1z}, j_1^2 - j_{1z}^2] + 2i\hbar j_{1y}j_{2x} - 2i\hbar j_{1x}j_{2y} \\
&= 2i\hbar(j_{1y}j_{2x} - j_{1x}j_{2y}).
\end{aligned} \tag{6.6.4}$$

The commutator is *not* zero, and so we *cannot* specify m_{j1} (or m_{j2}) if we specify j. Therefore we have to make a choice concerning the specification of the system. Either we use the *uncoupled representation* $|j_1 m_{j1}; j_2 m_{j2}\rangle$, which leaves the magnitude of the total angular momentum unspecified (and therefore, in effect, says nothing about the relative orientation of the two angular momenta), or we use the *coupled representation* $|j_1 j_2; jm_j\rangle$, which leaves the individual components unspecified. At this stage which choice we make is entirely arbitrary. Later, when we consider the energy of interaction between different angular momenta, we shall see that one representation is sometimes more natural than the other. At this stage the two representations are simply alternative ways of specifying, or thinking about, a composite system.

If we decide to use the coupled representation, the question arises as to the permissible values of j and m_j for given values of j_1 and j_2. We have seen that the commutation relations allow j to have the values $j = 0, \frac{1}{2}, 1, \ldots$, but it is plausible that the values that actually occur are restricted and depend on the values of j_1 and j_2 of the components that are contributing to the total. What, for example, are the possible values of the magnitude of the total angular momentum of a p-electron, an electron with $s = \frac{1}{2}$ and $l = 1$? Surely the total cannot exceed $j = \frac{3}{2}$?

In order to arrive at the allowed values first note that the eigenvalue of j_z corresponding to the state $|j_1 m_{j1}; j_2 m_{j2}\rangle$ is

$$\begin{aligned}
j_z\, |j_1 m_{j1}; j_2 m_{j2}\rangle &= (j_{1z} + j_{2z})\, |j_1 m_{j1}; j_2 m_{j2}\rangle \\
&= (m_{j1} + m_{j2})\hbar\, |j_1 m_{j1}; j_2 m_{j2}\rangle.
\end{aligned} \tag{6.6.5}$$

But eigenvalues of j_z are denoted $m_j\hbar$. Therefore

$$m_j = m_{j1} + m_{j2}. \tag{6.6.6}$$

Now the maximum possible value of m_j arises when m_{j1} and m_{j2} have their maximum values. These are j_1 and j_2. Hence the maximum value of m_j is j_1+j_2. But the maximum value of m_j is called j; hence the value of j that j_1 and j_2 give rise to is j_1+j_2. This, however, is only the maximum value of j. We can see that there must be other values, because there are $(2j_1+1)(2j_2+1)=2j_1+2j_2+1+4j_1j_2$ states available, but a system with $j=j_1+j_2$ can give rise to only $2j+1=2j_1+2j_2+1$ states (corresponding to the $2j+1$ values allowed to m_j). What are the other $4j_1j_2$ states?

A system with $j=j_1+j_2$ can be in any of the states $m_j=j_1+j_2$, $j_1+j_2-1, \ldots, -(j_1+j_2)$. The state with $m_j=j_1+j_2$ can arise in only one way: $m_{j1}=j_1$, $m_{2j}=j_2$.† On the other hand, the state with $m_j=j_1+j_2-1$ can arise in two ways: from $m_{j1}=j_1$, $m_{j2}=j_2-1$ and from $m_{j1}=j_1-1$, $m_{j2}=j_2$. A system with $j=j_1+j_2$ accounts for only one of these (or one of two possible linear combinations), and so there is a second state with $m_j=j_1+j_2-1$, a second maximum value of m_j, unaccounted for. It must correpond to a system with $j=j_1+j_2-1$. A system with this value of j accounts for a further $2j+1=2j_1+2j_2-1$ states, corresponding to the permitted m_j values. The process can be continued by considering the next lower value of m_j, $m_j=j_1+j_2-2$, which can be formed in three ways ($m_{j1}, m_{j2}=j_1, j_2-2$ or j_1-1, j_2-1 or j_1-2, j_2). $j=j_1+j_2$ needs one of them, $j=j_1+j_2-1$ needs a second, and so the third (or the third remaining independent linear combination) must arise from $j=j_1+j_2-2$. We have therefore established that the value of j may be $j_1+j_2, j_1+j_2-1, j_1+j_2-2, \ldots$. All the states are accounted for by the time j has fallen to $|j_1-j_2|$ (j has to be positive, hence the modulus signs). Therefore the permitted states of total angular momentum that can arise from a system composed of two sources of angular momentum are those given by the

Clebsch–Gordan series: $j=j_1+j_2, j_1+j_2-1, \ldots, |j_1-j_2|$, (6.6.7)

and $m_j=m_{j1}+m_{j2}$. For example, a p-electron can have $j=(1+\frac{1}{2})$, $(1+\frac{1}{2}-1), \ldots, |1-\frac{1}{2}|$: that is, $j=\frac{3}{2}$ or $\frac{1}{2}$.

The Clebsch–Gordan series can be expressed in a simple pictorial way. Suppose we are given rods of lengths j_1 units and j_2 units (with j_1 and j_2 either integers or half-integers) and were asked to list the possible lengths of the third side that could be used to form a triangle

† The alert reader should object that we are now specifying j, m_{j1}, and m_{j2} simultaneously, in conflict with the fact that $[j^2, j_{1z}] \neq 0$. This is a special case of the type anticipated in the footnote on p. 96 where we saw that although $[A, B]$ may be non-zero, $[A, B]\psi = 0$ because of the special form of ψ. It turns out (Problem 8.12) that $[j^2, j_{1z}]|j_1j_1; j_2j_2\rangle = 0$, and so j, m_{j1}, m_{j2} may be specified. The same applies to the state $|j_1, -j_1; j_2, -j_2\rangle$, corresponding to $m_j=-j$.

given that its length, j units, had to start at j_1+j_2 and could change only in unit steps. The answer is precisely the Clebsch–Gordan series, Fig. 6.4. The states of total angular momentum that can arise are therefore the ones for which j_1, j_2, and j form a triangle: this is called the *triangle condition*. For example, $j_1=1$, $j_2=1$ require rods of length 2, 1, 0 to form a triangle, Fig. 6.5, and so angular momenta $j_1=1$, $j_2=1$ give rise to $j=2, 1, 0$. Although the triangle condition is no more than a simple and sometimes helpful mnemonic, it does suggest that angular momenta in quantum mechanics to some extent behave like vectors, and that the total angular momentum can be regarded as the *resultant* of the addition of the contributing momenta. Exploration of this point leads to the *vector model* of coupled angular momenta.

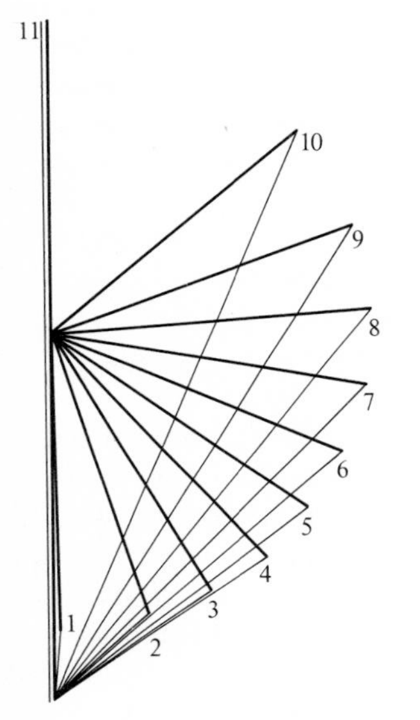

Fig. 6.4. The triangle condition for the addition $5+6=11, 10, \ldots 1$.

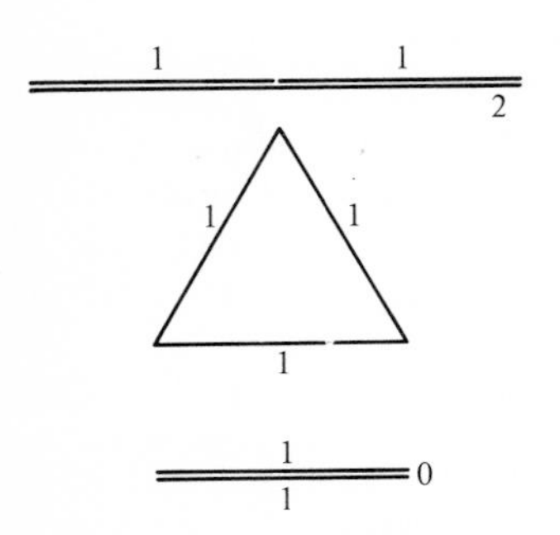

Fig. 6.5. The Clebsch–Gordan series $1+1=2, 1, 0$.

6.7 The vector model of coupled angular momenta

The implications of the commutation relations outlined in the last section allow us to extend the basic vector model of angular momentum to depict states of coupled systems. This gives some insight into the significance of the various coupling schemes and is often a helpful guide to the imagination: it puts flesh on the operator bones.

The features that any diagrams must express are as follows.

(1) The length of the total angular momentum vector is $\{j(j+1)\}^{\frac{1}{2}}$ units, with j one of the values permitted by the Clebsch–Gordan series.
(2) This vector must lie in an indeterminate angle on a cone about the z-axis (because j_x and j_y cannot be specified if j_z has been).
(3) The lengths of the contributing angular momentum vectors are $\{j_1(j_1+1)\}^{\frac{1}{2}}$ and $\{j_2(j_2+1)\}^{\frac{1}{2}}$ units; these have definite values even when j is specified.
(4) The projection of the total angular momentum on the z-axis is m_j units; in the coupled representation (where j is specified) the values of m_{j1} and m_{j2} are indefinite (except in the special case of $m_{j1}=j_1$; $m_{j2}=j_2$; see footnote p. 119), but their sum is equal to m_j.
(5) In the uncoupled representation, where j is not specified, the individual components m_{j1} and m_{j2} may be specified.

The diagrams in Figs. 6.6 and 6.7 capture these points. Figure 6.6(a) shows one of the states in the uncoupled representation: both m_{j1} and m_{j2} are specified, but there is no information about the relative orientation of j_1 and j_2 (apart from the fact that they lie on their respective cones), and so the total angular momentum is indeterminate. For example, it could be either of the resultants in Fig. 6.6(b) or (c), or something in between. Figure 6.7(a) shows one of the states in the coupled representation. Now the resultant, total angular momentum j has a well-defined magnitude and component on the z-axis, but while the magnitudes of the contributing momenta are well defined, their projections on the z-axis are indeterminate (we merely know that their sum has a definite value, namely m_j). For example, the arrangements in Figs. 6.7(b) and (c) have the same resultant and total

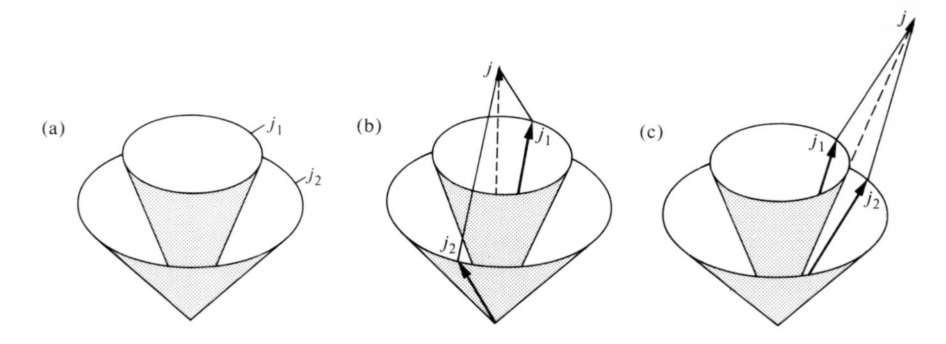

Fig. 6.6. (a) The uncoupled representation of $|j_1m_1; j_2m_2\rangle$; (b) and (c) show two possible values of total j.

projection but different individual projections. Still do not think of the vectors as precessing around the cones: at this stage the vector model is a display of possible but unspecifiable orientations: movement comes as a later feature of the model.

An important example, and one that we shall run into many times in Part 3, is the case of two particles with spin $s=\frac{1}{2}$, such as two electrons (or two protons), and the total angular momentum states they may have. For each electron $s=\frac{1}{2}$ and $m_s=\pm\frac{1}{2}$. We have already referred to the convention of denoting the state $|\frac{1}{2},\frac{1}{2}\rangle$ as α and $|\frac{1}{2},-\frac{1}{2}\rangle$ as β. In the *uncoupled representation* the electrons may be in any of the four states $|s_1m_{s1}; s_2m_{s2}\rangle$ with $m_{s1}=\pm\frac{1}{2}$ and $m_{s2}=\pm\frac{1}{2}$. These four are

$$\begin{aligned} &|\tfrac{1}{2},\tfrac{1}{2};\tfrac{1}{2},\tfrac{1}{2}\rangle=\alpha_1\alpha_2, \qquad |\tfrac{1}{2},\tfrac{1}{2};\tfrac{1}{2},-\tfrac{1}{2}\rangle=\alpha_1\beta_2,\\ &|\tfrac{1}{2},-\tfrac{1}{2};\tfrac{1}{2},\tfrac{1}{2}\rangle=\beta_1\alpha_2, \qquad |\tfrac{1}{2},-\tfrac{1}{2};\tfrac{1}{2},-\tfrac{1}{2}\rangle=\beta_1\beta_2. \end{aligned} \tag{6.7.1}$$

They are illustrated in Fig. 6.8. The individual angular momenta lie at unspecified positions on their cones, and the total angular momentum is indeterminate.

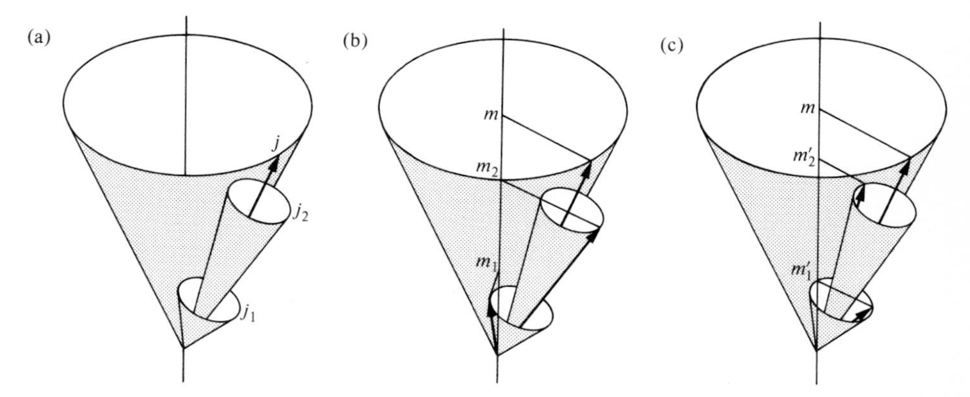

Fig. 6.7. (a) The coupled representation of $|j_1j_2jm\rangle$; (b) and (c) show how $|jm\rangle$ may arise from different projections m_1, m_2.

Now consider the *coupled representation* of the same two spins. Since $s_1 = \frac{1}{2}$ and $s_2 = \frac{1}{2}$ the triangle condition tells us that the total spin angular momentum S (capital letters are usually used to denote angular momenta of collections of particles), can take the values $S = 1, 0$:

$S = 1$ $s_2 = \frac{1}{2}$

$s_1 = \frac{1}{2}$ $s_1 = \frac{1}{2}$ $s_1 = \frac{1}{2}$ $S = 0.$

When $S = 0$, the only possible value of its z-component (now denoted M_S) is $M_S = 0$. There is only one state with $S = 0$, and so this arrangement is called a *singlet.* In the coupled representation notation $|s_1 s_2; SM_S\rangle$ it is denoted $|\frac{1}{2}, \frac{1}{2}; 0, 0\rangle$. When $S = 1$ the z-component of the spin may take the three values $M_S = 1, 0, -1$, and this arrangement is therefore called a *triplet.* The three states of the triplet are $|\frac{1}{2}, \frac{1}{2}; 1, M_S\rangle$ with $M_S = 1, 0, -1$.

The vector model of the triplet is shown in Fig. 6.9. We have drawn the vectors to scale (and much more carefully than is normally done; vector model diagrams are normally used only to indicate points, not to express detailed features), and several points are apparent. One is that in order to arrive at a resultant angular momentum corresponding to $S = 1$ (that is, to a vector of length $\sqrt{2}$) using component vectors of length $\{\frac{1}{2}(\frac{1}{2}+1)\}^{\frac{1}{2}} = \frac{1}{2}\sqrt{3}$, the individual vectors cannot lie at arbitrary relative orientations: if one is at some orientation on its cone, the other is at a definite angle on its. The absolute location of the vectors, however, remains unknown, and various possibilities are shown in Fig. 6.9. Another point is that when $M_S = +1$ we also know the individual projections even though we are specifying the total spin magnitude. This is the peculiar case referred to in the footnote on p. 119: note however, that the individual vectors make a definite angle to each other, and this is quite unlike the uncoupled picture, Fig. 6.8, where they are at arbitrary relative angles. In one there is a definite total angular momentum, in the other there is not: in the coupled representation there are definite *phase relations* between the vectors; in the uncoupled representation there are not.

The vector model of the singlet must represent a state in which the angular momenta have zero resultant: this is illustrated in Fig. 6.10. On account of this condition there is also a definite phase relation between the individual spin vectors, even though their absolute position in space is arbitrary.

We shall take this example up again in the next section.

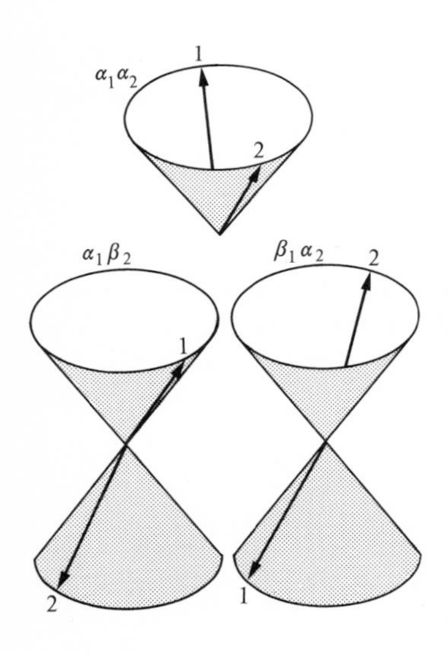

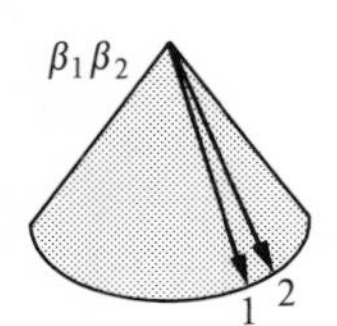

Fig. 6.8. Two spin-$\frac{1}{2}$ angular momenta in the uncoupled representation.

6.8 The relation between schemes

The state $|j_1 j_2; jm_j\rangle$ contains all values of m_{j1} and m_{j2} such that $m_{j1} + m_{j2} = m_j$. This suggests that it should be possible to express the coupled state as a sum over all the uncoupled states $|j_1 m_{j1}; j_2 m_{j2}\rangle$

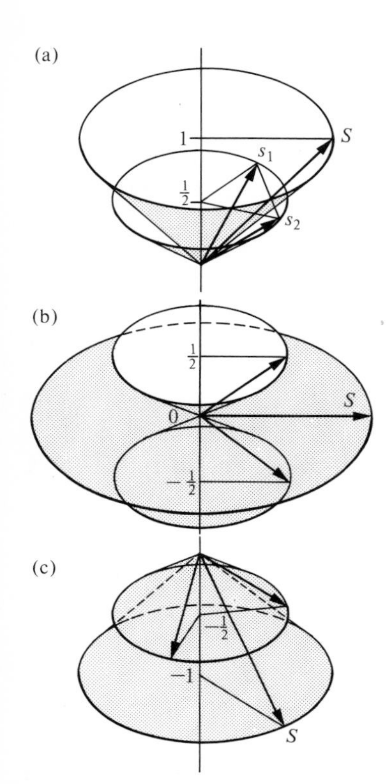

Fig. 6.9. Two spin-$\frac{1}{2}$ angular momenta in the coupled representation with $S=1$ and $M_S=$ (a) 1, (b) 0, (c) -1.

which conform to $m_j = m_{j1} + m_{j2}$. (This means that in the case of two spins we expect to be able to express the state $|\frac{1}{2}, \frac{1}{2}; 1, 0\rangle$ as a sum of $\alpha_1\beta_2$ and $\beta_1\alpha_2$ since each corresponds to $M_S = 0$.) This leads us to write

$$|j_1 j_2; j m_j\rangle = \sum_{m_{j1}} \sum_{m_{j2}} C_{m_{j1}m_{j2}} |j_1 m_{j1}; j_2 m_{j2}\rangle, \tag{6.8.1}$$

where the $C_{m_{j1}m_{j2}}$ are coefficients called the *vector coupling coefficients* (other common names are the *Clebsch–Gordan coefficients*, the *Wigner coefficients*, and, in a slightly modified form, the *3j-symbols*). In the case of the $M_S = 0$ state of the triplet of two spins, we expect to be able to write

$$|\tfrac{1}{2}, \tfrac{1}{2}; 1, 0\rangle = C_{\frac{1}{2},-\frac{1}{2}}\alpha_1\beta_2 + C_{-\frac{1}{2},\frac{1}{2}}\beta_1\alpha_2. \tag{6.8.2}$$

The vector coupling coefficients are tabulated for many individual angular momenta, and a sample is given in Table 6.1 and Appendix 9. Note that for the triplet, Table 6.1 gives

$$|\tfrac{1}{2}, \tfrac{1}{2}; 1, 0\rangle = (\tfrac{1}{2})^{\frac{1}{2}}\alpha_1\beta_2 + (\tfrac{1}{2})^{\frac{1}{2}}\beta_1\alpha_2 = (\tfrac{1}{2})^{\frac{1}{2}}(\alpha_1\beta_2 + \beta_1\alpha_2) \tag{6.8.3a}$$

while for the singlet (the same value of M_S)

$$|\tfrac{1}{2}, \tfrac{1}{2}; 0, 0\rangle = (\tfrac{1}{2})^{\frac{1}{2}}\alpha_1\beta_2 - (\tfrac{1}{2})^{\frac{1}{2}}\beta_1\alpha_2 = (\tfrac{1}{2})^{\frac{1}{2}}(\alpha_1\beta_2 - \beta_1\alpha_2). \tag{6.8.3b}$$

The difference of sign can be related to the vector diagram of the coupling: the + sign in the triplet state is interpreted as the two individual vectors having the same phase (i.e. lying at the same angle), while the − sign in the singlet is interpreted as meaning that the vectors have opposite phase (180° out of phase, corresponding to an arrangement in which their resultant is zero).

Table 6.1 Vector coupling coefficients for $s_1 = \frac{1}{2}$, $s_2 = \frac{1}{2}$

m_{s1}	m_{s2}	$\|1, 1\rangle$	$\|1, 0\rangle$	$\|0, 0\rangle$	$\|1, -1\rangle$
$\frac{1}{2}$	$\frac{1}{2}$	1	0	0	0
$\frac{1}{2}$	$-\frac{1}{2}$	0	$\sqrt{\frac{1}{2}}$	$\sqrt{\frac{1}{2}}$	0
$-\frac{1}{2}$	$\frac{1}{2}$	0	$\sqrt{\frac{1}{2}}$	$-\sqrt{\frac{1}{2}}$	0
$-\frac{1}{2}$	$-\frac{1}{2}$	0	0	0	1

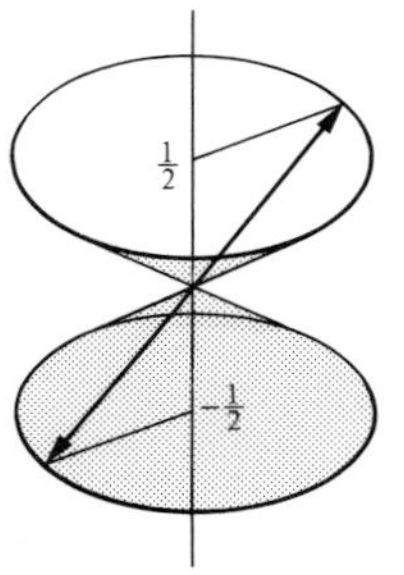

General expressions for the vector coupling coefficients can be derived, but they are very complicated, and the tables usually carry enough information. They can be derived in particular cases quite simply, and we shall illustrate the technique using the two spin-$\frac{1}{2}$ states, and then touch on a more complicated example. The general point worth noting about them is that they represent the overlap of one coupling scheme with another. For instance, on multiplying both sides of eqn (6.8.1) by $\langle j_1 m'_{j1}; j_2 m'_{j2}|$ we find

$$\langle j_1 m'_{j1}; j_2 m'_{j2} | j_1 j_2; j m_j\rangle = C_{m'_{j1}m'_{j2}}, \tag{6.8.4}$$

Fig. 6.10. As in Fig. 6.9, but with $S=0$, $M_S=0$.

and so the coefficient $C_{m'_{j1}m'_{j2}}$ can be interpreted as the degree of overlap (loosely the extent of resemblance) of the states $|j_1m'_{j1}; j_2m'_{j2}\rangle$ and $|j_1j_2; jm_j\rangle$.

The state $|\tfrac{1}{2}, \tfrac{1}{2}; 1, 1\rangle$, which from now on will be written $|1, 1\rangle$, must arise from $m_{s1} = \tfrac{1}{2}$, $m_{s2} = \tfrac{1}{2}$ as these are the only projections of the individual momenta that can combine to give $M_S = 1$. Therefore, without ambiguity, we can write

$$|1, 1\rangle = |\tfrac{1}{2}, \tfrac{1}{2}; \tfrac{1}{2}, \tfrac{1}{2}\rangle = \alpha_1\alpha_2. \qquad (6.8.5)$$

The effect of the lowering operator S^- on this state is given by eqn (6.3.20):

$$S^-|S, M_S\rangle = \{S(S+1) - M_S(M_S - 1)\}^{\frac{1}{2}}\hbar\,|S, M_S - 1\rangle \qquad (6.8.6)$$

(S_x, S_y, and S_z obey the angular momentum commutation rules). This gives

$$S^-|1, 1\rangle = 2^{\frac{1}{2}}\hbar\,|1, 0\rangle.$$

But we may also write $S^- = s_1^- + s_2^-$ (because $S_x = s_{1x} + s_{2x}$ etc.) and so the effect of S^- is also

$$S^-|1, 1\rangle = (s_1^- + s_2^-)\alpha_1\alpha_2 = \hbar\{\alpha_1\beta_2 + \beta_1\alpha_2\},$$

where eqn (6.5.2) has been used for the effects of S^- on α and β. Comparison of these equations gives

$$|1, 0\rangle = (\tfrac{1}{2})^{\frac{1}{2}}\{\alpha_1\beta_2 + \beta_1\alpha_2\}, \qquad (6.8.7)$$

as found from the tables. The third state of the triplet is obtained by repeating the procedure on this state. We obtain

$$\begin{aligned} S^-|1, 0\rangle &= 2^{\frac{1}{2}}\hbar\,|1, -1\rangle \\ &= (s_1^- + s_2^-)(\tfrac{1}{2})^{\frac{1}{2}}\{\alpha_1\beta_2 + \beta_1\alpha_2\} \\ &= 2^{\frac{1}{2}}\hbar\beta_1\beta_2. \end{aligned}$$

Therefore

$$|1, -1\rangle = \beta_1\beta_2. \qquad (6.8.8)$$

Only the singlet state $|0, 0\rangle$ remains to be found. As it must be constructed from $\alpha_1\beta_2$ and $\beta_1\alpha_2$ and yet be orthogonal to $|1, 0\rangle$ ($|1, 0\rangle$ and $|0, 0\rangle$ are eigenstates of a hermitian operator S^2 corresponding to different eigenvalues) we can write immediately that

$$|0, 0\rangle = (\tfrac{1}{2})^{\frac{1}{2}}\{\alpha_1\beta_2 - \beta_1\alpha_2\}, \qquad (6.8.9)$$

because $\langle\alpha_1 | \alpha_2\rangle = 1$ and $\langle\alpha_1 | \beta_1\rangle = 0$, etc.

Example. Confirm that the state in eqn (6.8.9) corresponds to $S = 0$, $M_S = 0$.

- *Method.* For M_S evaluate the eigenvalue of $S_z = s_{1z} + s_{2z}$. For S evaluate the eigenvalue of S^2. Note that $S^2 = (\mathbf{s}_1 + \mathbf{s}_2)^2 = s_1^2 + s_2^2 + 2\mathbf{s}_1 \cdot \mathbf{s}_2$, and so the effect of

S^2 can be expressed in terms of the operators for the individual spin-$\frac{1}{2}$ species. For the second part note that $s^2\alpha = \frac{3}{4}\hbar^2\alpha$, $s^2\beta = \frac{3}{4}\hbar^2\beta$ (eqn (6.5.1)) and $\mathbf{s}_1 \cdot \mathbf{s}_2 = s_{1z}s_{2z} + \frac{1}{2}(s_1^+ s_2^- + s_1^- s_2^+)$; then use eqn (6.5.2).

• *Answer.*

$$\begin{aligned} S_z\,|0, 0\rangle &= (\tfrac{1}{2})^{\frac{1}{2}}(s_{1z} + s_{2z})\{\alpha_1\beta_2 - \beta_1\alpha_2\} \\ &= (\tfrac{1}{2})^{\frac{1}{2}}\{s_{1z}\alpha_1\beta_2 - s_{1z}\beta_1\alpha_2 + \alpha_1 s_{2z}\beta_2 - \beta_1 s_{2z}\alpha_2\} \\ &= \hbar(\tfrac{1}{2})^{\frac{1}{2}}\{\tfrac{1}{2}\alpha_1\beta_2 + \tfrac{1}{2}\beta_1\alpha_2 - \tfrac{1}{2}\alpha_1\beta_2 - \tfrac{1}{2}\beta_1\alpha_2\} = 0. \end{aligned}$$

Hence $M_S = 0$.

$$\begin{aligned} S^2\,|0, 0\rangle &= (\tfrac{1}{2})^{\frac{1}{2}}\{s_1^2\alpha_1\beta_2 - s_1^2\beta_1\alpha_2 + \alpha_1 s_2^2\beta_2 - \beta_1 s_2^2\alpha_2 \\ &\quad + 2s_{1z}s_{2z}\alpha_1\beta_2 - 2s_{1z}s_{2z}\beta_1\alpha_2 \\ &\quad + s_1^+\alpha_1 s_2^-\beta_2 - s_1^+\beta_1 s_2^-\alpha_2 + s_1^-\alpha_1 s_2^+\beta_2 - s_1^-\beta_1 s_2^+\alpha_2\} \\ &= (\tfrac{1}{2})^{\frac{1}{2}}\hbar^2\{\tfrac{3}{4}\alpha_1\beta_2 - \tfrac{3}{4}\beta_1\alpha_2 + \tfrac{3}{4}\alpha_1\beta_2 - \tfrac{3}{4}\beta_1\alpha_2 - \tfrac{1}{2}\alpha_1\beta_2 + \tfrac{1}{2}\beta_1\alpha_2 - \alpha_1\beta_2 + \beta_1\alpha_2\} \\ &= 0. \end{aligned}$$

Hence $S = 0$.

• *Comment.* The same analysis may be applied to $\alpha_1\beta_2 + \beta_1\alpha_2$, when it turns out that $M_S = 0$ but $S = 1$.

As a second example, consider two electrons, each with orbital quantum numbers $l_1 = l_2 = 2$. This could correspond to two d-electrons in an atom. What are the possible values of the total orbital angular momentum of the system, and what are the states in the coupled representation in terms of the states in the uncoupled representation? The first question is easily answered by appealing to the triangle conditions. The orbital angular momentum quantum number for the combined system, L, can take the values $2+2, 2+2-1, \ldots, |2-2|$; that is, $L = 4, 3, 2, 1, 0$. With these states there are associated 9, 7, 5, 3, 1 values of M_L respectively, twenty-five in all, as there should be (from five m_{l1} values for $l_1 = 2$ and five for $l_2 = 2$). The state with $L = 4$ must have $M_L = 4$ as one of its components, and this can be formed only from the states with $m_{l1} = 2$ and $m_{l2} = 2$ in the uncoupled representation. Therefore

$$|2, 2; 4, 4\rangle = |2, 2; 2, 2\rangle, \tag{6.8.10}$$

where the left uses the notation $|l_1 l_2; LM_L\rangle$ and the right $|l_1 m_{l1}; l_2 m_{l2}\rangle$. This notation is very confusing, and so we shall modify it slightly. Just as in the hydrogen atom we called the orbitals with $l = 0, 1, 2, \ldots$ the s-, p-, d-orbitals, the states with $L = 0, 1, 2, 3, 4, \ldots$ we shall denote S, P, D, F, G, ... It is unfortunate that S occurs here as well as for spin, but the risk of confusion is slight. We shall omit the values of l_1 and l_2 from the kets in eqn (6.8.10), which therefore becomes

$$|G, 4\rangle = |2, 2\rangle. \tag{6.8.11}$$

We may now proceed to generate the remaining eight states with $L = 4$

and $M_L = 3, 2, \ldots, -4$ by applying the operator L^- to this equation, but using the form $l_1^- + l_2^-$ on the right-hand side.

$$L^- |G, 4\rangle = 8^{\frac{1}{2}}\hbar\, |G, 3\rangle \qquad \text{from eqn (6.3.20)}$$
$$= (l_1^- + l_2^-)\,|2, 2\rangle \qquad \text{from } L^- = l_1^- + l_2^-$$
$$= 4^{\frac{1}{2}}\hbar\{|1, 2\rangle + |2, 1\rangle\} \qquad \text{from eqn (6.3.20).}$$

Therefore

$$|G, 3\rangle = (\tfrac{1}{2})^{\frac{1}{2}}\{|1, 2\rangle + |2, 1\rangle\}, \tag{6.8.12}$$

where $|1, 2\rangle$ and $|2, 1\rangle$ mean $|2, 1; 2, 2\rangle$ and $|2, 2; 2, 1\rangle$ respectively in the notation $|l_1m_1; l_2m_2\rangle$. The remaining seven states of this set may be generated similarly. The state with $L = 3$, $M_L = 3$, an F state, also comes from the uncoupled states $|1, 2\rangle$ and $|2, 1\rangle$, for from them we have $M_L = m_{l1} + m_{l2} = 3$. This state must be orthogonal to $|G, 3\rangle$, and so $\langle F, 3 \mid G, 3\rangle = 0$. From this requirement and eqn (6.8.12) we conclude that

$$|F, 3\rangle = (\tfrac{1}{2})^{\frac{1}{2}}\{|1, 2\rangle - |2, 1\rangle\}. \tag{6.8.13}$$

The remaining six states of this set ($L = 3$; $M_L = 2, 1, \ldots, -3$) may then be generated from this state. The same argument may then be applied to find the D, P, and S states, and the table of coefficients constructed (Appendix 9).

Example. Construct the state with $j = \frac{3}{2}$ and $m_j = -\frac{1}{2}$ for a 2p-electron.

- *Method.* For a 2p-electron $l = 1$, $s = \frac{1}{2}$; the state with $j = \frac{3}{2}$, $m_j = -\frac{1}{2}$ is a linear combination of the states $|1m_l; \frac{1}{2}m_s\rangle$. Use Table 6.1 for the vector coupling coefficients. Only states with $m_l + m_s = -\frac{1}{2}$ contribute.
- *Answer.* Write the coupled state $|\frac{3}{2}, -\frac{1}{2}\rangle$ in the form

$$|\tfrac{3}{2}, -\tfrac{1}{2}\rangle = C_{0,-\frac{1}{2}}\,|10; \tfrac{1}{2} -\tfrac{1}{2}\rangle + C_{-1,\frac{1}{2}}\,|1 -1; \tfrac{1}{2}\tfrac{1}{2}\rangle$$
$$= (\tfrac{2}{3})^{\frac{1}{2}}\,|10; \tfrac{1}{2} -\tfrac{1}{2}\rangle + (\tfrac{1}{3})^{\frac{1}{2}}\,|1 -1; \tfrac{1}{2}\tfrac{1}{2}\rangle.$$

- *Comment.* A neater notation for the uncoupled states is $p_0\beta$ and $p_{-1}\alpha$ respectively; then $|\frac{3}{2}, -\frac{1}{2}\rangle = (\frac{2}{3})^{\frac{1}{2}}p_0\beta + (\frac{1}{3})^{\frac{1}{2}}p_{-1}\alpha$; p_0 can be interpreted as the $2p_z$-orbital and p_{-1} as a linear combination of the $2p_x$- and $2p_y$-orbitals

6.9 Coupling of several angular momenta

The final point we shall make in the present discussion of the coupling of momenta is concerned with the cases where there are three or more momenta to be coupled. We have the choice of first coupling $\mathbf{j}_1$ to $\mathbf{j}_2$ to give $\mathbf{j}_{12}$ and then coupling $\mathbf{j}_3$ to $\mathbf{j}_{12}$ to give the resultant $\mathbf{j}$, or of coupling $\mathbf{j}_1$ to $\mathbf{j}_3$, and then $\mathbf{j}_{13}$ to $\mathbf{j}_2$. The triangle conditions apply to each step in the coupling procedure, but the states obtained are different.

The angular momentum of a system with three or more sources of angular momentum can be established by coupling the momenta pairwise. First systems 1 and 2 are combined using the Clebsch–Gordan series. Then system 3 is combined with each possible outcome of the first pairing. For example, suppose we want the total orbital angular momentum of three p-electrons. The coupling of one pair gives $l_{12}=(1+1), (1+1-1), \ldots, 1-1=2, 1, 0$. Then the third couples with each of these in turn: $l_{12}=2$ gives rise to $L=(2+1)$, $(2+1-1), \ldots, (2-1)$, or 3, 2, 1; $l_{12}=1$ gives rise to 2, 1, 0; and $l_{12}=0$ gives rise to $L=1$ only. Therefore L may take the values 3, 2, 1, 0, with $L=2, 1$ obtained in two different ways.

The states which arise from different modes of coupling the components are related, and any one can be expressed as a linear combination of the states obtained by the other method of coupling. The coefficients which occur in the linear combinations are known as *Racah coefficients.* These too are tabulated, sometimes in terms of the related quantities, the *6j-symbols.* The question of alternative coupling schemes, and how to select the most appropriate ones, arises in discussions of atomic and molecular spectra, and we return to it in Part 3.

6.10 The conservation of angular momentum

In Chapter 5 we saw that the general expression for the rate of change of an expectation value is

$$(\mathrm{d}/\mathrm{d}t)\langle\Omega\rangle=(\mathrm{i}/\hbar)\langle[H, \Omega]\rangle. \tag{6.10.1}$$

If j^2 commutes with the hamiltonian for the system, then its expectation value remains the same, and it is a constant of the motion. Similarly, if j_z (or some other component) commutes with the hamiltonian, then its expectation value remains unchanged, and hence it too is a constant of the motion.

In the case of orbital angular momentum we have to consider the commutation properties of l^2 and l_z with the hamiltonian. This is most readily done by considering the position representation of the operators, using $l_z=(\hbar/\mathrm{i})(\partial/\partial\phi)$. If we write the hamiltonian in the form

$$H=(-\hbar^2/2m)\nabla^2+V,$$

its commutator with l_z has the following value (remember that we have to think of it as operating on some unwritten function):

$$\begin{aligned}[H, l_z]&=(\hbar/\mathrm{i})[(-\hbar^2/2m)\nabla^2, \partial/\partial\phi]+(\hbar/\mathrm{i})[V, \partial/\partial\phi]\\ &=0+(\hbar/\mathrm{i})\{V(\partial/\partial\phi)-(\partial/\partial\phi)V\}\\ &=(\hbar/\mathrm{i})\{V(\partial/\partial\phi)-(\partial V/\partial\phi)-V(\partial/\partial\phi)\}\\ &=-(\hbar/\mathrm{i})(\partial V/\partial\phi).\end{aligned} \tag{6.10.2}$$

Therefore, the rate of change of the expectation value of l_z is

$$\mathrm{d}\langle l_z\rangle/\mathrm{d}t=-\langle(\partial V/\partial\phi)\rangle. \tag{6.10.3}$$

In the hydrogen atom the potential energy of the electron is independent of the angle ϕ; therefore the right-hand side is zero, and $\langle l_z\rangle$ is independent of time. A similar argument applies to $\langle l^2\rangle$. Hence the

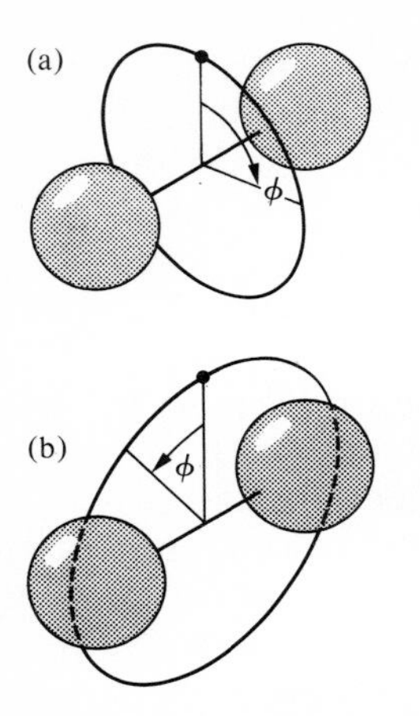

Fig. 6.11. (a) Conservation and (b) non-conservation of angular momentum in a linear system.

orbital angular momentum is conserved in the hydrogen atom. The same applies to the z-component of the orbital angular momentum in a linear molecule, Fig. 6.11(a) when z refers to the symmetry axis, because V is then independent of ϕ. On the other hand, if z denotes a direction perpendicular to the axis, the potential energy of an electron depends on ϕ, Fig. 6.11(b), therefore $\partial V/\partial\phi$ is not zero, and the component of orbital angular momentum about that axis is not well defined. Since the components are not constant in time, the magnitude is also not a constant of the motion. (As the electron bumps into the nuclei it causes the molecule to rotate; hence although the electronic orbital angular momentum is not conserved, the total angular momentum of electron + molecule may be.)

We can conclude that the conservation of angular momentum is an aspect of the *isotropy of the hamiltonian.* This points to a deep connection between the symmetry of a system and its states of angular momentum. It is therefore time to turn to aspects of symmetry.

Further reading

Background information and techniques

Elementary theory of angular momentum. M. E. Rose; Wiley, New York, 1957.

Angular momentum. D. M. Brink and G. R. Satchler; Oxford University Press, 1968.

Angular momentum in quantum mechanics. A. R. Edmonds; Princeton University Press, 1957.

Applications

Atomic structure. E. U. Condon and H. Odabaşi; Cambridge University Press, 1980.

Operator techniques in atomic spectroscopy. B. R. Judd; McGraw-Hill, New York, 1963.

Angular momentum theory for diatomic molecules. B. R. Judd; Academic Press, New York, 1975.

Tables of vector coupling coefficients

Group theory. V. Heine; Pergamon, Oxford, 1960.

The 3j- and 6j-symbols. M. Rotenberg, R. Bivins, N. Metropolis, and J. K. Wooten Jr; Technology Press, M.I.T., 1959.

The last title contains a bibliography. See also the books referred to under *Background information*, and also *Handbook of mathematical functions*, M. Abramowitz and I. A. Stegun; Dover, New York, 1965.

Problems

6.1. Evaluate the commutator $[l_x, l_y]$ in the position representation.

6.2. Evaluate the commutators $[l_y^2, l_x]$, $[l_y^2, l_x^2]$, and $[l_x, [l_x, l_y]]$. *Hint.* Use the basic commutators in eqn (6.1.6).

6.3. Verify that eqn (6.1.8) expresses the basic angular momentum commutation rules. *Hint.* Expand the left of eqn (6.1.8) and compare coefficients of the unit vectors. Be careful with the ordering of the vector components when expanding the determinant: the operators in the second row always precede those in the third.

6.4. Confirm that the *Pauli matrices*

$$\sigma_x = \begin{pmatrix} 0 & 1 \\ 1 & 0 \end{pmatrix}, \quad \sigma_y = \begin{pmatrix} 0 & -i \\ i & 0 \end{pmatrix}, \quad \sigma_z = \begin{pmatrix} 1 & 0 \\ 0 & -1 \end{pmatrix}$$

satisfy the angular momentum commutation relations when we write $s_x = \frac{1}{2}\hbar\sigma_x$, etc. and hence provide a matrix representation of angular momentum. Why does the representation correspond to $s = \frac{1}{2}$? *Hint.* For the last part, form the matrix representing s^2 and establish its eigenvalues.

6.5. Using the Pauli matrix representation, reduce the operators $s_x s_y$, $s_x s_y^2 s_z^2$, and $s_x^2 s_y^2 s_z^2$ to a single spin operator. *Hint.* On writing $s_x s_y = \frac{1}{4}\hbar^2 \sigma_x \sigma_y$ and evaluating the matrix product it turns out that $s_x s_y \propto s_z$, etc.

6.6. Suppose that in place of the actual angular momentum commutation rules, the operators obeyed $[l_x, l_y] = -i\hbar l_z$. What would be the roles of $l^{\pm}$?

6.7. Calculate the matrix elements $\langle 0, 0|\, l_z\, |0, 0\rangle$, $\langle 1, 1|\, l^+\, |0, 0\rangle$, $\langle 2, 1|\, l^+\, |2, 0\rangle$, $\langle 2, 2|\, l^{+2}\, |2, 0\rangle$, $\langle 2, 0|\, l^+ l^-\, |2, 0\rangle$, $\langle 2, 0|\, l^- l^+\, |2, 0\rangle$, and $\langle 2, 0|\, l^{-2} l_z l^{+2}\, |2, 0\rangle$. (The notation is $\langle l', m_l'|\, A\, |l, m_l\rangle$.)

6.8. Calculate the values of the matrix elements $\langle p_x|\, l_z\, |p_y\rangle$, $\langle p_x|\, l^+\, |p_y\rangle$, $\langle p_z|\, l_y\, |p_x\rangle$, $\langle p_z|\, l_x\, |p_y\rangle$, and $\langle p_z|\, l_x\, |p_x\rangle$.

6.9. Confirm that the spherical polar forms of the orbital angular momentum operators in eqn (6.4.1) satisfy the angular momentum commutation relation $[l_x, l_y] = i\hbar l_z$, and that the shift operators in eqn (6.4.2) satisfy $[l^+, l^-] = 2\hbar l_z$.

6.10. Verify that successive application of l^- to $\psi_{l,l}$ with $l = 2$ in eqn (6.4.3) generates the five normalized spherical harmonics $Y_{2,m}$ as set out in Table 4.1. *Hint.* Use (a) $l^-\psi_{l,m_l} = \hbar\{l(l+1) - m_l(m_l - 1)\}^{\frac{1}{2}}\psi_{l,m_l-1}$ and (b) l^- in the explicit form given in eqn (6.4.2b). The normalization constant N is specified in the *Example* following eqn (6.4.3).

6.11. Demonstrate that if $[j_{1q}, j_{2q'}] = 0$ for all q, q', then $\mathbf{j}_1 \wedge \mathbf{j}_2 = -\mathbf{j}_2 \wedge \mathbf{j}_1$. Go on to show that if $\mathbf{j}_1 \wedge \mathbf{j}_1 = i\hbar\mathbf{j}_1$, and $\mathbf{j}_2 \wedge \mathbf{j}_2 = i\hbar\mathbf{j}_2$, then $\mathbf{j} \wedge \mathbf{j} = i\hbar\mathbf{j}$ where $\mathbf{j} = \mathbf{j}_1 + \mathbf{j}_2$. Confirm that $\mathbf{j}_1 - \mathbf{j}_2$ is not an angular momentum.

6.12. In the footnote on p. 119 it is pointed out that in some cases m_{j1} and m_{j2} may be specified at the same time as j because although $[j^2, j_{1z}]$ is non-zero, the effect of $[j^2, j_{1z}]$ on the state with $m_1 = j_1$, $m_2 = j_2$ is zero. Confirm that $[j^2, j_{1z}]\,|j_1 j_1; j_2 j_2\rangle = 0$ and $[j^2, j_{1z}]\,|j_1, -j_1; j_2, -j_2\rangle = 0$.

6.13. Determine what total angular momenta may arise in the following composite systems: (a) $j_1 = 3$, $j_2 = 4$; (b) the orbital momenta of two electrons (i) both in p-orbitals, (ii) both in d-orbitals, (iii) the configuration pd; (c) the spin angular momenta of four electrons. *Hint.* Use the Clebsch–Gordan series, eqn (6.6.7); apply it successively in (c).

6.14. Construct the vector coupling coefficients for a system with $j_1 = 1$ and $j_2 = \frac{1}{2}$ and evaluate the matrix elements $\langle j'm'|\, j_{1z}\, |jm\rangle$. *Hint.* Proceed as in Section 6.8 and check the answer against the values in Appendix 9. For the matrix element, express the coupled states in the uncoupled representation, and then operate with j_{1z}.

6.15. Consider a system of two electrons (e.g. a biradical). The energy of the system depends on the relative orientation of their spins. Show that the operator $(J/\hbar)\mathbf{s}_1 \cdot \mathbf{s}_2$ distinguishes between singlet and triplet states. The system is now exposed to a magnetic field in the z-direction. Because the two electrons are in different environments, they experience different local fields and their interaction energy can be written $(\mu_B/\hbar)B(g_1 s_{1z} + g_2 s_{2z})$ with $g_1 \neq g_2$. Establish the matrix of the total hamiltonian, and demonstrate that when $\hbar J \gg \mu_B B$ the coupled representation is 'better', but that when $\mu_B B \gg \hbar J$ the uncoupled representation is 'better'. Find the eigenvalues and eigenstates of the system in each case.

6.16. What is the expectation value of the z-component of orbital momentum of electron 1 in the $|G, M_L\rangle$ state of the configuration d^2? *Hint.* Express $|G, M_L\rangle$ in terms of the uncoupled states, find $\langle G, M_L| l_{1z} |G, M_L\rangle$ in terms of the vector coupling coefficients, and evaluate it for $M_L = 4, 3, \ldots, -4$.

6.17. Prove that $\sum_{m_{j1}} \sum_{m_{j2}} |C_{m_{j1}m_{j2}}|^2 = 1$ for a given j_1, j_2, j. *Hint.* Use eqn (6.8.1) and form $\langle j_1 j_2; jm_j | j_1 j_2; jm_j\rangle$.

7 Group theory

THE subject treated in this chapter is one of the most remarkable in quantum mechanics. Not only does it simplify calculations, but it also reveals unexpected connections between apparently disparate phenomena. Whole regions of study are brought together in terms of its concepts. Angular momentum is a facet of group theory; so too are the properties of the harmonic oscillator. Turn the subject over and one finds conservation of momentum and energy. Group theory is used to classify the elementary particles and to discuss the structure of transition-metal complexes. The subject simply glitters with power and achievements.

What are its capabilities as far as quantum chemistry is concerned? First, it should be clear that by taking into account the full symmetry of a problem one stands a good chance of solving it with the least effort. It may be that a qualitative answer is sufficient for the purpose at hand, and that group theory furnishes this type of answer. We shall see that group theory is particularly useful for determining whether or not a quantity is zero; but it is usually unable to provide the magnitude of the quantity if it is not zero. This is because such an answer is normally expressed in terms of the fundamental constants or in terms of properties of the system that are extraneous to its symmetry. For example, we know that the magnitudes of physical observables are determined by integrals such as $\langle n| \Omega |n\rangle$, and we shall see in Chapter 8 that another type of observable, the probability of a transition between the states $|n\rangle$ and $|m\rangle$, is determined by the integral $\langle m| \Omega |n\rangle$; if either of these quantities can be said to be zero for some combinations of states $|m\rangle$ and $|n\rangle$ then we have a very powerful result. In Chapter 3 we encountered the condition known as degeneracy, and realized that it was an aspect of the symmetry of the system. Group theory lets us anticipate the occurrence of degeneracy, and to discuss its degree. In the construction of wavefunctions for electrons in molecules it is almost too trivial to remark that they must be constructed so as to reflect the symmetry of the nuclear framework. We shall see that group theory not only enables us to build molecular wavefunctions, but also to classify them in a useful way.

7.1 Symmetry operations

An *operation* applied to an object is the act of doing something to it, such as rotating it through some angle. A *symmetry operation* is an operation which, when applied to an object, leaves it apparently unchanged. For example, if the object is a sphere, any rotation about its centre leaves it apparently unchanged, and is therefore a symmetry operation. The translation of the function $\sin x$ through an interval $x = 2\pi$ leaves it apparently unchanged, and so it is a symmetry

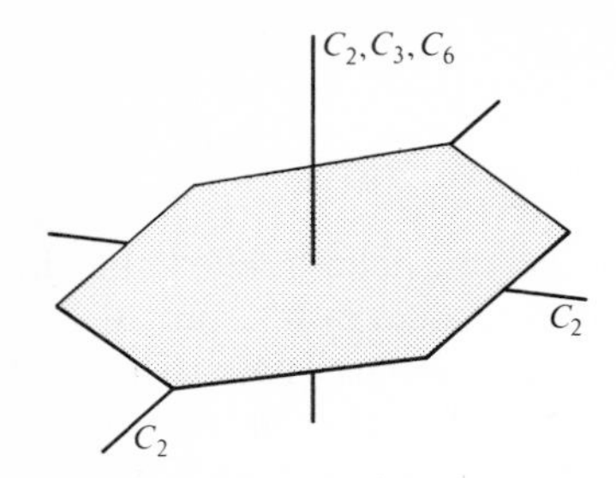

Fig. 7.1. Symmetry elements of a planar hexagon.

operation on the function. Not all operations are symmetry operations. The rotation of a rectangle through 90° is not a symmetry operation unless the rectangle happens to be a square; the translation of sin x by $x = \pi$ does not leave it apparently unchanged. Every object has at least one symmetry operation: the *identity operation*, the operation of doing nothing.

To each symmetry operation there corresponds a *symmetry element: the point, line, or plane with respect to which the operation is carried out.* For instance, there is a line, the *rotation axis*, corresponding to the symmetry operation of rotation of a sphere or the 90° rotation of a square. If we disregard the symmetry elements and operations corresponding to translational motion (that is, if we confine our attention to *point groups* as distinct from the *space groups* of interest when crystal symmetry is being discussed), there are five types of symmetry operation and their corresponding elements, and we list them below. The operation and the corresponding element are denoted by the same symbol.

E. The *identity*, which consists of doing nothing. Including it is more than just pedantry, as we shall see.

C_n. A *rotation* (the operation) about an *axis of symmetry* (the element). If a rotation through $360°/n$ gives rise to an indistinguishable orientation, the object has an n-fold axis. A hexagon (or a hexagonal molecule, such as benzene) has 2, 3, and 6-fold axes (C_2, C_3, C_6) perpendicular to the plane, Fig. 7.1, and several 2-fold axes (C_2) in the plane. For $n > 2$, the sense of rotation is significant: the n orientations are encountered in a different order depending on whether the rotation is clockwise (C_n^+) or anticlockwise (C_n^-) and so to a single *element* (the n-fold axis) there may correspond more than one *operation* (for this reason n-fold axes, $n > 2$, are sometimes called 'double-sided'). If an object has several rotation axes, the one with the greatest value of n is called the *principal axis.*

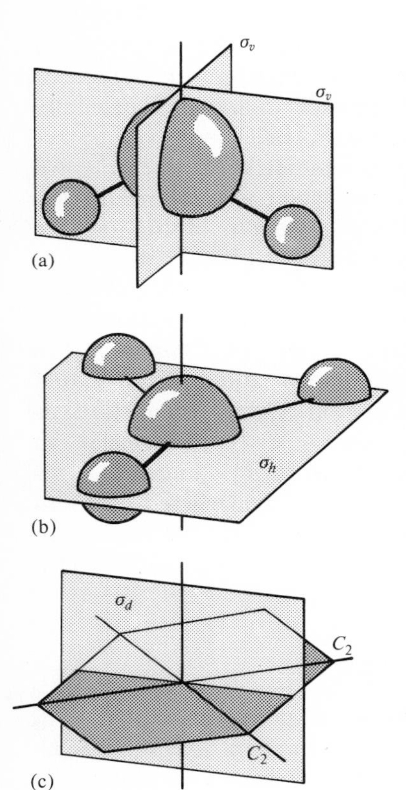

Fig. 7.2. (a) Vertical mirror planes, (b) horizontal mirror planes, (c) dihedral mirror planes.

σ. A *reflection* (the operation) in a *mirror plane* (the element). When the principal rotation axis lies in the mirror plane (as for H_2O) it is called a *vertical plane*, Fig. 7.2(a), and denoted σ_v. When the principal axis is perpendicular to the mirror plane, it is called a *horizontal plane*, σ_h, Fig. 7.2(b). A *dihedral plane*, σ_d, is a special case of a vertical plane, and occurs when the mirror plane bisects two C_2 axis that lie perpendicular to the principal axis, Fig. 7.2(c).

i. An *inversion* through a *centre of symmetry*. The inversion operation is the imaginary operation of taking each point of the object through its centre and out to an equal distance the other side, Fig. 7.3. If that operation leaves the object's appearance unchanged, it has *inversion symmetry*.

S_n. An *improper rotation* (or *rotary-reflection*) about an *axis of improper rotation* (or a *rotary-reflection axis*). This is a composite operation consisting of an n-fold rotation followed by a horizontal reflection. The methane molecule, CH_4, has three S_4 axes; one of them is shown in Fig. 7.4.

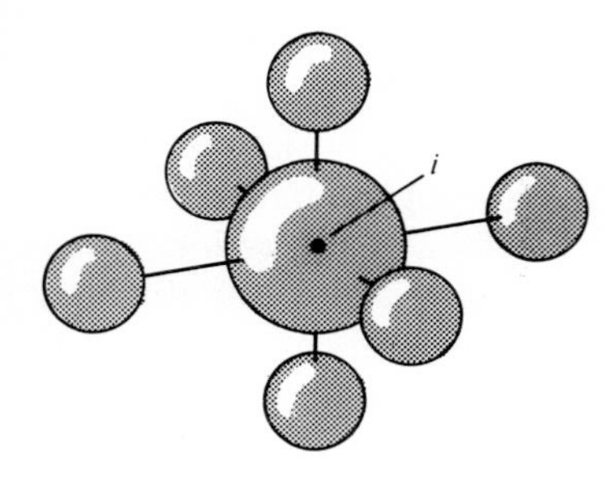

Fig. 7.3. A centre of inversion.

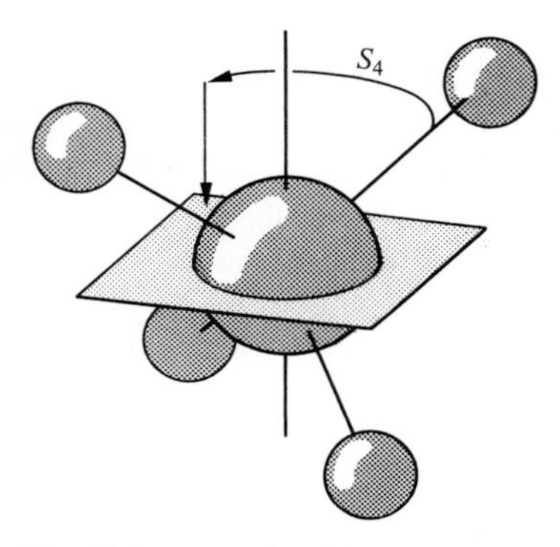

Fig. 7.4. An axis of improper rotation (S_4).

All these operations leave a single point of an object unaffected: this is the source of the name 'point group'.

7.2 The classification of molecules

The simplest application of these ideas is to the classification of molecules according to their symmetry. The procedure consists of listing the symmetry elements of a molecule, and then classifying the molecule according to the list. This has the effect of putting H_2O and H_2S into the same *group*, but into a different group from those occupied by CO_2 and C_6H_6. The advantage of such a classification is that molecules belonging to the same symmetry group have numerous properties in common, such as the possession of an electric dipole or optical activity, and are described by electronic wavefunctions of similar composition. Group classification is the first step in making full use of the symmetry of objects in order to reduce the labour involved in the calculation of their properties.

The list of symmetry elements of a molecule determines to which *point symmetry group* or *molecular point group* a molecule belongs. The name of the group is arrived at as follows. There are two systems of nomenclature the *Schoenflies* (the one we use) and the *International* (or *Hermann–Mauguin*) *systems*. A translation dictionary is given in Table 7.1 for some common groups.

C_s, C_i, and C_1 consist of the identity alone (C_1), or the identity and one other element, either a plane of symmetry (C_s) or a centre of inversion (C_i).

C_n consists of the identity and an n-fold axis of symmetry. (Note that C_n is used to label the symmetry element, the symmetry operation, and the group itself: this economy never gives rise to any real difficulty.)

C_{nv} consists of the identity, an n-fold rotation axis, and n vertical mirror planes. The molecule H_2O possesses the elements E, C_2, $2\sigma_v$,

Table 7.1 Notation for point groups (Schoenflies: International)
This table shows the relations between the Schoenflies (e.g. C_{2v}) and the International, Hermann–Mauguin (e.g. $2mm$) systems of nomenclature for the 32 crystallographic point groups. D_2 is sometimes denoted V and called the *Vierer group* (group of four)

$C_i:\bar{1}$	$C_s:m$			
$C_1:1$	$C_2:2$	$C_3:3$	$C_4:4$	$C_6:6$
	$C_{2v}:2mm$	$C_{3v}:3m$	$C_{4v}:4mm$	$C_{6v}:6mm$
	$C_{2h}:2/m$	$C_{3h}:\bar{6}$	$C_{4h}:4/m$	$C_{6h}:6/m$
	$D_2:222$	$D_3:32$	$D_4:422$	$D_6:622$
	$D_{2h}:mmm$	$D_{3h}:\bar{6}2m$	$D_{4h}:4/mmm$	$D_{6h}:6/mmm$
	$D_{2d}:\bar{4}2m$	$D_{3d}:\bar{3}m$	$S_4:\bar{4}$	$S_6:\bar{3}$
$T:23$	$T_d:\bar{4}3m$	$T_h:m3$	$O:43$	$O_h:m3m$

Further information: Group theory, M. Hamermesh; Addison-Wesley, Reading, 1962.

and so belongs to the group C_{2v}. The molecule NH_3 possesses E, C_3, $3\sigma_v$, and so belongs to C_{3v}.

C_{nh} consists of the identity, an n-fold axis, and a horizontal mirror plane. Note that the presence of some elements implies the presence of others. For instance, C_{2h} automatically possesses an inversion centre because a 180° rotation followed by a reflection in a horizontal mirror plane is the same as an inversion: the way that elements combine to produce other elements is a crucial aspect of the symmetry properties of molecules, and is developed in the remainder of the chapter.

D_n consists of the identity, an n-fold axis, and n 2-fold axes perpendicular to the n-fold axis.

D_{nh} consists of the elements involved in D_n together with a horizontal mirror plane. The C_6H_6 molecule belongs to D_{6h} because it has a principal 6-fold axis, six perpendicular 2-fold axes, a horizontal mirror plane, and several other elements that the possession of these elements implies. A uniform cylinder belongs to $D_{\infty h}$ while a cone, which lacks the horizontal mirror plane belongs to $C_{\infty v}$; homonuclear diatomic molecules belong to $D_{\infty h}$, heteronuclear diatomic molecules belong to $C_{\infty v}$.

D_{nd} is basically D_n, but it also possesses vertical mirror planes bisecting the angles between neighbouring C_2 axes. That is, D_{nd} possesses the elements of D_n plus n σ_d-planes, together with whatever other elements these imply.

The *cubic groups* arise when the object possesses more than one principal axis (i.e. when it has several C_n axes). These play a specially important role in chemistry because they include the group of the tetrahedron and the octahedron. T_d is the group of the regular tetrahedron (it has four C_3 axes, one along each C—H bond in CH_4, for instance). If the group lacks the reflection planes of the tetrahedron, it is called T, but if it contains a centre of inversion it is called T_h. The group of the regular octahedron is called O_h; if it lacks the mirror planes of the octahedron it is called O.

The *full rotation group* R_3 is the group of the sphere. Atoms belong to R_3, but no molecule does. The symmetry properties of R_3 turn out to be the properties of angular momentum. This is the deep link between this chapter and the last, and we explore it later.

Classification, while tidy, is not particularly helpful unless it leads to rationalization or prediction. The remainder of this chapter explores the consequences of the symmetries of molecules.

7.3 The calculus of symmetry

Symmetry operations may be performed consecutively. We shall use the convention that *the operation R followed by the operation S is the joint operation SR*. The order of the operations is important, and in general the result of the operation RS is not the same as the result of the operation SR. Inspection of the symmetry properties of objects leads to the conclusion that a joint symmetry operation is *always* equivalent to a single symmetry operation of the object. We have already glanced at this property when we saw that a rotation by 180° followed by a horizontal reflection is the same as an inversion. For an object belonging to a group possessing the elements C_2, σ_h, and i we could write that $\sigma_h C_2 = i$ for the corresponding operations. This type of relation applies to all the symmetry operations of an object, and it is always the case that we can write $RS = T$ where R, S, and T are three symmetry operations of the object. Another point is that the joint operation TSR, R followed by S followed by T, can be thought of as being either $T(SR)$ or $(TS)R$ as well as TSR itself. In other words $T(SR) = (TS)R$, and so the operations multiply *associatively*. These observations, and another couple of obvious ones, can be collected together as follows:

(1) The identity is a symmetry operation.
(2) Symmetry operations combine in accord with the associative law of multiplication.
(3) If R and S are symmetry operations, then RS is also a symmetry operation.
(4) The inverse of each symmetry operation is also a symmetry operation.

The remarkable point to notice is that in mathematics a set of entities called *elements* are said to form a *group* if they satisfy the following conditions:

(1) The identity is an element of the set.
(2) The elements multiply associatively.
(3) If R and S are elements, then RS is also an element of the set.
(4) The inverse of each element is a member of the set.

That is, the set of symmetry operations of an object fulfill conditions ensuring that they form a group in the mathematical sense. Consequently the mathematical theory of groups, *group theory*, may be applied to the study of the symmetry of objects. This is the justification of the title of this chapter.

Note the unfortunate double meaning of the term 'element'. We must distinguish *element*, meaning member of a group, from *symmetry element*, as defined earlier. A molecule is classified according to the symmetry elements it possesses; the group properties are then expressed in terms of the *operations*, which form the elements of the appropriate symmetry group.

7.4 Matrix representations

Relations such as $i = \sigma_h C_2$ look like ordinary algebraic expressions, but they are really only symbolic equations. We can move away from this *abstract group theory* by finding a way of representing symmetry operations by quantities which can be manipulated just like ordinary algebra. Since the order of operations is important, and we have already seen that matrices multiply together in an order-dependent

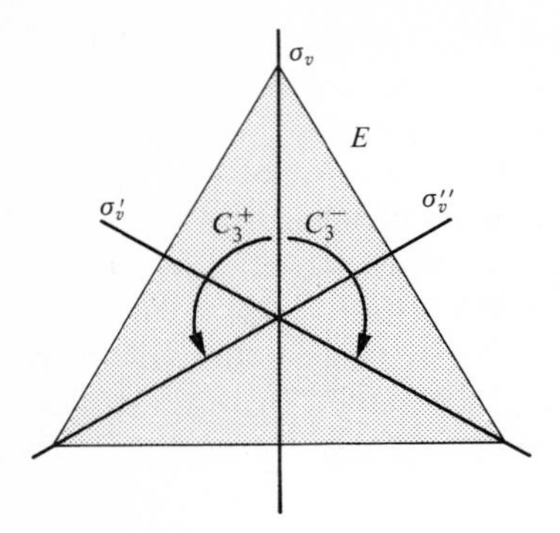

Fig. 7.5. Symmetry operations of the group C_{3v}.

manner, we are led to look for a *matrix representation* of the symmetry operations, a set of matrices which when multiplied together reproduce equations such as $i = \sigma_h C_2$. This leads to an immensely powerful form of group theory.

We shall establish the matrix representation of symmetry operations by considering the special case of the group C_{3v}. At each stage we generalize the expressions so that they can be applied to any symmetry group.

The symmetry elements and operations of the group C_{3v} are set out in Fig. 7.5: it is important to note that the elements remain unshifted as the symmetry operations are applied to the object. The mirror plane σ_v, for example, remains at the same orientation on the page at all times, including when rotations and other reflections have been performed.

There are six operations in the group and so we say that the *order* of the group is 6. The elements are the identity E, the two 3-fold rotations C_3^+ and C_3^-, and the three reflections σ_v, σ_v', and σ_v''. C_3^+ denotes a *clockwise rotation seen from below* and therefore an anticlockwise rotation seen as drawn on the page. This funny convention matches the right-handed screw rule for angular momentum. C_3^- is the anticlockwise rotation seen from below, and therefore is drawn with a clockwise sense on the page.

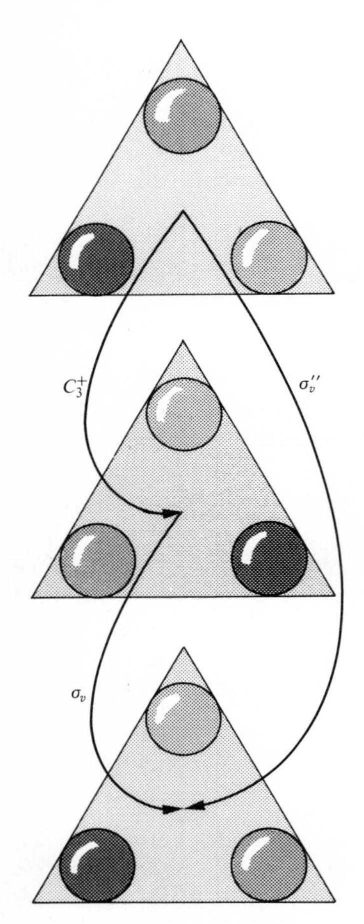

Fig. 7.6. The group multiplications $\sigma_v C_3^+ = \sigma_v''$.

A variety of relations can now be established by considering diagrams such as those in Fig. 7.6. We have, for instance, $C_3^- C_3^+ = E$, $\sigma_v C_3^+ = \sigma_v''$, and $C_3^- C_3^- = C_3^+$. The list of all $6^2 = 36$ multiplications lets us compile the *group multiplication table* in Table 7.2. For a *general group* of order h (i.e. a group of h elements), the group multiplication table consists of h^2 entries formed by taking all the products RS in turn.

Now we switch from the abstract group multiplication table to its representation in terms of matrices. There are an infinite number of ways of choosing matrices to represent the symmetry operations, and which representation we select is determined by its *basis*, a set of functions or labels attached to the object of interest. In order to be as concrete as possible, we take as a basis the four objects s_A, s_B, s_C, and s_N as drawn in Fig. 7.7: these can be thought of as the hydrogen

Table 7.2 The C_{3v} group multiplication table

First	E	C_3^+	C_3^-	σ_v	σ_v'	σ_v''
Second						
E	E	C_3^+	C_3^-	σ_v	σ_v'	σ_v''
C_3^+	C_3^+	C_3^-	E	σ_v'	σ_v''	σ_v
C_3^-	C_3^-	E	C_3^+	σ_v''	σ_v	σ_v'
σ_v	σ_v	σ_v''	σ_v'	E	C_3^-	C_3^+
σ_v'	σ_v'	σ_v	σ_v''	C_3^+	E	C_3^-
σ_v''	σ_v''	σ_v'	σ_v	C_3^-	C_3^+	E

1s-orbitals and the nitrogen 2s-orbital in NH_3. The *dimension* of this basis is 4. *In general* the dimension, d, of a basis $(f_1, f_2, \ldots, f_d)$ is the number of functions it contains. We write $(f_1, f_2, \ldots, f_d)$ collectively as the vector $\mathbf{f}$.

Under the operation σ_v the vector changes from (s_N, s_A, s_B, s_C) to $\sigma_v(s_N, s_A, s_B, s_C) = (s_N, s_A, s_C, s_B)$. This can be represented by a matrix multiplication:

$$\sigma_v(s_N, s_A, s_B, s_C) = (s_N, s_A, s_C, s_B) = (s_N, s_A, s_B, s_C)\begin{bmatrix} 1 & 0 & 0 & 0 \\ 0 & 1 & 0 & 0 \\ 0 & 0 & 0 & 1 \\ 0 & 0 & 1 & 0 \end{bmatrix}. \tag{7.4.1}$$

We call the matrix the *representative* of the operation σ_v and denote it $\mathbf{D}(\sigma_v)$. The matrix expression can be interpreted as follows: the effect of the operation σ_v on a member of the basis, s_B say, is to change it to s_C; hence we can write $\sigma_v s_B = s_C$. *In general*, for any component f_i the operation R converts it to Rf_i, where

$$Rf_i = \sum_j f_j D_{ji}(R), \tag{7.4.2}$$

$\mathbf{D}(R)$ being the matrix representative of the operation R for the basis. It is easy to check that in the present case, $\sigma_v s_B = s_C$, as required.

The representatives of the other operations of the group can be found in the same way. For instance, in the case of C_3^+ the effect of the operation is $C_3^+(s_N, s_A, s_B, s_C) = (s_N, s_B, s_C, s_A)$ which can be expressed as the matrix multiplication

$$C_3^+(s_N, s_A, s_B, s_C) = (s_N, s_B, s_C, s_A) = (s_N, s_A, s_B, s_C)\begin{bmatrix} 1 & 0 & 0 & 0 \\ 0 & 0 & 0 & 1 \\ 0 & 1 & 0 & 0 \\ 0 & 0 & 1 & 0 \end{bmatrix}. \tag{7.4.3}$$

This has the form

$$C_3^+ f_i = \sum_j f_j D_{ji}(C_3^+), \tag{7.4.4}$$

for each f_i, in accord with eqn (7.4.2). The full set of representatives for the six elements of the group are set out in Table 7.3.

Now we get to a crucially important point. Consider the effect of the consecutive operations C_3^+ followed by σ_v. Under the first operation we have the set turning from (s_N, s_A, s_B, s_C) to $C_3^+(s_N, s_A, s_B, s_C) = (s_N, s_B, s_C, s_A)$; we know that under σ_v the member f_2 (i.e. s_A) is unchanged but $f_3(s_B)$ and $f_4(s_C)$ are interchanged, and so the joint effect is $\sigma_v C_3^+(s_N, s_A, s_B, s_C) = (s_N, s_C, s_B, s_A)$, which is the same overall effect as $\sigma_v''(s_N, s_A, s_B, s_C) = (s_N, s_C, s_B, s_A)$. That is,

$$\sigma_v C_3^+(s_N, s_A, s_B, s_C) = \sigma_v''(s_N, s_A, s_B, s_C),$$

Table 7.3 The matrix representation of C_{3v} in the basis (s_N, s_A, s_B, s_C)

$\mathbf{D}(E)$	$\mathbf{D}(C_3^+)$	$\mathbf{D}(C_3^-)$
$\begin{bmatrix} 1 & 0 & 0 & 0 \\ 0 & 1 & 0 & 0 \\ 0 & 0 & 1 & 0 \\ 0 & 0 & 0 & 1 \end{bmatrix}$	$\begin{bmatrix} 1 & 0 & 0 & 0 \\ 0 & 0 & 0 & 1 \\ 0 & 1 & 0 & 0 \\ 0 & 0 & 1 & 0 \end{bmatrix}$	$\begin{bmatrix} 1 & 0 & 0 & 0 \\ 0 & 0 & 1 & 0 \\ 0 & 0 & 0 & 1 \\ 0 & 1 & 0 & 0 \end{bmatrix}$
$\chi(E)=4$	$\chi(C_3^+)=1$	$\chi(C_3^-)=1$
$\mathbf{D}(\sigma_v)$	$\mathbf{D}(\sigma_v')$	$\mathbf{D}(\sigma_v'')$
$\begin{bmatrix} 1 & 0 & 0 & 0 \\ 0 & 1 & 0 & 0 \\ 0 & 0 & 0 & 1 \\ 0 & 0 & 1 & 0 \end{bmatrix}$	$\begin{bmatrix} 1 & 0 & 0 & 0 \\ 0 & 0 & 1 & 0 \\ 0 & 1 & 0 & 0 \\ 0 & 0 & 0 & 1 \end{bmatrix}$	$\begin{bmatrix} 1 & 0 & 0 & 0 \\ 0 & 0 & 0 & 1 \\ 0 & 0 & 1 & 0 \\ 0 & 1 & 0 & 0 \end{bmatrix}$
$\chi(\sigma_v)=2$	$\chi(\sigma_v')=2$	$\chi(\sigma_v'')=2$

as the group multiplication table requires. Now consider this in terms of the matrix representatives. Take the two matrix representatives, eqns (7.4.1) and (7.4.3), and multiply them together using the rules of matrix multiplication:

$$\mathbf{D}(\sigma_v)\mathbf{D}(C_3^+) = \begin{bmatrix} 1 & 0 & 0 & 0 \\ 0 & 1 & 0 & 0 \\ 0 & 0 & 0 & 1 \\ 0 & 0 & 1 & 0 \end{bmatrix} \begin{bmatrix} 1 & 0 & 0 & 0 \\ 0 & 0 & 0 & 1 \\ 0 & 1 & 0 & 0 \\ 0 & 0 & 1 & 0 \end{bmatrix} = \begin{bmatrix} 1 & 0 & 0 & 0 \\ 0 & 0 & 0 & 1 \\ 0 & 0 & 1 & 0 \\ 0 & 1 & 0 & 0 \end{bmatrix}$$

$$= \mathbf{D}(\sigma_v'').$$

That is, the matrix representatives multiply together in exactly the same way as the operations of the group. The set of six matrices is therefore called a *representation* of the group.

The group property of the representatives can be expressed in more general terms as follows. Equation (7.4.4) gives the effect of C_3^+ on one member of the basis. If we act on this with σ_v using eqn (7.4.2) we have

$$\sigma_v C_3^+ f_i = \sigma_v \sum_j f_j D_{ji}(C_3^+) = \sum_j \sum_k f_k D_{kj}(\sigma_v) D_{ji}(C_3^+).$$

The sum over j of the product $D_{kj}D_{ji}$ is the definition of a matrix product (Appendix 8), and so

$$\sigma_v C_3^+ f_i = \sum_k f_k \{D(\sigma_v)D(C_3^+)\}_{ki}.$$

Since we know that $\sigma_v C_3^+ = \sigma_v''$, we can also write

$$\sigma_v C_3^+ f_i = \sigma_v'' f_i = \sum_k f_k D_{ki}(\sigma_v'').$$

Therefore, comparing the two, we have $\{D(\sigma_v)D(C_3^+)\}_{ki} = D_{ki}(\sigma_v'')$ for all elements, and so

$$\mathbf{D}(\sigma_v)\mathbf{D}(C_3^+) = \mathbf{D}(\sigma_v'').$$

Therefore the representatives multiply in the same way as the operations, as we set out to demonstrate. The general form of this expression is that if the representatives of the operations R, S, and T are $\mathbf{D}(R)$, $\mathbf{D}(S)$, and $\mathbf{D}(T)$ in a particular basis and $RS = T$, then

$$\mathbf{D}(R)\mathbf{D}(S) = \mathbf{D}(RS) = \mathbf{D}(T). \tag{7.4.5}$$

When the elements $R, S, T, \ldots$ of a group G and the elements $R', S', T', \ldots$ of another group G' satisfy the same multiplication table, so that if $RS = T$ then $R'S' = T'$, the two groups are said to be *homomorphous*. In the present case, the matrix representatives form a group which is homomorphous with the group of symmetry operations. When the basis is d-dimensional, the group is represented by matrices of dimension d (i.e. $d \times d$ matrices) and the representation is called *d-dimensional*. The specific example of C_{3v} we are considering consists of a 4-dimensional representation. Note that the precise form of the representation of the group depends on the basis, and in particular *the dimensionality of the representation is equal to the dimensionality of the chosen basis*. For example, had we included the three nitrogen 2p-orbitals in the basis, we should have arrived at a 7-dimensional representation, and the six 7×7 matrices would have formed another homomorphous representation of the symmetry group. The discovery of a matrix representation of a symmetry group means that instead of dealing with abstract symbols we can deal with *numbers*. This is an immense simplification, and leads to results of extraordinary power.

7.5 The properties of matrix representations

In order to develop the content of matrix representations, we need to develop some of their properties. In each case we shall use the (s_N, s_A, s_B, s_C) basis of C_{3v} to introduce an idea, and then strengthen it by generalizing to any group.

As a first step we introduce the idea of a *similarity transformation*. Suppose instead of the s-orbital basis, we choose to use some linear combination of it. One such set might be s_N, $s_1 = s_A + s_B + s_C$, $s_2 = 2s_A - s_B - s_C$, and $s_3 = s_B - s_C$ (the choice is arbitrary but this one will later be seen to be specially important). We should expect the matrix representation in this basis to be 'similar' to that in the original basis. This similarity can be given a formal definition as follows.

Since the new basis $\mathbf{f}' = (f_1', f_2', \ldots, f_d')$ is a linear combination of the original basis $\mathbf{f} = (f_1, f_2, \ldots, f_d)$, we can express any member as

$$f_i' = \sum_j f_j c_{ji} \tag{7.5.1}$$

where the c_{ji} are coefficients (e.g. the coefficients 0, 2, −1, −1 in the expression $0s_N + 2s_A - s_B - s_C$ for the linear combination s_2). This can be expressed as the matrix equation

$$\mathbf{f}' = \mathbf{fc} \tag{7.5.2}$$

where $\mathbf{c}$ is the matrix formed from the elements c_{ji}. Suppose that in the original basis the representative of the element R is $\mathbf{D}(R)$. This means that for some member f_i the effect of R is given by

$$Rf_i = \sum_k f_k D_{ki}(R), \quad \text{or} \quad R\mathbf{f} = \mathbf{fD}(R). \tag{7.5.3}$$

Likewise, the effect of the same operation on f_i' is given by

$$Rf_i' = \sum_k f_k' D_{ki}'(R), \quad \text{or} \quad R\mathbf{f}' = \mathbf{f}'\mathbf{D}'(R). \tag{7.5.4}$$

We are looking for the connection between the representatives $\mathbf{D}'(R)$ and the $\mathbf{D}(R)$, and we expect them to be 'similar'. The connection can be found by noticing that on substituting $\mathbf{f}' = \mathbf{fc}$ the second equation becomes

$$R\mathbf{fc} = \mathbf{fcD}'(R).$$

Then, since $\mathbf{cc}^{-1} = 1$, where $\mathbf{c}^{-1}$ is the reciprocal of $\mathbf{c}$ (Appendix 8), when this equation is multiplied from the right by $\mathbf{c}^{-1}$,

$$R\mathbf{f} = \mathbf{fcD}'(R)\mathbf{c}^{-1}.$$

Comparison of this with eqn (7.5.3) shows that the two representatives are connected by the

Similarity relation: $\mathbf{D}(R) = \mathbf{cD}'(R)\mathbf{c}^{-1}$. (7.5.5)

The inverse relation ($\mathbf{D}'$ in terms of $\mathbf{D}$) is obtained by multiplying by $\mathbf{c}^{-1} \ldots \mathbf{c}$:

$$\mathbf{D}'(R) = \mathbf{c}^{-1}\mathbf{D}(R)\mathbf{c}. \tag{7.5.6}$$

How this works in the case of C_{3v} can be illustrated as follows. The linear combinations (s_N, s_1, s_2, s_3) can be expressed in terms of the original set by

$$(s_N, s_1, s_2, s_3) = (s_N, s_A, s_B, s_C)\begin{bmatrix} 1 & 0 & 0 & 0 \\ 0 & 1 & 2 & 0 \\ 0 & 1 & -1 & 1 \\ 0 & 1 & -1 & -1 \end{bmatrix},$$

which determines the matrix $\mathbf{c}$. Its inverse $\mathbf{c}^{-1}$ can be obtained by the methods set out in Appendix 8, and is

$$\mathbf{c}^{-1} = \frac{1}{6}\begin{bmatrix} 6 & 0 & 0 & 0 \\ 0 & 2 & 2 & 2 \\ 0 & 2 & -1 & -1 \\ 0 & 0 & 3 & -3 \end{bmatrix}.$$

The representative of C_3^+ in the new basis is therefore

$$\mathbf{D}'(C_3^+) = \mathbf{c}^{-1}\mathbf{D}(C_3^+)\mathbf{c}$$

$$= \frac{1}{6}\begin{bmatrix} 6 & 0 & 0 & 0 \\ 0 & 2 & 2 & 2 \\ 0 & 2 & -1 & -1 \\ 0 & 0 & 3 & -3 \end{bmatrix}\begin{bmatrix} 1 & 0 & 0 & 0 \\ 0 & 0 & 0 & 1 \\ 0 & 1 & 0 & 0 \\ 0 & 0 & 1 & 0 \end{bmatrix}\begin{bmatrix} 1 & 0 & 0 & 0 \\ 0 & 1 & 2 & 0 \\ 0 & 1 & -1 & 1 \\ 0 & 1 & -1 & -1 \end{bmatrix}$$

$$= \frac{1}{6}\begin{bmatrix} 6 & 0 & 0 & 0 \\ 0 & 6 & 0 & 0 \\ 0 & 0 & -3 & -3 \\ 0 & 0 & 9 & -3 \end{bmatrix} = \begin{bmatrix} 1 & 0 & 0 & 0 \\ 0 & 1 & 0 & 0 \\ 0 & 0 & -\frac{1}{2} & -\frac{1}{2} \\ 0 & 0 & \frac{3}{2} & -\frac{1}{2} \end{bmatrix}.$$

The same technique may be applied to the other five elements of the group, with the results set out in Table 7.4. Apart from the important applications we shall shortly describe, this is an explicit and simple way of building the matrix representation for any basis that can be expressed as a linear combination of another basis.

There is one striking feature of the two representations. In Tables 7.3 and 7.4 we have listed the sum of the diagonal elements of each representative matrix. The first point to notice is that, for a given R, the diagonal sum is the same in both representations. In other words,

Table 7.4 The matrix representation of C_{3v} in the basis (s_N, s_1, s_2, s_3)

$\mathbf{D}(E)$	$\mathbf{D}(C_3^+)$	$\mathbf{D}(C_3^-)$
$\begin{bmatrix} 1 & 0 & 0 & 0 \\ 0 & 1 & 0 & 0 \\ 0 & 0 & 1 & 0 \\ 0 & 0 & 0 & 1 \end{bmatrix}$	$\begin{bmatrix} 1 & 0 & 0 & 0 \\ 0 & 1 & 0 & 0 \\ 0 & 0 & -\frac{1}{2} & -\frac{1}{2} \\ 0 & 0 & \frac{3}{2} & -\frac{1}{2} \end{bmatrix}$	$\begin{bmatrix} 1 & 0 & 0 & 0 \\ 0 & 1 & 0 & 0 \\ 0 & 0 & -\frac{1}{2} & \frac{1}{2} \\ 0 & 0 & -\frac{3}{2} & -\frac{1}{2} \end{bmatrix}$
$\chi(E)=4$	$\chi(C_3^+)=1$	$\chi(C_3^-)=1$
$\mathbf{D}(\sigma_v)$	$\mathbf{D}(\sigma_v')$	$\mathbf{D}(\sigma_v'')$
$\begin{bmatrix} 1 & 0 & 0 & 0 \\ 0 & 1 & 0 & 0 \\ 0 & 0 & 1 & 0 \\ 0 & 0 & 0 & -1 \end{bmatrix}$	$\begin{bmatrix} 1 & 0 & 0 & 0 \\ 0 & 1 & 0 & 0 \\ 0 & 0 & -\frac{1}{2} & \frac{1}{2} \\ 0 & 0 & \frac{3}{2} & \frac{1}{2} \end{bmatrix}$	$\begin{bmatrix} 1 & 0 & 0 & 0 \\ 0 & 1 & 0 & 0 \\ 0 & 0 & -\frac{1}{2} & -\frac{1}{2} \\ 0 & 0 & -\frac{3}{2} & \frac{1}{2} \end{bmatrix}$
$\chi(\sigma_v)=2$	$\chi(\sigma_v')=2$	$\chi(\sigma_v'')=2$

the diagonal sum is invariant under a similarity transformation, and in some sense the representatives are characterized by their diagonal sums. This is the source of the term *character* of a representative:

$$\textit{Character of } R\text{:}\quad \chi(R)=\sum_i D_{ii}(R)=\mathrm{tr}\,\mathbf{D}(R). \qquad (7.5.7)$$

The sum of the diagonal elements of a matrix is called its trace,† denoted tr. Different bases give rise to different characters, but if the bases are related by a similarity transformation, the characters are the same. In other words, *similar representatives of the same operation have the same character.* (The terms 'similar' and 'character' convey the common sense meanings of the words, but in fact have their precise technical meanings: notice here and elsewhere how the mathematics of group theory sharpens intuitively obvious concepts.)

The general proof of the invariance of the character is easy to find. It depends on the fact that *the trace of a product of matrices is invariant under cyclic permutation of their order:*

$$\mathrm{tr}\,\mathbf{ABC}=\mathrm{tr}\,\mathbf{BCA}=\mathrm{tr}\,\mathbf{CAB}. \qquad (7.5.8)$$

The proof of this invariance (which we use several times in the following sections) runs as follows. Express the trace as the diagonal sum

$$\mathrm{tr}\,\mathbf{ABC}=\sum_i (ABC)_{ii}$$

and expand the matrix product according to the usual rules:

$$\mathrm{tr}\,\mathbf{ABC}=\sum_i\sum_j\sum_k A_{ij}B_{jk}C_{ki}.$$

The matrix elements are simple numbers, and may be written in any order. If we permute them cyclically, neighbouring subscripts continue to match, and so the matrix sum may be done again but with the matrix elements in the new order:

$$\mathrm{tr}\,\mathbf{ABC}=\sum_i\sum_j\sum_k B_{jk}C_{ki}A_{ij}=\sum_j (BCA)_{jj}=\mathrm{tr}\,\mathbf{BCA},$$

as required. The equality of this to tr **CAB** follows in the same way.

Now we apply this result to prove the invariance of the character. Suppose the character of the representative $\mathbf{D}'(R)$ for the operation R in some basis is $\chi'(R)$, and that the basis is related to another by a similarity transformation. Then we know that the new representative is related to the old by eqn (7.5.5). The trace is the new basis is therefore

$$\begin{aligned}\chi(R)&=\mathrm{tr}\,\mathbf{D}(R)=\mathrm{tr}\,\mathbf{c}\mathbf{D}'(R)\mathbf{c}^{-1}\\&=\mathrm{tr}\,\mathbf{D}'(R)\mathbf{c}^{-1}\mathbf{c}=\mathrm{tr}\,\mathbf{D}'(R)=\chi'(R),\end{aligned} \qquad (7.5.9)$$

as we wanted to prove.

† The name trace is the same as the word for an animal's imprint—sometimes trace is called 'spur', from the German word *Spur* meaning spoor.

The next point to notice about the representatives is that they can be classified according to the values of their characters. The characters of C_3^+ and C_3^-, for instance, are both 1 (in either similar basis) while the three reflections have the character 2. This suggests that the concept of the 'class' of an operation can be sharpened and related to the characters of the corresponding representatives.

The formal definition of *class* of an operation runs as follows. Suppose we have two elements R and R' in a symmetry group, then they are defined as falling into the same class if there is some element S of the group such that

$$R' = S^{-1}RS. \tag{7.5.10}$$

Such elements (R and R') are then called *conjugate*. Conjugate elements fall into the same class.

The physical interpretation of the definition of conjugacy is that R and R' are the same kind of operation (e.g. rotation) but performed with respect to a new direction, the new direction being determined by the effect of the element S. For instance, consider the nature of the operation $\sigma_v^{-1}C_3^+\sigma_v$. From the group multiplication table we know that $C_3^+\sigma_v = \sigma_v'$, and as $\sigma_v^{-1} = \sigma_v$ (because the inverse of a reflection is the reflection; or more formally, the group multiplication table shows that in order to have $\sigma_v^{-1}\sigma_v = E$ we must take $\sigma_v^{-1} = \sigma_v$) it follows that

$$\sigma_v^{-1}C_3^+\sigma_v = \sigma_v^{-1}\sigma_v' = \sigma_v\sigma_v' = C_3^-.$$

That is, C_3^+ and C_3^- are conjugate elements, and so fall into the same class. Their conjugacy arises from the fact that a reflection of a rotation in one of the vertical planes changes its sense. In the same way $\sigma_v' = (C_3^+)^{-1}\sigma_v C_3^+$, showing that σ_v' and σ_v are conjugate and fall into the same class: in this case a C_3 rotation turns σ_v into σ_v'.

It is now a trivial matter to prove that *conjugate elements have the same character.* Using the cyclic invariance of a trace we have

$$\chi(R') = \operatorname{tr} R' = \operatorname{tr} S^{-1}RS = \operatorname{tr} RSS^{-1} = \operatorname{tr} R = \chi(R).$$

A glance at Tables 7.3 and 7.4 shows that this is the case for the two bases used to construct them. Note, however, that the converse is not true: elements with the same character are not necessarily members of the same class because non-conjugate operations may by chance have the same character.

The importance of the character is beginning to emerge. We shall see that although it contains only some of the information about the representatives (only the sum of their diagonal elements, not the complete matrix) this is often enough to draw far-reaching conclusions. The character will gradually move towards the centre of the stage in the following discussion, and we shall begin to see why *character tables*, the lists of characters for symmetry groups, play such an important part in quantum chemistry.

7.6 Irreducible representations

Inspection of the representation of the group C_{3v} in Table 7.3 shows that the s_N member is left unchanged by every operation. This is reflected in the form of the matrices, which are all of the

$$\textit{Block-diagonal form:}\quad \begin{bmatrix} 1 & 0 & 0 & 0 \\ 0 & & & \\ 0 & & \blacksquare & \\ 0 & & & \end{bmatrix}.$$

As a consequence, the original basis may be divided into two, one consisting of s_N alone and the other the three-dimensional basis (s_A, s_B, s_C). The 4-dimensional representation corresponding to the original basis therefore also separates into a 1-dimensional representation and a 3-dimensional representation:

E	C_3^+	C_3^-	σ_v	σ_v'	σ_v''	
1	1	1	1	1	1	basis: s_N
$\begin{bmatrix} 1 & 0 & 0 \\ 0 & 1 & 0 \\ 0 & 0 & 1 \end{bmatrix}$	$\begin{bmatrix} 0 & 0 & 1 \\ 1 & 0 & 0 \\ 0 & 1 & 0 \end{bmatrix}$	$\begin{bmatrix} 0 & 1 & 0 \\ 0 & 0 & 1 \\ 1 & 0 & 0 \end{bmatrix}$	$\begin{bmatrix} 1 & 0 & 0 \\ 0 & 0 & 1 \\ 0 & 1 & 0 \end{bmatrix}$	$\begin{bmatrix} 0 & 1 & 0 \\ 1 & 0 & 0 \\ 0 & 0 & 1 \end{bmatrix}$	$\begin{bmatrix} 0 & 0 & 1 \\ 0 & 1 & 0 \\ 1 & 0 & 0 \end{bmatrix}$	basis: (s_A, s_B, s_C)

We call this separation of a representation its *reduction.* In the present case the original 4-dimensional representation has been reduced to a 1-dimensional and a 3-dimensional representation. The formal way of saying this is that a 4-dimensional representation has been *reduced* to the *direct sum* of 1- and 3-dimensional representations, written

$$\mathbf{D}^{(4)} = \mathbf{D}^{(1)} + \mathbf{D}^{(3)}.$$

The + sign here is only formal: it does not mean that the 4-dimensional representation is a sum in the ordinary sense of 1- and 3-dimensional representations, but merely indicates how the original representation may be separated into the reduced representations. Sometimes the symbol $\oplus$ is used instead.

There are several points worth noting about this reduction. In the first place, the 1-dimensional representation consists of the six 1×1 matrices 1, 1, 1, 1, 1, 1. These represent the group trivially because $1\times 1=1$ for any selection of elements. Nevertheless, although 'trivial' we shall see that this representation is in some senses the most important of all. This representation satisfies the rule that conjugate elements have the same character: the characters are all equal to unity and hence the rule is satisfied trivially. In this case the characters even of non-conjugate elements are also all equal, hence they do not characterize different operations. For this reason the representation is called the *unfaithful representation* (the characters are unfaithful to their class).

The 3-dimensional representation of the example is not unfaithful: different classes have different characters. The question that arises, though, is whether this 3-dimensional representation can itself be reduced. The representation listed in Table 7.4 shows that it is reducible: if the basis is changed to (s_1, s_2, s_3) then the representation that results from the corresponding symmetry transformation consists of matrices that are all in the block-diagonal form

$$\begin{bmatrix} 1 & 0 & 0 \\ 0 & \blacksquare & \\ 0 & & \end{bmatrix}.$$

That is, a suitable linear combination of the members of the basis leads to representatives that are *simultaneously* in block-diagonal form, and therefore corresponds to the reduction of the entire 3-dimensional representation to the direct sum of 1-dimensional and 2-dimensional representations:

$$\mathbf{D}^{(3)} = \mathbf{D}^{(1)} + \mathbf{D}^{(2)}.$$

Note that the rule about the characters of classes is obeyed, and that the 1-dimensional representation is the same, unfaithful, representation of the group as we had before.

The question that immediately arises is whether the 2-dimensional representation can be reduced further by forming a suitable linear combination of the members of the basis (s_2, s_3). Group theory shows that this is impossible. There is no choice of linear combination of s_2 and s_3 that leads to a reduction of $\mathbf{D}^{(2)}$: $\mathbf{D}^{(2)}$ is an *irreducible representation* (or an *irrep*) of the group. The unfaithful representation is also trivially an irreducible representation (because a 1-dimensional representation cannot be reduced to a smaller dimension). That is, we find that the original 4-dimensional representation can be reduced to the irreps $\mathbf{D}^{(2)} + 2\mathbf{D}^{(1)}$, and can be reduced no further.

A glance at Fig. 7.7 shows the physical basis of this reduction: the s_N orbital has the 'same symmetry' as the linear combination s_1, but the pair of linear combinations s_2 and s_3 have a 'different symmetry'. Group theory has provided a quantitative assessment of the judgements 'same' and 'different'. 'Same' means 'are a basis for the same irreducible representation'; 'different' means 'are bases for different irreducible representations'.

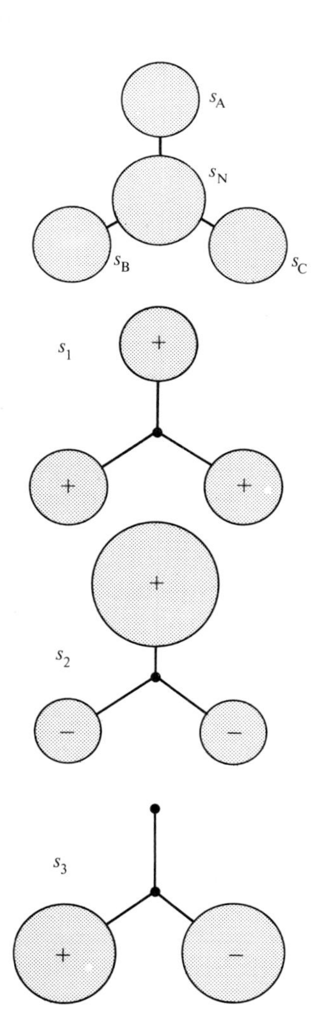

Fig. 7.7. The bases (s_N, s_A, s_B, s_C) and (s_1, s_2, s_3).

Now that we are discovering various irreps of a group we need some way of labelling them. Each irrep can be classified according to its *symmetry species*, the list of its characters for the operations of the group. Thus the unfaithful irrep has characters $(1, 1, 1, 1, 1, 1)$ and belongs to one symmetry species while the 2-dimensional irrep encountered here has characters $(2, -1, -1, 0, 0, 0)$ and belongs to a different symmetry species. We use the labels Γ_l for the symmetry species, and these two irreps belong to Γ_1 and Γ_3 respectively (we encounter Γ_2 soon). The notation A_1 and E is also widely used, and the convention is described in Appendix 10. We shall use both systems: Γ_l will be used for general expressions, and A/E, used in particular cases. If a set of functions is a basis for an irrep of symmetry species Γ_l, then we say it *spans* that irrep. Thus (s_2, s_3) spans the irrep of symmetry species Γ_3 (or E) in C_{3v} (or simply '(s_2, s_3) spans E').

There are now three jobs. One is to discover what symmetry species of irreps may occur in a group. The second is to find what irreps an arbitrary representation reduces to (or what symmetry species of irreps an arbitrary basis spans). The third is to find the linear combination of the members of a basis that reduces the representation. This requires powerful machinery, which the next section provides.

7.7 The great orthogonality theorem

The quantitative development of group theory is based on the *Great Orthogonality Theorem* (GOT) which asserts the following. Consider a group of order h, and let $\mathbf{D}^{(l)}(R)$ be the representative of the operation R in a d-dimensional irreducible representation of symmetry species Γ_l of the group. Then

$$GOT: \sum_R D_{ij}^{(l)}(R)^* D_{i'j'}^{(l')}(R) = (h/d_l)\delta_{ll'}\delta_{ii'}\delta_{jj'}. \tag{7.7.1}$$

Although this may look fearsome, it is simple to apply. In words it states that if you select any position in a matrix of one irreducible representation, and any position in a matrix of a second irreducible representation, multiply together the numbers you find there, and then sum the product over all the elements of the group, the answer is zero unless you choose not only the same symmetry species in the two cases, but also the same location to take the numbers from in both sets of matrices, in which case the answer is the number h/d_l. Examples are better than words. In the case of C_{3v}, which has $h=6$, take the 1-dimensional irreducible representation A_1 $(d=1)$ in which the matrices are the numbers 1, 1, 1, 1, 1, 1. The sum on the left of the theorem for this irrep taken with itself is

$$1\cdot 1+1\cdot 1+1\cdot 1+1\cdot 1+1\cdot 1+1\cdot 1=6,$$

which is also equal to $h/d=6/1$, as required by the theorem. Now consider two different locations in the 2-dimensional $(d=2)$ irrep of symmetry species E set out in Table 7.4; for example, the 34 and 33 locations of the matrices:

$$\begin{aligned}\sum_R D_{34}^{(E)}(R)^* D_{33}^{(E)}(R) &= D_{34}^{(E)}(E)^* D_{33}^{(E)}(E) + D_{34}^{(E)}(C_3^+)^* D_{33}^{(E)}(C_3^+) + \ldots \\ &= 0\cdot 1 + (-\tfrac{1}{2})\cdot(-\tfrac{1}{2}) + (\tfrac{1}{2})\cdot(-\tfrac{1}{2}) + 0\cdot 1 \\ &\quad + (\tfrac{1}{2})\cdot(-\tfrac{1}{2}) + (-\tfrac{1}{2})\cdot(-\tfrac{1}{2}) \\ &= 0,\end{aligned}$$

which is also in accord with the theorem.

GOT is almost too powerful for our purposes. Suppose we set $i=j$ and $i'=j'$, and then sum over these diagonal elements. This leads to a simpler version in terms of the characters of the irreducible representations: we shall call it the *little orthogonality theorem* (LOT). Specifically, on the left:

$$\sum_i \sum_{i'} \sum_R D_{ii}^{(l)}(R)^* D_{i'i'}^{(l')}(R) = \sum_R \left\{\sum_i D_{ii}^{(l)}(R)^*\right\}\left\{\sum_{i'} D_{i'i'}^{(l')}(R)\right\}$$
$$= \sum_R \chi^{(l)}(R)^* \chi^{(l')}(R),$$

and on the right

$$\sum_i \sum_{i'} (h/d_l)\delta_{ll'}\delta_{ii'}\delta_{ii'} = (h/d_l)\delta_{ll'} \sum_i \delta_{ii}.$$

There are d_l values of the index i in a matrix of dimension d_l, and so the sum of the right is the sum of 1 taken d_l times, or d_l itself. Hence, by combining the two parts of the calculation, we arrive at the explicit form

$$\textit{LOT}: \sum_R \chi^{(l)}(R)^* \chi^{(l')}(R) = h\delta_{ll'}. \tag{7.7.2}$$

This expression can be simplified still further by drawing on the result that the characters of all members of a class are the same. Suppose the number of times an operation occurs in a class c is $g(c)$, so that $g(C_3)=2$ and $g(\sigma_v)=3$ in C_{3v}, then

$$\textit{LOT}: \sum_c g(c)\chi^{(l)}(c)^* \chi^{(l')}(c) = h\delta_{ll'} \tag{7.7.3}$$

where now the sum is over the classes.

The form of LOT suggests the following analogy. Suppose we interpret the quantity $v_c^{(l)} = g(c)^{\frac{1}{2}}\chi^{(l)}(c)$ as a component of a vector $\mathbf{v}^{(l)}$, individual components being distinguished by the labels c, then LOT can be written

$$\sum_c v_c^{(l)*} v_c^{(l')} = \mathbf{v}^{(l)*} \cdot \mathbf{v}^{(l')} = h\delta_{ll'}$$

and is equivalent to the orthogonality of two different vectors $\mathbf{v}^{(l)}$ and $\mathbf{v}^{(l')}$. But the number of orthogonal vectors in a space of dimension N cannot exceed N (think of the three orthogonal vectors in ordinary space). In the present case the dimensionality of the 'space' inhabited by the vectors is equal to the number of components each has; this is the number of values of c, which in turn is the *number of classes* of the group. Therefore, the number of values of the index l which distinguishes different orthogonal vectors cannot exceed the number of classes of the group. Since l labels the symmetry species of the irreps of the group, it follows that the number of symmetry species cannot exceed the number of classes. In fact it follows from a more detailed analysis of GOT (as distinct from LOT) that these two numbers are equal. Hence we arrive at the following important restriction on the structure of a group:

The number of symmetry species = The number of classes. (7.7.4)

This vector interpretation can be applied to GOT itself. The number $D_{ij}^{(l)}(R)$ can be interpreted as the Rth component of some vector $\mathbf{v}$ identified by the three indices l, i, j. The orthogonality condition

$$\mathbf{v}^{(l',i',j')*} \cdot \mathbf{v}^{(l,i,j)} = (h/d_l)\delta_{ll'}\delta_{ii'}\delta_{jj'}$$

then implies that any pair of vectors with different labels are orthogonal. The orthogonality condition is expressed in terms of a sum over all h elements of the group, and so the vectors are h-dimensional. The total number of vectors of a given irrep is d_l^2 because the labels i and j can each take d_l values in a $d_l \times d_l$ matrix. The total dimensionality of the 'space' is therefore the sum of d_l^2 over all the symmetry species. This number ($\sum_l d_l^2$) cannot exceed the dimension (h) of the space the vectors inhabit, and it may be shown that the numbers are equal. Therefore we have the following restriction on the structure of the group:

$$\sum_l d_l^2 = h. \qquad (7.7.5)$$

We are now at the point where we can illustrate how these important restrictions affect the properties of groups. Once again we consider the group C_{3v}, which, as we know, has $h = 6$ and three classes of elements. Two symmetry species of irrep have already been found: there is the unfaithful representation of symmetry species A_1 and the 2-dimensional irrep of symmetry species E. The character table for the two symmetry species of irrep is given in Table 7.5. Are there any more? Since there are three classes there are three symmetry species; hence one (*and only one*) has not yet been found. The condition in eqn (7.7.5) takes the form $1^2 + 2^2 + d^2 = 6$, where d is the dimension of the remaining irrep. The only solution is $d = 1$; hence the missing irrep is 1-dimensional: we call it A_2. In order to complete the character table we use eqn (7.7.3). This breaks down into three equations for the

Table 7.5 The C_{3v} character table

C_{3v}	E	$2C_3$	$3\sigma_v$
A_1	1	1	1
A_2	1	1	−1
E	2	−1	0

characters of A_2. We can set $l = l' = A_2$, and obtain (anticipating that the characters are real)

$$\chi^{(A_2)}(E)^2 + 2\chi^{(A_2)}(C_3)^2 + 3\chi^{(A_2)}(\sigma_v)^2 = 6$$

or set $l = A_2$ and $l' = A_1$ or E to obtain two more relations:

$$\chi^{(A_2)}(E)\chi^{(A_1)}(E) + 2\chi^{(A_2)}(C_3)\chi^{(A_1)}(C_3) + 3\chi^{(A_2)}(\sigma_v)\chi^{(A_1)}(\sigma_v) = 0$$
$$\chi^{(A_2)}(E)\chi^{(E)}(E) + 2\chi^{(A_2)}(C_3)\chi^{(E)}(C_3) + 3\chi^{(A_2)}(\sigma_v)\chi^{(E)}(\sigma_v) = 0,$$

which, when the known values of the characters of A_1 and E are substituted, turn into

$$\chi^{(A_2)}(E) + 2\chi^{(A_2)}(C_3) + 3\chi^{(A_2)}(\sigma_v) = 0$$
$$2\chi^{(A_2)}(E) - 2\chi^{(A_2)}(C_3) = 0.$$

These three equations are enough to determine the three unknowns completely, and we find $\chi^{(A_2)}(E) = 1$ (which we know independently, because A_2 is 1-dimensional), $\chi^{(A_2)}(C_3) = 1$, and $\chi^{(A_2)}(\sigma_v) = -1$. Hence the complete C_{3v} character table with the characters for all possible species of irrep can now be completed: it is shown in Table 7.5.

The character tables for any symmetry group may be constructed in this way, and a selection of them is given in Appendix 10. Now we have to see how to use them.

7.8 Reducing representations

If we are given a general set of basis functions, how do we find which symmetry species of irreps they span? Often we are interested not in the irrep itself (i.e. not the set of matrices) but simply in its symmetry species. In the case of the C_{3v} example, for instance, it turns out to be sufficient to know that s_N and s_1 span the same species of irrep while (s_2, s_3) spans another. Only for detailed considerations does the representation itself need to be determined. Reduction of some representation with matrices $\mathbf{D}(R)$ consists of finding the similarity transformation which leads to the expression of all the $\mathbf{D}(R)$ as block-diagonal matrices. Formally we write that $\mathbf{D}(R)$ is reduced to the direct sum

$$\mathbf{D}(R) = \sum_l a_l \mathbf{D}^{(l)}(R). \tag{7.8.1}$$

Pictorially this means

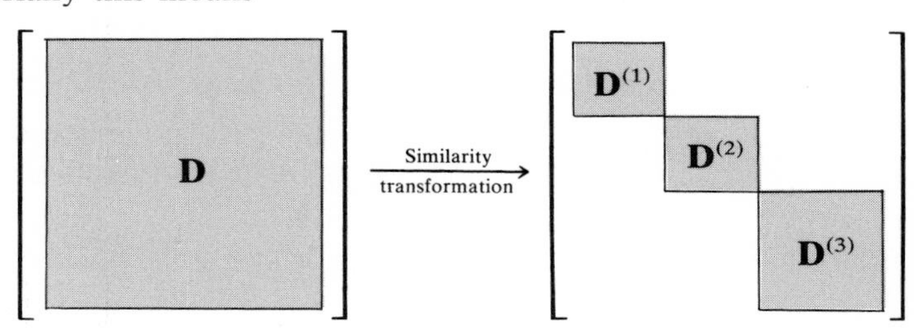

simultaneously for all R. For example in the reduction of the representation based on (s_N, s_A, s_B, s_C) we found $\mathbf{D}=2\mathbf{D}^{(A_1)}+\mathbf{D}^{(E)}$ for each R, and so $a_{A_1}=2$, $a_{A_2}=0$, $a_E=1$. This can be expressed more succinctly in terms of the symmetry species of the irreps as

$$\Gamma=\sum_l a_l\Gamma_l \tag{7.8.2}$$

which in the present example reads $\Gamma=2A_1+E$. The advantage of this notation is that it focuses attention on the symmetry species and away from the irreps themselves.

Our job is to find the *reduction coefficients* a_l. We have seen that a similarity transformation leaves the trace of a representative matrix unchanged, and so the character of any given operation is the same after the reduction as it was before:

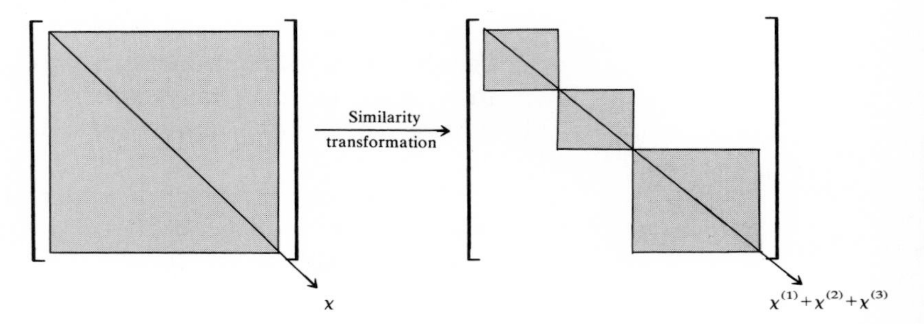

Therefore the character of the original representative is equal to the sum of the characters of the representatives of the reduced representations:

$$\chi(R)=\sum_l a_l\chi^{(l)}(R). \tag{7.8.3}$$

The coefficients can be found from a simple application of LOT. If eqn (7.8.3) is multiplied on both sides by $\chi^{(l')}(R)^*$ and then summed over the elements of the group, we find

$$\sum_R \chi^{(l')}(R)^*\chi(R)=\sum_R\sum_l \chi^{(l')}(R)^*\chi^{(l)}(R)a_l$$
$$=\sum_l h\delta_{ll'}a_l=ha_{l'}.$$

That is, the coefficients are given by the rule

$$a_l=(1/h)\sum_R \chi^{(l)}(R)^*\chi(R). \tag{7.8.4}$$

Since the characters of members of the same class of operation are the same, we can express this in terms of the characters of classes:

$$a_l=(1/h)\sum_c g(c)\chi^{(l)}(c)^*\chi(c). \tag{7.8.5}$$

Although this is the formally precise way of finding the reduction coefficients, in many cases it is possible to find them by inspection. For example, in the original basis for the C_{3v} problem the characters are 4, 1, 2 for the classes E, $2C_3$, $3\sigma_v$. By inspection of the character table we immediately conclude that the reduction is to $2A_1+E$. The same result is obtained from the mechanical application of eqn (7.8.5). The use of eqn (7.8.5) is almost essential in more complicated groups as the following *Example* shows.

Example. What symmetry species do the four 1s-orbitals of the hydrogen atoms of methane span?

- *Method.* Methane, CH_4, belongs to the group T_d; the character table is in Appendix 10. The character of each operation in the basis (H_a, H_b, H_c, H_d) can be obtained by noting the number (N) of members left in their original location after the operation: a 1 occurs on the diagonal of the representative in each case, and so the character is the sum of 1 taken N times. Then analyse the collection of characters using eqn (7.8.5).
- *Answer.* The numbers of unchanged basis members under the operations E, C_3, C_2, S_4, σ_d are 4, 1, 0, 0, 2. For the group T_d, $h=24$. Then

$$\begin{aligned}
a(A_1)&=(\tfrac{1}{24})\{(4\times 1)+8(1\times 1)+3(0\times 1)+6(0\times 1)+6(2\times 1)\}=1\\
a(A_2)&=(\tfrac{1}{24})\{(4\times 1)+8(1\times 1)+3(0\times 1)-6(0\times 1)-6(2\times 1)\}=0\\
a(E)&=(\tfrac{1}{24})\{(4\times 2)-8(1\times 1)+3(0\times 2)+6(0\times 0)+6(0\times 0)\}=0\\
a(T_1)&=(\tfrac{1}{24})\{(4\times 3)+8(1\times 0)-3(0\times 1)+6(0\times 1)-6(2\times 1)\}=0\\
a(T_2)&=(\tfrac{1}{24})\{(4\times 3)+8(1\times 0)-3(0\times 1)-6(0\times 1)+6(2\times 1)\}=1.
\end{aligned}$$

  Hence, the four orbitals span A_1+T_2.
- *Comment.* In some cases an operation changes the sign of a member of the basis without moving its location (e.g. O2p_x in H_2O under the operation C_2 of the group C_{2v}). This gives rise to -1 on the diagonal of the representative, and hence to a contribution of -1 to the character. Sometimes (e.g. for the basis p_x, p_y in C_{3v}) fractional terms arise on the diagonal: this will be seen in Section 7.10.

7.9 Symmetry-adapted bases

We now establish how to find the linear combinations of the members of a basis that span an irreducible representation of given symmetry, this is called finding the *symmetry-adapted basis.* The next page or so will bristle with subscripts: the reader who does not wish to pick through the thicket will probably find it possible to use the final result, eqn (7.9.7), and we shall meet again there.

Consider the set of functions $\mathbf{f}^{l'}=(f_1^{l'}, f_2^{l'}, \ldots, f_d^{l'})$ that form a basis for a $d_{l'}$-dimensional irrep $\mathbf{D}^{(l')}$ of symmetry species $\Gamma_{l'}$ of a group of order h. This means that we can express the effect of any operation as

$$Rf_{j'}^{(l')}=\sum_{i'} f_{i'}^{(l')}D_{i'j'}^{(l')}(R).$$

GOT may now be invoked. Multiply by some element $D_{ij}^{(l)}(R)^*$ of a representative of the same operation and sum over the elements R

using GOT:

$$\begin{aligned}\sum_R D_{ij}^{(l)}(R)^* R f_{j'}^{(l')} &= \sum_R \sum_{i'} D_{ij}^{(l)}(R)^* f_{i'}^{(l')} D_{i'j'}^{(l')}(R) \\ &= \sum_{i'} f_{i'}^{(l')} \Big\{ \sum_R D_{ij}^{(l)}(R)^* D_{i'j'}^{(l')}(R) \Big\} \\ &= \sum_{i'} f_{i'}^{(l')} (h/d_{l'}) \delta_{ll'} \delta_{ii'} \delta_{jj'} \\ &= (h/d_{l'}) f_i^{(l')} \delta_{ll'} \delta_{jj'} = (h/d_l) f_i^{(l)} \delta_{ll'} \delta_{jj'}.\end{aligned}$$

We then define the

Projection operator: $P_{ij}^{(l)} = (d_l/h) \sum_R D_{ij}^{(l)}(R)^* R.$ (7.9.1)

It can be thought of as a mixed operation, a linear combination of the operations of the group (e.g. some linear combination of the identity, the rotations, and the three reflections of C_{3v}, the combination coefficients being the various elements of the representatives). Then the equation becomes

$$P_{ij}^{(l)} f_{j'}^{(l')} = f_i^{(l)} \delta_{ll'} \delta_{jj'}. \tag{7.9.2}$$

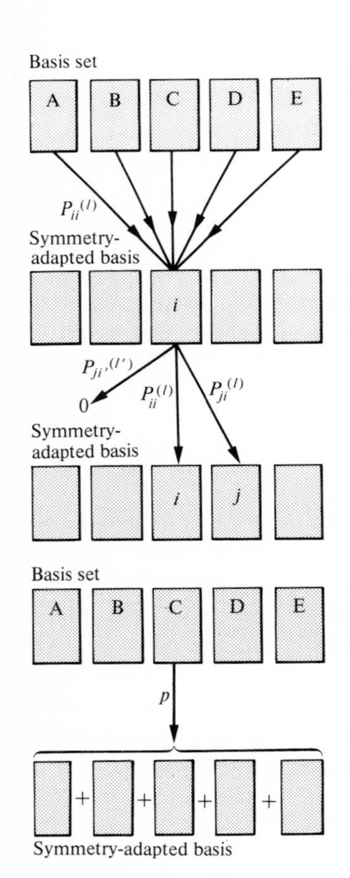

Fig. 7.8. The operation of the projection operators.

The reason why P is called a projection operator can now be made clear. In the first place, suppose either $l \neq l'$ or $j \neq j'$; then when it acts on some member $f_{j'}^{(l')}$ it gives zero. That is, when $P_{ij}^{(l)}$ acts on a function that is not a member of a set that spans Γ_l, or, if it is a member, is not at location j in the set, then it gives zero. On the other hand, if the member is at the location j in the basis of Γ_l, then when P_{ij} operates it generates from f_j the function f_i, the member of the set standing at location i. P *projects* a member at one location to another location, Fig. 7.8. The importance of this result is that *if we know only one member of a basis of a representation, then all the other members can be projected from it.*

In the special case of $l' = l$ and $i = j$, the effect of the projection operator, which is now $P_{ii}^{(l)}$, on some member of the basis $f_j^{(l)}$ is

$$P_{ii}^{(l)} f_j^{(l)} = f_i^{(l)} \delta_{ij}, \tag{7.9.3}$$

and so P then generates either 0 (if $i \neq j$), or regenerates the original function (if $i = j$). The significance of this special case will become apparent soon.

Now suppose that we are given an *arbitrary* set of functions $\mathbf{f} = (f_1, f_2, \ldots)$, such as the original s_N, s_A, s_B, s_C set of the C_{3v} example. What is the effect of the projection operator $P_{ii}^{(l)}$ on any member? Just as any member of the symmetry-adapted basis can be expressed as the appropriate linear combination of the members of the arbitrary basis we can express any f_j as a linear combination of all the $f_{j'}^{(l')}$:

$$f_j = \sum_{l'} \sum_{j'} f_{j'}^{(l')}. \tag{7.9.4}$$

(The expansion coefficients have been absorbed into the $f_{j'}^{(l')}$.) Now operate on this expression with the projection operator $P_{ii}^{(l)}$:

$$P_{ii}^{(l)} f_j = \sum_{l'} \sum_{j'} P_{ii}^{(l)} f_{j'}^{(l')} = \sum_{l'} \sum_{j'} f_{j'}^{(l')} \delta_{ll'} \delta_{ij'} = f_i^{(l)}. \qquad (7.9.5)$$

That is, when $P^{(l)}$ operates on *any* member of the arbitrary initial basis, *it generates the ith member of the basis for the irrep of symmetry species Γ_l*. Having obtained that member, we can act on it with $P_{ji}^{(l)}$ and so construct the jth member of the same symmetry-adapted basis. Therefore the problem of finding a symmetry-adapted basis has been solved.

The problem with the method is that the nature of the projection operator depends on the elements of the representatives of all the operations of the group (via its definition in eqn (7.9.1)). Often we know only the characters (e.g. from the character tables of the group). Nevertheless, these still convey some information, as may be seen in the following. Consider the projection operator $p^{(l)}$ formed by summing over the diagonal elements of $P^{(l)}$:

$$p^{(l)} = \sum_i P_{ii}^{(l)} = \sum_i \sum_R (d_l/h) D_{ii}^{(l)}(R)^* R.$$

That is,

$$p^{(l)} = (d_l/h) \sum_R \chi^{(l)}(R)^* R. \qquad (7.9.6)$$

This operator can be constructed from the character tables alone, but what is its effect? When it is applied to the member f_i of the arbitrary initial basis we find the following:

$$p^{(l)} f_j = \sum_i P_{ii}^{(l)} f_j = \sum_i f_i^{(l)}. \qquad (7.9.7)$$

Therefore it generates a *sum* of the members of the basis spanning the irrep of symmetry species Γ_l. This is no complication in the case of 1-dimensional irreps because there is then only one member of the basis set and $p^{(l)}$ is unambiguous in its effect. In the case of higher dimensionality irreps, $p^{(l)}$ gives the sum of the members of symmetry adapted bases, and so the information is incomplete. Nevertheless since we are often concerned only with low-dimensional irreps (such as the 2-dimensional E-irrep of C_{3v}) this is rarely a severe complication. The way in which the technique is used to generate the symmetry adapted basis from (s_N, s_A, s_B, s_C) in C_{3v} is described in the following *Example*, which also shows how to resolve the ambiguity. The information in the character tables in Appendix 10 can be used in the same way to establish the symmetry-adapted linear combinations for members of an arbitrary basis for any symmetry group. This is of particular importance when we come to discuss molecular wavefunctions.

Example. Construct the symmetry-adapted orbitals for the group C_{3v} using the basis (s_N, s_A, s_B, s_C).

- *Method.* We have seen already that the basis spans $2A_1+E$; therefore we use eqn (7.9.6) to construct the appropriate symmetry-adapted orbitals. The character table is in Appendix 10 and the effects of the operations were established in Section 7.4. The best way of using eqn (7.9.6) is to follow this recipe:

  (a) Draw up a table headed by the basis and showing in the columns the effects of the operations. (A given column is headed by f_i, and an entry in it shows Rf_i.)
  (b) Multiply each member of the column by the character of the corresponding operation. (This produces $\chi(R)Rf_i$ at each location.)
  (c) Add the entries within each column. (This produces $\sum_R \chi(R)Rf_i$; for a given f_i.)
  (d) Multiply by dimension/order. (This produces pf_i.)

  For C_{3v}, $h=6$.

- *Answer.*

Original set:	s_N	s_A	s_B	s_C
Under E	s_N	s_A	s_B	s_C
C_3^+	s_N	s_B	s_C	s_A
C_3^-	s_N	s_C	s_A	s_B
σ_v	s_N	s_A	s_C	s_B
σ_v'	s_N	s_B	s_A	s_C
σ_v''	s_N	s_C	s_B	s_A

For A_1, $d=1$ and all $\chi(R)=1$. Hence the first column gives $(\frac{1}{6})\times(s_N+s_N+s_N+s_N+s_N+s_N)=s_N$. The second gives $(\frac{1}{6})(s_A+s_B+s_C+s_A+s_B+s_C)=\frac{1}{3}(s_A+s_B+s_C)$. The remaining two columns give the same. For E, $d=2$ and $\chi=(2,-1.-1,0,0,0)$ for the six operations. The first column gives $(\frac{2}{6})\times(s_N-1s_N-1s_N+0+0+0)=0$. The second column gives $(\frac{2}{6})(2s_A-s_B-s_C+0+0+0)=\frac{1}{3}(2s_A-s_B-s_C)$. The remaining columns produce $\frac{1}{3}(2s_B-s_C-s_A)$ and $\frac{1}{3}(2s_C-s_A-s_B)$. We cannot have three independent linear combinations. It is sensible to choose a linear combination of the second two which is orthogonal to the first. If we take $(2s_B-s_C-s_A)-(2s_C-s_A-s_B)=3(s_B-s_C)$, this is orthogonal to $2s_A-s_B-s_C$. Note also that the two orbitals $s_2=2s_A-s_B-s_C$ and $s_3=s_B-s_C$ have different parity under σ_v.

- *Comment.* In this example we have taken the characters to be real. This is true for all characters commonly encountered, but in advanced applications complex χs arise. The recipe for generating symmetry-adapted orbitals is almost in the form of a computer algorithm. You may care to explore the possibility of writing and using a program.

7.10 Symmetry properties of functions

We now turn to the consideration of the transformation properties of functions defined more generally than hitherto. As a first step, consider the problem of discovering how the three p-orbitals of the nitrogen atom in NH_3 transform under the operations of the group C_{3v}. The basis for the representation of the group is now (p_x, p_y, p_z). Intuitively we expect this basis to be reducible into an irrep spanned by p_z (but is it of symmetry species A_1 or A_2?) and a 2-dimensional irrep spanned by $(\mathrm{p}_x, \mathrm{p}_y)$ which must be of symmetry species E. But suppose we allowed for the possibility of d-orbitals on the central atom: what symmetry species does the 5-dimensional basis of d-orbitals span in this group? In order to answer questions like these we need a systematic procedure that can be applied even when the results are not obvious. This is the subject we tackle here. The procedure is essentially the same as that already described, but it differs in detail largely because the basis, and therefore the representatives, are different. The point of doing the calculation is that its results can be used in many applications, as we shall soon see.

Consider the basis $(\mathrm{p}_x, \mathrm{p}_y, \mathrm{p}_z)$ for the group C_{3v}. The p-orbitals have the form $\mathrm{p}_x = r \sin\theta \cos\phi f(r)$, $\mathrm{p}_y = r \sin\theta \sin\phi f(r)$, $\mathrm{p}_z = r\cos\theta f(r)$. But in spherical polar coordinates we have $x = r\sin\theta\cos\phi$, etc. and so we can also express the orbitals as $\mathrm{p}_x = xf(r)$, $\mathrm{p}_y = yf(r)$, and $\mathrm{p}_z = zf(r)$. All the operations of the point group C_{3v} (as for any point group) leave distances from the origin unchanged, and so the transformations of the basis are fully captured by considering the simpler basis (x, y, z). How does this basis transform in C_{3v} (or any group)?

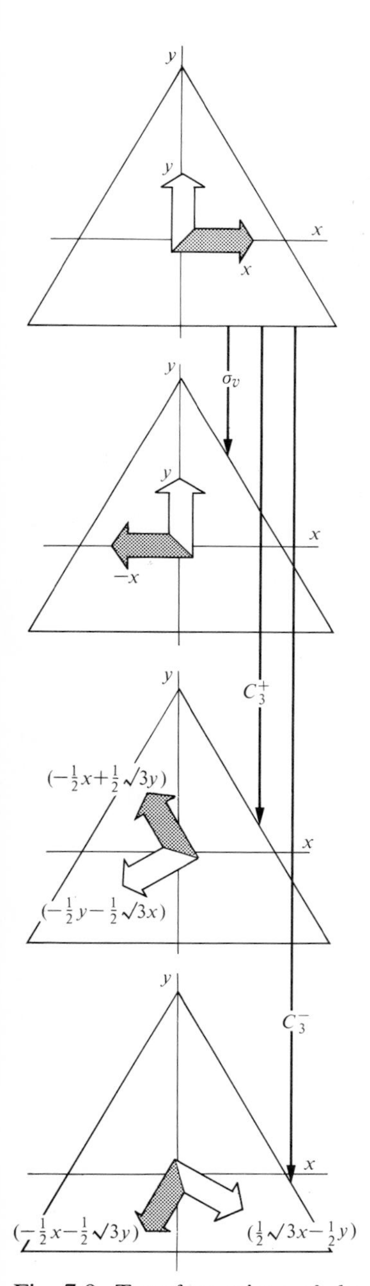

Fig. 7.9. Transformations of the vector $\mathbf{r} = (x, y, z)$.

Refer to Fig. 7.9 which shows the way that the functions (x, y, z) change when the symmetry operations of the group are performed on the object in which they reside.

The effect of the reflection σ_v on the basis is $\sigma_v(x, y, z) = (-x, y, z)$. This can be expressed in matrix notation as

$$\sigma_v(x, y, z) = (-x, y, z) = (x, y, z)\begin{bmatrix} -1 & 0 & 0 \\ 0 & 1 & 0 \\ 0 & 0 & 1 \end{bmatrix}, \tag{7.10.1}$$

which therefore leads to the identification of the representative $\mathbf{D}(\sigma_v)$ for this basis. Under the rotation C_3^+ the basis transforms into

$$\begin{aligned} C_3^+(x, y, z) &= (-\tfrac{1}{2}x + \tfrac{1}{2}\sqrt{3}y, -\tfrac{1}{2}\sqrt{3}x - \tfrac{1}{2}y, z) \\ &= (x, y, z)\begin{bmatrix} -\frac{1}{2} & -\frac{1}{2}\sqrt{3} & 0 \\ \frac{1}{2}\sqrt{3} & -\frac{1}{2} & 0 \\ 0 & 0 & 1 \end{bmatrix}, \end{aligned} \tag{7.10.2}$$

which leads to the identification of $\mathbf{D}(C_3^+)$ for the basis. The complete set of representatives can be found in this way, and are set out in Table 7.6 together with their characters.

The characters of the operations E, $2C_3$, $3\sigma_v$ in the basis (x, y, z) are 3, 0, 1 respectively. This corresponds to the reduction $A_1 + E$, and so the basis spans irreps of symmetry species A_1 and E. The function z is a basis for A_1 (the representation is in a block-diagonal form), and the pair (x, y) span an irrep of symmetry species E. We therefore also know that the three p-orbitals span $A_1 + E$, and that p_z is a basis for A_1 and $(\mathrm{p}_x, \mathrm{p}_y)$ a basis for E.

The identity of the symmetry species of the irreps spanned by x, y, and z are so important that they are normally given specifically in the

Table 7.6 The matrix representation of C_{3v} in the basis (x, y, z)

$\mathbf{D}(E)$	$\mathbf{D}(C_3^+)$	$\mathbf{D}(C_3^-)$
$\begin{bmatrix} 1 & 0 & 0 \\ 0 & 1 & 0 \\ 0 & 0 & 1 \end{bmatrix}$	$\begin{bmatrix} -\frac{1}{2} & -\frac{1}{2}\sqrt{3} & 0 \\ \frac{1}{2}\sqrt{3} & -\frac{1}{2} & 0 \\ 0 & 0 & 1 \end{bmatrix}$	$\begin{bmatrix} -\frac{1}{2} & \frac{1}{2}\sqrt{3} & 0 \\ -\frac{1}{2}\sqrt{3} & -\frac{1}{2} & 0 \\ 0 & 0 & 1 \end{bmatrix}$
$\chi(E)=3$	$\chi(C_3^-)=0$	$\chi(C_3^+)=0$
$\mathbf{D}(\sigma_v)$	$\mathbf{D}(\sigma_v')$	$\mathbf{D}(\sigma_v'')$
$\begin{bmatrix} -1 & 0 & 0 \\ 0 & 1 & 0 \\ 0 & 0 & 1 \end{bmatrix}$	$\begin{bmatrix} \frac{1}{2} & -\frac{1}{2}\sqrt{3} & 0 \\ -\frac{1}{2}\sqrt{3} & -\frac{1}{2} & 0 \\ 0 & 0 & 1 \end{bmatrix}$	$\begin{bmatrix} \frac{1}{2} & \frac{1}{2}\sqrt{3} & 0 \\ \frac{1}{2}\sqrt{3} & -\frac{1}{2} & 0 \\ 0 & 0 & 1 \end{bmatrix}$
$\chi(\sigma_v)=1$	$\chi(\sigma_v')=1$	$\chi(\sigma_v'')=1$

character tables. This has been done in the tables in Appendix 10. Exactly the same procedure may be applied to the quadratic forms x^2, xy, etc. that arise when the d-orbitals are expressed in cartesian coordinates ($\mathrm{d}_{xy}=xyf(r)$, $\mathrm{d}_{yz}=yzf(r)$, $\mathrm{d}_{zx}=zxf(r)$, $\mathrm{d}_{x^2-y^2}=(x^2-y^2)f(r)$, $\mathrm{d}_{z^2}=(3z^2-r^2)f(r)$), and the symmetry species these span (A_2+2E in C_{3v}) are also normally reported.

The question that arises at this stage is stimulated by the observation that the d-orbital basis is expressed in terms of products of x, y, and z. Is it possible to find the symmetry species of products such as xy from the known symmetry species of x and y without having to go through the business of establishing the transformation properties all over again? In more general terms, if we know what symmetry species a set of functions $(f_1, f_2, \ldots)$ span, can we say what symmetry species their products $(f_1^2, f_1f_2, \ldots)$ span? This information is also included in the character tables, as we now show.

7.11 The decomposition of direct-product bases

Suppose that $f_i^{(l)}$ is a member of a basis for an irrep of symmetry species Γ_l of dimension d_l and $f_{i'}^{(l')}$ is a member of a basis of symmetry species $\Gamma_{l'}$ of dimension $d_{l'}$. This means that under an operation R of the group they transform as follows:

$$Rf_i^{(l)}=\sum_j f_j^{(l)}D_{ji}(R); \qquad Rf_{i'}^{(l')}=\sum_{j'} f_{j'}^{(l')}D_{j'i'}^{(l')}(R).$$

It follows that their product transforms into

$$R(f_i^{(l)}f_{i'}^{(l')})=\sum_j\sum_{j'} f_j^{(l)}f_{j'}^{(l')}D_{ji}^{(l)}(R)D_{j'i'}^{(l')}(R), \tag{7.11.1}$$

which is a linear combination of products $f_j^{(l)}f_{j'}^{(l')}$. This means that the products also constitute a basis for a representation. But is this *direct-product representation*, of dimension $d_l d_{l'}$, reducible, and, if so, into what? We can answer this by establishing its characters.

The matrix representative of R in the product basis is $D_{ji}^{(l)}(R)D_{j'i'}^{(l')}(R)$ where the pair of indices jj' label the row of the matrix and the pair ii' label the column. The diagonal elements are the ones with $j=i$ and $j'=i'$, and so the character of the operation R is the sum

$$\chi(R)=\sum_i\sum_{i'} D_{ii}^{(l)}(R)D_{i'i'}^{(l')}(R)$$
$$=\left\{\sum_i D_{ii}^{(l)}(R)\right\}\left\{\sum_{i'} D_{i'i'}^{(l')}(R)\right\}=\chi^{(l)}(R)\chi^{(l')}(R). \tag{7.11.2}$$

This is a very nice result, and easy to use. It simply says that the characters of the operations in the product basis are the products of the corresponding characters for the original bases. In the case of the product basis of (x, y, z) taken with itself, in C_{3v} where the characters are 3, 0, 1 (in the usual order), the nine quadratic products $x^2, xy, \ldots, z^2$ span a representation with characters 9, 0, 1.

In general the direct-product basis is reducible. This is clearly the case for the C_{3v} example, because 9, 0, 1 does not correspond to any single-symmetry species. Inspection shows that the direct sum $2A_1+A_2+3E$ has the same set of characters, and so we know that the set of nine quadratic forms span irreps of these symmetry species. Inspection is sometimes difficult (as in this case); then it is better to find the decomposition of the representation using eqn (7.8.5).

The symmetry species spanned by (xz, yz) alone can be investigated by the same technique, because this basis is the product of the bases spanning $A_1(z)$ and $E(x, y)$. The characters of these two symmetry species are 1, 1, 1 and 2, -1, 0 respectively, and so the product basis spans a representation with characters 2, -1, 0. This is an irreducible representation because it has the same characters as an irrep of species E. Therefore the product basis (xz, yz) is a basis for E. We write this result formally as $A_1\times E=E$, where the multiplication sign signifies a *direct product* (it is sometimes written $\otimes$).

The symmetry species spanned by the product of the basis (x, y) with itself is obtained by forming the product of the characters for the symmetry species E with itself. Since E has characters 2, -1, 0 the product representation has characters $E\times E=4, 1, 0$. This is reducible to A_1+A_2+E, and so we write $E\times E=A_1+A_2+E$. Hence the four product basis functions (x^2, xy, yx, y^2) span these three irreps.

Tables of decomposition of products like $E\times E=A_1+A_2+E$ are called *direct-product tables*. They can be worked out once and for all, and some are listed in Appendix 11. We shall see that they are often more important than the character tables themselves!

The discussion so far has ignored the fact that terms like xy and yx are being encountered, whereas they are clearly not independent. The new aspect which resolves the point is to note that the *symmetrized* and *antisymmetrized products* $\frac{1}{2}\{f_i^{(l)}f_j^{(l)}+f_j^{(l)}f_i^{(l)}\}$ and $\frac{1}{2}\{f_i^{(l)}f_j^{(l)}-f_j^{(l)}f_i^{(l)}\}$ of a basis taken with itself also constitute a basis for the direct product group. Clearly the latter vanish identically: $xy - yx = 0$. Therefore we must establish which irreps the antisymmetrized product basis spans, and discard them from $A_1 + A_2 + E$. The characters of the symmetrized and antisymmetrized products are given by

$$\chi^+(R) = \tfrac{1}{2}\{\chi^{(l)}(R)^2 + \chi^{(l)}(R^2)\}; \qquad \chi^-(R) = \tfrac{1}{2}\{\chi^{(l)}(R)^2 - \chi^{(l)}(R^2)\},$$

and in the direct-product tables the symmetry species of the antisymmetrized product is denoted $[\Gamma]$. (The fact that it is recorded at all signifies that is has some use: we shall encounter it again in Chapter 10.) In the present case $E \times E = A_1 + [A_2] + E$, and so (x^2, xy, y^2) spans $A_1 + E$.

7.12 Direct-product groups

We can now consider another example of using group theory to build up information from existing results. This time, instead of building up the transformation properties of functions from the behaviour of x, y, and z, we build up the properties of the groups themselves by showing how to cement together the character tables for small groups into the character tables for bigger ones.

Suppose there exists the group G of order h with elements $R_1, R_2, \ldots, R_h$, and another group G' of order h' with elements $R'_1, R'_2 \ldots R'_{h'}$. Let the groups satisfy the conditions that (a) *the only element they have in common is the identity*, and (b) *the elements of G commute with all the elements of G'* ('commutation' means the same here as in the chapter on operators: operations commute if their order is unimportant: $RR' = R'R$). Examples of two such groups are C_{3v} and C_s (the group of order 2 consisting of the elements E and σ_h). Then the product RR' of each element of G with each element of G' forms a group called the *direct-product group $G'' = G \times G'$*. That G'' is a group can be proved by checking that the group property is obeyed for every pair of elements. Since in G, $R_iR_j = R_k$ (because G is a group) and in G', $R'_rR'_s = R'_t$, for the same reason, in G'' with elements $R_iR'_r$, etc.

$$(R_iR'_r)(R_jR'_s) = R_iR'_rR_jR'_s = R_iR_jR'_rR'_s = R_kR'_t, \quad \text{a member of } G'',$$

hence the group property is satisfied (but notice that R and R' have to commute in order to get the second equality). The order of G'' is hh' (and so the order of the group $C_{3v} \times C_s$ is 12).

The direct-product group can be identified by constructing its elements ($C_{3v} \times C_s$ will turn out to be D_{3h}) and the character table constructed from the character tables of the component groups. We do this as follows. Let $(f_1, f_2, \ldots)$ be a basis for an irrep of G and $(f'_1, f'_2, \ldots)$ be a basis for an irrep of G'. It follows that we can write

$$Rf_i = \sum_j f_jD_{ji}(R) \quad \text{and} \quad R'f'_r = \sum_s f'_sD_{sr}(R')$$

for the effects of R on $\mathbf{f}$ and R' on $\mathbf{f}'$. Then the effect of the operation

RR' on the direct-product basis $\mathbf{ff'}$ is

$$RR'f_if'_r = (Rf_i)(R'f'_r) = \sum_j \sum_s f_jf'_sD_{ji}(R)D_{sr}(R'). \tag{7.12.1}$$

The character of the operation RR' is, as usual, the sum of the diagonal elements of the representative in the product basis, and so

$$\chi(RR') = \sum_i \sum_r D_{ii}(R)D_{rr}(R') = \chi(R)\chi(R'). \tag{7.12.2}$$

Therefore the character table of the direct-product group can be written down simply by multiplying together the appropriate characters of the two contributing groups.

As an example, consider the direct-product group $C_{3v} \times C_s$ of order 12. C_{3v} is a group of three classes and C_s is a group of two classes. The total number of classes in the direct-product group is therefore six. It follows that the irreps of the group fall into six symmetry species. The six classes arise as follows. On formation of the direct-product group, each of the three classes of C_{3v} is multiplied by the identity E of C_s: this leads to the same three classes of the new group. Each of the three classes of C_{3v} is also multiplied by σ_h. The operation $E\sigma_h$ is the same as σ_h itself. The two operations $C_3^+\sigma_h$ and $C_3^-\sigma_h$ are the rotary-reflections S_3^+ and S_3^- (p. 132). The three operations $\sigma_v\sigma_h$, etc. are the same as 2-fold rotations about the bisectors of the angles of the triangle (Fig. 7.10) and so are denoted C_2. The direct-product group is therefore formed as follows:

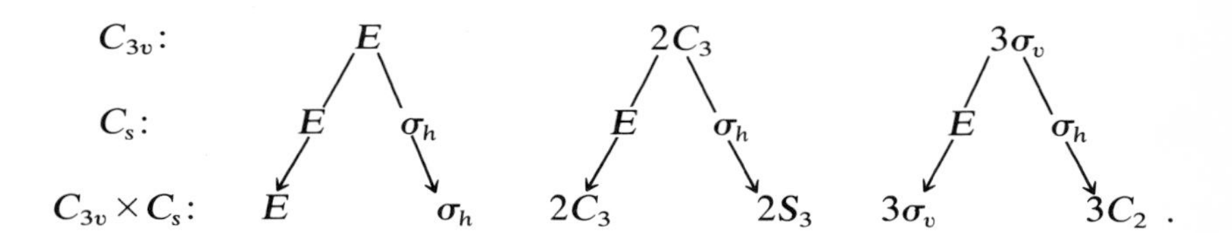

According to the system of nomenclature outlined in Section 7.2, this is the group D_{3h}. We can now use the rule about characters to construct its character table. The two component group character tables are

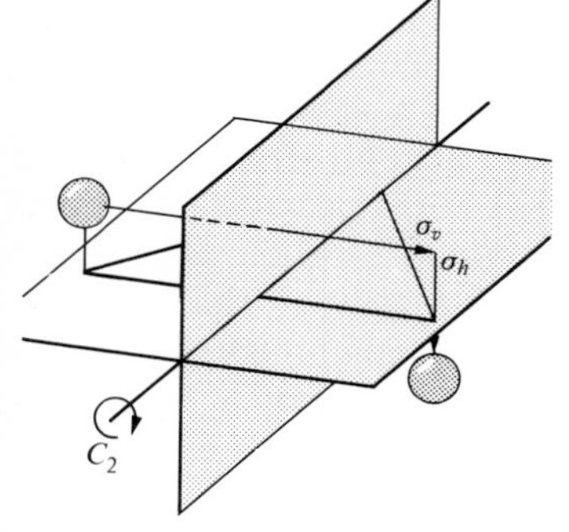

Fig. 7.10. The implication of C_2 by σ_v and σ_h.

C_{3v}	E	$2C_3$	$3\sigma_v$
A_1	1	1	1
A_2	1	1	−1
E	2	−1	0

C_s	E	σ_h
A'	1	1
A''	1	−1

Taking all the appropriate products leads to

	E		$2C_3$		$3\sigma_v$	
	E	σ_h	E	σ_h	E	σ_h
	E	σ_h	$2C_3$	$2S_3$	$3\sigma_v$	$3C_2$
A_1' $(=A_1A')$	1	1	1	1	1	1
A_1'' $(=A_1A'')$	1	−1	1	−1	1	−1
A_2' $(=A_2A')$	1	1	1	1	−1	−1
A_2'' $(=A_2A'')$	1	−1	1	−1	−1	1
E' $(=EA')$	2	2	−1	−1	0	0
E'' $(=EA'')$	2	−2	−1	1	0	0

which is the character table given in Appendix 10 for this group. This is extremely important and an easy way of building up more complex groups (e.g. D_{6h} from $D_6 \times C_i$ and O_h from $O \times C_i$).

7.13 Vanishing integrals

One of the most important applications of group theory is to the problem of deciding when integrals are necessarily zero on account of the symmetry of the problem. This can be illustrated in a simple way by considering two functions $f(x)$ and $g(x)$, and the integral over a symmetric range around $x=0$ of each one separately and of their squares and products.

Let $f(x)$ be a function that is antisymmetric with respect to the replacement of x by $-x$: $f(-x)=-f(x)$. The integral of this function from $x=-a$ to $x=a$ is zero, Fig. 7.11(a). Now consider the function $g(x)$ which is symmetric with respect to $x \rightarrow -x$: $g(-x)=g(x)$. The integral of this function over the same range is not necessarily zero, Fig. 7.11(b). Note that it may, by accident, be zero, as in Fig. 7.11(c), but that is because of a numerical coincidence; on the other hand, the integral over $f(x)$ is necessarily zero by symmetry. Now consider another way of looking at the difference between the two functions. The range $(-a, a)$ is an 'object' with two symmetry elements: the identity and a mirror plane perpendicular to the x-axis, Fig. 7.12. The object has the symmetry of the group C_s. A representation of the group with $f(x)$ as a basis has the characters $(1, -1)$ because $Ef=f$ and $\sigma_h f=-f$. It spans the irrep of symmetry species A''. On the other hand the function $g(x)$ spans A' because $Eg=g$ and $\sigma_h g=g$. That is, *if the integrand is not a basis for the totally symmetric irrep of the group* (A') *it is necessarily zero.* If it is a basis for the totally symmetrical irrep, it is not necessarily zero (but it may be zero by accident). This is a feature that generalizes to all groups: the value of the integral (e.g. the volume occupied by a molecule) cannot depend on the orientation of the object (e.g. the molecule) being integrated over, and so it must be

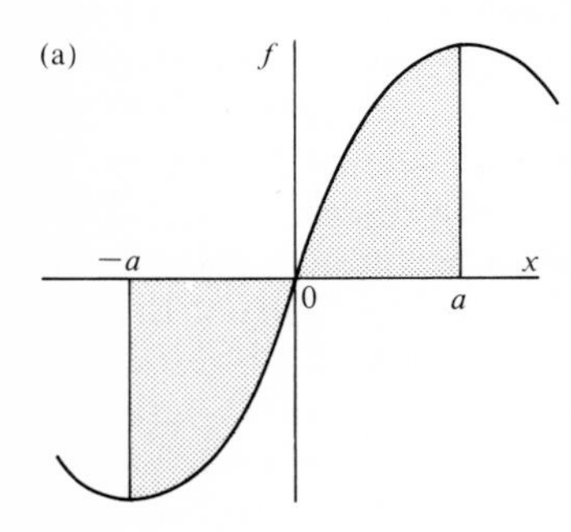

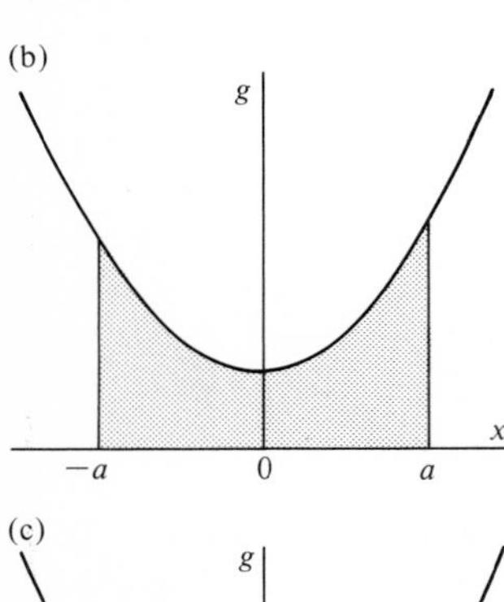

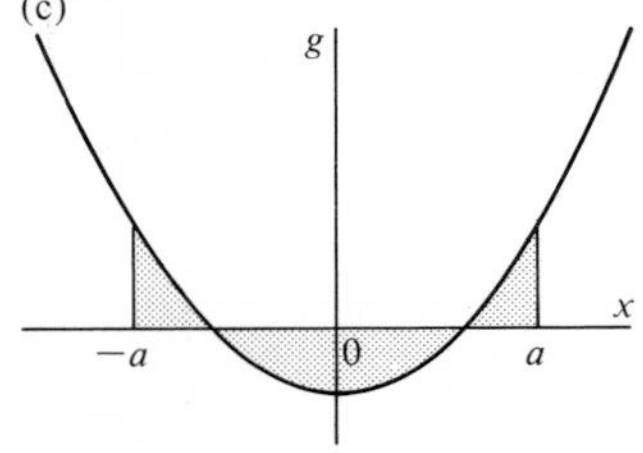

Fig. 7.11. (a) $\int f\,dx=0$, (b) $\int g\,dx \neq 0$, (c) $\int g\,dx=0$, with integrations over a symmetric range.

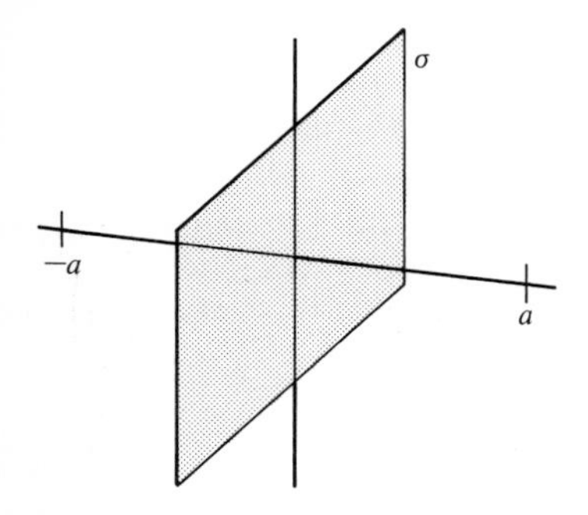

Fig. 7.12. Symmetry elements of a domain of integration.

invariant under all symmetry operations of the group. Therefore only if the integrand is a basis of the totally symmetric representation is its integral over a symmetric range not necessarily zero.

This simple example also introduces a further point which generalizes to all groups. Consider the integrals (over the same range as before) of f^2: it is not necessarily zero. Likewise the integral over g^2 is not necessarily zero. On the other hand the integral over the product is necessarily zero. There are two ways of looking at this result in terms of the symmetry of the group C_s. (1) The product $f^2 = f \times f$ is a basis for $A'' \times A'' = A'$, and hence its integral is not necessarily zero. The same is true of g^2 which spans $A' \times A' = A'$. But fg spans $A'' \times A' = A''$, which is not the totally symmetric irrep, and so the integral is necessarily zero. (2) The function f is a basis of an irrep of one symmetry species, g is a basis for an irrep of a different symmetry species. The vanishing of the integral over their product is equivalent to saying that *basis functions spanning irreps of different symmetry species are orthogonal.* In more detail: if $f_i^{(l)}$ is the ith member of a basis that spans an irreducible representation of symmetry species Γ_l in some group, and $f_j^{(l')}$ the jth member of a basis for the irrep of symmetry species $\Gamma_{l'}$, then, for a symmetric range of integration,

$$\int f_i^{(l)*} f_j^{(l')} \,\mathrm{d}\tau \propto \delta_{ll'} \delta_{ij}. \tag{7.13.1}$$

The proof of this result is based on GOT, and is given in Appendix 12. Note that the integral *may* be zero even though $l = l'$ and $i = j$ because it says nothing about the value of the proportionality constant.

We have now arrived at one of the most important results of group theory. We can summarize and extend the result into a useful rule of thumb as follows.

An integral $\int f^{(l)*} f^{(l')} \,\mathrm{d}\tau$ over a symmetric range is necessarily zero unless the integrand is a basis for the totally symmetric irrep of the group.

Since $f^{(l)}$ is a basis for Γ_l and $f^{(l')}$ is a basis for $\Gamma_{l'}$ we can assess whether the integrand is a basis for the totally symmetric irrep by forming the direct product $\Gamma_l \times \Gamma_{l'}$ and seeing whether it contains A_1. We have already seen how to do this on p. 150: it depends on knowing the character tables for the group, and nothing more.

Example. Determine which orbitals of nitrogen in ammonia may have non-vanishing overlap with the hydrogen 1s-orbitals.

- *Method.* The overlap integral has the form $\int \psi_i^* \psi_j \,\mathrm{d}\tau$; hence it is non-vanishing only if $\Gamma_i \times \Gamma_j$ includes A_1. Identify the symmetry species of the nitrogen 2s- and 2p-orbitals (character table in Appendix 10) on the basis that $p_x \propto x$, etc. and decide which can have non-vanishing overlap with s_1, s_2, s_3 (which span $A_1 + E$). Use the direct-product tables in Appendix 11.

- *Answer.* In C_{3v} the 2p-orbitals span $A_1(\mathrm{p}_z)$ and $E(\mathrm{p}_x, \mathrm{p}_y)$. Since $A_1 \times A_1 = A_1$ and $E \times E = A_1 + A_2 + E$, the p_z-orbital (A_1) may have non-vanishing overlap with $s_1(A_1)$ and the $\mathrm{p}_x, \mathrm{p}_y$-orbitals ($E$) with s_2, s_3 (E). The nitrogen 2s-orbital also spans A_1, and so it may also overlap with s_1.

- *Comment.* Note that whether the s_1-orbital-combination has non-vanishing overlap with $\mathrm{N2p}_z$ depends on the bond angle: when the molecule is flat s_1 lies in the node of $\mathrm{N2p}_z$ and the overlap is zero. (This is another example of group theory merely allowing an integral to be non-zero.)

An integral $\int f^{(l)*} f^{(l')} f^{(l'')}\, \mathrm{d}\tau$ is also necessarily zero unless the integrand is a basis for A_1. In order to assess whether the integrand does span A_1 we form $\Gamma_l \times \Gamma_{l'}$ and expand it in the usual way. Then we take each Γ_k in the expansion and form the direct product $\Gamma_k \times \Gamma_{l''}$, and expand them. If A_1 occurs *nowhere* in all the terms so produced, then the integral is necessarily zero.

The connection of group theory with the work in Chapter 5 can at last be made. There we saw that quantum mechanics involved the matrix elements of operators. These are integrals of the form $\langle a|\, \Omega\, |b\rangle$ or $\int \psi_a^* \Omega \psi_b\, \mathrm{d}\tau$, and have exactly the same form as the integral we have just encountered. Therefore *group theory can tell us when matrix elements are necessarily zero.* This is an immense simplification.

Example. Does the integral $\int \mathrm{d}_{xy} z\, \mathrm{d}_{x^2-y^2}\, \mathrm{d}\tau$ vanish in a C_{4v} molecule? Does $\int \mathrm{d}_{xy} l_z\, \mathrm{d}_{x^2-y^2}\, \mathrm{d}\tau$?

- *Method.* Assess whether $\Gamma_i \times \Gamma_j \times \Gamma_k$ contains A_1; use the character table in Appendix 10 to identify the transformation properties. Angular momenta transform as rotations (l_z as R_z in the table). Use Appendix 11 for the direct product decompositions.

- *Answer.* In C_{4v}, d_{xy} and $\mathrm{d}_{x^2-y^2}$ span B_2 and B_1 respectively, z spans A_1 and l_z spans A_2. The integrand $\mathrm{d}_{xy} z\, \mathrm{d}_{x^2-y^2}$ therefore spans $B_2 \times A_1 \times B_1 = B_2 \times B_1 = A_2$; hence the integral must vanish. The integrand $\mathrm{d}_{xy} l_z \mathrm{d}_{x^2-y^2}$ spans $B_2 \times A_2 \times B_1 = B_2 \times B_2 = A_1$; hence the integrand need not vanish (but may).

- *Comment.* Integrals of this kind are particularly important for discussing electronic spectra: we shall see that group theory is the basis of selection rules which govern which transitions may occur.

The symmetry of an object also governs its quantum-mechanical properties as follows. The hamiltonian for any system must be invariant under every symmetry operation of the symmetry group of the object: $(RH) = H$. One example of this is the hamiltonian for the harmonic oscillator: the kinetic energy operator $\mathrm{d}^2/\mathrm{d}x^2$ and the potential energy $\frac{1}{2}kx^2$ are both invariant under $x \rightarrow -x$: the hamiltonian therefore spans the totally symmetric irrep of the group C_s. This invariance has two consequences. Suppose we have solved the Schrödinger equation and have found a wavefunction ψ_i corresponding to the energy E, then

$$H\psi_i = E\psi_i.$$

Now consider the effect of some symmetry operation R. When R is applied to this equation we have

$$RH\psi_i = RE\psi_i, \quad \text{or} \quad HR\psi_i = ER\psi_i, \tag{7.13.2}$$

because $RH\psi_i = (RH)(R\psi_i) = HR\psi_i$. This shows two things. One is that any operation of the group commutes with the hamiltonian ($RH = HR$): in fact this can be taken as the definition of a symmetry operation: *a transformation is a symmetry operation if it commutes with the system's hamiltonian.* The other point is that the functions ψ_i and $R\psi_i$ correspond to the same energy E. Therefore we conclude that functions that can be generated from each other by symmetry operations of the system have the same energy. That is, *eigenfunctions related by symmetry operations are degenerate.* This was the case for the square square-well eigenfunctions discussed in Section 3.4.

We can go on to find a rule for the maximum degree of degeneracy that can occur in a system of given symmetry. Consider some member ψ_j of a basis for an irrep of dimension d of the symmetry group of the system. Let it be an eigenfunction of the hamiltonian corresponding to the energy E. We have seen that all the other members of the basis can be generated by acting on this function with the projection operator P_{ij} as defined in eqn (7.9.2). But P_{ij} is simply a linear combination of the symmetry operations of the group, and it therefore commutes with the hamiltonian. Therefore

$$P_{ij}H\psi_j = HP_{ij}\psi_j = H\psi_i \quad \text{and} \quad P_{ij}H\psi_j = P_{ij}E\psi_j = E\psi_i\,;$$

hence $H\psi_i = E\psi_i$, and ψ_i belongs to the same eigenvalue. But we can generate all d members of the d-dimensional basis by choosing i appropriately, and so all d basis functions correspond to the same energy. Therefore *the degree of degeneracy of a set of functions is equal to the dimension of the irreducible representation they span.* (The dimension is always given by $\chi(E)$, the character of the identity.) In the harmonic oscillator, symmetry group C_s, the only irreps are 1-dimensional and so there can be no degeneracies. In the case of the square square-well (symmetry group C_{4v}), 2-dimensional irreps are allowed, and so energy levels may be doubly degenerate. In an object of C_{3v} symmetry there are 2-dimensional irreps, and so doubly degenerate levels can also occur. Triply degenerate levels enter the scene only with the cubic groups, for these are the only ones that have 3-dimensional irreps. The full rotation group, R_3, has irreps of arbitrarily high dimension, and so degeneracies of any degree are permitted. This *infinite group* is so important (and different from the *finite point groups* we have been considering so far) that it deserves a special section.

7.14 The full rotation group

We shall consider the full rotation groups in two and three dimensions and discover the deep connection between group theory and the quantum mechanics of angular momentum. The techniques are no different in principle from those made familiar by the rest of the chapter, but there are some interesting points of detail.

Consider first the full rotation group in two dimensions, $C_{\infty v}$. This is the group of the circle, Fig. 7.13. Bear in mind the analogous diagram for the group of the equilateral triangle, Fig. 7.5, to see the analogies between the treatment of the infinite and finite rotation groups. The difference between C_{3v} and $C_{\infty v}$ is that a rotation through *any* angle is a symmetry operation of the circle. In particular, rotations through infinitesimal angles are symmetry operations. Because this aspect is so striking we shall build on it, for once we can discuss the effect of infinitesimal rotations we can string enough of them together to obtain a finite rotation.

Consider the effect of an infinitesimal anticlockwise rotation through an angle $\delta\phi$ about the z-axis on the basis (x, y, z). Under the operation (which we call $C_{\delta\phi}$) the basis $(r \sin\theta \cos\phi,\ r \sin\theta \sin\phi,\ r\cos\theta)$ changes to

$$\begin{aligned}C_{\delta\phi}(x, y, z) &= \{r\sin\theta\cos(\phi-\delta\phi),\ r\sin\theta\sin(\phi-\delta\phi),\ r\cos\theta\}\\ &= \{r\sin\theta\cos\phi\cos\delta\phi + r\sin\theta\sin\phi\sin\delta\phi,\\ &\qquad r\sin\theta\sin\phi\cos\delta\phi - r\sin\theta\cos\phi\sin\delta\phi,\ r\cos\theta\}\\ &= \{r\sin\theta\cos\phi + r\,\delta\phi\sin\theta\sin\phi,\\ &\qquad r\sin\theta\sin\phi - r\,\delta\phi\sin\theta\cos\phi,\ r\cos\theta\}\\ &= (x + y\,\delta\phi,\ y - x\,\delta\phi,\ z) = (x, y, z) - (-y, x, 0)\,\delta\phi,\end{aligned} \tag{7.14.1}$$

where we have expanded $\sin\delta\phi = \delta\phi + \ldots$ and $\cos\delta\phi = 1 - \tfrac{1}{2}\delta\phi^2 + \ldots$ and kept only terms first order in the infinitesimal $\delta\phi$. The calculation shows that under an infinitesimal rotation the basis differs from the original only by an infinitesimal amount. That, of course, is hardly unexpected. But we can go further. Consider the effect of the derivative operator $x(\partial/\partial y) - y(\partial/\partial x)$ on the basis:

$$\{x(\partial/\partial y) - y(\partial/\partial x)\}(x, y, z) = (-y, x, 0).$$

Therefore if we define the operator

$$l_z = (\hbar/\mathrm{i})\{x(\partial/\partial y) - y(\partial/\partial x)\}, \tag{7.14.2}$$

where the $\hbar/\mathrm{i}$ factor has been introduced for future convenience (and on the basis of the hindsight provided by Chapter 6!) the effect of the operator $C_{\delta\phi}$ on the basis is

$$C_{\delta\phi}(x, y, z) = \{1 - (\mathrm{i}/\hbar)\,\delta\phi\, l_z\}(x, y, z). \tag{7.14.3}$$

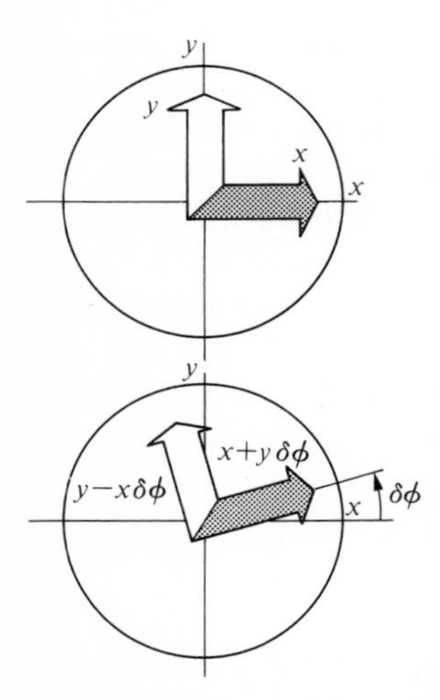

Fig. 7.13. Elements of the axial rotation group.

The infinitesimal rotation operator therefore differs from the identity by a quantity determined by the effect of the operator l_z, which is called the *generator* of the infinitesimal rotation about the z-axis. It is not hard to believe that had we considered infinitesimal rotations around the y- and x-axes, as would be permitted as symmetry operations in the full rotation group R_3, then the generators would have been the operators l_y and l_x obtained from l_z by cyclic permutation of the labels z, y, x and identifiable as the angular momentum operators of Chapter 6, eqn (6.1.4).

We know that the angular momentum operators satisfy a set of commutation relations. These can be seen in a different light as follows. The effect of rotations about different axes depends on their order, (Fig. 7.14). Consider the effects of (a) a rotation by $\delta\alpha$ about x followed by a rotation by $\delta\beta$ about y, (b) a rotation by $\delta\beta$ about y followed by a rotation of $\delta\alpha$ about x, both $\delta\alpha$ and $\delta\beta$ being infinitesimal. The basis (x, y, z) changes as follows:

$$\text{(a)}\quad C^y_{\delta\beta}C^x_{\delta\alpha}(x, y, z) = \{1-(\mathrm{i}/\hbar)\,\delta\beta l_y\}\{1-(\mathrm{i}/\hbar)\,\delta\alpha l_x\}(x, y, z)$$
$$= \{1-(\mathrm{i}/\hbar)(\delta\beta l_y+\delta\alpha l_x)+(\mathrm{i}/\hbar)^2\,\delta\beta\,\delta\alpha l_y l_x\}(x, y, z).$$

$$\text{(b)}\quad C^x_{\delta\alpha}C^y_{\delta\beta}(x, y, z) = \{1-(\mathrm{i}/\hbar)\,\delta\alpha l_x\}\{1-(\mathrm{i}/\hbar)\,\delta\beta l_y\}(x, y, z)$$
$$= \{1-(\mathrm{i}/\hbar)(\delta\alpha l_x+\delta\beta l_y)+(\mathrm{i}/\hbar)^2\,\delta\alpha\,\delta\beta l_x l_y\}(x, y, z).$$

The difference between these two operations is

$$C^y_{\delta\beta}C^x_{\delta\alpha} - C^x_{\delta\alpha}C^y_{\delta\beta} = (\mathrm{i}/\hbar)^2\,\delta\alpha\,\delta\beta(l_y l_x - l_x l_y) = (\mathrm{i}/\hbar)\,\delta\alpha\,\delta\beta l_z \qquad (7.14.4)$$

because we have already encountered the commutator $[l_x, l_y] = \mathrm{i}\hbar$. That is, the difference of the two rotations is equivalent to an infinitesimal rotation through $-\delta\alpha\,\delta\beta$ about the z-axis, as is geometrically plausible, Fig. 7.14. The *reverse* argument, that it is geometrically obvious that the difference between the two composite rotations is another rotation, implies that $[l_x, l_y] = \mathrm{i}\hbar l_z$; hence the angular momentum commutation relations can be regarded as a direct consequence of the geometrical properties of composite rotations.

We shall now look for an irreducible matrix representation of the full rotation group. As a starting point we note that the spherical harmonics $Y_{l,m_l}(\theta, \phi)$ of a given l transform into linear combinations of each other when rotated. (For instance, p-orbitals rotate into each other, d-orbitals rotate into each other, and so on; p-orbitals do not rotate into d-orbitals.) Therefore the functions $Y_{l,l}, Y_{l,l-1}, \ldots Y_{l,m_l}, \ldots Y_{l,-l}$ constitute a basis for a $(2l+1)$-dimensional, and it turns out irreducible, representation of the group. Each one has the form $\mathrm{e}^{\mathrm{i}m_l\phi}P(\theta)$, and so as a result of a rotation by α around the z-axis each one transforms to $\mathrm{e}^{\mathrm{i}m_l(\phi-\alpha)}P(\theta)$, and so the entire basis transforms into

$$C^z_\alpha(Y_{l,l}, Y_{l,l-1}, \ldots, Y_{l,-l}) = \{\mathrm{e}^{\mathrm{i}l(\phi-\alpha)}P, \mathrm{e}^{\mathrm{i}(l-1)(\phi-\alpha)}P, \ldots\}$$

$$= (Y_{l,l}, Y_{l,l-1}, \ldots, Y_{l,-1})\begin{bmatrix} \mathrm{e}^{-\mathrm{i}l\alpha} & 0 & 0\ldots0 \\ 0 & \mathrm{e}^{-\mathrm{i}(l-1)\alpha} & 0\ldots0 \\ 0 & 0 & \ddots & \vdots \\ \vdots & \vdots & & \\ 0 & 0\ldots & \ldots & \mathrm{e}^{\mathrm{i}l\alpha} \end{bmatrix}$$

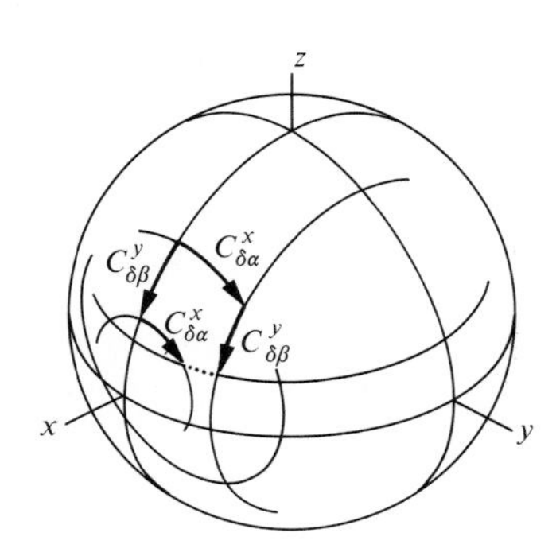

Fig. 7.14. The composition of infinitesimal rotations.

and hence we can recognize the matrix representation of the rotation in the basis.

The character of the rotation through the angle α about the z-axis (and therefore about any axis, because all rotations through the same angle but about different axes are conjugate and fall into the same class) is the sum

$$\chi(C_\alpha) = e^{il\alpha} + e^{i(l-1)\alpha} + \ldots + e^{-il\alpha}$$
$$= 1 + 2\cos\alpha + 2\cos 2\alpha + \ldots + 2\cos l\alpha. \qquad (7.14.5)$$

(The 1 comes from the central term with $m_l = 0$, and we have used $e^{im_l\alpha} + e^{-im_l\alpha} = 2\cos m_l\alpha$.) This simple expression can be used to establish the character of any rotation for a $(2l+1)$-dimensional basis. An even simpler version is obtained by recognizing that the first line is a geometric series† from $e^{-il\alpha}$ to $e^{il\alpha}$ in steps of $e^{i\alpha}$ and hence has the sum

$$\chi(C_\alpha) = \sum_{m_l=-l}^{l} e^{im_l\alpha} = \frac{e^{-il\alpha}(e^{i(2l+1)\alpha} - 1)}{e^{i\alpha} - 1},$$

which is readily manipulated into

$$\chi(C_\alpha) = \frac{\sin(l+\frac{1}{2})\alpha}{\sin\frac{1}{2}\alpha}. \qquad (7.14.6)$$

When $\alpha = 0$ the character is $2l+1$ (take the limit as $\alpha \to 0$), and so the levels with angular momentum quantum number l are $(2l+1)$-fold degenerate in a spherical system (because $\chi(C_0) = \chi(E)$, and on p. 151 we saw that $\chi(E)$ is equal to the dimension of the representation).

Example. An atom is in a configuration giving rise to an F term. What symmetry species would it span in an octahedral environment?

- *Method.* Identify which rotations are common to R_3 and O and calculate their characters from eqn (7.14.6) with $l = 3$. Use the O character table in Appendix 10 and eqn (7.8.5) to identify the symmetry species spanned.
- *Answer.* The characters for the group O correspond to $\alpha = 0(E)$, $2\pi/3(C_3)$, $\pi(C_2)$, $\pi/2(C_4)$ and $\pi(C_2')$. Since $\chi(C_\alpha) = \sin(7\alpha/2)/\sin(\alpha/2)$ we find $\chi = (7, 1, -1, -1, -1)$ for (E, C_3, C_2, C_4, C_2'). Use of eqn (7.8.5) with $h = 24$ gives $a(A_2) = 1$, $a(T_1) = 1$, and $a(T_2) = 1$. Therefore $F \to A_2 + T_1 + T_2$.
- *Comment.* The step from one group to a subgroup is called *descent in symmetry*. It is a particularly important technique in the theory of the structure and spectra of transition-metal ions, and we re-encounter it in Chapter 10.

We now explore the group-theoretical description of the coupling of two angular momenta. We suppose that we have two sets of functions which are bases for the irreducible representations $\Gamma^{(j_1)}$ and $\Gamma^{(j_2)}$ of the full rotation group. We call these functions $f_{m_{j1}}^{(j_1)}$ and $f_{m_{j2}}^{(j_2)}$ respectively.

† The sum of a geometric series $a + ar + ar^2 + \ldots ar^n$ is $a(r^{n+1} - 1)/(r-1)$. In the present case the series is $e^{-il\alpha}\{1 + e^{i\alpha} + e^{2i\alpha} + \ldots + e^{2il\alpha}\}$ and so $r = e^{i\alpha}$, $n = 2l$, and $a = e^{-il\alpha}$.

The products $f_{m_{j1}}^{(j_1)} f_{m_{j2}}^{(j_2)}$ provide a basis for the direct product representation $\Gamma^{(j_1)} \times \Gamma^{(j_2)}$. This representation is in general reducible, and the manner of its reduction is given by eqns (7.8.2–5). That is, we seek the reduction

$$\Gamma^{(j_1)} \times \Gamma^{(j_2)} = \sum_j a_j \Gamma^{(j)}, \tag{7.14.7}$$

and obtain it by considering the characters for the rotation C_α:

$$\chi(C_\alpha) = \chi^{(j_1)}(C_\alpha)\chi^{(j_2)}(C_\alpha)$$
$$= \sum_{m_{j1}=-j_1}^{j_1} \sum_{m_{j2}=-j_2}^{j_2} e^{-i(m_{j1}+m_{j2})\alpha} = \sum_j a_j \sum_{m_j=-j}^{j} e^{-im_j\alpha}. \tag{7.14.8}$$

Since $|m_{j_1} + m_{j_2}| \leq j_1 + j_2$ it follows that $|m_j| \leq j_1 + j_2$, and so $j \leq j_1 + j_2$. Therefore $a_j = 0$ if $j > j_1 + j_2$. The maximum value of m_j may be obtained from m_{j_1} and m_{j_2} in only one way: when $m_{j_1} = j_1$ and $m_{j_2} = j_2$. Therefore $a_{j_1+j_2} = 1$. $m_{j_{\max}} - 1$ may be obtained in two ways, when $m_{j_1} = j_1$ and $m_{j_2} = j_2 - 1$, or when $m_{j_1} = j_1 - 1$ and $m_{j_2} = j_2$; one of these ways also contributes to the values of m_j from $j = j_1 + j_2$ and so $a_{j_1+j_2-1} = 1$. The argument may be continued down to $j = |j_1 - j_2|$ and so eqn (7.14.8) is equivalent to

$$\chi(C_\alpha) = \sum_{j=|j_1-j_2|}^{j_1+j_2} \chi^{(j)}(C_\alpha).$$

Therefore we conclude that the direct product decomposes as follows:

$$\Gamma^{(j_1)} \times \Gamma^{(j_2)} = \Gamma^{(j_1+j_2)} + \Gamma^{(j_1+j_2-1)} + \ldots \Gamma^{(|j_1-j_2|)}, \tag{7.14.9}$$

which is nothing other than the Clebsch–Gordan series (eqn (6.6.7)). This shows, in effect, that the whole of angular momentum theory can be regarded as an aspect of group theory and the symmetry properties of rotations.

7.15 Applications

There are numerous applications of group theory, both explicit and implicit, and we shall encounter many in the following pages. We shall not give more examples here, but merely indicate the types of application to be encountered, and where. The application of the rotation groups R_3, $D_{\infty h}$, $C_{\infty v}$ to spherical systems (atoms, rotating molecules) and linear systems (diatomic molecules) will appear whenever we discuss their angular momentum (Chapters 9, 10, and 11). Finite groups play an important role in the discussion of molecular structure. When an atom is exposed to a crystal field in a transition-metal complex, the symmetry of its environment is reduced from spherical, and the degeneracy of its states may be removed; we shall examine this in Chapter 10. Molecular orbitals must be constructed in conformity with the symmetry of the molecule, and so too must hybrid orbitals

(Chapter 10). We have seen that spectroscopy draws on group theory to determine which transitions are possible: in Chapters 11 and 12 we apply the techniques of group theory to electronic, vibrational, and rotational transitions. Group theory is able to dismiss integrals on the basis of their symmetry and the symmetries of the components of the integrand: this will prove to be of great value in the discussion of molecular properties (Chapters 13 and 14) for these are usually calculated on the basis of perturbation theory (Chapter 8), in which integrals occur of just the kind we have been considering. The following chapters will confirm that group theory pervades the whole of quantum chemistry.

Further reading

Background information and techniques

Symmetry in chemistry. H. H. Jaffe and M. Orchin; Wiley, New York, 1965.
Chemical applications of group theory. F. A. Cotton; Wiley, New York, 2nd edition, 1971.
Group theory and chemistry. D. Bishop; Oxford University Press, 1973.
Group theory and quantum mechanics. M. Tinkham; McGraw-Hill, New York, 1964.
Group theory. M. Hamermesh; Addison-Wesley, Reading, Mass., 1962.
Group theory. E. P. Wigner; Academic Press, New York, 1959.

Chemical applications

Physical chemistry. P. W. Atkins; Oxford University Press and W. H. Freeman and Company, San Francisco, 2nd edition, 1982.
Symmetry and spectroscopy. D. C. Harris and M. D. Bertolucci; Oxford University Press, New York, 1978.
Introduction to the theory of molecular vibrations and vibrational spectroscopy. L. A. Woodward; Oxford University Press, 1972.
Symmetry properties of molecules. G. S. Ezra; Springer, Berlin, 1982.

See also the first three titles under *Background information.*

Character tables

Tables for group theory. P. W. Atkins, M. S. Child, and C. S. G. Phillips; Oxford University Press, 1970.

Problems

7.1. Classify the following molecules according to their point symmetry group: (a) H_2O, (b) CO_2, (c) C_2H_4, (d) *cis*-ClHC=CHCl, (e) *trans*-ClCH=CHCl, (f) benzene, (g) naphthalene, (h) CHClFBr, (i) $B(OH)_3$.

7.2. Which of the molecules listed above may possess a permanent electric dipole moment? *Hint.* Decide on the criterion for the non-vanishing of $\boldsymbol{\mu}_e = \int \psi^* \boldsymbol{\mu} \psi \, d\tau$, and refer to the tables in Appendix 10; $\boldsymbol{\mu}$ transforms as $\mathbf{r} = \{x, y, z\}$.

7.3. Find the representatives of the operations of the group C_{2v} using as a basis the valence orbitals of H and O in H_2O (i.e. $H1s_A$, $H1s_B$, O2s, O2p). *Hint.* The group is of order 4 and so there are four 6-dimensional matrices to find.

7.4. Confirm that the representatives established in Problem 3 reproduce the group multiplications $C_2^2 = E$, $\sigma_v C_2 = \sigma_v'$.

7.5. Determine which symmetry species are spanned by the six orbitals. Find the symmetry-adapted linear combinations, and confirm that the representatives are in block-diagonal form. *Hint.* Decompose the representation established in Problem 7.3 by analysing the characters. Use the projection operator in eqn (7.9.6) to establish the symmetry-adapted bases (using the elements of the representatives established in Problem 7.3), form the matrix of eqn (7.5.1), and use eqn (7.5.6) to construct the irreps.

7.6. Find the representatives of the operations of the group T_d using as a basis four 1s-orbitals, one at each apex of a regular tetrahedron (as in CH_4). *Hint.* The $C_3^{\pm}$ operations can be constructed like eqn (7.4.3). The basis is 4-dimensional; the order of the group is 24, and so there are twenty-four matrices to find.

7.7. Confirm that the representations established in Problem 7.6 reproduce the group multiplications $C_3^+C_3^- = E$, $S_4C_3 = S_4'$, and $S_4C_3 = \sigma_d$.

7.8. Determine which irreducible representations are spanned by the four 1s-orbitals in methane. Find the symmetry-adapted linear combinations, and confirm that the representatives for C_3^+ and S_4 are in block-diagonal form. *Hint.* Decompose the representation into irreps by analysing the characters. Use the projection operator in eqn (7.9.6) to establish the symmetry-adapted bases, form the matrix $\mathbf{c}$ of eqn (7.5.1), and use eqn (7.5.6) to construct the irreps.

7.9. Analyse the following direct products into the symmetry species they span: (a) C_{2v}: $A_2 \times B_1 \times B_2$, (b) C_{3v}: $A_1 \times A_2 \times E$, (c) C_{6v}: $B_2 \times E_1$, (d) $C_{\infty v}$: E_1^2, (e) O: $T_1 \times T_2 \times E$.

7.10. Show that $3x^2y - y^3$ is a basis for an A_1 irrep of C_{3v}. *Hint.* Use eqn (7.10.2) to show that $C_3^+(3x^2y - y^3) \propto 3x^2y - y^3$; likewise for the other elements of the group.

7.11. A function $f(x, y, z)$ was found to be a basis for a representation of C_{2v}, the characters being (4, 0, 0, 0). What symmetry species of irrep does it span? *Hint.* Proceed by inspection to find the a_l in eqn (7.8.3), or use eqn (7.8.5).

7.12. Find the components of the function $f(x, y, z)$ acting as a basis for each irrep it spans. *Hint.* Use eqn (7.9.6). The basis for A_1, for example, turns out to be $\frac{1}{4}\{f(x, y, z) + f(-x, -y, -z) + f(x, -y, -z) + f(-x, y, z)\}$.

7.13. Regard the naphthalene molecule as having C_{2v} symmetry (C_2 perpendicular to the plane), which is a subgroup of its full symmetry group. Consider the π-orbitals on each carbon as a basis. What symmetry species do they span? Construct the symmetry-adapted bases. *Hint.* Proceed as in the *Example* following eqn (7.9.7).

7.14. Repeat the process for benzene, using the subgroup C_{6v} of the full symmetry group. After constructing the symmetry-adapted linear combinations, refer to the D_{6h} character table to label them according to the full group.

7.15. Show that in an octahedral array, hydrogen 1s-orbitals span $A_{1g} + E_g + T_{1u}$ of the group O_h.

7.16. Classify the terms that may arise from the following configurations: (a) C_{2v}: $a_1^2b_1b_2$; (b) C_{3v}: a_2e, e^2; (c) T_d: a_2e, et_1, t_1t_2, t_1^2, t_2^2; (d) O: e^2, et_1, t_2^2. *Hint.* Use the direct product tables; triplet terms have antisymmetric spatial functions.

7.17. Construct the character tables for the groups O_h and D_{6h}. *Hint.* Use $D_{6h} = D_6 \times C_i$ and $O_h = O \times C_i$ and the procedure in Section 7.12.

7.18. Demonstrate that there are no non-zero integrals of the form $\int \psi' H \psi \, d\tau$ when ψ' and ψ belong to different symmetry species.

7.19. The ground states of the C_{2v} molecules NO_2 and ClO_2 are $^2A_1$ and $^2B_1$

respectively; the ground state of O_2 is $^3\Sigma_g^-$. To what states may (a) electric-dipole, (b) magnetic-dipole transitions take place. *Hint.* The electric-dipole operator transforms as translations, the magnetic as rotations.

7.20. What is the maximum degeneracy of the energy levels of a particle confined to the interior of a regular tetrahedron?

7.21. Demonstrate that the linear momentum operator $p = (\hbar/i)(d/dx)$ is the generator of infinitesimal translations. *Hint.* Proceed as in eqns (7.14.1–3).

7.22. An atom bearing a single p-electron is trapped in an environment with C_{3v} symmetry. What symmetry species does it span? *Hint.* Use eqn (7.14.6) with $\alpha = 120°$.

8 Techniques of approximation

THIS is a sad but necessary chapter. It is sad because we have reached the point where the hope of finding exact solutions to problems is set aside, and we begin to look for methods of approximation. It is necessary, because most of the problems of quantum chemistry cannot be solved exactly, and so we must learn how to tackle them. There are very few problems for which the Schrödinger equation can be solved exactly, and the examples in Chapters 3 and 4 almost exhaust the list. As soon as the shape of the potential is distorted from the forms considered there, or one takes into account the interactions between particles, the equation cannot be solved exactly. There are three major ways of making progress. The first is to try to guess the shape of the wavefunction for the system. Even people with profound insight need a criterion of success, and this is provided by the *variation principle.* It is useful to be able to be guided to the form of the function by a knowledge of the distortion of the system induced by the complicating aspects of the potential or the interactions. For example, the exact solutions for a system that resembles the true system may be known and we can be led to the solutions for the real system by observing how the hamiltonians for the two system differ. This is the province of *perturbation theory.* Both techniques are powerful, and may be applied to the calculation of molecular properties. Perturbation theory is particularly useful when we are interested in responses of atoms and molecules to electric and magnetic fields. When these fields change with time (as in a light wave) we have to deal with *time-dependent perturbation theory.* This is the case when we calculate transition probabilities. The third approximation method is that of *self-consistent fields,* which is an iterative method of solving the Schrödinger equation for systems of many particles. This method is discussed in the next chapter.

8.1 Time-independent perturbation theory

Our first concern is with time-independent perturbation theory, when the hamiltonians for the true system and the simple model system differ by a quantity that is independent of the time.

Consider a system in which there are only two energy levels. We write the hamiltonian of the true system as H and that of a simpler system which it resembles as $H^{(0)}$. H and $H^{(0)}$ differ by an amount represented by the *perturbation* $H^{(1)}$, so that $H = H^{(0)} + H^{(1)}$. We suppose that the two eigenfunctions of $H^{(0)}$ are known, and call them ψ_1 and ψ_2. They correspond to the energies E_1 and E_2, and so

$$H^{(0)}\psi_m = E_m\psi_m; \qquad m = 1, 2. \tag{8.1.1}$$

The wavefunctions and energies of the true system differ only slightly

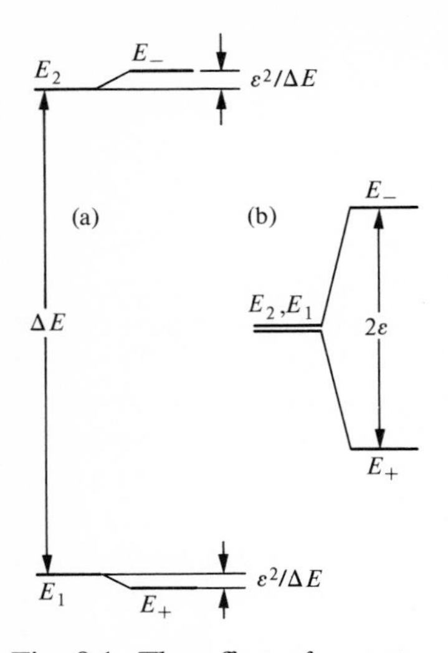

Fig. 8.1. The effect of a perturbation on (a) a non-degenerate system, (b) a degenerate system.

from these, and so we shall attempt to solve the eigenvalue equation

$$H\psi = E\psi \tag{8.1.2}$$

in terms of them by writing

$$\psi = a_1\psi_1 + a_2\psi_2, \tag{8.1.3}$$

where a_1 and a_2 are constants. It proves useful to use a ket notation for ψ_1 and ψ_2, and so we denote them $|1\rangle$ and $|2\rangle$ respectively. If this equation is inserted in the preceding one we obtain

$$a_1(H-E)\,|1\rangle + a_2(H-E)\,|2\rangle = 0.$$

When this is multiplied from the left by the bras $\langle 1|$ and $\langle 2|$ in turn, and use made of the orthonormality of the functions ψ_1 and ψ_2 ($\langle m\mid n\rangle = \delta_{mn}$), we obtain

$$a_1(H_{11}-E) + a_2H_{12} = 0; \qquad a_1H_{21} + a_2(H_{22}-E) = 0, \tag{8.1.4}$$

where $H_{mn} = \langle m|\,H\,|n\rangle$. The condition for the existence of non-trivial solutions of this pair of equations is that the determinant of the coefficients of a_1 and a_2 should disappear:

$$\begin{vmatrix} H_{11}-E & H_{12} \\ H_{21} & H_{22}-E \end{vmatrix} = 0,$$

and this is satisfied for the following values of E:

$$E_\pm = \tfrac{1}{2}(H_{11}+H_{22}) \pm \tfrac{1}{2}\{(H_{11}-H_{22})^2 + 4H_{12}H_{21}\}^{\frac{1}{2}}. \tag{8.1.5}$$

In the special (but common) case of a perturbation for which $H^{(1)}_{11} = H^{(1)}_{22} = 0$, the diagonal elements H_{mm} are equal to $H^{(0)}_{mm} = E_m$. Therefore, since $H_{12} = H^{(1)}_{12}$ (because $H^{(0)}$ is diagonal in its own eigenstates),

$$\begin{aligned} E_\pm &= \tfrac{1}{2}(E_1+E_2) \pm \tfrac{1}{2}\{(E_1-E_2)^2 + 4H^{(1)}_{12}H^{(1)}_{21}\}^{\frac{1}{2}} \\ &= \tfrac{1}{2}(E_1+E_2) \pm \tfrac{1}{2}\{(E_1-E_2)^2 + 4\varepsilon^2\}^{\frac{1}{2}} \end{aligned} \tag{8.1.6}$$

where $\varepsilon^2 = H^{(1)}_{12}H^{(1)}_{21}$. The energies of the perturbed system, E, are therefore shifted from E_1 and E_2, Fig. 8.1, and it is important to note that *the upper energy level rises in energy and the lower level falls.* The perturbation causes a kind of repulsion between them. Furthermore, *the closer E_1 and E_2 are in the unperturbed system, the greater the effect of a given perturbation.* Finally, when $E_1 = E_2$, so that the perturbation is applied to a degenerate system, $E_+ - E_- = 2\varepsilon$, and so *the stronger the perturbation the greater the effective repulsion of the levels.* The behaviour of the energy of the perturbed system for different values of $\Delta E = E_2 - E_1$ is shown in more detail in Fig. 8.2.

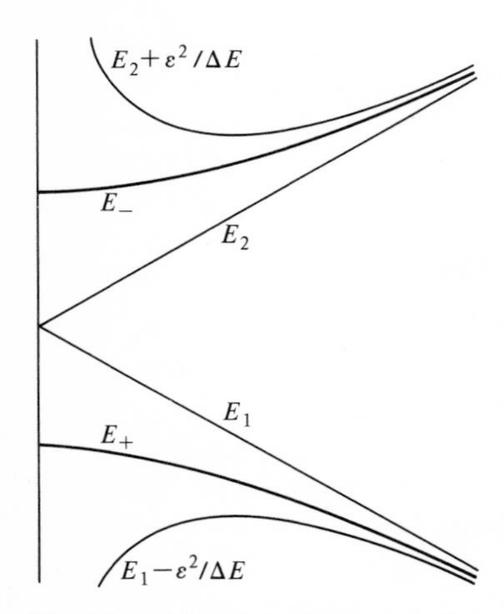

Fig. 8.2. Unperturbed energies (E_1, E_2), exact perturbed energies (E_+, E_-), and the perturbation approximation for $\varepsilon/\Delta E \ll 1$.

How the perturbed energies emerge from the unperturbed can be deduced by considering the values of $E_\pm$ when the perturbation is much smaller than the energy separation, so that $\varepsilon/\Delta E \ll 1$. In this case eqn (8.1.6) can be expanded using $(1+x)^{\frac{1}{2}} = 1 + \frac{1}{2}x + \ldots$; then

$$E_\pm = \tfrac{1}{2}(E_1+E_2) \pm \tfrac{1}{2}(E_1-E_2)\{1 + 2\varepsilon^2/\Delta E^2 + \ldots\}.$$

Then to second order in ε we have

$$E_+ \approx E_1 - \varepsilon^2/\Delta E; \qquad E_- \approx E_2 + \varepsilon^2/\Delta E. \tag{8.1.7}$$

The way these agree with the exact solution when $\varepsilon/\Delta E$ is small is shown in Fig. 8.2; when $\varepsilon \approx \Delta E$ the approximate expressions diverge markedly from their true values. Nevertheless, the energy corrections $\pm\varepsilon^2/\Delta E$ in these approximate expressions are easy to calculate, and a reasonable guide to the true behaviour when the perturbations are weak and the energy separations large. We shall see their form emerging from more complicated expressions in due course.

The perturbed wavefunctions are obtained by solving eqns (8.1.4) for the coefficients using $E = E_+$ (to get ψ_+) and $E = E_-$ (to get ψ_-). A convenient way of writing them (p. 100) is

$$\psi_+ = \psi_1 \cos\beta + \psi_2 \sin\beta; \qquad \psi_- = -\psi_1 \sin\beta + \psi_2 \cos\beta \tag{8.1.8}$$

where $\tan 2\beta = 2H_{12}^{(1)}/(E_1 - E_2)$. In the case of degeneracy in the initial system ($E_1 = E_2$), $\tan 2\beta = \infty$ corresponding to $\beta = \pi/4$. In this case the perturbed wavefunctions are

$$\psi_+ = (\psi_1 + \psi_2)/\surd 2; \qquad \psi_- = (-\psi_1 + \psi_2)/\surd 2 \tag{8.1.9}$$

and so each state is a 50 per cent mixture of the two original states. In the case of a weak perturbation acting on states widely separated in energy $H_{12}^{(1)}/\Delta E \ll 1$ and so $\tan 2\beta \approx 2\beta = -2H_{12}/\Delta E$. Furthermore, because $\beta \ll 1$ in this case, $\sin\beta \approx \beta$ and $\cos\beta \approx 1$. Therefore

$$\psi_+ \approx \psi_1 - (H_{12}^{(1)}/\Delta E)\psi_2; \qquad \psi_- \approx \psi_2 + (H_{12}^{(1)}/\Delta E)\psi_1 \tag{8.1.10}$$

and we see that each state is weakly contaminated by an admixture of the other.

Now we have to generalize these results to the case of a system consisting of numerous levels. Special precautions have to be taken if the state of interest is degenerate, and so we discuss that possibility later.

We suppose that we know all the eigenfunctions and eigenvalues of a model system with hamiltonian $H^{(0)}$ which differs from the hamiltonian for the true system to a small extent. An example of an actual system might be an anharmonic oscillator or a molecule in an electric field; the simple model systems would then be the harmonic oscillator itself or the molecule itself. We therefore suppose that we have solved the equations

$$H^{(0)}\psi_n = E_n\psi_n, \qquad n = 0, 1, 2, \ldots \tag{8.1.11}$$

and that we have the complete set of solutions. Whenever convenient we replace the unperturbed functions ψ_n by the kets $|n\rangle$. Any state whose perturbed form and energy we are calculating will be denoted $|0\rangle$, its unperturbed wavefunction written ψ_0 and its unperturbed energy E_0. When $|0\rangle$ is perturbed its wavefunction changes from ψ_0 to ψ and its energy from E_0 to E. $|0\rangle$ is not necessarily the ground state, but it often is.

As the hamiltonian for the actual system of interest is so similar to that for the model system we may express it as

$$H = H^{(0)} + H^{(1)} + H^{(2)} + \dots \tag{8.1.12}$$

where $H^{(1)}$ is a hamiltonian first order in some small quantity (e.g. the electric field strength), $H^{(2)}$ is second order, and so on. So as to keep the orders of the terms in mind we write

$$H = H^{(0)} + \lambda H^{(1)} + \lambda^2 H^{(2)} + \dots \tag{8.1.13}$$

where λ signifies the order of magnitude of the perturbation. At the end of the calculation, when we have collected together all terms of the same order of magnitude, we erase all the λs. Because the actual and model systems are so similar, we can write the function for the level of interest of the perturbed system, ψ, as

$$\psi = \psi_0 + \lambda \psi_0^{(1)} + \lambda^2 \psi_0^{(2)} + \dots \tag{8.1.14}$$

Likewise, because the energies of the actual system are similar to those of the model,

$$E = E_0 + \lambda E_0^{(1)} + \lambda^2 E_0^{(2)} + \dots \tag{8.1.15}$$

$E_0^{(1)}$ is the *first-order correction* to the energy, $E_0^{(2)}$ the *second-order correction*, and so on.

The equation to solve is

$$H\psi = E\psi. \tag{8.1.16}$$

On inserting the preceding three equations into this, and then collecting powers of λ,

$$\lambda^0\{H^{(0)}\psi_0 - E_0\psi_0\} + \lambda^1\{H^{(0)}\psi_0^{(1)} + H^{(1)}\psi_0 - E_0\psi_0^{(1)} - E_0^{(1)}\psi_0\}$$
$$+ \lambda^2\{H^{(0)}\psi_0^{(2)} + H^{(1)}\psi_0^{(1)} + H^{(2)}\psi_0 - E_0\psi^{(2)} - E^{(1)}\psi^{(1)} - E_0^{(2)}\psi_0\} + \dots = 0.$$

As λ is an arbitrary parameter, the coefficients of each λ^n must equal zero separately, and so we have the following equations:

$$H^{(0)}\psi_0 = E_0\psi_0 \tag{8.1.17a}$$
$$\{H^{(0)} - E_0\}\psi_0^{(1)} = \{E_0^{(1)} - H^{(1)}\}\psi_0, \tag{8.1.17b}$$
$$\{H^{(0)} - E_0\}\psi_0^{(2)} = \{E_0^{(2)} - H^{(2)}\}\psi_0 + \{E_0^{(1)} - H^{(1)}\}\psi_0^{(1)}, \tag{8.1.17c}$$

etc.

The solution of the first of these is assumed known (it is just eqn (8.1.11)). We try to find the first-order correction to the wavefunction, $\psi_0^{(1)}$, in the same way as in the initial example, by writing it as a linear combination of the known functions

$$\psi_0^{(1)} = \sum_n a_n \psi_n. \tag{8.1.18}$$

The sum is over all functions of the model system, including those belonging to the continuum of energies, if that exists. When this expansion is inserted into eqn (8.1.17b) and the ket notation introduced for the ψ_n, we find

$$\sum_n a_n\{H^{(0)}-E_0\}\,|n\rangle=\sum_n a_n\{E_n-E_0\}\,|n\rangle=\{E_0^{(1)}-H^{(1)}\}\,|0\rangle. \qquad (8.1.19)$$

When this is multiplied from the left by the bra $\langle 0|$, because the ψ_m are orthonormal,

$$\sum_n a_n\{E_n-E_0\}\delta_{0n}=E_0^{(1)}-\langle 0|\,H^{(1)}\,|0\rangle.$$

The left-hand side of this equation is zero, and so we conclude that the first-order correction to the energy of the state $|0\rangle$ is

First-order energy correction: $E_0^{(1)}=\langle 0|\,H^{(1)}\,|0\rangle=H_{00}^{(1)}$. (8.1.20)

The matrix element $H_{00}^{(1)}=\int \psi_0^* H^{(1)}\psi_0\,\mathrm{d}\tau$, is a kind of average of the first-order perturbation over the system. A mechanical analogy is that the first-order shift in the frequency of a violin string when small weights are added along its length is an average of the effects of the weights, those at the antinodes affecting the motion the most, those near the nodes the least. This is illustrated by the following *Example.*

Example. A small step in the potential energy was introduced into the one-dimensional square well problem such that the potential energy was zero from $x=0$ to $x=\frac{1}{2}(L-a)$, ε from $x=\frac{1}{2}(L-a)$ to $x=\frac{1}{2}(L+a)$, and then zero from there to $x=L$. Find the first-order correction to the energy of a particle confined to the well, and evaluate the expression for $a=L/10$ (so that the blip in the potential occupies the central 10 per cent) and $n=1$, $n=2$.

- *Method.* Evaluate eqn (8.1.20) using $H^{(1)}=\varepsilon$ for $\frac{1}{2}(L-a)\le x\le\frac{1}{2}(L+a)$ and zero elsewhere. Use the wavefunctions in eqn (3.3.5). Use

$$\int \sin^2 kx\,\mathrm{d}x=\tfrac{1}{2}x-(1/4k)\sin(2kx).$$

- *Answer.*

$$E^{(1)}=(2/L)\varepsilon\int_{\frac{1}{2}(L-a)}^{\frac{1}{2}(L+a)}\sin^2(n\pi x/L)\,\mathrm{d}x$$
$$=\varepsilon\{(a/L)-(-1)^n(1/n\pi)\sin(n\pi a/L)\}.$$

When $n=1$, $E^{(1)}=0.1984\varepsilon$; when $n=2$, $E^{(1)}=0.0065\varepsilon$.

- *Comment.* The central 10 per cent of the box lies at the peak (antinode) of ψ_1 and so the energy correction is large. On the other hand, it lies at the node of ψ_2, and its effect on the energy is correspondingly slight. When $n\sim\infty$, $E^{(1)}\sim(a/L)\varepsilon$ and is independent of n. This is because the distribution of the particle is then uniform and is found in the region of length a for a fraction a/L of the time. Note that if $\varepsilon>0$ the energy is increased from $E^{(0)}$.

Now we look for the first-order correction to the wavefunction. To find it, eqn (8.1.19) is multiplied from the left by the bra $\langle k|$, where

$k \neq 0$. Using the orthonormality of the functions again gives

$$\sum_n a_n\{E_n - E_0\}\delta_{kn} = a_k\{E_k - E_0\}$$
$$= E_0^{(1)}\langle k \mid 0\rangle - \langle k|\, H^{(1)}\,|0\rangle$$
$$= -\langle k|\, H^{(1)}\,|0\rangle = -H_{k0}^{(1)}. \tag{8.1.21}$$

Because we have restricted the discussion to a non-degenerate level, $|0\rangle$, all the differences $E_k - E_0$ are non-zero for $k \neq 0$; therefore, the coefficients are given by

$$a_k = H_{k0}^{(1)}/(E_0 - E_k). \tag{8.1.22}$$

The wavefunction corrected to first-order in the perturbation is therefore

$$\textit{First-order wavefunction:}\ \psi \approx \psi_0 + \sum_k{}' \left\{\frac{H_{k0}^{(1)}}{E_0 - E_k}\right\}\psi_k, \tag{8.1.23}$$

where the prime on the summation means that the state with $k = 0$ should be omitted. This is an important equation (and the generalization of eqn (8.1.10)), for we see how the perturbation is guiding us towards the true function. The distortion of the wavefunction is described by mixing the other states of the system into the one of interest, and a loose way of speaking is to say that the perturbation 'induces transitions' to these other states. However, the transitions are not real, they are *virtual*.

The next problem is to find the second-order correction to the energy, and to extract this information from eqn (8.1.17c) we use the same approach as before. The second-order correction to the function is written as

$$\psi_0^{(2)} = \sum_n b_n\psi_n, \tag{8.1.24}$$

and this is inserted into eqn (8.1.17c), which in a ket notation is

$$\sum_n b_n^{(2)}\{E_n - E_0\}\,|n\rangle = \{E_0^{(2)} - H^{(2)}\}\,|0\rangle + \sum_n a_n\{E_0^{(1)} - H^{(1)}\}\,|n\rangle.$$

This equation is now multiplied from the left by $\langle 0|$, and yields

$$\sum_n b_n\{E_n - E_0\}\delta_{0n} = E_0^{(2)} - \langle 0|\, H^{(2)}\,|0\rangle + \sum_n a_n\{E_0^{(1)}\,\delta_{0n} - \langle 0|\, H^{(1)}\,|n\rangle\}.$$

The left-hand side is zero, and so

$$E_0^{(2)} = H_{00}^{(2)} - \sum_n a_n\{E_0^{(1)}\,\delta_{0n} - H_{0n}^{(1)}\} = H_{00}^{(2)} + \sum_n{}' a_n H_{0n}^{(1)}$$

because when $n = 0$ in the sum the quantity in brackets is zero (by eqn (8.1.20)), and when $n \neq 0$ $E_0^{(1)}\delta_{0n}$ disappears. We use eqn (8.1.22) for

a_n. Then

Second-order energy correction:

$$E_0^{(2)} = H_{00}^{(2)} + \sum_n{}' \left\{ \frac{H_{0n}^{(1)} H_{n0}^{(1)}}{E_0 - E_n} \right\}. \tag{8.1.25}$$

The prime signifies the omission from the sum of the state with $n = 0$.

This is a very important result and we shall use it frequently. It is a generalization of the approximate form of the solutions for the 2-level problem. The correction consists of two parts. One, $H_{00}^{(2)}$, is the same kind of average that occurs in the first-order correction, but this time it is the second-order hamiltonian that is averaged over the undistorted wavefunctions. The other part is more involved but may be interpreted as an average of the first-order perturbation taking into account the distortions of the original wavefunctions. The latter are of first order and so the overall contribution is of second order.

Example. The square-well potential was modified by the addition of a contribution of the form $-\varepsilon \sin(\pi x/L)$, resembling a dip of depth ε hanging between $x = 0$ and $x = L$. Find the second-order correction to the energy of the $n = 1$ state by numerical evaluation of the perturbation sum.

- *Method.* Evaluate the matrix elements H_{n0} analytically using the wavefunctions in eqn (3.3.5). Tabulate their numerical values. The denominator is obtained from the energy expression in eqn (3.2.5) and is proportional to $1 - n^2$. Evaluate the terms in the perturbation sum up to about $n = 9$. By symmetry, only odd values of n contribute: this also emerges directly from the analysis.

- *Answer.*

$$H_{n0} = -\varepsilon(2/L) \int_0^L \sin(n\pi x/L)\sin(\pi x/L)\sin(\pi x/L)\,dx$$

$$= (\varepsilon/\pi)\{1/n - 1/2(n+2) - 1/2(n-2)\}\{(-1)^n - 1\}.$$

$$E_1 - E_n = (1 - n^2)(h^2/8mL^2).$$

$$E^{(2)} = -\sum_n{}' \{|H_{n0}|^2/(n^2 - 1)(h^2/8mL^2)\}.$$

$n =$	3	5	7	9	...
$H_{n0}/\varepsilon =$	0.1698	0.0243	0.0081	0.0037	...
$(H_{n0}/\varepsilon)^2/(n^2-1) =$	3.60×10^{-3}	2.45×10^{-5}	1.36×10^{-6}	1.69×10^{-7}	...

Therefore $E^{(2)} = -3.63 \times 10^{-3} \varepsilon^2/(h^2/8mL^2)$.

- *Comment.* The energy of the ground state is lowered: this is a general feature of perturbation theory. The distorted wavefunction may be calculated from eqn (8.1.23) with the matrix elements listed above: we find

$$\psi \approx \psi_1 - \{\varepsilon/(h^2/8mL^2)\}\{2.12 \times 10^{-2}\psi_3 + 1.01 \times 10^{-3}\psi_5 + 1.68 \times 10^{-4}\psi_7 + \ldots\},$$

corresponding (as expected) to an extra accumulation of amplitude in the middle of the well (ψ_3 is negative at $x = \tfrac{1}{2}L$).

We could go on to find the second-order correction to the functions, and then the third-order corrections to the energy. Such corrections are rarely needed. If they are, it suggests that the model system should be chosen to resemble the true system more closely. Furthermore, a useful theorem states that in order to know the energy of a system correct to the $(2n+1)$-order in a perturbation, it is necessary to know only the wavefunctions correct to the nth order. Thus, from the first-order wavefunctions we may calculate the energy correct to third order. A final technical problem ought also to be mentioned: does the perturbation series converge? This is answered affirmatively for most common cases by a theorem due to Rellich and Kato, but normally one just assumes that it does. The bibliography suggests places where this delicate question may be pursued.

The practical difficulty with eqn (8.1.25) is that we do not normally have information about the states and energies that occur in the sum. The sum extends, for instance, over all the states of the unperturbed system, and therefore includes the states of the continuum, if that exists. There are, happily, several aspects of quantum mechanics that alleviate this problem.

In the first place, the contribution of states that differ very much in energy from the one of interest will normally be small, on account of the appearance of the energy differences in the denominator. Other things being equal, only energetically nearby states contribute appreciably to the sum, and the continuum states are often so high in energy (they correspond in atoms and molecules to ionized states) that they can safely be ignored. The difficulty, though, is that although the high energy of states may diminish their importance, there may be a very large number of them, and many small terms in the sum may contribute as much as one or two big ones. In the case of the hydrogen atom, for instance, the degeneracy of levels increases as n^2 (p. 77), and so at $n \approx 1000$ there are 10^6 states of the same energy, each one possibly contributing a little bit to the sum.

In the second place, it may turn out that the matrix elements in the numerators of the sum vanish identically for many states, even those nearby in energy. In the case of a hydrogen atom in a uniform electric field it turns out that for each n all the matrix elements except one vanish identically. Therefore, although at $n \approx 1000$ there are 10^6 states queuing to be included in the sum, only one of them is selected. The vanishing of matrix elements depends on the symmetry properties of the system, and this is where group theory plays such a striking role. The matrix elements $H^{(1)}_{0n}$ are actually integrals of the form $\int \psi_0^* H^{(1)} \psi_n \, d\tau$, and we saw in Chapter 7 that such integrals are necessarily zero unless a particular symmetry condition is satisfied (p. 161). No matrix element contributes to the perturbation sum unless the direct product $\Gamma^{(0)} \times \Gamma^{(\text{pert.})} \times \Gamma^{(n)}$ contains A_1, where $\Gamma^{(0)}$, $\Gamma^{(\text{pert.})}$, and $\Gamma^{(n)}$ are the symmetry species. The physical basis of this important rule (and you can see how important it is if it discards all but one of 10^6 states for one value of n alone) can be understood by considering the distortion in the wavefunction induced by the perturbation. Suppose the state of interest $|0\rangle$ in the unperturbed molecule is totally symmetric (e.g. the 1s-orbital of the hydrogen atom), then the matrix element vanishes unless $A_1 \times \Gamma^{(\text{pert.})} \times \Gamma^{(n)} = \Gamma^{(\text{pert.})} \times \Gamma^{(n)}$ contains A_1; but it does so only if $\Gamma^{(n)} = \Gamma^{(\text{pert.})}$. Therefore, according to eqn (8.1.23), the only states mixed into the original state are those with the same symmetry as the perturbation. In other words, *the distortion picked up by the system has the same symmetry as the disturbing perturbation.* The second-order energy correction takes into account the distortion induced by the perturbation, and so it is natural that it involves contributions only of an appropriate symmetry type. This is the reason why only one state contributes for each n in the hydrogen atom: the perturbation distorts the atom by stretching it in the z-direction, and only p_z-orbitals have the appropriate symmetry to model the effect.

It is often the case that the quantum chemist wants a 'back of the envelope' assessment of the magnitude of a property, but the perturbation expressions seem too complicated for them to be used in this way. So they are; but they themselves can also be approximated. Suppose the arrangement of states is more or less as depicted in Fig. 8.3. We make the approximation that the actual energy differences $E_n - E_0$ in the perturbation expression can be replaced by some kind of mean energy difference ΔE. Then the expression for the second-order correction to the energy becomes

$$E_0^{(2)} \approx H_{00}^{(2)} - (1/\Delta E) \sum_n{}' H_{0n}^{(1)} H_{n0}^{(1)}.$$

The sum is almost in the form of a matrix product. It would be such a product if the sum extended over all n, including the value $n=0$. So we extend the sum, but subtract the term that should not be there:

$$\begin{aligned} E_0^{(2)} &\approx H_{00}^{(2)} - (1/\Delta E) \sum_n H_{0n}^{(1)} H_{n0}^{(1)} - (1/\Delta E) H_{00}^{(1)} H_{00}^{(1)} \\ &\approx H_{00}^{(2)} - (1/\Delta E)(H^{(1)} H^{(1)})_{00} - (1/\Delta E) H_{00}^{(1)} H_{00}^{(1)}. \end{aligned} \tag{8.1.26}$$

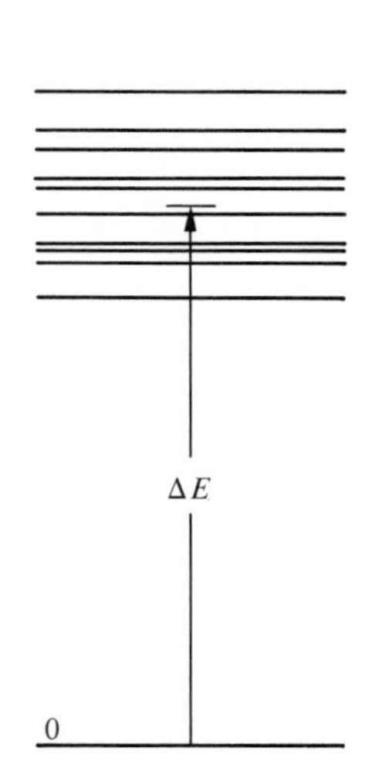

Fig. 8.3. The basis of the closure approximation.

The energy correction is now expressed solely in terms of integrals involving the wavefunction ψ_0, and so we need no information about the excited states. This *closure approximation* expression is very approximate, but it does give an idea of magnitudes. If we write

$$\Delta\varepsilon^2 = \langle 0|\, H^{(1)2}\, |0\rangle - \langle 0|\, H^{(1)}\, |0\rangle^2, \tag{8.1.27}$$

a *mean square deviation of the perturbation*, the second-order energy is simply

$$\textit{Closure approximation:}\quad E_0^{(2)} \approx H_{00}^{(2)} - \Delta\varepsilon^2/\Delta E, \tag{8.1.28}$$

an expression that we shall use several times in Part 4.†

† If ΔE is identified with the *minimum* excitation energy of the system, then $\Delta\varepsilon^2/\Delta E$ is larger than the true correction term. Therefore $H_{00}^{(2)} - \Delta\varepsilon^2/\Delta E$ is then a *lower bound* to the true energy correction. This is *Unsöld's theorem*. Since the variation method leads to an upper bound, in combination they locate the true energy to within a range.

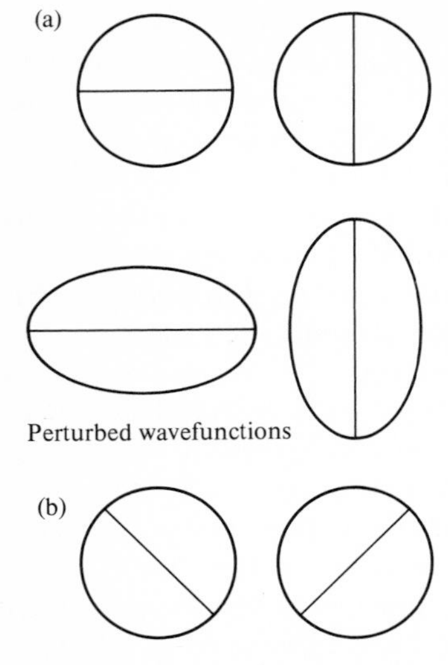

Fig. 8.4. (a) Good, (b) poor choices of a degenerate-basis linear combination for a perturbation leading to the wavefunctions shown in the middle.

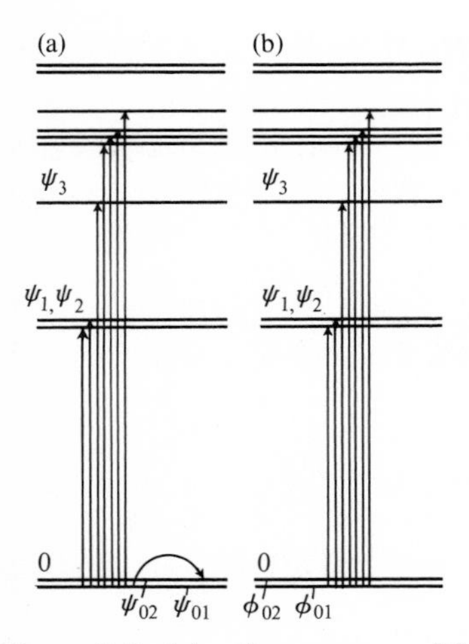

Fig. 8.5. (a) Improper, (b) proper choices of linear combinations of degenerate states.

8.2 Perturbation theory for degenerate states

Figure 8.3 has warned us that we shall get totally wrong answers from perturbation theory for systems in which the level of interest is degenerate, unless, that is, we are circumspect and avoid energy denominators going to zero. If we want to apply perturbation theory we must start with wavefunctions that most closely resemble the function in the distorted system, otherwise we have to deal with large changes and should not expect first- or second-order corrections to be adequate. For example, Fig. 8.4(a) shows the two linear combinations of a degenerate pair that provides a sensible set to perturb in the manner shown. Any linear combination of degenerate functions is an eigenfunction of the same energy (p. 85), and so the starting point for the perturbation calculation could, in principle, be any linear combination of the functions depicted in Fig. 8.4(a). But a pair of linear combinations like those shown in Fig. 8.4(b) would be quite unsuitable, for under the perturbation they change a great deal. There are therefore two points to watch. One is that we avoid energy differences of the two components of a degenerate level occurring in the denominator of any expressions, Fig. 8.5, and the other is to ensure that we start with sensible linear combinations, Fig. 8.4.

We suppose that the level of interest is r-fold degenerate and that the functions corresponding to the energy E_0 are ψ_{0l} $(l=1, 2, \ldots, r)$. We use the ket notation $|0l\rangle$ for the states when convenient. All r-functions satisfy the equations

$$H^{(0)}\psi_{0l} = E_0\psi_{0l}, \qquad l=1, 2, \ldots, r. \tag{8.2.1}$$

In the perturbed system the hamiltonian is H and the functions ψ_i; these might be non-degenerate. We shall suppose that the correct initial combinations of the degenerate unperturbed functions, the ones which (a) most closely resemble the perturbed functions and (b) which give identically zero matrix elements $\langle 0l'|\, H^{(1)}\, |0l\rangle$ among themselves, are ϕ_{0i}:

$$\phi_{0i} = \sum_{l=1}^{r} c_{il}\psi_{0l}, \quad \text{all of energy } E_0. \tag{8.2.2}$$

When the perturbation is applied we suppose that the combination ϕ_{0i} is distorted into ψ_i, which it closely resembles, and its energy changes from E_0 to a similar value, E_i. Observe that we now need the index i on the energy because the degeneracy might be lifted by the perturbation. As in Section 8.1 we write

$$\psi_i = \phi_{0i} + \lambda\psi_{0i}^{(1)} + \ldots \tag{8.2.3}$$

$$E_i = E_0 + \lambda E_{0i}^{(1)} + \ldots \tag{8.2.4}$$

If the previous procedure of substituting these into the equation

$$H\psi_i = E_i\psi_i \tag{8.2.5}$$

is repeated and powers of λ collected we obtain

$$H^{(0)}\phi_{0i} = E_0\phi_{0i}, \tag{8.2.6a}$$

$$\{H^{(0)} - E_0\}\psi_{0i}^{(1)} = \{E_{0i}^{(1)} - H^{(1)}\}\phi_{0i}. \tag{8.2.6b}$$

As before, we attempt to express the first-order correction to the wavefunction as a sum over all functions. The simplest procedure is to divide the sum into two parts, one being a sum over the members of the degenerate set ψ_{0l} and the other the sum over all other states (which may be degenerate among themselves):

$$\psi_{0i}^{(1)} = \sum_l a_l\psi_{0l} + \sum_n{}' a_n\psi_n. \tag{8.2.7}$$

This should now be inserted into eqn (8.2.6b), and changing to a ket notation we obtain

$$\sum_l a_l\{E_0 - E_0\}\,|0l\rangle + \sum_n{}' a_n\{E_n - E_0\}\,|n\rangle = \sum_l c_{il}\{E_{0i}^{(1)} - H^{(1)}\}\,|0l\rangle.$$

The first term is plainly zero. On multiplying the remaining terms from the left by the bra $\langle 0k|$ we obtain zero on the left ($\langle 0k \mid n\rangle = 0$ because $|0k\rangle$ and $|n\rangle$ are orthogonal), and hence

$$\sum_l c_{il}\{E_{0i}^{(1)}\langle 0k \mid 0l\rangle - \langle 0k|\,H^{(1)}\,|0l\rangle\} = 0.$$

The degenerate functions need not be orthogonal (but can always be chosen to be so). We therefore write

$$S_{kl} \equiv \langle 0k \mid 0l\rangle = \int \psi_{0k}^{*}\psi_{0l}\,\mathrm{d}\tau \tag{8.2.8}$$

and so obtain the

Secular equations: $\sum_l c_{il}\{E_{0i}^{(1)}S_{kl} - \langle 0k|\,H^{(1)}\,|0l\rangle\} = 0.$ (8.2.9)

We call the matrix element $\langle 0k|\,H^{(1)}\,|0l\rangle$ simply $H_{kl}^{(1)}$ and observe that the last equation is really a set of r ($i = 1, 2, \ldots, r$) simultaneous equations for the coefficients c_{il}, and that the set has a non-trivial solution if the determinant of the coefficients of the c_{il} vanishes, that is if the

Secular determinant: $\det|H_{kl}^{(1)} - E_{0i}^{(1)}S_{kl}| = 0.$ (8.2.10)

The solution of this equation gives the energies $E_{0i}^{(1)}$. This result may be compared with eqn (8.1.20) into which it reduces when the level 0 is non-degenerate. It is always possible to choose the functions in a degenerate set to be mutually orthogonal, and if that is done the quantities S_{kl} may be replaced by δ_{kl}.

As a simple example, consider the case when the level of interest is doubly degenerate: the effect of a perturbation on the energy is then given by

$$\begin{vmatrix} H_{11}^{(1)} - E_{0i}^{(1)} & H_{12}^{(1)} \\ H_{21}^{(1)} & H_{22}^{(1)} - E_{0i}^{(1)} \end{vmatrix} = 0.$$

This is the same as eqn (8.1.5) in the case when $E_1 = E_2$, and so the two energies $E_{0i}^{(1)}$ ($i = 1, 2$ or $\pm$ in the earlier notation) are given by the same expression, and the degeneracy is lifted as in that example.

The coefficients c_{il} may be determined from eqn (8.2.9), using the energies obtained by solving the secular determinant. The effect of the mixing of the functions corresponding to *different energies* may be found by applying non-degenerate perturbation theory in the manner already described.

8.3 Variation theory

Variation theory is a way of checking and improving guesses about the forms of wavefunctions in complicated systems. In the first place it provides a criterion of excellence for judging the quality of the guess (which is called, more formally, the *trial function* ψ_{trial}) and also indicates how to optimize it. Suppose the system is described by a hamiltonian H, and that the lowest eigenvalue (i.e. the energy of the ground state of the true system) is E_0. Evaluate the

Rayleigh ratio: $\mathscr{E} = \int \psi_{\text{trial}}^* H \psi_{\text{trial}}\, d\tau \Big/ \int \psi_{\text{trial}}^* \psi_{\text{trial}}\, d\tau.$ (8.3.1)

(This is just a matter of evaluating the two integrals; that is, it is a *quadrature*, it doesn't involve solving any differential equations.) Then:

The variation theorem: $\mathscr{E} \geq E_0$ for any ψ_{trial}. (8.3.2)

The equality holds only if ψ_{trial} is the true ground state wavefunction ψ_0.

The proof of the theorem runs as follows. The trial function can be written as a linear combination of the true eigenfunctions of the hamiltonian (they form a complete set, p. 84):

$$\psi_{\text{trial}} = \sum_n c_n \psi_n; \qquad H\psi_n = E_n \psi_n.$$

Now consider the integral

$$\begin{aligned} \int \psi_{\text{trial}}^* (H - E_0) \psi_{\text{trial}}\, d\tau &= \sum_n \sum_{n'} c_n^* c_{n'} \int \psi_n^* (H - E_0) \psi_{n'}\, d\tau \\ &= \sum_n \sum_{n'} c_n^* c_{n'} (E_{n'} - E_0) \int \psi_n^* \psi_{n'}\, d\tau \\ &= \sum_n c_n^* c_n (E_n - E_0) \geq 0. \end{aligned}$$

This expression is non-negative because $E_n \geq E_0$ and $|c_n|^2$ is non-negative. Therefore

$$\int \psi^*_{\text{trial}}(H - E_0)\psi_{\text{trial}}\, d\tau \geq 0,$$

which is then readily reorganized into $\mathscr{E} \geq E_0$, completing the proof.

The significance of the variation theorem is that whatever trial function we take, the 'energy' calculated from it (i.e. the Rayleigh ratio $\mathscr{E}$) is never less than the true energy of the ground state of the system. This is normally interpreted as meaning that the smaller the value of $\mathscr{E}$, the closer it is to the true ground-state energy, and therefore the more closely ψ_{trial} resembles the true ground-state wavefunction. Therefore, in order to arrive at the wavefunction of some system, a trial function is set up in a form that is sufficiently flexible—it might be expressed in terms of several parameters, such as p_1 and p_2 in the function $x^{p_1}e^{-p_2x}$, and then the parameters are varied until the Rayleigh ratio takes its minimum value, Fig. 8.6. These values of the parameters are then used to construct the 'best' wavefunction of that form. The variation of the parameters and the search for the values corresponding to a minimum of $\mathscr{E}$ is clearly a job for the differential calculus, and the procedure involves solving the set of equations $(\partial\mathscr{E}/\partial p_1) = 0$, $(\partial\mathscr{E}/\partial p_2) = 0, \ldots,$ where the p_i are the parameters. This procedure is illustrated in the following *Example*.

Example. Find the optimum form and the corresponding energy of a trial function e^{-kr} for the ground state of a hydrogen-like atom of atomic number Z.

- *Method.* The hamiltonian is $-(\hbar^2/2\mu)\nabla^2 - Ze^2/4\pi\varepsilon_0 r$; since the trial function is spherically symmetrical we can replace ∇^2 by its radial part, eqn (4.2.3). Evaluate the Rayleigh ratio, and find its minimum value by solving $d\mathscr{E}/dk = 0$.
- *Answer.*

$$\int \psi^*_{\text{trial}}\psi_{\text{trial}}\, d\tau = \int_0^{2\pi} d\phi \int_0^{\pi} \sin\theta\, d\theta \int_0^{\infty} e^{-2kr}r^2\, dr = (\pi/k^3).$$

$$\int \psi^*_{\text{trial}}(1/r)\psi_{\text{trial}}\, d\tau = \int_0^{2\pi} d\phi \int_0^{\pi} \sin\theta\, d\theta \int_0^{\infty} e^{-2kr}(1/r)r^2\, dr = \pi/k^2.$$

$$\int \psi^*_{\text{trial}}\nabla^2\psi_{\text{trial}}\, d\tau = \int \psi^*_{\text{trial}}\{(1/r)(d^2/dr^2)re^{-kr}\}\, d\tau$$

$$= \int \psi^*_{\text{trial}}\{k^2 - (2k/r)\}\psi_{\text{trial}}\, d\tau = \pi/k - 2\pi/k = -\pi/k.$$

Therefore,

$$\int \psi^*_{\text{trial}}H\psi_{\text{trial}}\, d\tau = (\hbar^2/2\mu)(\pi/k) - (Ze^2/4\pi\varepsilon_0)(\pi/k^2).$$

$$\mathscr{E} = \frac{(\hbar^2/2\mu)(\pi/k) - (Ze^2/4\pi\varepsilon_0)(\pi/k^2)}{(\pi/k^3)} = (\hbar^2/2\mu)k^2 - (Ze^2/4\pi\varepsilon_0)k.$$

$$d\mathscr{E}/dk = 2(\hbar^2/2\mu)k - (Ze^2/4\pi\varepsilon_0)$$

$$= 0 \text{ when } k = Ze^2\mu/4\pi\varepsilon_0\hbar^2.$$

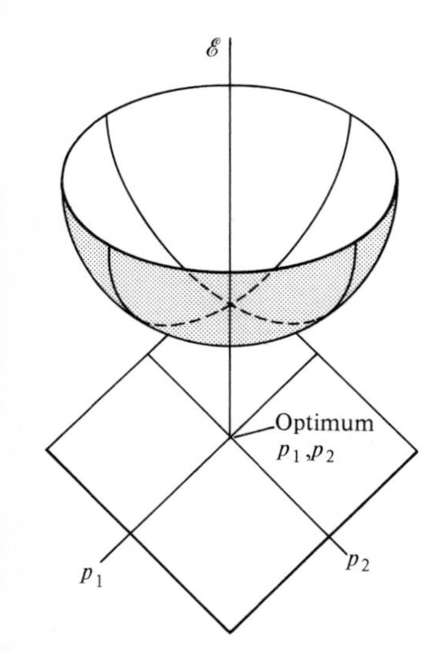

Fig. 8.6. Two-parameter optimization.

The best value of $\mathscr{E}$ is therefore $\mathscr{E}=-Z^2e^4\mu/32\pi^2\varepsilon_0^2\hbar^2$ and the optimum form of the wavefunction is $\exp(-Ze^2\mu r/4\pi\varepsilon_0\hbar^2)$.

• *Comment.* The 'optimum' energy turns out to be the exact energy in this case. Had we chosen $\psi_{\text{trial}}=\mathrm{e}^{-kr^2}$ (see Problems) $\mathscr{E}$ would have been less negative and the trial function 'less good'.

The basic variation procedure is due to Rayleigh (like so many things). The modification of using a trial function expressed as a linear combination of functions $\psi_{\text{trial}}=c_1\psi_1+c_2\psi_2+\ldots$ in which the combination coefficients are the parameters to be varied is due to Ritz and is called the *Rayleigh–Ritz method.* The trial function is taken to be

$$\psi_{\text{trial}}=\sum_i c_i\psi_i \tag{8.3.3}$$

where the functions $\psi_1, \psi_2, \ldots$ are frozen throughout the calculation, and constitute the *basis set.* The Rayleigh ratio is

$$\mathscr{E}=\frac{\int\psi^*_{\text{trial}}H\psi_{\text{trial}}\,\mathrm{d}\tau}{\int\psi^*_{\text{trial}}\psi_{\text{trial}}\,\mathrm{d}\tau}=\frac{\sum_i\sum_j c_ic_j\int\psi_iH\psi_j\,\mathrm{d}\tau}{\sum_i\sum_j c_ic_j\int\psi_i\psi_j\,\mathrm{d}\tau}=\frac{\sum_i\sum_j c_ic_jH_{ij}}{\sum_i\sum_j c_ic_jS_{ij}},$$

where we have assumed that the coefficients and basis functions are real. In order to find the minimum value of $\mathscr{E}$ we differentiate with respect to each coefficient in turn and set $(\partial\mathscr{E}/\partial c_k)=0$ in each case. Thus:

$$(\partial\mathscr{E}/\partial c_k)=\frac{\sum_j c_jH_{kj}+\sum_i c_iH_{ik}}{\sum_i\sum_j c_ic_jS_{ij}}-\frac{\left\{\sum_j c_jS_{kj}+\sum_i c_iS_{ik}\right\}\sum_i\sum_j c_ic_jH_{ij}}{\left\{\sum_i\sum_j c_ic_jS_{ij}\right\}^2}$$

$$=\frac{\sum_j c_j\{H_{kj}-\mathscr{E}S_{kj}\}}{\sum_i\sum_j c_ic_jS_{ij}}+\frac{\sum_i c_i\{H_{ik}-\mathscr{E}S_{ik}\}}{\sum_i\sum_j c_ic_jS_{ij}}=0.$$

This is satisfied if the numerators each vanish. Therefore we must solve the secular equations

$$\sum_i c_i(H_{ik}-\mathscr{E}S_{ik})=0, \tag{8.3.4}$$

a set of simultaneous equations for the coefficients c_i. The condition for existence of solutions is that the secular determinant should be zero:

$$\det|H_{ik}-\mathscr{E}S_{ik}|=0. \tag{8.3.5}$$

Solving this determinant leads to a set of values of $\mathscr{E}$ (for example, in a basis of two members, the determinant expands into a quadratic equation, and there are two values of $\mathscr{E}$, its two roots): the lowest value of $\mathscr{E}$ is the best value of the energy of the ground state of the system with the basis set in the selected form. The coefficients of the functions corresponding to this best energy are then found by solving the set of secular equations with this value of $\mathscr{E}$. The procedure is illustrated in the following *Example.*

Example. How does the mass of the nucleus affect the distribution of the electron in the hydrogen atom? If the mass of the nucleus were infinite the 1s-orbital would be $\psi_1=(1/\pi a_0^3)^{\frac{1}{2}}e^{-r/a_0}$ with $a_0=4\pi\varepsilon_0\hbar^2/m_e e^2$ and the 2s-orbital would be $(1/32\pi a_0^3)^{\frac{1}{2}}(2-r/a_0)e^{-r/2a_0}$. Express the true ground state wavefunctions as $c_1\psi_1+c_2\psi_2$ and find the optimum values of c_1 and c_2.

- *Method.* The basis functions are orthonormal, and so $S_{11}=S_{22}=1$, $S_{12}=S_{21}=0$. The hamiltonian is $-(\hbar^2/2\mu)\nabla^2-(e^2/4\pi\varepsilon_0)(1/r)$. Since ψ_i are spherically symmetrical, use $\nabla^2=(1/r)(\mathrm{d}^2/\mathrm{d}r^2)r$. Express energies in terms of $hcR_\infty=\hbar^2/2a_0^2m_e$. Evaluate H_{11}, H_{22}, H_{12}, H_{21} $(=H_{12})$; set up the 2×2 secular determinant; find the lower energy. Solve the secular equations for c_2 using eqn (8.3.4) and $c_1^2+c_2^2=1$.

- *Answer.*

$$H_{11}/hcR_\infty=\gamma-1,\qquad H_{22}/hcR_\infty=\tfrac{1}{4}(\gamma-1),$$
$$H_{12}/hcR_\infty=16\gamma/27\surd2;\qquad \gamma=m_e/m_p.$$

The secular determinant expands as follows:

$$\begin{vmatrix} H_{11}-\mathscr{E}S_{11} & H_{12}-\mathscr{E}S_{12} \\ H_{21}-\mathscr{E}S_{21} & H_{22}-\mathscr{E}S_{22}\end{vmatrix}=\begin{vmatrix} H_{11}-\mathscr{E} & H_{12} \\ H_{21} & H_{22}-\mathscr{E}\end{vmatrix}$$
$$=\mathscr{E}^2-(H_{11}+H_{22})\mathscr{E}+(H_{11}H_{22}-H_{12}H_{21})=0.$$

Substitution of the matrix elements gives the lower root

$$\mathscr{E}/hcR_\infty=\tfrac{1}{8}(\gamma-1)\{5+3(1+2\Gamma^2)^{\frac{1}{2}}\},\qquad \Gamma=(2^6/3^4)\gamma/(\gamma-1)=-0.000\,43.$$

Hence $\mathscr{E}/hcR_\infty=-0.999\,46$. The secular equations are

$$c_1(H_{11}-\mathscr{E})+c_2H_{21}=0;\qquad c_1H_{12}+c_2(H_{22}-\mathscr{E})=0,$$

and with the lower value of $\mathscr{E}$ (as above) and $c_1^2+c_2^2=1$,

$$c_2=-0.000\,54,\qquad c_2^2=3.0\times10^{-7}.$$

- *Comment.* The wavefunction has a 3.0×10^{-5} per cent admixture of 2s-orbital and the phase is negative. Since $\psi_1(0)=(1/\pi a_0^3)^{\frac{1}{2}}$ and $\psi_2(0)=(1/32\pi a_0^3)^{\frac{1}{2}}$, this represents a slight reduction of the amplitude of the electron at the nucleus from $(1/\pi a_0^3)^{\frac{1}{2}}$ if the nucleus were infinite, to $(1/\pi a_0^3)^{\frac{1}{2}}+c_2(1/32\pi a_0^3)^{\frac{1}{2}}$ when it is finite. Since $\mu<m_e$ this represents the greater freedom of the less massive 'effective' particle.

Note that the variation principle gives an *upper bound* to the energy of the ground state (E_0 is not greater than $\mathscr{E}$). There are also variational techniques for finding lower bounds, and so the true energy can be sandwiched from above and below and located more definitely. These calculations are often quite difficult because they involve integrals over the square of the hamiltonian. Finally, a word of caution. Although the variation method may give a good value for the energy, there is no guarantee that the optimum trial wavefunction will give a good value for some other property of the system (e.g. its dipole moment).

8.4 The Hellmann–Feynman theorem

The variation theorem takes a fixed hamiltonian and investigates how the Rayleigh ratio varies as the form of the trial function is modified. The Hellmann–Feynman theorem investigates how the energy of a system varies when the *hamiltonian* changes.

Consider a system characterized by a hamiltonian that depends on a parameter P. This parameter might be the internuclear distance in a molecule, or the strength of the electric field to which the molecule is exposed. The exact (not trial) wavefunction describing the system is a solution of the Schrödinger equation, and so it depends on the value of the parameter P. The energy of the system also depends on P. The question arises as to how the energy varies when the parameter P is modified: is it necessary to solve the Schrödinger equation for each value of P, or can the response of the system to changes of P be expressed more simply?

Consider how the expectation value of the hamiltonian varies with P. We suppose that the wavefunctions are normalized to unity for all values of P; then $E(P)=\int \psi^* H\psi \,\mathrm{d}\tau$ with H, ψ^*, and ψ all depending on P. The derivative of E with respect to P is

$$\begin{aligned}
\mathrm{d}E/\mathrm{d}P &= \int (\partial\psi^*/\partial P)H\psi\,\mathrm{d}\tau + \int \psi^*(\partial H/\partial P)\psi\,\mathrm{d}\tau + \int \psi^* H(\partial\psi/\partial P)\,\mathrm{d}\tau \\
&= E\int (\partial\psi^*/\partial P)\psi\,\mathrm{d}\tau + \int \psi^*(\partial H/\partial P)\psi\,\mathrm{d}\tau + E\int \psi^*(\partial\psi/\partial P)\,\mathrm{d}\tau \\
&= E(\mathrm{d}/\mathrm{d}P)\int \psi^*\psi\,\mathrm{d}\tau + \int \psi^*(\partial H/\partial P)\psi\,\mathrm{d}\tau.
\end{aligned}$$

The first term is zero because the integral is equal to 1 and so its derivative is zero. The second term is the expectation value of $(\partial H/\partial P)$. Therefore we have arrived at

The Hellmann–Feynman theorem: $\mathrm{d}E/\mathrm{d}P=\langle(\partial H/\partial P)\rangle.$ (8.4.1)

The great advantage of this theorem is that the operator $(\partial H/\partial P)$ might be extremely simple in form. For instance, if the total hamiltonian has the form $H=H^{(0)}+Px$, then $(\partial H/\partial P)=x$, and there is no mention of $H^{(0)}$, which might be a complicated molecular hamiltonian. In this example, $\mathrm{d}E/\mathrm{d}P$ is equal to the expectation value of x itself, which is a trivial computational problem.

There is (as always) a complication. The proof of the theorem has supposed that the wavefunctions are the *exact* eigenfunctions of the *total* hamiltonian. Therefore, in order to work out the expectation value of even a simple operator like x we need to have solved the Schrödinger equation for the complete, complicated hamiltonian. Nevertheless, we can use the perturbation theory outlined earlier in the chapter to arrive at successively better approximations to the true wavefunctions, and therefore at successively better approximations to the value of $\mathrm{d}E/\mathrm{d}P$, the response of the system to changes in the hamiltonian. We shall use this technique in Part 4 when we examine the behaviour of molecules in electric and magnetic fields.

8.5 Time-dependent perturbation theory

Just about every perturbation in chemistry is time-dependent, even those that appear to be stationary. This is because perturbations almost always have to be turned on: samples are inserted into electric and magnetic fields, the shapes of vessels are changed, and so on. The reason why time-independent perturbation theory can be applied successfully, is that the molecular response is so rapid that for all practical purposes the systems forget that they were ever unperturbed, and settle down into their perturbed states as though the switching had never occurred. Nevertheless, if we really want to understand quantum mechanics, we should see how the systems respond to a newly imposed perturbation, and then settle down to a stationary state after it has been present for long enough.

But there is a much more important reason for studying time-dependent perturbation theory. This is because many important perturbations never 'settle down' to a constant value: a molecule exposed to electromagnetic radiation is an especially important example. In this case the molecule experiences an electric field that oscillates for as long as the light is present. Time-dependent perturbation theory has to be used to treat such problems and time-independent theory is totally inadequate. The kind of problem that arises in this connection is the discussion of transition probabilities between states, the rates of transitions, and the intensities of spectral lines. The topic is therefore central to the whole of spectroscopy.

We shall use the same approach as for time-independent perturbations. First we consider a two-level system, and then generalize that to systems of arbitrary complexity.

The total hamiltonian of the system is $H = H^{(0)} + H^{(1)}(t)$, the difference between this and the earlier case being that the perturbation is time-dependent (it might be of the form $H^{(1)}(t) = A \cos \omega t$ if the perturbation is one that oscillates at a frequency ω). Since the hamiltonian is time-dependent, we have to deal with the solutions of the time-dependent Schrödinger equation

$$H\Psi = \mathrm{i}\hbar(\partial \Psi/\partial t). \tag{8.5.1}$$

Consider a two-level unperturbed system. As in the earlier part of the chapter, we denote its two energies as E_1 and E_2 and the corresponding eigenfunctions as ψ_1 and ψ_2. These are the solutions of $H^{(0)}\psi_n = E_n\psi_n$. We saw in Chapter 2 how to form the time-dependent form of these functions:

$$\Psi_n(t) = \psi_n \mathrm{e}^{-\mathrm{i}E_n t/\hbar}. \tag{8.5.2}$$

In the presence of the perturbation $H^{(1)}(t)$ the state of the system is described by a linear combination of the basis functions:

$$\Psi(t) = a_1(t)\Psi_1(t) + a_2(t)\Psi_2(t). \tag{8.5.3}$$

Notice that the coefficients are also time-dependent, and so the overall time-dependence of Ψ arises both from the oscillating time-dependence of the basis functions and from the time-dependence of the coefficients. The latter represents the fact that in general the system evolves under the influence of the perturbation. If it starts as pure state 1 its complexion may change, and in due course it may become pure state 2. The probability that at any instant the system is in state 2 is $|a_2(t)|^2$ (and the probability that it remains in state 1 is $|a_1(t)|^2 = 1 - |a_2(t)|^2$).

Substitution of the linear combination into the Schrödinger equation leads to the following expression:

$$\begin{aligned} H\Psi &= a_1 H^{(0)}\Psi_1 + a_1 H^{(1)}(t)\Psi_1 + a_2 H^{(0)}\Psi_2 + a_2 H^{(1)}(t)\Psi_2 \\ &= \mathrm{i}\hbar(\partial/\partial t)(a_1\Psi_1 + a_2\Psi_2) \\ &= \mathrm{i}\hbar a_1(\partial\Psi_1/\partial t) + \mathrm{i}\hbar(\mathrm{d}a_1/\mathrm{d}t)\Psi_1 + \mathrm{i}\hbar a_2(\partial\Psi_2/\partial t) + \mathrm{i}\hbar(\mathrm{d}a_2/\mathrm{d}t)\Psi_2. \end{aligned}$$

Each basis function satisfies

$$H^{(0)}\Psi_n = \mathrm{i}\hbar(\partial\Psi_n/\partial t),$$

and so the last equation simplifies to

$$a_1 H^{(1)}(t)\Psi_1 + a_2 H^{(1)}(t)\Psi_2 = \mathrm{i}\hbar\dot{a}_1\Psi_1 + \mathrm{i}\hbar\dot{a}_2\Psi_2, \tag{8.5.4}$$

where $\dot{a} = \mathrm{d}a/\mathrm{d}t$.

The next job is to extract from this equation, equations for the rates of change of the individual coefficients. We use the orthonormality of the basis functions. First, express the equation with its time-dependent and time-independent factors written explicitly:

$$\begin{aligned} a_1 H^{(1)}(t)\psi_1 \mathrm{e}^{-\mathrm{i}E_1 t/\hbar} &+ a_2 H^{(1)}(t)\psi_2 \mathrm{e}^{-\mathrm{i}E_2 t/\hbar} \\ &= \mathrm{i}\hbar\dot{a}_1\psi_1\mathrm{e}^{-\mathrm{i}E_1 t/\hbar} + \mathrm{i}\hbar\dot{a}_2\psi_2\mathrm{e}^{-\mathrm{i}E_2 t/\hbar}. \end{aligned}$$

Then multiply through by ψ_1^* and integrate over all space. Since ψ_1 and ψ_2 are orthonormal

$$a_1 H_{11}^{(1)}(t)\mathrm{e}^{-E_1 t/\hbar} + a_2 H_{12}^{(1)}(t)\mathrm{e}^{-\mathrm{i}E_2 t/\hbar} = \mathrm{i}\hbar\dot{a}_1\mathrm{e}^{-\mathrm{i}E_1 t/\hbar},$$

where

$$H_{ij}^{(1)}(t) = \int \psi_i^* H^{(1)}(t)\psi_j \, \mathrm{d}\tau.$$

This expression can be tidied up in two ways. In the first place the exponential on the right can be moved to the left; the term $E_2 - E_1$ that then appears as an exponent will be written $\hbar\omega_0$. Secondly, as in the time-independent calculation, we suppose that the perturbation has no diagonal elements, and set $H_{11}^{(1)}(t) = H_{22}^{(1)}(t) = 0$ at all times. (This condition is satisfied by many types of common perturbation.) The equation for $\dot{a}_1$ now becomes

$$\dot{a}_1 = (1/\mathrm{i}\hbar)a_2 H_{12}^{(1)}(t)\mathrm{e}^{-\mathrm{i}\omega_0 t}. \tag{8.5.5a}$$

The solution of this equation depends on the time-dependence of the coefficient a_2; therefore we need its equation too. Repeating the procedure but multiplying through by ψ_2^* instead of ψ_1^* leads to

$$\dot{a}_2 = (1/\mathrm{i}\hbar)a_1 H_{21}^{(1)}(t)\mathrm{e}^{\mathrm{i}\omega_0 t}. \tag{8.5.5b}$$

In the case where the perturbation is absent, so that its matrix elements are zero, we have $\dot{a}_1 = 0$ and $\dot{a}_2 = 0$. Hence, whatever the initial state of the system (i.e. whatever the initial values of a_1 and a_2),

$$\Psi = a_1\psi_1\mathrm{e}^{-\mathrm{i}E_1 t/\hbar} + a_2\psi_2\mathrm{e}^{-\mathrm{i}E_2 t/\hbar}; \qquad a_1, a_2 \text{ constant}, \tag{8.5.6}$$

and although Ψ oscillates, the probability of finding the system in one state or the other remains constant (e.g. the probability of it being found in ψ_1 is $|a_1\mathrm{e}^{-\mathrm{i}E_1 t/\hbar}|^2 = |a_1|^2$, a constant).

Now consider the case where a *constant perturbation* is switched on at $t = 0$. It has constant, non-zero matrix elements $H_{12}^{(1)}$ and $H_{21}^{(1)}$ between the two states for all times $t \geq 0$ until it is switched off. Before the perturbation is switched on, $\dot{a}_1 = \dot{a}_2 = 0$. After $t = 0$

$$\dot{a}_1 = (1/\mathrm{i}\hbar)a_2 H_{12}^{(1)}\mathrm{e}^{-\mathrm{i}\omega_0 t}, \qquad \dot{a}_2 = (1/\mathrm{i}\hbar)a_1 H_{21}^{(1)}\mathrm{e}^{\mathrm{i}\omega_0 t}. \tag{8.5.7}$$

There are several ways of solving coupled differential equations. The most elementary involves substituting one into the other (the best method is to use Laplace transforms). By differentiating $\dot{a}_2$ and using the expression for $\dot{a}_1$ we find:

$$\begin{aligned}\ddot{a}_2 &= (1/\mathrm{i}\hbar)\dot{a}_1 H_{21}^{(1)}\mathrm{e}^{\mathrm{i}\omega_0 t} + \mathrm{i}\omega_0(1/\mathrm{i}\hbar)a_1 H_{21}^{(1)}\mathrm{e}^{\mathrm{i}\omega_0 t}\\ &= (1/\mathrm{i}\hbar)^2 a_2 H_{12}^{(1)}H_{21}^{(1)} + \mathrm{i}\omega_0\dot{a}_2.\end{aligned}$$

For simplicity we write $H_{12}^{(1)}H_{21}^{(1)} = \hbar^2 V^2$, then

$$\ddot{a}_2 = -V^2 a_2 + \mathrm{i}\omega_0\dot{a}_2.$$

(The corresponding expression for $\ddot{a}_1$ is obtained by differentiating $\dot{a}_1$ a second time.) The solutions are

$$a_2 = (A\mathrm{e}^{\mathrm{i}\Omega t} + B\mathrm{e}^{-\mathrm{i}\Omega t})\mathrm{e}^{\frac{1}{2}\mathrm{i}\omega_0 t}, \qquad \Omega = \tfrac{1}{2}(\omega_0^2 + 4V^2)^{\frac{1}{2}} \tag{8.5.8}$$

where A and B are constants determined by the initial conditions. A similar expression holds for a_1. Now suppose that at $t = 0$ the system was definitely in state 1; then $a_1(0) = 1$ and $a_2(0) = 0$. These initial conditions are sufficient to determine A and B. After some straightforward algebra this leads to

$$\begin{aligned}a_1 &= \{\cos\Omega t - \mathrm{i}(\omega_0/2\Omega)\sin\Omega t\}\mathrm{e}^{\frac{1}{2}\mathrm{i}\omega_0 t};\\ a_2 &= -\mathrm{i}(V/\Omega)\sin\Omega t\,\mathrm{e}^{\frac{1}{2}\mathrm{i}\omega_0 t},\end{aligned} \tag{8.5.9}$$

which lets us calculate the coefficients at any time after the perturbation was switched on.

We are interested in the probability of finding the system in one of

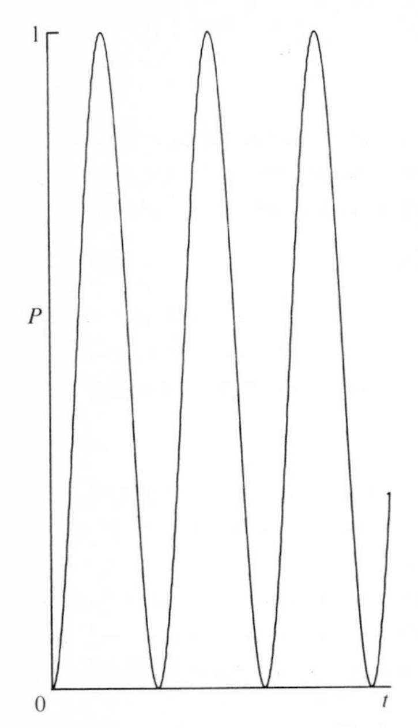

Fig. 8.7. Probability oscillation for degenerate states.

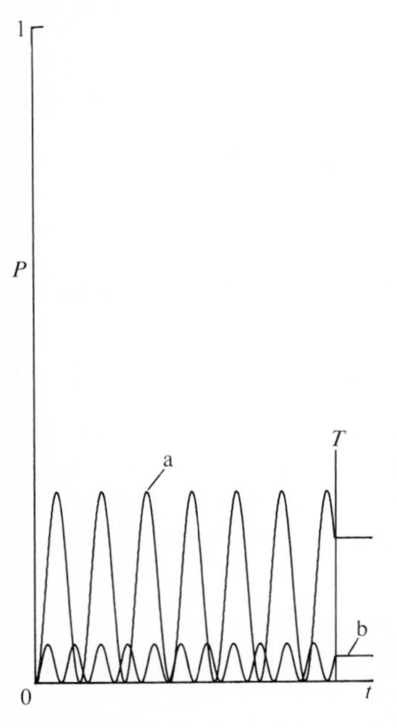

Fig. 8.8. Probability oscillation for non-degenerate states: (a) small energy separation; (b) larger energy separation.

the states as a function of time. These probabilities are given by $P_1 = |a_1|^2$ and $P_2 = |a_2|^2$ (and, because the system must be in one of the two states, $|a_2|^2 = 1 - |a_1|^2$). For state 2, the initially unoccupied state, we find the

$$\textit{Rabi formula}: \; P_2 = |a_2|^2 = \left(\frac{4V^2}{\omega_0^2 + 4V^2}\right)\sin^2\tfrac{1}{2}(\omega_0^2 + 4V^2)^{\frac{1}{2}}t, \quad (8.5.10)$$

an expression at the centre of the following discussion.

Consider first the case of a system in which $E_1 = E_2$ so that $\omega_0 = 0$. The probability that the system is in state 2 if initially it was certainly in state 1 is

$$P_2 = \sin^2 Vt. \quad (8.5.11)$$

This function is plotted in Fig. 8.7. The system oscillates between the two states, and periodically may be found with certainty in state 2. The frequency with which it reappears in state 2 is $2V$: strong perturbations drive the system between the states more quickly than weak perturbations. The fact that the system can make the transition completely to the second state, however weak the perturbation ($\sin^2 Vt$ will reach 1 if we wait until $t = \pi/2V$), is a special property of degenerate systems. Degenerate systems are 'loose', and *populations* (i.e. the occupation probabilities) can be driven completely between degenerate states with even the weakest perturbations.

Now consider the other extreme, when the separation of energy levels is large compared to the strength of the perturbation, in the sense $\omega_0^2 \gg 4V^2$. Then we can neglect $4V^2$ in comparison with ω_0^2 in both the denominator and the sine function, and obtain

$$P_2 \approx (2V/\omega_0)^2 \sin^2\tfrac{1}{2}\omega_0 t. \quad (8.5.12)$$

The behaviour is now quite different. The population still oscillates, but its maximum value is only $4V^2/\omega_0^2 \ll 1$. There is only a small probability that the weak perturbation will drive the system into state 2. The frequency of the oscillation is now determined solely by the energy separation: as the separation increases the frequency increases, Fig. 8.8, but the maximum value of P_2 decreases. In the limit of the perturbation being effectively zero (i.e. in the limit $4V^2/\omega_0^2 = 0$) the population of state 2 never rises from zero. The important criterion for the effectiveness of the perturbation is not its absolute strength but *its strength relative to the energy separation of the states of the system.* This is an exceptionally important point.

Finally, we can see how it is possible to prepare systems in *mixed states* (e.g. 50 per cent state 1, 50 per cent state 2). For simplicity consider the case of doubly degenerate levels; then eqn (8.5.11) shows that if we wait for a time $t = \pi/4V$, $P_2 = \frac{1}{2}$ (and $P_1 = \frac{1}{2}$). If the perturbation is immediately extinguished, this 50:50 mixed state persists, because now $\dot{a}_1 = \dot{a}_2 = 0$. Similar behaviour is shown in Fig. 8.8. An inspection of the composition at any later instant will give the outcome 'state 1' 50 per cent of the time and 'state 2' for the other 50 per cent. This is a general way of preparing mixed states, and is widely used in nuclear magnetic resonance, where nuclei are exposed to magnetic fields for exactly specified lengths of time to cause them to evolve from a pure state (spin α for instance) into a superposition of α and β states. This is the underlying quantum mechanics of 'pulse methods' in n.m.r.

8.6 General perturbations: the variation of constants

The previous section reveals two points. One is that even very simple systems lead to complicated differential equations. For a two-level system we had to solve a second-order differential equation; for three levels we have a third-order differential equation, and so on. There is no hope of discovering general solutions for such systems. The other point concerns the simplicity of the perturbation: a constant when it was on. The differential equation was bad enough then; it becomes very much more complicated when the perturbation has any more complicated time-dependence. Even the case of $\cos \omega t$ is very difficult. What we need is a way of handling systems of many states exposed to perturbations of arbitrary time-dependence. The price to pay will be that we can expect to obtain only approximate solutions.

We shall describe the technique invented by Dirac and known (agreeably paradoxically) as *variation of constants*. It is a generalization of the two-level problem, and that should be held in mind in the course of the calculation. We shall see how closely they are related when we have obtained the final expressions.

As before, the hamiltonian is taken to have the form $H = H^{(0)} + H^{(1)}(t)$. The eigenfunctions of $H^{(0)}$ are the functions ψ_n corresponding to the states $|n\rangle$. Their eigenvalues are E_n, and their full time-dependent forms are

$$\Psi_n = \psi_n e^{-iE_n t/\hbar}, \qquad H^{(0)}\Psi_n = i\hbar(\partial \Psi_n/\partial t). \tag{8.6.1}$$

The state of the perturbed system is Ψ, and, as before, we express it as a time-dependent linear combination of the time-dependent basis functions; this time, however, there are several basis states (and perhaps an infinite number):

$$\Psi = \sum_n a_n \Psi_n = \sum_n a_n \psi_n e^{-E_n t/\hbar}, \qquad H\Psi = i\hbar(\partial \Psi/\partial t). \tag{8.6.2}$$

Our problem is to discover how the linear combination evolves with time: this involves finding and then solving a differential equation for the mixing coefficients a_n.

We proceed as before. Substituting the linear combination into the Schrödinger equation it has to satisfy and then using the Schrödinger equations for the basis functions leads to the following expression:

$$\begin{aligned} H\Psi &= \sum_n a_n H^{(0)}\Psi_n + \sum_n a_n H^{(1)}\Psi_n \\ \mathrm{i}\hbar(\partial\Psi/\partial t) &= \sum_n a_n \mathrm{i}\hbar(\partial\Psi_n/\mathrm{d}t) + \sum_n \mathrm{i}\hbar\dot{a}_n\Psi_n, \end{aligned}$$

where the terms in the left-hand box are equal by eqn (8.6.2), and those in the right-hand box are equal by eqn (8.6.1). Therefore

$$\sum_n a_n H^{(1)}\psi_n \mathrm{e}^{-\mathrm{i}E_n t/\hbar} = \mathrm{i}\hbar \sum_n \dot{a}_n\psi_n \mathrm{e}^{-\mathrm{i}E_n t/\hbar}.$$

Now we have to extract one of the $\dot{a}_n$ on the right: as before we use the orthonormality of the basis and multiply both sides by ψ_k^* and integrate over all space. This gives

$$\sum_n a_n H_{kn}^{(1)}(t)\mathrm{e}^{-\mathrm{i}E_n t/\hbar} = \mathrm{i}\hbar\dot{a}_k \mathrm{e}^{-\mathrm{i}E_k t/\hbar}$$

because only one term survives on the right, the one with $n = k$. This expression can be tidied up as before by expressing the energy differences as frèquencies. This time there are many different frequencies, not just the single ω_0 we had before. Therefore we define $\hbar\omega_{kn} = E_k - E_n$. The final expresion for the rate of change of any coefficient a_k is therefore

$$\dot{a}_k = (1/\mathrm{i}\hbar) \sum_n a_n H_{kn}^{(1)}(t)\mathrm{e}^{\mathrm{i}\omega_{kn}t}. \tag{8.6.3}$$

It is easy to check that this reduces to eqn (8.5.5) in the case of a two-level system.

The last equation is exact. We now move towards finding approximate solutions, and so we diverge from the previous method of solution. The way of solving a differential equation is to integrate it. Therefore we integrate the last equation from some initial time $t = 0$ when the state of the system was described by a linear combination with coefficients having the values $a_n(0)$, to the time t of interest, where they have the values $a_n(t)$:

$$a_k(t) - a_k(0) = (1/\mathrm{i}\hbar) \sum_n \int_0^t a_n(t)H_{kn}^{(1)}(t)\mathrm{e}^{\mathrm{i}\omega_{kn}t}\,\mathrm{d}t. \tag{8.6.4}$$

This is still exact.

The trouble with eqn (8.6.4) is that although it appears to give an expression for any coefficient a_k, it does so in terms of all the other coefficients (and possibly including a_k itself if the perturbation has diagonal elements). These other coefficients are unknown, and have to be found from equations of the same kind. The only way to burst out of this circle is to make an approximation. At this point we shall draw, for the first time, on the fact that the perturbation is weak, and that the times for which it is applied are not long. This means that the mixture retains something close to its initial composition for all the times of

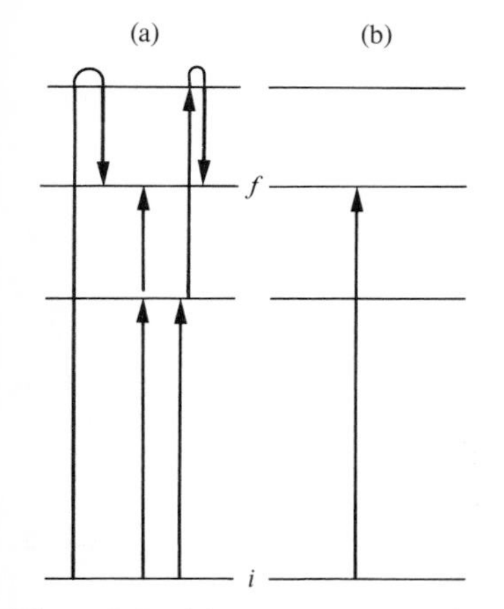

Fig. 8.9. (a) Indirect (high-order), (b) direct (first-order) routes for excitation.

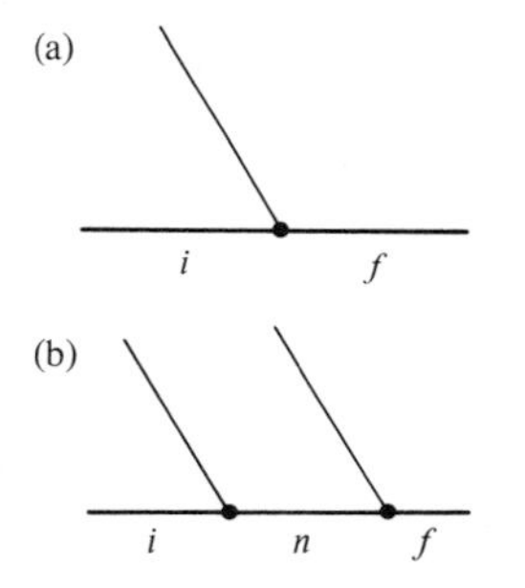

Fig. 8.10. Feynman diagrams for (a) a first-order process, (b) a second-order process.

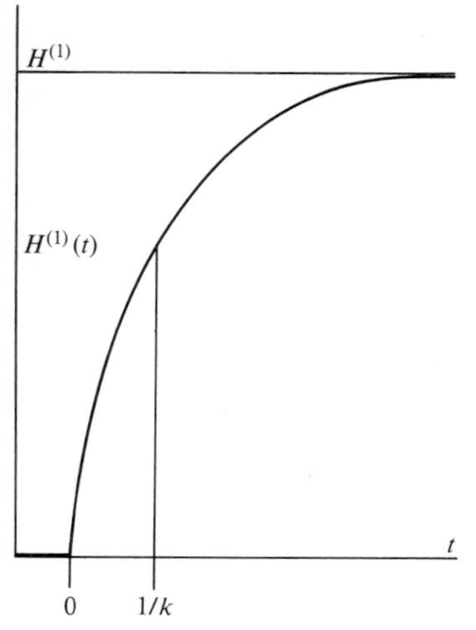

Fig. 8.11. An exponentially switched perturbation.

interest. We shall also suppose that at $t=0$ the system is known to be in the state i, its initial state, and that we are interested in knowing the probability of it being in some state f where $\mathrm{f}\neq\mathrm{i}$. We therefore need the value of $a_{\mathrm{f}}(t)$. At $t=0$ all the coefficients except a_{i} are zero, and $a_{\mathrm{i}}(0)=1$. Now we make the approximation. We suppose that *the probability that the system is in any state other than* i *is always so low that all the terms in the sum on the right can be set equal to zero, with the exception of the term with* $n=\mathrm{i}$. The right-hand side of eqn (8.6.4) then reduces to a single term involving an integral over a_{i}:

$$a_{\mathrm{f}}(t)-a_{\mathrm{f}}(0)=(1/\mathrm{i}\hbar)\int_0^t a_{\mathrm{i}}(t)H^{(1)}_{\mathrm{fi}}(t)\mathrm{e}^{\mathrm{i}\omega_{\mathrm{fi}}t}\,\mathrm{d}t. \tag{8.6.5}$$

But, on the same grounds as before, the coefficient a_{i} barely changes from its initial value (1) over the period for which the perturbation is active. Therefore we set $a_{\mathrm{i}}(t)=1$ in the integrand. Since $a_{\mathrm{f}}(0)=0$ we have

$$a_{\mathrm{f}}(t)=(1/\mathrm{i}\hbar)\int_0^t H^{(1)}_{\mathrm{fi}}(t)\mathrm{e}^{\mathrm{i}\omega_{\mathrm{fi}}t}\,\mathrm{d}t, \tag{8.6.6}$$

an explicit expression for the value of the coefficient of a state that was unoccupied initially.

The nature of the approximation can now be seen. We are ignoring the possibility that the perturbation can take the system from its initial state i to some selected state f by an *indirect* route, Fig. 8.9, in which it induces several transitions in sequence. The act of neglecting all coefficients except a_{i} is equivalent to supposing that the probability of the system being in the other states never rises to a point where they can provide a starting point for a transition to f. Put another way, we are allowing the perturbation to act only once; we are dealing with *first-order time-dependent perturbation theory*. In modern books you will often see a diagram like Fig. 8.10(a). This expresses the content of eqn (8.6.6) by showing how the perturbation (the sloping line) acts once to drive the system from state i to state f. *Diagrammatic perturbation theory* has been richly developed, and many calculations can be performed on the basis of the properties of the diagrams themselves. In the present case, the diagram in Fig. 8.10(a) should be regarded as a succinct way of expressing the right-hand side of eqn (8.6.6). Second-order time-dependent perturbation theory would give rise to a diagram like that in Fig. 8.10(b). This is the basis of the *Feynman diagrams* now widely used in quantum theory.

As the first example of the application of the general expression, consider a perturbation that rises from zero very slowly, and reaches a constant value, Fig. 8.11. This can be expressed as

$$H^{(1)}(t)=H^{(1)}(1-\mathrm{e}^{-kt}),\qquad t\geq 0 \tag{8.6.7}$$

where k is very small and positive. What is the coefficient a_{f} long after the hamiltonian has reached its final magnitude? That is, what is the value of $a_f(t)$ for $t \gg 1/k$? The answer is provided by eqn (8.6.6):

$$a_f(t) = (H_{fi}^{(1)}/i\hbar)\int_0^t (1-e^{-kt})e^{i\omega_{fi}t}\,dt$$
$$= (H_{fi}^{(1)}/i\hbar)\left\{\left(\frac{e^{i\omega_{fi}t}-1}{i\omega_{fi}}\right)+\left(\frac{e^{-(k-i\omega_{fi})t}-1}{(k-i\omega_{fi})}\right)\right\}. \quad (8.6.8)$$

This expression can be simplified first by making use of the facts that $t \gg 1/k$ (so that $e^{-kt} \approx 0$) and the rise of the perturbation is slow (so that $k \ll \omega_{fi}$, which makes definite what we mean by 'slow'). Then with these approximations we find

$$a_f(t) \approx (H_{fi}^{(1)}/i\hbar)(1/i\omega_{fi})e^{i\omega_{n}t} = -(H_{fi}^{(1)}/\hbar\omega_{fi})e^{i\omega_{fi}t}. \quad (8.6.9)$$

This is exactly of the form of the time-independent expression for the coefficients in the first-order correction to the wavefunction, eqn (8.1.22), apart from the time-dependent phase factor.†

We can now see the reason why time-independent perturbation theory can be used for most problems of chemical interest except where the perturbation continues to change. When a constant perturbation is simply 'switched on' in the laboratory, it is done extremely slowly in comparison with the frequencies associated with transitions in atoms and molecules ($k \approx 1/10^{-3}\ s = 10^3\ s^{-1}$; $\omega_{fi} \approx 10^{15}\ s^{-1}$). Furthermore, usually we are interested in the system's properties at times long after the switching is complete ($t \gg 10^{-3}$ s). These are the conditions under which time-dependent perturbation theory has settled down into time-independent theory: all the transients stimulated by the switching have subsided, and populations are steady.

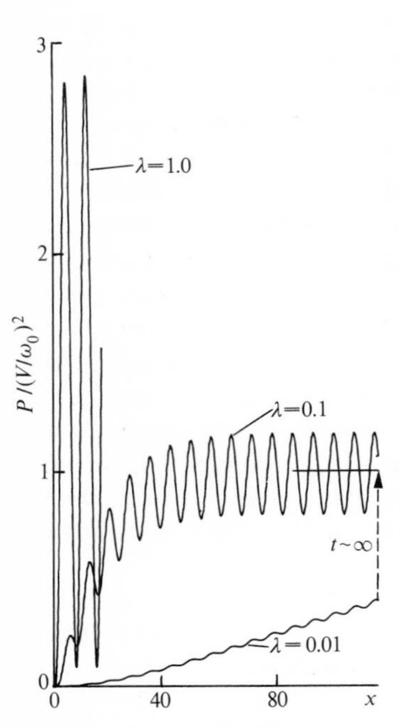

Fig. 8.12. Time-dependent probabilities for various switching rates.

Example. A constant perturbation was turned on exponentially starting at $t = 0$. Evaluate the probability of a two-level system being found in state 2 if initially it was in state 1, and examine the role of transients.

- *Method.* The perturbation is given by eqn (8.6.7). The solution for the amplitude is eqn (8.6.8). Find $P = |a_2|^2$ for general k and plot P as a function of t. For example plots take $\lambda = k/\omega$ and draw $P/(V/\omega_0)^2$ for $\lambda = 0.01, 0.1, 1.0$.
- *Answer.* From eqn (8.6.8) with $\lambda = k/\omega$ and $x = \omega t$,

$$P/(V^2/\omega_0^2) = \left(\frac{1}{1+\lambda^2}\right)\{1+2\lambda^2-[2\lambda^2\cos x+(2-e^{-\lambda x})e^{-\lambda x}$$
$$+2\lambda(1-e^{-\lambda x})\sin x]\}.$$

This is plotted in Fig. 8.12 for $\lambda = 0.01, 0.1, 1.0$ (corresponding to increasing switching rates).

† The phase factor disappears on forming the wavefunction. On substitution into eqn (8.6.2) the wavefunction takes the form

$$\Psi = \sum_f (-H_{fi}^{(1)}/\hbar\omega_{fi})\psi_f e^{-iE_it/\hbar} = \left\{\sum_f \left(\frac{H_{fi}^{(1)}}{E_i-E_f}\right)\psi_f\right\}e^{-iE_it/\hbar},$$

and so the time-independent part of the perturbed wavefunction, in this first-order approximation, is exactly the one given by eqn (8.1.23).

• *Comment.* Notice how slow switching ($\lambda = 0.01$) generates hardly any transient behaviour while rapid switching ($\lambda = 1.0$) is like an impulsive shock to the system, and causes the population to oscillate between the two states. For very rapid switching ($\lambda \gg 1$), $P/(V^2/\omega_0^2)$ varies as $2(1-\cos x)$, and so oscillates between 0 and 4 with an average value of 2. Very rapid switching is like a hammer blow.

Now consider the other important case of a system exposed to an *oscillating perturbation,* such as an atom experiences when it is exposed to electromagnetic radiation in a spectrometer or in sunlight. Once we can deal with oscillating perturbations we can treat any perturbation, because a general time-dependent function can always be regarded as a superposition of harmonically oscillating functions.

A perturbation oscillating with a frequency $\omega = 2\pi\nu$ and turned on at $t = 0$ has the form

$$H^{(1)}(t) = 2H^{(1)} \cos \omega t = H^{(1)}(e^{i\omega t} + e^{-i\omega t}), \qquad t \geq 0. \tag{8.6.10}$$

If this is inserted into eqn (8.6.6) we obtain

$$\begin{aligned} a_f(t) &= (H_{fi}^{(1)}/i\hbar) \int_0^t (e^{i\omega t} + e^{-i\omega t})e^{i\omega_{fi} t}\, dt \\ &= (H_{fi}^{(1)}/i\hbar) \left\{ \frac{e^{i(\omega_{fi}+\omega)t} - 1}{i(\omega_{fi}+\omega)} + \frac{e^{i(\omega_{fi}-\omega)t} - 1}{i(\omega_{fi}-\omega)} \right\}. \end{aligned} \tag{8.6.11}$$

As it stands this expression is fairly obscure, but it can be simplified so as to bring out its principal content by taking note of the conditions under which it is normally used. In applications to optical spectra the frequencies ω_{fi} and ω are of the order of $10^{15}\,s^{-1}$ and more. Even in nuclear magnetic resonance they are greater than $10^6\,s^{-1}$. The exponential in the numerator of the terms in brackets is never greater than 1 however large the frequencies (because $e^{ix} = \cos x + i \sin x$); the denominator of the first term is of the order of the frequencies, and so the first term is not larger than about 10^{-15} s for optical transitions and 10^{-6} s for n.m.r. The denominator in the second term, however, can become arbitrarily close to zero as the external perturbation approaches a transition frequency. Therefore the second term is normally larger than the first, and overwhelms it completely as the frequencies match. Consequently we can be confident about dropping the negligible first term. The probability of finding the system in state f after a time t if initially it was in state i, $P_f = |a_f|^2$, is therefore

$$P_f(t) = \frac{4H_{if}^{(1)}H_{fi}^{(1)} \sin^2\frac{1}{2}(\omega_{fi}-\omega)t}{\hbar^2(\omega_{fi}-\omega)^2}. \tag{8.6.12}$$

Once again we write $H_{if}^{(1)}H_{fi}^{(1)} = \hbar^2 V_{fi}^2$; then

$$P_f(t) = \left\{ \frac{4V_{fi}^2}{(\omega_{fi}-\omega)^2} \right\} \sin^2\tfrac{1}{2}(\omega_{fi}-\omega)t. \tag{8.6.13}$$

The last expression should be familiar. Apart from a small but

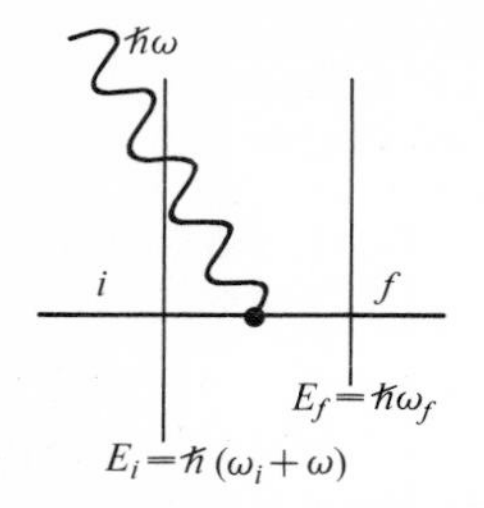

Fig. 8.13. The overall energy difference in the first-order process.

significant modification, it is exactly the same as eqn (8.5.12), the expression for a static perturbation applied to a two-level system in the limit of the perturbation being weak. The significant difference is that instead of the actual difference ω_{fi} appearing in the expression, it is replaced by $\omega_{fi} - \omega$. This can be thought of as an effective shift in the energy differences involved in exciting the system as a result of the presence of the photons associated with the oscillating field. As depicted in Fig. 8.13, where the wavy lines now represent an oscillating perturbation, the energy difference $E_f - E_i$ should really be thought of as

$$\begin{aligned} E_f - E_i &= E(\text{excited molecule, no photon}) \\ &\quad - E(\text{ground state, photon } \hbar\omega) \\ &= \hbar(\omega_{fi} - \omega). \end{aligned}$$

It follows that the transition probability may approach unity as the applied frequency approaches a transition frequency, because then the effective energy difference approaches zero, and the system becomes, in effect, a loose, degenerate system which can be nudged fully from state to state by gentle perturbations.

8.7 Transition rates

The time-dependence of the probability of being found in state f depends on the *frequency offset* $\omega_{fi} - \omega$, Fig. 8.14. When the frequency offset is zero the field and the system are in *resonance*, and the transition probability increases most sharply with time. In order to obtain the quantitative form of the time-dependence of the probability

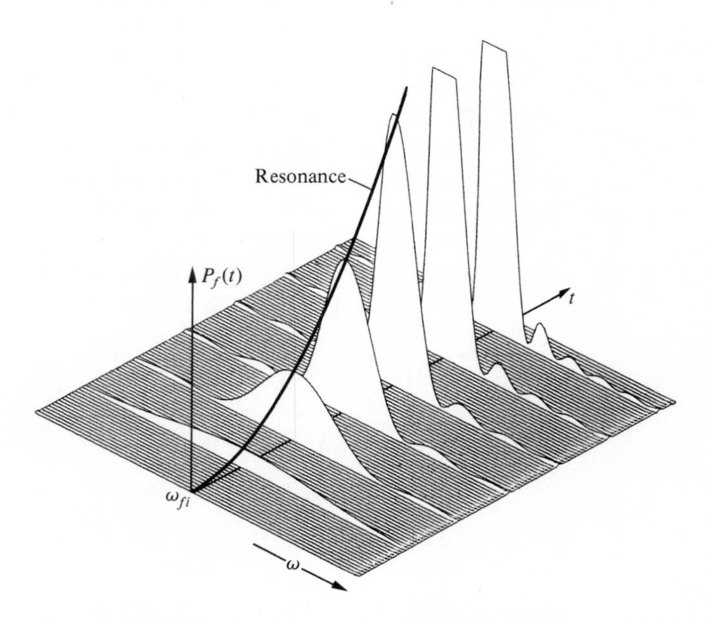

Fig. 8.14. Time and frequency dependence of P_f.

at resonance we set $x=\frac{1}{2}(\omega_{fi}-\omega)t$ and take the limit

$$\lim_{x\to 0}\left(\frac{\sin x}{x}\right)^2=\lim_{x\to 0}\left(\frac{x-\frac{1}{6}x^3+\dots}{x}\right)^2=1. \qquad (8.7.1)$$

Then

$$P_f(t)=V^2t^2 \text{ at resonance,} \qquad (8.7.2)$$

and the probability increases quadratically with time. Although this appears to predict that if we wait long enough ($t>1/V$) the probability will exceed unity, the approximations used in the deduction of eqn (8.6.13) mean that it is not applicable to such long intervals.

The *transition rate* is the rate of change of the probability of being in the initially empty state:

Transition rate: $W_{i\to f}=dP_f/dt.$ (8.7.3)

Spectral intensities are proportional to transition rates, because they depend on the rate at which energy is pumped into the electromagnetic field or absorbed from it. It follows that at resonance we predict a transition rate $W_{i\to f}=2V^2t$; this expression, while pleasantly simple is also ridiculous. It predicts that spectral lines should grow more intense as time passes.

The feature omitted from the discussion so far is the fact that in any real spectroscopic arrangement the incident light is not strictly monochromatic but has components covering at least a narrow band of frequencies. Furthermore, molecules often have numerous states of a similar energy, and so an absorption normally takes place to a band of states covering a narrow band of energies. In either case the transition probability is the *sum* of all the transition probabilities over the range covered, Fig. 8.15, in other words, the total probability is related to an

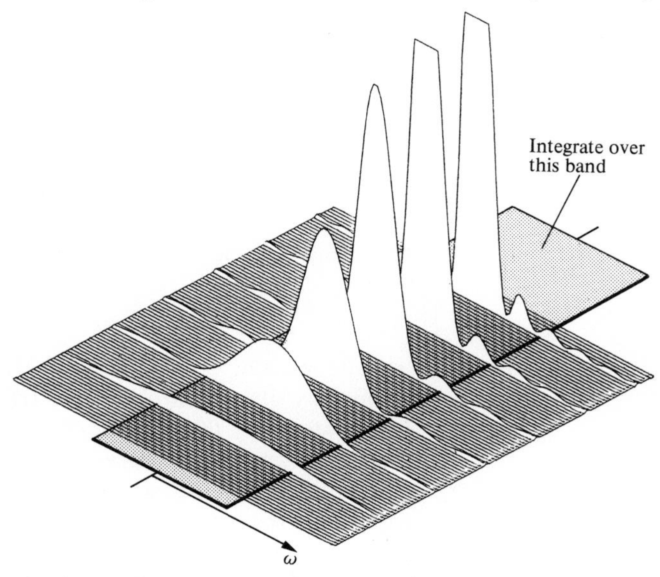

Fig. 8.15. The total transition probability for a range of perturbing frequencies.

integral of the area under the $\sin^2 x/x^2$ curve. This integral increases linearly with time, as we shall now show (you can see from Fig. 8.15 that although the height of the centre peak increases as t^2, the width decreases, and so the area increases less quickly than t^2 and in fact as t).

Suppose that the number of states in the energy range E_f to $E_\mathrm{f}+\mathrm{d}E$ is $\rho_N(E_\mathrm{f})\,\mathrm{d}E$, where $\rho_N(E)$ is called the *density of states*, Fig. 8.16. Then the total transition probability to all the states lying in a band near E_f is a sum (integral) of expressions like eqn (8.6.13) weighted according to the number of states:

$$P(t)=\sum_\mathrm{f} P_\mathrm{f}(t)=\int_\mathrm{band}\left\{\frac{4V_\mathrm{fi}^2\sin^2\frac{1}{2}(\omega_\mathrm{fi}-\omega)t}{(\omega_\mathrm{fi}-\omega)^2}\right\}\rho_N(E_\mathrm{f})\,\mathrm{d}E_\mathrm{f}.$$

We can write $\mathrm{d}E_\mathrm{f}=\hbar\,\mathrm{d}\omega_\mathrm{fi}$ to turn this into an integral over the transition frequencies.

The $\sin^2 x/x^2$ function falls to zero as soon as ω_fi moves far away from ω, and so to a good approximation we may assume that the density of states $\rho_N(E_\mathrm{f})$ can be approximated by its value at the centre of the band, $\rho_N(\bar{E}_\mathrm{f})$, and that the matrix elements V_fi are all very similar. This leads to

$$P(t)\approx\hbar V_\mathrm{fi}^2\rho_N(\bar{E}_\mathrm{f})t^2\int\left\{\frac{\sin^2\frac{1}{2}(\omega_\mathrm{fi}-\omega)t}{[\frac{1}{2}(\omega_\mathrm{fi}-\omega)t]^2}\right\}\mathrm{d}\omega_\mathrm{fi}.$$

The substitution $x=\frac{1}{2}(\omega_\mathrm{fi}-\omega)t$ turns this into

$$P(t)=2\hbar V_\mathrm{fi}^2\rho_N(\bar{E}_\mathrm{f})t\int\left(\frac{\sin x}{x}\right)^2\mathrm{d}x.$$

The integral is over the range of x corresponding to the range of energies in the band. But the integrand $\sin^2 x/x^2$ falls off so sharply that there is little error introduced by extending the limits of integration to $-\infty$ and ∞. The advantage is that the integral is then standard (i.e. listed) and has the value π. Hence the final expression for the transition probability to the band is

$$P(t)=2\pi\hbar V_\mathrm{fi}^2\rho_N(\bar{E}_\mathrm{f})t. \tag{8.7.4}$$

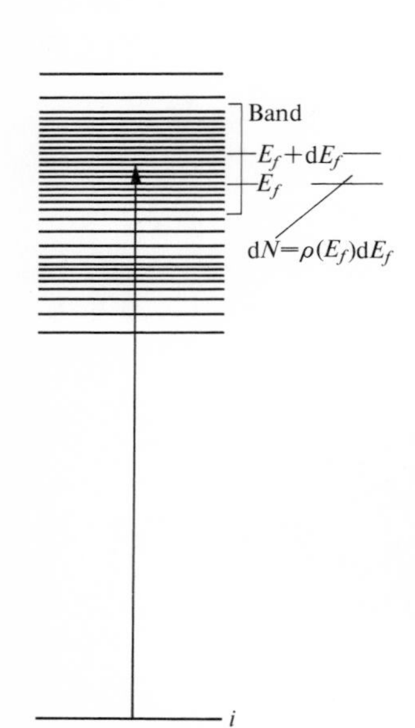

Fig. 8.16. The density of states and the transition to a band.

The transition rate to the band is therefore given by

> *Fermi's golden rule*: $W_{\mathrm{i}\to\mathrm{f}}=2\pi\hbar V_\mathrm{fi}^2\rho_N(\bar{E}_\mathrm{f})=(2\pi/\hbar)\,|H_\mathrm{fi}^{(1)}|^2\rho_N(\bar{E}_\mathrm{f})$, (8.7.5)

and is independent of the time. Therefore, to calculate the transition rate, multiply the square of the transition matrix element by the density of states of the band involved (and include the factor $2\pi/\hbar$).

When the system is exposed to a perturbation covering a band of frequencies, exactly the same procedure may be followed. Now we suppose that the perturbation is a sum (integral) of different frequency components, the number of components with frequencies in the range ν to $\nu+\mathrm{d}\nu$ being $\hat{\rho}_N(\nu)\,\mathrm{d}\nu$. Here $\hat{\rho}_N(\nu)$ is the *frequency density of states*.

The total transition probability is therefore

$$P_f(t) = \int P_f(t)\hat{\rho}_N(\nu)\,d\nu \approx V_{fi}^2\hat{\rho}_N(\nu_{fi})t \tag{8.7.6}$$

where we have made the same approximations as before, but supposed that $\hat{\rho}_N(\nu)$ may be approximated by $\hat{\rho}_N(\nu_{fi})$, where $\nu_{fi} = \omega_{fi}/2\pi$, because the $\sin^2 x/x^2$ integral is non-zero only when ω is near ω_{fi}. The transition rate is then

$$W_{i\to f} = (1/\hbar^2)\,|H_{if}^{(1)}|^2\hat{\rho}_N(\nu_{fi}), \tag{8.7.7}$$

another version of the golden rule. An example of how this expression is used is given in Appendix 13.

8.8 The Einstein transition probabilities

From now on we concentrate on the second form of the golden rule. In the case of electric dipole transitions, it turns out (Appendix 13) that $W \propto \mathscr{E}(\nu)\hat{\rho}_N/V$, where $\mathscr{E}(\nu)$ is the energy of the field at the frequency ν and V is the volume occupied by the field; therefore $\{\mathscr{E}(\nu)\hat{\rho}_N/V\}\,d\nu$ is the energy per unit volume in the frequency range ν to $\nu + d\nu$. We shall write this $\rho(\nu)\,d\nu$, where $\rho(\nu) = \mathscr{E}(\nu)\hat{\rho}_N/V$ is the *energy density of states* at the frequency ν. Then, since $W \propto \rho$ we can write

$$W_{i\to f} = B_{if}\rho(\nu_{fi}). \tag{8.8.1}$$

From now on we shall write ρ in place of $\rho(\nu_{fi})$, and remember that it has to be evaluated at the transition frequency. Einstein studied the equilibrium between matter and radiation, and distinguished several processes. For the present we shall suppose too that the state $|f\rangle$ lies higher in energy than $|i\rangle$. The rate of absorption of electromagnetic radiation is written

$$\textit{Stimulated absorption: } W_{f\leftarrow i} = B_{if}\rho \tag{8.8.2}$$

and the coefficient B_{fi} is known as *Einstein's coefficient of stimulated absorption.* The transition $f \to i$ corresponds to an emission, and Einstein introduced another coefficient, the *coefficient of stimulated emission* and wrote the rate as

$$\textit{Stimulated emission: } W_{f\to i} = B_{fi}\rho. \tag{8.8.3}$$

Since $|H_{fi}^{(1)}|^2$ is unchanged when i and f exchange places, the coefficient B_{fi} must be equal to the coefficient of stimulated absorption B_{if}. Hence *the rates of stimulated emission and absorption are equal.*

Einstein was next able to show that these two coefficients alone were unable to account for the existence of thermal equilibrium between matter and radiation. There must be another process in operation which allows emission to occur even in the absence of any stimulating radiation. The process is alled *spontaneous emission* and its rate is written A_{fi}, a quantity called *Einstein's coefficient of spontaneous emission*:

$$\textit{Spontaneous emission:}\quad W_{f\to i} = A_{fi}. \tag{8.8.4}$$

The total rate of emission is $A_{fi}+B_{fi}\rho$; the total rate of absorption is simply $B_{if}\rho$ because there is no spontaneous absorption process.

The argument that leads to the existence of A_{fi} and to an expression for it runs as follows. Consider a system of molecules in thermal equilibrium with radiation. Let there be N_i in the state $|i\rangle$ and N_f in the state $|f\rangle$ at equilibrium, and write the equilibrium density of radiation at the transition frequency as ρ_e. The total rate of absorption by all the molecules is $N_i W_{f\leftarrow i}$, and the total rate of emission is $N_f W_{f\to i}$. At equilibrium the total rates must be equal, and so we require that

$$N_f\{A_{fi}+B_{fi}\rho_e\} = N_i B_{if}\rho_e.$$

But at equilibrium at a temperature T the populations are in a ratio given by the Boltzmann distribution:

$$N_i/N_f = \exp(h\nu_{fi}/kT),$$

and so the equation rearranges to

$$\rho_e = (A_{fi}/B_{fi})\{e^{h\nu_{fi}/kT}-1\}^{-1}. \tag{8.8.5}$$

because $B_{if}=B_{fi}$. Clearly $A_{fi}\neq 0$, for otherwise the radiation density would be zero. This confirms the necessity of the spontaneous emission process.

Now note that at thermal equilibrium the radiation density of states is also given by Planck's distribution law (eqn (1.1.11)):

$$\rho_e = (8\pi h\nu_{fi}^3/c^3)\{e^{h\nu_{fi}/kT}-1\}^{-1}. \tag{8.8.6}$$

Comparison of eqn (8.8.5) with this expression leads to

$$A_{fi}/B_{fi} = 8\pi h(\nu_{fi}/c)^3. \tag{8.8.7}$$

The important point to note is that *spontaneous emission increases in relative importance as the cube of the transition frequency*, and is very important at very high frequencies. This is one reason why X-ray lasers are so elusive: highly excited populations are difficult to maintain, and simply throw off their energy at random instead of cooperating in a stimulated emission process.

Example. Calculate the rates of stimulated and spontaneous emission of the $3p\to 2s$ transition in hydrogen (the H_α radiation, responsible for its red glow) when it is inside a cavity at a temperature of 1000 K.

- *Method.* Use eqn (8.8.3) for the rate of stimulated emission, calculating B from the expression in Appendix 10 and using the Planck expression for the density of states, eqn (8.8.6). Calculate the transition moment from the hydrogen orbitals listed in Tables 4.1 and 4.2. For the energy of the transition, use eqn (4.1.7). For the rate of spontaneous emission, use eqn (8.8.7).

• *Answer.* First calculate the transition dipole moment. For the $3p_z \rightarrow 2s$ transition

$$\mu_z = -e\int \psi^*_{3p_z} z\psi_{2s}\,d\tau = -(3^3\times 2^{10}/5^6)ea_0 = -1.769ea_0$$
$$= -1.500\times 10^{-29}\ \text{C m}$$
$$\mu^2 = \mu_x^2+\mu_y^2+\mu_z^2 = 9.393e^2a_0^2 = 6.752\times 10^{-58}\ \text{C}^2\ \text{m}^2.$$
$$B = \mu^2/6\varepsilon_0\hbar^2 = 1.143\times 10^{21}\ \text{J}^{-1}\ \text{m}^3\ \text{s}^{-1}$$
$$\nu(3p \rightarrow 2s) = (hcR/h)\{\tfrac{1}{4}-\tfrac{1}{9}\} = (\tfrac{5}{36})(cR) = 4.567\times 10^{14}\ \text{Hz}.$$
$$A = 8\pi h(\nu/c)^3 B = 6.728\times 10^7\ \text{s}^{-1}.$$

At 1000 K and $\nu = 4.567\times 10^{14}$ Hz,

$$\rho = (8\pi h\nu^3/c^3)\{e^{h\nu/kT}-1\}^{-1} = 1.782\times 10^{-23}\ \text{J Hz}^{-1}\ \text{m}^{-3}.$$

Hence

$$W^{\text{stimulated}} = \rho B = 2.036\times 10^{-2}\ \text{s}^{-1};\quad W^{\text{spontaneous}} = A = 6.728\times 10^7\ \text{s}^{-1}.$$

• *Comment.* The dipole moment of the transition is equivalent to an electron moving through a distance of $-1.769a_0 \approx 94$ pm. The 'spontaneous lifetime' is $1/A \approx 1.49\times 10^{-8}$ s; the stimulated lifetime is 49 s.

The presence of the spontaneous process may be viewed as a manifestation of the zero-point fluctuations of the electromagnetic field. Since the field can be thought of as a collection of oscillators, it has zero-point energy even when the oscillators are in their lowest energy states (i.e. when no photons are present). The 'spontaneous' process may be regarded as emission stimulated by these fluctuating fields. Things that fluctuate also dissipate.

8.9 Lifetime and energy uncertainty

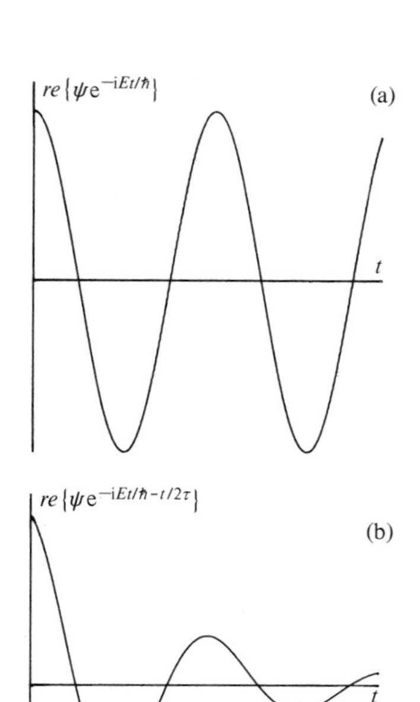

Fig. 8.17. (a) The time-dependence of a stationary state; (b) the amplitude of a state with finite lifetime.

We are now in a position to discover the relation betwen the lifetimes of states and the ranges of energies they may possess. We have seen that if a state possesses a definite energy it has a time-dependent wavefunction of the form $\Psi = \psi e^{-iEt/\hbar}$ and $|\Psi|^2 = \psi^2$. But suppose a state is described by a wavefunction that is decaying in amplitude, perhaps because the system is making a transition to another state. Suppose that the amplitude of the state is not oscillating for ever, Fig. 8.17(a), but is decaying exponentially too, Fig. 8.17(b), so that $\Psi = \psi e^{-iEt/\hbar - t/2\tau}$ and $|\Psi|^2 = \psi^2 e^{-t/\tau}$, where τ is the time constant for the decay. What is its energy?

The only way of arriving at a clear-cut answer is to try to express the function in the form $e^{-iE't/\hbar}$, but it is obvious that one such function cannot model the decaying exponential. On the other hand, the decaying function can be modelled by a superposition of oscillating functions. In fact, we can use the techniques of Fourier transforms (which essentially is the mathematical technique for discussing any function in terms of oscillating functions) to write

$$e^{-iEt/\hbar - t/2\tau} = \int g(E')e^{-iE't/\hbar}\,dE';\qquad g(E') = \frac{(\hbar/\tau)}{(E-E')^2+(\hbar/2\tau)^2}. \tag{8.9.1}$$

This shows that the decaying wavefunction corresponds to a range of energies (all the values of E' in the integral). Therefore, *any state with a finite lifetime has to be regarded as having an imprecise energy.*

We can arrive at a quantitative relation between lifetime and energy range by considering the shape of the weighting function $g(E')$, Fig. 8.18. The width at half-height is readily shown to be equal to $\hbar/2\tau$, and can be taken as a rough guide to the range of energies the state possesses. Denoting it δE, we have $\delta E \approx \hbar/2\tau$, or

Lifetime broadening: $\tau\,\delta E \approx \hbar/2$ (8.9.2)

which is reminiscent of the uncertainty relations encountered in Chapter 5 (but we see how different the derivation has been). The expression is known as the *lifetime broadening relation* because it shows how the spread of energy is related to the lifetime of the state, in this case the characteristic time for the state to decay.

The shorter the lifetime of the state, the less precise its energy. When a state has zero lifetime we can say nothing about its energy. When the lifetime is infinite, the energy can be specified precisely.

Further reading

General accounts of perturbation theory

Quantum chemistry. W. Kauzmann; Academic Press, New York, 1957.

Introduction to quantum mechanics. L. Pauling and E. B. Wilson; McGraw-Hill, New York, 1935.

Recent developments in perturbation theory. J. O. Hirschfelder, W. Byers Brown, and S. T. Epstein; *Adv. Quantum Chem.*, **1**, 256, 1964.

Perturbation theory and its applications in quantum mechanics. C. H. Wilcox (ed.); Wiley, New York, 1966.

Diagrammatic techniques

A guide to Feynman diagrams in the many-body problem. R. D. Mattuck; McGraw-Hill, New York, 2nd edition, 1976.

Elements of advanced quantum theory. J. M. Ziman; Cambridge University Press, 1969.

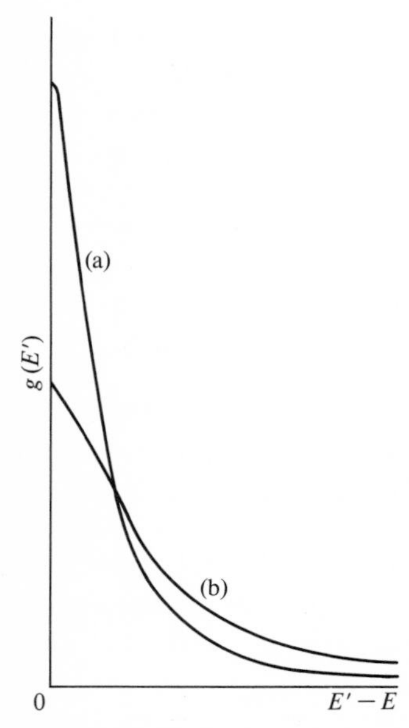

Fig. 8.18. The shape function for (a) τ long, (b) τ short.

Problems

8.1. One excited state of the sodium atom lies at 25 739.86 cm^{-1} above the ground state, another lies at 50 266.88 cm^{-1}. Suppose they are connected by a perturbation equivalent in energy to (a) 100 cm^{-1}, (b) 1000 cm^{-1}, (c) 5000 cm^{-1}. Calculate the energies and composition of the states of the perturbed system. *Hint*. Use eqn (8.1.6) for the energies and eqn (8.1.8) for the states, and express the composition as the contribution of the unperturbed states (as percentages).

8.2. A simple calculation of the energy of the helium atom supposes that each electron occupies the same hydrogen-like 1s-orbital (but with $Z = 2$). The electron–electron interaction is regarded as a perturbation, and calculation gives

$$\int \psi_{1s}^2(r_1)(e^2/4\pi\varepsilon_0 r_{12})\psi_{1s}^2(r_2)\,d\tau = (\tfrac{5}{4})(e^2/4\pi\varepsilon_0 a_0);$$

see below, p. 225. Estimate (a) the binding energy of helium, (b) its first ionization energy. *Hint.* Use eqn (8.1.6) with $E_1 = E_2 = E_{1s}$. Be careful not to count the electron–electron interaction energy twice.

8.3. Show that the energy of the perturbed levels is related to the mean energy of the unperturbed levels $\bar{E} = \frac{1}{2}(E_1 + E_2)$ by $E_\pm - \bar{E} = \pm\frac{1}{2}(E_1 - E_2)\sec 2\beta$, where β is the parameter in eqn (8.1.8). Devise a diagrammatic method of showing how $E_\pm - \bar{E}$ depends on $E_1 - E_2$ and ε. *Hint.* use eqn (8.1.6).

8.4. We normally think of the one-dimensional well as being horizontal. Suppose it is vertical; then the particle's potential energy depends on x because of the presence of the gravitational field. Calculate the first-order correction to the zero-point energy, and evaluate it for an electron in a box on the surface of the earth. Account for the result. *Hint.* The energy of the particle depends on its height as mgx where $g = 9.81 \text{ m s}^{-2}$. Use eqn (8.1.20) with $\psi_1(x)$ given in eqn (3.3.5). Since g is so small the energy correction is tiny; but it would be significant if the box were on the surface of a neutron star.

8.5. Calculate the second-order correction to the energy for the same problem and the distortion of the ground-state wavefunction induced by the gravitational field. Account for the shape of the distortion. *Hint.* Use eqn (8.1.25) for the energy and eqn (8.1.23) for the wavefunction. The integral involved is of the form

$$\int x \sin ax \sin bx \, dx = -(d/da) \int \cos ax \sin bx \, dx$$

$$\int \cos ax \sin bx \, dx = \frac{\cos(ab)x}{2(a-b)} - \frac{\cos(a+b)x}{2(a+b)}.$$

Evaluate the sum over n numerically.

8.6. In the free-electron molecular orbital method (Problem 3.14) the potential energy may be made slightly more realistic by supposing that it varies sinusoidally along the polyene chain. Select a potential energy with suitable periodicity, and calculate the first-order correction to the wavelength of the lowest energy transition.

8.7. Show group-theoretically that when a perturbation of the form $H^{(1)} = az$ is applied to a hydrogen atom, the 1s-orbital is contaminated by the admixture of np_z-orbitals. Deduce which orbitals contaminate (a) $2p_x$-orbitals, (b) $2p_z$-orbitals, (c) $3d_{xy}$-orbitals.

8.8. The symmetry of the ground electronic state of the water molecule is A_1. (a) An electric field, (b) a magnetic field is applied perpendicular to the molecular plane. What symmetry species of excited states may be mixed in to the ground state by the perturbations? *Hint.* The electric interaction has the form $H^{(1)} = ax$; the magnetic interaction has the form $H^{(1)} = bl_x$.

8.9. Repeat Problem 8.5, but find the second-order energy correction using the closure approximation. Compare the two calculations and deduce the appropriate value of ΔE. *Hint.* Use eqns (8.1.27) and (8.1.28).

8.10. Calculate the second-order energy correction to the ground state for a perturbation of the form $H^{(1)} = -\varepsilon \sin(\pi x/L)$ using the closure approximation. Infer a value of ΔE by comparison with the numerical calculation in the *Example* on p. 177. These two problems show that the parameter ΔE depends on the perturbation and is not simply a characteristic of the system itself.

8.11. The potential energy of a particle on a ring depended on the angle ϕ as $H^{(1)} = \varepsilon \sin^2\phi$. Calculate the first-order corrections to the energy of the degenerate $m_l = \pm 1$ states, and find the correct linear combinations for the perturbation calculation. Find the first- and second-order corrections to the energy. *Hint.* This is an example of degenerate-state perturbation theory, and so find the correct linear combinations by solving eqn (8.2.9) after deducing the energies from the roots of the secular determinant in eqn (8.2.10). Compare your results with Fig. 8.4. For the matrix elements, express $\sin\phi$ as $(1/2i)(e^{i\phi}-e^{-i\phi})$. Don't forget the $m_l=0$ state lying beneath the degenerate pair when evaluating eqn (8.1.25). Energies go as $m_l^2\hbar^2/2mr^2$; use $\psi_{m_l}=(1/2\pi)^{\frac{1}{2}}e^{im_l\phi}$ for the unperturbed states.

8.12. A trial function of the form $\psi=Ne^{kr^2}$ was taken for the ground-state wavefunction of the hydrogen atom. Find the optimum value of k and the corresponding energy.

8.13. A particle of mass m is confined to a one-dimensional square well of the type treated in Chapter 3. Choose trial functions of the form (a) $\sin kx$, (b) $(x-x^2/L)+k(x-x^2/L)^2$, (c) $e^{-k(x-\frac{1}{2}L)}-e^{-\frac{1}{2}kL}$ for $x\geq\frac{1}{2}L$ and $e^{k(x-\frac{1}{2}L)}-e^{-\frac{1}{2}kL}$ for $x\leq\frac{1}{2}L$. Find the optimum values of k and the corresponding energies.

8.14. Consider the linear H_3 molecule. The wavefunctions may be modelled by expressing them as $\psi=c_As_A+c_Bs_B+c_Cs_C$, the s_i denoting hydrogen 1s-orbitals of the relevant atom. Use the Rayleigh–Ritz method to find the optimum values of the coefficients and the energies of the orbitals. Make the approximations $H_{ss}=\alpha$, $H_{ss'}=\beta$ but $H_{s_As_C}=0$, $S_{ss}=1$, $S_{ss'}=0$. *Hint.* Although the basis can be used as it stands, it leads to a 3×3 determinant, and hence to a cubic equation for the energies. A better procedure is to set up symmetry-adapted combinations, and then to use the vanishing of $\int\psi_iH\psi_j\,d\tau$ unless $\Gamma_i=\Gamma_j$.

8.15. Repeat the last problem but set $H_{s_As_C}=\gamma$ and $S_{ss'}\neq0$. Evaluate the overlap integrals between 1s-orbitals on centres separated by R; use

$$S=\{1+(R/a_0)+\tfrac{1}{3}(R/a_0)^2\}e^{-R/a_0},$$

(a result deduced on p. 255). Suppose that $\beta/\gamma=S_{s_As_B}/S_{s_As_C}$. For a numerical result, take $R=80$ pm; $a_0=53$ pm.

8.16. A hydrogen atom in a 2s-configuration passes into a region where it experiences an electric field in the z-direction for a time τ. What is its electric dipole moment during its exposure and after it emerges? *Hint.* Use eqn (8.5.9) with $\omega_0=0$; the dipole moment is the expectation value of $-ez$. The integral $\int\psi_{2s}z\psi_{2p_z}d\tau$ is equal to $3a_0$.

8.17. A biradical is prepared with its two electrons in a singlet state. A magnetic field is present, and because the two electrons are in different environments their interaction with the field is $(\mu_B/\hbar)B(g_1s_{z1}+g_2s_{z2})$ with $g_1\neq g_2$. Evaluate the time-dependence of the probability that the electron spins will take on a triplet configuration (i.e. that the $S=1$, $M_s=0$ state will be populated). Examine the role of the energy separation $\hbar J$ of the singlet state and the $M_s=0$ state of the triplet. Suppose $g_1-g_2\approx10^{-3}$ and $J\approx0$; how long does it take for the triplet state to emerge in a field of 1 T? *Hint.* Use eqn (8.5.10); take $|0,0\rangle=(1/\sqrt{2})(\alpha\beta-\beta\alpha)$ and $|1,0\rangle=(1/\sqrt{2})(\alpha\beta+\beta\alpha)$.

8.18. An electric field in the z-direction is increased linearly from zero. What is the probability that a hydrogen atom, initially in the ground state, will be found with its electron in a $2p_z$-orbital at a time t? *Hint.* Use eqn (8.6.6) with $H_{fi}^{(1)}(t)\propto t$.

8.19. At $t=\frac{1}{2}T$ the strength of the field used in Problem 8.18 begins to decrease linearly. What is the probability that the electron is in the $2p_z$-orbital at $t=T$? What would the probability be if initially the electron was in a 2s-orbital?

8.20. Instead of the perturbation being switched linearly it was switched on and off exponentially and extremely slowly, the switching off commencing long after the switching on was complete. Calculate the probabilities, long after the perturbation has been extinguished, of the $2p_z$-orbital being occupied, the initial states being as in Problem 8.18. *Hint.* Take $H^{(1)}\propto1-e^{-kt}$ for $0\leq t\leq T$ and $H^{(1)}\propto e^{-k(t-T)}$ for $t\geq T$, interpret 'slow' as $k\ll\omega$, and 'long after' as both

$kT \gg 1$ (for 'long after switching on') and $k(t-T) \gg 1$ (for 'long after switching off').

8.21. Calculate the rates of stimulated and spontaneous emission for the $2p \rightarrow 1s$ transition in hydrogen when it is inside a cavity at a temperature of 1000 K. *Hint.* Follow the *Example* on p. 200.

8.22. Find the complete atomic-number dependence of the A and B coefficients for the $2p \rightarrow 1s$ transitions of hydrogen-like one-electron atoms. Calculate how the stimulated emission rate depends on Z when the atom is exposed to black-body radiation at 1000 K. *Hint.* The relevant density of states also depends on Z.

8.23. Examine how the A and B coefficients depend on the length of a one-dimensional square well for the transition $n+1 \rightarrow n$.

Part 3 Applications

THE basic principles of quantum mechanics, as far as they are necessary for quantum chemistry, are with one exception now at our fingertips. The exception concerns wavefunctions for many-particle systems, and the remarkable rule (the Pauli principle) that they have to satisfy. The experimental basis for the rule emerges as we move through the field of atomic spectra. Then, with it established, we shall have enough quantum mechanics at our disposal to enable us to discuss most aspects of atomic and molecular structures and properties. In this part of the book we look at the straightforward applications of quantum mechanics to basic properties of chemical interest. We look in some detail at atomic structure, and then extend the experience that provides to the discussion of molecular structure (but since that is such a richly developed field, we shall only sketch its basis and scope). Then we examine how transitions occur between the energy levels of molecules, and see the quantum mechanical basis of the extremely important techniques of spectroscopy. We shall see that the principles established in the first two parts let us discuss how molecules undergo changes in their states, and see the basis of the relation between spectral transitions and molecular structure.

9 Atomic spectra and atomic structure

A GREAT deal of chemically interesting information can be obtained by interpreting the *line spectra* of atoms. For one thing, we can establish the structures of the atoms themselves, and check the predictions of quantum mechanics. Atomic spectra led, in fact, to the discovery of a major principle of quantum mechanics, the Pauli principle, without which it is impossible to understand atomic structure, chemical periodicity, and molecular structure. We shall see the evidence for this principle, and then explore its implications for wavefunctions of many-electron species.

In order to discuss molecular structure we need to know the energy levels of electrons in atoms. This information comes from atomic spectra. We shall see that the ionization energies and the spin–orbit coupling constants of atoms are central to an understanding of molecular structure and properties, and that their values come quite directly from atomic spectra. The effects of electric and magnetic fields on the positions of spectral lines gives additional information. The former is particularly useful when we discuss the structures of transition metal complexes, because the effect of the ligands on the central ion can be modelled as the effect of an electric field on the central ion's electrons. There are also the straightforward applications of atomic spectra to the identification of elements in samples, the assessment of energy resources in reactions, and the detection and identification of excited atoms produced in photochemical processes and under extreme conditions such as in plasma, rocket exhausts, and stellar atmospheres.

We begin with the spectrum of atomic hydrogen. We have already established the atom's structure (in broad outline at least), and we should now test and refine the description.

9.1 The spectrum of atomic hydrogen

So long as we ignore electron spin, the state of an electron in an atom of hydrogen is specified by the three quantum numbers n, l, and m_l. The wavefunction corresponding to the state $|nlm_l\rangle$ is ψ_{nlm_l}, and has the form described in Chapter 4. These wavefunctions are referred to as *atomic orbitals*. The energy of the atom in the state $|nlm_l\rangle$ is given by eqn (4.3.7) as

$$E = -(\mu e^4/32\pi^2\varepsilon_0^2\hbar^2)(1/n^2), \qquad n = 1, 2, \ldots \qquad (9.1.1)$$

and is independent of the values of l $(=0, 1, \ldots, n-1)$ and m_l $(=l, l-1, \ldots, -l)$. It is normally written in the form

$$E_n = -hcR_{\mathrm{H}}/n^2, \qquad R_{\mathrm{H}} = \mu e^4/8\varepsilon_0^2 h^3 c, \qquad (9.1.2)$$

where R_{H} is the *Rydberg constant for hydrogen*. It has the dimensions of length^{-1} and is normally expressed in cm^{-1}. Substitution of the values of the fundamental constants gives $R_H = 109\,678\ \text{cm}^{-1}$. The *energy level diagram* for the atom is shown in Fig. 9.1.

As can be seen from its definition, R_H depends on the reduced mass of the proton and the electron, $\mu = m_e m_p/(m_e + m_p)$; sometimes it is convenient to use a form of the constant called R_∞ which involves the mass only of the electron:

$$R_\infty = m_e e^4/8\varepsilon_0^2 h^3 c = 109\,737.31\ \text{cm}^{-1}. \tag{9.1.3}$$

The two constants are related by $R_H = R_\infty/(1 + m_e/m_p)$. The energy levels with R_∞ would be the ones obtained if the proton had infinite mass and took no part in the motion. Expressing the energy using R_∞ makes it particularly easy to adapt to the isotopes of hydrogen and to hydrogen-like, one-electron, highly ionized atoms of other elements simply by using the appropriate mass in the correction factor $1/(1 + m_e/m)$ (and the appropriate value of the nuclear charge Z; $E \propto Z^2$).

The spectrum of atomic hydrogen arises from transitions between states of different energy, the excess energy being emitted as radiation (or absorbed if the transition is upwards in energy). The energy involved in the transition from a state $|n_1 l_1 m_{l1}\rangle$ to a state $|n_2 l_2 m_{l2}\rangle$ is predicted to be

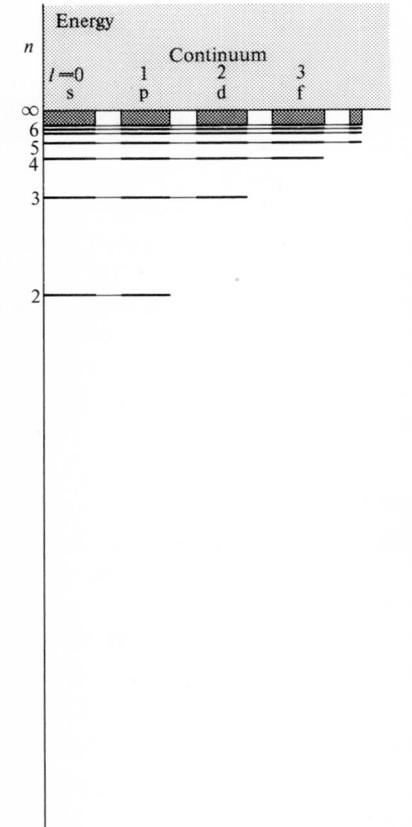

Fig. 9.1. Hydrogen atom energy levels.

$$\Delta E = hcR_H\left\{\frac{1}{n_1^2} - \frac{1}{n_2^2}\right\} \tag{9.1.4}$$

irrespective of the values of l and m_l of the states. This energy is carried away as a photon of energy ΔE and therefore of frequency $\nu = \Delta E/h$. Spectra are often reported in terms of the *wavenumbers* of the lines $\tilde{\nu} = \nu/c$. Since $\tilde{\nu} = (c/\lambda)/c = 1/\lambda$, we can imagine the wavenumber as being the number of wavelengths per unit length. Wavenumbers are normally expressed in cm^{-1}, and so they are the number of wavelengths needed to span 1 cm. In terms of wavenumbers the transitions are predicted to lie at

$$\tilde{\nu} = \Delta E/hc = R_H\left\{\frac{1}{n_1^2} - \frac{1}{n_2^2}\right\} \tag{9.1.5}$$

(which is why the Rydberg constant was defined with a dangling hc).

Consider the absorptions that occur as n_1 takes the values $1, 2, \ldots$ successively. For each vaue of n_1, transitions may take the atom to a state with $n_2 = n_1 + 1,\ n_1 + 2, \ldots,$ and hence give rise to a *series* of absorptions (and the corresponding series of emissions). These series have been named after their principal investigators, and are the *Lyman* ($n_1 = 1$; $n_2 = 2, 3, \ldots$), *Balmer* ($n_1 = 2$; $n_2 = 3, 4, \ldots$), *Paschen* ($n_1 = 3$), *Brackett* ($n_1 = 4$), *Pfund* ($n_1 = 5$), and *Humphreys* ($n_1 = 6$) *series*. The Lyman series occurs in the ultraviolet, the Balmer in the visible (which is why it was first to be studied), and the rest in the infrared and the far infrared (which is why they were studied much later, because detectors became available only relatively recently). The *limit* of each series is the wavenumber corresponding to the transition that involves just enough energy to ionize the atom. The limits correspond to $n_2 = \infty$, and hence

$$\textit{Series limit}: \ \tilde{\nu} = R_H/n_1^2. \tag{9.1.6}$$

The *ionization energy*, I, of the atom is the minimum energy required to ionize it from its ground state. This is readily determined from the series limit with $n_1 = 1$ and hence we can predict (and find) that $I = hcR_H$, or 2.179×10^{-18} J (corresponding to 1319 kJ mol^{-1}, 13.67 eV).

9.2 Selection rules

All is well, so far. The question that now arises recalls that a level corresponding to the quantum number n is n^2-fold degenerate (p. 77). The transitions shown in Fig. 9.1 suggest that they occur indiscriminately between states of any l, m_l values. This is not so. There are *selection rules* that govern which transitions may occur (are *allowed*) and which may not (are *forbidden*).

An indication of the basis of the existence of selection rules comes from noting that *a photon has spin*, an intrinsic, unremovable angular momentum. The evidence for this is the existence of left- and right-circularly polarized light, corresponding to two states of spin with respect to the direction of propagation.† The interpretation of this evidence is not quite straightforward because two components of angular momentum with respect to some direction would normally signify $s = \frac{1}{2}$. In the case of a particle travelling at the speed of light (like photons, and perhaps also like neutrinos and gravitons) relativity forbids any component perpendicular to the propagation direction, and so whatever the spin, only two components are allowed. Therefore, all we can conclude from the existence of polarized light is that the spin of the photon is non-zero. A deeper analysis, taking note of the fact that the electromagnetic field is a *vector field* (i.e. **E** and **B** are vectors varying periodically through space) shows that photons have $s = 1$. The two components correspond to $m_s = 1$ (*left-circularly polarized light*, according to a historically based convention) and $m_s = -1$ (*right-circularly polarized light*).

When a photon plunges into an atom and is absorbed it transfers all its angular momentum to the electrons. When a photon is emitted, it carries away unit angular momentum. Therefore in either case, so long as we consider only the spin of the photon, the angular momentum of the electron cannot change by more than one unit. It follows that an electron in a d-orbital cannot make a transition to an s-orbital. This

† So too is the observation that if a circularly polarized beam of light is passed through as optically birefringent crystal hanging on a thread, the crystal acquires an angular momentum (R. A. Beth, *Phys. Rev.*, **50,** 115 (1936)).

can be summarized by the following selection rule:

Electric dipole transitions: $\Delta l = \pm 1$. (9.2.1)

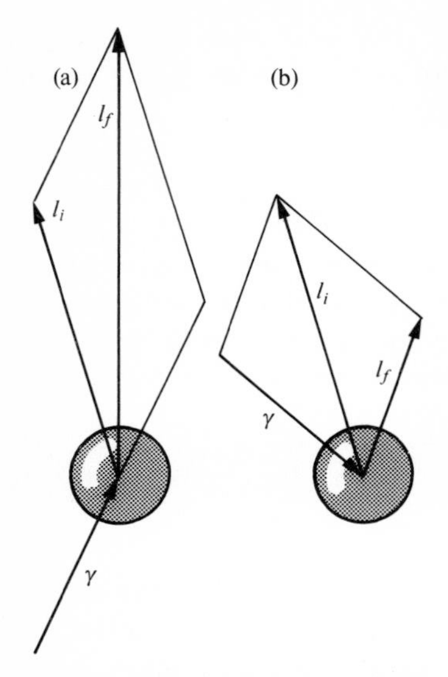

Fig. 9.2. Conservation of orbital angular momentum on absorption of a photon (γ).

There are various points hidden in this remark. One concerns the name 'electric dipole transition'. This is the simplest (and strongest; i.e. most probable) type of transition that may occur, and we examine it in more detail shortly. As expressed, the selection rule is equivalent to supposing that the angular momentum conservation may be discussed solely in terms of the spin of the photon: we have been careful to emphasize that the spin is only the *minimum* angular momentum of a photon. A photon can also possess orbital angular momentum, and so its total angular momentum may exceed unity. Therefore transitions with greater values of Δl may also occur. These higher-order transitions are normally much weaker because the photon has to be given an orbital 'flip' as it is generated. More on this below. Note too that the + and − signs in $\Delta l = \pm 1$ do not denote emission and absorption: $\Delta l = +1$ (e.g. s→p) and $\Delta l = -1$ (e.g. p→s) may apply in either process. The angular momentum of an electron may increase or decrease during emission or absorption, because satisfying the angular momentum conservation law depends on the relative directions of the angular momenta, Fig. 9.2.

It is quite easy to extend these pictures to obtain selection rules for Δm_l. Suppose Δm_l refers to the component of angular momentum of the electron with respect to the propagation direction of the light; then an absorption of a left-handed circularly polarized photon ($m_s = 1$) results in $\Delta m_l = +1$, while its emission results in $\Delta m_l = -1$, Fig. 9.3. The opposite holds for right-circularly polarized photons. *Plane polarized photons* are linear superpositions of circularly polarized photons, Fig. 9.4, and when they are absorbed give rise to linear combinations of orbitals. Thus, x-polarized light (a superposition of $m_s = +1$ and $m_s = -1$) generates the superposition $p_{+1} + p_{-1}$, which is a p_x-orbital, Fig. 9.4. This example also shows clearly the connection between the shift of charge in an atom and the polarization of the wave it requires or produces. Transmitting antennae used for radio and television are macroscopic 'classical' versions of the same phenomenon.

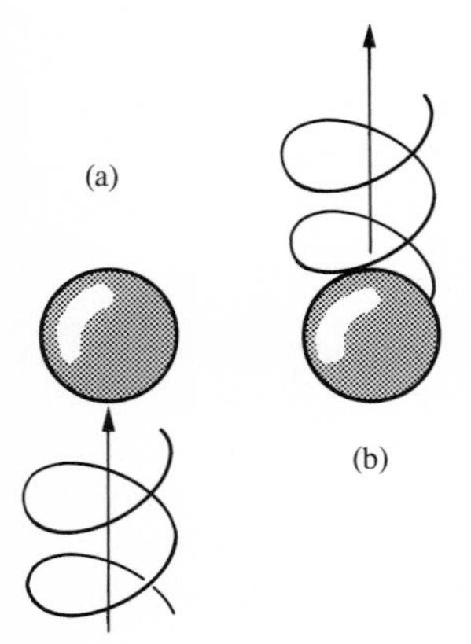

Fig. 9.3. (a) Absorption, (b) emission of (left-hand) circularly polarized components.

These remarks can be put on a more formal basis by setting up the quantum mechanical description of the transitions and then arguing on the basis of symmetry. *Selection rules are aspects of symmetry*, as we shall now see.

Time-dependent perturbation theory (Chapter 8) shows that the transition rate (and therefore the spectral intensity) depends on the square of the matrix element of the perturbation. The simplest form of the interaction between the electromagnetic field $\mathbf{E}(t)$ and an atom (or any system of charges) is a

Dipolar interaction: $H^{(1)}(t) = -\boldsymbol{\mu} \cdot \mathbf{E}(t)$. (9.2.2)

$\boldsymbol{\mu}$ is the electric dipole moment operator, which for a single electron of charge $-e$ is $\boldsymbol{\mu}=-e\mathbf{r}$, $\mathbf{r}$ being its location relative to the nucleus. The intensity of a transition from some state $|n_1l_1m_{l1}\rangle$ to another state $|n_2l_2m_{l2}\rangle$ is proportional to the square of the *transition matrix element* $\langle n_2l_2m_{l2}|\,H^{(1)}\,|n_1l_1m_{l1}\rangle$ and therefore to the *transition dipole moment* $\langle n_2l_2m_{l2}|\,\boldsymbol{\mu}\,|n_1l_1m_{l1}\rangle$. The transition moment can be regarded as a measure of the size of the jolt the system gives to the electromagnetic field when the charge redistribution occurs: big charge shifts through large distances deliver strong impulses, and give rise to intense lines. Finding the condition under which the transition dipole moment is non-zero is equivalent to finding the selection rules for the transitions.

Group theory tells us that *the transition dipole moment is zero unless it is totally symmetrical under the symmetry operations of the system.* The easiest operation to consider is the inversion. Under the inversion $\mathbf{r}\rightarrow-\mathbf{r}$; wavefunctions corresponding to l even (s, d, . . .) do not change sign (have *even parity*); wavefunctions with l odd (p, f, . . .) do change sign (have *odd parity*). The structure of the integrand is therefore $(-1)^{l_2}(-1)(-1)^{l_1}$, and is even overall only when one l is even and the other is odd. This is the *Laporte selection rule*: *the only allowed electric dipole transitions are those involving a change of parity.* Even $\rightarrow$ even and odd $\rightarrow$ odd are forbidden; even $\leftrightarrow$ odd are allowed. Notice how this includes yet goes beyond the angular momentum conservation rule. According to that argument a p $\rightarrow$ p (or in general an $l\rightarrow l$ transition) is allowed, because, for example, a spin-1 photon can be emitted in a direction such as to allow a p-electron to switch into another p-electron (the triangle condition for angular momenta can be satisfied by three unit momenta); the Laporte selection rule, however, forbids this transition, because it involves no change of parity. The angular momentum selection rule also goes beyond the Laporte rule, for the latter allows f $\rightarrow$ s transitions, and other odd–even transitions; but these are forbidden by angular momentum conservation. Always be aware of the interplay of selection rules, for different rules may forbid different transitions.

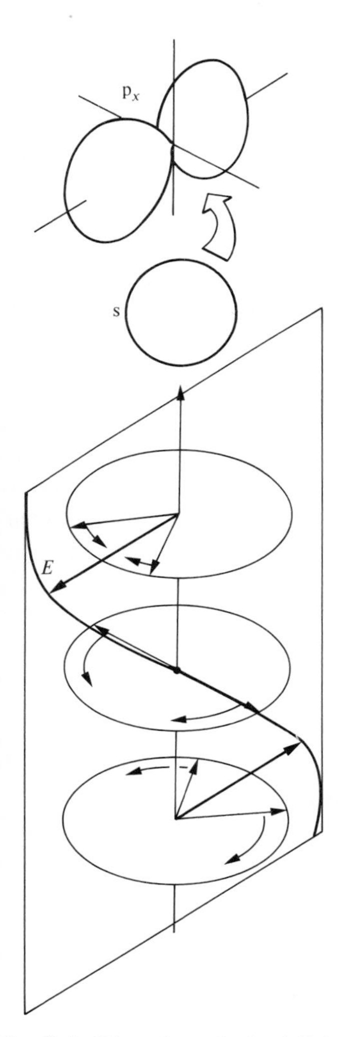

Fig. 9.4. Linearly polarized light as a superposition of circularly polarized components.

The form of the transition dipole moment lets us arrive at the selection rule for Δl and Δm_l algebraically. The latter is easier, and we consider it here. Consider plane polarized radiation with the electric field pointing in the z-direction. The interaction hamiltonian has the form $-\mu_zE(t)=ezE(t)=er\cos\theta E(t)$. The interaction dipole moment is an integral over the electron's coordinates r, θ, and ϕ. The ϕ integral has the form

$$\int_0^{2\pi}\mathrm{e}^{-\mathrm{i}m_{l2}\phi}(er\cos\theta)\mathrm{e}^{\mathrm{i}m_{l1}\phi}\,\mathrm{d}\phi=er\cos\theta\int_0^{2\pi}\mathrm{e}^{-\mathrm{i}(m_{l2}-m_{l1})\phi}\,\mathrm{d}\phi.$$

This is zero unless $m_{l2}=m_{l1}$. Therefore, for z-polarized radiation the selection rule is $\Delta m_l=0$. The rules $\Delta m_l=\pm1$ arise in a similar way for light polarized in the xy-plane.

Example. Estimate the electric dipole transition moment for a $2p_z\rightarrow 2s$ transition of an electron between Slater orbitals.

- *Method.* The forms of the orbitals are set out on p. 234. Evaluate the integral $\int \psi_{2p_z}\mu_z\psi_{2s}\,d\tau$ with $\mu_z = -ez = -er\cos\theta$.
- *Answer.*

$$\psi_{2p_z} = (Z^{*5}/32\pi a_0^5)^{\frac{1}{2}} r\cos\theta e^{-Z^*r/2a_0},$$

$$\psi_{2s} = (Z^{*5}/96\pi a_0^5)^{\frac{1}{2}} r e^{-Z^*r/2a_0}.$$

$$\mu_z = -(e/\sqrt{3})(Z^{*5}/32\pi a_0^5)\int_0^{2\pi} d\phi \int_0^{\pi} \cos^2\theta \sin\theta\, d\theta \int_0^{\infty} r^5 e^{-Z^*r/a_0}\, dr$$

$$= -(5/Z^*\sqrt{3})ea_0.$$

- *Comment.* In the case of a transition involving an oxygen atom, $Z^* = 4.55$ and so $\mu_z = -0.634ea_0$, corresponding to the movement of an electron through a distance of $0.634a_0 = 33.6$ pm.

At this stage we have three rules for electric dipole transitions:

$$\Delta n\text{: unrestricted; } \Delta l = \pm 1;\ \Delta m_l = 0, \pm 1 \text{ (polarization dependent)} \tag{9.2.3}$$

This suggests that a more precise way of depicting the transitions that contribute to the spectrum of atomic hydrogen is to separate the contributions corresponding to different values of *l*. The resulting *Grotrian diagram* is shown in Fig. 9.5.

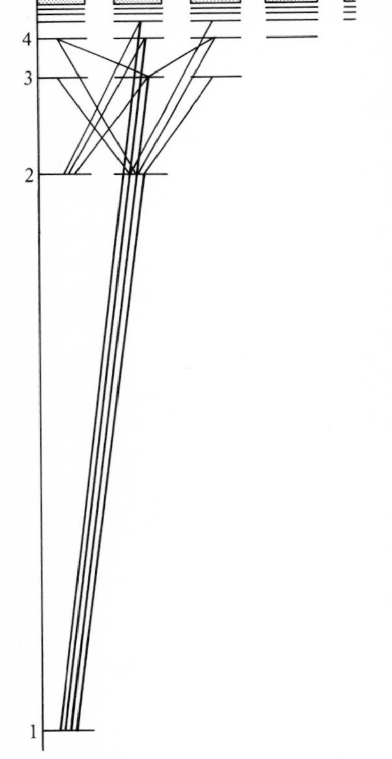

Fig. 9.5. Grotrian diagram for the transitions of atomic hydrogen.

Electric dipole transitions are not the only types of transition that may occur. Light is an electro*magnetic* phenomenon, and the perturbation arising from the effect of the oscillating magnetic component of the light beam can induce *magnetic dipole transitions*. Such transitions are generally much weaker (about 10^5 times weaker) than the electric dipole transition, but because they satisfy different selection rules they may give rise to lines where the electric dipole transitions are forbidden. Another type of transition is the *electric quadrupole transition*. This is weaker still, having only 10^{-8} times an electric dipole transition's strength. Its selection rule is $\Delta l = 0, \pm 2$, the large angular momentum change arising from the fact that the electric quadrupole transition imparts an orbital angular momentum to the photon as it is generated, and this can augment its spin to produce a greater total angular momentum. Both magnetic dipole and electric quadrupole transitions are weaker than electric dipole transitions because while the latter depends on the electric field itself, they depend on its variation over the extent of the molecule, and since the wavelength is much longer than the molecular diameter the variation is only slight. In some cases a system can generate two photons by an electric dipole mechanism more efficiently than it can generate one by a magnetic dipole transition. An example of such *multiple-quantum dipole transitions* is provided by the excited ^1S state of the helium atom, where the two-photon process governs the radiative lifetime because the magnetic dipole transition probability is so low.

9.3 Spin: fine structure and spin–orbit coupling

So far we have ignored the spin of the electron. Now we consider its effects on the spectrum of the hydrogen atom, and therefore on the energy levels. While its effect is not very pronounced in hydrogen (which is why we have been able to ignore it until now) in the atoms of the heavier elements it is of great importance, and we shall see how the ideas we are about to introduce can have a dominating influence on their spectra. In hydrogen the role of the spin is so small that, it does little more than blur the lines except under high resolution, where structure—*fine structure*—is seen. This blurring is the source of the nomenclature *sharp, principal, diffuse,* and *fundamental* which now survives as the names of the s-, p-, d-, and f-orbitals (which are involved in the respective series of transitions).

An electron is a charged particle, and so if it has angular momentum it should also have a magnetic moment. But an electron in an atom may possess two kinds of angular momentum, its orbital angular momentum and its spin. Therefore an electron should give rise to two magnetic moments. We should anticipate that they interact (like any pair of magnetic dipoles) and that the energy of interaction affects the atom's energy levels and therefore the appearance of its spectrum. As early as 1887 Michelson and Morley had reported a doublet structure in the H_α spectral line, and so there is obviously something to explain.

First, consider the magnetic moment associated with the electron's orbital angular momentum. We deduce its magnitude by using a classical analogy, but there is a 'quantum mechanical' way of arriving at the same result, and it is described in Chapter 14. If a charge circulates in an orbit of radius r at a speed v, the magnitude of the current, the magnitude of the charge per unit time passing some point of the orbit, is $I = e(v/2\pi r)$. This current gives rise to a magnetic dipole of magnitude IA, where A is the area enclosed by the orbit. Since $A = \pi r^2$ for a circular orbit, the magnetic moment is $(ev/2\pi r)\pi r^2 = (e/2)rv = (e/2m_e)m_e rv$. The factor $m_e rv$ is the magnitude of the angular momentum of a body of mass m_e, which in quantum mechanics is $\{l(l+1)\}^{\frac{1}{2}}\hbar$. Therefore the magnitude of the orbital magnetic moment is $(e/2m_e)\{l(l+1)\}^{\frac{1}{2}}\hbar$, and so the magnitude of the moment is proportional to the magnitude of the momentum. This suggests that the magnetic moment itself (the vector) is proportional to the angular momentum vector. Since the electron is negatively charged, its magnetic moment is opposite in direction to its angular momentum, Fig. 9.6. Therefore we write

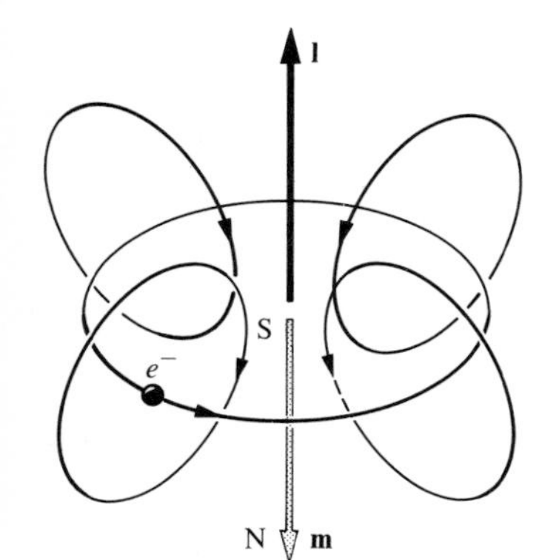

Fig. 9.6. The orbital magnetic moment of an electron.

Orbital magnetic moment: $\mathbf{m} = \gamma_e \mathbf{l}, \qquad \gamma_e = -e/2m_e.$ (9.3.1)

γ_e is called the *magnetogyric ratio.*

The properties of the orbital magnetic moment follow from the properties of the angular momentum. Its magnitude is quantized, and its z-component is restricted to the values

$$m_z = m_l \gamma_e \hbar. \tag{9.3.2}$$

The positive quantity $-\gamma_e \hbar = e\hbar/2m_e$ is called the *Bohr magneton* and denoted μ_B; $\mu_B = 9.274 \times 10^{-24}$ J T^{-1}. It can be regarded as an elementary unit of magnetism (the *nuclear magneton*, $\mu_N = e\hbar/2m_p$, plays the same role for the magnetism arising from nuclei).

Now turn to the magnetic moment arising from the electron spin. By analogy we might expect the spin and the magnetic moment to be related by $\mathbf{m} = \gamma_e \mathbf{s}$, but this turns out not to be the case. This should not be too surprising, because the spin has no classical analogy, yet we have tried to argue by analogy to arrive at its magnetic moment. The experimental demonstration of the existence of the spin magnetic moment was by means of the *Stern–Gerlach experiment*. In this experiment a beam of silver atoms (which have a single outermost electron, and so the rest of the atom acts as a kind of platform for the electron spin) are shot through an inhomogeneous magnetic field. This field separates the beam into two components, one corresponding to the α ($m_s = \frac{1}{2}$) state of the electron, and the other to the β ($m_s = -\frac{1}{2}$) state. The separation of the detected peaks give the magnitude of the dipole moment, and it is found to be twice that suggested by the classical analogy. Modern experiments have greatly refined the measurement of the electron magnetic moment, and it is now known with great precision. We now write

$$\mathbf{m}_s = g_e \gamma_e \mathbf{s}, \qquad g_e = 2.002\,319\,314. \tag{9.3.3}$$

We shall often take $g_e = 2$ because it greatly simplifies expressions and calculations. The quantity g_e is called the *g-factor* of the electron. The value $g_e = 2$ can be deduced from Dirac's theory of the electron but the remaining 0.002 3 . . . comes from the more sophisticated theory of *quantum electrodynamics*, which accounts for it fully.†

The danger of arguing by classical analogy is also shown up when we turn to the next aspect of the problem, the calculation of the interaction energy between the orbital and spin magnetic momenta of a single electron. The classical calculation runs as follows.

A body moving with a velocity $\mathbf{v}$ in an electric field $\mathbf{E}$ experiences a magnetic field $\mathbf{B} = \mathbf{E} \wedge \mathbf{v}/c^2$. If the field is due to an isotropic potential ϕ the field $\mathbf{E}$ is the gradient $\mathbf{E} = -(\mathbf{r}/r)\,\mathrm{d}\phi/\mathrm{d}r$. Therefore $\mathbf{B} = -(1/c^2 r)(\mathrm{d}\phi/\mathrm{d}r)\mathbf{r} \wedge \mathbf{v}$. The angular momentum of the particle (an electron of mass m_e) is $\mathbf{l} = \mathbf{r} \wedge \mathbf{p} = \mathbf{r} \wedge m_e \mathbf{v}$, and so $\mathbf{B} = (-1/m_e c^2 r)(\mathrm{d}\phi/\mathrm{d}r)\mathbf{l}$. The energy of interaction between a field $\mathbf{B}$ and a magnetic dipole $\mathbf{m}_s$ is equal to $-\mathbf{m}_s \cdot \mathbf{B}$. Therefore we expect the spin–orbit coupling hamiltonian to be $(1/m_e c^2 r)(\mathrm{d}\phi/\mathrm{d}r)\mathbf{m}_s \cdot \mathbf{l}$, or $-(e/m_e^2 c^2 r)(\mathrm{d}\phi/\mathrm{d}r)\mathbf{l} \cdot \mathbf{s}$, where we have used $g_e = 2$. This is twice the result obtained from the solution

† Quantum electrodynamics takes the electromagnetic field and quantizes it in terms of a collection of harmonic oscillators. We know that an harmonic oscillator is never completely at rest because it has a zero-point energy; therefore, even in the absence of photons, the 'vacuum' consists of fluctuating electric and magnetic fields. An electron, being charged, experiences these vacuum fluctuations, and instead of moving smoothly it jitter-bugs ('*Zitterbewegung*'). As it 'spins' we may also imagine that it wobbles because of the fluctuating fields, and one may imagine a wobble in its equatorial plane that increases the magnitude of its magnetic moment.

of the Dirac equation. The error is the implicit assumption that one can step from the stationary nucleus to the moving electron without treating the change of viewpoint relativistically. When the correct relativistic calculation is carried through, an extra factor of $\frac{1}{2}$ appears.† The correct result is that the hamiltonian for the interaction takes the form

Spin-orbit interaction: $H_{so} = \xi(r)\mathbf{s}\cdot\mathbf{l},\ \xi(r) = -(e/2m_e^2c^2)(1/r)(\mathrm{d}\phi/\mathrm{d}r).$ (9.3.4)

The radial average of this hamiltonian is important, and gives the average interaction energy of an electron in some orbital with its own spin. Writing the wavefunction for the state $|nlm_l\rangle$ as the product $R_{nl}(r)Y_{lm_l}(\theta, \phi)$, we have the

Spin-orbit coupling constant: $hc\zeta_{nl} = \hbar^2 \int_0^\infty \xi(r)R_{nl}^2(r)r^2\,\mathrm{d}r.$ (9.3.5)

(Defined in this way the ζ has the dimensions of a wavenumber.) In the case of the hydrogen-like atom with a nucleus of atomic number Z, $\phi(r)$ is the Coulomb potential $Ze/4\pi\varepsilon_0 r$ (remember that ϕ is the potential, not the potential energy V; $V = -e\phi$). Therefore

$$\xi(r) = (Ze^2/8\pi\varepsilon_0 m_e^2c^2)(1/r^3) \tag{9.3.6}$$

and the evaluation of the radial integral using the appropriate hydrogenic orbitals (i.e. the hydrogen orbitals for a general value of Z, Table 4.2) and $l > 0$ leads to

$$\langle 1/r^3\rangle_{nl} = \frac{(Z/a_0)^3}{n^3 l(l+\frac{1}{2})(l+1)}; \qquad a_0 = 4\pi\varepsilon_0\hbar^2/m_e e^4. \tag{9.3.7}$$

(a_0 is the Bohr radius.) Therefore, for an nl-electron in a hydrogen-like atom the spin–orbit coupling constant is

$$\zeta_{nl} = \frac{(Ze^2/8\pi\varepsilon_0 m_e^2c^2)(Z/a_0)^3}{n^3 l(l+\frac{1}{2})(l+1)hc} = \frac{\alpha^2 R_\infty Z^4}{n^3 l(l+\frac{1}{2})(l+1)}. \tag{9.3.8}$$

The dimensionless constant α is called the *fine-structure constant*, and

† The phenomenon that gives rise to the factor $\frac{1}{2}$ is *Thomas precession*. The electron moves in its orbital with velocities close to the velocity of light, and one has to take account of this on stepping from the coordinate system fixed on the nucleus to the coordinate system fixed on the electron; that is necessary if one is going to calculate the interaction energy of the spin and orbital moments. To an observer on the nucleus, the coordinate system on the electron seems to rotate in the plane of motion, and the electron moves with just the right velocity so that its coordinate system seems to have rotated by 180° when it has completed one circuit of the nucleus. It is spinning within its own frame, and so it appears to the observer on the nucleus that it is spinning with only one-half its rate if it were static: this effectively reduces the magnetic moment for the interaction to $\frac{1}{2}g_e\gamma_e\mathbf{s}$.

has the value

$$\alpha = e^2/4\pi\varepsilon_0\hbar c = 1/137.036\,04. \tag{9.3.9}$$

Although we have encountered it in this special application (which is from where it takes its name) it is of extraordinarily broad significance because it is a fundamental constant for the strength of the coupling of charge to the electromagnetic field.

Now we revert to the special case of hydrogen itself, and take $Z = 1$. The magnitude of the coupling constant in the case $n = 2$, $l = 1$ (i.e. for a 2p-electron) is $\alpha^2 R_\infty/24 \approx R_\infty/4.51 \times 10^5$. Energy level separations and transition wavenumbers are of the order of R_∞ itself, and so spin–orbit interaction energies are about 5×10^5 times smaller. Since $R_\infty \approx 10^5\ \text{cm}^{-1}$, we should expect effects of only fractions of cm^{-1} to appear in the spectrum. Since the doublet splitting of the H_α line reported by Michelson and Morley was $0.3\ \text{cm}^{-1}$, we are on the right track for finding an explanation. In passing, note that $\zeta \propto Z^4$, and so as the atomic number of the element increases, spin–orbit coupling effects can become very large. What may seem to be a niggling problem for hydrogen can become of dominating importance for heavy elements, and what we learn now will be of crucial importance for them.

What effect does the spin–orbit interaction have on the energy levels and the spectrum of the hydrogen atom? It is easy to get a qualitative idea of the effect by considering the diagram in Fig. 9.7.

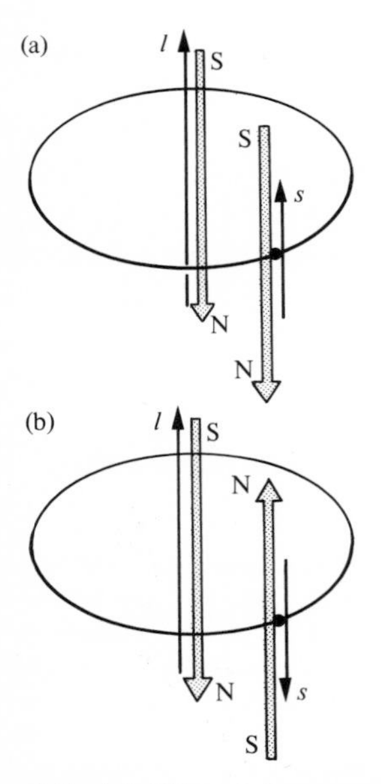

Fig. 9.7. (a) High j, low energy and (b) low j, high energy orientations of orbital and spin momenta.

When the spin and the orbital momenta are in the same direction, so that the total angular momentum j takes its largest value ($j = \frac{3}{2}$ for $l = 1$ and $s = \frac{1}{2}$, and $j = l + \frac{1}{2}$ in general when the orbital quantum number is l), the angular momenta have a relative orientation roughly as indicated in Fig. 9.7(a). The corresponding magnetic moments are then also parallel. This is an energetically unfavourable arrangement for two magnetic dipoles. When j takes its lowest value ($l - \frac{1}{2}$ in general, and $j = \frac{1}{2}$ in the case of a 2p-electron) the momenta are roughly antiparallel, and so the two magnetic moments are also antiparallel. This is an energetically favourable arrangement. We conclude that the energy of the level with $j = l + \frac{1}{2}$ should lie above that of the level with $j = l - \frac{1}{2}$, and that the magnitude of the energy separation should be of the order of the spin–orbit coupling constant (the magnetic interaction energy). This splitting of energy levels gives rise to the fine structure of the spectrum.

Now consider the quantitative treatment of the effect. Since the interaction is so weak in comparison with the energy level separations of the atom, we can use first-order perturbation theory to assess the effects. The first-order correction to the energy of a state $|nls; jm_j\rangle$ is $\langle nls; jm_j|\, H_{\text{so}}\, |nls; jm_j\rangle$. (Note that we are using the coupled representation of the states of the atom: this is the natural choice because it leads to the simplest description, in the sense that H_{so} is then diagonal.) The spin–orbit coupling energy for this state is

$$E_{\text{so}} = \langle nls; jm_j|\, H_{\text{so}}\, |nls; jm_j\rangle = \langle nls; jm_j|\, \xi(r)\mathbf{l}\cdot\mathbf{s}\, |nls; jm_j\rangle. \tag{9.3.10}$$

The matrix element of a scalar product of two operators can be evaluated very simply by noting that

$$j^2 = |(\mathbf{l}+\mathbf{s})|^2 = l^2 + s^2 + 2\mathbf{l}\cdot\mathbf{s}; \tag{9.3.11}$$

therefore

$$\begin{aligned}\mathbf{l}\cdot\mathbf{s}\,|nls;jm_j\rangle &= \tfrac{1}{2}(j^2 - l^2 - s^2)\,|nls;jm_j\rangle \\ &= \tfrac{1}{2}\hbar^2\{j(j+1) - l(l+1) - s(s+1)\}\,|nls;jm_j\rangle,\end{aligned}$$

and so the interaction energy is

$$\begin{aligned}E_{\mathrm{so}} &= \tfrac{1}{2}\hbar^2\{j(j+1) - l(l+1) - s(s+1)\}\langle nls;jm_j|\,\xi(r)\,|nls;jm_j\rangle \\ &= \tfrac{1}{2}hc\zeta_{nl}\{j(j+1) - l(l+1) - s(s+1)\} \\ &= \alpha^2 hcR_\infty\left\{\frac{j(j+1) - l(l+1) - s(s+1)}{2n^3 l(l+\frac{1}{2})(l+1)}\right\}.\end{aligned} \tag{9.3.12}$$

In the case of s-orbitals, the spin–orbit interaction energy is zero (there is no orbital magnetic moment). In the case of p-orbitals the separation between the $j=\frac{3}{2}$ and $j=\frac{1}{2}$ levels is $\alpha^2 hcR_\infty/2n^3$, and so it rapidly becomes negligible as the principal quantum number increases. For a 2p-electron the fine-structure splitting is $(\alpha^2/16)R_\infty = 0.365\ \mathrm{cm}^{-1}$, Fig. 9.8. Note that the energy is independent of the value of m_j (i.e. independent of the orientation of the total angular momentum in space, as is physically plausible in the absence of external fields), and so a level with quantum number j has $2j+1$ degenerate states. Note too that the 'centre of gravity' of the levels is the same before and after the spin–orbit interaction has been taken into account.

9.4 Term symbols and spectral details

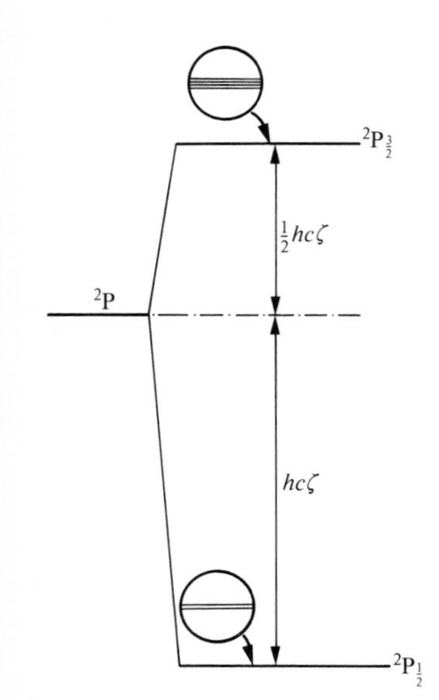

Fig. 9.8. Level structure of a ^{2}P term.

We now have the detailed energy level diagram for the hydrogen atom, but in order to simplify the discussion we need some more notation. Spectral transition lines arise from transitions between *terms*, the wavenumber of a line being expressed as the difference between two terms: $\tilde{\nu} = T' - T$. Any transition can be denoted $T' \rightarrow T$ or $T' \leftarrow T$, and we shall adopt the convention that *the first term is higher in energy than the second.* Therefore $T' \leftarrow T$ is an absorption and $T' \rightarrow T$ is an emission.

The *configuration* of an atom is the specification of the orbitals the electrons occupy. In the case of the hydrogen atom there is only one electron and so we speak of the configuration 1s, or the configuration 2p, and so on. A single configuration (e.g. 2p) may give rise to several terms. In the case of the hydrogen atom every term with $l>0$ is a *doublet* in the sense that each one has two *levels* distinguished by two values of j. For example, each np configuration gives rise to the two levels $j=\frac{3}{2}$, $j=\frac{1}{2}$; each nd configuration gives rise to the levels $j=\frac{5}{2}$, $j=\frac{3}{2}$. The name level always applies to terms with a specified value of j. Each level may be degenerate because each has $2j+1$ *states* distinguished by the quantum number m_j. In summary, we have the hierarchy: *configuration–term–level–state.*

The *term symbol* is a label constructed as follows.

Multiplicity → $2S+1$
$\{L\}$ ← Orbital angular momentum
J ← Level

In the case of the hydrogen atom, the total orbital angular momentum is the same as the orbital angular momentum of the single electron, and so $L=l$. The value of L is denoted by the letters S, P, D, F, G, . . . according to $L=0, 1, 2, 3, 4, \ldots$. Therefore the 1s configuration gives rise to an S term; the 2p configuration gives rise to a P term, and so on. The *multiplicity* of a term is the number of levels it can give rise to. When $L \geq S$ the Clebsch–Gordan series has $2S+1$ members, and so there are $2S+1$ levels. The left superscript on the term symbol is the value of $2S+1$, and therefore, when $L \geq S$, this is equal to the multiplicity of the term. When $L < S$ there are only $2L+1$ entries in the Clebsch–Gordan series, and so the multiplicity is only $2L+1$. Therefore, although the left superscript is always the value of $2S+1$, and is also loosely called the multiplicity, it is in fact the multiplicity only if $L \geq S$. In the case of the hydrogen atom, where $S=s=\frac{1}{2}$, $2S+1=2$ and so all the terms are called doublets and denoted $^2\{L\}$; but ^2S terms have only one level ($J=j=\frac{1}{2}$).

The right subscript of the term symbol denotes the *level*. The ground term of hydrogen is therefore $^2S_{\frac{1}{2}}$. The 2p configuration gives rise to the doublet term ^2P with its two levels $^2P_{\frac{3}{2}}$ and $^2P_{\frac{1}{2}}$. In the rare cases where a *state* needs to be specified, the term symbol carries the value of M_J ($=m_j$ for hydrogen) as a right superscript. Thus the $m_j=\frac{1}{2}$ state of the $\frac{3}{2}$ level arising from the 2p configuration would be denoted $^2P^{\frac{1}{2}}_{\frac{3}{2}}$. The configuration is also sometimes written in front of the term symbol. Then the fully dressed specification of a state might be 2p $^2P^{\frac{1}{2}}_{\frac{3}{2}}$.

Example. Construct the term symbols that may arise from the ground configurations of the neutral carbon and fluorine atoms.

- *Method.* The ground configurations are C: $1s^2 2s^2 2p^2$ and F: $1s^2 2s^2 2p^5$. The former consists of a p^2-configuration outside a closed shell; the latter should be regarded as a single hole in an otherwise closed shell. A hole may be treated as a particle. First construct L, then S, then J for the configurations using the Clebsch–Gordan series. The triplet term of carbon has an antisymmetric spatial component: see Appendix 11.

- *Answer.* For carbon, p^2 gives rise to $L=2$, [1], 0, where [ ] denotes the antisymmetric contribution. Two electrons ($s_1=s_2=\frac{1}{2}$) give $S=1, 0$. The terms are therefore $^1D+{}^3P+{}^1S$. ^1D has $J=2$ (because $L=0$, $S=0$); ^3P has $J=2, 1, 0$ (because $L=1$, $S=1$); ^1S has $J=0$ (because $L=0$, $S=0$). Hence the configuration gives rise to $^1D_2$; $^3P_2$, $^3P_1$, $^3P_0$; $^1S_0$. For fluorine p^5 is equivalent to p^1, and so $L=l=1$, $S=s=\frac{1}{2}$. Therefore only a ^2P term arises, its levels being $^2P_{\frac{3}{2}}$, $^2P_{\frac{1}{2}}$.

- *Comment.* The treatment of holes and particles as identical (for angular momentum considerations) amounts to the *particle–hole equivalence*. We shall see soon that the order of the levels of the two complementary configurations (i.e. p^n and p^{-n}) are opposite (p. 241).

The term symbols have been written on the energy level diagram for the hydrogen atom in Fig. 9.9. In order to predict the appearance of the spectrum we need to express the selection rules in terms of them. Electric dipole transitions must satisfy

$$\Delta J = 0, \pm 1 (J = 0 \nleftrightarrow J = 0); \qquad \Delta L = \pm 1, 0 \qquad \Delta l = \pm 1; \qquad \Delta S = 0. \tag{9.4.1}$$

The ΔJ and ΔL rules express the general point about the conservation of angular momentum. The Δl rule reflects this conservation for a single electron and goes on to include the Laporte selection rule (relating to the parity change that must occur). The ΔS rule reflects the point that the electric component of the electromagnetic field can have no effect on the internal spin angular momentum of the electron.

Now consider the application of these rules to the energy level diagram. Consider the transitions that can contribute to the H_α line of the Balmer series (the red light in the spectrum). The upper terms have $n = 3$ and the lower have $n = 2$. The configuration 3s gives rise to the ^2S term with the single level $^2S_{\frac{1}{2}}$. The configuration 3p gives rise to $^2P_{\frac{1}{2}}$ and $^2P_{\frac{3}{2}}$, and although these levels are split by the spin–orbit interaction, the splitting is very small and almost negligible. The 3d configuration gives rise to the ^2D term with the two levels $^2D_{\frac{5}{2}}$ and $^2D_{\frac{3}{2}}$. Now consider the lower energy terms. The configuration 2s gives rise to $^2S_{\frac{1}{2}}$. The other configuration, 2p, gives rise to the doublet $^2P_{\frac{1}{2}}$ and $^2P_{\frac{3}{2}}$ split by the spin–orbit interaction by 0.36 cm^{-1}, as we saw before. The allowed transitions between the terms from the upper configuration to those of the lower are marked on Fig. 9.9. Because the only appreciable splitting occurs in the lower ^2P term, the transitions fall into two groups separated by 0.36 cm^{-1}. The doublet structure in the spectrum is therefore a *compound doublet* arising from two almost coincident groups of transitions.

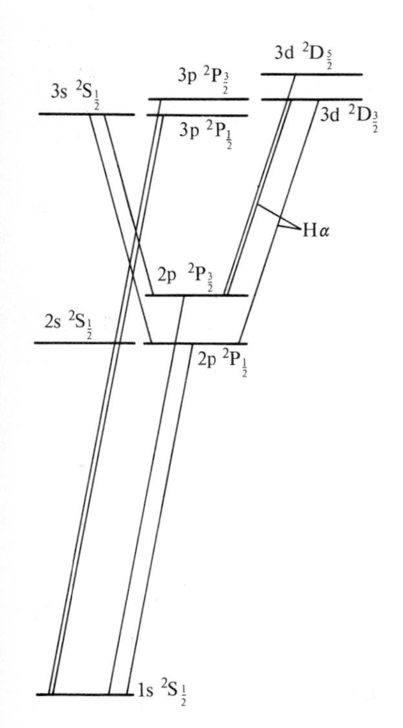

Fig. 9.9. Hydrogen atom fine structure.

We shall now leave hydrogen and apply similar reasoning to more complicated atoms. Nevertheless, there are still features of the spectrum to explain, but they depend on more advanced techniques, and are of less relevance to chemistry.†

† One thing that may confuse is that the energy of the term $^2S_{\frac{1}{2}}$ is the same as the energy of the level $^2P_{\frac{1}{2}}$ arising from the same value of n: at least that is what the Dirac equation predicts. One way to view this is that we have neglected to take into account the relativistic increase of the mass of the electron with its velocity: when this is done properly it gives rise to a contribution to the energy of the same order of magnitude as the spin–orbit coupling, and is of such a form that it makes levels with the same value of j but different l degenerate. Nevertheless, there is a small difference in energy between the levels $^2S_{\frac{1}{2}}$ and $^2P_{\frac{1}{2}}$ which is known as the *Lamb shift*, and which cannot be accounted for by the Dirac equation. We must draw on quantum electrodynamics again in order to explain it successfully, and a crude picture of the reason for the slight raising of $^2S_{\frac{1}{2}}$ over $^2P_{\frac{1}{2}}$ may be obtained by considering the wobbling electron again (p. 216 footnote). This effectively smears the electron over a small region of space, and the effect of this is particularly noticeable for s-electrons because they spend time close to the nucleus: any smearing tends to reduce the effectiveness of this clustering at the nucleus and the energy of the configuration rises. A p-electron does not cluster so close to the nucleus and the effect of smearing the electron over a small region is less.

9.5 The spectra of alkali metal atoms

The alkali metal atoms have one electron in an s-orbital outside a core of closed-shell electrons. (This is known from elementary accounts of atomic structure and may be taken as given for the sake of this discussion; we return to the matter shortly and justify it more rigorously.) So long as we confine attention to the relatively loosely bound outermost electron, we can expect the spectra of the atoms to resemble the spectrum of atomic hydrogen. There will be both resemblances and differences. The resemblances stem from the fact that the terms are all doublets, and so the pattern of possible terms resembles the pattern shown by hydrogen. The differences arise from the presence not only of the higher nuclear charge but also of the *core electrons*. Now the outermost *valence electron* no longer experiences a Coulomb potential due to a single point source at the centre of the atom, but a sum of the potentials due to the nucleus and all the core electrons surrounding it. This potential is *non-central* and varies in a complicated way with distance. It can be simplified by regarding it, on the average, as a central field. This is achieved as follows.

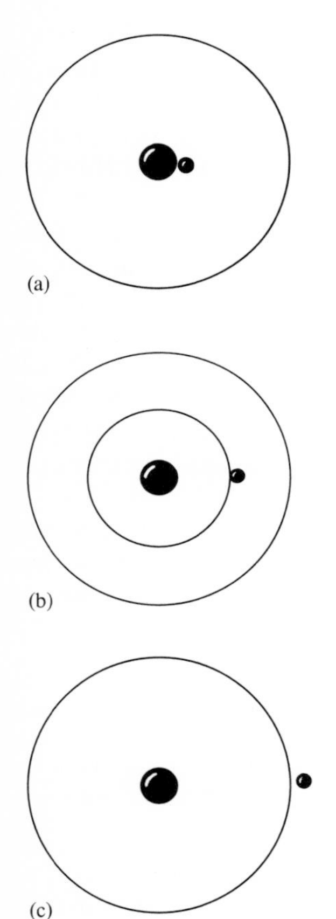

Fig. 9.10. An electron experiences (a) full nuclear charge Z, (b) Z shielded by electrons within sphere, (c) Z shielded by all core electrons.

Suppose we regard the core electrons as a spherically symmetrical haze of charge. When the electron is outside any spherical region of charge it experiences a net force equivalent to a single charge at the nucleus, Fig. 9.10, but the charge outside makes no contribution (this is a result from elementary electrostatics). Therefore the further the electron is from the nucleus, the greater the net repulsion it experiences from the electrons in the closed shell. When it is at the nucleus it experiences the full potential of its charge, Ze. When it is completely outside the closed shell it experiences a charge of only $1e$ on account of the shielding effect of the $Z-1$ electrons of the core. Hence the spherically averaged charge distribution gives rise to a *shielded potential* that falls off more rapidly with distance than the pure Coulomb potential.

There are two principal consequences of this rapid falling off of the shielded nuclear potential. One is that the energy of an electron is no longer independent of the angular momentum quantum number l: the effectively non-Coulombic potential experienced by the electron remove the degeneracy of orbitals of different values of l. The order of the orbital energies can be predicted quite simply on the basis of their *penetration* of the core, their average probability of being found close to the nucleus and therefore inside the shielding core. s-orbitals have non-zero amplitude at the nucleus, and so penetrate the core more completely than other orbitals (which have a node there). They therefore, on the average, experience a greater nuclear charge, and so their energies are lower than the p- and d-orbitals of the same principal quantum number. For a given value of n the average distance of the electron from the nucleus increases as l increases (in a classical sense, the centrifugal effect of the angular momentum flings the electron away to greater distances) and so, on the basis of penetration through a

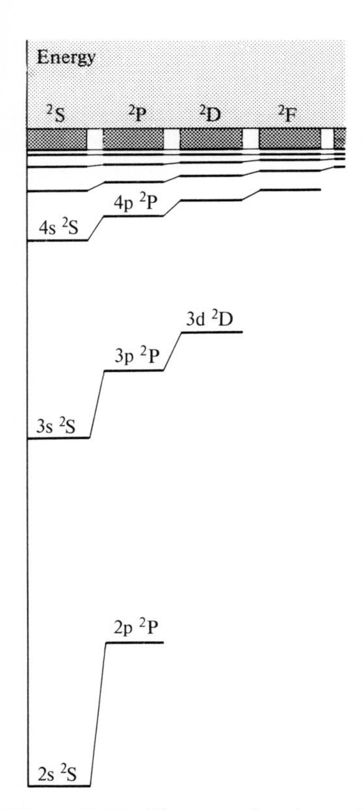

Fig. 9.11. Energy levels of atomic lithium.

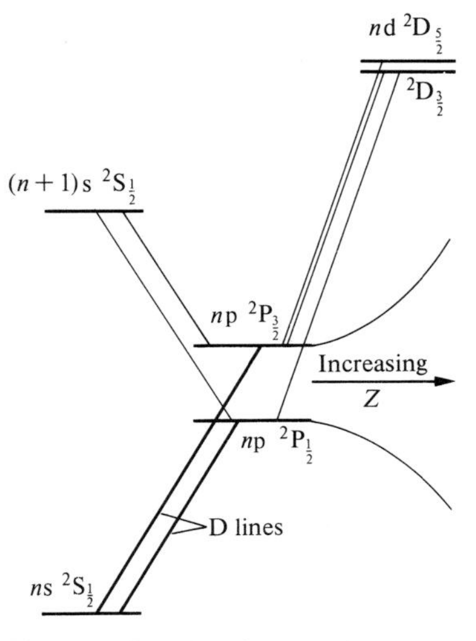

Fig. 9.12. Alkali metal atom fine structure.

shielding core, we expect the energies of orbitals to increase in the order s<p<d...

The other effect of increasing Z is the strong increase of the spin–orbit coupling constant, roughly as Z^4. There are complicating features arising from the other electrons, but the principal effect is a strong increase in spin–orbit coupling constant as the gradient of the potential increases ($\xi \propto \mathrm{d}\phi/\mathrm{d}r$, eqn (9.3.4)) on increasing Z.

These effects can be seen in the spectra. Consider first the spectrum of the lithium atom. It can be interpreted in terms of the energy level diagram shown in Fig. 9.11. The absence of the degeneracy of the terms arising from a given configuration is marked, and in accord with the penetration argument: the ^{2}S terms lie lower in energy than the ^{2}P and ^{2}D terms of the same principal quantum number. Notice that the degeneracy appears to be returning in configurations having higher values of n. This is because at high principal quantum numbers the electron is so far from the nucleus on average that differences of penetration have little effect on its energy. Since the electron then experiences an effective nuclear charge of $+e$ irrespective of its orbital angular momentum quantum number, the energy levels become increasingly hydrogen-like. They are sometimes expressed in the form $R/(n-a)^2$ in order to emphasize the resemblance; n is an integer, a is a correction called the *quantum defect* of the orbital.

The effect of the increasing spin–orbit coupling constant is to lead to a more marked fine structure in the spectrum. The splitting of the ^{2}P term is greater than in the hydrogen atom, and it increases down the Group I atoms, as anticipated. The splitting of the ^{2}P term implies that the $^2\mathrm{P} \rightarrow {}^2\mathrm{S}$ transitions produces doublets. This is the source of the famous pair of *D-lines* of the sodium atom (where 'D' has no relation to $L=2$), which corresponds to the transitions $3\mathrm{p}\,^2\mathrm{P}_{\frac{3}{2}} \rightarrow 3\mathrm{s}\,^2\mathrm{S}_{\frac{1}{2}}$ and $3\mathrm{p}\,^2\mathrm{P}_{\frac{1}{2}} \rightarrow 3\mathrm{s}\,^2\mathrm{S}_{\frac{1}{2}}$. The $^2\mathrm{D} \rightarrow {}^2\mathrm{P}$ transitions correspond to three lines, Fig. 9.12, but as the levels are almost unsplit in ^{2}D and under low resolution they form a compound doublet, much as we found in the case of hydrogen. The analysis of the fine structure of the spectrum gives the values of the spin–orbit coupling constant for the atoms (and its dependence on the atom's state), and the analysis of series limits gives ionization energies.

9.6 The structure of the helium atom

The structure of the hydrogen atom was a basis for a discussion of atoms that behaved as though they had only one electron. We now investigate the structure of the helium atom in the same spirit, and use it as a basis for the discussion of atoms containing more than one electron.

The helium atom is built from two electrons and a nucleus of charge $2e$. The hamiltonian is therefore

$$H = -(\hbar^2/2m_\mathrm{e})(\nabla_1^2+\nabla_2^2) - (2e^2/4\pi\varepsilon_0 r_1) - (2e^2/4\pi\varepsilon_0 r_2) + (e^2/4\pi\varepsilon_0 r_{12}). \quad (9.6.1)$$

∇_1^2 and ∇_2^2 are the second derivatives with respect to the coordinates of the two electrons, and are the operators for their kinetic energies. The next two terms are the Coulombic potential energies for each electron in the field of the nucleus. The final term is the contribution to the potential energy from the repulsive interaction between the two electrons. We have not troubled to include the reduced mass because the calculation is going to be too crude to warrant it.

The Schrödinger equation has the form $H\psi(\mathbf{r}_1, \mathbf{r}_2) = E\psi(\mathbf{r}_1, \mathbf{r}_2)$, and the wavefunction depends on the coordinates of both electrons. It appears to be impossible to find analytical solutions of such a complicated partial differential equation, and most work has been directed to finding increasingly refined numerical solutions. The simplest version of these approximate solutions is based on a perturbation approach, and this is the line we shall take here. The obvious candidate to use as the perturbation is the electron–electron repulsion, but as it is not particularly small with respect to the other terms in the hamiltonian we should not expect striking agreement with experiment.

The unperturbed part of the hamiltonian is the sum of two hydrogen-like hamiltonians:

$$H^{(0)} = H_1 + H_2, \qquad H_i = -(\hbar^2/2m_e)\nabla_i^2 - 2e^2/4\pi\varepsilon_0 r_i. \tag{9.6.2}$$

Whenever a hamiltonian can be written as the sum of two independent terms, the wavefunction can be expressed as a product of two factors, each factor being the wavefunction for one of the electrons. That is, for a hamiltonian of the form $H_1 + H_2$, the wavefunction is of the form $\psi(\mathbf{r}_1, \mathbf{r}_2) = \psi(\mathbf{r}_1)\psi(\mathbf{r}_2)$. This is proved as follows. Let $H_i\psi(\mathbf{r}_i) = E_i\psi(\mathbf{r}_i)$; then

$$\begin{aligned}(H_1 + H_2)\psi(\mathbf{r}_1, \mathbf{r}_2) &= (H_1 + H_2)\psi(\mathbf{r}_1)\psi(\mathbf{r}_2)\\ &= H_1\psi(\mathbf{r}_1)\psi(\mathbf{r}_2) + \psi(\mathbf{r}_1)H_2\psi(\mathbf{r}_2)\\ &= E_1\psi(\mathbf{r}_1)\psi(\mathbf{r}_2) + E_2\psi(\mathbf{r}_1)\psi(\mathbf{r}_2) = (E_1 + E_2)\psi(\mathbf{r}_1, \mathbf{r}_2).\end{aligned}$$

Therefore the product function satisfies the Schrödinger equation, and corresponds to an energy which is the sum of the eigenvalues of the two component equations. Since the hamiltonians are hydrogen-like with $Z = 2$, it follows that the 'unperturbed' wavefunctions of the helium atom are of the form

$$\psi(\mathbf{r}_1, \mathbf{r}_2) = \psi_{n_1 l_1 m_{l1}}(\mathbf{r}_1)\psi_{n_2 l_2 m_{l2}}(\mathbf{r}_2) \tag{9.6.3}$$

and that the energies of the atom are

$$E = -4hcR_\infty\{(1/n_1^2) + (1/n_2^2)\}, \qquad n_1, n_2 = 1, 2, \ldots \tag{9.6.4}$$

Now consider the influence of the electron–electron repulsion. The first-order correction to the energy is

$$E^{(1)} = \langle n_1 l_1 m_{l1}; n_2 l_2 m_{l2}|\,(e^2/4\pi\varepsilon_0 r_{12})\,|n_1 l_1 m_{l1}; n_2 l_2 m_{l2}\rangle = J$$

where J is the

Coulomb integral:

$$J=(e^2/4\pi\varepsilon_0)\int|\psi_{n_1l_1m_{l1}}(\mathbf{r}_1)|^2\,(1/r_{12})\,|\psi_{n_2l_2m_{l2}}(\mathbf{r}_2)|^2\,\mathrm{d}\tau_1\,\mathrm{d}\tau_2. \qquad (9.6.5)$$

J has a very simple interpretation, Fig. 9.13. $|\psi_{n_1l_1m_{l1}}(\mathbf{r}_1)|^2\,\mathrm{d}\tau_1$ is the probability of finding the electron in the volume element $\mathrm{d}\tau_1$, and so when multiplied by $-e$ it is the charge associated with that region. Likewise, $-e\,|\psi_{n_2l_2m_{l2}}(\mathbf{r}_2)|^2\,\mathrm{d}\tau$ is the charge associated with the element $\mathrm{d}\tau_2$. The integrand is therefore the electrostatic potential energy of interaction of these two charge elements. The integral, J, is therefore the total average potential energy of interaction of one electron in a hydrogen-like $n_1l_1m_{l1}$-orbital with another electron in an $n_2l_2m_{l2}$-orbital. This integral can be evaluated analytically (see below). When both electrons occupy the same 1s-orbital, so that the helium is in the configuration $1s^2$, the total energy is

$$E=2E_{1s}+J;\qquad J=(e^2/4\pi\varepsilon_0)\int|\psi_{1s}(r_1)|^2\,(1/r_{12})\,|\psi_{1s}(r_2)|^2\,\mathrm{d}\tau_1\,\mathrm{d}\tau_2, \qquad (9.6.6)$$

J being the Coulomb integral for the interaction of two 1s-electrons. In the case $Z=2$, as for helium $J=(\frac{5}{4})e^2/4\pi\varepsilon_0a_0=5.45\times10^{-18}\,\mathrm{J}$ (corresponding to $3280\,\mathrm{kJ\,mol^{-1}}$). Since $E_{1s}=-4hcR_\infty=-8.72\times10^{-18}\,\mathrm{J}$ (corresponding to $-5251\,\mathrm{kJ\,mol^{-1}}$), the total energy of the atom (the energy relative to the infinite separation of its components) is

$$E=2(-8.72\times10^{-18}\,\mathrm{J})+5.45\times10^{-18}\,\mathrm{J}=-1.20\times10^{-17}\,\mathrm{J} \qquad (9.6.7)$$

(corresponding to $-7220\,\mathrm{kJ\,mol^{-1}}$). The experimental value (the sum of the first and second ionization energies) is $-7619\,\mathrm{kJ\,mol^{-1}}$. The agreement is not brilliant, but we are obviously on the right track. One of the deficiencies of the calculation is that the perturbation is not small: J is $\frac{5}{16}$ the value of $2E_{1s}$, and so first-order perturbation theory cannot be expected to lead to an adequate result. Nevertheless we shall continue to use perturbation theory since it captures the main features of the structure of the atom without getting us involved in detailed technicalities.

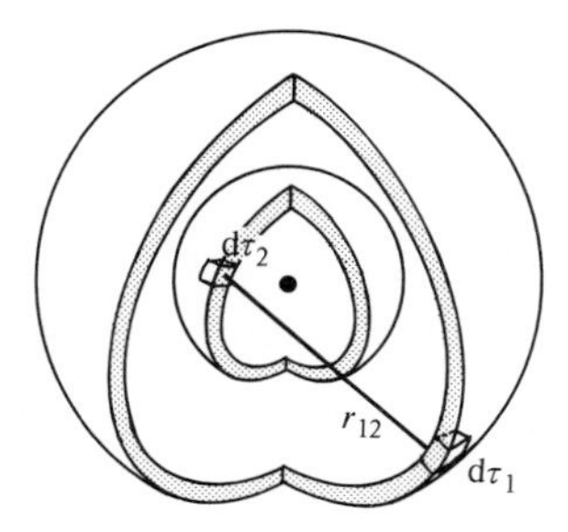

Fig. 9.13. Contributions to the electron–electron repulsion.

Example. Evaluate the Coulombic interaction energy between two electrons in a hydrogen-like 1s-orbital.

• *Method.* Evaluate eqn (9.6.5) with $\psi(r_1)=(Z^3/\pi a_0^3)^{\frac{1}{2}}\mathrm{e}^{-Zr_1/a_0}$ and $\psi(r_2)=(Z^3/\pi a_0^3)^{\frac{1}{2}}\mathrm{e}^{-Zr_2/a_0}$. The term in $1/r_{12}$ is the complicating feature. The central result needed is the expansion of $1/r_{12}$ in terms of r_1 and r_2. When $r_1>r_2$

$$1/r_{12}=(1/r_1)\sum_{l=0}^{\infty}\sum_{m_l=1}^{l}\left(\frac{4\pi}{2l+1}\right)(r_2/r_1)^lY_{lm_l}(\theta_1,\phi_1)Y_{lm_l}(\theta_2,\phi_2),$$

and when $r_2>r_1$ interchange r_1 and r_2. The wavefunctions are independent of θ and ϕ, and so in the integral J the angular integrations eliminate all terms in this sum (the average values of $Y_{lm_l}(\theta_1,\phi_1)$ are zero) except the one with

$l = m_l = 0$. For the surviving term use $Y_{00} = (1/4\pi)^{\frac{1}{2}}$; and hence

$$1/r_{12} = 1/r_1 \text{ when } r_1 > r_2; \qquad 1/r_{12} = 1/r_2 \text{ when } r_2 > r_1.$$

(The relations are meaningful only inside the integral, of course.) Divide the radial integrations into two parts corresponding to $r_1 > r_2$ and $r_2 > r_1$.

• *Answer.*

$$J = (e^2/4\pi\varepsilon_0)(Z^3/\pi a_0^3)^2 \int_0^{2\pi} d\phi_1 \int_0^{2\pi} d\phi_2 \int_0^{\pi} d\theta_1 \sin\theta_1 \int_0^{\pi} d\theta_2 \sin\theta_2$$

$$\times \int r_1^2 r_2^2 (1/r_{12}) e^{-2Z(r_1+r_2)/a_0} dr_1 dr_2$$

$$= (e^2/4\pi\varepsilon_0)(Z^3/\pi a_0^3)^2 (2\pi)^2 Z^2$$

$$\times \int_0^{\infty} r_2^2 e^{-2Zr_2/a_0} \left\{ \int_0^{r_2} r_1^2 (1/r_2) e^{-2Zr_1/a_0} dr_1 + \int_{r_2}^{\infty} r_1^2 (1/r_1) e^{-2Zr_1/a_0} dr_1 \right\} dr_2$$

$$= \tfrac{5}{8}(e^2/4\pi\varepsilon_0)(Z/a_0).$$

• *Comment.* The radial integral has the value $(5/2^7)(a_0/Z)^5$. The same analysis may be carried through for (ns, n's) interactions, but when the wavefunctions have an angular dependence (as for (2p, 2s) interactions), different terms survive in the expansion of $1/r_{12}$. Use $Z = 2$ for helium.

A new feature arises when we consider the excited states of the atom. The configuration in which one electron occupies the orbital $n_1 l_1 m_{l1}$ and the other occupies a different orbital $n_2 l_2 m_{l2}$ is described by an unperturbed wavefunction which may be either $\psi_{n_1 l_1 m_{l1}}(\mathbf{r}_1)\psi_{n_2 l_2 m_{l2}}(\mathbf{r}_2)$ or $\psi_{n_2 l_2 m_{l2}}(\mathbf{r}_1)\psi_{n_1 l_1 m_{l1}}(\mathbf{r}_2)$. Both arrangements have the same energy, and so they form a degenerate pair of functions. For simplicity we denote them a(1)b(2) and b(1)a(2) respectively, their unperturbed energies each being $E_a + E_b$. In order to calculate the perturbed energy we must use the degenerate form of first-order perturbation theory as described in Chapter 8. In order to use the recipe in eqn (8.2.10) we set up the secular determinant using the following matrix elements of the perturbation:

$$H_{11} = \int a(1)b(2)\{H_1 + H_2 + (e^2/4\pi\varepsilon_0 r_{12})\}a(1)b(2)\, d\tau_1\, d\tau_2$$

$$= E_a + E_b + J, \tag{9.6.8a}$$

$$H_{22} = E_a + E_b + J, \tag{9.6.8b}$$

$$H_{12} = \int a(1)b(2)\{H_1 + H_2 + (e^2/4\pi\varepsilon_0 r_{12})\}a(2)b(1)\, d\tau_1\, d\tau_2$$

$$= (E_a + E_b) \int a(1)b(2)b(1)a(2)\, d\tau_1\, d\tau_2$$

$$+ (e^2/4\pi\varepsilon_0) \int a(1)b(2)(1/r_{12})b(1)a(2)\, d\tau_1\, d\tau_2. \tag{9.6.8c}$$

The first of these integrals is zero because the orbitals a and b are orthogonal, and

$$\int a(1)b(2)b(1)a(2)\, d\tau_1\, d\tau_2 = \int a(1)b(1)\, d\tau_1 \int b(2)a(2)\, d\tau_2 = 0.$$

The second integral is not necessarily zero. It is called the

Exchange integral:

$$K=(e^2/4\pi\varepsilon_0)\int\{a(1)b(1)\}(1/r_{12})\{a(2)b(2)\}\,d\tau_1\,d\tau_2. \tag{9.6.9}$$

Therefore $H_{12}=K$, and H_{21} has the same value. The secular determinant is therefore

$$\begin{vmatrix} H_{11}-ES_{11} & H_{12}-ES_{12} \\ H_{21}-ES_{21} & H_{22}-ES_{22}\end{vmatrix} = \begin{vmatrix} H_{11}-E & H_{12} \\ H_{21} & H_{22}-E\end{vmatrix}$$
$$= \begin{vmatrix} E_a+E_b+J-E & K \\ K & E_a+E_b+J-E\end{vmatrix}=0.$$

Its solutions are

$$E=E_a+E_b+J\pm K \tag{9.6.10}$$

and the corresponding wavefunctions are

$$\psi_\pm(1,2)=(\tfrac{1}{2})^{\frac{1}{2}}\{a(1)b(2)\pm b(1)a(2)\}, \tag{9.6.11a}$$

or, in more detail,

$$\psi_\pm(\mathbf{r}_1,\mathbf{r}_2)=(\tfrac{1}{2})^{\frac{1}{2}}\{\psi_{n_1l_1m_{l1}}(\mathbf{r}_1)\,\psi_{n_2l_2m_{l2}}(\mathbf{r}_2)\pm\psi_{n_2l_2m_{l2}}(\mathbf{r}_1)\,\psi_{n_1l_1m_{l1}}(\mathbf{r}_2)\}, \tag{9.6.11b}$$

the individual functions being hydrogen-like atomic orbitals for $Z=2$.

The striking feature of this result is that the degeneracy of the two product functions is removed by the presence of the electron–electron repulsion terms, the separation being $2K$. The integral K, and the effect giving rise to it, has no classical counterpart, and should be regarded as a quantum mechanical correction to the Coulomb integral J. In order to come to some kind of interpretation of its source, consider the behaviour of the wavefunctions $\psi_\pm(\mathbf{r}_1,\mathbf{r}_2)$ as one of the electrons approaches the other. We can ask the question: what is the probability of finding both electrons in the same region of space? In order to find an answer, consider the amplitude of the wavefunction ψ_- when $\mathbf{r}_2=\mathbf{r}_1$: it vanishes. On the other hand, the amplitude ψ_+ does not necessarily vanish as $\mathbf{r}_1$ approaches $\mathbf{r}_2$. The difference in behaviour of the probability density is sketched in Fig. 9.14. The answer to the question is that there is *zero probability* of finding the electrons at the same point of space if they occupy the state described by ψ_-, but there is no such restriction when they occupy ψ_+ (in fact there is a slight tendency for one electron to be found near the other one). The dip in the probability density $|\psi_-|^2$ wherever $\mathbf{r}_1=\mathbf{r}_2$ is called a *Fermi hole*. Its existence is purely quantum mechanical, and has nothing to do with the charges of the electrons. Even 'uncharged electrons' would show the phenomenon: if they are in a state described by the wavefunction ψ_-, then they have an intrinsic tendency to avoid each other.

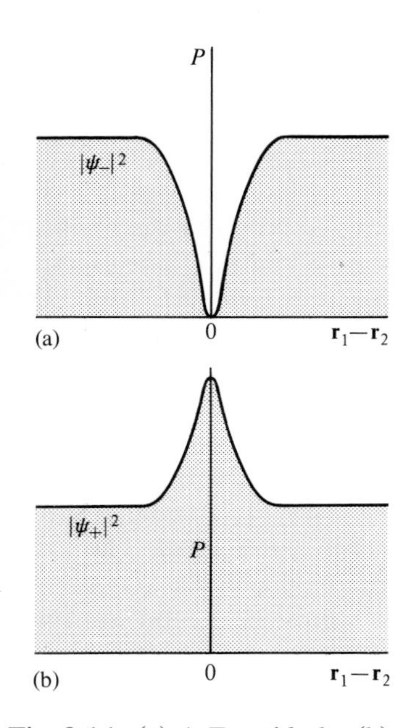

Fig. 9.14. (a) A Fermi hole, (b) a Fermi heap.

Having established that feature of the wavefunction, we can investigate its influence on the energy. The first-order energy correction is the average of the perturbation over the undistorted wavefunction; that is, the average of the repulsive energy $H^{(1)}$ over the wavefunctions ψ_+ or ψ_-. Clearly, since the electrons have an intrinsic tendency to avoid each other in ψ_-, their average repulsion must be less than in ψ_+. The total potential energy of the former is $J-K$, the exchange integral reducing the total repulsive interaction, and in ψ_+ it is $J+K$, the extra contribution reflecting the tendency of the electrons to congregate.

In passing, note the difference in the *permutation symmetry* of ψ_+ and ψ_-. The combination ψ_+ is *symmetric* under particle-label interchange: it does not change sign when $\mathbf{r}_1$ and $\mathbf{r}_2$, the coordinates of the electrons, are interchanged. The combination ψ_- is *antisymmetric* under particle-label interchange, because when 1 and 2 are interchanged a(1)b(2) − b(1)a(2) changes sign. This aspect of the functions turns out to be of crucial importance, as we shall soon see.

At this stage we have seen that when both electrons are in the same orbital (as in the ground state of the atom) the configuration gives rise to a single term with the energy $2E+J$, both E and J depending on which orbital is doubly occupied. When the electrons occupy different orbitals the configuration gives rise to two terms, one at $E+E'+J-K$ and the other at $E+E'+J+K$. The separation of the terms by $2K$ should be detectable in the spectrum. In order to see what to expect, we now consider helium's energy levels in slightly more detail.

The ground configuration is $1s^2$; its total orbital angular momentum is zero (because $l_1=l_2=0$), and so $L=0$ is the only possible total angular momentum. It is therefore an S term. The only excited configurations we need consider in practice are those involving the excitation of a single electron, and therefore of the form 1snl. (The energy of a helium atom with both electrons excited exceeds the ionization energy, and so He^+ is formed.) The configuration 1snl gives rise to terms with $L=1$ because only one of the electrons has any orbital angular momentum. Therefore we have to consider the energies of the terms 1s2s S, 1s2p P, 1s3s S, and so on. When only one electron is involved the selection rule $\Delta l=\pm1$ translates into $\Delta L=\pm1$. We therefore expect to see transitions between $1s^2$ S and 1s2p P, and such like. But what of the distinction between the symmetrical and antisymmetrical states represented by the symmetrical (ψ_+) and antisymmetrical (ψ_-) wavefunctions? All the configurations that interest us (apart from $1s^2$) have such wavefunctions. The appropriate selection rule is that *symmetrical terms make transitions only to symmetrical, and antisymmetrical terms make transitions only to antisymmetrical.* The technical expression for 'makes a transition' is *combine with,* and so this selection rule is *symmetrical combines with symmetrical,* and *antisymmetrical combines with antisymmetrical.*

The basis of the permutation-symmetry selection rule is the vanishing of the transition dipole moment when the two states have different symmetry. The argument runs as follows. The electric dipole moment operator for the two-electron system is $-e(\mathbf{r}_1+\mathbf{r}_2)$, which is obviously symmetrical under particle interchange. The transition dipole for a transition from a symmetric to an antisymmetric state is the integral $-e\int \psi_+(\mathbf{r}_1, \mathbf{r}_2)(\mathbf{r}_1+\mathbf{r}_2)\psi_-(\mathbf{r}_1, \mathbf{r}_2)\,\mathrm{d}\tau_1\,\mathrm{d}\tau_2$. Under particle-label interchange the integrand changes sign; the value of the integral cannot depend on the names of the electrons; therefore it must be independent of the labels 1 and 2, and so be symmetrical under particle interchange. The only way of satisfying both criteria is for the integral to vanish. Hence there can be no electric dipole transition between the symmetric and the antisymmetric sets of terms.

A final piece of information concerns the multiplicity of the terms. We are considering two electrons, each with spin $s=\frac{1}{2}$. Their total spin angular momentum may be either $S=0$ or $S=1$. The values of $2S+1$ are 1 and 3 respectively, and so we expect singlet and triplet terms. For a given value of L the levels are distinguished by the values of the total angular momentum quantum number J. For the singlet term only $J=L$ can occur. For the triplet term, the Clebsch–Gordan series gives $J=L+1$, L, and $L-1$ as the three levels of each triplet for $L\geq 1$. For $L=0$ we have $J=1$ only. This means that we should expect levels such as $^1P_1$ and $^3P_2$, $^3P_1$, $^3P_0$ to come from the configuration 1s2p and to be split by the spin–orbit interaction. We can also expect each of these four P terms to occur as both the symmetric and the antisymmetric forms. Hence, we expect eight terms to arise from 1s2p. From 1s2s we expect $^1S_0$ and $^3S_1$, each of which may be symmetric or antisymmetric, giving four terms in all.

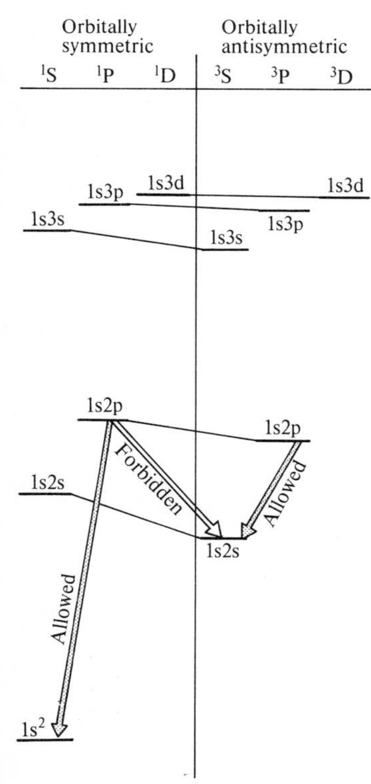

Fig. 9.15. Energy levels of atomic helium.

9.7 The spectrum of helium and the Pauli principle

The spectrum of helium is consistent with this scheme. Each 1snl configuration gives rise to two types of term, Fig. 9.15, one symmetric and the other antisymmetric: we can decide which is which, because only the symmetric terms have appreciable transition intensity to the ground state (which we know is symmetric). Furthermore, whenever both terms are observed, the antisymmetric one turns out to be the lower.† There is, however, an extraordinary feature. Analysis of the spectrum shows that *all the symmetric states are singlets*, and *all the antisymmetric states are triplets.* There are no symmetrical triplets, and there are no antisymmetrical singlets. There are only four terms from each 1snp configuration, not eight. In fact, half of all possible terms appear to be excluded.

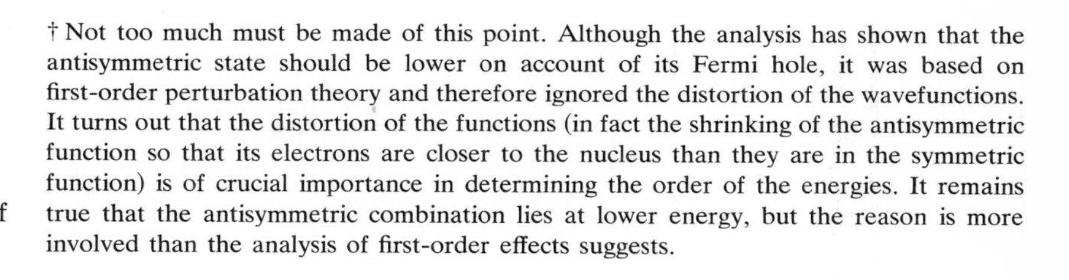
† Not too much must be made of this point. Although the analysis has shown that the antisymmetric state should be lower on account of its Fermi hole, it was based on first-order perturbation theory and therefore ignored the distortion of the wavefunctions. It turns out that the distortion of the functions (in fact the shrinking of the antisymmetric function so that its electrons are closer to the nucleus than they are in the symmetric function) is of crucial importance in determining the order of the energies. It remains true that the antisymmetric combination lies at lower energy, but the reason is more involved than the analysis of first-order effects suggests.

Pauli proposed the solution. We can move towards it by reflecting first on the form the wavefunction takes when we include the spin of the two electrons. In Chapter 6 we saw that the spin state of two electrons coupled together so that $S=0$ is $\alpha_1\beta_2-\beta_1\alpha_2$ (if we ignore the $(\frac{1}{2})^{\frac{1}{2}}$ normalization constant), which is antisymmetric under label interchange, while the three states corresponding to $S=1$ are $\alpha_1\alpha_2$, $(\alpha_1\beta_2+\beta_1\alpha_2)$, and $\beta_1\beta_2$, which are all symmetric under interchange. We can list all possible total wavefunctions that might occur:

$\psi_-(1,2)(\alpha_1\beta_2-\beta_1\alpha_2)$	$\psi_+(1,2)(\alpha_1\beta_2-\beta_1\alpha_2)$
$\psi_-(1,2)\alpha_1\alpha_2$ $\psi_-(1,2)(\alpha_1\beta_2+\beta_1\alpha_2)$ $\psi_-(1,2)\beta_1\beta_2$	$\psi_+(1,2)\alpha_1\alpha_2$ $\psi_+(1,2)(\alpha_1\beta_2+\beta_1\alpha_2)$ $\psi_+(1,2)\beta_1\beta_2.$

The excluded states have been printed without the tinted background. It is clear that there is a common feature in the admissible combinations: *overall* the wavefunction is antisymmetrical with respect to label interchange. This observation is raised into a general law of nature:

The Pauli principle: *the total wavefunction* (*including the spin*) *must be antisymmetric with respect to the interchange of any pair of electrons* (*or identical fermions*).

(The 'opposite' principle applies to identical bosons: their overall wavefunctions must be symmetrical under interchange.) We shall regard the Pauli principle as one more postulate of quantum mechanics, but it can be rationalized using relativistic arguments and the requirement that the energy of the universe be positive. For us it is a succinct, subtle summary of experience (the spectrum of helium) which, we shall see, is fully vindicated by the success it has in explaining the spectra of other species (as well as the whole of the theory of molecular structure).

As a consequence of the Pauli principle there is a restriction on electrons occupying the same state. This is

The Pauli exclusion principle: *no two electrons can have the same set of quantum numbers.*

In its simplest form the argument linking this to the Pauli principle runs as follows. Suppose the spin states of the two electrons are the same. We can always choose the z-direction such that their joint state is described as $\alpha_1\alpha_2$. This is symmetric. According to the Pauli principle the corresponding spatial part of the wavefunction must be antisymmetric, and therefore of the form $\mathrm{a}(1)\mathrm{b}(2)-\mathrm{b}(1)\mathrm{a}(2)$. But if a is the same orbital as b this amplitude is zero for all values of $\mathbf{r}_1$ and $\mathbf{r}_2$. Therefore the amplitude of the state for two electrons with the same spin in the same orbital is identically zero: no such state exists. Therefore, in order for electrons to occupy the same spatial orbital they must *pair* (i.e. they must have opposed spins).

Another way of looking at the relation between the two principles is as follows. Consider first the following way of writing the ground-state wavefunction of the helium atom. The total wavefunction is

$$\begin{aligned}\psi(1,2) &= \psi_{1s}(\mathbf{r}_1)\psi_{1s}(\mathbf{r}_2)\{(\tfrac{1}{2})^{\frac{1}{2}}(\alpha_1\beta_2-\beta_1\alpha_2)\}\\ &= (\tfrac{1}{2})^{\frac{1}{2}}\begin{vmatrix}\psi_{1s}(\mathbf{r}_1)\alpha_1 & \psi_{1s}(\mathbf{r}_1)\beta_1\\ \psi_{1s}(\mathbf{r}_2)\alpha_2 & \psi_{1s}(\mathbf{r}_2)\beta_2\end{vmatrix},\end{aligned} \tag{9.7.1}$$

because the expansion of the determinant yields the preceding line. We now introduce the *spin-orbital* $\psi_{1s}^{\alpha}(1)=\psi_{1s}(\mathbf{r}_1)\alpha_1$, and likewise for $\psi_{1s}^{\beta}(1)$. The ground-state wavefunction can then be written as a

Slater determinant: $$\psi(1,2)=(\tfrac{1}{2})^{\frac{1}{2}}\begin{vmatrix}\psi_{1s}^{\alpha}(1) & \psi_{1s}^{\beta}(1)\\ \psi_{1s}^{\alpha}(2) & \psi_{1s}^{\beta}(2)\end{vmatrix}. \tag{9.7.2}$$

A Slater determinant displays the overall antisymmetry of the state in a very neat manner, because if the labels 1 and 2 are interchanged the rows of the determinant are exchanged, and as a consequence it changes sign.

Now suppose that the electrons had the same spin and occupied the same orbitals. The Slater determinant for the state would be

$$\psi(1,2)=(\tfrac{1}{2})^{\frac{1}{2}}\begin{vmatrix}\psi_{1s}^{\alpha}(1) & \psi_{1s}^{\alpha}(1)\\ \psi_{1s}^{\alpha}(2) & \psi_{1s}^{\alpha}(2)\end{vmatrix},$$

and this vanishes identically, as is true for any determinant having two rows or two columns in common. Hence the state is excluded.

The general expression for an *antisymmetrized wavefunction* composed of spin-orbitals $\varphi_a, \varphi_b, \ldots$ and accommodating N electrons is

$$\psi(1,2,\ldots,N)=(1/N!)^{\frac{1}{2}}\begin{vmatrix}\varphi_a(1) & \varphi_b(1) & \cdots & \varphi_z(1)\\ \varphi_a(2) & \varphi_b(2) & \cdots & \varphi_z(2)\\ \vdots & \vdots & & \vdots\\ \varphi_a(N) & \varphi_b(N) & \cdots & \varphi_z(N)\end{vmatrix}.$$

It is fully antisymmetric under the interchange of any pair of electrons because that interchanges a pair of rows, which causes the determinant to change sign. Furthermore, if any two spin orbitals are the same, the determinant vanishes because it then has two columns in common. Instead of writing the determinant out in full, which is tiresome, it is normally denoted by its principal diagonal. The wavefunction in the last expression would therefore then be written $(1/N!)^{\frac{1}{2}}\det|\varphi_a(1)\varphi_b(2)\ldots\varphi_z(N)|$, or sometimes $\|\varphi_a(1)\varphi_b(2)\ldots\varphi_z(N)\|$.

We are now in a position to return to the helium spectrum. We have seen that two electrons tend to avoid each other if they are described by an antisymmetrical *spatial* wavefunction. When the electrons have this state their spins must be in a symmetrical state and correspond to $S=1$ which, as explained on p. 122, means that they are 'parallel'. Therefore we may summarize the effect by saying that *parallel spins tend to avoid each other.* This is called *spin correlation,* but the preceding discussion has shown that it only *indirectly* depends on spin (via the Pauli principle and the way that governs what spin states may go with what spatial functions). That is, if the spins are parallel then the Pauli principle requires them to have an antisymmetric spatial wavefunction; that being so, they cannot be found at the same point simultaneously.

A consequence of the spin-correlation effect is, as we have seen, that triplet states lie lower in energy than singlet states. Mark the point, however, that the difference in energy of states of different multiplicity is only an indirect consequence of the relative orientations of their spins, and that it should not be thought of as some kind of interaction between spins. The difference in energy between states of different multiplicity is a purely *electrostatic* effect, and reflects the influence that spin correlation has on the relative spatial distribution of the electrons.

9.8 The periodic table

We have seen that a fairly crude description of the ground state of the helium atom is $1s^2$, with the 1s-orbitals hydrogen-like but with $Z=2$, (or, in slightly more sophisticated versions, with Z about 1.3 so as to take into account in a rough way the shielding of the nucleus). This approach to atomic structure can be extended to the other elements, because the average effect of the electrons in a many-electron atom gives rise to a shielded central potential. Therefore, even in a many-electron atom we think of atomic orbitals as being roughly hydrogen-like but with a radial dependence distorted by the shielding effect and the consequent complicated radial dependence of the potential. The *orbital approximation* is based on the view that since the average electron–electron interactions give rise to a shielded central potential, the electrons in the atom can be regarded as occupying a set of atomic orbitals centred on the nucleus. There are going to be profound distortions of this picture on account of the true form of the electron–electron repulsion terms, which are strongly non-central, Fig. 9.16, but at least it is somewhere to start.

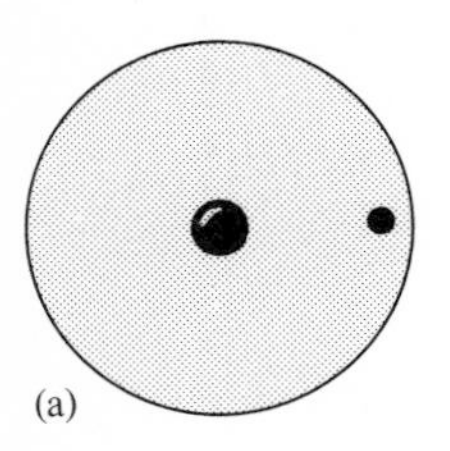

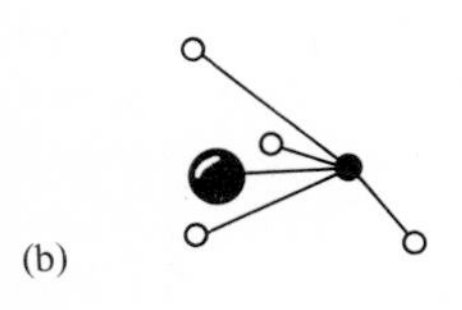

Fig. 9.16. (a) The central field approximation; (b) the actual non-central field.

The energies of orbitals must reflect a number of features. The closer they are to the nucleus the more favourable the potential energy. This orders them very roughly into *shells* according to the principal quantum number n. Thus, we have the *K-shell* ($n=1$), *L-shell* ($n=2$), *M-shell* ($n=3$), and so on. The order of the orbitals within the shells can be anticipated on the basis of shielding and penetration effects, and so we expect *subshells* in the order s, p, d, . . . with the s-subshell lying lowest on account of its greatest penetration into the vicinity of the nucleus. We should not be surprised if the energy differences between components of subshells is so great that it upsets the ordering of the shells themselves, and it is indeed found to be the case that the 4s-subshell lies lower in energy than the 3d. A rough indication of the general order of energies is 1s; 2s, 2p; 3s, 3p; 4s, 3d, 4p; . . .

The Pauli principle is the key to the problem of accommodating the electrons in these orbitals. It had long been a vexing problem as to why all the electrons in an atom did not simply enter the orbital of lowest energy. The Pauli principle limits the occupancy to two to each orbital, and in order for two to be in a single orbital they must pair. The *building-up principle* (or *Aufbau principle*) then runs as follows: (i) set up the orbitals; (ii) decide how many electrons have to be accommodated (Z for neutral atoms); (iii) feed them successively into the lowest lying available energy orbital, allowing no more than two to occupy any orbital.

The K-shell can accommodate only two electrons. Therefore, when we turn to lithium ($Z=3$) the third electron must enter the 2s-orbital of the L-shell. This shell can accommodate up to eight electrons and accounts for the *first short period* of the periodic table. The next shell has to be started at sodium ($Z=11$), and can accommodate a further eight. Electron number 19 enters the 4s-orbital (potassium, $Z=19$) and the subshell is complete with calcium ($Z=20$). At this point the 3d-subshell is accessible, and can accommodate up to ten electrons. The sequential occupation of these gives rise to the *transition elements* and accounts for the *first long period* of the table. When the 3d-subshell is full the 4p-orbitals become available, and the corresponding elements resemble their partners in the preceding short periods. This process is continued, and similar electron configurations are encountered periodically. More transition series are encountered as the 4d- and 5d-orbitals become available, and at $Z=57$ the 4f-orbitals are next in line to be filled. Their sequential filling gives rise to the *lanthanides*. The pattern of periodic repetition of electron configuration, and therefore periodic repetition of chemical characteristics, is continued until about $Z=108$, when the table of known elements fizzles out. Modern techniques, however, continue to lead to the synthesis of new elements, and there is some expectation (which is founded on views about the shell structure of *nuclei*) of finding an *island of stability*, a collection of fairly stable nuclides with Z considerably higher than 100 (around 120 and around 160).

9.9 Ionization energies

This is not a textbook of inorganic chemistry, but it is worth reflecting on an interesting feature of the elements in the first short period of the periodic table, from lithium to neon. One of the most important properties is the *ionization energy* of the atom, the minimum energy required for the process $E(g) \rightarrow E^+(g) + e^-(g)$ for atoms of the element. (This is the *first ionization energy*; the second, third, etc. ionization energies are the minimum energies required for ionization of the succession of ions left after each ionization.) Ionization energies may be calculated, but they are normally found by detecting the series limits of the atom's spectrum.

Across the periodic table there is a general rise in ionization energy with increasing Z, Fig. 9.17. This is because the effect of the nuclear attraction on the outermost electron increases more rapidly than the repulsion arising from the more dispersed electrons of the core. The lowest ionization energy is that for lithium ($Z=3$) because its electron is attracted by a weakly charged nucleus which the two electrons of the K-shell shield effectively. The drop in ionization energy between beryllium and boron is because in the latter the outermost electron occupies a 2p-orbital, which is more distant on average from the nucleus and so is more effectively shielded than the 2s-orbital. The drop between nitrogen and oxygen is because in nitrogen the configuration is $K2s^2 2p_x 2p_y 2p_z$ but in oxygen it is $K2s^2 2p_x^2 2p_y 2p_z$, and so two electrons have to enter the same orbital, and there is a greater electron–electron repulsion. The dramatic drop between neon and sodium corresponds to the need for the outermost electron to occupy a new, distant shell where it can interact only weakly with the nucleus.

9.10 Approximate atomic orbitals

We have spoken of 'distorted hydrogen-like atomic orbitals' as being the basis of the orbital description of atoms, but what are their shapes? No definitive answer can be given because the orbital approximation is very crude. Nevertheless it is useful to have available a set of orbitals which roughly model the actual electron distributions found using the more sophisticated techniques we describe shortly. These orbitals have to take into account the shielding effects of the other electrons present, and so we need rules for deciding on the *effective atomic number*, the average nuclear charge experienced by a given electron. One quite useful set of approximate orbitals are the *Slater atomic orbitals*. They are constructed as follows.

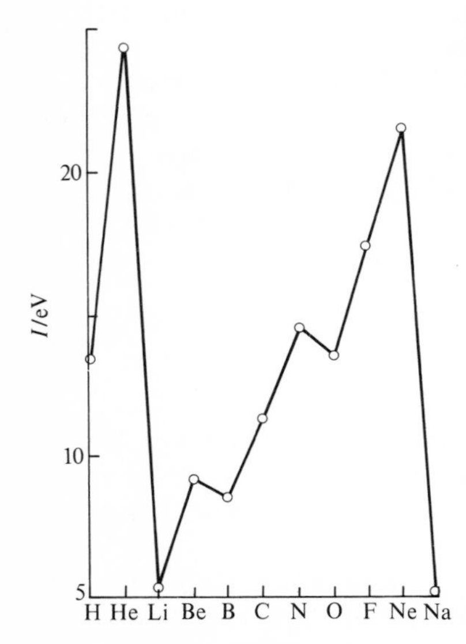

Fig. 9.17. Ionization energies of light atoms.

(1) An orbital with quantum numbers n, l, m_l on a nucleus of atomic number Z is written

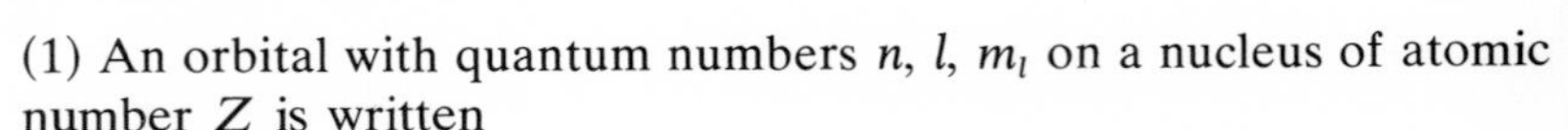

$$\psi_{nlm_l}(r, \theta, \phi) = N r^{n^*-1} e^{-Z^*\rho/n^*} Y_{lm_l}(\theta, \phi)$$

where N is a normalizing constant, Y_{lm_l} is a spherical harmonic (Table 4.1), and $\rho = r/a_0$.

(2) The *effective principal quantum number* n^* is related to n as follows:

n	1	2	3	4	5	6
n^*	1	2	3	3.7	4.0	4.2

(3) The *effective atomic number* Z^* is related to the true atomic number by $Z^* = Z - \sigma$ where σ is the *shielding constant*. The latter is calculated by dividing the orbitals into the following groups:

1s; 2s, 2p; 3s, 3p, 3d; 4s, 4p; 4d; 4f; 5s, 5p; 5d

and summing the following contributions to obtain the value of σ for a given orbital.

(a) For an electron in a group outside the one of interest: 0;
(b) from each other electron in the group of interest: 0.35 but if the electron is 1s: 0.30 instead;
(c) from each electron from groups inside the one of interest: 1.00 if the one of interest is d or f; 0.85 from each electron in the next inner group if the electron of interest is s or p; 1.00 for all deeper electrons.

Typical (and improved) values of Z^* are given in Table 9.1.

Example. Find expressions for the Slater type 2s-, $2p_z$-, and $2p_x$-orbitals of nitrogen.

- *Method.* In each case $Z^* = Z - \sigma$ has the same value. Use $Z = 7$ and calculate σ for the configuration $1s^2 2s^2 2p^2$ using the rules expressed above. For the $2p_z$-orbital use Y_{10} from Table 4.1. For the $2p_x$-orbital, use the linear combinations $(Y_{1-1} - Y_{11})/\sqrt{2}$ corresponding to $p_x = (p_+ + p_-)\sqrt{2}$. Normalize the orbitals to unity.
- *Answer.* $\sigma = 4\times 0.35 + 2\times 0.85 = 3.10$, and so $Z^* = 3.90$. Use $Y_{00} = (1/4\pi)^{\frac{1}{2}}$, so

Table 9.1. Values of $Z^* = Z - \sigma$ for neutral ground-state atoms

	H							He
1s	1 (1.000)							1.70 (1.6875)
	Li	Be	B	C	N	O	F	Ne
1s	2.70 (2.6906)	3.70 (3.6848)	4.70 (4.6795)	5.70 (5.6727)	6.70 (6.6651)	7.70 (7.6579)	8.70 (8.6501)	9.70 (9.6421)
2s	1.30 (1.2792)	1.95 (1.9120)	2.60 (2.5762)	3.25 (3.2166)	3.90 (3.8474)	4.55 (4.4916)	5.20 (5.1276)	5.85 (5.7584)
2p			2.60 (2.4214)	3.25 (3.1358)	3.90 (3.8340)	4.55 (4.4532)	5.20 (5.1000)	5.85 (5.7584)
	Na	Mg	Al	Si	P	S	Cl	Ar
1s	10.70 (10.6259)	11.70 (11.6089)	12.70 (12.5910)	13.70 (13.5745)	14.70 (14.5578)	15.70 (15.5409)	16.70 (16.5239)	17.70 (17.5075)
2s	6.85 (6.5714)	7.85 (7.3920)	8.85 (8.2136)	9.85 (9.0200)	10.85 (9.8250)	11.85 (10.6288)	12.85 (11.4304)	13.85 (12.2304)
2p	6.85 (6.8018)	7.85 (7.8258)	8.85 (8.9634)	9.85 (9.9450)	10.85 (10.9612)	11.85 (11.9770)	12.85 (12.9932)	13.85 (14.0082)
3s	2.20 (2.5074)	2.85 (3.3075)	3.50 (4.1172)	4.15 (4.9032)	4.80 (5.6418)	5.45 (6.3669)	6.10 (7.0683)	6.75 (7.7568)
3p			3.50 (4.0656)	4.15 (4.2852)	4.80 (4.8864)	5.45 (5.4819)	6.10 (6.1161)	6.75 (6.7641)

Figures in normal type are values of Z^* obtained on the basis of the Slater rules. The figures in parentheses are improved values of Z^* suggested by Clementi and Raimondi; see *Further reading*, p. 247.

that

$$\psi_{2s} = N(1/4\pi)^{\frac{1}{2}} r e^{-Z^* r/2a_0} = (Z^{*5}/96\pi a_0^5)^{\frac{1}{2}} r e^{-Z^* r/2a_0}$$

when it is normalized to unity. In the case of N(2s):

$$\psi_{2s} = (1.73/a_0^{\frac{5}{2}}) r e^{-1.95 r/a_0}$$

For $2p_z$ use $Y_{10} = (3/4\pi)^{\frac{1}{2}} \cos\theta$, so that

$$\psi_{2p_z} = N(3/4\pi)^{\frac{1}{2}} r \cos\theta e^{-Z^* r/2a_0} = (Z^{*5}/32\pi a_0^5)^{\frac{1}{2}} r \cos\theta e^{-Z^* r/2a_0}.$$

In the case of $N(2p_z)$:

$$\psi_{2p_z} = (3.00/a_0^{\frac{5}{2}}) r \cos\theta e^{-1.95 r/a_0} = (3.00/a_0^{\frac{5}{2}}) z e^{-1.95 r/a_0}$$

For $2p_x$ use $(Y_{1-1} - Y_{11})/\sqrt{2} = (3/4\pi)^{\frac{1}{2}} \sin\theta \cos\phi$, so that

$$\psi_{2p_x} = N(3/4\pi)^{\frac{1}{2}} r \sin\theta \cos\phi e^{-Z^* r/2a_0} = (Z^{*5}/32\pi a_0^5)^{\frac{1}{2}} r \sin\theta \cos\phi e^{-Z^* r/2a_0}.$$

In the case of $N(2p_x)$:

$$\psi_{2p_x} = (3.00/a_0^{\frac{5}{2}}) r \sin\theta \cos\theta e^{-1.95 r/a_0} = (3.00/a_0^{\frac{5}{2}}) x e^{-1.95 r/a_0}.$$

• *Comment.* The 2s-orbital does not have the correct form at $r = 0$: it vanishes there. Nor is it orthogonal to the 1s-orbital. It can be corrected by modifying to $a\psi_{2s} + b\psi_{1s}$, where a and b are chosen so that the modified orbital is normalized and orthogonal to ψ_{1s}. This is called *Schmidt orthogonalization.*

9.11 Self-consistent fields

The Slater orbitals are only rough approximations and are totally inadequate for accurate calculations. In order to arrive at better descriptions of atoms, their energies, and their electron distributions, we have to find a way of solving the Schrödinger equation numerically, because there is no hope of finding exact, analytical solutions. The original way of doing this was introduced by Hartree and is known as the method of *self-consistent fields* (SCF). The technique was then modified by Fock and by Slater to include the effect of electron exchange, and the orbitals obtained by the method are called *Hartree–Fock self-consistent field atomic orbitals* (HF-SCF-AOs) or just *Hartree–Fock orbitals.*

The assumption behind the technique is that any one electron moves in a potential which is a spherical average of the potential due to all the other electrons. Then the Schrödinger equation is integrated numerically for that electron and that average potential. This supposes that the wavefunctions for all the other electrons are already known so that their average potential can be evaluated. This is in general not the case, and so the calculation is started by guessing the form of their wavefunctions (e.g. by supposing that they are Slater type orbitals). The Schrödinger equation for the electron of interest is then solved, and the wavefunction so found is used in the calculation of the potential experienced by one of the other electrons. The latter's Schrödinger equation is then solved, and in turn is used to refine the average potential experienced by another electron. This is repeated for all the electrons in the atom, and the potential experienced by the first electron can be recalculated. Usually this refined potential differs from the original guess, and so the whole cycle is repeated until the solutions for all the electrons are unchanged in a cycle of calculation—then the orbitals are *self-consistent*.

The Hartree–Fock equations are a bit involved to derive (see *Further reading*, p. 247) but quite easy to interpret. Suppose we express the structure of the (closed-shell) atom in the orbital approximation as the Slater determinant $\|a(1)\bar{a}(2)b(3)\bar{b}(4)\ldots\bar{z}(N)\|$ where a is a spin-orbital with α spin, $a=\psi_a^\alpha$ and $\bar{a}$ is one with β-spin, ψ_a^β. The hamiltonian for the atom has the form

$$H=\sum_i H_i+\tfrac{1}{2}\sum_i\sum_j{}'(e^2/4\pi\varepsilon_0 r_{ij}) \tag{9.11.1}$$

where H_i is the 'hydrogen-like' hamiltonian for electron i in the field of a bare nucleus of charge Ze: this is called the *core hamiltonian*. The factor $\frac{1}{2}$ in the repulsion term is there to avoid double-counting of interactions in the double sum: $1/r_{12}$, for instance, appears twice in the uncorrected sum but it should occur only once in the potential energy. The prime on the summation excludes $j=i$. The hamiltonian is a general form of the one used for helium, eqn (9.6.1). The Hartree–Fock equation for some space orbital ψ_s occupied by electron 1 in the determinant is

Hartree–Fock equation: $$\left\{H_1+\sum_r(2J_r-K_r)\right\}\psi_s(1)=\varepsilon_s\psi_s(1) \tag{9.11.2}$$

where J_r and K_r are *operators* defined in terms of their effects on $\psi_s(1)$:

Coulomb operator: $$J_r\psi_s(1)=\left\{\int\psi_r^*(2)(e^2/4\pi\varepsilon_0 r_{12})\psi_r(2)\,\mathrm{d}\tau_2\right\}\psi_s(1), \tag{9.11.3}$$

which is the average Coulombic potential energy of electron 1 when it is at some point where its amplitude is $\psi_s(1)$ and interacting with another electron in orbital ψ_r.

Exchange operator: $$K_r\psi_s(1)=\left\{\int\psi_r^*(2)(e^2/4\pi\varepsilon_0 r_{12})\psi_s(2)\,\mathrm{d}\tau_2\right\}\psi_r(1), \tag{9.11.4}$$

which is the quantum mechanical correction term that takes account of the effects of spin correlation. The quantity ε_s is the *one-electron orbital energy*. The sum on the left of eqn (9.11.2) runs over all the occupied space orbitals ($r=$ a, b, . . . , z). The equation shows very clearly that in order to find the orbital ψ_s, whichever one that may be, it is necessary to know all the other wavefunctions in order to set up the operators J and K.

When the final, self-consistent form of the orbitals has been established, we can find the orbital energies by multiplying both sides of eqn (9.11.2) by $\psi_s^*(1)$ and integrating over all space. The right-hand side is simply ε_s and so

$$\textit{One-electron energy:}\ \varepsilon_s = \int \psi_s^*(1) H_1 \psi_s(1)\,\mathrm{d}\tau_1 + \sum_r (2J_{sr} - K_{sr}) \tag{9.11.5}$$

where

$$J_{sr} = \int \psi_s^*(1) J_r \psi_s(1)\,\mathrm{d}\tau_1$$
$$= (e^2/4\pi\varepsilon_0) \int \psi_s^*(1)\psi_r(2)(1/r_{12})\psi_r^*(2)\psi_s(1)\,\mathrm{d}\tau_1\,\mathrm{d}\tau_2 \tag{9.11.6}$$

which, after reorganizing the integrand a little, is seen to be the *Coulomb integral* encountered in the helium problem, the average electrostatic interaction betwen two charge distributions r and s, and where

$$K_{rs} = \int \psi_s^*(1) K_r \psi_s(1)\,\mathrm{d}\tau_1$$
$$= (e^2/4\pi\varepsilon_0) \int \psi_s^*(1)\psi_r(2)(1/r_{12})\psi_s^*(2)\psi_r(1)\,\mathrm{d}\tau_1\,\mathrm{d}\tau_2, \tag{9.11.7}$$

which, after reorganizing the integrand, is seen to be the *exchange integral*, the correction to the Coulomb integral needed in order to allow for the effects of spin correlation. In passing, note that $J_{rr} = K_{rr}$.

The sum of the orbital energies is not the total energy of the atom, because simply adding together all the ε_s counts all the interactions twice. Therefore, in order to arrive at the total energy it is necessary to eliminate the double counting:

$$\textit{Total energy:}\ E = 2\sum_s \varepsilon_s - \sum_r \sum_s (2J_{rs} - K_{rs}), \tag{9.11.8}$$

where the sum is over the occupied orbitals a, b, . . . , z (each one being doubly occupied). We can check how this works for the ground state of helium with the configuration $1s^2$. The single-electron orbital energy is

$$\varepsilon_{1s} = E_{1s} + (2J_{1s,1s} - K_{1s,1s}) = E_{1s} + J_{1s,1s}$$

and the total energy is

$$E = 2\varepsilon_{1s} - (2J_{1s,1s} - K_{1s,1s}) = 2(E_{1s} + J_{1s,1s}) - J_{1s,1s} = 2E_{1s} + J_{1s,1s},$$

exactly as before.

When some electron is removed from an orbital ψ_r and we *assume* that the atom's electronic structure does not adjust, the energy required is the one-electron energy ε_r. Therefore we may equate the one-electron orbital energy with the ionization energy of the electron from that specified orbital. When r denotes the outermost orbital, this ionization energy is the normal first ionization energy. This identification of I_r and ε_r is the content of *Koopmans' theorem*. Note however, that it is an approximation, because in a real atom the electrons do readjust when one of their number is removed: the atom does *relax*, and so its energy is not changed by exactly ε_r.

The solutions of the Hartree–Fock equations are generally given either in numerical tables, or fitted to sets of simple functions. Once they are available the total electron density in an atom may be calculated very simply by summing the squares of the amplitudes for each electron. These electron densities are quite revealing because they have a series of maxima, Fig. 9.18, corresponding to the shell structure of the atom; the *total* electron density has inflections but no maxima—it falls off monotonically.

HF-SCF-AOs are by no means the end of the story. They remain tied to the orbital approximation, and therefore to the approximate central field form of the potential. The true wavefunctions, whatever they are, depend explicitly on the separations of the electrons, not merely on their distances from the nucleus. The incorporation of the separations r_{ij} into the wavefunction is the background to the *correlation problem*, which is at the centre of much modern work.

9.12 Many-electron atoms: term symbols and term energies

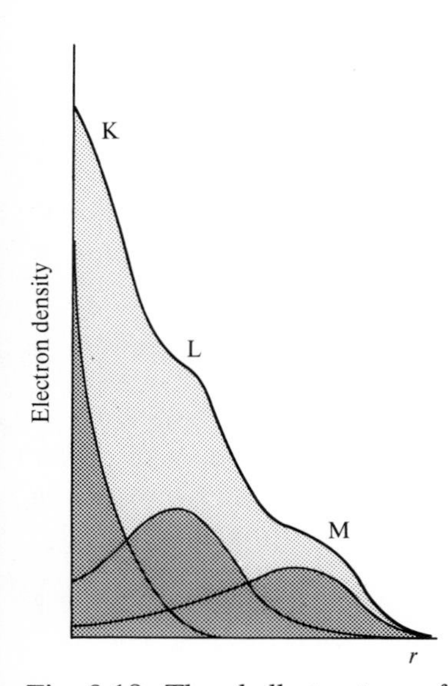

Fig. 9.18. The shell structure of a many electron atom (the electron density not the radial distribution function).

We have already met the basic nomenclature for the states of atoms. The *configuration* is the specification of the orbital occupancy, such as $1s^2 2s^2 2p^2$ for carbon (or more briefly, $K2s^2 2p^2$, K denoting the complete K shell). A variety of *terms* can arise from any given configuration, and are labelled with the appropriate term symbol, a p^2 configuration ($l_1 = l_2 = 1$) gives rise to $L = 2, 1, 0$ and hence to D, P, and S terms. Since p^2 consists of two electrons outside a closed shell the total spin is $S = 1$ or 0; hence the terms may be either *triplet* (and, for $L \geq 1$, have three *levels* distinguished by the value of J) or *singlet* (and have only a single level, with $J = L$). At this stage, therefore, it looks as though p^2 gives rise to $^3D_3$, $^3D_2$, $^3D_1$, $^3P_2$, $^3P_1$, $^3P_0$, $^3S_1$, $^1D_2$, $^1P_1$, $^1S_1$ as possible terms (the J values have been constructed from the Clebsch–Gordan series $J = L + S,\ L + S - 1, \ldots, |L - S|$ in each case). This would be the case if the configuration were 2p3p, but when it is $2p^2$ and $3p^2$ etc., the Pauli principle excludes many of the terms. It is easy to see that ^3D must be excluded: for it to occur, both electrons must have the same angular momentum component ($m_{l1} = 1$, $m_{l2} = 1$ to give $M_L = 2$) and the same spin orientation (to give the $\alpha_1\alpha_2$ state of the triplet): but two electrons cannot have the same set of quantum numbers, and so this term does not arise.

The Pauli principle allows the occurrence of only ^1D, ^3P, and ^1S from the $2p^2$ configuration. This can be seen as follows. Suppose we denote the spin-orbital occupied by an electron as m_l if it is nlm_l with

Table 9.2. The microstates of p^2

M_L	$M_S =$ 1	0	−1
2		$(1, \bar{1})$	
1	$(1, 0)$	$(1, \bar{0}), (\bar{1}, 0)$	$(\bar{1}, \bar{0})$
0	$(1, -1)$	$(1, -\bar{1}), (\bar{1}, -1), (0, \bar{0})$	$(\bar{1}, -\bar{1})$
−1	$(-1, 0)$	$(-1, \bar{0}), (-\bar{1}, 0)$	$(-\bar{1}, \bar{0})$
−2		$(-1, -\bar{1})$	

$m_s = \frac{1}{2}$ and as $\bar{m}_l$ if it has $m_s = -\frac{1}{2}$. A *microstate* of both electrons will be denoted $(1, \bar{1})$ if one occupies $m_l = 1$ with α-spin and the other occupies $m_l = 1$ with β-spin, and so on. This lets us draw up Table 9.2 of possible microstates, and to classify them as contributing to states of given $M_L = m_{l1} + m_{l2}$ and $M_S = m_{s1} + m_{s2}$. States such as $(1, 1)$ are excluded by the Pauli principle and have been omitted. The only $M_L = 2$ state has $M_S = 0$, and so must come from a ^{1}D term. There are five *states* with $L = 2$, and so one linear combination of the states in each row of the $M_S = 0$ column has been accounted for. The next row shows that there is a state with $M_L = +1$, $M_S = -1$. This must arise from $L = 1$, $S = 1$ and therefore from a ^{3}P term. There are nine states in a ^{3}P term, $(2S + 1) \times (2L + 1) = 3 \times 3$, and so we can strike out nine of the microstates in the table. This leaves only one microstate unaccounted for: it corresponds to $M_L = 0$ and $M_S = 0$ and so comes from a ^{1}S term. Therefore the terms ^{1}D, ^{3}P, ^{1}S account for all the permitted microstates of the system, as we set out to demonstrate. The same kind of analysis can be applied to other configurations involving electrons in equivalent orbitals.

Having arrived at the existence of the three terms of the p^2 configuration, the question arises as to their relative energies. Hund devised a set of rules which, while not infallible, are a useful guide even though the actual order has to be established either by detailed calculation or from the analysis of the spectrum of the atom.

Rule 1: *The term with the maximum multiplicity lies lowest in energy*. Therefore, in the configuration p^2 we expect the order $^3\text{P} < (^1\text{D}, {}^1\text{S})$. The explanation of the rule lies in the effects of spin correlation. On account of the Fermi hole, orbitals containing electrons of like spin can congregate more closely round the nucleus and lower their potential energy without the increased electron–electron repulsion overcoming the advantage.

Rule 2: *For a given multiplicity, the term with the largest value of L lies lowest in energy*. Therefore, in p^2 we expect $^3\text{P} < {}^1\text{D} < {}^1\text{S}$. The basis for this rule is essentially that if the electrons are orbiting in the same direction (and so have a large total angular momentum) they meet less often than when they orbit in opposite directions. Hence their repulsion is less on average when L is large.

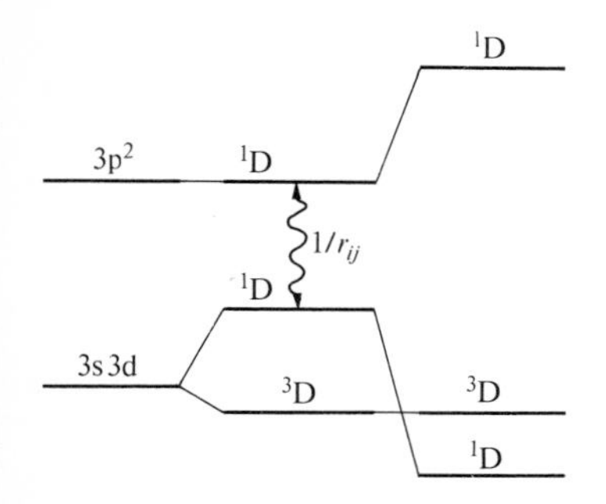

Fig. 9.19. Configuration interaction between ^{1}D terms.

Rule 3: *For atoms with less than half-filled shells, the level with the lowest value of J lies lowest in energy.* Since p^2 is less than half-filled, the three levels of ^{3}P are predicted to lie in the order $^3P_0 < {}^3P_1 < {}^3P_2$. When the shell is more than half full, the opposite rule holds (highest J lies lowest). The basis of the rule is the spin–orbit coupling, and is the same as that discussed for hydrogen on p. 218. Formally we see that ξ is positive ($d\phi/dr$ is positive because ϕ increases towards zero as r increases); the scalar product $\mathbf{l}\cdot\mathbf{s}$ is negative if $\mathbf{l}$ and $\mathbf{s}$ point in opposite directions; hence $\xi\mathbf{l}\cdot\mathbf{s}$ is negative when $\mathbf{l}$ and $\mathbf{s}$ are antiparallel and j is small.

These rules are quite simple-minded, and may fail for a variety of reasons. One possibility is that the structure of the atom is inaccurately described by a single configuration of electrons distributed among orbitals. It may be more accurately described as a mixture of several configurations; that is, in terms of *configuration interaction.* An example of this comes from the excited states of magnesium. In the configuration KL3s3d we expect ^{3}D to lie beneath ^{1}D: the opposite is found to be the case. A possible explanation is that the ^{1}D term is really a mixture of about 75 per cent 3s3d and 25 per cent 3p^2 (which can also give rise to a ^{1}D term). If the configurations 3s3d and 3p^2 are not too far apart in energy, the $1/r_{12}$ interaction perturbs them, and like any two-level system, they tend to move apart, Fig. 9.19 (recall Fig. 8.1). The lower combination is pressed down, and as a result may drop below the ^{3}D term (which is not depressed, because there is no 3p^2 ^{3}D term to interact with it). The lesson to draw is that a quoted configuration may signify only the major component and should not be taken at face value.

9.13 Alternative coupling schemes

Even the term symbols themselves should not be taken too literally. They imply that L and S take definite values, but this may be far from the truth when the atom has strong spin–orbit coupling, which is the case for atoms of high atomic number ('heavy atoms').

Consider the case of an atom with two electrons outside a closed shell. Let their orbital angular momentum quantum numbers be l_1 and l_2 (and, of course $s_1 = s_2 = \frac{1}{2}$). When the spin–orbit coupling is weak the electrostatic interaction between electrons dominates, and the orbital angular momenta couple together to form a resultant $\mathbf{L}$. The weak spin–orbit interaction now couples the spins to the combined orbital angular momenta, and so the total angular momentum is the resultant of $\mathbf{L}$ and $\mathbf{S}$: $\mathbf{J} = \mathbf{L} + \mathbf{S}$. A vector diagram of this coupling scheme, which is called *Russell–Saunders coupling* or *LS-coupling* is illustrated in Fig. 9.20: this is based on the vector diagrams introduced in Chapter 6, the cones representing the possible but unpredictable locations of the vectors. Since we also want to indicate the strength of the coupling (this is a new feature) we now imagine the component vectors as

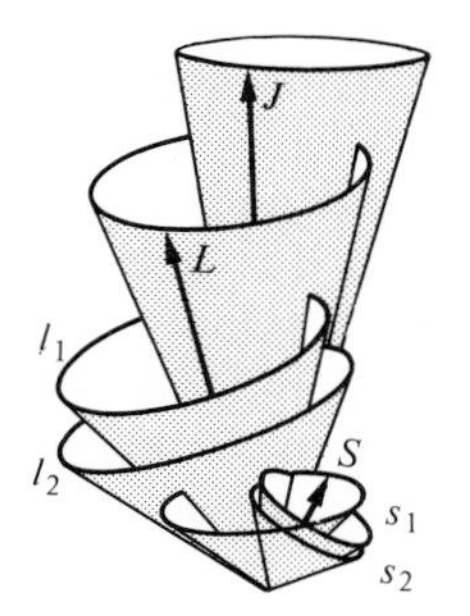

Fig. 9.20. Russell–Saunders, *LS*-coupling.

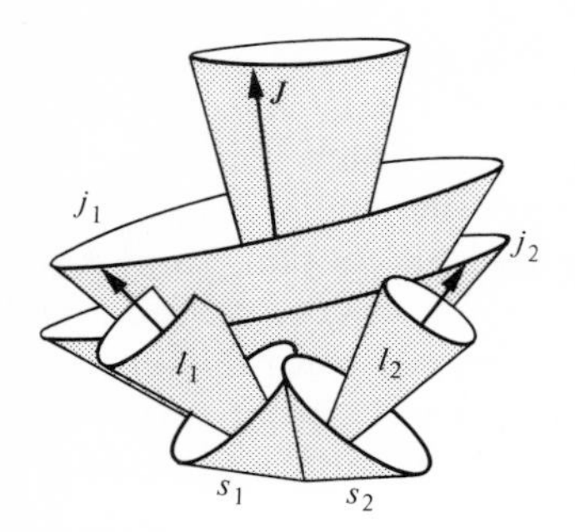

Fig. 9.21. *jj*-coupling.

sweeping round their cones. This motion is called *precession.* Fast precession denotes strong coupling; slow precession denotes weak coupling. For a light atom where *LS*-coupling is appropriate, the individual orbital momenta precess rapidly ($\mathbf{l}_1$ and $\mathbf{l}_2$ are strongly coupled electrostatically), the resultants **L** and **S** precess slowly around **J**. **J** itself is not (at this stage) precessing around the z-direction but is simply distributed about it.

Now consider the other extreme case, when the spin-orbit coupling is strong. Now the orbital and spin momenta of individual electrons couple together effectively, and give rise to resultant angular momenta $\mathbf{j}_1$ and $\mathbf{j}_2$. These resultants interact weakly via electrostatic coupling between the electron distributions they represent, and so form a resultant **J**. The vector coupling diagram for this *jj-coupling scheme* is shown in Fig. 9.21: now $\mathbf{l}_1$ and $\mathbf{s}_1$ precess rapidly around $\mathbf{j}_1$; $\mathbf{l}_2$ and $\mathbf{s}_2$ precess rapidly around $\mathbf{j}_2$, and $\mathbf{j}_1$ and $\mathbf{j}_2$ precess slowly around their resultant **J** lying static somewhere on its cone. In this scheme L and S are not specified, hence the term symbol loses its significance.

Although the significance of the term symbol is lost when the spin–orbit coupling is large, it can still be used to label the terms. This can be understood by considering the terms that can arise from the $n\text{p}^2$ configurations of the Group IV elements, where n is the quantum number of the valence shell. In the Russell–Saunders scheme we expect ^1S, ^1D, and ^3P terms, the last possessing the levels $^3\text{P}_2$, $^3\text{P}_1$, and $^3\text{P}_0$ as seen above. On the other hand, if *jj*-coupling is used, each p-electron is allowed either $j=\frac{1}{2}$ or $j=\frac{3}{2}$. When $j_1=j_2=\frac{1}{2}$ we may have $J=1, 0$; when $j_1=\frac{1}{2}$ and $j_2=\frac{3}{2}$ (or vice versa) we can have $J=2, 1$; and when $j_1=j_2=\frac{3}{2}$ we can have $J=3, 2, 1, 0$. If the Russell–Saunders scheme is appropriate the ^1D and ^3P terms are much further apart than the separation of the levels of ^3P because the fine structure is due to the weak spin–orbit coupling. If the *jj*-scheme is appropriate, we expect the states with $(\frac{1}{2}, \frac{1}{2})$ to be well below the higher energy couplings $(\frac{1}{2}, \frac{3}{2})$ and $(\frac{3}{2}, \frac{3}{2})$ where individual spins and orbital momenta are less advantageously aligned. These considerations let us construct the left and right sides of the diagram in Fig. 9.22, where on the left we have the array of terms for pure LS-coupling, and on the right the typical array for pure *jj*-coupling. The important feature is that the two sets of terms can be correlated because J is a well-defined quantum number for both coupling schemes. Furthermore, we have seen that states do not cross when perturbations are present (recall Fig. 8.2). Therefore the correlation of the terms is unambiguous, and even though the Russell–Saunders scheme is inappropriate it can still be used to construct labels for the terms. Hence term symbols based on the scheme may be used for any type of coupling, but when the spin–orbit coupling is not negligible the term symbol cannot be interpreted as indicating the values of L and S. The way this works for the Group IV elements is indicated in Fig. 9.22: the $^1\text{S}_0$ term of Pb is unambiguously labelled, even though S and L are ill-defined in this heavy atom.

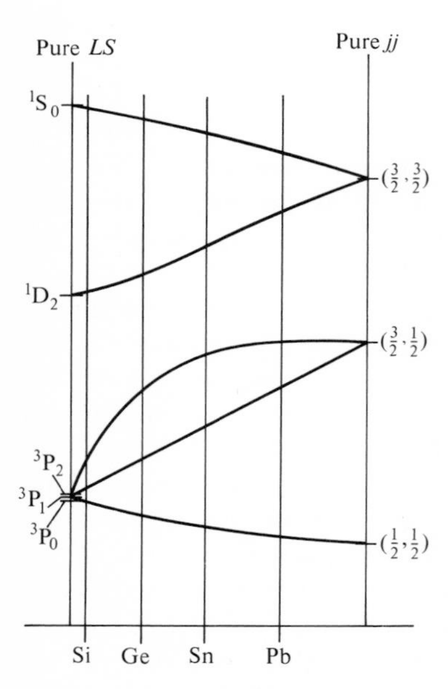

Fig. 9.22. *LS–jj* correlation for Group IV atoms.

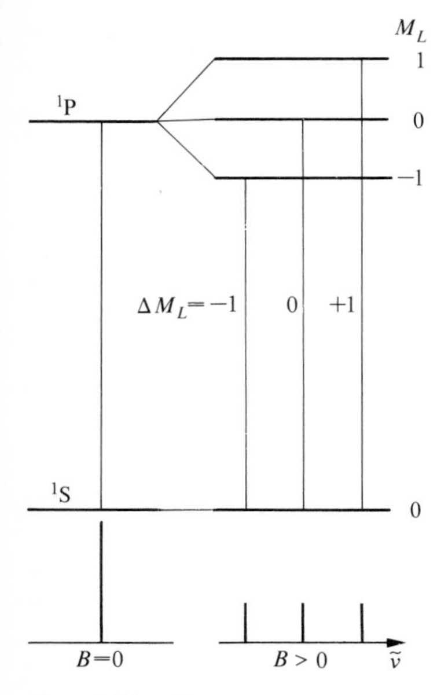

Fig. 9.23. The normal Zeeman effect for the ^{1}P–^{1}S transition.

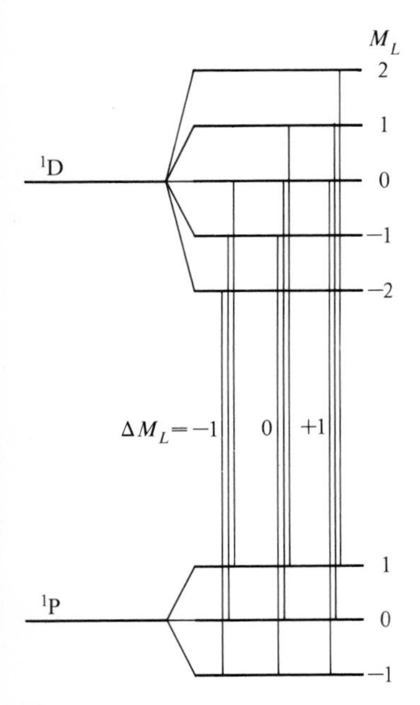

Fig. 9.24. The normal Zeeman effect for the ^{1}D–^{1}P transition.

9.14 The Zeeman effect

Electrons possess magnetic moments as a consequence of both their orbital and spin angular momenta. A magnetic field affects the energies of magnetic moments, and so when an atom is in a magnetic field we should expect its spectrum to change. The *Zeeman effect* is the response of the spectrum to the application of a magnetic field.

Consider first the effect of a magnetic field on a singlet term, ^{1}P for instance. Since $S=0$ and $L=1$, the magnetic moment arises solely from the orbital motion. Since in a magnetic field $\mathbf{B}$ a magnetic dipole $\mathbf{m}=\gamma_e\mathbf{l}$ has an energy $-\mathbf{m}\cdot\mathbf{B}$, the hamiltonian for the interaction is

$$H^{(1)} = -\mathbf{m}\cdot\mathbf{B} = -\gamma_e\mathbf{l}\cdot\mathbf{B} \tag{9.14.1}$$

for a single electron, and the sum of such terms for several electrons:

$$H^{(1)} = -\gamma_e(\mathbf{l}_1+\mathbf{l}_2)\cdot\mathbf{B} = -\gamma_e\mathbf{L}\cdot\mathbf{B}. \tag{9.14.2}$$

The ^{1}P term has the states $M_L=-1,0,+1$, and so the first-order correction to the energy when the field lies in the z-direction is

$$\begin{aligned} E^{(1)} &= \langle {}^1\mathrm{P}^{M_L}|\,H^{(1)}\,|{}^1\mathrm{P}^{M_L}\rangle = -\gamma_e\langle {}^1\mathrm{P}^{M_L}|\,L_z\,|{}^1\mathrm{P}^{M_L}\rangle B \\ &= -\gamma_e\hbar M_L B = \mu_B M_L B; \qquad M_L=-1,0,+1. \end{aligned} \tag{9.14.3}$$

(recall that $-\gamma_e\hbar=\mu_B$, the Bohr magneton). If the term is ^{1}S it has neither orbital nor spin angular momentum, and so the magnetic field leaves it unaffected. It follows that in a transition $^1\mathrm{P}\rightarrow{}^1\mathrm{S}$ we should expect to observe *three* lines in place of the one present originally, their wavenumbers corresponding to the energies ΔE (for $M_L=0$), $\Delta E\pm\mu_B B$ (for $M_L=\pm1$). This is illustrated in Fig. 9.23. A 1 T magnetic field splits lines by only about 0.5 cm^{-1}, and so the effect is very small.

We have already seen that transitions corresponding to different values of ΔM_L correspond to different polarizations of the emitted light. In the present case an observer perpendicular to the field should see that the outer lines ($\Delta M_L=\pm1$) of the trio (the *σ-lines*) are circularly polarized in opposite senses while the central line ($\Delta M_L=0$) should be linearly polarized parallel to the field (it is called the *π-line*). These are the polarizations observed.

For all allowed transitions between singlet states the effect of the magnetic field is to split the original spectral line into three. This is called the *normal Zeeman effect.* It occurs even for transitions of the type $^1\mathrm{D}\rightarrow{}^1\mathrm{P}$ where there are five states in the first term and three in the second. This is because the splittings are the same in both terms and the selection rules restrict the transitions to $\Delta M_L=0,\pm1$. This has the effect of leading to three groups of coincident lines, Fig. 9.24.

More common than the normal effect is the *anomalous Zeeman effect.* This is observed when the transitions involve states other than singlets. In the anomalous effect, the spectral lines are split into more than three components when the field is applied. The basic reason is that the splitting is unequal in the two terms involved in the transition, and hence the neat coincidences of the normal effect do not occur. The ultimate reason for the anomalous effect is the anomalous g-value of the electron.

The reason is as follows. If the g-value were 1 and not 2 the total angular momentum **J** and the total magnetic moment of the electron would be collinear, Fig. 9.25(a). But in fact $g=2$, not 1, and so the magnetic moment arising from the spin should be drawn twice as long as it is drawn in Fig. 9.25(a). As a result the total magnetic moment is not collinear with the total angular momentum, Fig. 9.25(b). The **S** and **L** vectors precess about their resultant (as a consequence of the spin–orbit coupling), and drag their magnetic moments round with them. The only component of the magnetic moment that does not average to zero is the one collinear with the direction of **J**. This projection is what we refer to as the 'magnetic moment' of the system, $\mathbf{m}_J$. The surviving component is parallel to **J**, and so we can still write $\mathbf{m}_J \propto \mathbf{J}$, but the proportionality constant depends on the values of S, L, and J because of the way that the vectors of different lengths lead to different resultants. Therefore the magnetic moment of the system depends on the values of these quantum numbers, and so different terms are split to different extents.

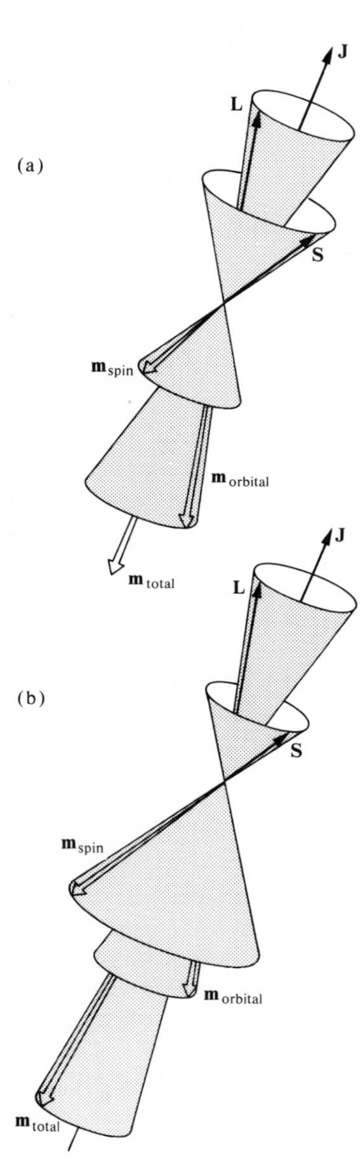

Fig. 9.25. (a) Hypothetical case for $g_e=1$; (b) actual case of $g_e=2$.

The calculation of the magnetic moment proceeds as follows. The hamiltonian for the interaction of a magnetic field with orbital and spin angular momenta is

$$H^{(1)}=-\mathbf{m}_{\text{orbital}}\cdot\mathbf{B}-\mathbf{m}_{\text{spin}}\cdot\mathbf{B}=-\gamma_e(\mathbf{L}+2\mathbf{S})\cdot\mathbf{B} \tag{9.14.4}$$

where we have used 2 in place of g_e. If the field is along the z-direction the hamiltonian becomes

$$H^{(1)}=-\gamma_e(L_z+2S_z)B, \tag{9.14.5}$$

but this form is inconvenient because when J is well-defined, the individual values of the z-components of the component momenta are not well-defined (Chapter 6). The expression would be more useful if it could be expressed in the form

$$H^{(1)}=-g_J\gamma_e J_z B \tag{9.14.6}$$

because the z-component of **J** is well-defined and has the eigenvalues $M_J\hbar$. In this expression g_J is a constant whose value we need to find so that this expression for the interaction hamiltonian is equivalent to the 'true' version in eqn (9.14.5). For a general direction of the field it follows that we are looking for g_J such that eqn (9.14.4) is equivalent to

$$H^{(1)}=-g_J\gamma_e\mathbf{J}\cdot\mathbf{B}. \tag{9.14.7}$$

In fact they are not equivalent; but so long as we confine ourselves to *first-order effects*, and ignore all terms arising from the off-diagonal elements of the hamiltonians, we can find a suitable expression for g_J.

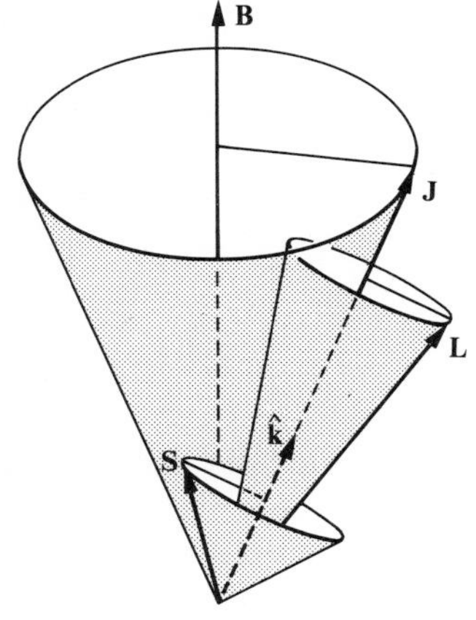

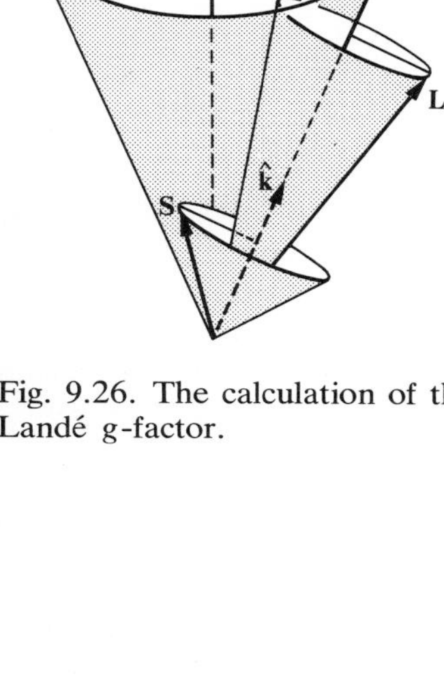

Fig. 9.26. The calculation of the Landé g-factor.

Consider Fig. 9.26. There are now three precessional motions: **S** about **J**, **L** about **J**, and now **J** about **B** (**J** is now coupled to the external field, and so we can signify the strength of its coupling by a rate of precession). The effective magnetic moment can be found by projecting **L** on to **J** and then **J** on to **B**, and then doing the same for **S**. The precessional motion averages to zero all the components perpendicular to this motion (this 'averaging to zero' is the classical equivalent of neglecting off-diagonal terms of the hamiltonian). Let $\hat{\mathbf{k}}$ be a unit vector in the direction of **J**, $\mathbf{J}=|J|\,\hat{\mathbf{k}}$. Then

$$\mathbf{L}\cdot\mathbf{B}=\mathbf{L}\cdot\hat{\mathbf{k}}\hat{\mathbf{k}}\cdot\mathbf{B}=\mathbf{L}\cdot\mathbf{J}\mathbf{J}\cdot\mathbf{B}/|J|^2;\ \mathbf{S}\cdot\mathbf{B}=\mathbf{S}\cdot\hat{\mathbf{k}}\hat{\mathbf{k}}\cdot\mathbf{B}=\mathbf{S}\cdot\mathbf{J}\mathbf{J}\cdot\mathbf{B}/|J|^2$$

(which is valid if the perpendicular components are ignored). Therefore

$$(\mathbf{L}+2\mathbf{S})\cdot\mathbf{B}=(\mathbf{L}\cdot\mathbf{J}+2\mathbf{S}\cdot\mathbf{J})\mathbf{J}\cdot\mathbf{B}/|J|^2.$$

Since $\mathbf{L}+\mathbf{S}=\mathbf{J}$ we have

$$2\mathbf{L}\cdot\mathbf{J}=J^2+L^2-|\mathbf{J}-\mathbf{L}|^2=J^2+L^2-S^2$$
$$2\mathbf{S}\cdot\mathbf{J}=J^2+S^2-|\mathbf{J}-\mathbf{S}|^2=J^2+S^2-L^2.$$

If these quantities are inserted into eqn (9.14.4) and the quantum mechanical versions of the magnitudes, $J(J+1)\hbar^2$, are used in place of J^2 etc., we find

$$H^{(1)}=-\gamma_e(\mathbf{L}+2\mathbf{S})\cdot\mathbf{B}=-\gamma_e\left\{1+\frac{J(J+1)+S(S+1)-L(L+1)}{2J(J+1)}\right\}\mathbf{J}\cdot\mathbf{B},$$

which is the form we require, eqn (9.14.7). The g_J-factor can now be identified as

The Landé g-factor: $$g_J=1+\frac{J(J+1)+S(S+1)-L(L+1)}{2J(J+1)}. \quad (9.14.8)$$

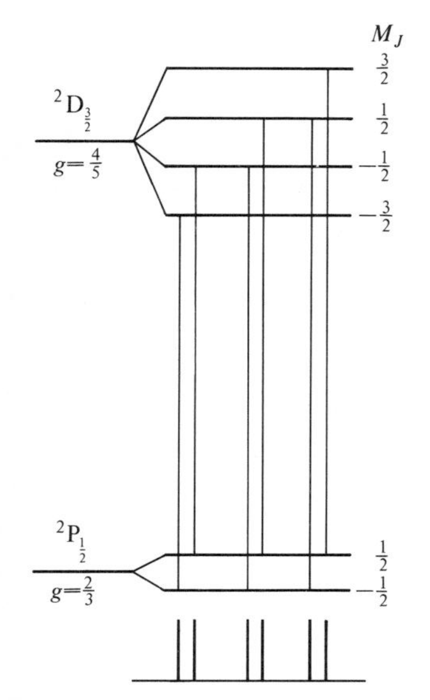

Fig. 9.27. The anomalous Zeeman effect for $^2D_{\frac{3}{2}}$–$^2P_{\frac{1}{2}}$.

When $S=0$, $g_J=1$ because then J must equal L. In this case the magnetic moment is independent of L, and so all terms are split to the same extent. Hence we regain the normal Zeeman effect. When $S\neq 0$ the value of g_J depends on the values of L and S, and so different terms are split to different extents. This is illustrated in Fig. 9.27: note that the selection rule $\Delta M_J=0,\pm 1$ continues to limit the transitions that contribute to the spectrum, but the lines no longer coincide and form three groups.

When the applied field is very strong, the coupling between **S** and **L** may be broken in favour of their direct coupling to the magnetic field. (This was the problem treated in Appendix 9 where the full significance of the recoupling was seen to be the search for the representation that gave matrices with the smallest off-diagonal elements: the vector recoupling diagram is a pictorial representation of the effect.) The momenta, and therefore the magnetic moments, now precess

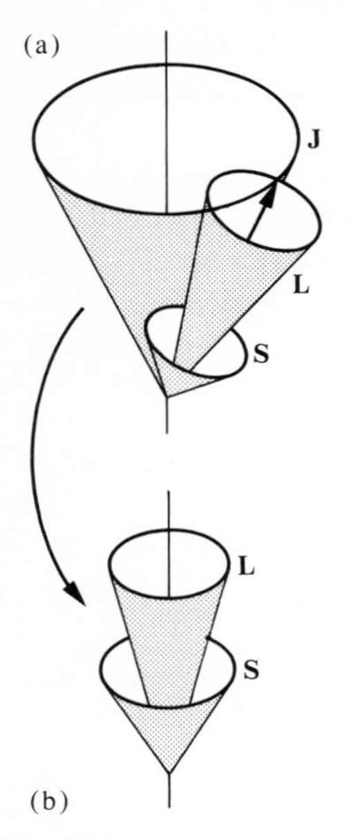

Fig. 9.28. Angular momentum coupling at (a) low fields and (b) the Paschen–Back effect at high fields.

independently about the field direction, Fig. 9.28. As the electromagnetic field induces electric dipole transitions by coupling with the orbital distribution of the electrons the selection rule obeyed is $\Delta S=0$; therefore the presence of the spin does not show up in the transitions. As a consequence the normal Zeeman effect is observed even though the states might not be singlets. This switch from anomalous to normal is called the *Paschen–Back effect.*

9.15 The Stark effect

The effect produced on a spectrum when an electric field is applied is called the *Stark effect.* There are two varieties.

The *linear Stark effect* is a modification of the spectrum linear in the strength of the applied field. It arises on account of the degeneracy typical of hydrogen-like atoms (where the energy is independent of the quantum number l). The hamiltonian for the interaction of the field and the atom is

$$H^{(1)}=-\boldsymbol{\mu}\cdot\mathbf{E}=-\mu_z E=ezE \tag{9.15.1}$$

when the field lies in the z-direction ($\boldsymbol{\mu}$ is the dipole moment operator, $\boldsymbol{\mu}=-e\mathbf{r}$). This operator has matrix elements between orbitals that differ in l by unity and in m_l by zero (p. 213). Therefore it may mix the 2s-orbital of hydrogen with the degenerate $2p_z$-orbital. The magnitude of the matrix element is $\langle 2p_z|\,H^{(1)}\,|2s\rangle=3ea_0E$, and so from Fig. 8.1 (or eqn (8.1.6)) we know that 2s and $2p_z$ mix and give rise to two states separated by $6ea_0E$. The two perturbed functions are $(\frac{1}{2})^{\frac{1}{2}}(2s\pm 2p_z)$, and their forms are sketched in Fig. 9.29. It is easy to see that they correspond to a shift of charge density into and out of the direction of the field, and hence the electron has an energy that depends on which state it occupies. The splitting of the 2s- and $2p_z$-orbitals, which we see is indeed linear in the field strength, shows up in the spectrum for transitions involving the 2s- and or 2p-orbitals, but even for fields of the order of $10^6\ \mathrm{V\,m^{-1}}$ the splitting is equivalent to a wavenumber $6ea_0E/hc$ of only $2.6\ \mathrm{cm^{-1}}$.

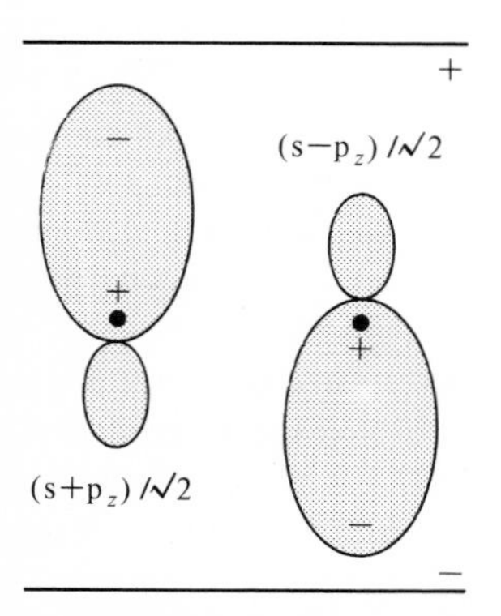

Fig. 9.29. The linear Stark effect in atomic hydrogen.

The linear Stark effect depends on the peculiar degeneracy of the Coulomb problem, and it does not occur in the spectra of many-electron atoms (where the degeneracy does not exist). In these it is replaced by the *quadratic Stark effect,* which is much weaker still. The basis of the effect is the same, but now the distortion of the charge distribution occurs only in higher orders of perturbation theory and the energy shifts are of second-order in the strength of the perturbation, and therefore proportional to E^2. The field has to distort a non-degenerate (and therefore 'tight' system) and then interact with the dipole it has induced. Hence the quadratic nature of the effect.

At very high field strengths the H_α line is seen to broaden and its intensity to decrease. This is traced to the tunnelling of the electron. In high fields the form of the potential experienced by an electron is shown in Fig. 9.30. The tails of the orbitals seep through the region of high potential, and penetrate into the external region where the potential can strip them away from the atom. This results in fewer of them being available for the radiative emission step, and so the intensity falls. Furthermore, the upper state now has a finite lifetime, and hence its energy undergoes lifetime broadening, and the transition becomes more diffuse.

Further reading

Atomic spectra

Structure and spectra of atoms. W. G. Richards and P. R. Scott; Wiley, London, 1976.

Atomic spectra and atomic structure. G. Herzberg; Dover, New York, 1944.

Atomic spectra. H. Kuhn; Longmans, London, 1962.

Atomic structure. E. U. Condon and H. Odabaşi; Cambridge University Press, 1980.

Atomic data

American Institute of Physics Handbook. D. E. Gray (ed.); McGraw-Hill, New York, 1972.

Atomic energy levels (3 volumes). C. E. Moore; NBS-Circ. 467, Washington, 1949, 1952, 1958.

Tables of spectral lines of neutral and ionized atoms. A. R. Striganov and N. S. Sventitskii; Plenum, New York, 1968.

Atomic energy levels and Grotrian diagrams. S. Bashkin and J. O. Stoner Jr.; North-Holland, Amsterdam, 1975 *et seq.*

Atomic screening constants from SCF functions. E. Clementi and D. L. Raimondi; IBM Res. Note NJ-27, 1963.

Introduction to applied quantum chemistry (Appendix B). S. P. McGlynn, L. G. Vanquickenborne, M. Kinoshita, and D. G. Carroll; Holt, Rinehart, and Winston, 1972.

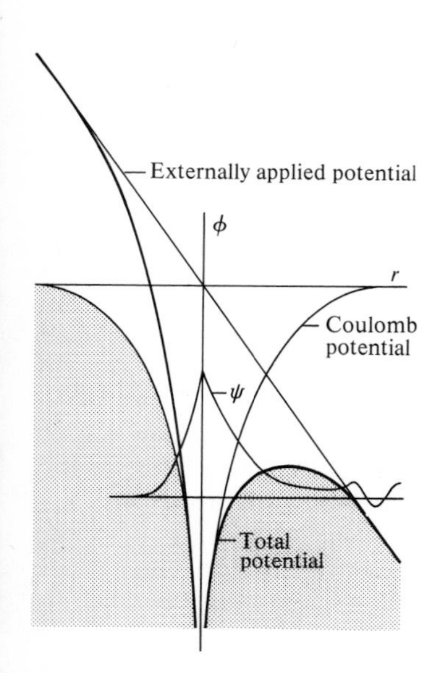

Fig. 9.30. Tunnelling at high field strengths.

Problems

9.1. Calculate the wavenumbers of the transitions of He^+ for the analogue of the Balmer series of hydrogen. *Hint.* Use eqn (9.1.5) with the Rydberg constant modified to account for the mass and charge differences.

9.2. Predict the form of the spectrum of the muonic atom formed from a proton in association with a μ-meson ($m_\mu = 207m_e$, charge $-e$).

9.3. Which of the following transitions are electric-dipole allowed: $1s \rightarrow 2s$, $1s \rightarrow 2p$, $2p \rightarrow 3d$, $3s \rightarrow 5d$, $3s \rightarrow 5p$?

9.4. The spectrum of a one-electron ion of an element showed that its ns-orbitals were at 0, 2 057 972 cm^{-1}, 2 439 156 cm^{-1}, and 2 572 563 cm^{-1} for $n = 1, 2, 3, 4$ respectively. Identify the species and predict its ionization energy.

9.5. Demonstrate that for one-electron atoms the selection rules are $\Delta l = \pm 1$, $\Delta m_l = 0, \pm 1$, and Δn unlimited. *Hint.* Evaluate the electric-dipole transition moment $\langle n'l'm_l'|\,\boldsymbol{\mu}\,|nlm_l\rangle$ using $\mu_x = -er \sin\theta \cos\phi$, $\mu_y = -er \sin\theta \sin\phi$, and $\mu_z = -er \cos\theta$. The easiest way of evaluating the angular integrals is to recognize that the components just listed are proportional to Y_{lm} with $l = 1$, and to analyse the resulting integral group-theoretically.

9.6. Confirm that in hydrogen-like atoms, the spin–orbit coupling constant depends on n and l as in eqn (9.3.8).

9.7. Calculate the spin–orbit coupling constant for a 2p-electron in a Slater type atomic orbital, and evaluate it for the neutral atoms of the first short period of the periodic table (from B to F).

9.8. Deduce the *Landé interval rule*, that for a given l and s the energy difference between two levels differing in j by unity is proportional to j. *Hint.* Evaluate E_{so} in eqn (9.3.12) for j and $j-1$; use the second line in the equation (in terms of ζ).

9.9. The ground configuration of iron is $3d^6 4s^2$, and the ^5D term has five levels ($J=4, 3, \ldots, 0$) at relative wavenumbers 0, 415.9, 704.0, 888.1, and 978.1 cm^{-1}. Investigate how well the Landé interval rule is obeyed. Deduce a value of ζ.

9.10. Calculate the energy difference between the levels with the greatest and smallest values of j for given l and s. Each term of a level is $(2j+1)$-fold degenerate. Demonstrate that the centre of gravity of the energy of a term is the same as the energy in the absence of spin–orbit coupling. *Hint.* Weight each level with $2j+1$ and sum the energies given in eqn (9.3.12; second line) from $j=|l-s|$ to $j=l+s$. Use the relations

$$\sum_{s=0}^{n} s=\tfrac{1}{2}n(n+1), \qquad \sum_{s=0}^{n} s^2=\tfrac{1}{6}n(n+1)(2n+1); \qquad \sum_{s=0}^{n} s^3=\tfrac{1}{4}n^2(n+1)^2.$$

9.11. Deduce what terms may arise from the ground configurations of the atoms of elements of the first short period, and suggest the order of their energies. *Hint.* Construct the term symbols as explained in Section 9.4, and use Hund's rules to arive at their relative orders. Recall the hole–particle rule explained in the *Example* on p. 220.

9.12. Find the first-order corrections to the energies of the hydrogen atom that result from the relativistic mass increase of the electron. *Hint.* The energy is related to the momentum by $E=(p^2c^2+m^2c^4)^{\frac{1}{2}}+V$. When $p^2c^2 \ll m^2c^4$, $E \approx \mu c^2+p^2/2\mu+V-p^4/8\mu^3c^2$ (where the reduced mass has replaced m). Ignore the rest energy μc^2, which simply fixes the zero. The term $-p^4/8\mu^3c^2$ is a perturbation; hence calculate $\langle nlm|\,H^{(1)}\,|nlm\rangle=-(1/2\mu c^2)\langle nlm|\,(p^2/2\mu)^2\,|nlm\rangle=-(1/2\mu c^2)\langle nlm|\,(E_{nlm}-V)^2\,|nlm\rangle$. We know E_{nlm}; therefore calculate the matrix elements of $V=-e^2/4\pi\varepsilon_0 r$ and V^2.

9.13. Write the hamiltonian for the lithium atom ($Z=3$) and confirm that, if electron–electron repulsions are neglected, the wavefunction can be written as a product $\psi(1)\psi(2)\psi(3)$ of hydrogen-like atomic orbitals, and the energy is a sum of the corresponding energies.

9.14. Take a trial function for the helium atom as $\psi=\psi(1)\psi(2)$, with $\psi(1)=(\zeta^3/\pi)^{\frac{1}{2}}e^{-\zeta r_1}$ and $\psi(2)=(\zeta^3/\pi)^{\frac{1}{2}}e^{-\zeta r_2}$, ζ being a parameter, and find the best ground-state energy for a function of this form, and the corresponding value of ζ. Calculate the first and second ionization energies. *Hint.* Use the variation theorem. All the integrals are standard, the electron repulsion term is calculated in the *Example* on p. 225. Interpret Z in terms of a shielding constant. The experimental ionization energies correspond to 24.850 eV and 54.40 eV.

9.15. On the basis of the same kind of calculation as in Problem 14, but for general Z, account for the first ionization energies of the ions Li^+, Be^{2+}, B^{3+}, C^{4+}. The experimental values are 73.5, 153, 258, and 389 eV respectively.

9.16. Consider a one-dimensional square well containing two electrons. One occupies ψ_1, the other ψ_2. Plot a two-dimensional contour diagram of the probability distribution of the electrons when their spins are (a) parallel, (b) antiparallel. Devise a measure of the radius of the Fermi hole. *Hint.* Recall the discussion in Section 9.7. When the spins are parallel (e.g. $\alpha\alpha$) the antisymmetric space combination $\psi_1(1)\psi_2(2)-\psi_2(1)\psi_1(2)$ must be taken, and when antiparallel, the symmetric combination. In each case plot ψ^2 against axes labelled x_1, x_2. Computer graphics may be used to obtain striking diagrams, but a sketch is sufficient.

9.17. The first few S terms of helium lie at the following wavenumbers: $1s^2$ ^1S: 0; 1s2s ^1S: 166 272 cm^{-1}; 1s2s ^3S: 159 850 cm^{-1}; 1s3s ^1S: 184 859 cm^{-1}; 1s3s ^3S: 183 231 cm^{-1}. What are the values of K in the 1s2s and 1s3s configurations?

9.18. What levels may arise from the following terms: ^1S, ^2P, ^3P, ^3D, ^2D, ^1D, ^4D? Arrange in order of increasing energy the terms that may arise from the following configurations: 1s2p, 2p3p, 3p3d. What terms may arise from (a) a d^2 configuration, (b) an f^2 configuration?

9.19. Write down the Slater determinant for the ground term of the beryllium atom, and find an expression for its energy in terms of Coulomb and exchange integrals. Find expressions for the energy in terms of the Hartree–Fock expression, eqn (9.11.8). *Hint.* Use eqn (9.7.3) for the configuration $1s^22s^2$; evaluate the expectation value $\langle\psi| H |\psi\rangle$.

9.20. Calculate the magnetic field required to produce a splitting of 1 cm^{-1} between the states of a ^1P term. Calculate the Landé g-factor for (a) a term in which J has its maximum value for a given L, S, (b) a term in which J has its minimum value.

9.21. Transitions are observed and ascribed to $^1F \rightarrow {}^1D$. How many lines will be observed in a 4 T magnetic field? Calculate the form of the spectrum for the Zeeman effect on a $^3P \rightarrow {}^3S$ transition.

10 Molecular structure

Now we come to the heart of chemistry. If we can understand what holds atoms together as molecules we may also start to understand why, under certain conditions, old arrangements change in favour of new ones. We shall understand structure, and through structure, the mechanism of change. The aim of this chapter is to select a few points to illustrate the main features of *valence theory*, the theory of molecular structure. The subject has been so richly developed in recent years, particularly in the sophistication of numerical methods for the computation of electron distributions and energies, that there is little hope of covering it except in broad outline.

10.1 The Born–Oppenheimer approximation

It is unfortunate that, having arrived in sight of the promised land, we realize that as even the simplest molecule must contain at least three particles, its Schrödinger equation is not solvable analytically. We have to make an approximation at the outset. The *Born–Oppenheimer approximation* is physically plausible and mathematically justifiable; but like all approximations, its range of validity is sometimes overstepped, and deviations modify the appearance of spectra. We shall see some of them in the next chapters.

The Born–Oppenheimer approximation is based on the great difference of masses of the electrons and the nuclei in a molecule. When nuclei move, the electrons can almost instantaneously adjust to their new positions. Therefore, instead of trying to solve the Schrödinger equation for a collection of mobile electrons and nuclei, we regard the nuclei as frozen in a single arrangement (the *molecular conformation*) and solve the Schrödinger equation for the electrons moving in the stationary potential they generate. Different conformations of the molecule may then be taken, and the Schrödinger equation solved for each one. This lets us construct a curve showing the dependence of the energy of the molecule on its conformation (e.g. the bond length of a diatomic molecule) which is called a *molecular potential energy curve* (or *surface* for a polyatomic molecule where the geometry depends on several parameters). The *equilibrium conformation* of the molecule corresponds to the minimum of the curve or surface.

10.2 The hydrogen molecule-ion

Even if the Born–Oppenheimer approximation is used there is still only one molecule for which the Schrödinger equation can be solved exactly. This is the hydrogen molecule-ion, H_2^+. As it has only one electron, it has a status in the theory of molecules similar to that of the hydrogen atom in the theory of atoms. Whereas the equation for the atom is separable in spherical polar coordinates, the equation for the molecule-ion is separable in *ellipsoidal coordinates* in which the two nuclei are the foci of ellipses. Just as the atom has solutions called atomic orbitals, so the molecule-ion has solutions called *molecular orbitals* which spread over both nuclei. We used the atomic orbitals as a basis of the description of many-electron atoms. Here we use the molecular orbitals of the diatomic molecule as a basis for the discussion not only of many-electron diatomic molecules, but also of all molecules, including polyatomics as big as DNA.

As a first step we look at the form of the first few molecular orbitals of H_2^+ obtained from the exact solution of its Schrödinger equation. By analysing the electron density distribution we shall see what holds the molecule together.

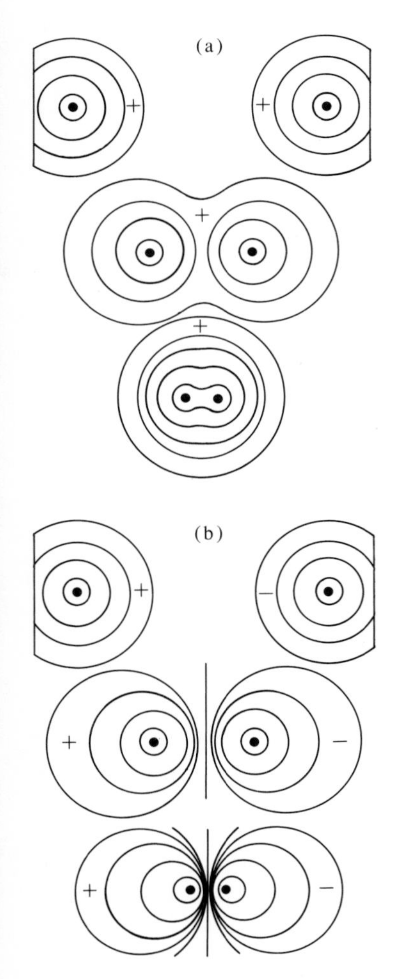

Fig. 10.1. (a) Bonding and (b) antibonding orbitals of H_2^+.

The two lowest energy orbitals are drawn in Fig. 10.1 and shown at various stages as the internuclear distance is changed. The striking difference between them is that the lower orbital has no node while the upper has a node slicing through the internuclear region. There is therefore a much greater probability of finding the electron between the nuclei when it occupies the lower orbital than when it occupies the upper. The energy level diagram in Fig. 10.2 shows how the orbital energies change with distance (it is an example of a *molecular potential energy curve*). We see that an electron occupying the lower orbital gives rise to a structure having a lower energy than the infinitely separated $H \ldots H^+$. Hence the molecular orbital is called a *bonding orbital*. In contrast, when an electron occupies the upper orbital, the system has an energy that is greater than the separated $H \ldots H^+$, and hence it is called an *antibonding orbital*. The difference between the two is clearly related to the presence of the internuclear nodal plane.

The contributions to the energy of the molecule that account for the distance-dependence of the orbital energies are harder to identify than might be thought. The rise of energy at short distances is largely due to the repulsion between the two nuclei (which is proportional to $1/\boldsymbol{R}$, where $\boldsymbol{R}$ is the internuclear distance). This repulsion is largely responsible for the molecule not collapsing down to a point (and forming the *united atom*, an isotope of He^+ in this case). The energy reduction associated with the bonding orbital at intermediate values of $\boldsymbol{R}$ is *conventionally* interpreted as arising from the favourable electrostatic potential of the electron sandwiched between the two nuclei. The heightened energy of the antibonding orbital is conventionally ascribed to the unfavourable electrostatic energy it is forced to have on account of its exclusion from the internuclear region. However, the total energy of the molecule arises from a variety of sources, including the kinetic energy of the electron and the changes in its average distance from the nuclei when the molecular orbital is formed. What appears to happen is that when the bonding orbital is formed, the wavefunction distorts by shrinking more closely into the region of the nuclei, which enhances the probability of finding the electron in a favourable potential. The formation of the antibonding orbital, on the other hand, seems to lead

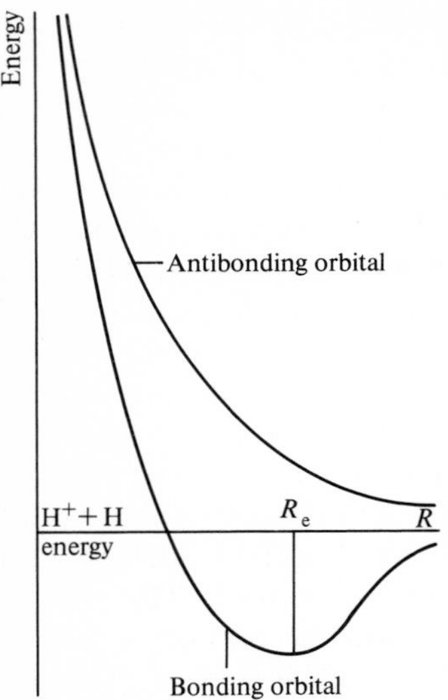

Fig. 10.2. The molecular potential energy curves for H_2^+.

to an expansion of the wavefunction in the region of the nuclei, and so in this more diffuse arrangement it samples a less favourable potential. In other words, it is not the act of shifting the electron into the internuclear region itself that lowers the molecule's energy, but the freedom this gives for the wavefunctions to distort and cluster more closely around the nuclei. Therefore although the bonding character accompanies enhanced internuclear overlap, it seems that the effect actually arises from the shrinkage of the orbitals then allowed (in H_2^+ at least).

In what follows we shall anticipate bonding orbitals whenever there is an enhanced probability density in the internuclear region, and antibonding orbitals whenever there is exclusion from these regions, and we shall not enquire too closely about the source of the energy changes. *Accumulation signifies bonding; exclusion signifies antibonding.*

Once we turn from H_2^+ to the hydrogen molecule itself, we make the same transition in complexity as in the step from the hydrogen atom to helium. The hamiltonian for the molecule with a specified, static internuclear separation R is

$$H=(-\hbar^2/2m_e)(\nabla_1^2+\nabla_2^2)$$
$$+(e^2/4\pi\varepsilon_0)\left\{-\frac{1}{r_{1a}}-\frac{1}{r_{1b}}-\frac{1}{r_{2a}}-\frac{1}{r_{2b}}+\frac{1}{r_{12}}+\frac{1}{R}\right\}. \tag{10.2.1}$$

The labels 1, 2 signify the electrons and a, b the nuclei. We have ignored the question of reduced mass and used m_e in the two kinetic energy terms. The four $-1/r$ terms are the attractive potential energy terms for the interaction of the two electrons and the two nuclei, Fig. 10.3. The $+1/r_{12}$ term is the electron–electron repulsion, and the $+1/R$ term is the internuclear repulsion. There is clearly no hope of finding analytical solutions of the corresponding Schrödinger equation, and we have to make another approximation. There are two basic approaches: the molecular orbital method, and the valence bond method.

10.3 The molecular orbital method

The *molecular orbital* (MO) *method* supposes that the individual electrons of a molecule can each be thought of as occupying an orbital that spreads throughout the nuclear framework. In the case of hydrogen they resemble the molecular orbitals found on the basis of the exact solution of the Schrödinger equation for the hydrogen molecule-ion; in the case of heavier molecules the orbitals are still recognizable as being similar, but differ on account of the different nuclear charge and the effects of the numerous electron–electron interactions. Solving the exact Schrödinger equation for the bare nuclei is not a very good starting point when we already have at our disposal quite good many-electron atomic orbitals which already take into account many of the electron–electron repulsions (e.g. HF-SCF-AOs or even Slater orbitals in less sophisticated treatments). Therefore we first look for a

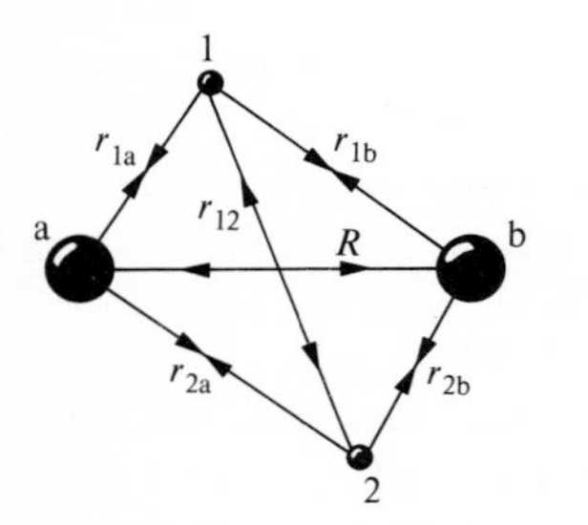

Fig. 10.3. The contributions to the potential energy of H_2.

way of modelling the exact molecular orbitals of the hydrogen molecule-ion (those of Fig. 10.1) using the atomic orbitals of the hydrogen atom, and then use the same modelling procedure to construct molecular orbitals for many-electron molecules using the orbitals of the parent atoms in an analogous fashion. The procedure is called the *linear combination of atomic orbitals* (LCAO) *molecular orbital* (MO) *method* because the true molecular orbitals are modelled by expressing them as linear combinations of the atomic orbitals of the parent atoms.

We return to the hydrogen molecule-ion in order to establish the modelling procedure. Its hamiltonian is

$$H = -(\hbar^2/2m_e)\nabla^2 + (e^2/4\pi\varepsilon_0)\left\{-\frac{1}{r_a} - \frac{1}{r_b} + \frac{1}{R}\right\} \tag{10.3.1}$$

where r_a and r_b are the distances of the single electron from the two nuclei a and b separated by R. The lowest energy solution, Fig. 10.1, can be modelled as the linear combination $\psi_a + \psi_b$ of the hydrogen 1s-orbitals centred on each nucleus, Fig. 10.4. The antibonding orbital can also be modelled as $\psi_a - \psi_b$ using the same two atomic orbitals, but this time superimposed with opposing amplitudes, Fig. 10.4(b). In the first case there is a *constructive interference* where the two wavefunctions overlap and their amplitudes augment each other: this models the accumulation of amplitude in the internuclear region which we know is an important feature of the bonding process. In the antibonding combination the amplitudes tend to cancel each other, and there is exact cancellation to form a nodal plane half way between the two nuclei, just as in the exact solution. The *destructive interference* where the two wavefunctions overlap with opposite phase models the annihilation of the amplitude in the internuclear region.

The same conclusions can be reached in a slightly more formal way which has the advantage that it is readily extended to other types of molecule. We set out by choosing a *basis set* of atomic orbitals, in this case the two hydrogen atomic orbitals. For simplicity we write them a and b. Then we express the molecular orbitals of the system as the LCAO

$$\psi = c_a a + c_b b, \tag{10.3.2}$$

where the coefficients have to be determined. Their optimum values can be found using the variation principle (Chapter 8) and solving the secular equations

$$\sum_r c_r(H_{rs} - ES_{rs}) = 0, \tag{10.3.3}$$

where H_{rs} is a matrix element of the hamiltonian and S_{rs} is an overlap matrix element. These equations have solutions when the secular determinant $|H_{rs} - ES_{rs}|$ vanishes. Therefore, in order to find the best combination coefficients (and the energy of the system) we have to evaluate the relevant matrix elements. We introduce the following

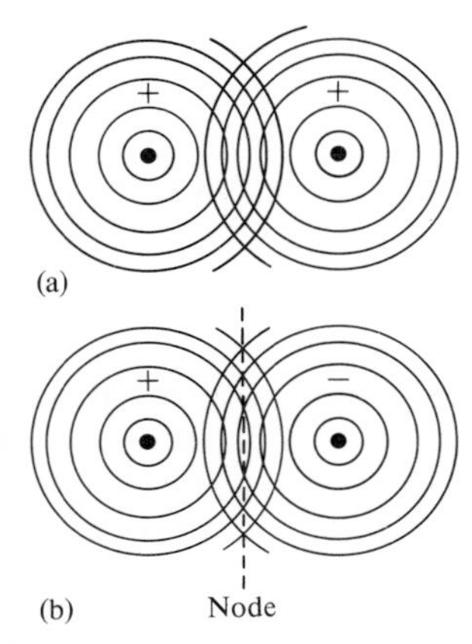

Fig. 10.4. (a) Constructive, (b) destructive overlap between 1s-orbitals.

integrals (using the fact that a and b are real):

$S_{aa} = \int a^2 \, d\tau = 1$ (ψ_a is already normalized to unity);

$S_{bb} = 1$ (likewise);

$S_{ab} = \int ab \, d\tau = S$ (the *overlap integral*, a measure of the extent to which the two orbitals located on different centres overlap);

$H_{aa} = \int aHa \, d\tau = \alpha$ (the *Coulomb integral;* we find its explicit form shortly; $H_{aa} = H_{bb}$ for a homonuclear diatomic);

$H_{ab} = \int aHb \, d\tau = \beta$ (the *resonance integral:* its explicit form and its significance will be seen soon).

The secular determinant is then

$$\begin{vmatrix} \alpha - E & \beta - ES \\ \beta - ES & \alpha - E \end{vmatrix} = 0.$$

This equation (a quadratic equation in E when the determinant is expanded) has the solutions

$$E_{\pm} = \frac{\alpha \pm \beta}{1 \pm S} \qquad (10.3.4)$$

and the corresponding values of the coefficients are

$$c_a = c_b, \qquad c_a = 1/\{2(1+S)\}^{\frac{1}{2}} \quad \text{for} \quad E_+ = (\alpha+\beta)/(1+S) \qquad (10.3.5a)$$

$$c_a = -c_b, \qquad c_a = 1/\{2(1-S)\}^{\frac{1}{2}} \quad \text{for} \quad E_- = (\alpha-\beta)/(1-S). \qquad (10.3.5b)$$

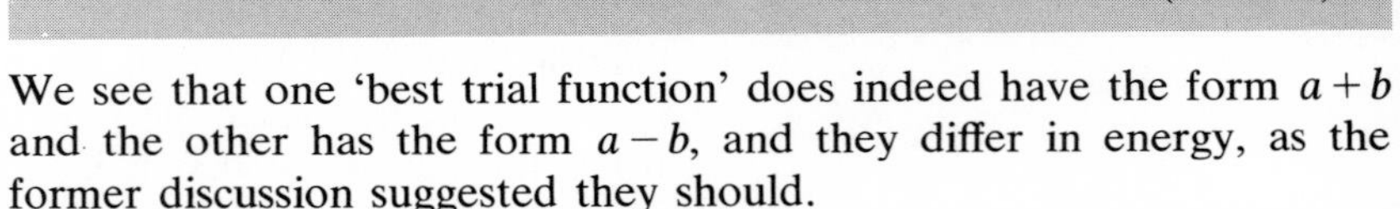

We see that one 'best trial function' does indeed have the form $a+b$ and the other has the form $a-b$, and they differ in energy, as the former discussion suggested they should.

The detailed form of the Coulomb and resonance integrals can be established as follows. On inserting the explicit form of the hamiltonian into their definitions we find

$$\alpha = \int \psi_a\{(-\hbar^2/2m_e)\nabla^2 - (e^2/4\pi\varepsilon_0 r_a) - (e^2/4\pi\varepsilon_0 r_b) + (e^2/4\pi\varepsilon_0 R)\}\psi_a \, d\tau$$

$$= E_{1s} - (e^2/4\pi\varepsilon_0)\int \psi_a(1/r_b)\psi_a \, d\tau + (e^2/4\pi\varepsilon_0 R). \qquad (10.3.6)$$

The first term follows because ψ_a is an eigenfunction of the hydrogen atom hamiltonian, which we have separated out. The second term corresponds to the total electrostatic energy of interaction between a charge distribution with a probability density ψ_a^2 and the *other* nucleus b. This is illustrated in Fig. 10.5(a). We denote it j':

$$j' = (e^2/4\pi\varepsilon_0)\int \psi_a^2(1/r_b) \, d\tau \qquad (10.3.7)$$

(a)

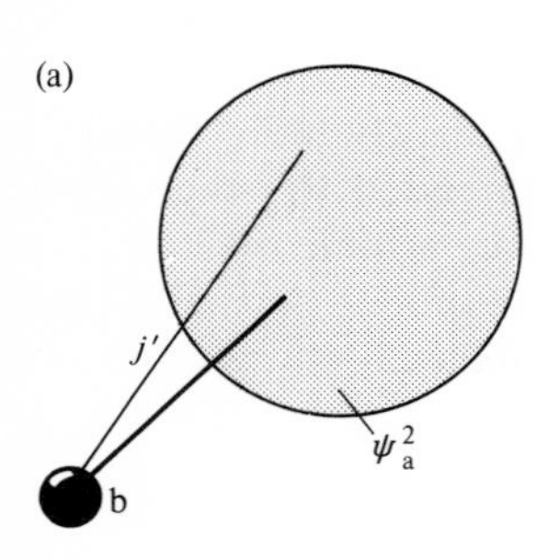

(b)

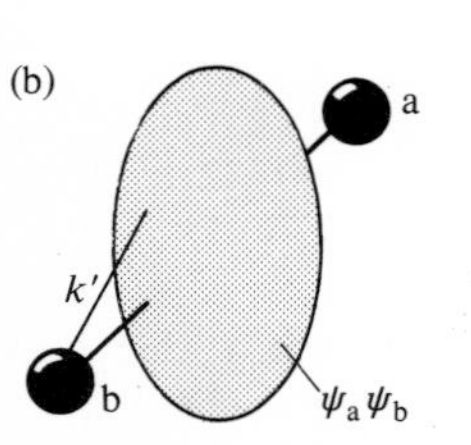

Fig. 10.5. Contributions to the integrals (a) j', (b) k'.

and note that it is positive. The Coulomb integral is therefore

$$\alpha = E_{1s} - j' + (e^2/4\pi\varepsilon_0 R). \tag{10.3.8}$$

Example. Evaluate (a) the overlap integral and (b) the integral j' for an electron in a molecular orbital composed of two hydrogen-like 1s-orbitals.

- *Method.* The natural coordinates are the *ellipsoidal coordinates* $\mu = (r_a + r_b)/R$, $\nu = (r_a - r_b)/R$, ϕ. The volume element is $d\tau = \frac{1}{8}R^3(\mu^2 - \nu^2)\, d\mu\, d\nu\, d\phi$ and $1 \leq \mu \leq \infty$, $-1 \leq \nu \leq 1$, $0 \leq \phi \leq 2\pi$. Use $\psi_a = (Z^3/\pi a_0^3)^{\frac{1}{2}} e^{-Zr_a/a_0}$ and $\psi_b = (Z^3/\pi a_0^3)^{\frac{1}{2}} e^{-Zr_b/a_0}$. For j', note that $r_b = \frac{1}{2}R(\mu - \nu)$ and $r_a = \frac{1}{2}R(\mu + \nu)$. All the integrations are then elementary. Write $j_0 = e^2/4\pi\varepsilon_0$.
- *Answer.*

$$S = \int \psi_a \psi_b\, d\tau = (Z^3/\pi a_0^3) \int_0^{2\pi} d\phi \int_1^{\infty} d\mu \int_{-1}^{1} d\nu \tfrac{1}{8}R^3(\mu^2 - \nu^2) e^{-Z\mu R/a_0}$$
$$= \{1 + (ZR/a_0) + \tfrac{1}{3}(ZR/a_0)^2\} e^{-ZR/a_0}.$$

$$j'/j_0 = \int \psi_a^2 (1/r_b)\, d\tau = (Z^3/\pi a_0^3) \int_0^{2\pi} d\phi \int_1^{\infty} d\mu \int_{-1}^{1} d\nu \left\{ \frac{\tfrac{1}{8}R^3(\mu^2 - \nu^2) e^{-\frac{1}{2}(\mu+\nu)ZR/a_0}}{\tfrac{1}{2}R(\mu - \nu)} \right\}$$
$$= \tfrac{1}{2}(ZR/a_0)^3 \int_1^{\infty} d\mu \int_{-1}^{1} d\nu (\mu + \nu) e^{-\frac{1}{2}(\mu+\nu)ZR/a_0}$$
$$= (1/R)\{1 - (1 + ZR/a_0) e^{-2ZR/a_0}\}.$$

- *Comment.* Both S and j' decrease as Z increases because the orbitals shrink down on to their nuclei. A more detailed account of the calculation of molecular integrals is given in S. P. McGlynn, L. G. Vanquickenborne, M. Kinoshita, and D. G. Carroll, *Introduction to applied quantum chemistry*, Holt, Rinehart, and Winston (1972), and a few results are listed in Appendix 14. The lines of constant μ are ellipses and the lines of constant ν are hyperbolas orthogonal to the ellipses.

In the case of the resonance integral we use the fact that ψ_b is an eigenfunction of hydrogen atom b's hamiltonian with eigenvalue E_{1s}, and write

$$\beta = \int \psi_a \{-(\hbar^2/2m_e)\nabla^2 - (e^2/4\pi\varepsilon_0 r_b) - (e^2/4\pi\varepsilon_0 r_a) + (e^2/4\pi\varepsilon_0 R)\} \psi_b\, d\tau$$
$$= E_{1s} \int \psi_a \psi_b\, d\tau - (e^2/4\pi\varepsilon_0) \int \psi_a (1/r_a) \psi_b\, d\tau + (e^2/4\pi\varepsilon_0 R) \int \psi_a \psi_b\, d\tau$$
$$= \{E_{1s} + (e^2/4\pi\varepsilon_0 R)\} S - k', \tag{10.3.9}$$

$$k' = (e^2/4\pi\varepsilon_0) \int \psi_a (1/r_a) \psi_b\, d\tau. \tag{10.3.10}$$

The k' integral (which is positive in this case) has no classical analogue, but we can think of it as representing the interaction of the *overlap charge density*, $-e\psi_a\psi_b$, with nucleus a, Fig. 10.5(b) (its interaction with the other nucleus has the same value, by symmetry). It follows that the

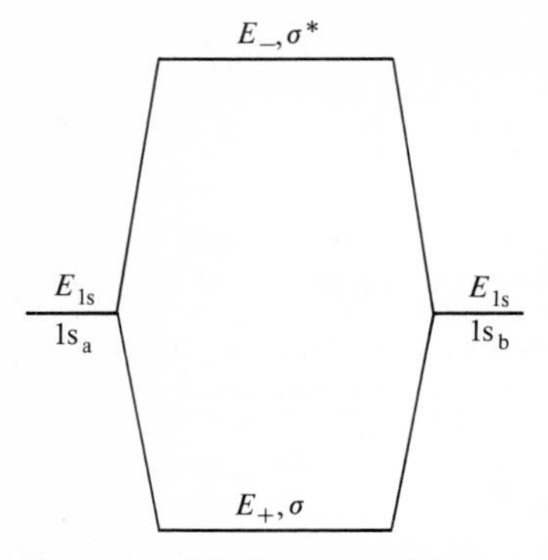

Fig. 10.6. Molecular orbital energy level diagram for H_2.

energies of the two LCAO-MOs are

$$E_+ = E_{1s} + (e^2/4\pi\varepsilon_0 R) - \left(\frac{j' + k'}{1+S}\right) \tag{10.3.11a}$$

$$E_- = E_{1s} + (e^2/4\pi\varepsilon_0 R) - \left(\frac{j' - k'}{1-S}\right). \tag{10.3.11b}$$

The integrals j' and k' are both positive, and so the lower of the two energies is E_+, and the corresponding function is the one with constructive interference in the internuclear region. The *energy level diagram* for H_2^+ (the ladder of orbital energies corresponding to the equilibrium nuclear conformation) is shown in Fig. 10.6. Note that the antibonding orbital lies further above E_{1s} than the bonding orbital lies below it: this is largely because the nucleus–nucleus repulsion pushes both levels upwards. In other words, *an antibonding orbital is more antibonding than a bonding orbital is bonding.*

The integrals in the expression for the energy can all be evaluated analytically, and so the molecular potential energy curve can be plotted. It turns out to have a minimum at $R = 130$ pm and the *dissociation energy* $(E_{1s} - E_+)$ is then $170\ \text{kJ mol}^{-1}$. The experimental values are 106 pm and $251\ \text{kJ mol}^{-1}$, and so the agreement is not spectacular. The principal deficiency of the calculation is that the trial function has been taken to have the form of two hydrogen 1s-orbitals, whereas in fact we saw from the exact solution that a major contribution to the bonding comes from the shrinkage of the orbitals on to the nuclei. This feature cannot be captured in this model, but could be incorporated by allowing the basis functions to have a measure of flexibility.

At this stage it is convenient to introduce a nomenclature for the molecular orbitals of homonuclear diatomics. The sausage-shaped molecular orbitals have rotational symmetry around the internuclear axis, and hence are called *σ-orbitals* (by analogy with the spherically symmetrical s-orbitals of atoms). An antibonding orbital is denoted by *; hence we have the σ bonding orbital and the σ^* antibonding orbital. The molecule also possesses inversion symmetry. Inverting the molecule through the centre of inversion, Fig. 10.7, leaves the bonding orbital apparently unaffected: this *even parity* is denoted by a subscript g (for *gerade*, even). The antibonding orbital changes sign under inversion, and so its *odd parity* is denoted by a subscript u (for *ungerade*). The two orbitals are therefore denoted σ_g and σ_u^*. If their parentage needs to be emphasized they are written $1s\sigma_g$ and $1s\sigma_u^*$.

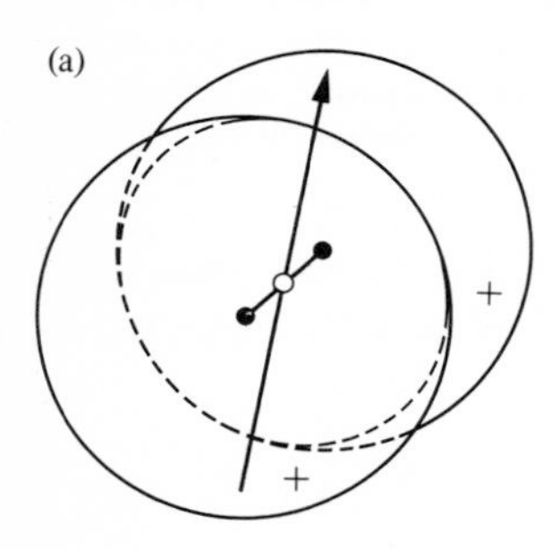

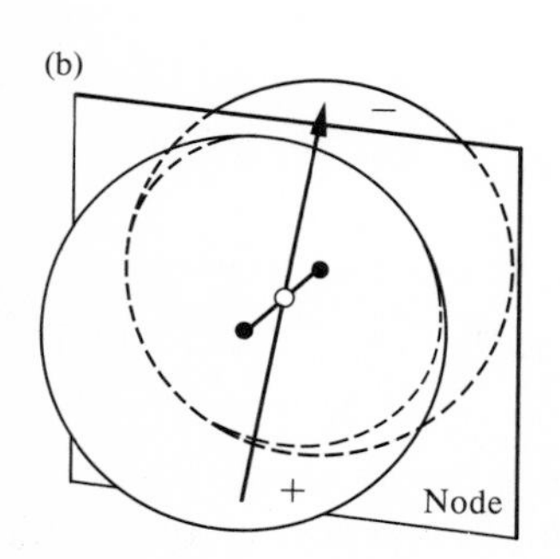

Fig. 10.7. (a) Gerade, g and (b) ungerade, u parity.

Now we turn to the hydrogen molecule. Just as we went from the hydrogen atom to the helium atom by adding an electron to a hydrogen-like $(Z = 2)$ atomic orbital, so we model the structure of the hydrogen molecule simply by inserting one more electron into the σ_g-orbital of H_2^+. The orbital description is therefore $\sigma_g(1)\sigma_g(2)$ where 1 and 2 are short for $\mathbf{r}_1$ and $\mathbf{r}_2$, the coordinates of the two electrons. This *spatial* wavefunction is symmetric under particle interchange;

therefore the spin component of the total wavefunction must be $\alpha_1\beta_2-\beta_1\alpha_2$ to ensure that the total wavefunction is antisymmetrical. Therefore, when the electrons enter the molecular orbital, they do so with paired spins. *Spin pairing in bond formation is seen not to be an end in itself, but the way that electrons have to arrange themselves in order to pack into the lowest energy orbital.*

The full form of the H_2 wavefunction is

$$\psi(1,2)=\{1/2(1+S)\}\{a(1)+b(1)\}\{a(2)+b(2)\}(\tfrac{1}{2})^{\frac{1}{2}}\{\alpha_1\beta_2-\beta_1\alpha_2\} \tag{10.3.12}$$

where we have also included the normalized spin function. Its energy is found by evaluating the expectation value of the hamiltonian. This gives rise to several integrals (see Appendix 14) which represent the interaction of various charge distributions with each other. For instance, there is an integral like j' representing the interaction of the charge density of a with nucleus b, an integral like k' representing the contribution from the overlap charge, and several integrals representing the interactions between the two electrons when they occupy a and b, when they are both in a, and so on. They may all be evaluated analytically (although it is quite a heavy job) and the molecular potential energy curve calculated. It is found to have a minimum at $R=74$ pm and the dissociation energy (the energy difference between H_2 and two widely separated H atoms) turns out to be 350 kJ mol^{-1}. The experimental values are 74.2 pm and 432 kJ mol^{-1}, and so although there is a fair measure of agreement there is room for improvement (e.g. by allowing for a more flexible basis function, using the self-consistent field procedure, and investigating electron correlation effects).

The molecular orbital method is easily extended qualitatively to diatomic molecules of heavier atoms, and we turn to that shortly. First, though, we consider the alternative approximate procedure.

10.4 The valence bond method

The valence bond (VB) method takes the view that the natural way to look at a bond is to concentrate on the *electron pair*, for the early theoretical chemists such as Lewis had emphasized its importance.

Consider first the description of a pair of hydrogen atoms separated by a large distance. The total wavefunction is the product $\psi_a(\mathbf{r}_1)\psi_b(\mathbf{r}_2)$, or $a(1)b(2)$ for short, where a is the 1s-orbital on a and b the 1s-orbital on b. The total energy is $2E_{1s}$. When the atoms are at their bonding distance this wavefunction may also be a fair description of the molecule. Therefore, as a start, we set $\psi\approx a(1)b(2)$ and calculate its energy by evaluating the expectation value of the hamiltonian. In a sense we are supposing that the function $a(1)b(2)$ can be regarded as the wavefunction for an unperturbed system, and then calculating the first-order correction to the energy using as a perturbation the interaction between the two electrons and their interactions with the 'wrong' nucleus. The total energy of the system is

$$E = \int a(1)b(2)Ha(1)b(2)\,\mathrm{d}\tau_1\,\mathrm{d}\tau_2$$

$$= 2E_{1s} + (e^2/4\pi\varepsilon_0 R) + (e^2/4\pi\varepsilon_0)\int a(1)^2(1/r_{12})b(2)^2\,\mathrm{d}\tau_1\,\mathrm{d}\tau_2$$

$$-\int a(1)^2(1/r_{1b})\,\mathrm{d}\tau_1 - \int b(2)^2(1/r_{2a})\,\mathrm{d}\tau_2$$

$$= 2E_{1s} + (e^2/4\pi\varepsilon_0 R) + j - 2j'. \qquad (10.4.1)$$

The integral j' we have met already; the integral j is

$$j = (e^2/4\pi\varepsilon_0)\int a(1)^2(1/r_{12})b(2)^2\,\mathrm{d}\tau_1\,\mathrm{d}\tau_2, \qquad (10.4.2)$$

and is the total Coulomb interaction between an electron in the orbital a and another in orbital b. Note that j turns into the integral J of the helium problem as R approaches zero and the two atomic orbitals coincide. All the integrals may be evaluated analytically and a minimum in the molecular potential energy curve is found at $R \approx$ 90 pm, corresponding to a dissociation energy of only about 25 kJ mol^{-1}. Something is obviously dreadfully wrong.

The elementary error in the calculation is the supposition that electron 1 remains in orbital a even when the atoms are so close that they overlap. As soon as the atoms are within bonding distance, electron 1 can move on to atom b, and vice versa, and so the description of the molecule as $b(1)a(2)$ is expected to be just as valid as the description $a(1)b(2)$. Therefore we ought to take the linear combination $a(1)b(2) \pm b(1)a(2)$, in which the alternative states appear with equal weight (so that 50 per cent of the time we should expect to find the arrangement $a(1)b(2)$ if the molecule is examined, and 50 per cent $b(1)a(2)$). Which sign leads to the lower energy depends on calculation, but it is not hard to anticipate that the + combination will be the lower. An alternative approach (which leads to the same conclusion) is to regard the problem as one of degenerate perturbation theory. In the absence of terms such as $1/r_{12}$ in the hamiltonian (i.e. when the atoms are far apart) the states $a(1)b(2)$ and $b(1)a(2)$ are degenerate. When the atoms are close together the 'atomic' hamiltonians are augmented by the terms like $1/r_{12}$ which act as a perturbation. As is common in degenerate perturbation theory, the effect of the perturbation is to mix the levels to give equal weights of each and causing one of the mixtures to rise and the other to fall (recall Fig. 8.1 again). Therefore, either we can set up the secular determinant and find its roots, or we can anticipate that the solutions are of the form $a(1)b(2) \pm b(1)a(2)$ and simply evaluate the expectation value of the total hamiltonian. We do the latter.

First normalize the two combinations. We require

$$\int \psi_{\pm}^2 \, d\tau = N_{\pm}^2 \int \{a(1)b(2) \pm b(1)a(2)\}^2 \, d\tau_1 \, d\tau_2$$

$$= N_{\pm}^2 \int \{a(1)^2 b(2)^2 + b(1)^2 a(2)^2 \pm 2a(1)b(1)a(2)b(2)\} \, d\tau_1 \, d\tau_2$$

$$= N_{\pm}^2 (1 + 1 \pm 2S^2) = 1; \quad \text{or} \quad N_{\pm} = 1/\{2(1 \pm S^2)\}^{\frac{1}{2}}. \qquad (10.4.3)$$

Now evaluate the expectation value of the hamiltonian in eqn (10.2.1). The outcome (Appendix 14) is

$$E_{\pm} = 2E_{1s} + (e^2/4\pi\varepsilon_0 R) + \left(\frac{J \pm K}{1 \pm S^2}\right) \qquad (10.4.4)$$

where $J = j - 2j'$, $K = k - 2k'S$; j, j', and k' we have already encountered and interpreted. The remaining integral is

$$k = (e^2/4\pi\varepsilon_0) \int a(1)b(1)(1/r_{12})a(2)b(2) \, d\tau_1 \, d\tau_2, \qquad (10.4.5)$$

which can be interpreted as the interaction between the overlap charge distribution $a(1)b(1)$ and the other $a(2)b(2)$: it reduces to the exchange integral of the helium problem when $R = 0$. J is called the *Coulomb integral* (yet another!) and K is called the *exchange integral*, although they are distinctly different from the integrals of the same name encountered for atoms.

Both J and K are *negative* ($2j' > j$ and $2Sk' > k$). Therefore E_+ is the lower energy. This corresponds to the + combination of degenerate states in the linear combination, and so we can conclude that that describes the ground state of the system. The energy may be calculated by evaluating the integrals. All can be treated analytically (with some effort), and the dissociation energy ($2E_{1s} - E_+$) turns out to have the value 303 kJ mol^{-1}, an order of magnitude greater than before. Not only is this better than the first calculation (and by the variation theorem, closer to the true value) it is greater than (i.e. the energy of the molecule calculated in this way is lower than) that calculated by the molecular orbital method. Therefore, by the same token, the valence bond description gives a better description of H_2 than the molecular orbital wavefunction.

10.5 Comparison of the methods

We can identify the differences between the methods by comparing the wavefunctions for the ground state of hydrogen. Disregarding the normalization constants, for the valence bond description we have

$$\psi^{VB} = a(1)b(2) + b(1)a(2) \qquad (10.5.1)$$

while in the molecular orbital description

$$\psi^{\mathrm{MO}} = a(1)b(2) + b(1)a(2) + a(1)a(2) + b(1)b(2) \tag{10.5.2}$$

(which is obtained by expanding the spatial function in eqn (10.3.12)). The second has two extra terms which represent the probability of finding both electrons in the same atomic orbital. If we think of the valence bond structure as representing a *pure covalent structure* H—H in which both electrons contribute equally to the bond and are never found on the same atom, the molecular orbital function represents a mixture where the covalent structure occurs with the same weight as the *ionic structures* H^-H^+ or H^+H^-. Put another way, the VB wavefunction suggests that when the molecule dissociates it forms the neutral atoms, but the MO wavefunction suggests that when the molecule dissociates there are equal probabilities of forming the neutral atoms and the ions H^+ and H^-. These predictions are both too extreme. There must be some chance that both electrons will be found in the same atomic orbital, but not with as high a probability as the MO function allows. The avoidance of one electron by another on electrostatic grounds is called *charge correlation* (and is to be distinguished from spin correlation discussed in Chapter 9). Therefore the MO wavefunction contains too much ionic character (it underestimates charge correlation), while the VB function contains too little (it overestimates charge correlation). Furthermore, the MO description seems to need more improvement than the VB because it lies higher in energy.

Because the VB function excludes ionic terms, we can improve it by adding to it components that represent the probability that both electrons will be found in the same orbital. These extra components are the two extra terms in the MO description. Therefore we modify the basic covalent function to

$$\begin{aligned}\psi^{\mathrm{VB}} &= a(1)b(2) + b(1)a(2) + \lambda\{a(1)a(2) + b(1)b(2)\} \\ &= \psi^{\mathrm{VB}}_{\mathrm{cov.}} + \lambda\psi^{\mathrm{VB}}_{\mathrm{ion.}}\end{aligned} \tag{10.5.3}$$

where $\psi^{\mathrm{VB}}_{\mathrm{cov.}}$ is the covalent description of the bond and $\psi^{\mathrm{VB}}_{\mathrm{ion.}}$ is the ionic. This admixture is called *ionic–covalent resonance.* λ is a parameter which can be optimized by using the variation theorem and looking for a value that minimizes the energy. The optimum value appears to be about $\frac{1}{6}$ (i.e. $\frac{1}{36}$ or 3 per cent admixture of the ionic component as distinct from the 50 per cent admixture tacitly assumed in the MO method). The energy of the molecule drops by a further 93 kJ mol^{-1} below the VB value (to give a dissociation energy of 396 kJ mol^{-1} (the experimental value is 432 kJ mol^{-1}).

Greater subtlety is needed to improve the MO description, but the method (like the MO method itself) is widely used and we should see it in action in a simple case. We have to return to the basic description of the molecule in terms of the occupation of the available H_2^+-like molecular orbitals. Two orbitals can be constructed from the $1s_a$, $1s_b$

basis set; one is the bonding orbital σ_g and the other is the antibonding orbital σ_u^*. There are four ways of accommodating the two electrons:

$$\sigma_g(1)\sigma_g(2); \qquad \sigma_g(1)\sigma_u^*(2); \qquad \sigma_u^*(1)\sigma_g(2); \qquad \sigma_u^*(1)\sigma_u^*(2).$$

By now we know that a correct description of the second and third (degenerate) states which allows for the indistinguishability of the electrons (or, alternatively, allows for the effect of a mixing perturbation) is as the two linear combinations $\sigma_g(1)\sigma_u^*(2) \pm \sigma_u^*(1)\sigma_g(2)$. Three of the wavefunctions are symmetric under particle interchange and must therefore be multiplied by an antisymmetric (singlet) spin function. The remaining function (the linear combination with $-$) is antisymmetric, and so must be multiplied by a symmetric (triplet) spin function. Therefore, the states allowed by the Pauli principle are as follows:

$$^1\Sigma_g\text{:}\ \ \sigma_g(1)\sigma_g(2)(\alpha_1\beta_2-\beta_1\alpha_2);$$
$$^1\Sigma_u\text{:}\ \ \{\sigma_g(1)\sigma_u^*(2)+\sigma_u^*(1)\sigma_g(2)\}(\alpha_1\beta_2-\beta_1\alpha_2);$$
$$^1\Sigma_g\text{:}\ \ \sigma_u(1)\sigma_u^*(2)(\alpha_1\beta_2-\beta_1\alpha_2);$$
$$^3\Sigma_u\text{:}\ \ \{\sigma_g(1)\sigma_u^*(2)-\sigma_u^*(1)\sigma_g(2)\}\begin{cases}\alpha_1\alpha_2\\ (\alpha_1\beta_2+\beta_1\alpha_2)\,.\\ \beta_1\beta_2\end{cases}$$

The labels are the term symbols for the states. We explain how they are constructed in Appendix 15: all we need to know here is that they denote the overall symmetries of the states.

The MO description so far has considered only the lowest energy $^1\Sigma_g$ term. If the energies of all the terms are calculated (all the integrals are available) we get the picture shown in Fig. 10.8. The important feature is that there are two terms with the same symmetry, and they converge as the bond length increases. This means that they approach degeneracy at large distances, and are then very responsive to any perturbation that can mix them. States of the same symmetry always avoid crossings because the hamiltonian of the problem always has non-vanishing matrix elements between them (H is totally symmetric under all the symmetry operations of the system and so matrix elements $\langle\Gamma|\,H\,|\Gamma\rangle$ transform as $\Gamma\times A_1\times\Gamma=A_1$). Therefore *configuration interaction* (CI, the mixing of configurations of the same overall symmetry) occurs, and the two terms move apart and do not cross. This is shown in Fig. 10.8. The configuration lowers the energy of the lower term (because the interaction pushes them apart, just as in the by-now-famous Fig. 8.1), which is exactly what is needed to lower the energy of the ground state. The final wavefunction after CI is the mixture $\sigma_g(1)\sigma_g(2)+\lambda'\sigma_u^*(1)\sigma_u^*(2)$, and when this is expanded in terms of the component atomic orbitals we find

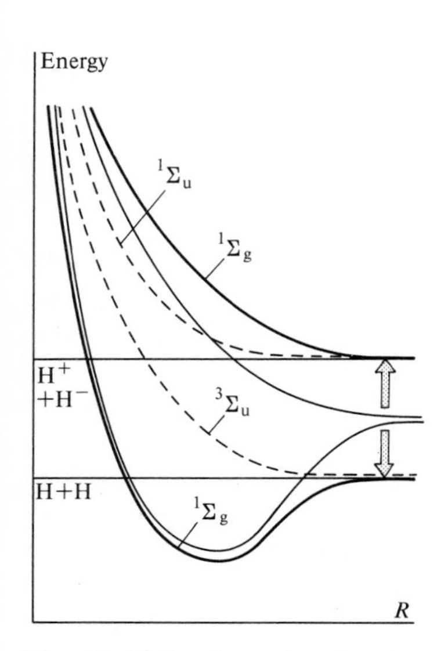

Fig. 10.8. Configuration interaction in H_2.

$$\psi_{CI}^{MO}=a(1)b(2)+a(2)b(1)+\lambda''\{a(1)a(2)+b(1)b(2)\} \tag{10.5.4}$$

where λ'' is some constant, which is exactly the form of the improved VB wavefunction, eqn (10.5.3).

The similarities and differences of the two approaches may be summarized as follows:

(1) Both emphasize the importance of the displacement of charge into the internuclear region. The lowest energy wavefunction in each case is the one where *constructive interference occurs in the bonding region.*

(2) The lowest energy is achieved with both electrons in the same symmetric spatial function (either $\sigma_g(1)\sigma_g(2)$ or $a(1)b(2)+b(1)a(2)$) and so the spin function has to be the antisymmetric singlet: *electrons pair on bond formation.*

(3) The VB method *overestimates charge correlation*, the MO method *underestimates* it. The former is improved by *ionic–covalent resonance*, and the latter by *configuration interaction.*

(4) The VB theory emphasizes the bond as a distinct atom–atom entity in the molecule, and concentrates on *pairs of electrons* from the outset; the MO method views the molecule as a continuous electron distribution and less as a collection of identifiable bonds. This is an advantage when the spectroscopic properties of molecules are being considered.

(5) The MO method has turned out to be easier to apply to more complicated molecules, and has received much more attention. It has been refined almost to a precision technique with the use of computers. For that reason from now on we shall concentrate on it alone.

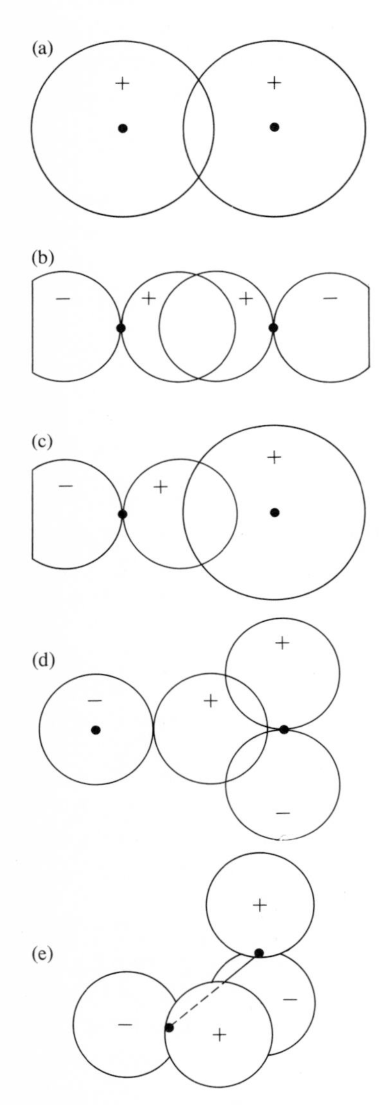

Fig. 10.9. Non-zero overlap in (a)–(c); zero overlap for (d) and (e).

10.6 The structures of diatomic molecules

The electronic configurations of the homonuclear diatomics can be established using the molecular version of the building-up principle. We decide what orbitals are available, judge the order of their energies, and then feed in the appropriate number of electrons into the lowest available orbitals consistent with the Pauli exclusion principle. First, though, we have to extend the molecular orbital description of diatomic species to the cases where MOs can be built from a variety of s- and p-orbitals on the participating atoms. The following points give us the necessary guidance.

In order for orbitals to overlap and participate in bonding they must have *the same symmetry with respect to rotations about the internuclear axis.* A measure of the extent of overlap is the overlap integral S first introduced on p. 254. Two adjacent s-orbitals ((1s, 1s), (1s, 2s), (2s, 2s), etc.) have non-zero overlap, Fig. 10.9. Two adjacent p_z-orbitals (with z the internuclear axis) also overlap; so does an s-orbital on one atom and a p_z-orbital on the other. But a p_x-orbital has zero overlap with an s-orbital and with a p_z-orbital on the neighbouring atom. They 'overlap' in the colloquial sense, but the net amount of overlap, which is what the overlap integral measures, is zero because positive regions of the product $\psi_a\psi_b$ are cancelled by negative regions, Fig. 10.10. The importance of overlap in bond formation has already been stressed. In simple terms we see that if the overlap integral is zero there are equal

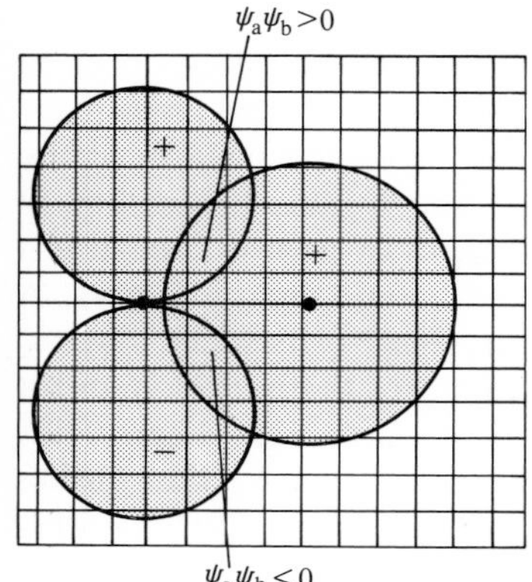

Fig. 10.10. Cancelling contributions to the overlap integral.

amounts of constructive and destructive interference between the participating atomic orbitals, and so there is no net shift of electron density into the bonding region. Hence s- and p_z-orbitals may participate in bonding and give rise to cylindrically symmetric σ-orbitals. On the other hand, the perpendicular p_x- and p_y-orbitals may overlap in a broadside sense, Fig. 10.9 and give rise to *π-orbitals*. Since the overlap occurs away from the prime bonding region in π-orbitals we might expect them to be less strongly bonding or antibonding than σ-orbitals between the same atoms. This is only partially true, as we shall see.

In order to participate in bond formation, *atomic orbitals must be neither too diffuse nor too compact*. In either case there is only feeble constructive or destructive overlap, and consequently only feeble bonds result. This suggests that 1s–1s overlap in the first short period (Li, . . . Ne) is much less important than 2s–2s overlap because the 1s-orbitals are packed closely round the nucleus.

The *energies of the participating orbitals must be similar*. In order to see why this is so, consider the following secular determinant for the bond between two different atoms (it is the determinant in Section 10.3 generalized slightly)

$$\begin{vmatrix} \alpha_a - E & \beta - ES \\ \beta - ES & \alpha_b - E \end{vmatrix} = 0.$$

The roots may be found by solving the quadratic equation for the energy, and when the difference $\alpha_a - \alpha_b$ is large

$$E_+ \approx \alpha_a - \beta^2/(\alpha_b - \alpha_a), \qquad E_- \approx \alpha_b + \beta^2/(\alpha_b - \alpha_a).$$

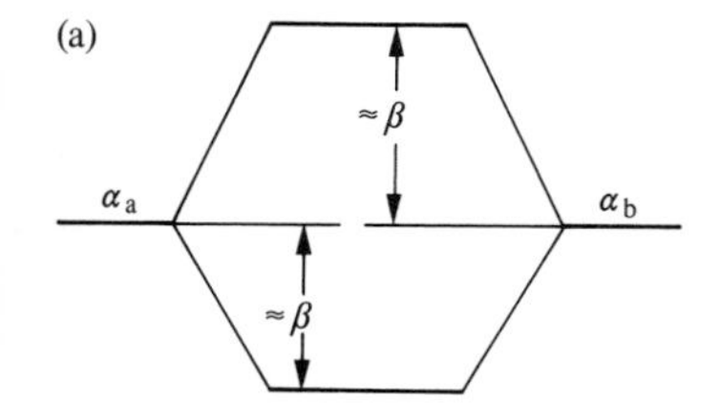

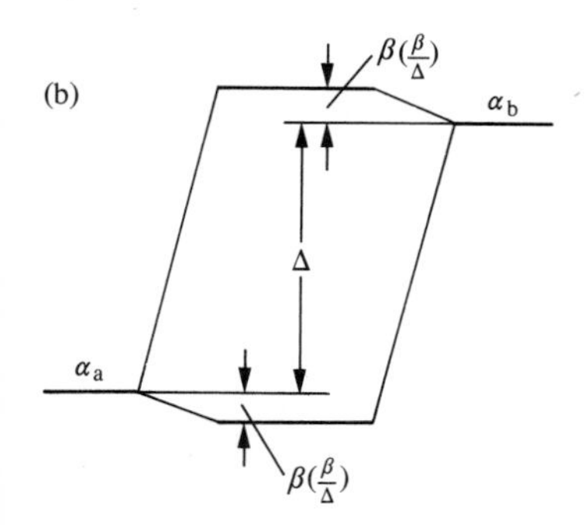

Fig. 10.11. (a) Strong interaction between degenerate orbitals and (b) weak interaction between orbitals widely different in energy.

This signifies, Fig. 10.11, that the molecular orbital energies are shifted from the atomic orbital energies (α_a and α_b effectively) by only a small amount when α_a and α_b are very different. This means that in homonuclear diatomics the atomic orbitals of identical energy on each atom dominate the bonding, and that there is little bonding arising from the overlap of 1s- and 2s-orbitals on different centres and, if we stretch a point, little bonding from (2s, $2p_z$) overlap (in species other than molecular hydrogen). The last point is only very approximately valid, and we shall avoid the assumption shortly.

With these rules in mind we can begin to set up the *molecular orbital energy level diagram* for the first row homonuclear diatomics. First we draw the atomic orbital energy levels for the two participating atoms. This is shown in Fig. 10.12; the actual 1s–2s separation is much greater than that drawn. Since the 1s-orbitals are so compact they have relatively little effect on the bonding, and so the bonding and antibonding σ-orbitals they give rise to lie at almost the same energy. On the basis of the energy matching criterion we expect the two neighbouring 2s-orbitals to interact and form a strong σ-bond and antibond, the $2p_z$-orbitals to do likewise, and the perpendicular $2p_x$- and $2p_y$-orbitals to form two π- and two π^*-orbitals. On the basis of the degree of overlap we expect the π-orbitals to give rise to weaker effects than the σ-orbitals formed from the p_z-atomic orbitals. The scheme this suggests is set out in Fig. 10.12(a). More detailed calculations (e.g. using an SCF procedure) confirm this expected order in general, with the exception of the order of the $2p\pi$-orbitals, which are found to lie beneath the $2p_z\sigma$-orbital. Apart from the naïvety of the approach, there are several explanations. One is that the extent of overlap is not the sole criterion for the strength of a bond, because it depends on a variety of potential energy and kinetic energy factors. A more constructive point, however, is that the $2s\sigma^*$-orbital lies close to the $2p_z\sigma$-orbital; as they have the same symmetry the hamiltonian for the system has non-vanishing matrix elements between them. They therefore perturb each other and move apart as they mix. This pushes the upper orbital above the π-orbitals, as required. The resulting energy level diagram is therefore as shown in Fig. 10.12(b), and this is the one we shall use. Note that the interacting σ and σ^* are no longer purely s- or p-orbitals: they are *hybrid orbitals* and possess both s and p character.

Now we use the molecular orbital energy level diagram as a basis for the building-up principle. We use the same diagram for each homonuclear diatomic (it expresses only qualitative information, and qualitatively their molecular orbitals are similar) and feed in the appropriate number of electrons ($2Z$ in the case of neutral diatomics formed from atoms of atomic number Z).

Take as an example nitrogen, N_2. Its 14 electrons enter the orbitals two by two (note that the $2p\pi$-level consists of two orbitals and so can accommodate up to four electrons). This results in the configuration

$$N_2\text{: } 1s\sigma_g^2 1s\sigma_u^{*2} 2s\sigma_g^2 2s\sigma_u^{*2} 2p\pi_u^2 2p\sigma_g^2, \qquad {}^1\Sigma_g.$$

The net number of bonds is $1-1+1-1+2+1=3$ if we allow for the cancellation of bonds by their corresponding antibonds. We can therefore report that the N_2 molecule has a *triple bond* composed of one σ-bond and two π-bonds. The term symbol (Appendix 15) shows that the ground state is a singlet ($S=0$) because all the electrons must be paired, that the overall parity is g (use the rule $g\times u=u$, $g\times g=g$, $u\times u=g$ for all the electrons), and that there is no net orbital angular momentum about the internuclear axis (there is none from electrons in σ-orbitals; the two π-orbitals have angular momenta in opposite senses around the axes, and since they are fully occupied their net momentum is zero; hence Σ for the term symbol).

The configuration of O_2 is obtained using 16 electrons, two more than possessed by N_2; therefore it is

$$O_2\text{: } 1s\sigma_g^2 1s\sigma_u^{*2} 2s\sigma_g^2 2s\sigma_u^{*2} 2p\pi_u^4 2p\sigma_g^2 2p\pi_g^{*2}.$$

This configuration gives rise to a number of terms because the two electrons in the π^*-orbitals can be accommodated in a variety of ways. For example, if they both enter the same orbital the total orbital angular momentum about the axis is $\pm 2\hbar$, which gives rise to a Δ term. In this case, though, the spins must be paired, and so the Δ term is

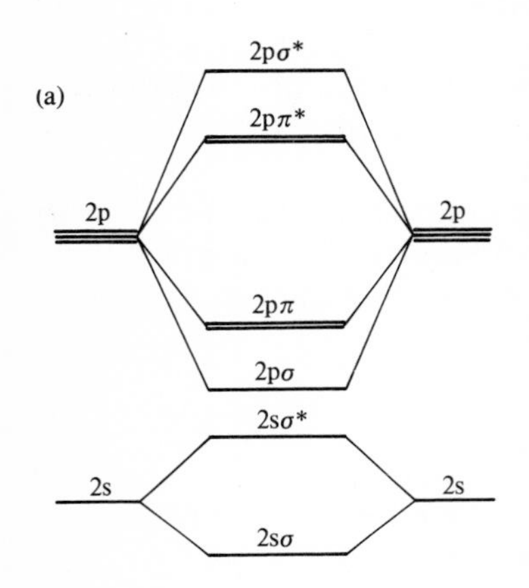

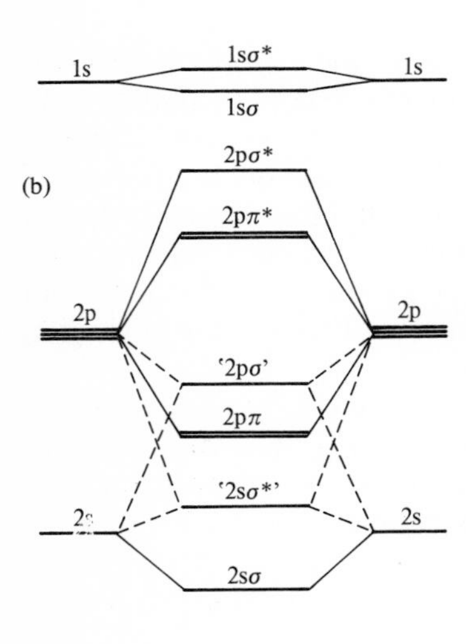

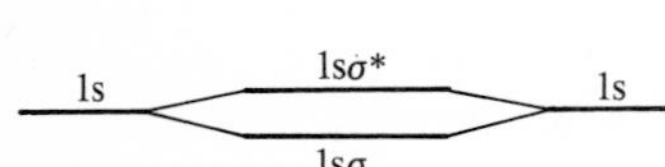

Fig. 10.12. (a) Simple and (b) more accurate molecular orbital energy level diagrams.

necessarily a singlet ($S=0$) and denoted $^1\Delta$. Alternatively, one electron may enter one π^*-orbital and one the other. The total orbital angular momentum is then zero, and so this arrangement gives rise to a Σ term. Since the electrons are in different orbitals their interaction is less than in the Δ term, and so this Σ term lies lower in energy than the Δ. Because the electrons are now in different orbitals there is no restriction on their relative spin orientations: they may be paired (to give a $^1\Sigma$ term) or parallel (to give a $^3\Sigma$ term). If we use Hund's rule of maximum multiplicity (p. 240), and suppose that it applies to molecules as well as to atoms) we expect the triplet term to lie beneath the singlet. Therefore we predict a $^3\Sigma$ ground state for the molecule. This is observed. The net bond character of the molecule is 2 (we count the $2p_x\pi^*2p_y\pi^*$ configuration as a single antibond), and so O_2 is a *doubly-bonded* $^3\Sigma$ species.

The molecule F_2 has the configuration $\ldots 2p\pi_u^4 2p\sigma_g^2 2p\pi_g^{*4}$, $^1\Sigma_g$ and so, as a result of the cancellation of all the π-bonding, can be regarded as held together by a single σ-bond. This accounts for its very low dissociation energy (about 154 kJ mol^{-1}; compare this to 942 kJ mol^{-1} for N_2). The addition of two further electrons to produce Ne_2 leads to the configuration $\ldots 2p\pi_g^{*4} 2p\sigma_u^{*2}$ and therefore to the complete cancellation of all bonding character. Since antibonding effects slightly outweigh bonding effects, the molecule is predicted to be unstable. Nevertheless, it is known to have some stability: its dissociation energy is equivalent to 50 cm^{-1}, and it has about four spectroscopically detectable vibrational levels. Clearly, simple MO theory is inadequate at this level of description.

The case of C_2 is instructive. The straightforward application of the building-up principle suggests the configuration $1s\sigma_g^2 1s\sigma_u^{*2}$-$2s\sigma_g^2 2s\sigma_u^{*2} 2p\pi_u^4$, $^1\Sigma_g$. The ground state is in fact $^3\Pi_u$. The reason is that there are alternative ways of arriving at the lowest energy configuration. On the one hand, the electrons can enter the orbitals as we have suggested, Fig. 10.13(a), but the disadvantage is that the final electron has to be accommodated in an orbital that is already half full, and because it has to enter it with paired spin, there are no advantageous spin correlation effects to lessen the electron–electron repulsion. On the other hand, it could enter the next higher orbital giving the molecule the configuration $\ldots 2p\pi_u^3 2p\sigma_g$, Fig. 10.13(b). The disadvantage of entering the higher orbital may be offset by the lessening of the electron–electron repulsions because the electron is now in a different region of space, and, furthermore, spin-correlation is able to play a helpful role. So long as the orbital energies are not too different, the latter configuration may turn out to have the lower total energy. This is the case for C_2, and accounts for its $^3\Pi_u$ ground term.

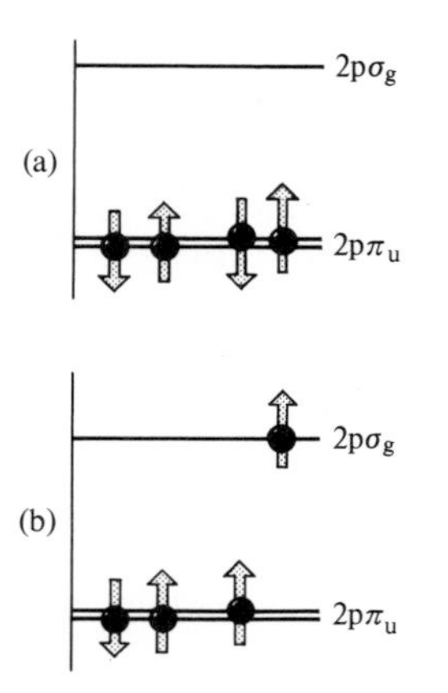

Fig. 10.13. Competition between (a) low orbital energy and (b) reduced electron repulsion.

The same style of discussion may be applied to the other rows of the periodic table and to heteronuclear diatomics. It can be turned into a quantitative form by using a development of the Hartree–Fock SCF procedure introduced for atoms. The difficulty with this procedure is that whereas in atoms there is the simplifying feature of spherical symmetry (and only one nucleus), this is lacking in molecules, and the calculations are very much more difficult. As a consequence many of these *ab initio calculations* (i.e. calculations attempting a direct solution without drawing on empirical data to simplify some of the integrals) generally use an LCAO approach, and model the true wavefunction in terms of a linear combination of a basis set of orbitals. The Hartree–Fock equations then turn into the *Roothaan equations* for the coefficients. The Roothaan method would coincide with the pure HF-SCF procedure if the basis set were infinite, for it could then model every nuance of the electron distribution. The energies and wavefunctions would then be in the *Hartree–Fock limit*, but still not perfect because *electron correlation* has still been neglected, just as in HF-SCF calculations on atoms. The study of correlation effects in molecules, the attempt to build even more realistic many-electron wavefunctions and to push the calculated energies down beneath the Hartree–Fock limit and to reproduce the true energy of the molecule, is a principal subject of modern research.

10.7 The structures of polyatomic molecules

The ideas introduced on the basis of the diatomic molecules, and in particular the role of the overlap of atomic orbitals, is a basis for the discussion of polyatomic molecules. We are now in a position to explain why the water molecule is triangular, the ammonia molecule pyramidal, and the methane molecule tetrahedral.

At its simplest level we look at the atomic orbitals of the central atom, see how they are disposed in space, and then attach hydrogen atoms so as to obtain the optimum overlap in each bond. For example, the ground configuration of the oxygen atom is $K2s^2 2p_z^2 2p_y 2p_x$, and so there are two half-filled p-orbitals. These can form bonds with two hydrogen atoms. Since the p-orbitals are mutually perpendicular, a 90° H—O—H bond angle is predicted. Likewise, the configuration of the nitrogen atom is $2s^2 2p^3$, and on the same grounds we expect three mutually perpendicular bonds to be formed. H_2O is bent, but its bond angle is 104.5° not 90°; NH_3 is a triangular pyramid, but its bond angle is 107°, not 90°. The description is a good first shot; but it must obviously be refined.

Refinement of the description requires us to go back to the basic LCAO procedure for constructing molecular orbitals and not to start in the middle of the problem, as we did in the last paragraph. If we proceed systematically, then the correct molecular orbital description will emerge, and we shall see that the bond angles of species can be explained. In order to implement the LCAO-MO procedure, we need to know which atomic orbitals can be combined to form molecular orbitals. In linear molecules it is obvious that s- and p_z-orbitals contribute to σ-orbitals and that p_x- and p_y-orbitals contribute to π-orbitals; but as the complexity of the molecule increases so the forms of the appropriate combinations become less clear. Now we have to resort to group theory in order to identify and construct the appropriate combinations. The H_2O molecule is happily on the border of having a sufficiently simple structure for its symmetry to be discussed in common sense terms, but it is also just complex enough to provide an example of the application of more formal group theoretical techniques of the kind introduced in Chapter 7. One further reason for using a group theoretical language is that we shall be able to classify the orbitals and the molecular states they give rise to in terms of the symmetry group of the molecule. This is a generalization of the procedure for diatomic molecules where we referred to σ and π orbitals and to Σ and Π terms; they are in fact symmetry classifications relevant to the $D_{\infty h}$ symmetry group of homonuclear diatomics.

Consider the H_2O molecule. It belongs to the symmetry group C_{2v}. The basis orbitals for the LCAO procedure are the 2s- and 2p-orbitals of oxygen and the two 1s-orbitals of the hydrogen atoms, Fig. 10.14. The 2s- and the $2p_z$-orbitals have the same symmetry (A_1); the p_x and p_y have different symmetries in the group (B_1 and B_2) and the linear combination $H1s_a + H1s_b$ and $H1s_a - H1s_b$ have A_1 and B_2 symmetries. These can all be seen by inspection and a quick glance at the character table in Appendix 10. The more formal procedure is the one set out in Chapter 7. We carry it through as the following *Example*.

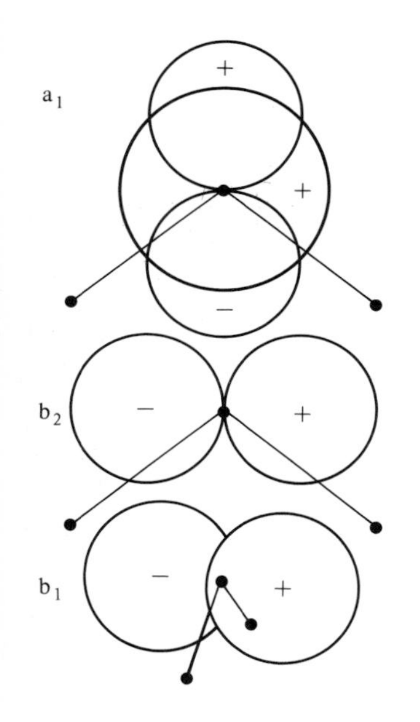

Example. Construct symmetry adapted linear combinations of H1s, O2s, O2p orbitals for H_2O.

- *Method.* Follow the method established in the *Example* on p. 154. Use the basis (O2s, $O2p_x$, $O2p_y$, $O2p_z$, $H1s_A$, $H1s_B$) in the group C_{2v}; $h = 4$.
- *Answer.*

Original set	O2s	$O2p_x$	$O2p_y$	$O2p_z$	$H1s_A$	$H1s_B$
Under E	O2s	$O2p_x$	$O2p_y$	$O2p_z$	$H1s_A$	$H1s_B$
C_2	O2s	$-O2p_x$	$-O2p_y$	$O2p_z$	$H1s_B$	$H1s_A$
σ_v	O2s	$-O2p_x$	$O2p_y$	$O2p_z$	$H1s_A$	$H1s_B$
σ_v'	O2s	$O2p_x$	$-O2p_y$	$O2p_z$	$H1s_B$	$H1s_A$

For A_1 $d = 1$ and all $\chi(R) = 1$. Hence column 1 gives O2s, column 2 and 3 give zero, column 4 gives $O2p_z$, and columns 5 and 6 give $\frac{1}{2}(H1s_A + H1s_B)$; hence $a_1 = c_1 O2s + c_2 O2p_z + c_3(H1s_A + H1s_B)$. For A_2 with characters 1, 1, −1, −1 no column survives. For $B_1(1, -1, -1, 1)$ column 2 gives $O2p_x$ and all other columns give zero. Hence $b_1 = O2p_x$, a non-bonding orbital. For $B_2(1, -1, 1, -1)$ column 3 gives $O2p_y$ and columns 5 or 6 give $\frac{1}{2}(H1s_A - H1s_B)$; hence $b_2 = c_1'' O2p_y + c_2''(H1s_A - H1s_B)$.

- *Comment.* If there were d-orbitals available (as in H_2S) the d_{z^2} and $d_{x^2-y^2}$-orbitals would contribute to a_1, d_{yz} to b_2, and d_{xz} and d_{xy} would be non-bonding.

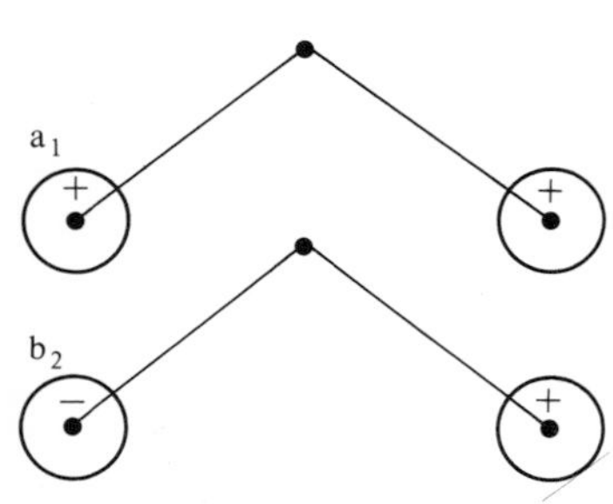

Fig. 10.14. Symmetry-adapted bases for H_2O.

Only orbitals of the same symmetry have net overlap ($S = \int \psi_a \psi_b \, d\tau = 0$ unless $\Gamma_a \times \Gamma_b$ contains A_1, which requires $\Gamma_a = \Gamma_b$), and so it follows that the molecular orbitals are of the following three

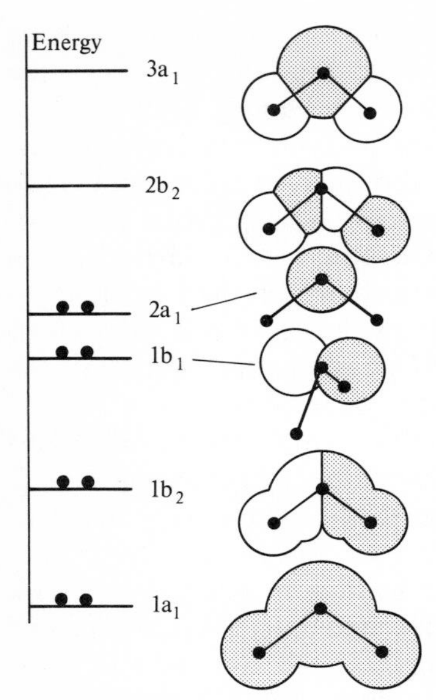

Fig. 10.15. Molecular orbitals and approximate orbital energies for H_2O.

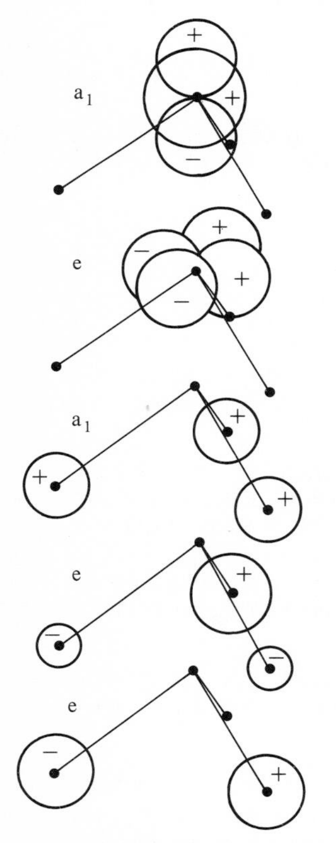

Fig. 10.16. Symmetry-adapted bases for NH_3.

symmetry types:

$$A_1\colon\ a_1 = c_1(\mathrm{H1s_a} + \mathrm{H1s_b}) + c_2(\mathrm{O2p}_z) + c_3(\mathrm{O2s});$$

$$B_1\colon\ b_1 = \mathrm{O2p}_x;$$

$$B_2\colon\ b_2 = c_1'(\mathrm{H1s_a} - \mathrm{H1s_b}) + c_2'(\mathrm{O2p_y}).$$

The coefficients are found by solving the secular equations arising from the variation principle, but because the hamiltonian has zero matrix elements between orbitals of different symmetry types (H spans A_1, hence $H_{ab} = \int \psi_a H \psi_b \, d\tau = 0$ unless $\Gamma_a \times A_1 \times \Gamma_b = A_1$, which requires $\Gamma_a = \Gamma_b$), the 6×6 secular determinant factorizes into 3×3, 1×1, and 2×2 determinants, each of which can be solved separately. In this way we arrive at three a_1 orbitals (called $1a_1$, $2a_1$, $3a_1$ in order of increasing energy), one b_1 orbital, which is simply the pure $\mathrm{O2p_x}$ orbital (it is quite clear that there are no hydrogen orbitals that can form a bond with it), and two b_2 orbitals, $1b_2$ and $2b_2$. The shapes of these orbitals and their approximate energies are set out in Fig. 10.15. Since there are 8 electrons to be accommodated, the configuration of the water molecule is predicted to be $1a_1^2 2a_1^2 1b_2^2 1b_1^2$, ${}^1\mathrm{A}_1$, the term symbol coming from the direct product of the symmetry species of the occupied orbitals (Appendix 15). Calculation shows that the lowest energy is obtained when the bond angle is a little greater than 90°, and if HF-SCF procedures are used to refine the calculation by taking electron–electron interactions into account in a better way, the experimentally observed bond angle can be predicted. (Computational techniques are now so advanced that in some cases it is possible to predict molecular geometries with a greater precision than they have been measured.) Note that all the occupied orbitals of H_2O are those with a net bonding effect; hence the molecule is a particularly stable species (to our advantage).

The same technique may be applied to NH_3. Now the *minimum basis set* (the basis set using only the valence orbitals) consists of the nitrogen 2s- and 2p-orbitals and the three hydrogen 1s-orbitals; the molecular point group is C_{3v}. Intuitively we expect 2s and $2p_z$ to belong to one symmetry type and p_x and p_y to belong to another. This can be seen with a glance at the character table in Appendix 10, because s and p_z on the central atom span A_1 and p_x and p_y (which transform as x and y) span E. The symmetry types of the three hydrogen orbitals were established in Chapter 7, and we know that the linear combinations there called s_1, s_2, s_3 form a basis for $A_1 + E$. These points can be seen by reference to Fig. 10.16 or by repeating the work set out in Chapter 7. The complete 7×7 secular determinant coming from the variation method applied to this basis set therefore factorizes into a 3×3 (corresponding to the three members of the basis with A_1 symmetry) and a 4×4 (corresponding to the four members, s_2, s_3, p_x, p_y, with E symmetry). The molecular orbitals of ammonia are therefore of the form

$$A_1\colon\ a_1 = c_1 s_1 + c_2(\mathrm{N2s}) + c_3(\mathrm{N2p}_z)$$

$$E\colon\ e = c_1' s_2 + c_2'(\mathrm{N2p_x}); \qquad e = c_1'' s_3 + c_2''(\mathrm{N2p_y}).$$

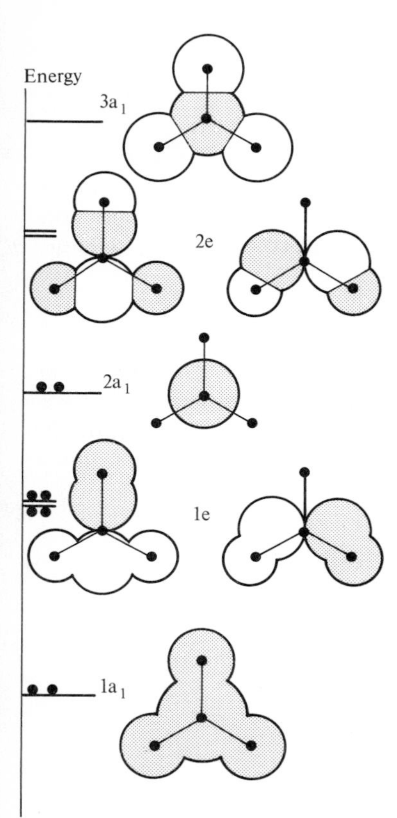

Fig. 10.17. Molecular orbitals (viewed from above) and orbital energies for NH_3.

(The two sets of e-orbitals are distinguished by their reflection symmetry.) The secular determinant is solved for the energy levels of the orbitals, Fig. 10.17. There are 8 electrons to accommodate, and so the configuration of the ground state is expected to be $1a_1^2 2a_1^2 1e^4$, ${}^1A_1$. The total energy reaches a minimum when the bond angle is greater than 90°, and HF-SCF type calculations give excellent agreement with the experimental value.

10.8 Hybridization and bond angles

The direct LCAO-MO procedure applied to H_2O and NH_3 does not give much insight into the reason why the two molecules adopt their particular bond angles. This must lie in the inclusion of the 2s-orbital of the central atom, because when it is excluded from the basis set we revert to the description of the molecule with which we began, and that leads to predictions of 90° for each. Analysing the source of the bending is not an essential part of the calculation of molecular structure, and many publications on polyatomic molecules simply quote the energies of states and the molecular geometry without looking for deeper reasons. Nevertheless, chemists do find it useful to build up rules of thumb which let them visualize what is happening inside molecules, and they like to try to relate sophisticated accounts of molecular structure to familiar, simpler versions. How, then, can we relate bond angles to the involvement of s-orbitals in molecular orbitals, especially when those orbitals appear to spread throughout the molecule? Is there some way of thinking of the structure of the molecule in terms of *individual bonds* identifiable with links between pairs of atoms?

The answer lies in the modification of the basis set to be used in the molecular orbital description. Instead of working with the 2s- and the 2p-orbitals themselves, we can form mixtures of them called *hybrid orbitals*, and then build the LCAO molecular orbitals from them. We shall see that the hybrids point in definite directions, just like the original set of p-orbitals in our initial description of H_2O and NH_3, and so we can see very plainly why the molecules adopt their characteristic shapes.

Consider the hybrid orbitals that may be formed from the 2s-, $2p_z$-, and $2p_y$-orbitals lying in the plane of an H_2O molecule. We shall show that from them it is possible to construct two *equivalent* hybrids pointing along the directions of the bonds and a third hybrid pointing along the exterior bisector of the bond angle, and that the composition of the hybrids depends on the bond angle. The two criteria we use are (a) the orthogonality of the three hybrids and (b) the equivalence of the two pointing along the bonding directions (so that the C_{2v} symmetry is preserved). The axes are set out in Fig. 10.18.

Consider first the combination of p_y and p_z corresponding to a p-orbital pointing along one of the bonds. Since $p_y \propto y$ and $p_z \propto z$ it follows that

$$p = p_z \cos \tfrac{1}{2}\Theta + p_y \sin \tfrac{1}{2}\Theta. \qquad (10.8.1a)$$

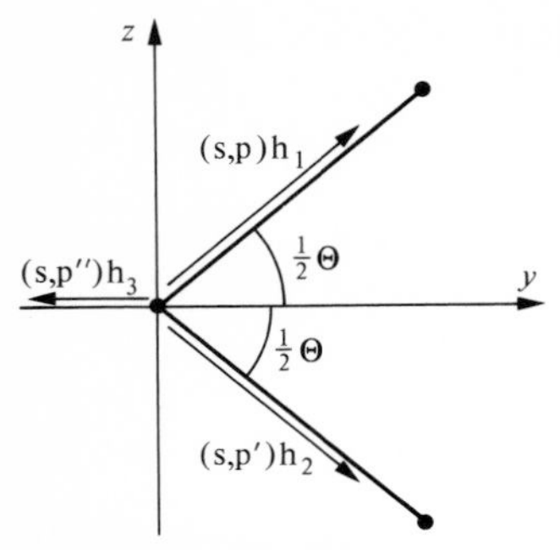

Fig. 10.18. The formation of hybrid orbitals in a C_{2v} system.

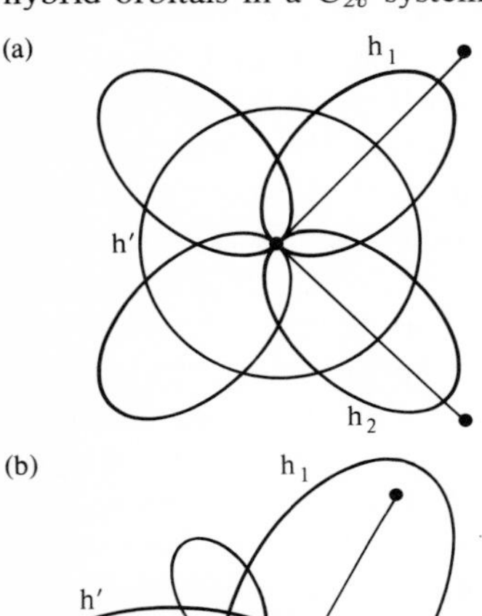

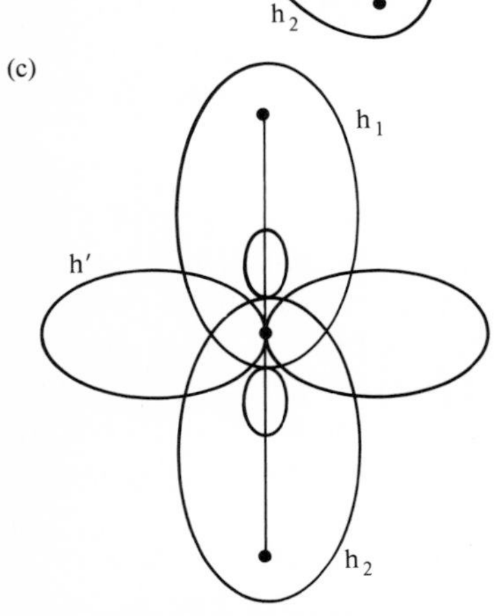

Fig. 10.19. Hybrid orbitals for $\theta =$ (a) 90°, (b) 120°, (c) 180°.

The equivalent p-orbital pointing along the other bond direction is

$$p' = p_z \cos\tfrac{1}{2}\Theta - p_y \sin\tfrac{1}{2}\Theta. \tag{10.8.1b}$$

These two p-orbitals are not orthogonal (their overlap integral is $\cos^2\frac{1}{2}\Theta - \sin^2\frac{1}{2}\Theta = \cos\Theta$). They are made orthogonal by mixing in an appropriate amount of s-character (we shall see that the proportion required is $\cos\Theta$). We therefore write the two orthogonal, equivalent hybrids as

$$h_1 = as + bp, \qquad h_2 = as + bp'. \tag{10.8.2}$$

Both hybrids should be constructed so that they are normalized to unity (we have seen that this always simplifies expressions), and so we require $a^2 + b^2 = 1$ (as s and p are individually normalized and mutually orthogonal). The orthogonality requirement translates into the following condition on the coefficients:

$$\int h_1 h_2\, d\tau = \int (as + bp_z \cos\tfrac{1}{2}\Theta + bp_y \sin\tfrac{1}{2}\Theta)$$
$$\times (as + bp_z \cos\tfrac{1}{2}\Theta - bp_y \sin\tfrac{1}{2}\Theta)\, d\tau$$
$$= a^2 + b^2 \cos^2\tfrac{1}{2}\Theta - b^2 \sin^2\tfrac{1}{2}\Theta = a^2 + b^2 \cos\Theta = 0.$$

Therefore the two conditions give

$$\cos\Theta = -a^2/(1-a^2), \quad \text{or} \quad a^2 = \cos\Theta/(\cos\Theta - 1). \tag{10.8.3}$$

As the angle changes from 90° to 180° the amount of s-character in each hybrid increases from zero (pure p-orbitals when the bond angle is 90°) to 50 per cent s-character when the molecule is linear. In the case of 104.45°, the experimental value for water, $a^2 = 0.20$, or 20 per cent s-character, and the hybrids are formed with $a = 0.45$ and $b = 0.89$. The third orthogonal hybrid that may be formed from the three basis orbitals is also orthogonal to the two just formed (but it is not required to be equivalent to the first pair). It can be expressed in the form $a's + b'p_z$, and little effort is needed to deduce that in this case

$$a'^2 = (1 + \cos\Theta)/(1 - \cos\Theta). \tag{10.8.4}$$

Therefore it changes from pure s when $\Theta = 90°$ to pure p when the bonds are collinear ($\Theta = 180°$). The shapes of the hybrid orbitals as Θ changes are shown in Fig. 10.19: notice how their amplitudes are projected strongly into one direction. Notice too the special case $\Theta = 120°$ when all three hybrids are equivalent and have the composition $a^2 = \frac{1}{3}$, $b^2 = \frac{2}{3}$; these are called *sp^2-hybrids*, the name reflecting the s:p composition ratio.

Now consider the configuration an oxygen atom needs in order to form a molecule of bond angle Θ. In general it is $2p_x^2 h'^2 h_1 h_2$. When the angle is 90° the hybrid h' is pure 2s and the pair of mutually equivalent hybrids are pure 2p; hence the configuration in this 'state of preparation' for bond formation, or *valence state*, to give a 90° molecule is $2s^2 2p_x^2 2p_1 2p_2$ and so overall the configuration is $2s^2 2p^4$.

Now take the other extreme of possibilities, when the atom is in the valence state appropriate to forming a linear molecule. Since $\Theta = 180°$, h' is pure p_z while h_1 and h_2 are 50:50 mixtures of 2s and $2p_y$. The configuration is therefore $2p_x^2 2p_z^2 (2p_y^{\frac{1}{2}}, 2s^{\frac{1}{2}})(2p_y^{\frac{1}{2}}, 2s^{\frac{1}{2}})$ and so overall the configuration is $2s2p^5$. The point that immediately appears is that in order to attain the valence state of a linear molecule, an electron has to be *promoted* from an s-orbital to a p-orbital. This is an energy disadvantage. When the bonds with the hydrogen atoms are formed, some of this promotion energy will be regained because Fig. 10.19 shows very clearly that as the s-character increases the distribution of amplitude increasingly favours strong overlap in the internuclear region. Therefore the strength of the resulting bond can be expected to increase as the hybridization increases the proportion of s-orbital.

Example. Find an expression for the valence configuration and extent of promotion of an s^2p^4 atom needed to achieve two equivalent bonds making an angle Θ.

- *Method.* Find an expression of the form $s^m p^n$ on the basis of eqns (10.8.3) and (10.8.4). The fraction of electron promoted is $n-4$.
- *Answer.* One bond hybrid is $s^{a^2}p^{b^2}$, the other is the same, and so together they account for $s^{2a^2}p^{2b^2}$ in the valence configuration. The lone pair of electrons accounts for $s^{2a'^2}p^{2b'^2}$. Another lone pair occupies p_x^2. The overall valence configuration is therefore $s^{2a^2+2a'^2}p^{2b'^2+2}$. Since $b^2 = 1 - a^2$ and $b'^2 = 1 - a'^2$ the configuration is $s^{2(a^2+a'^2)}p^{2(3-a^2-a'^2)}$. From eqns (10.8.3–4), $a^2 + a'^2 = 1/(1-\cos\Theta)$, and so the valence configuration for a bond angle Θ is $s^{2/(1-\cos\Theta)}p^{2(2-3\cos\Theta)/(1-\cos\Theta)}$. The extent of promotion is $\{(2-3\cos\Theta)/(1-\cos\Theta)\} - 4 = 2\cos\Theta/(\cos\Theta - 1)$.
- *Comment.* The extent of promotion changes from 0 (at $\Theta = \pi/2$) to 1 (at $\Theta = \pi$), the configurations being s^2p^4 and s^1p^5 respectively. The *hybridization ratio* of the bond hybrids is $b^2/a^2 = -1/\cos\Theta = -\sec\Theta$, and so they are $sp^{-\sec\Theta}$-hybrids (e.g. sp-hybrids when $\Theta = \pi$).

We can now see there is a competition. On the one hand it is energetically disadvantageous to promote the atom to its valence state. On the other hand, when it is promoted it is able to form stronger bonds on account of the improved overlap of hybrid orbitals. There is also an advantageous decrease in the electrostatic bond–bond repulsions when they move apart from 90°. It is then easy to see that there must be a compromise: there is a degree of promotion, not complete, such as to bring about the greatest net lowering of the energy of the molecule, and the molecular geometry reflects its extent. In the case of H_2O the optimum energy is obtained when there is the equivalent of 20 per cent promotion of an s-electron; in which case the hybridization corresponds to a bond angle of just over 104°.

The structure of NH_3 can be accounted for in the same way. The configuration of the unpromoted nitrogen atom is $2s^2 2p^3$. If we impose on the atom the requirement that the bonds should be built from three

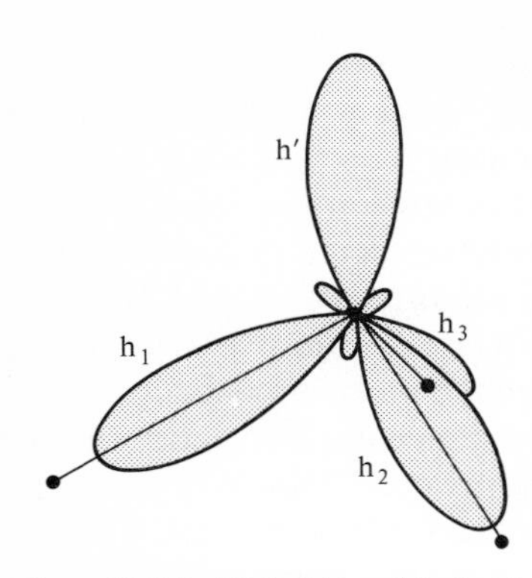

Fig. 10.20. Hybrid orbitals in NH_3; h_1, h_2, and h_3 are equivalent.

equivalent hybrid orbitals pointing towards the corners of an equilateral triangle and a fourth directed along the 3-fold axis, Fig. 10.20, then in order to achieve the valence state $h'^2h_1h_2h_3$ some degree of promotion is necessary, and it depends on the angle between the hybrids. When the molecule is flat its valence state has the configuration $2s2p^4$, corresponding to the complete promotion of an s-electron. Partial promotion of an s-electron results in the formation of hybrid orbitals forming a triangular pyramid. A fairly straightforward calculation shows that when the bond angle is 107° (the experimental angle for NH_3) the valence configuration is $2s^{1.22}2p^{3.78}$, corresponding to 78 per cent promotion. This compromise value represents the competition between promotion energy investment and bonding energy gain as a consequence both of the enhanced overlap, and of the modification of the repulsive interactions between the bonding and non-bonding electrons in the molecule (the three N—H bonds are now further apart, and so interact less strongly).

The most striking example of hybridization is the carbon atom, which in its ground state has the configuration $2s^22p^2$. In the case of methane, CH_4, we expect four equivalent orbitals pointing towards the corners of a regular tetrahedron. This is a special case of a triangular pyramid (as in NH_3), because now all four hybrids are equivalent and the inter-bond angles are all 109.47° ($=\arccos(-\frac{1}{3})$). The same techniques as before (mutual orthogonality but identical compositions) leads to the following forms of the hybrids in this special case:

$$\begin{aligned} h_1 &= s+p_x+p_y+p_z, & h_2 &= s+p_x-p_y-p_z \\ h_2 &= s-p_x+p_y-p_z, & h_3 &= s-p_x-p_y+p_z. \end{aligned} \qquad (10.8.5)$$

These hybrids are illustrated in Fig. 10.21; because they consist of 1 part s-orbital to 3 parts p-orbital they are called *sp³-hybrids*.

An atom of carbon in its tetrahedral valence state has the configuration $h_1h_2h_3h_4$, and since each hybrid is $\frac{1}{4}$ s-electron and $\frac{3}{4}$ p-electron, overall this configuration is sp^3, representing a promotion of an *entire* s-electron from the ground configuration. In the case of carbon there are special reasons why the promotion of an electron is energetically less disadvantageous than in nitrogen and oxygen. In the first place, the promotion can be regarded as taking place from an s-orbital into an *empty* p-orbital. In the case of oxygen ($2s^22p^4$) and nitrogen ($2s^22p^3$) the promoted electron has to enter an already half full orbital and suffer greater electron–electron repulsion. Therefore less promotion energy is required in carbon than in either nitrogen or oxygen. Next, *four* bonds (of enhanced strength) may be formed from the $2s2p^3$ configuration, whereas only three may be formed in the case of nitrogen or two for oxygen. Furthermore, the electrostatic repulsions between electrons in four different bonds are minimized when they adopt a tetrahedral arrangement. Therefore there is a greater gain to offset the promotion energy in the case of carbon, and it occurs readily. This accounts for the huge array of tetrahedral carbon compounds.

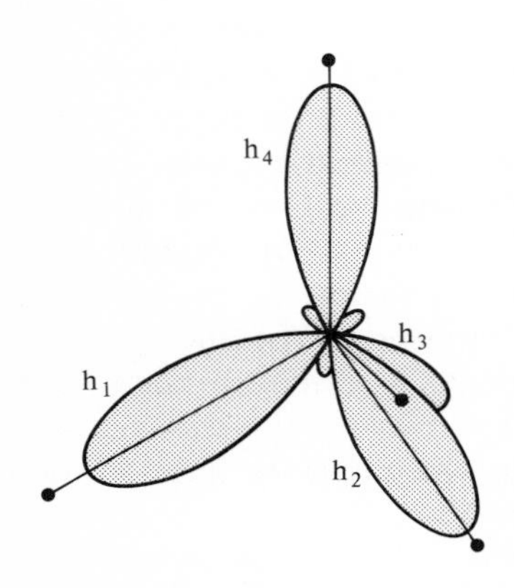

Fig. 10.21. sp^3-hybrid orbitals; h_1, h_2, h_3, and h_4 are equivalent.

When thinking about the promotion of an atom to its valence state do not think of the atom as 'undergoing' first promotion and then bonding. The two occur together: as the hydrogen atoms approach, the entire system evolves towards its state of lowest energy. Indeed, the valence state is not even a spectroscopic state. This is because the four electrons (in sp^3) have spin directions that are random with respect to each other, and the overall angular momentum of the atom is not well defined. The valence state can be regarded as a linear combination of spectroscopic states, and for the carbon atom its description is

$$(\tfrac{1}{8})\{2.5^{\frac{1}{2}}(\mathrm{sp}^3\,{}^5\mathrm{S})+3(\mathrm{p}^4\,{}^3\mathrm{P})+3.2^{\frac{1}{2}}(\mathrm{sp}^3\,{}^3\mathrm{D})-3(\mathrm{s}^2\mathrm{p}^2\,{}^3\mathrm{P})$$
$$-3^{\frac{1}{2}}(\mathrm{p}^4\,{}^1\mathrm{D})-2^{\frac{1}{2}}(\mathrm{sp}^3\,{}^1\mathrm{D})+3^{\frac{1}{2}}(\mathrm{s}^2\mathrm{p}^2\,{}^1\mathrm{D})\}.$$

Furthermore, hybridization is not a *necessary* component of the description of the structures of polyatomic molecules. The structure of the methane molecule, for instance, can be described in LCAO-MO terms using the basis functions 2s, $2p_x$, $2p_y$, $2p_z$ on the carbon atom. It is only when we want to think in terms of the various contributions to the energy that it is sometimes convenient to use an alternative basis expressed as a linear combination of the primitive basis. The final expressions, energies, and so on, are entirely equivalent, but one form may be intuitively more appealing (and easier to analyse) than the other.

10.9 Conjugated π-systems

A special class of polyatomic molecules are those containing π-bonded atoms, especially the conjugated polyenes and aromatic molecules based on benzene. They fall into a unique class because the σ-bonds and the π-bonds can (to some extent) be discussed individually. One reason is that the electrons in π-bonds are generally less strongly bound than those in the σ-bonds, and so to some extent can be discussed separately. Another is that since π-bonds are often found in extensive planar molecules, they have symmetry properties different from the σ-bonds, and therefore span different irreducible representations of the molecular point group; as a consequence the secular equations factorize and they can be discussed separately.

The simplest π-system is that in the ethene molecule. Ethene can be regarded as the combination of two CH_2 fragments with sp^2-hybrid bonds forming the σ-framework and the remaining p-orbitals of the carbon valence shell overlapping to form a π-bond. The maximum overlap between the two 2p-orbitals in the π-bond is achieved when the planes of the CH_2 groups are parallel. Any deviation from this conformation involves raising the energy of the molecule, and so we have a very simple explanation of the *torsional rigidity* of π-bonded molecules.

When the π-system is *conjugated* (i.e. there are p-orbitals on several neighbouring atoms capable of forming an extended system of π-orbitals, as in benzene) the simplest description of the bonding is in terms of the *Hückel molecular orbital method.* This takes full account of the symmetry of the network of atoms through which the π-bonds extend (e.g. the planar hexagonal symmetry of benzene), and leads to expressions for the relative energies of the different molecular orbitals in terms of a single parameter. Once the orbitals have been found and their energies roughly assessed, electrons are accommodated in them in accord with the usual rules of the building-up principle.

As an example, consider the π-system of butadiene, a molecule with the σ-framework H_2C—CH—CH—CH_2. Each carbon atom carries a 2p-orbital perpendicular to the plane defined by the σ-framework of bonds. We denote these orbitals p_1, p_2, p_3, p_4. The molecular orbitals of the π-system are linear combinations of these four atomic orbitals. The appropriate coefficients can be obtained from the variation theory, which involves solving the secular equations

$$\sum_i c_i\{H_{ij} - ES_{ij}\} = 0 \qquad i = 1, 2, 3, 4, \tag{10.9.1}$$

and hence finding the energies by solving

$$\det|H_{ij} - ES_{ij}| = 0. \tag{10.9.2}$$

For butadiene its full form is

$$\begin{vmatrix} H_{11}-ES_{11} & H_{12}-ES_{12} & H_{13}-ES_{13} & H_{14}-ES_{14} \\ H_{21}-ES_{21} & H_{22}-ES_{22} & H_{23}-ES_{23} & H_{24}-ES_{24} \\ H_{31}-ES_{31} & H_{32}-ES_{32} & H_{33}-ES_{33} & H_{34}-ES_{34} \\ H_{41}-ES_{41} & H_{42}-ES_{42} & H_{43}-ES_{43} & H_{44}-ES_{44} \end{vmatrix} = 0.$$

The Hückel method then goes on to make the following drastic simplifications:

(1) *All overlap integrals are set to zero:* $S_{ij} = 0$ unless $i = j$ when $S_{ii} = 1$. This is an awful approximation because typical values of the overlap integral for neighbouring atoms are $S \approx 0.20$–0.25. Nevertheless, when the rule is relaxed the energies are shifted in a simple way and their relative order is not greatly disturbed.

(2) *All the diagonal elements of the hamiltonian* (these correspond to the Coulomb integrals introduced on p. 254) *are assumed to be the same:* $H_{ii} = \alpha$. This is reasonable in the case of species not containing hetero-atoms because it is based on the view that all the atoms are electronically equivalent. Some justification for this comes from the *Coulson–Rushbrooke theorem* which states that in alternant† hydrocarbons the charge density on every atom is the same (and unity).

(3) *All the off-diagonal elements* (the resonance integrals of the discussion on p. 254) *are set equal to zero except those on neighbouring atoms, all of which are taken to be equal:* $H_{ij} = 0$ if i, j are not

† An *alternant* hydrocarbon is one in which the atoms can be divided into two groups by putting a star (for instance) on alternate atoms and if, when all the stars are present, there are no neighbouring stars. Benzene is alternant. Azulene is non-alternant.

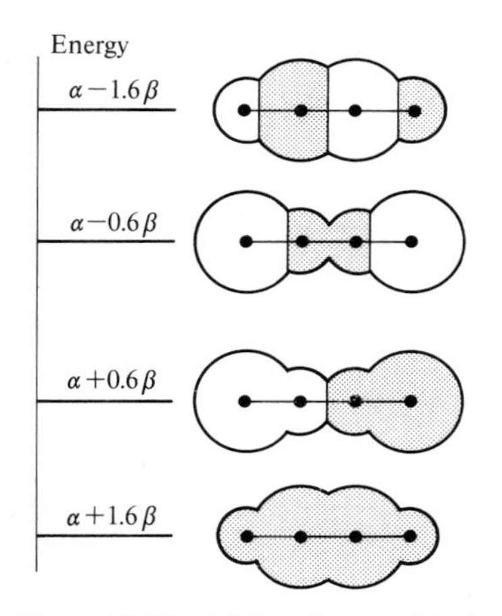

Fig. 10.22. Molecular orbitals (viewed from above) and Hückel energies for butadiene.

neighbours and $H_{ij}=\beta$ if they are neighbours. The parameter β is the single parameter characteristic of Hückel theory (α turns out to be redundant).

The imposition of these simplifications on the secular determinant turns it into the following form:

$$\begin{vmatrix} \alpha-E & \beta & 0 & 0 \\ \beta & \alpha-E & \beta & 0 \\ 0 & \beta & \alpha-E & \beta \\ 0 & 0 & \beta & \alpha-E \end{vmatrix}=0.$$

This can be solved very easily, and gives the four roots $E=\alpha\pm1.6\beta$ and $E=\alpha\pm0.6\beta$ (the numbers are $\frac{1}{2}(\sqrt{5}\pm1)$ in fact). Both α and β are negative (p. 254) and so the energy level diagram is as shown in Fig. 10.22. There are four p-electron to be accommodated, and so they complete the two lower orbitals. If we neglect the electron–electron repulsions that arise when the electrons occupy the orbitals, the energy of the ground state of the molecule is $2(\alpha+1.6\beta)+2(\alpha+0.6\beta)=4\alpha+4.4\beta$. The energy of a single unconjugated π-orbital is $\alpha+\beta$, and so if the picture of the molecule had been taken to be two simple, unconjugated bonds, the total energy of the four electrons would have been $4\alpha+4\beta$. There is therefore an additional stabilization of the molecule as a result of allowing the electron orbitals to be spread throughout the molecule. The additional energy, 0.4β in this case, is called the *delocalization energy* of the molecule. The delocalization energy is independent of the value taken for α. Furthermore the coefficients of the four orbitals (Fig. 10.23) are also independent of α, and it is in this sense that α is a redundant parameter.

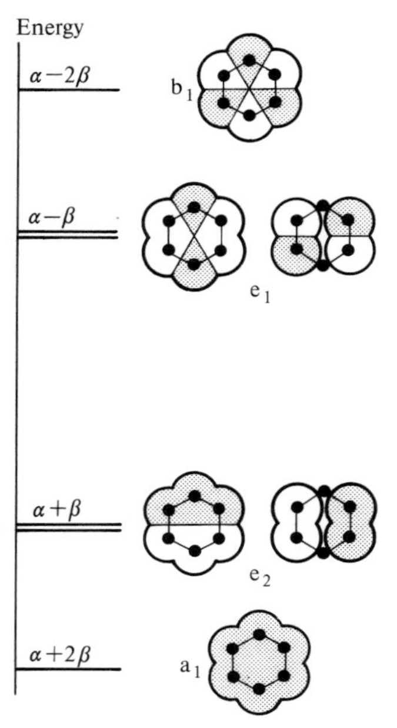

Fig. 10.23. Molecular orbitals (viewed from above) and Hückel energies for benzene.

The Hückel scheme still leads to secular determinants of large dimension. Happily the techniques of group theory let us factorize them into smaller, simpler-to-solve, components. So far we have used only the planar symmetry of a conjugated system to separate the σ-system from the π-system of orbitals. We can also use its symmetry to classify and distinguish the π-orbitals themselves. This follows from the fact that the hamiltonian transforms totally symmetrically in the molecular point group (by definition), and so it has vanishing matrix elements between orbitals spanning irreducible representations of different symmetry species. In the case of benzene, for instance, the 6×6 secular determinant (giving a sixth-order equation to solve for the energies) can be reduced dramatically by using the molecule's D_{6h} symmetry and setting up symmetry-adapted linear combinations of the six 2p-orbitals. In fact we do not even need to use the full point group symmetry: we know that all the p-orbitals change sign on reflection in the molecular plane, and so the subgroup C_{6v} carries enough information.

When the standard procedures for setting up the symmetry-adapted combinations are applied to the basis set of six p-orbitals we find that they span $A_1+B_1+E_1+E_2$ and take the form:

$$A_1\colon\ a_1=(\tfrac{1}{6})^{\frac{1}{2}}(p_1+p_2+p_3+p_4+p_5+p_6)$$
$$B_1\colon\ b_1=(\tfrac{1}{6})^{\frac{1}{2}}(p_1-p_2+p_3-p_4+p_5-p_6)$$
$$E_1\colon\ e_{1a}=(\tfrac{1}{12})^{\frac{1}{2}}(2p_1-p_2-p_3+2p_4-p_5-p_6)$$
$$e_{1b}=(\tfrac{1}{2})(p_2-p_3+p_5-p_6)$$
$$E_2\colon\ e_{2a}=(\tfrac{1}{12})^{\frac{1}{2}}(2p_1+p_2-p_3-2p_4-p_5+p_6)$$
$$e_{2b}=(\tfrac{1}{2})(p_2+p_3-p_5-p_6).$$

These are sketched in Fig. 10.23. Note that the orbitals are determined solely by the symmetry of the molecule and make no reference to the values of α and β.

Example. Find expressions for the energy levels of the π-orbitals of benzene.

- *Method.* The symmetry-adapted π-orbitals are specified above. Since $H_{rs}=\int\psi_r H\psi_s\,d\tau$ vanishes unless ψ_r and ψ_s are of the same symmetry species (because H is a basis for A_1) there are no off-diagonal elements between orbitals of different species. Hence the 6×6 determinant factorizes. Evaluate the matrix elements H_{rs} on the basis of the Hückel rules, using $\int p_i H p_j\,d\tau=0$ unless either (a) $i=j$, when it equals α or (b) i and j are neighbours, when it equals β. Find the eigenvalues.
- *Answer.*

For A_1: $$H_{a_1a_1}=\frac{1}{6}\int(p_1+\ldots+p_6)H(p_1+\ldots+p_6)\,d\tau=\alpha+2\beta$$

For B_1: $$H_{b_1b_1}=\frac{1}{6}\int(p_1-\ldots-p_6)H(p_1-\ldots-p_6)\,d\tau=\alpha-2\beta$$

For E_1: $$H_{e_{1a}e_{1a}}=\alpha+(\tfrac{1}{12})(-2\beta-2\beta-2\beta+\beta+\beta-2\beta-2\beta-2\beta-2\beta+\beta+\beta-2\beta)$$
$$=\alpha-\beta$$
$$H_{e_{1b}e_{1b}}=\alpha+(\tfrac{1}{4})(-\beta-\beta-\beta-\beta)=\alpha-\beta$$
$$H_{e_{1a}e_{1b}}=H_{e_{1b}e_{1a}}=0.$$

Therefore the 2×2 E_1-determinant is diagonal and the eigenvalues are both $\alpha-\beta$.

For E_2: $$H_{e_{2a}e_{2a}}=\alpha+(\tfrac{1}{12})(2\beta+2\beta+2\beta-\beta-\beta+2\beta+2\beta+2\beta+2\beta-\beta-\beta+2\beta)$$
$$=\alpha+\beta$$
$$H_{e_{2b}e_{2b}}=H_{e_{2a}e_{2a}};\qquad H_{e_{2a}e_{2b}}=H_{e_{2b}e_{2a}}=0.$$

Therefore the E_2-determinant is diagonal, the two eigenvalues being $\alpha+\beta$.

- *Comment.* The use of group theory greatly reduces the effort needed to find solutions; note that choice of orthogonal linear combinations of the e-orbitals even factorize the 2×2 determinants. The C_{6v} symmetry of benzene is particularly favourable to the factorization. Go on to try naphthalene (use C_{2v}) and pyridine (distinguish α_N from α_C and β_{CC} from β_{NC}).

When the symmetry-adapted basis is used the secular determinant factorizes into two 1×1 determinants (which are solvable trivially) and two 2×2 determinants which give rise to two sets of equations for E. It is therefore a simple matter to arrive at the energy level diagram in Fig. 10.23. Since there are six electrons to be accommodated, they enter the lowest three orbitals and give rise to the configuration $a_1^2e_2^2$, ${}^1A_1$. (In the full symmetry group D_{6h} the orbitals are called $a_{1u}e_{2g}$, and so the symmetry of the ground state is ${}^1A_{1g}$.) The approximate total π-bond energy is $2(\alpha+2\beta)+4(\alpha+\beta)=6\alpha+8\beta$. In contrast, three localized double bonds with the same bond length would have a total energy of $6\alpha+6\beta$. The delocalization energy is therefore 2β. Notice that the electrons just fill the MOs with a net bonding effect and leave unfilled the antibonding orbitals. This is a major contribution to the stability of the aromatic ring. Furthermore, note the symmetrical nature of the energy level array—to every bonding level there occurs a symmetrically disposed antibonding level. This symmetry is a characteristic feature of alternant hydrocarbons, and can be traced to the topological properties of the molecules. Indeed, many of the results of Hückel theory can be established on the basis of *graph theory*, the branch of topology concerned with the properties of networks.

Hückel theory, which virtually hijacks the disagreeable integrals, is only the most primitive stage of discussing the structures of π-molecules. The improvements that have to be made are quite easy to see. Within the LCAO-MO scheme the obvious first step involves retaining the overlap terms in the secular determinant. Then the matrix elements of the hamiltonian between the atomic orbitals of the basis set have to be retained and dealt with with more respect. Some of the integrals that arise can be related quite easily to spectroscopic quantities, and the methods that draw on data of this kind are called *semi-empirical.* Each semi-empirical method has its own name, and techniques called the *Pariser–Parr–Pople* (PPP) method, the *complete neglect of different overlap* (CNDO) method, *modified intermediate neglect of differential overlap* (MINDO) method, and many others, have been developed, where integrals of the form $\int a(1)b(1)(1/r_{12})c(2)d(2)\,d\tau_1\,d\tau_2$ are systematically neglected or approximated. The thrust of much of the work, however, is towards the direct numerical evaluation of these difficult (or at least time-consuming) integrals, and to move towards the refinement of the *ab initio techniques*, where no empirical data are used. These have a greater intrinsic reliability, and computing power is now sufficient to make them feasible even for quite large molecules.

10.10 Transition metal ions: ligand field theory

The success of the Hückel method is rooted in the fact that the orbitals themselves are determined by the symmetry of the system (so long as all the atoms are the same and do not attract electron density to different extents). These symmetry-determined orbitals are then put into an order of energies, essentially by assessing the number and importance of their nodes. The energy differences between orbitals are so large that the coarseness of this procedure does not unduly misrepresent their order. A similar situation occurs in the complexes of the transition metal ions. These consist of a central positive ion surrounded by a three-dimensional array of ligands. The orbitals of the complex are largely determined by the symmetry of the environment, and a single parameter can be used to give a rough indication of the order of the energies of the molecular orbitals of the complex. Ligand field theory is a kind of three-dimensional Hückel theory, where the orbitals of interest are clustered on the single, central ion.

The simplest version of ligand field theory is *crystal field theory*. In this approximation the complex is modelled as a central metal ion (M) surrounded by ligands (L) which act only as a source of electric potential. In other words, they are not regarded as supplying atomic orbitals from which molecular orbitals spreading over the entire complex may be formed. The ligands produce a potential which removes the degeneracy of the five d-orbitals of the central ion, and the structure of the complex can be discussed in terms of the building-up principle applied to the set of energy levels produced in this way.

The first step involves classifying the d-orbitals of the central ion according to the irreps they span in the molecular point group of the complex. If the complex is a regular octahedron the appropriate point group is O_h (a regular tetrahedron is T_d and a square planar complex is D_{4h}). A glance at the character table in Appendix 10 shows that the d-orbitals (which transform as $d_{xy} \propto xy$ etc.) span $E_g + T_{2g}$, and so we can immediately conclude that the five-fold degeneracy of the atom is removed by the octahedral field and replaced by a two-fold degenerate pair (d_{xy}, $d_{x^2-y^2}$) and a triply degenerate set (d_{xy}, d_{yz}, d_{xz}). Group theory tells us nothing about which set of orbitals is the higher, but we can decide on their relative order on the basis of a simple electrostatic argument.

Let there be a single d-electron and consider the ligands to be point negative charges. This represents for instance, the lone pairs of electrons on the NH_3 molecule when it is acting as a ligand. The energy of the entire system is reduced as the six ligands approach the central ion, but if the electron is in one of the t_{2g}-orbitals (which point between the positions occupied by the ligands) its energy is reduced more than when it occupies one of the two e_g-orbitals, (which point directly at the ligand sites and therefore experience a stronger repulsion). Therefore the t_{2g}-orbitals correspond to the lower level in an octahedral environment, Fig. 10.24. The energy separation between the two sets of orbitals is called $10Dq$; the e_g-orbitals lie at $6Dq$ above the mean and the three t_{2g}-orbitals each lie at $4Dq$ beneath the mean.

When there is only a single electron in the ion, as in Ti^{3+}, the ground configuration is t_{2g}^1 and the term symbol is $^2T_{2g}$. The complex in water (where the ligands are H_2O molecules) has an absorption peaking at $20\,000\ cm^{-1}$ which can be interpreted as $e_g \leftarrow t_{2g}$. Therefore we may identify the energy equivalent of $20\,000\ cm^{-1}$ with the magnitude of $10Dq$ ($20\,000\ cm^{-1}$ corresponds to $2340\ kJ\ mol^{-1}$, or $2.5\ eV$).

When there are several d-electrons present there are three cases to

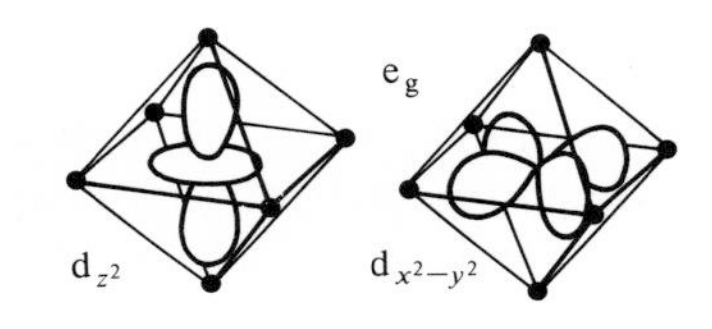

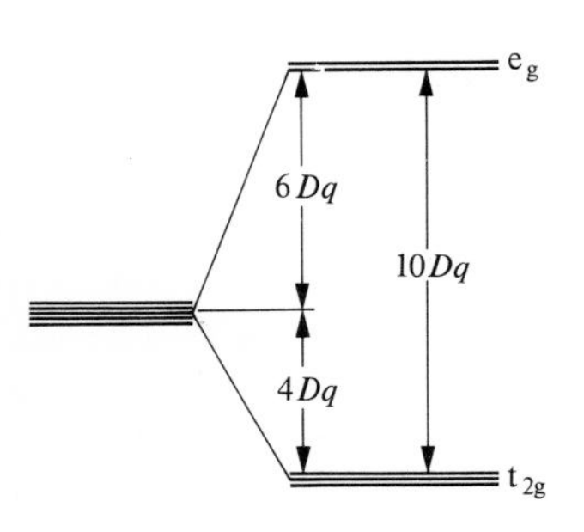

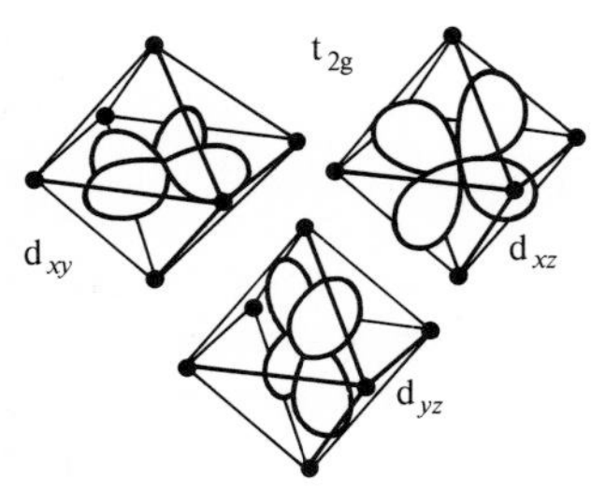

Fig. 10.24. Removal of d-orbital degeneracy in an octahedral complex.

consider:

(1) The *strong field case*, in which the potential energy due to the crystal field is much stronger than the electron–electron repulsion energy.
(2) The *weak field* in which the crystal field is weak in comparison with the repulsion energy but still strong in comparison with the spin–orbit coupling energy.
(3) The *very weak field case* in which the crystal field is weaker than both the electron–electron repulsion energy and the spin–orbit interaction.

The *very weak field* case is usually found in the lanthanides because their f-electrons are embedded deeply in the atom and experience the potential of the surrounding ligands only very weakly, and we shall not consider it further.

The *strong field case* uses the sets of orbitals just established as a basis for the applications of the building-up principle to ions with several d-electrons. We have to be careful about the relative orders of magnitude of the electron–electron repulsion terms and the splitting $10Dq$ because the latter is not necessarily dominantly large. Exactly the same occurred in the discussion of the C_2 molecule, where the lowest energy term involved the use of higher energy orbitals. There may be an energy advantage in octahedral complexes in making use of the e_g-orbitals before the t_{2g}-orbitals are full. This is because when the electron enters empty orbitals it can adopt the most advantageous spin, and the lowering of energy due to spin correlation may be strong enough to overcome the relatively small $t_{2g} \leftarrow e_g$ promotion energy.

The second d-electron enters one of the t_{2g}-orbitals, and can do so with parallel spin (which has the advantage of reducing its repulsive interaction with the electron already there). The configuration is therefore t_{2g}^2. The third electron can also occupy a t_{2g}-orbital with parallel spin, and so a d^3 ion is likely to have the configuration t_{2g}^3 in an octahedral complex. The fourth electron has to resolve a competition. It may enter the lower set and result in the configuration t_{2g}^4; but it then must have a spin opposite to the electrons already present and there is no advantageous spin correlation effects to reduce the electron–electron repulsions. On the other hand it could enter the upper e_g-level with its spin parallel to the first three electrons. The energy disadvantage would be of the order of $10Dq$, but there might be a net energy gain if the spin-correlation effect reduces the electron–electron repulsion term sufficiently. The configuration t_{2g}^4 will therefore be adopted if $10Dq$ is large, but $t_{2g}^3e_g$ if it is small. The total spin of the complex is greater when the configuration is $t_{2g}^3e_g$, and hence this is called a *high-spin complex* in contrast to the *low-spin* t_{2g}^4 configuration. It follows that we expect high-spin complexes when $10Dq$ is small and low-spin complexes when $10Dq$ is large. A similar competition occurs in d^5, d^6, and d^7 complexes, but when we reach d^8 and d^9 the configurations must be $t_{2g}^6e_g^2$ and $t_{2g}^6e_g^3$ because the additional electron can enter the orbitals only by pairing with electrons already present, and there is no energy advantage in using the upper orbitals before the lower.

The *weak field case* is quite tricky to deal with because the crystal field is no more than a perturbation which slightly affects the energies of the free atom states, and the configurations $t_{2g}^n e_g^{n'}$ are not really applicable because they are so strongly mixed by the $1/r_{ij}$ electron repulsion terms in the hamiltonian. The contrast between this and the strong field case, where the $t_{2g}^n e_g^{n'}$ configurations are meaningful, can be appreciated by considering a d^2-configuration and its classification in the two cases.

In the free ion a d^2 configuration can give rise to the terms ^3F, ^3P, ^1D, ^1G, and ^1S. When the crystal field is weak it merely reduces the crystal field from spherical to octahedral (from R_3 to O_h) without affecting the energies much. Concentrate on the lower ^3F and ^3P terms, which have $L=3$ and $L=1$ respectively. In order to see which symmetry species of irrep they span in O_h we apply eqn (7.14.6) with $L=3$ or 1 and identify the symmetry species from the character table in Appendix 10. We find

$$^3P \rightarrow {}^3T_{1g}; \qquad {}^3F \rightarrow {}^3A_{2g} + {}^3T_{1g} + {}^3T_{2g}.$$

The separation of the terms stemming from ^3F increases as the perturbation becomes stronger, as shown on the left of Fig. 10.25. In the strong field case we can speak of the three configurations e_g^2, t_{2g}^2, and $t_{2g}e_g$. If we confine our attention to triplet terms these configurations give rise to $e_g^2\,{}^3A_{2g}$, $t_{2g}^2\,{}^3T_{1g}$, and $t_{2g}e_g\,{}^3T_{1g}$ or $t_{2g}e_g\,{}^3T_{2g}$ (deciding which terms arise depends on considering the implication of the Pauli principle much as we did for the p^2 configuration of the free atom described in Chapter 9). The previous discussion of orbital energies lets us put these three configurations in order of increasing energy, the right of Fig. 10.25, and therefore to draw the complete correlation diagram for the triplet states of the ion for all relative strengths of the crystal field and the electron repulsions. The actual state of the complex corresponds to some intermediate point on the diagram, and its location can be explored spectroscopically. The details of the procedures for discussing these intermediate cases are set out in the books referred to in the bibliography.

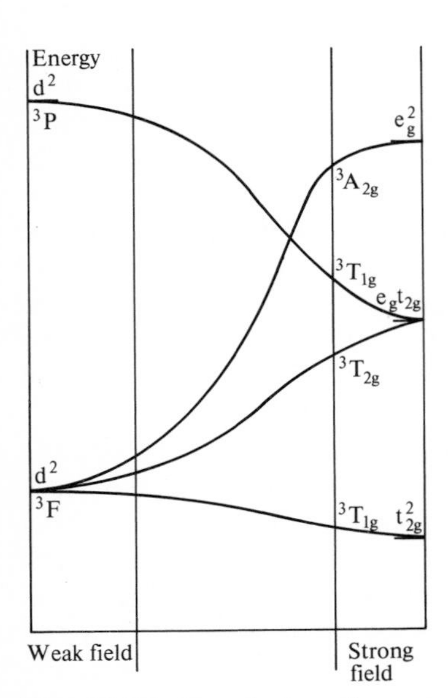

Fig. 10.25. Correlation diagram for d^2 in an octahedral complex.

10.11 The Jahn–Teller theorem

There is one final point about the predictions of the crystal field theory concerning the structure of complexes. The *Jahn–Teller theorem* states that *in any non-linear system there exists some vibrational mode that removes the degeneracy of an orbitally degenerate state.* We may see this in connection with a d^9-ion in an octahedral field. The configuration of such an ion is $t_{2g}^6e_g^3$, and this is orbitally degenerate because the 'hole' in the d^{10} configuration may be in either the d_{z^2}- or the $d_{x^2-y^2}$-orbital of the e_g-set. If the octahedron of ligands were to distort so that the complex lengthened along one C_4 direction, the energy of the d_{z^2}-orbital may fall below that of $d_{x^2-y^2}$ (Fig. 10.26), and so the energy of the complex is least if the configuration $d_{z^2}^2 d_{x^2-y^2}$ is adopted. If the axis were to shorten, or the equatorial bonds lengthen, the configuration $d_{x^2-y^2}^2 d_{z^2}$ would be adopted. In either case the system remains in its distorted conformation because it then has lower energy. While this prediction is quite clear, the distortion is often difficult to distinguish from that caused by the packing in a crystal.

An interesting case is the d^8 configuration, for then there is a competition between filling the lower of the e_g-pair (but having the disadvantage of a high electron repulsion energy), or half-filling both (and achieving a reduced repulsion energy but gaining nothing from the effect of the distortion). In this case distortion occurs only when the energy difference it induces is greater than the electron repulsion energy. Therefore, if a distortion is observed, it is likely to be a large one. d^7 configurations resemble the behaviour of d^9 because the competition no longer occurs.

10.12 Molecular orbital theory of transition metal complexes

The crystal field theory must be incomplete because it is known that the electrons of the central metal ion do drift on to the ligands. The evidence for this comes from electron spin resonance spectroscopy (see *Further reading*, p. 283). It is therefore natural to investigate whether the structure of complexes can be explained successfully in terms of a molecular orbital scheme, for then the electrons would acquire some of the character of the ligands.

In order to construct molecular orbitals we first consider orbitals (λ) on the ligands (i) that are σ-type with respect to the particular M–L axis, and label them $\lambda_i^{(\sigma)}$, with $i = 1, 2, \ldots, 6$ for an octahedral complex. We already know that the d-orbitals of the central ion can be labelled t_{2g} and e_g, and so if there are linear combinations of the ligand orbitals that span T_{2g} and E_g of the group O_h we may form molecular orbitals from these and the corresponding d-orbitals. The standard techniques of group theory show that the six ligand σ-orbitals span $A_{1g} + E_g + T_{1u}$ of the group, and the explicit linear combinations are as follows:

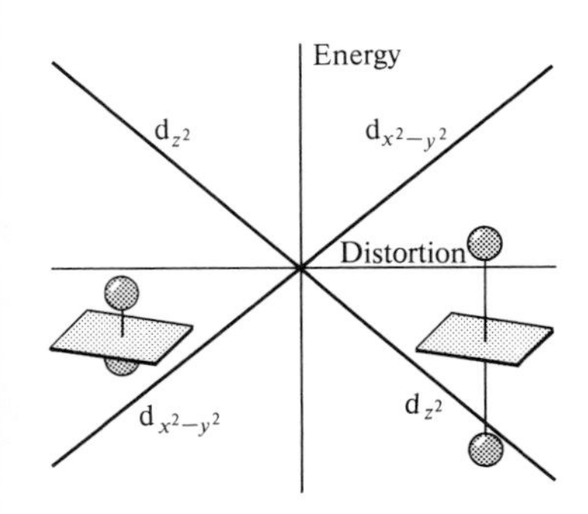

Fig. 10.26. The Jahn–Teller effect in an octahedral complex.

$$A_{1g}:\ a_{1g} = (\tfrac{1}{6})^{\frac{1}{2}}(\lambda_1^{(\sigma)} + \lambda_2^{(\sigma)} + \lambda_3^{(\sigma)} + \lambda_4^{(\sigma)} + \lambda_5^{(\sigma)} + \lambda_6^{(\sigma)}),$$

$$E_g:\ e_g = \begin{cases} \tfrac{1}{2}(\tfrac{1}{3})^{\frac{1}{2}}(2\lambda_5^{(\sigma)} + 2\lambda_6^{(\sigma)} - \lambda_1^{(\sigma)} - \lambda_2^{(\sigma)} - \lambda_3^{(\sigma)} - \lambda_4^{(\sigma)}) \\ \tfrac{1}{2}(\lambda_1^{(\sigma)} + \lambda_2^{(\sigma)} - \lambda_3^{(\sigma)} - \lambda_4^{(\sigma)}), \end{cases}$$

$$T_{1u}:\ t_{iu} = \begin{cases} (\tfrac{1}{2})^{\frac{1}{2}}(\lambda_1^{(\sigma)} - \lambda_2^{(\sigma)}) \\ (\tfrac{1}{2})^{\frac{1}{2}}(\lambda_3^{(\sigma)} - \lambda_4^{(\sigma)}) \\ (\tfrac{1}{2})^{\frac{1}{2}}(\lambda_5^{(\sigma)} - \lambda_6^{(\sigma)}). \end{cases}$$

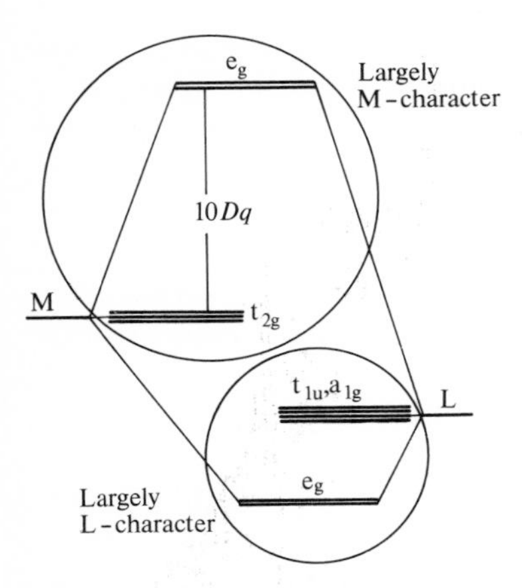

Fig. 10.27. Molecular orbital diagram for an octahedral complex.

There is no T_{2g} combination. It follows that only the e_g(d)-orbitals can overlap the ligand σ-orbitals with non-vanishing overlap, and that the t_{2g}-orbitals remain confined to the nucleus (and therefore are *non-bonding*). The energy level scheme this suggests is shown in Fig. 10.27. The building-up principle can be applied to this orbital scheme in the usual way and so we need to know how many electrons to accommodate. Generally the ligand σ-orbitals approach the ion already filled (for example, they may represent the lone-pair of electrons in the ammonia molecule, or the p-electrons in the closed-shell fluoride ion). This being the case, there are twelve ligand electrons and n d-electrons from the metal ion to add to the molecular orbitals. The first twelve electrons enter the six e_g-, t_{1u}-, and a_{1g}-orbitals. These are all predominantly confined to the ligands. The e_g-orbitals do spread a little into the metal system, but in the present scheme (in which the s- and p-orbitals of the metal are neglected) the single a_{1g}-orbital and the three degenerate t_{1u}-orbitals are totally confined to the ligands, because no d-orbitals transform suitably to have a net overlap with them. The remaining n electrons then enter either the non-bonding t_{2g}-orbitals or the antibonding e_g-orbitals, depending on the difference in energy of the two sets and the strength of the electron repulsion energy. As the antibonding e_g-orbitals have predominantly metal-ion d-orbital character, the description is essentially the same as in the crystal field theory, only the source of the splitting is different. The upper e_g-orbitals are only predominantly composed of metal-ion orbitals and in fact contain some ligand orbital character, as we wanted. The extent of delocalization depends on the relative energies of the ligand and metal orbitals, and as the energies approach each other the delocalization increases.

The molecular orbital theory enables the description of the bonding to be extended to cases where π-bonding can occur between the metal and ligands. We suppose that on each of the six ligands there are orbitals with π-symmetry with respect to the particular M–L axis. They span $T_{1u}+T_{2u}+T_{1g}+T_{2g}$ and only the last can combine with the t_{2g}-orbitals on the ion. The explicit combinations of the π-orbitals are as follows:

$$t_{1u}=\begin{cases}\frac{1}{2}(p_{3x}+p_{4x}+p_{5x}+p_{6x}),\\ \frac{1}{2}(p_{1y}+p_{2y}+p_{5y}+p_{6y}),\\ \frac{1}{2}(p_{1z}+p_{2z}+p_{3z}+p_{4z});\end{cases}\qquad t_{1g}=\begin{cases}\frac{1}{2}(p_{5x}-p_{6x}-p_{1z}+p_{2z}),\\ \frac{1}{2}(p_{5y}-p_{6y}-p_{3z}+p_{4z}),\\ \frac{1}{2}(p_{1y}-p_{3x}-p_{2y}+p_{4x});\end{cases}$$

$$t_{2g}=\begin{cases}\frac{1}{2}(p_{1y}-p_{2y}+p_{3y}-p_{4y}),\\ \frac{1}{2}(p_{1z}-p_{2z}+p_{5x}-p_{6z}),\\ \frac{1}{2}(p_{3z}-p_{4z}+p_{5y}-p_{6y});\end{cases}\qquad t_{2u}=\begin{cases}\frac{1}{2}(p_{5y}+p_{6y}-p_{1y}-p_{2y}),\\ \frac{1}{2}(p_{1z}+p_{2z}-p_{3z}-p_{4z}),\\ \frac{1}{2}(p_{5x}+p_{6x}-p_{3x}-p_{4x}).\end{cases}$$

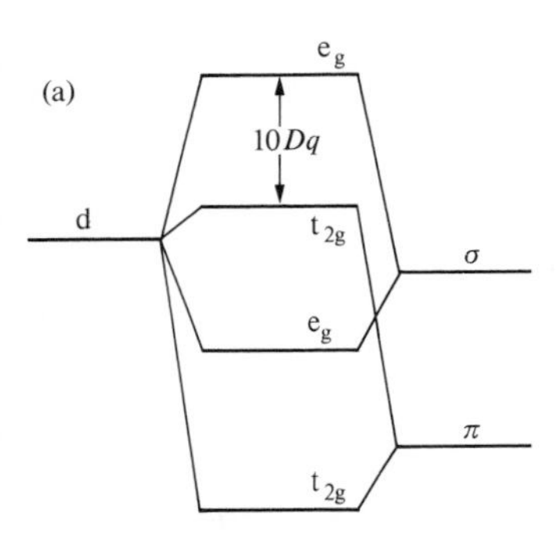

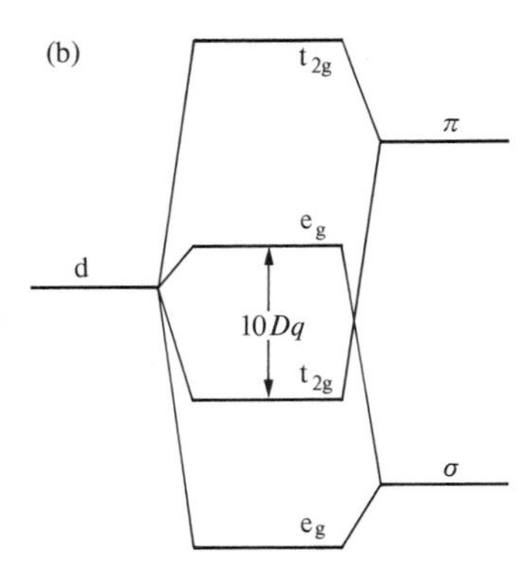

Fig. 10.28. Effect of ligand π-orbitals when they lie (a) below, (b) above the ligand σ-orbitals.

Two cases may be distinguished, and are illustrated in Fig. 10.28. In the first, Fig. 10.28(a), the π-orbitals of the ligands are full and the electrons from the ligands fill all the lower e_g- and t_{2g}-orbitals and the remaining n electrons must be distributed among the upper e_g- and

t_{2g}-orbitals. The important point is that the value of $10Dq$ is *reduced* as a result of the π-bonding because the originally non-bonding orbitals on the metal become slightly antibonding. Therefore, not only does the orbital splitting decrease but the metal–ligand bonds are weakened. In the second case, where the π-orbitals of the ligands are initially empty and therefore lie above the σ-orbitals, the effect on the metal t_{2g}-orbitals is to make them slightly bonding. Therefore the orbital splitting is increased, the bonding is enhanced, and the metal electrons are delocalized even more on to the ligands. This is observed when the ligand is NO, and since the delocalization occurs into π^*-orbitals of the ligand it may be detected as a weakening of the N—O bond (Fig. 10.29).

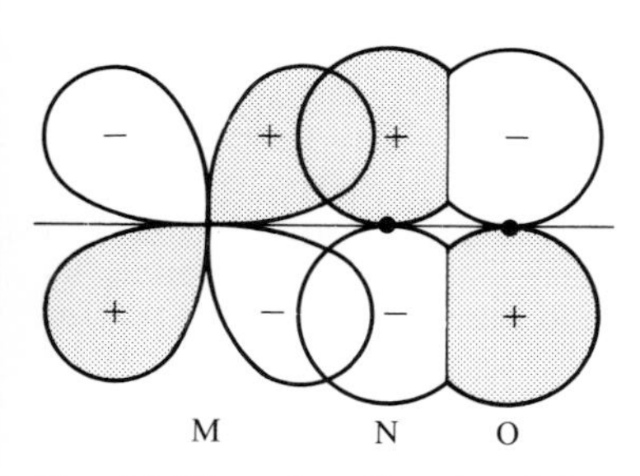

Fig. 10.29. π-bonding involving a ligand π^*-orbital (of NO).

Further reading

General valence theory

The shape and structure of molecules. C. A. Coulson, revised R. McWeeny; Oxford University Press, 2nd edition, 1982.
Coulson's Valence. R. McWeeny; Oxford University Press, 3rd edition, 1979.
The chemical bond. J. N. Murrell, S. F. A. Kettle, and J. M. Tedder; Wiley, Chichester, 1978.
Valence theory. J. N. Murrell, S. F. A. Kettle, and J. M. Tedder; Wiley, Chichester, 2nd edition, 1974.
Quantum theory. J. P. Lowe; Academic Press, New York, 1978.
Introduction to applied quantum chemistry. S. P. McGlynn, L. G. Vanquickenborne, M. Kinoshita, and D. G. Carroll; Holt, Rinehart, and Winston, 1972.

Organic molecules; semiempirical techniques

Molecular orbital theory for organic chemists. A. Streitweiser; Wiley, New York, 1961.
The molecular orbital theory of conjugated systems. L. Salem; Benjamin, New York, 1966.
The organic chemist's book of orbitals. W. L. Jorgensen and L. Salem; Academic Press, New York, 1973.
Approximate molecular orbital theory. J. A. Pople and D. L. Beveridge; McGraw-Hill, New York, 1970.
Semi-empirical self-consistent-field molecular orbital theory of molecules. J. N. Murrell and A. J. Harget; Wiley–Interscience, London, 1971.

Ab initio calculations

Molecular wavefunctions. E. Steiner; Cambridge University Press, 1976.
Ab initio molecular orbital calculations for chemists. W. G. Richards and D. L. Cooper; Oxford University Press, 2nd edition, 1982.
A bibliography of ab initio molecular wavefunctions. W. G. Richards, T. E. H. Walker, and R. K. Hinkley; Oxford University Press, 1971 and Supplements, 1974, 1978, and 1981.

Transition metal complexes

An introduction to transition metal chemistry. L. E. Orgel; Methuen, London, 1966.
Introduction to ligand field theory. C. Ballhausen; McGraw-Hill, New York, 1962.
The theory of transition metal ions. J. S. Griffith; Cambridge University Press, 1964.

Solids

Band theory of metals. S. L. Altmann; Pergamon, Oxford, 1970.
Electronic structure and the properties of solids. W. A. Harrison; W. H. Freeman, San Francisco, 1980.

Problems

10.1. Consider a one-electron, one-nucleus 'molecule' in one dimension, and show that the electronic and nuclear motions are approximately separable. (This is a primitive version of the Born–Oppenheimer approximation: the general case is proved in the same way.) *Hint.* This is a heavy beginning to the Problems: the rest are easier! The hamiltonian is $-(\hbar^2/2M)(\partial^2/\partial X^2)-(\hbar^2/2m_e)(\partial^2/\partial x^2)+V(X,x)$. Show that $\psi=\psi_{el}(x,X)\psi_{vib}(X)$ is an approximate solution, and would be exact if $(\hbar^2/2M)\{2(\partial\psi_{el}/\partial X)(\partial\psi_{vib}/\partial X)+(\partial^2\psi_{el}/\partial X^2)\psi_{vib}\}$ were zero. Appreciate that $(\hbar^2/2m_e)(\partial^2\psi_{el}/\partial X^2)\approx(\hbar^2/2m_e)(\partial^2\psi_{el}/\partial x^2)$ because $\psi_{el}(x,X)$ is largely a function of $x-X$, and $(\hbar^2/2m_e)(\partial^2\psi_{el}/\partial x^2)\approx$ electron energy. But M stands in place of m_e in the correction term.

10.2. The R-dependences of the molecular integrals j', k', and S for the hydrogen molecule ion are specified in eqns (A14.1–3). Plot E_+ and E_- against R and identify the equilibrium bond length and the dissociation energy of the molecule. Explore the R-dependences of α and β similarly.

10.3. Evaluate the probability density of the electron in H_2^+ at the mid-point of the bond, and plot it as a function of R. Evaluate the difference density $\rho=\psi_\pm^2-\frac{1}{2}(\psi_a^2+\psi_b^2)$ at points along the line joining the two nuclei (including the regions outside the nuclei) for $R=130$ pm. The difference density shows the modification to the electron distribution brought about by constructive (or destructive) overlap. *Hint.* Use the $\psi_\pm$ in eqn (10.3.5); S is given in Appendix 14. If you have access to a computer graphics system, consider plotting ρ throughout a plane containing the two nuclei. Repeat the calculation for several values of R.

10.4. Take the hydrogen molecule wavefunction in eqn (10.3.12) and find an expression for the expectation value of the hamiltonian (eqn (10.2.1)) in terms of molecular integrals. *Hint.* The outcome of this calculation is eqn (A14.4).

10.5. All the integrals involved in the H_2 molecular orbital calculation are listed in Appendix 14. Write and run a program to calculate $E-2E_{1s}$ as a function of R, and identify the equilibrium bond length and the dissociation energy.

10.6. Evaluate the probability density for a single electron at a point on a line running between the two nuclei in H_2, and plot the difference density $\rho_1-\frac{1}{2}(\psi_a^2+\psi_b^2)$ for $R\approx 74$ pm. *Hint.* Use eqn (10.3.12). The probability density of electron 1, ρ_1, is obtained from $\psi^2(1,2)$ by integrating over all locations of electron 2, because the latter's position is irrelevant. Therefore, begin by forming $\rho_1=\int\psi^2(1,2)\,d\tau_2$.

10.7. Calculate the valence bond E_+, eqn (10.4.4), for H_2 as a function of R, and locate the equilibrium bond length and identify the dissociation energy.

10.8. Normalize the wavefunctions in eqns (10.5.3) and (10.5.4) to unity. Relate λ' and λ'' in eqn (10.5.4). Find expressions in terms of molecular integrals for λ and λ'' which optimize the modified valence bond and molecular orbital descriptions of H_2. *Hint.* Use the variation theorem: form $\mathscr{E}_{VB}(\lambda)$ and $\mathscr{E}_{MO}(\lambda'')$ and seek the values of λ and λ'' for which $d\mathscr{E}_{VB}(\lambda)/d\lambda=0$ and for which $d\mathscr{E}_{MO}(\lambda'')/d\lambda''=0$.

10.9. Predict the ground configuration of the molecules C_2, C_2^+, C_2^-, N_2^+, N_2^-, O_2, O_2^+, CO, NO, F_2^+, and Ne_2^+. Decide which terms can arise in each case, and suggest which lies lowest.

10.10. Using a minimum basis set for the MO description of the molecule H_2O, show that the secular determinant factorizes into $(1\times1)+(2\times2)+(3\times3)$ determinants. Set up the secular determinant, denoting the Coulomb integrals α_H, α'_O, α_O for H1s, O2s, and O2p respectively, and writing the (O2p, H1s) and (O2s, H1s) resonance integrals as β and β'. Neglect overlap. First, neglect the 2s-orbital, and find expressions for the energies of the molecular orbitals for a bond angle of $\Theta=90°$.

10.11. Now develop the calculation by taking into account the O2s-orbital. Use the hybrid orbitals of eqn (10.8.2) to construct molecular orbitals of the appropriate symmetry species, and set up the secular determinant with Θ as a parameter. Find expressions for the energies of the molecular orbitals and of the entire molecule. As a first step in analysing the expressions, set $\alpha_H\approx\alpha_O\approx\alpha'_O$ and $\beta\approx\beta'$. Can you devise improvements to the values of the Coulomb integrals on the basis of atomic spectral data?

10.12. Show that three equivalent bonds in a planar, equilateral triangular molecule may be formed from sp^2-hybrid orbitals. Go on to show that three equivalent bonds may also be constructed with the hybridizations dp^2, sd^2, and d^3, and establish which d-orbitals are involved in each case. *Hint.* Use the group D_{3h}. For the first part show that the bonds σ_1, σ_2, σ_3 span A'_1+E'; so too do s, p_x, p_y.

10.13. Show that four equivalent tetrahedral hybrids may be formed from the hybrids sp^3 and sd^3. Show that four equivalent tetragonal planar bonds arise from dsp^2- and d^2p^2-hybridizations, that five pentagonal bipyramidal bonds arise from dsp^3- and d^3sp-hybridization, and that six equivalent octahedral bonds arise from d^2sp^3-hybridization.

10.14. Set up and solve the secular determinants for (a) butadiene, (b) square planar *cyclo*butadiene in the Hückel π-electron scheme; find the energy levels and molecular orbitals, and estimate the delocalization energy. *Hint.* Use the groups C_2 and C_{4v} to factorize the determinants.

10.15. Confirm that the symmetry-adapted linear combinations of $2p\pi$-orbitals for benzene are those set out on p. 282. Find the corresponding combinations for naphthalene.

10.16. Heterocyclic molecules may be incorporated into the Hückel scheme by modifying the Coulomb integral of the atom concerned and the resonance integrals to which it contributes. Consider pyridine, its symmetry group being C_{2v}. Construct and solve the Hückel secular determinant with $\beta_{CC}\approx\beta_{CN}$ and $\alpha_N\approx\alpha_C+\frac{1}{2}\beta$. Estimate the π-electron energy and the delocalization energy. *Hint.* The roots of the 4×4 determinant, which is one of the factors of the 6×6 determinant, are in the vicinity of $(E-\alpha)/\beta\approx2.11$, 1.17, -0.84, and -1.93. These are best found numerically (e.g. by successive approximation or, if to hand, on a computer by finding the eigenvalues of the appropriate matrix).

10.17. Explore the role of π-orbital overlap in π-electron calculations. Take the *cyclo*-butadiene secular determinant, but construct it without neglect of overlap between neighbouring atoms. Show that in place of $x=(\alpha-E)/\beta$ and 1 the elements of the determinant become $w=(\alpha-E)/(\beta-ES)$ and 1 respectively. Hence the roots in terms of w are the same as the roots in terms of x. Solve for E. Typically $S\approx0.25$.

10.18. Find the effect of including neighbouring atom overlap on the π-electron energy levels of benzene. If you have a computer available, explore how the energies depend on the bond lengths, using $\beta\propto S$ and

$$S(2p\pi,2p\pi)=\{1+s+2s^2/5+s^3/15\}e^{-s},\qquad s=Z^*R/na_0.$$

Consider the difference in resonance energy between the cases where the molecule has six equivalent C—C bond lengths of 140 pm (the experimental value) and where it has alternating lengths of 133 pm and 153 pm (typical C=C and C—C lengths respectively).

10.19. A one-dimensional metal can be modelled as a line of N equally spaced atoms, and treated by Hückel theory (but we may be considering σ-bonds if the atoms carry s-orbitals). Show that the Hückel orbitals occur at the energies $(E-\alpha)/\beta = -2\cos\{n\pi/(N+1)\}$ with $n = 1, 2, \ldots, N$. Plot the energies for $N = 1, 2, 3, 4, \ldots$, and show that they lie in a band of finite width even when $N \sim \infty$. Find an expression for the energy density of states and plot it across the band. *Hint*. The energy density of states is dn/dE.

10.20. Determine which symmetry species are spanned by d-orbitals in a tetrahedral complex.

10.21. An ion with the configuration f^2 enters an environment of octahedral symmetry. What terms arise in the free ion, and which terms do they correlate with in the complex? *Hint*. Follow the discussion at the end of Section 10.10.

10.22. In the strong field case, the d^2-configuration gives rise to e_g^2, $t_{2g}e_g$, and t_{2g}^2. What terms may arise? How do the singlet terms of the complex correlate with the singlet terms of the free ion? What configurations arise in a tetrahedral complex, and what are the correlations.

10.23. Find the symmetry-adapted linear combinations of (a) σ-orbitals, (b) π-orbitals on the ligands of an octahedral complex. *Hint*. Set cartesian axes on each ligand site, with z pointing towards the central ion, determine how the orbitals are transformed under the operations of the group O, and use the procedures for establishing symmetry-adapted orbitals as described in Chapter 7.

10.24. Repeat Problem 10.23, but for a tetrahedral complex. What is the role of π-bonding in such complexes?

11 Molecular rotational and vibrational transitions

MOLECULAR spectra are both more complex and more useful than atomic spectra. Their complexity arises from the complicated structures of molecules, for while the spectra of atoms are due only to their electronic transitions, the spectra of molecules arise from electronic, vibrational, and rotational transitions. These modes are not independent of each other, and the complexity of the spectra is made richer by the possibility of interactions between them. If they can be analysed, the information they yield should be correspondingly great, and we shall see that it is possible to extract details of molecular dimensions, the strengths of bonds, and the shapes of molecular potential energy curves, as well as to understand the basic processes involved in photochemistry.

The energy associated with rotational transitions is usually less than that involved in vibrational transitions, and that is usually less than in electronic transitions. As a consequence, although it is possible to observe pure rotational spectra (in the *microwave region*), when a vibration is excited it is normally accompanied by rotational transitions, and so a vibrational spectrum (in the *infrared region*) has superimposed on it a structure due to simultaneous rotational transitions. In an electronic transition both vibrational and rotational transitions are stimulated and the spectrum (in the *visible* and *ultraviolet regions*) contains information on all of them. Because of this hierarchy of transitions and complexity we shall deal with the transitions in order of size of the associated quanta.

11.1 Types of spectral transition

We saw in Chapter 9 that the most intense transitions are induced by the interaction of the electric field of the radiation with an electric dipole, and that the intensity of the transition is proportional to the square of the *transition dipole moment*, the matrix element of the electric dipole moment operator between the two states involved. This moment can be interpreted as a measure of the dipole associated with the migration of electric charge during the transition, Fig. 11.1. We shall be concerned almost exclusively with electric dipole transitions because they are the most intense, but other types may also occur, as discussed in Chapter 9. Molecular collisions do not respect electric dipole selection rules, and hence may cause all kinds of transitions. Their effect is to establish thermal equilibrium populations of rotational, vibrational, and electronic states, and as a consequence to affect the spectra indirectly (because the appearance of a spectrum depends on the populations of the initial states in the sample).

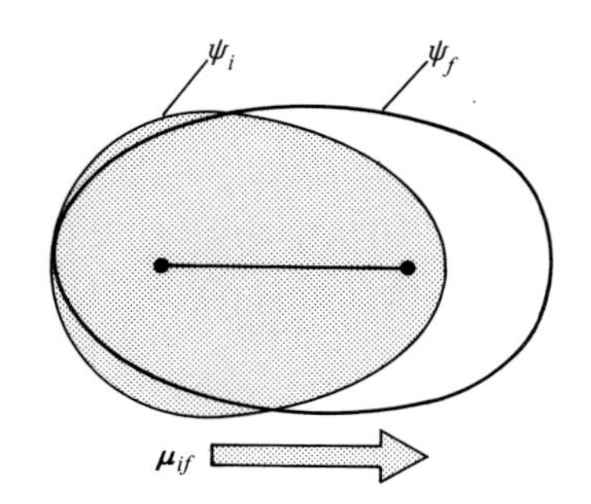

Fig. 11.1. The transition dipole for $\psi_i \rightarrow \psi_f$.

A special class of transitions gives rise to *Raman spectra*. The

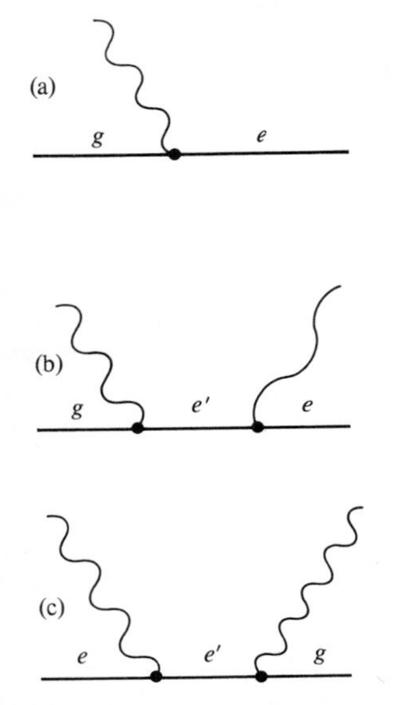

Fig. 11.2. Feynman diagrams for (a) absorption, (b) Stokes-Raman scattering, (c) anti-Stokes-Raman scattering (g is a state lower in energy than e in each case).

process involves the *inelastic scattering* of a photon. When a photon collides with a molecule and is scattered it may lose some of its energy to the molecule, Fig. 11.2. Since the molecule may accept energy only in quantized amounts, the photons in the scattered radiation differ in frequency to extents that depend on how much energy they have lost. The scattered radiation consists of a series of components on the low-frequency side of the incident radiation, and is seen as the *Stokes lines* in the spectrum. If the molecule is already in an excited state before the collision it may transfer some energy to the photon, which therefore emerges from the encounter with a higher frequency. These components appear in the spectrum as the *anti-Stokes lines.*

The classical interpretation of the observation of several frequency components in the scattered spectrum must be that the molecule has an electric dipole that oscillates with a superposition of frequencies. We can see by the following argument that under some circumstances the natural motion of the molecule (e.g. its rotation) 'beats' with the incident frequencies, Fig. 11.3, and gives rise to sum and difference frequencies in the dipole moment; these generate the corresponding radiation, and are detected as Stokes and anti-Stokes lines. When a molecule is exposed to an electric field it is *polarized* and acquires a dipole moment as a result of the distortion. We can write $\mu=\alpha E$, where the constant of proportionality is the *polarizability* (of which much more in Chapter 13). The polarizability may change periodically with time if the molecule is rotating or vibrating, and so we may write $\alpha=\alpha_0+\Delta\alpha\cos\omega_{\text{int}}t$ where α_0 is the mean polarizability and ω_{int} is some internal frequency (e.g. a vibrational frequency) of the molecule. Since the electric field has the time-dependence $E_0\cos\omega t$, the induced dipole moment of the molecule is

$$\begin{aligned}\mu(t)&=(\alpha_0+\Delta\alpha\cos\omega_{\text{int}}t)E_0\cos\omega t\\&=\alpha_0E_0\cos\omega t+\tfrac{1}{2}\Delta\alpha E_0\{\cos(\omega+\omega_{\text{int}})t+\cos(\omega-\omega_{\text{int}})t\},\end{aligned}\qquad(11.1.1)$$

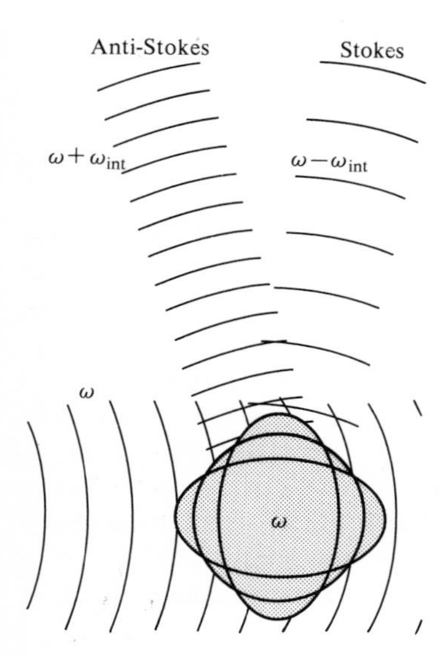

Fig. 11.3. The classical version of Raman scattering.

and so it has a time-dependence that can be resolved into three components. There is one component at the incident frequency, which therefore generates light of unshifted frequency and gives rise to the *Rayleigh radiation* in the scattered light. There are also two other components at $\omega+\omega_{\text{int}}$ and $\omega-\omega_{\text{int}}$, corresponding to the anti-Stokes and the Stokes components respectively.

It is clear from this classical analysis that the Raman radiation appears only if the polarizability oscillates, for the intensities of the lines depend on $\Delta\alpha$. For rotational transitions this requires the polarizability to vary as the molecule rotates; in other words, *for rotational Raman spectra to appear, the molecular polarizability must be anisotropic. For vibrational Raman transitions to occur, the polarizability must change as the molecule swells and contracts, or as parts of it bend.* The question of Raman activity will be taken up again when we consider specific cases.

11.2 The rotation of molecules: the general case

We shall adopt the strategy of deciding on the energy levels that molecules may possess, and then relating the levels to the spectra by deciding what transitions are allowed by the selection rules.

In the case of the rotation of molecules it is a simple matter to arrive at expressions for their energy levels. In classical physics the kinetic energy of rotation of a body of moment of inertia I_{qq} about an axis q with an angular velocity ω_q is $\frac{1}{2}I_{qq}\omega_q^2$. The total kinetic energy of a body able to rotate in three dimensions is

$$\begin{aligned} T &= \tfrac{1}{2}I_{xx}\omega_x^2 + \tfrac{1}{2}I_{yy}\omega_y^2 + \tfrac{1}{2}I_{zz}\omega_z^2 \\ &= J_x^2/2I_{xx} + J_y^2/2I_{yy} + J_z^2/2I_{zz} \end{aligned} \tag{11.2.1}$$

where we have introduced the classical expressions for the components of the angular momentum, $J_q = I_{qq}\omega_q$. The hamiltonian for molecular rotation therefore is the same expression, but with the angular momentum components interpreted as operators (there is no potential energy term).

Consider first a *symmetric top*, a molecule having an axis of at least three-fold rotation. Examples include NH_3 and CH_3Cl. Two of the moments of inertia are equal, and different from the third; therefore we write $I_{xx} = I_{yy} = I_\perp$ and $I_{zz} = I_\parallel$. The hamiltonian is then

$$H = (1/2I_\perp)(J_x^2 + J_y^2) + (1/2I_\parallel)J_z^2.$$

It can be expressed in terms of the total magnitude of the angular momentum $J^2 = J_x^2 + J_y^2 + J_z^2$ as

$$H = (1/2I_\perp)J^2 + \{(1/2I_\parallel) - (1/2I_\perp)\}J_z^2 \tag{11.2.2}$$

and so we can relate the eigenvalues of H to the eigenvalues of the angular momentum operators as established in Chapter 6. We know that the square of the magnitude has the values $J(J+1)\hbar^2$, $J = 0, 1, 2, \ldots$, and the z-component (in this case the component of angular momentum about the *figure axis*, the molecule's principal axis) has the values $K\hbar$, $K = J, J-1, \ldots, -J$ (the quantum number K is used in place of M to denote the component on an *internal* axis of the molecule). Therefore the energy levels of the symmetric top are

$$E = (\hbar^2/2I_\perp)J(J+1) + \{(\hbar^2/2I_\parallel) - (\hbar^2/2I_\perp)\}K^2. \tag{11.2.3}$$

It is conventional, and it will turn out to be convenient, to express the energies in wavenumbers. Therefore we introduce the

Rotation constants: $A = \hbar/4\pi cI_\parallel$, $\quad B = \hbar/4\pi cI_\perp$, (11.2.4)

and obtain

$$E/hc = BJ(J+1) + (A-B)K^2. \tag{11.2.5}$$

There is a hidden degeneracy. K denotes the component of the angular momentum about the figure axis of the molecule. The component of angular momentum about the laboratory Z-axis may also be specified (its operator J_Z commutes with J_z and J^2) and, like any component, is allowed to have the values $M_J\hbar$, $M_J = J, J-1, \ldots, -J$. The vector diagram representing the state of rotation of a cylindrical molecule is therefore as shown in Fig. 11.4. Note that when $K \approx J$, Fig. 11.4(b), the molecule is rotating fast around its figure axis and only slowly about a perpendicular axis, and the vector $\mathbf{J}$ is lying somewhere (not precessing: there are no external fields present yet) on its cone corresponding to the value of M_J. When $K=0$ there is no rotation around the figure axis, and so the entire angular momentum is end-over-end rotation, Fig. 11.4(a). In the absence of an external field all the M_J states have the same rotational energy (for a given J, K) because the orientation of the angular momentum in space is immaterial. Hence the full specification of the rotational energy and state is as follows:

Symmetric tops

$$\text{State: } |JKM_J\rangle \begin{cases} J = 0, 1, 2, \ldots \\ K = J, J-1, \ldots, -J; \quad 2J+1 \text{ values} \\ M_J = J, J-1, \ldots, -J; \quad 2J+1 \text{ values.} \end{cases}$$

$$\text{Energy: } E_{JKM_J}/hc = BJ(J+1) + (A-B)K^2. \tag{11.2.6}$$

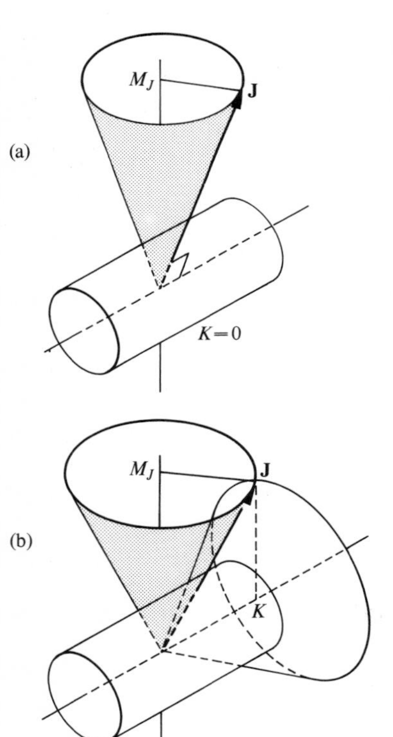

Fig. 11.4. The rotation of a symmetric top: (a) $K=0$, no rotation about the figure axis; (b) $K \approx J$, most rotation about the figure axis.

Since E_{JKM_J} is independent of the value of M_J, an energy level with specified J, K values is $(2J+1)$-fold degenerate.

The dependence of the energies on the value of K depends on the relative sizes of A and B. A *prolate top* has $I_\| < I_\perp$ (hence $A > B$) and is a cigar-shaped species. An *oblate top* has $I_\| > I_\perp$ (hence $A < B$), and is disc-shaped. The magnitude of K determines how much of the angular momentum arises from motion about the figure axis, and therefore how much of the energy depends on the moment of inertia about that axis. When $K=0$ the motion is entirely about the perpendicular axis, and the energy expression depends only on $I_\perp$ When $K \approx J$ the molecule is rotating principally about its figure axis, and now the energies are determined principally by $I_\|$ (when $J=K$, $E/hc = AJ^2 + BJ$, which is dominated by AJ^2 when J is large). All this is in accord with the physical picture of rotation. Furthermore, since the energy depends on K as K^2 we also have the physically expected result that the direction of rotation about the figure axis is irrelevant. The energy levels and degeneracies of the symmetric top are illustrated in Fig. 11.5.

There are two important special cases of the symmetric top. One is when $A = B$: such molecules are called *spherical tops*, and include CH_4 and SF_6. In group theoretical terms, spherical tops are molecules belonging to one of the cubic groups (e.g. regular tetrahedra and octahedra). When $A = B$ there is an additional degeneracy because the energy levels are independent of K. From eqn (11.2.6) with $A = B$ we have

> *Spherical tops*
>
> *State*: $|JKM_J\rangle$, J, K, M_J as above
>
> *Energy*: $E_{JKM_J}/hc = BJ(J+1)$. (11.2.7)

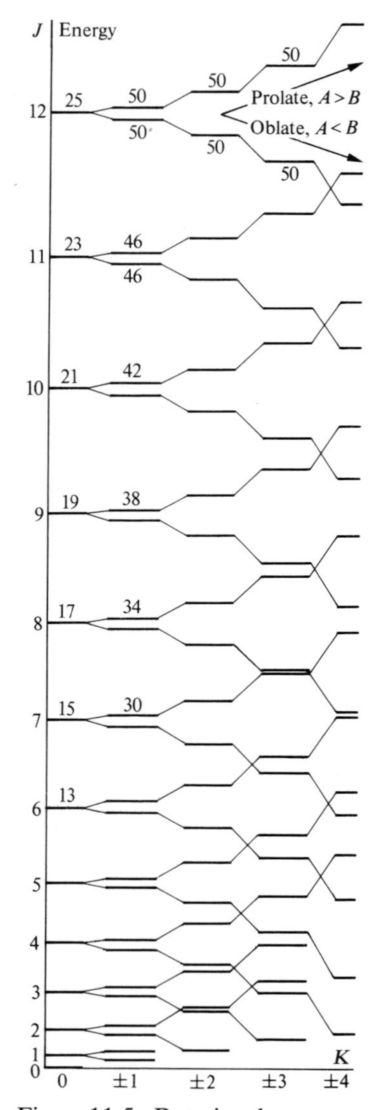

Fig. 11.5. Rotational energy levels of a symmetric top (the numbers on the levels denote the degeneracies).

Since K and M_J may each take $2J+1$ values for a given value of J, the energy levels are $(2J+1)^2$-fold degenerate, Fig. 11.6. Physically this K-degeneracy arises because it is immaterial about which internal axis a spherical top rotates.

The other special case of a symmetric top is the *linear rotor*, of which examples are CO_2 and all the diatomic molecules. A linear rotor cannot rotate about its figure axis, and so the angular momentum vector is necessarily perpendicular. Consider, for example, a diatomic molecule. The location of its two atoms can be specified completely by quoting $2\times3=6$ coordinates. Instead of using the cartesian coordinates of each atom, an equivalent procedure is to specify the location of the centre of mass (3 coordinates), the separation of the atoms (1), and the orientation of their line of centres (2). Rotation corresponds to the change in this orientational degree of freedom, and hence is a motion perpendicular to the axis, Fig. 11.7(a). (In non-linear molecules, in contrast, the orientation of the molecule is fully specified only if a third angle is quoted, Fig. 11.7(b); hence in non-linear molecules there are three degrees of rotational freedom.) We are still able to use the symmetric top energy level expression, but now we have to impose the constraint that the angular momentum is necessarily perpendicular to the axis; that is, $K\equiv0$. The specification is therefore

> *Linear rotor*
>
> *State*: $|JM_J\rangle, \begin{cases} J=0,1,2,\ldots, \\ M_J=J, J-1, \ldots, -J \end{cases}$
>
> *Energy*: $E_{JM_J}/hc = BJ(J+1)$. (11.2.8)

Since the energy is still independent of M_J, each level is $(2J+1)$-fold degenerate, Fig. 11.8. Note that while the energy levels of linear rotors and spherical tops are given by the same expressions, their degeneracies are different, and for a given value of J a spherical top has many more rotational states† (compare Figs. 11.6 and 11.8).

The case of *asymmetric tops*, when all three moments of inertia are

† The difference of degeneracy is the quantum mechanical basis of the difference between the rotational heat capacities of linear and non-linear molecules. Classical equipartition theory gives R for the former and $(3/2)R$ for the latter. Heat capacities depend on the availability of thermally accessible states: there are many more in spherical tops than in linear molecules because of the degeneracy differences; hence spherical tops have a greater heat capacity even though they have the same expression for the energy levels.

different, is too difficult to treat by elementary methods, and we shall not consider it.

11.3 The rotational selection rules

The intensities of rotational transitions depend on the magnitudes of the transition dipoles between the states. Consider the case of a linear molecule in an electronic state labelled ε and in a rotational state $|JM_J\rangle$. The transition dipole moment for an electric dipole transition from this state to a different rotational state of the same electronic state is $\langle \varepsilon J'M_J'|\,\boldsymbol{\mu}\,|\varepsilon JM_J\rangle$, where $\boldsymbol{\mu}$ is the electric dipole moment operator. According to the Born–Oppenheimer approximation we can separate the rotational motion of the molecule from the motion of the electrons themselves, and so write the wavefunction as $\psi = \psi^{\mathrm{el}}\psi^{\mathrm{rot}}$. The transition matrix element therefore factorizes into an integral over the electronic wavefunction and an integral over the orientation of the molecule:

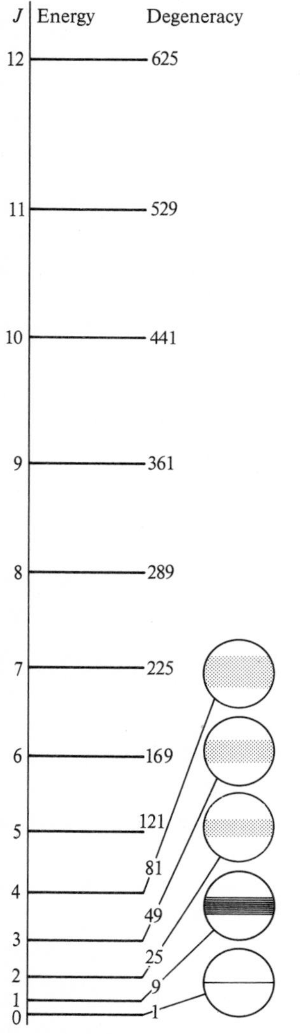

Fig. 11.6. Rotational energy levels and degeneracies of a spherical top.

$$\langle \varepsilon J'M'|\,\boldsymbol{\mu}\,|\varepsilon JM\rangle = \int \psi_\varepsilon^{\mathrm{el}*}\psi_{J'M_J'}^{\mathrm{rot}*}\boldsymbol{\mu}\psi_\varepsilon^{\mathrm{el}}\psi_{JM_J}^{\mathrm{rot}}\,\mathrm{d}\tau_{\mathrm{el}}\,\mathrm{d}\tau_{\mathrm{rot}}.$$

The *permanent electric dipole moment* of the molecule in some electronic state ε is the expectation value of the operator $\boldsymbol{\mu}$ over the wavefunctions of the state; but this is exactly the integral that appears in the last expression. Therefore, denoting the permanent moment as $\boldsymbol{\mu}_0$, where $\boldsymbol{\mu}_0 = \int \psi_\varepsilon^{\mathrm{el}*}\boldsymbol{\mu}\psi_\varepsilon^{\mathrm{el}}\,\mathrm{d}\tau_{\mathrm{el}}$, the transition matrix element is seen to be the matrix element of the permanent dipole moment between the two rotational states, $\langle J'M_J'|\,\boldsymbol{\mu}_0\,|JM_J\rangle$. The first conclusion, therefore, is that the necessary condition for the appearance of a pure electric dipole rotational transition is that *the molecule has a non-vanishing permanent electric dipole moment.* This is called a *gross selection rule.*

In order to find the specific values of the quantum numbers of the states between which transitions may occur, that is, in order to find the *specific selection rules*, we have to examine the conditions for the integral $\langle J'M_J'|\,\boldsymbol{\mu}_0\,|JM_J\rangle$ to be non-vanishing. In the case of a linear molecule the rotational wavefunctions are the spherical harmonics $Y_{JM_J}(\theta, \phi)$, where θ, ϕ is the orientation of the molecular axis. The permanent electric dipole moment is a vector with components μ_{0X}, μ_{0Y}, and μ_{0Z} in the laboratory frame, Fig. 11.9. We can establish the conditions for the non-vanishing of the integral in two ways. One consists of examining the integrals $\int Y^*_{J'M'}(\theta, \phi)\mu_{0X}Y_{JM}(\theta, \phi)\,\mathrm{d}\tau_{\mathrm{rot}}$ with $\mu_{0X} = \mu \sin\theta \cos\phi$, and likewise for the Y and Z components (and using $\mathrm{d}\tau_{\mathrm{rot}} = \sin\theta\,\mathrm{d}\theta\,\mathrm{d}\phi$, the area element in spherical polar coordinates); this leads to the result that $\Delta J = \pm 1$, $\Delta M_J = 0, \pm 1$. The other way is to use group theory, using the fact that the spherical harmonics are a basis for irreps $\Gamma^{(J)}$ of the full rotation group and a vector is a basis for $\Gamma^{(1)}$. The direct product $\Gamma^{(J')} \times \Gamma^{(1)} \times \Gamma^{(J)}$ contains the totally symmetric representation only if $J' = J, J \pm 1$. The case $J' = J$ can be ruled out by the same parity argument as we used in Chapter 8. (The

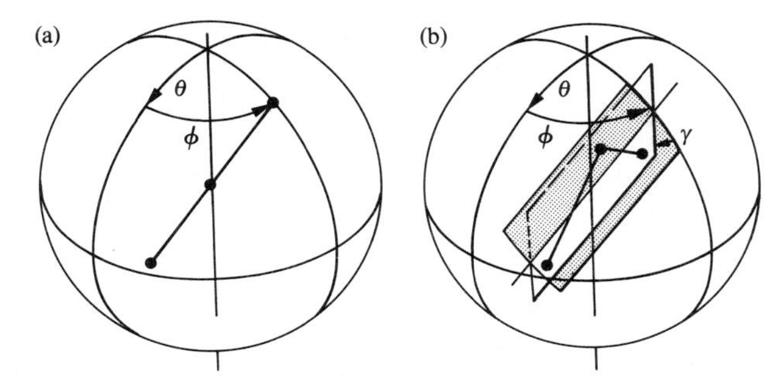

Fig. 11.7. (a) Two orientation angles are required to specify the orientation of a linear molecule, but (b) three are required for a non-linear molecule.

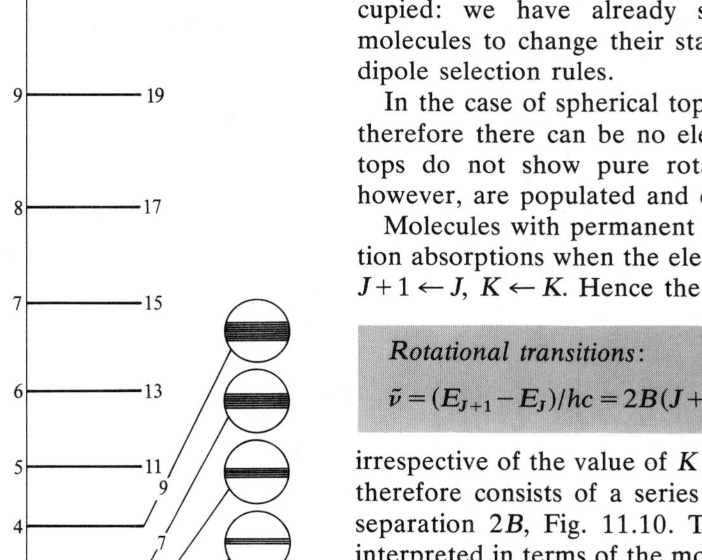

Fig. 11.8. Rotational energy levels and degeneracies of a linear rotor.

parity of Y_{JM_J} is $(-1)^J$, and since the parity of the electric dipole moment is -1, the parity overall is $(-1)^{J'+1+J}$ which is odd if $J'=J$; hence the integral is then zero.)

In the case of symmetric top molecules other than linear rotors, the question of the change in the value of K during an electronic dipole transition must be considered. Since in a symmetric top the electric dipole moment must lie parallel to the molecular axis, there is no component perpendicular to the axis. Hence an electric dipole transition cannot accelerate the rotation of the molecule about that axis: there is no 'handle' for the field to grip. The selection rule is therefore $\Delta K=0$. This does not mean that in a sample the K-states are unoccupied: we have already stressed that collisions can cause the molecules to change their states of rotation regardless of the electric dipole selection rules.

In the case of spherical tops there is no permanent dipole moment; therefore there can be no electric dipole transitions. Hence spherical tops do not show pure rotational spectra. Their rotational states, however, are populated and change as a result of collisions.

Molecules with permanent electric dipole moments show pure rotation absorptions when the electromagnetic field induces the absorption $J+1 \leftarrow J$, $K \leftarrow K$. Hence the wavenumbers of the absorptions are

Rotational transitions:

$$\tilde{\nu}=(E_{J+1}-E_J)/hc=2B(J+1), \qquad J=0,1,2,\ldots \qquad (11.3.1)$$

irrespective of the value of K and M_J in the initial state. The spectrum therefore consists of a series of equally spaced absorption lines with separation $2B$, Fig. 11.10. This separation, when measured, can be interpreted in terms of the moment of inertia $I_{\perp}$, and hence in terms of the dimensions of the molecule. This is illustrated in the following *Example*.

Example. The microwave spectrum of $^{14}NH_3$ gave a series of lines at wavenumbers $\tilde{\nu}/cm^{-1}=19.95, 39.91, 59.86, \ldots$. Deduce the moment of inertia of the molecule perpendicular to its figure axis, and infer what you can about its shape. $^{14}ND_3$ gave a series of absorptions at $\tilde{\nu}/cm^{-1}=10.32$, 20.63,

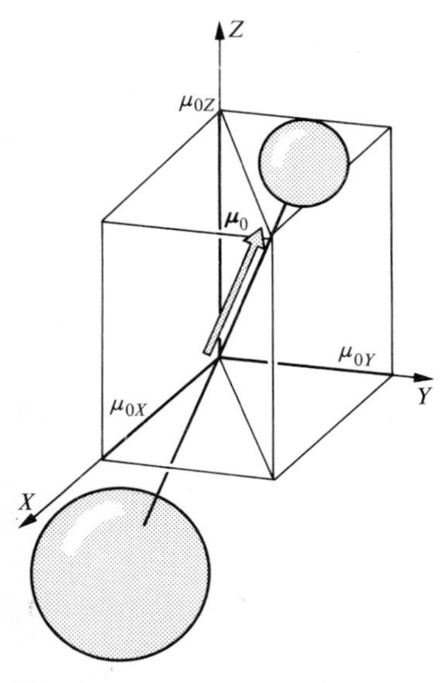

Fig. 11.9. The components of a molecular dipole in the laboratory frame.

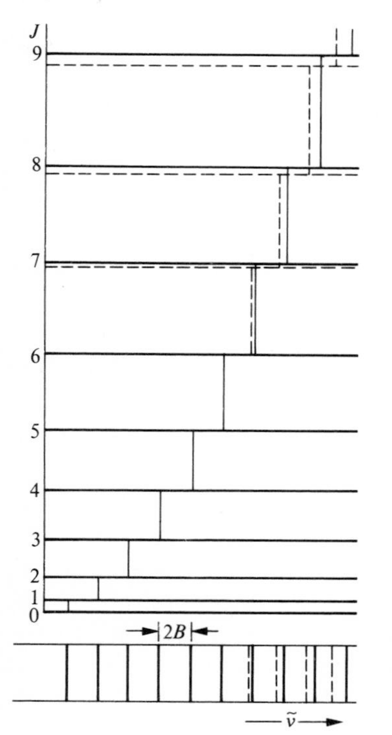

Fig. 11.10. Transitions contributing to a pure rotation spectrum; the broken lines denote the effect of centrifugal distortion.

30.95, ... Combine the two pieces of information and obtain the bond angle and bond length.

- *Method.* Ammonia is a symmetric top, and the transitions are given by eqn (11.3.1). Hence find B for the two isotropic species. Use eqn (11.2.40) to find $I_\perp$. A pyramidal AB_3 molecule has a moment of inertia $I_\perp = m_B R^2(1-\cos\theta)+(m_A m_B/m)R^2(1+2\cos\theta)$, where m_A and m_B are the masses of the respective atoms, $m = m_A + 3m_B$, R the A—B bond length, and Θ the BAB angle. The value of $I_\perp$ alone is insufficient to determine R and Θ, but if we assume the geometry of the molecule is unaffected by substitution the two values $I_\perp(NH_3)$ and $I_\perp(ND_3)$ can be interpreted in terms of Θ and R.
- *Answer.* Use $m(^1H) = 1.6735\times10^{-27}$ kg, $m(^2H) = 3.3445\times10^{-27}$ kg, and $m(^{14}N) = 2.3253\times10^{-26}$ kg. From the data, identifying the successive lines as arising from $J = 0, 1, 2, \ldots$ we find $B(NH_3) = 9.977\ \text{cm}^{-1}$, $B(ND_3) = 5.158\ \text{cm}^{-1}$ Hence $I_\perp(NH_3) = 2.806\times10^{-47}\ \text{kg m}^2$, $I_\perp(ND_3) = 5.427\times10^{-47}\ \text{kg m}^2$. The two equations for $I_\perp$ are then

$$2.806\times10^{-47} = 1.674\times10^{-27}(R/\text{m})^2(1-\cos\theta)+1.376\times10^{-27}(R/\text{m})^2(1+2\cos\theta)$$

$$5.427\times10^{-47} = 3.345\times10^{-27}(R/\text{m})^2(1-\cos\theta)+2.336\times10^{-27}(R/\text{m})^2(1+2\cos\theta).$$

Hence $\theta = 106.7°$ and $R = 101.2$ pm.

- *Comment.* Isotopic substitution is often used to elicit molecular geometry. Careful studies have shown that molecular geometry is largely insensitive to substitution, except for small effects.

Molecules with anisotropic polarizabilities can show pure rotational Raman lines. The selection rules are now $\Delta J = \pm2$ and $\Delta J = \pm1$ (except for $K=0 \to K=0$) and $\Delta K = 0$. There are several ways of understanding why 2 appears in place of 1. In the first place the classical explanation as embodied in eqn (11.1.4), should be interpreted with $\omega_{\text{int}} = 2\omega_{\text{rot}}$. This is because the polarizability of a molecule reverts to its initial value after a rotation of π, Fig. 11.11, rather than 2π in the case of $\boldsymbol{\mu}_0$. Hence the polarizability oscillates at twice the rotational frequency. The presence of rotational angular momentum about the figure axis upsets this simple analysis and allows $\Delta J = \pm1$. The more formal procedure is to use the fact that the anisotropy of α transforms as a *second rank tensor* (i.e. as the components $XX, XY, \ldots, ZZ$ as distinct from the first rank tensor behaviour of the dipole moment with its X-, Y-, and Z-components). In place of $\Gamma^{(1)}$ in the transition element $\Gamma^{(J')}\times\Gamma^{(1)}\times\Gamma^{(J)}$ we must have $\Gamma^{(2)}$. Hence $J' = J\pm2, J\pm1, 0$. It follows that for a linear molecule ($K\equiv0$) Raman lines appear at

$$\textit{Anti-Stokes lines } (\Delta J=-2):\ \tilde{\nu} = \tilde{\nu}_0 + 4B(J-\tfrac{1}{2}), \quad J = 2, 3, \ldots \tag{11.3.2a}$$

$$\textit{Stokes lines } (\Delta J=2):\ \tilde{\nu} = \tilde{\nu}_0 - 4B(J+\tfrac{3}{2}), \quad J = 0, 1, \ldots \tag{11.3.2b}$$

where $\tilde{\nu}_0$ is the wavenumber of the incident light, Fig. 11.12.

There are two major complicating features. The first is that molecules are not perfectly rigid, and may deform under the stress of rotation. The nuclei respond to the centrifugal force due to the rotation and bonds are extended and angles change. This increases the moment of inertia, and reduces the rotational constant. As a result, the separation between the lines diminishes, Fig. 11.10. The effect is normally expressed by writing the energy levels as

$$E/hc = BJ(J+1) - DJ^2(J+1)^2 - \ldots \tag{11.3.2}$$

where D is the *centrifugal distortion constant.* It can be related to the rigidity of the bond and therefore to the shape of the molecular potential energy curve.

The second major complicating feature concerns the intensities of the lines and the effect of nuclear spin. This is so important that it deserves a section to itself.

11.4 Nuclear statistics

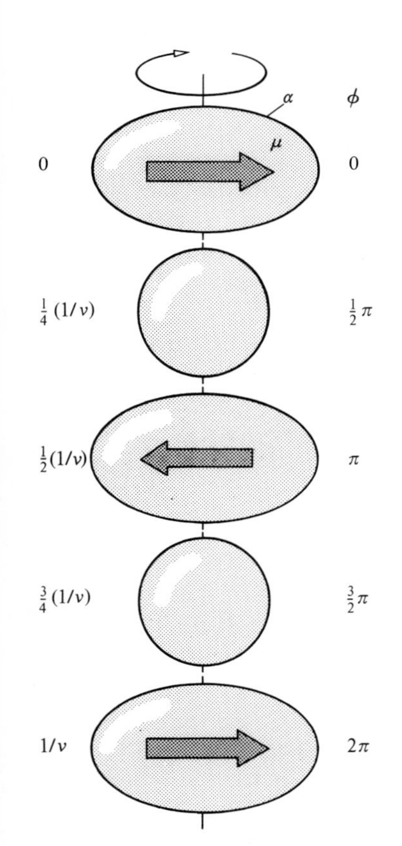

Fig. 11.11. The polarizability oscillates at twice the frequency of the dipole moment as the molecule rotates.

Consider the Raman spectrum expected from molecular hydrogen. The intensities of the lines depend on the number of molecules present in the sample and having the appropriate initial J value. At first glance we might expect the populations to follow a simple Boltzmann distribution, the number present with energy $hcBJ(J+1)$ being proportional to the product of the degeneracy of the level, $(2J+1)$, and the exponential factor $e^{-hcBJ(J+1)/kT}$. This suggests that the populations should depend on J as indicated in Fig. 11.13, the maximum population arising at $J \approx \frac{1}{2}\{(2kT/hcB)^{\frac{1}{2}} - 1\}$, and the intensitities passing through a maximum at about the same value. (The intensities do not follow the populations exactly because the transition moment depends on J in a complicated way, but the populations do account for the broad features of the intensities.) In contrast to this prediction, when the spectrum of hydrogen is inspected it is seen to consist of lines that *alternate* in intensity, Fig. 11.14. It turns out that in order to account for this behaviour we have to examine the influence of the *spin states of the nuclei* on the permissible rotational states of the molecule.

Rotation of a homonuclear diatomic molecule (and any symmetrical linear molecule) interchanges two identical particles. The Pauli principle (p. 230) imposes conditions on the acceptability of wavefunctions under particle label interchange, and requires fermion wavefunctions to change sign and boson wavefunctions to retain their sign. This has profound effects on the populations of rotational states, as we shall now see.

When the hydrogen molecule rotates through 180° more happens than simply a relabelling of the nuclei, Fig. 11.15. The electronic wavefunctions are also rotated, and the states of the nuclei as well as their labels change places. Therefore, in order to see how the wavefunction changes when the nuclei are relabelled, the operation P, and there is no other change, involves taking account of all these effects. This can be done as follows: watch Fig. 11.15. First, the overall wavefunction for the molecule is written $\psi = \psi^{el}\psi^{vib}\psi^{rot}\psi^{nuc}$ in an obvious notation (and within the Born–Oppenheimer approximation). When the molecule as a whole is rotated by 180° the rotational wavefunction changes sign by $(-1)^J$ (this follows from the behaviour of the spherical harmonics when θ is increased by π, Chapter 4). The electrons are now moved back into their original location first by an inversion (i_{el}) of the electronic wavefunction and then by its reflection (σ_{el}) in a plane perpendicular to the initial rotation. This joint operation produces $\sigma_{el} i_{el} \psi^{el}$ out of the electronic wavefunction. The states of the nuclei must now be brought back to their original arrangement (for instance, if one is α and the other β, the overall rotation will have interchanged both the labels and the states, Fig. 11.15. Let the operator achieving this *state permutation* be denoted p_{nuc}; then when it operates on ψ^{nuc} it produces $p_{nuc}\psi^{nuc}$. At this stage we have undone all the unwanted effects of the original molecular rotation, and the outcome is equivalent to a simple relabelling of the two nuclei: $P\psi = (\sigma_{el} i_{el} \psi^{el})\psi^{vib}(C_2\psi^{rot})(p_{nuc}\psi^{nuc})$.

The Pauli principle requires the overall wavefunction to satisfy $P\psi = -\psi$, where P is the label permutation operator, if the nuclei are identical fermions (half-integer spin particles) but $P\psi = +\psi$ if they are bosons. In H_2 the nuclei are protons, hence $P\psi = -\psi$.

The term symbol for H_2 is ${}^1\Sigma_g^+$. The g signifies that under i_{el} the electronic wavefunction changes to $i_{el}\psi^{el} = \psi^{el}$, while the + signifies that $\sigma_{el}\psi^{el} = \psi^{el}$. Hence, $\sigma_{el} i_{el} \psi^{el} = \psi^{el}$. Therefore $P\psi = (-1)^J \psi^{el}\psi^{vib}\psi^{rot} p_{nuc}\psi^{nuc}$. We therefore have to establish the effect of p_{nuc}. Since the protons are spin-$\frac{1}{2}$ particles they may be in the spin states α or β. The possible nuclear spin states of two protons are therefore similar to the α, β combinations we encountered when discussing the coupling of two electron spins (Section 6.8). That is, there is a *nuclear spin triplet* (spins parallel) and a *nuclear spin singlet* (spins antiparallel):

$$^3\psi^{nuc} = \begin{cases} \alpha_1\alpha_2 \\ \alpha_1\beta_2 + \beta_1\alpha_2, \\ \beta_1\beta_2 \end{cases} \qquad {}^1\psi^{nuc} = \alpha_1\beta_2 - \beta_1\alpha_2.$$

The former are symmetric, the latter antisymmetric ($p_{nuc}{}^1\psi^{nuc} = -{}^1\psi^{nuc}$). Therefore, if the protons are described by the triplet nuclear spin state (i.e. if their spins are parallel), $p_{nuc}\psi^{nuc} = \psi^{nuc}$ and so overall $P\psi = (-1)^J\psi$. In order for this to conform to the Pauli principle it follows that J must be odd. On the other hand, if the nuclear spins are antiparallel, they are described by the nuclear singlet spin function, and hence $p_{nuc}\psi^{nuc} = -\psi^{nuc}$. Therefore $P\psi = (-1)^{J+1}\psi$, and so, in order to conform to the Pauli principle, J must be even.

This discussion leads to the following remarkable conclusion. Molecular hydrogen consists of two types of molecule. One, in which the proton spins are parallel, is called *ortho-hydrogen*, and it can exist only in rotational states with odd values of J ($J = 1, 3, \ldots$). The other has its proton spins opposed and is called *para-hydrogen*; it can exist only in even-J rotational states ($J = 0, 2, \ldots$). Since there are three

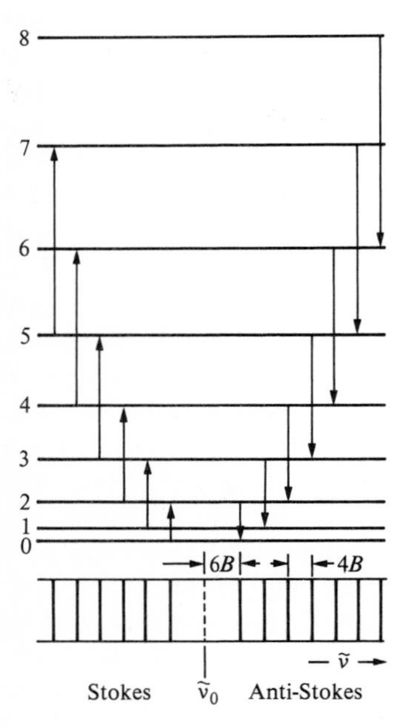

Fig. 11.12. Transitions contributing to a rotational Raman spectrum.

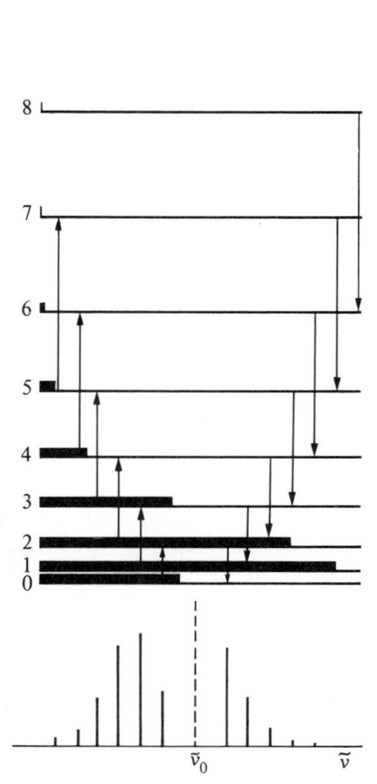

Fig. 11.13. The effect of populations on transition intensities.

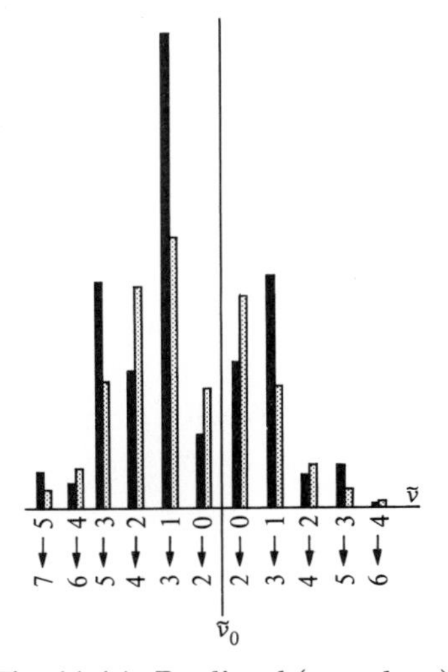

Fig. 11.14. Predicted (open bars) and observed (full lines) rotational Raman intensities for H_2.

symmetric nuclear spin states but only one antisymmetric state, we should expect *ortho*-hydrogen to be three times as abundant as *para*-hydrogen, and therefore Raman lines due to transitions starting at odd values of J should be three times as intense as those starting at even values. This accounts for the intensity variation in the spectrum as shown in Fig. 11.14.

An interesting point arises in connection with the cooling of hydrogen. The lowest energy level is the rotational state with $J=0$. In order to enter this state the proton spins must be antiparallel, but the switch from parallel to antiparallel is very slow because it involves the relative reorientation of the nuclear spins, and nuclear spins interact only very weakly (via their weak magnetic moments) with the environment. Therefore, when hydrogen is cooled it forms a mixture of *ortho*- and *para*- forms and therefore a mixture of $J=0$ and $J=1$ rotational states. The nuclear spins do reorientate very slowly (for example, if one proton experiences a slightly stronger magnetic field than its companion it precesses to new relative orientation), and so the non-equilibrium mixture slowly settles down into the $J=0$ state. The rotational energy given out heats the sample, and this causes the liquid to evaporate. It is for this reason that in the commercial production of liquid hydrogen steps are taken to speed the *ortho-para* interconversion. One method is to introduce a catalyst: the hydrogen dissociates and recombines at random, and so the nuclear spin equilibrium is attained; another is to include a paramagnetic complex or molecule (e.g. NO), which gives rise to different magnetic fields at the two nuclei, and causes them to precess at different rates, and therefore to switch from parallel to antiparallel.

Similar arguments can be applied to molecules with nuclei of other spins. For instance, if the nuclei of a homonuclear diatomic have spin I there are $(2I+1)^2$ product functions (corresponding to the four product functions when $I=\frac{1}{2}$). Of these products, $2I+1$ have $m_{I1}=m_{I2}$ and so are symmetric (these are the two states $\alpha_1\alpha_2$ and $\beta_1\beta_2$ for $I=\frac{1}{2}$). Of the remaining $(2I+1)^2-(2I+1)=2I(2I+1)$ states, half are symmetric ($|m_{I1}m'_{I2}\rangle+|m'_{I1}m_{I2}\rangle$, such as $\alpha_1\beta_2+\beta_1\alpha_2$) and half are antisymmetric. Therefore the total numbers of symmetric and antisymmeric functions are $N_+=(2I+1)(I+1)$ and $N_-=(2I+1)I$, and the ratio of types is

$$N_+/N_-=(I+1)/I. \qquad (11.4.1)$$

In the case of $^1H_2$, where $I=\frac{1}{2}$, the ratio is 3, as we have already found. In the case of $^2H_2$, where $I=1$, the ratio is 2. Note that the deuteron is a boson, and so the acceptable states of $^2H_2$ must be symmetrical under particle interchange; hence symmetrical spin states entail even-J values. The Raman spectrum of $^2H_2$ shows an alternation of intensities with even-J lines having twice the intensity of odd-J lines.

These arguments may be applied to molecules containing more than two identical nuclei (e.g. NH_3 and CH_4), but the arguments rapidly become very complicated. Nevertheless, their correct investigation is crucial to a full interpretation of spectra and to a proper calculation of the populations of the rotational states of molecules.

11.5 The vibrations of molecules

The energy of a diatomic molecule increases if the nuclei are disturbed from their equilibrium separation. When the distortion $\xi = R - R_e$ is small we can express the potential energy of the molecule as the first few terms of a Taylor expansion:

$$V(\xi) = V(0) + (dV/d\xi)_0\xi + \tfrac{1}{2}(d^2V/d\xi^2)_0\xi^2 + \left(\frac{1}{3!}\right)(d^3V/d\xi^3)_0\xi^3 + \dots \tag{11.5.1}$$

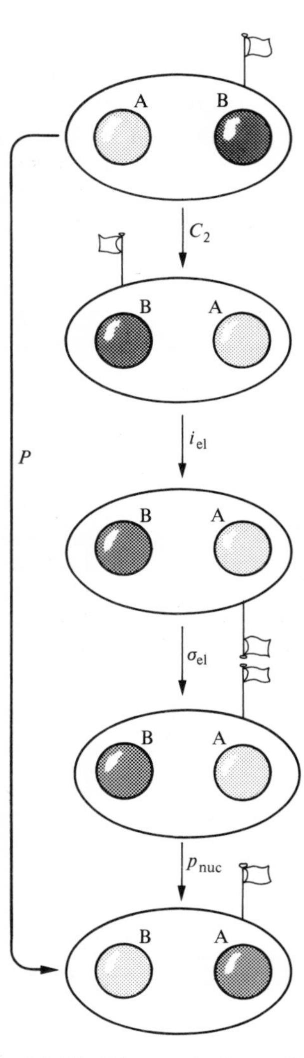

Fig. 11.15. The equivalence $P \equiv p_{nuc}\sigma_{el}i_{el}C_2$ for a homonuclear diatomic molecule.

where the subscript 0 indicates that the derivatives should be evaluated at the equilibrium bond length, when $\xi = 0$. We are not interested in the absolute value of the energy of the molecule, and so $V(0)$ may be set equal to zero. The first derivative $dV/d\xi$ is zero at the minimum. The contribution of the third derivative is proportional to ξ^3 and for small displacements may be neglected so long as $(d^3V/d\xi^3)_0$ is not enormous. This leaves only the second derivative. Therefore the potential energy for displacements from equilibrium may be written

$$V(\xi) = \tfrac{1}{2}k\xi^2, \qquad k = (d^2V/d\xi^2)_0. \tag{11.5.2}$$

A parabolic potential energy is characteristic of a harmonic oscillator, and so we may immediately conclude that the energy levels and wavefunctions of the vibrating molecule are the same as those calculated for the harmonic oscillator in Chapter 3 with the displacement ξ in place of x. What, though, is the value of the mass to use in the expression for the frequency $\omega = (k/m)^{\frac{1}{2}}$? It is clearly not the total mass of the molecule. In a diatomic molecule both atoms move during the vibration, but when one atom is much lighter than the other it dominates because it moves most (think of a 1 g mass attached by a spring to a brick). In Appendix 6 we show that the appropriate quantity is the

Reduced mass: $\mu = m_a m_b/(m_a + m_b)$, or $1/\mu = 1/m_a + 1/m_b$. (11.5.3)

Therefore the energy levels available to a diatomic molecule composed of atoms of masses m_a and m_b are

$$E_v = (v + \tfrac{1}{2})\hbar\omega, \quad \omega = (k/\mu)^{\frac{1}{2}}; \qquad v = 0, 1, 2, \dots. \tag{11.5.4}$$

The wavefunctions are those of the harmonic oscillator and are set out in Chapter 3.

The harmonic oscillator energy levels lie in a uniform ladder with rungs separated by $\hbar\omega$. In order to predict the form of the spectrum we need the electric dipole selection rules. The transition dipole moment is $\langle\varepsilon, v'|\boldsymbol{\mu}|\varepsilon, v\rangle$ because (at this stage) we are interested only in transitions between vibrational states of the same electronic state. The integration over the electronic coordinates can be carried out as before (on the basis of the Born–Oppenheimer approximation the wavefunction is factorizable into $\psi^{\text{el}}\psi^{\text{vib}}$), and so the transition moment is $\langle v'|\boldsymbol{\mu}|v\rangle$, where $\boldsymbol{\mu}$ is the dipole moment of the molecule when it is in the electronic state ε and has some bond length R. The dipole moment depends on the bond length, and so we can write

$$\boldsymbol{\mu}=\boldsymbol{\mu}_0+(\mathrm{d}\boldsymbol{\mu}/\mathrm{d}\xi)_0\xi+\ldots \tag{11.5.5}$$

where $\boldsymbol{\mu}_0$ is the dipole moment when the displacement is zero. If the dipole arose from charges fixed to the nuclei (e.g. the molecule consisted of two point charges) then the dipole moment would depend linearly on the displacement and the higher terms in the series would be zero, Fig. 11.16. In practice the charges redistribute themselves as the bond length changes, and so the dipole moment is not linear in ξ, but to a reasonable first approximation (which we refine later) we may ignore the higher order terms in the series. Then the transition matrix element becomes

$$\langle\varepsilon, v'|\boldsymbol{\mu}|\varepsilon v\rangle=\boldsymbol{\mu}_0\langle v'\mid v\rangle+(\mathrm{d}\boldsymbol{\mu}/\mathrm{d}\xi)_0\langle v'|\xi|v\rangle+\ldots. \tag{11.5.6}$$

The first term is zero on account of the orthogonality of the states. The second term is non-zero so long as the dipole moment depends on the displacement (i.e. $(\mathrm{d}\boldsymbol{\mu}/\mathrm{d}\xi)_0\neq 0$). Therefore, we have the *gross selection rule* that in order for a molecule to show vibrational transitions, *its dipole moment must change with displacement.* This means that homonuclear diatomic molecules do not show a pure vibrational spectrum, while heteronuclear diatomics do. The *specific selection rule* can be established by evaluating the integral $\langle v'|\xi|v\rangle$ using the form of the vibrational wavefunctions set out in Table 3.1. The integral vanishes unless $v'=v\pm 1$. A cleverer way of establishing the same result is shown in the following *Example.*

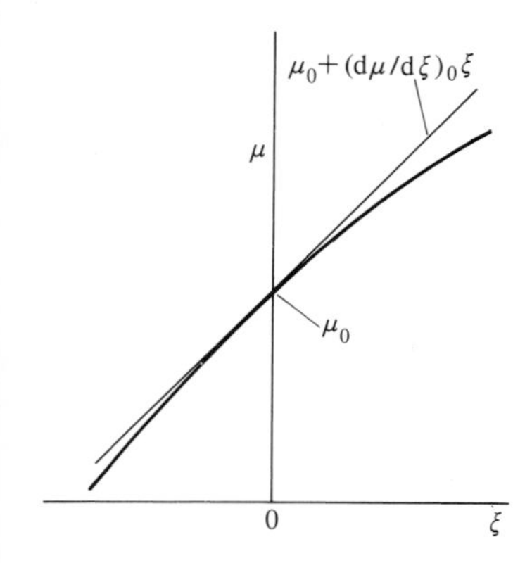

Fig. 11.16. The dependence of electric dipole moment on bond length.

Example. Use the creation and annihilation operators as described in Appendix 4 to establish the selection rules for the electric dipole transitions of a harmonic oscillator.

- *Method.* The displacement from equilibrium ξ ($\hat{=}\,x$) can be expressed in terms of a and a^+ using eqn (A4.2). The matrix elements of a and a^+ are indicated by the discussion following eqn (A4.6): $a|v\rangle\propto|v-1\rangle$; $a^+|v\rangle\propto|v+1\rangle$. Hence $\Delta v=\pm 1$ is established trivially. The precise magnitude of the transition dipole is obtained by finding the values of the matrix elements $\langle v+1|a^+|v\rangle$ and $\langle v-1|a|v\rangle$. In order to do so, find values of the elements that reproduce the commutation relations $[a, a^+]=-2$.

- *Answer.* Since $\xi=\frac{1}{2}(\hbar/m\omega)(a-a^+)$, we have (using eqn (11.5.6) and specializing to a one-dimensional system):

$$\langle\varepsilon, v'|\mu|\varepsilon, v\rangle=\tfrac{1}{2}(\hbar/m\omega)^{\frac{1}{2}}(\mathrm{d}\mu/\mathrm{d}\xi)_0\langle v'|a-a^+|v\rangle$$
$$=0 \text{ unless } v'=v\pm 1;\text{ hence } \Delta v=\pm 1.$$

The rule $[a, a^+] = -2$ is reproduced by setting

$$a\,|v\rangle = \{2v\}^{\frac{1}{2}}\,|v-1\rangle; \qquad a^+\,|v\rangle = -\{2(v+1)\}^{\frac{1}{2}}\,|v+1\rangle$$

because

$$\begin{aligned}[a, a^+]\,|v\rangle &= aa^+\,|v\rangle - a^+a\,|v\rangle \\ &= -\{2(v+1)\}^{\frac{1}{2}}a\,|v+1\rangle - \{2v\}^{\frac{1}{2}}a^+\,|v-1\rangle \\ &= -\{2(v+1)\}^{\frac{1}{2}}\{2(v+1)\}^{\frac{1}{2}}\,|v\rangle + \{2v\}^{\frac{1}{2}}\{2v\}^{\frac{1}{2}}\,|v\rangle \\ &= -2\,|v\rangle, \text{ in accord with } [a, a^+] = -2.\end{aligned}$$

Therefore $\langle \varepsilon, v+1|\, \mu\, |\varepsilon, v\rangle = \frac{1}{2}(\hbar/m\omega)^{\frac{1}{2}}\{2(v+1)\}^{\frac{1}{2}}(\mathrm{d}\mu/\mathrm{d}\xi)_0$;

$$\langle \varepsilon, v-1|\, \mu\, |\varepsilon, v\rangle = \tfrac{1}{2}(\hbar/m\omega)^{\frac{1}{2}}\{2v\}^{\frac{1}{2}}(\mathrm{d}\mu/\mathrm{d}\xi)_0.$$

• *Comment.* This technique is very readily extended to the evaluation of matrix elements of higher powers of the displacement (which occur in expressions we meet below). Note that the transition moment increases as $v^{\frac{1}{2}}$: this is another case of the populations not being the sole determinants of the spectral intensities.

From these considerations we deduce that vibrational absorption occurs at the wavenumbers

$$\tilde{\nu} = (E_{v+1} - E_v)/hc = \hbar\omega/hc = \omega/2\pi c, \qquad (11.5.7)$$

and that the spectrum should therefore consist of a single line. Knowing the reduced mass of the molecule then lets us deduce the force constant k, and hence the curvature $(\mathrm{d}^2V/\mathrm{d}\xi^2)_0$ of the molecular potential energy curve in the vicinity of the equilibrium conformation.

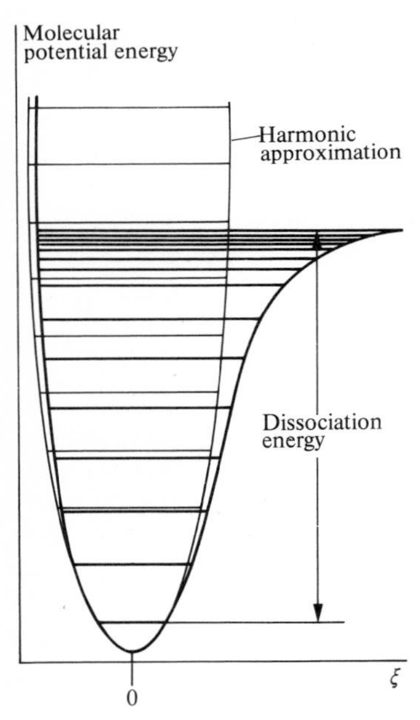

Fig. 11.17. The harmonic approximation and the actual molecular potential energy.

These expectations are not entirely fulfilled. In the first place the electric dipole moment may vary with displacement in such a way that the second and higher derivatives $\mathrm{d}^n\boldsymbol{\mu}/\mathrm{d}\xi^n$ are non-zero. This is called *electrical anharmonicity*. The dipole moment occurring in the transition matrix elements therefore has terms proportional to ξ^2, ξ^3, etc., and as a consequence transitions corresponding to $\Delta v = 2, 3$, etc., become allowed. These transitions occur as weak *overtones* or *harmonics* (the first overtone is the second harmonic, etc.) in the spectrum. In the second place, there may be true *mechanical anharmonicity*, because truncating the Taylor expansion of the molecular potential energy curve at the ξ^2 term is only an approximation, and in a real molecule the higher terms may be significant. The typical form of the potential is shown in Fig. 11.17, and as it is less confining than the parabolic potential, the energy levels can be expected to converge as the quantum number increases. Not only do the energy levels change, but the wavefunctions also change from those of a harmonic oscillator; therefore the selection rules, which are based on integrals over the wavefunctions, are also modified. Transitions with $\Delta v = \pm 2, \pm 3, \ldots$ then appear in the spectrum.

Anharmonicities are always important in high resolution spectroscopy, and even in low resolution work they are significant when the displacement of the nuclei reaches values far from equilibrium: this occurs when the vibration is excited to high quantum numbers. In pure vibrational spectra at normal temperatures only the lowest vibrational state is occupied initially because the Boltzmann expression is almost zero for higher states (because $\hbar\omega \gg kT$). Therefore the infrared spectrum normally consists of a *progression* of transitions consisting of the *fundamental* ($v = 1 \leftarrow v = 0$, normally denoted 1–0), and the *overtones* (2–0), (3–0), etc., with diminishing intensities. The energies of the progression are normally fitted to an expression of the form

$$E_v = (v + \tfrac{1}{2})\hbar\omega - (v + \tfrac{1}{2})^2\hbar\omega x + \ldots \qquad (11.5.8)$$

where the extra terms reflect the decreasing separations between neighbouring levels on account of the anharmonicity. The quantity ωx, the *anharmonicity constant*, can be related to the dissociation energy of the bond. One procedure is to suppose that the molecular potential energy curve has the form of the

$$\textit{Morse potential}: \; V(R) = D_e(1 - e^{-a\xi})^2, \qquad (11.5.9)$$

as drawn in Fig. 11.18. D_e is the depth of the potential minimum and $a = (k/2D_e)^{\frac{1}{2}}$. The Schrödinger equation can be solved exactly for this potential, the eigenvalues being

$$E_v = (v + \tfrac{1}{2})\hbar\omega - (v + \tfrac{1}{2})^2\hbar\omega x \qquad (11.5.10)$$

with $x = \hbar a^2/2\mu\omega$. Hence, if x is measured, a and therefore D_e may be determined. The *dissociation energy* is $D_e - E_0$, because the lowest state open to the molecule is in the level $v = 0$ with its zero-point energy $E_0 = \frac{1}{2}\hbar\omega - \frac{1}{4}\hbar\omega x$. Although the Morse potential has the advantages of simplicity and priority, it does not give a particularly good representation of the true curve, and the dissociation energy calculated from it is often only a poor approximation. Modern computer based methods for arriving at dissociation energies from spectral data are described in the books in *Further reading*, p. 316.

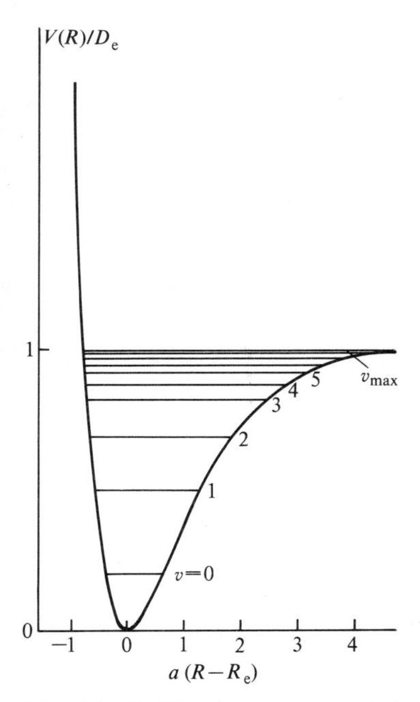

Fig. 11.18. The Morse potential energy and its eigenvalues.

11.6 Rotation–vibration spectra of diatomic molecules

The next complication in the vibrational spectra of diatomic molecules is that a vibrational transition may be accompanied by a simultaneous rotational transition in which J changes by ± 1. The total energy change in the transition then depends on the values of the rotational constant and the initial value of J. Furthermore, the magnitude of the rotational constant depends on the molecule's vibrational state (because vibrations effect its size, and so we must attach a label to B and write it B_v).

The energy of a rotating, vibrating molecule is

$$E(v, J) = (v + 1)\hbar\omega - (v + \tfrac{1}{2})^2\hbar\omega x + \ldots + hcB_vJ(J + 1) - hcD_vJ^2(J + 1)^2 + \ldots \qquad (11.6.1)$$

where both anharmonicity and centrifugal distortion effects have been included. The transitions in which $\Delta J = +1$ give rise to the lines called the *R-branch* of the spectrum. Their wavenumbers are

$$\tilde{\nu}^{R}(v, J) = \{E(v+1, J+1) - E(v, J)\}/hc$$
$$= \tilde{\nu} - 2(v+1)x\tilde{\nu} + \ldots + 2B_{v+1} + (3B_{v+1} - B_v)J$$
$$+ (B_{v+1} - B_v)J^2 + \ldots \quad (11.6.2)$$

A series of lines is obtained because in the initial state many rotational levels are thermally populated, and so J takes a wide range of values. When the rotational constants are the same in the upper and lower vibrational states ($B_{v+1} = B_v = B$) the wavenumbers of the branch lines are given by

$$\tilde{\nu}^{R}(v, J) = \tilde{\nu} + 2B(J+1), \qquad J = 0, 1, 2, \ldots \quad (11.6.3)$$

(we have also disregarded the anharmonicity effect). Hence the branch consists of a series of lines separated by $2B$ with an intensity distribution that mirrors the thermal population of the rotational states, Fig. 11.19.

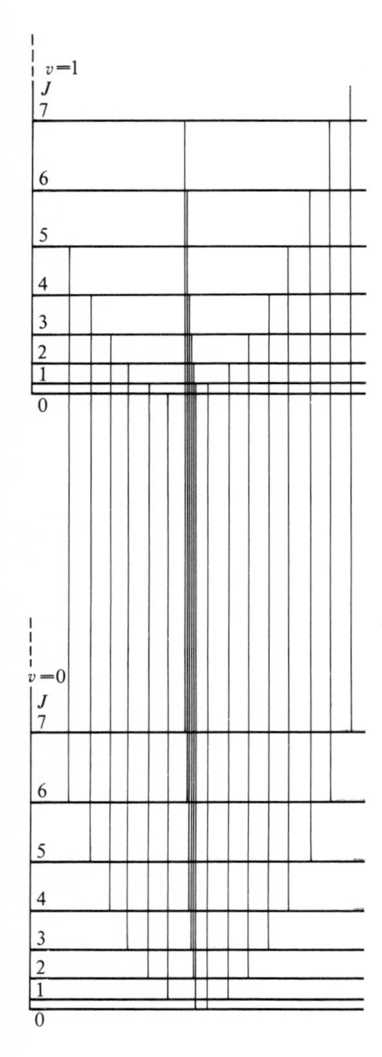

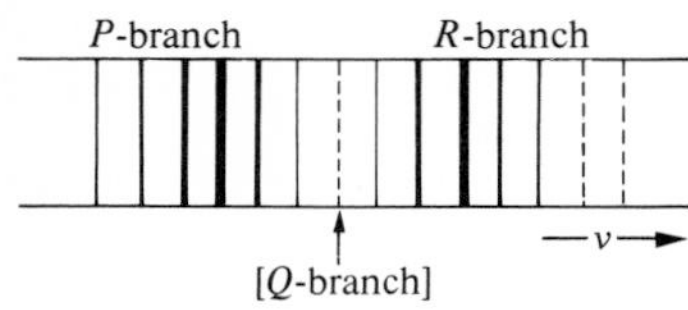

Fig. 11.19. The formation of P-, Q-, and R-branches.

When $\Delta J = -1$ the transitions constitute the *P-branch*, the wavenumbers being given by

$$\tilde{\nu}^{P}(v, J) = \tilde{\nu} - 2(v+1)x\tilde{\nu} + \ldots$$
$$- (B_{v+1} + B_v)J + (B_{v+1} - B_v)J^2 + \ldots \quad (11.6.4)$$

in the general case, and by

$$\tilde{\nu}^{P}(v, J) = \tilde{\nu} - 2BJ, \qquad J = 1, 2, \ldots \quad (11.6.5)$$

when the rotational constants are the same and we neglect anharmonicity. The branch therefore consist of an array of lines on the low-wavenumber side of the pure vibrational transition, and the intensities of the lines reflects the thermal populations of the initial rotational states.

When the rotational constants are markedly different in the two states, the spacing within the branches is no longer regular, and one of the branches may start to converge. If at high values of J the quantity $(B_{v+1} - B_v)J^2$ becomes large enough it may dominate the term linear in J and the branch may pass through a *head* and successive lines approach the origin instead of moving away from it. The effect is much more pronounced when a change of electronic state is also involved, and we shall encounter it again in Chapter 12.

The *Q-branch* consists of the lines corresponding to $\Delta J = 0$. It is allowed only when the molecule possesses angular momentum parallel to the internuclear axis (this is because the angular momentum of the molecule + field, which lies at the root of the selection rules of rotation–vibration transitions, can be satisfied by $\Delta J = 0$ for linear molecules only if they have this parallel component) and therefore a diatomic molecule may show a Q-branch only if it has orbital angular momentum about its axis ($\Lambda \neq 0$). The wavenumbers of the Q-branch lines are

$$\tilde{\nu}^{Q}(v, J) = \tilde{\nu} + (B_{v+1} - B_v)J + (B_{v+1} - B_v)J^2 + \ldots. \quad (11.6.6)$$

Since $B_{v+1} \approx B_v$ almost always, the Q-branch consists of a very narrow band of lines at the wavenumber of the vibration. What structure it has gives direct information on the difference $B_{v+1} - B_v$.

Vibrational Raman spectra of diatomic molecules can also be observed, the *gross selection rule* being the requirement that *the polarizability should change during a molecular vibration.* This means that all diatomic molecules are vibrationally Raman active. The *specific selection rule* is $\Delta v = \pm 1$, the $\Delta v = +1$ giving rise to the Stokes lines and the $\Delta v = -1$ giving rise to the anti-Stokes lines. Note that the selection rule is the same as for infrared transitions because the polarizability returns to its initial value once during each oscillation, Fig. 11.20, and so the polarizability oscillates at the same frequency as the bond ($\omega_{\text{int}} = \omega_{\text{vib}}$ in the language of eqn (11.1.4)).

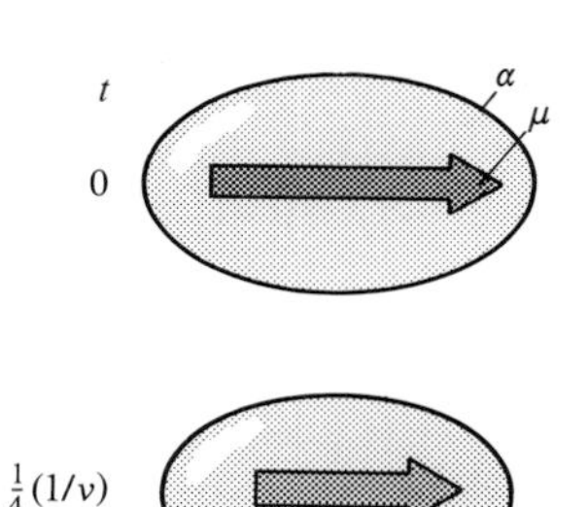

Example. Deduce the selection rule for vibrational Raman transitions, and evaluate the transition moment using the results obtained in the *Example* on p. 299.

- *Method.* Raman transitions depend on the polarizability α. Consider a one-dimensional system, and allow the polarizability to depend on displacement as $\alpha = \alpha_0 + (\mathrm{d}\alpha/\mathrm{d}\xi)_0 \xi + \ldots$. Express the matrix elements of α in terms of the annihilation and creation operator matrix elements established in the earlier *Example.*

- *Answer.* $\langle \varepsilon, v' | \alpha | \varepsilon, v \rangle = \langle v' | \alpha^{\varepsilon} | v \rangle$, where $\alpha^{\varepsilon} = \langle \varepsilon | \alpha | \varepsilon \rangle$, the polarizability in the electronic state ε. Then, with $\alpha^{\varepsilon} = \alpha_0^{\varepsilon} + (\mathrm{d}\alpha^{\varepsilon}/\mathrm{d}\xi)_0 \xi + \ldots$,

$$\langle \varepsilon, v' | \alpha | \varepsilon, v \rangle = \alpha_0^{\varepsilon} \langle v' | v \rangle + (\mathrm{d}\alpha^{\varepsilon}/\mathrm{d}\xi)_0 \langle v' | \xi | v \rangle + \ldots$$
$$= 0 \text{ unless } v' = v \text{ or } v' = v \pm 1.$$
$$\langle \varepsilon, v+1 | \alpha | \varepsilon, v \rangle = \tfrac{1}{2}(\hbar/m\omega)^{\frac{1}{2}} \{2(v+1)\}^{\frac{1}{2}} (\mathrm{d}\alpha^{\varepsilon}/\mathrm{d}\xi)_0;$$
$$\langle \varepsilon, v-1 | \alpha | \varepsilon, v \rangle = \tfrac{1}{2}(\hbar/m\omega)^{\frac{1}{2}} \{2v\}^{\frac{1}{2}} (\mathrm{d}\alpha^{\varepsilon}/\mathrm{d}\xi)_0.$$

- *Comment.* Note that the $\Delta v = \pm 1$ transitions may occur only if α^{ε} depends on the displacement ξ. As in the case of pure vibrational transitions, the transition moments increase as $v^{\frac{1}{2}}$.

In the gas phase both the Stokes and the anti-Stokes lines of the vibrational spectrum show branch structure. Now the rotational selection rule $\Delta J = \pm 2$ (and $\Delta J = \pm 1$ for $K = 0 \nrightarrow K = 0$) is applicable, and hence the lines form the *O-branch* ($\Delta J = -2$) and the *S-branch* ($\Delta J = +2$). When the selection rule $\Delta J = 0$ is admissible there is a Q-branch. The Raman spectra of homonuclear species are particularly important, for they are inactive in the infrared and microwave spectra.

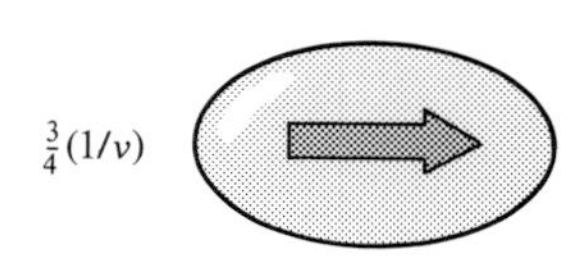

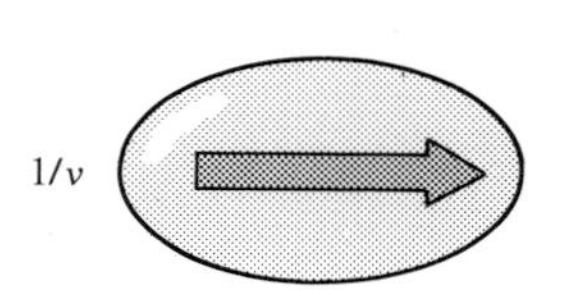

Fig. 11.20. The polarizability oscillates at the same frequency as the dipole moment as the molecule vibrates.

11.7 The vibrations of polyatomic molecules

The vibrations of polyatomics involve the entire molecule. Thus, if one bond of a triatomic molecule is vibrationally excited, the energy is rapidly transferred to the other bond through the motion of the central atom. The potential energy of a polyatomic molecule depends on all the displacements of all the atoms from their equilibrium locations. Therefore, to express the actual potential energy of the molecule in terms of all the displacements ξ_i from equilibrium we write

$$V = V(0) + \sum_i (\partial V/\partial \xi_i)_0 \xi_i + \tfrac{1}{2} \sum_{i,j} (\partial^2 V/\partial \xi_i\, \partial \xi_j)_0 \xi_i \xi_j + \ldots \qquad (11.7.1)$$

as a generalization of eqn (11.5.1), the subscript 0 denoting that the derivatives are evaluated at the equilibrium conformation. As in the case of the diatomic molecule, the constant $V(0)$ may be set equal to zero, and the first derivatives are all zero at $\xi = 0$. Therefore

$$V = \tfrac{1}{2} \sum_{i,j} k_{ij} \xi_i \xi_j, \qquad k_{ij} = (\partial^2 V/\partial \xi_i\, \partial \xi_j)_0. \qquad (11.7.2)$$

where k_{ij} is a *generalized force constant.* When there is only one distortion coordinate this expression reduces to eqn (11.5.2).

Not all the displacements ξ_i correspond to vibrations. In a polyatomic molecule there are $3N$ independent displacements of the N atoms it is built from. Three of these displacements (or three linear combinations of them) correspond to displacements of the centre of mass of the molecule. Three more correspond to displacements of the orientation of the molecule as a whole around its centre of mass (two in the case of a linear molecule) and so correspond to rotational displacements. Therefore of the $3N$ displacements overall, only $3N-6$ (or $3N-5$ for linear species) correspond to vibrational displacements, where one part of the molecule moves relative to some other part. Therefore six (or five) displacements in eqn (11.7.2) leave the potential energy of the molecule unchanged: the translations and the rotations have zero for their generalized force constants. The form of the potential in eqn (11.7.2) also reflects the way that a vibration in one part of the molecule (e.g. displacement i) may affect the vibration elsewhere (e.g. displacement j) because the potential depends on both displacements: k_{ij} links displacements and so couples together vibrations.

As a first step in the simplification of the problem we introduce the

Mass-weighted coordinates: $q_i = m_i^{\frac{1}{2}} \xi_i$, (11.7.3)

where m_i is the mass of the atom undergoing the displacement ξ_i. The potential energy of the molecule becomes,

$$V = \tfrac{1}{2} \sum_{i,j} K_{ij} q_i q_j, \qquad K_{ij} = (1/m_i m_j)^{\frac{1}{2}} k_{ij} = (\partial^2 V/\partial q_i\, \partial q_j)_0, \qquad (11.7.4)$$

the kinetic energy

$$T = \tfrac{1}{2} \sum_i m_i \dot{\xi}_i^2 = \tfrac{1}{2} \sum_i \dot{q}_i^2, \qquad (11.7.5)$$

and the total energy

$$E = \tfrac{1}{2} \sum_i \dot{q}_i^2 + \tfrac{1}{2} \sum_{i,j} K_{ij} q_i q_j. \qquad (11.7.6)$$

The difficult quantities are the cross terms ($i \neq j$) in the potential energy. The question therefore arises as to whether it is possible to find linear combinations Q_i of the mass-weighted coordinates q_i such that the total enegy can be written in the form

$$E = \tfrac{1}{2}\sum_i \dot{Q}_i^2 + \tfrac{1}{2}\sum_i \kappa_i Q_i^2 \tag{11.7.7}$$

where there are no cross terms. Linear combinations of this type, the *normal coordinates*, do exist. We can suspect that they do, because an alternative picture of the two stretching modes of a molecule like CO_2 is as the sum and difference of the displacements, Fig. 11.21. When the symmetrical mode is excited, for instance, the central atom is buffetted simultaneously on both sides, and the antisymmetric mode remains unexcited. We therefore expect the combination $q_1 - q_3$ to be one of the normal modes of CO_2. The formal procedure for finding the normal modes is described in Appendix 16. When that technique is applied to a linear triatomic AB_2 (like CO_2) we find the following expressions for the three normal modes arising from displacements parallel to the molecular axis:

$$\begin{aligned} Q_1 &= (\tfrac{1}{2})^{\frac{1}{2}}(q_1 - q_3), & \kappa_1 &= k/m_B \\ Q_2 &= (1/2M)^{\frac{1}{2}}(m_A^{\frac{1}{2}} q_1 - 2m_B^{\frac{1}{2}} q_2 + m_A^{\frac{1}{2}} q_3), & \kappa_2 &= kM/m_A m_B \\ Q_3 &= (1/M)^{\frac{1}{2}}(m_B^{\frac{1}{2}} q_1 + m_A^{\frac{1}{2}} q_2 + m_B^{\frac{1}{2}} q_3), & \kappa_3 &= 0. \end{aligned} \tag{11.7.8}$$

where $M = m_A + 2m_B$, the mass of the molecule. These normal coordinates are illustrated in Fig. 11.21. Note that for Q_3 the force constant $\kappa_3 = 0$, and so this coordinate corresponds to the displacement of the molecule as a whole; that Q_1 corresponds to the *symmetric mode* of displacement of the nuclei; and that Q_2 corresponds to the *antisymmetric mode*. As the mass of the central atom is increased relative to the masses of the outer atoms the Q_1 coordinate remains unchanged, and so does its force constant. On the other hand, the coordinate Q_2 approaches the form $(1/2)^{\frac{1}{2}}(q_1 + q_3)$ in which the central atom makes no contribution to the vibration (but it dominates the translation) and its force constant κ_2 changes to $2k/m_B$. This is the same as if the molecule consisted of two small masses attached by springs to opposite sides of a brick. The important point to note is that the relative masses of the atoms govern both the forms of the normal coordinates and, through the values of their effective force constants, their frequencies.

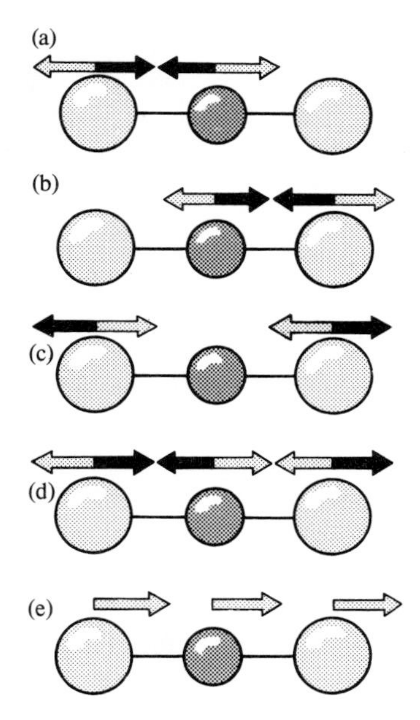

Fig. 11.21. (a) and (b) show two of the bond vibrations; (c) and (d) show two of their linear combinations corresponding to normal modes; (e) is a translational displacement.

Since the classical expression for the total energy consists of a sum of terms, the hamiltonian is constructed in the same way:

$$H = \sum_i H_i, \qquad H_i = -(\hbar^2/2)(\partial^2/\partial Q_i^2) + \tfrac{1}{2}\kappa_i Q_i^2. \tag{11.7.9}$$

The wavefunction for the entire molecule is therefore also factorizable into a product of wavefunctions for each mode:

$$\psi = \psi_{v_1}(Q_1)\psi_{v_2}(Q_2)\ldots, \tag{11.7.10}$$

there being $3N - 6$ factors for a polyatomic molecule composed of N

atoms (but $3N-5$ if it is linear). Each factor satisfies the Schrödinger equation $H_i\psi(Q_i)=E\psi(Q_i)$, or

$$-(\hbar^2/2)(\mathrm{d}^2\psi/\mathrm{d}Q_i^2)+\tfrac{1}{2}\kappa_i Q_i^2\psi=E\psi, \tag{11.7.11}$$

which is nothing other than the Schrödinger equation for a harmonic oscillator of unit mass undergoing vibration along the coordinate Q_i. We can therefore use the solutions in Chapter 3 and conclude that for each mode the energies and eigenfunctions are

$$E=(v_i+\tfrac{1}{2})\hbar\omega_i \qquad \omega_i=\kappa^{\frac{1}{2}}; \qquad v_i=0,1,2,\ldots$$
$$\psi_i=C_{v_i}H_{v_i}(y_i)\mathrm{e}^{-\frac{1}{2}y_i^2} \qquad y_i=(\omega_i/\hbar)^{\frac{1}{2}}Q_i, \tag{11.7.12}$$

C_{v_i} being a normalization constant (Table 3.1). It follows that the total vibrational energy of the molecule is

$$E=\sum_i(v_i+\tfrac{1}{2})\hbar\omega_i, \qquad i=1,2,\ldots,3N-6 \text{ or } 3N-5 \tag{11.7.13}$$

and the total wavefunction is a product of the wavefunctions in eqn (11.7.12). It follows that the *minimum vibrational energy* of a molecule is the sum of the zero-point energies of all the modes,

$$E_0=\tfrac{1}{2}\hbar\sum_i\omega_i. \tag{11.7.14}$$

For a medium–large molecule containing 50 atoms there are 144 modes of vibration; if we take $\omega/2\pi c\approx 300\ \mathrm{cm}^{-1}$ for each of them, the total zero-point vibrational energy is about $258\ \mathrm{kJ\,mol}^{-1}$, a sizeable store. It also follows that the wavefunction for the molecule in its lowest vibrational state is the product of gaussian functions (because $H_0(y)=1$, Table 3.1):

$$\psi_0=C_{10}\mathrm{e}^{-\frac{1}{2}y_1^2}C_{20}\mathrm{e}^{-\frac{1}{2}y_2^2}\ldots=C\mathrm{e}^{-\frac{1}{2}(y_1^2+y_2^2+\ldots)}=C\mathrm{e}^{-\frac{1}{2}y^2}. \tag{11.7.15}$$

The important feature is that the normal coordinates appear symmetrically and as their squares, and so *the ground-state wavefunction is totally symmetric under any symmetry transformation of the molecule*: the ground-state wavefunction spans A_1 in the symmetry group of the molecule. The significance of this point will become clear when we turn to a group-theoretical discussion of the normal coordinates.

The remaining point concerns the selection rules for the electric dipole transitions of an harmonic oscillator. We know that $\Delta v=\pm 1$ is a direct consequence of the form of the oscillator wavefunctions, and therefore it follows that each normal mode satisfies the same selection rule independently. We also know that in a diatomic molecule the vibration can be stimulated by an electric dipole interaction with the electromagnetic field only if the dipole moment changes with the nuclear separation. In that case the displacement of the nuclei from the equilibrium bond length separation is the only vibrational normal coordinate of the molecule; we generalize to the case of polyatomic species by saying that *only those modes leading to a change in the electric dipole moment of the molecule are active in the infrared.* This can be formulated quantitatively by generalizing the argument that led to eqn (11.5.5). The dipole moment at an arbitrary value of the normal coordinates (i.e. for an arbitrary distortion of the molecule) is

$$\boldsymbol{\mu} = \boldsymbol{\mu}_0 + \sum_i (\partial\boldsymbol{\mu}/\partial Q_i)_0 Q_i + \dots, \tag{11.7.16}$$

and so the transition dipole matrix element for the excitation of one of the modes (e.g. mode k) is

$$\langle v_1, v_2, \dots, v_k', \dots | \boldsymbol{\mu} | v_1, v_2, \dots, v_k, \dots \rangle = (\partial\boldsymbol{\mu}/\partial Q_k)_0 \langle v_k' | \, Q_k \, | v_k \rangle. \tag{11.7.17}$$

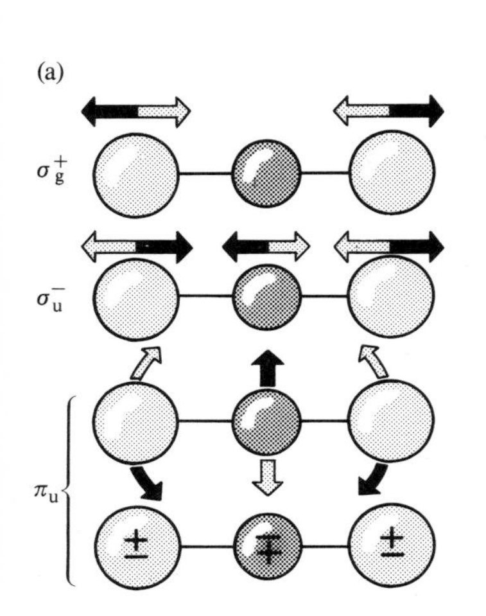

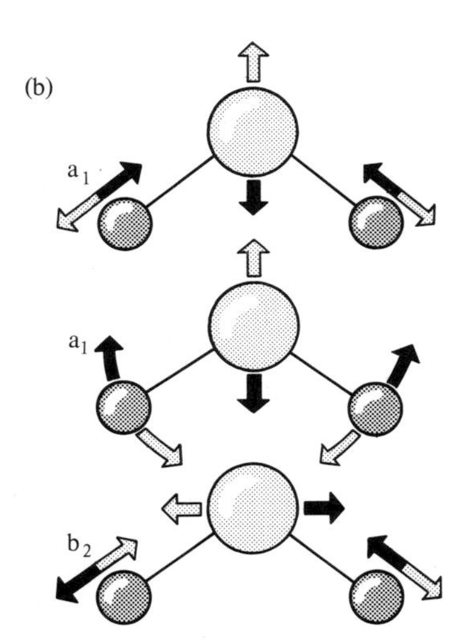

Fig. 11.22. Normal modes of (a) linear, (b) non-linear triatomic molecules.

Consequently $v_k' = v_k \pm 1$, and the transition has non-vanishing intensity only if $(\partial\boldsymbol{\mu}/\partial Q_k)_0 \neq 0$; that is, if the electric dipole moment changes when the molecule is distorted along the normal coordinate Q_k. In the case of CO_2, for example, it is clear that the symmetric stretch leaves the dipole moment unchanged at zero, and so this mode is *infrared inactive*; the antisymmetric stretch, on the other hand, distorts the molecule in such a way that it acquires a dipole moment (pointing in opposite directions at each end of its swing), and hence this mode is *infrared active.* In more complicated molecules it is obviously a complicated job, or involves inspired guesswork, to decide on the activities of modes. The problem is rationalized by group theory, which gives simple rules for deciding on mode activity, and so we must consider the group-theoretical description of molecular vibrations.

Molecular vibrations may also be active in Raman spectroscopy. The specific selection rule remains $\Delta v_i = \pm 1$ for each mode. The gross selection rule depends on the variation of the polarizability of the molecule as it oscillates along the normal coordinate. *If the polarizability changes, then the mode is Raman active.* It is often much harder to judge when the polarizability is going to change: in CO_2 for instance, the polarizability does change during the symmetric stretching vibration, but it does not change during the antisymmetric stretch. Very crudely we can see that this may be related to the fact that the molecule changes its overall size during the former stretching, but does not during the latter stretching, and polarizability reflects the sizes of molecules. At this point a group-theoretical analysis of the problem becomes essential.

11.8 Group theory and molecular vibrations

The detailed form of the normal coordinates does not need to be known in order to arrive at selection rules, and normally it is sufficient to deal with their symmetries. For instance, the normal coordinates of AB_2 molecules depend on the masses of the atoms, and different species have different normal coordinates; the *symmetries* of the normal coordinates—the shapes sketched in Fig. 11.22—remain the same irrespective of the details of the masses of the atoms, and the activity of the mode is an aspect of the symmetry of the displacement.

The first problem that arises is the symmetry species of the displacement coordinates (or of the mass-weighted coordinates: they are proportional). The procedure has already been described in Chapter 7. The q_i are regarded as a basis, and the characters of the reducible representation they span are determined by considering how they change into each other under the operations of the molecular point group. That set of characters is then used to find the symmetry species spanned by the basis. Three of the symmetry-adapted basis combinations correspond to translations, and three (or two) correspond to rotations: their symmetry species may be picked out from the character table for the group and subtracted. The remainder gives the symmetry species of the displacements spanned by the displacements of the atoms.

Consider the example of H_2O, which belongs to the point group C_{2v}. There are $3N=9$ displacement coordinates, Fig. 11.23, and so collectively they span a nine-dimensional reducible representation of the group. The characters of the representation can be calculated very simply *by counting* 1 *whenever a displacement is left unchanged by a symmetry operation,* -1 *when it is changed into the negative of itself, and* 0 *when the symmetry operation changes it into some entirely different displacement.* These rules are justified by the fact that the character is the sum of the *diagonal* elements of the representative of the operation in the basis, and the diagonal elements have the values 1 (if q is unchanged), -1 (if $q\rightarrow -q$), and 0 (if q moves elsewhere); recall the *Example* on p. 151. For example, under C_2

$$C_2(q_1, q_2, \ldots q_9) = (-q_1, -q_2, q_3, -q_7, -q_8, q_9, -q_4, -q_5, q_6)$$

$$= (q_1, q_2, \ldots, q_9)\begin{pmatrix} -1 & 0 & 0 & 0 & 0 & 0 & 0 & 0 & 0 \\ 0 & -1 & 0 & 0 & 0 & 0 & 0 & 0 & 0 \\ 0 & 0 & 1 & 0 & 0 & 0 & 0 & 0 & 0 \\ 0 & 0 & 0 & 0 & 0 & 0 & -1 & 0 & 0 \\ 0 & 0 & 0 & 0 & 0 & 0 & 0 & -1 & 0 \\ 0 & 0 & 0 & 0 & 0 & 0 & 0 & 0 & 1 \\ 0 & 0 & 0 & -1 & 0 & 0 & 0 & 0 & 0 \\ 0 & 0 & 0 & 0 & -1 & 0 & 0 & 0 & 0 \\ 0 & 0 & 0 & 0 & 0 & 1 & 0 & 0 & 0 \end{pmatrix},$$

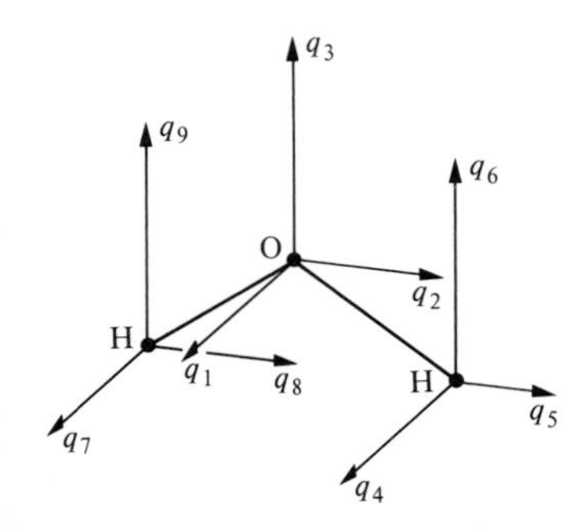

Fig. 11.23. Displacement coordinates for H_2O.

and so $\chi(C_2)=-1$. The rule is clearly much easier to apply than the business of setting up the explicit representative for each operation.

In the case of H_2O the four operations of the group give representatives with the characters $9, -1, 3, 1$, which decompose (using eqn (7.8.5)) into $3A_1 + A_2 + 2B_1 + 3B_2$. The translations of the molecule transform like the displacements x, y, and z, and so (from the character table, Appendix 10) span $A_1+B_1+B_2$. The rotations transform as $A_2+B_1+B_2$. When these two sets are subtracted we are left with $2A_1+B_2$ as the symmetry species of the vibrations of the water molecule H_2O. A sketch of their forms is given in Fig. 11.22.

We may do a great deal with a knowledge of the symmetries of the normal coordinates. First, we already know that *the ground-state vibrational wavefunction of the molecule is totally symmetric.* This can be demonstrated more formally as follows. The wavefunction for the ground state is a function of the Q_i^2; but in the group C_{2v} any symmetry operation simply multiplies the symmetry adapted basis by $+1$ or -1 (all the irreps are one-dimensional); therefore Q_i^2 remains unchanged under all the operations of the group, and therefore so too does the ground-state wavefunction. Therefore this function spans A_1. The fundamental transitions of the molecule are those in which a single quantum of any mode is excited. The wavefunctions for these $v=1$ states are proportional to $H_1(y_i)e^{-y^2/2} \propto y_i e^{-y^2/2} \propto Q_i e^{-\frac{1}{2}y^2}$ (because the Hermite polynomial $H_1(y)=2y$, Table 3.1). Therefore *the symmetry of the first vibrationally excited state of the molecule is the same as the normal coordinate of the mode.*

The last point provides a powerful method of determining the allowed electric dipole transitions. The transition moment between the ground state ψ_0 and the singly excited state of the mode k is $\int \psi_{1_k}^* \boldsymbol{\mu} \psi_0 \, d\tau$ where ψ_0 is the wavefunction for the ground state $|0, 0, \ldots, 0, 0, 0, \ldots\rangle$ and ψ_{1_k} that for the state $|0, 0, \ldots, 0, 1_k, 0, \ldots\rangle$. The integral is zero unless the direct product of the irreducible representations spanned by ψ_{1_k}, $\boldsymbol{\mu}$, and ψ_0 contains the totally symmetric irrep A_1. But ψ_0 is a basis for A_1. Therefore, ψ_{1_k} and $\boldsymbol{\mu}$ must span the same irrep if their product is to contain A_1. The electric dipole moment is a vector which has three components transforming in the same way as translations, and which therefore in C_{2v} spans $A_1+B_1+B_2$. Since the first excited state of each of the normal mode vibrations spans the same irrep as its normal coordinate, the three ψ_{1_k} span $A_1+A_1+B_2$. All three first excited states are therefore accessible by the electric dipole transitions; hence all three species of vibration are active.

Now consider the case of CO_2, which is also a triatomic molecule but belongs to the group $D_{\infty h}$. Four of its normal coordinates correspond to vibrations. Using the same techniques as for H_2O we find that the vibrational displacements span $\Sigma_g^+ + \Sigma_u^+ + \Pi_u$, the last being doubly degenerate. All four are depicted in Fig. 11.21. In $D_{\infty h}$ translations, and therefore the dipole moment components, transform as $\Sigma_u^+ + \Pi_u$, and so transitions from the ground state to the first excited states of both Σ_u^+ and Π_u modes are active. The Σ_g^+ mode is inactive, and a glance at Fig. 11.21 confirms that this is the mode that causes no change in the electric dipole moment of the molecule.

Symmetry arguments may also be applied in the same way to the transitions involved in the vibrational Raman effect, the selection rules now depending on the symmetry properties of the polarizability. The polarizability of a molecule transforms in the same way as the quadratic forms x^2, xy, etc. (the reason will become clear when we have formulated expressions for the polarizability in Chapter 13). The symmetry species these functions span are also listed in the character tables, and so exactly the same procedure may be followed. In the group C_{2v} the polarizability components span all the symmetry species of the group, and so all three fundamental transitions are Raman active. In $D_{\infty h}$ the quadratic forms, and therefore the components of the polarizability, span $\Sigma_g^+ + \Pi_g + \Delta_g$, and so the only fundamental transition allowed is from the ground state to the first excited state of the mode of symmetry Σ_g^+. Hence only that mode is Raman active.

The *exclusion rule* is a generalization of the last remark. If the molecule contains a centre of inversion, the concept of parity is applicable. The electric dipole moment has odd parity; the polarizability (being a quadratic form) has even parity. Therefore *in a molecule with a centre of inversion, a mode which is infrared active cannot be Raman active, and vice versa,* because the final state in the transition moment $\langle f|\,\Omega\,|i\rangle$ cannot simultaneously have both even and odd inversion symmetry. This accounts for the behaviour of CO_2 (which has an inversion centre), because the Σ_g^+ mode is Raman but not infrared active, and the other two types of mode are infrared but not Raman active. (Some transition modes may be both Raman and infrared inactive; that is not eliminated by the exclusion rule.) In H_2O the exclusion rule does not apply because the molecule has no centre of symmetry; we have already seen that its modes are both Raman and infrared active.

Example. Establish the symmetry species of the vibrations of CH_4 and decide which are infrared active and which are Raman active.

- *Method.* Proceed as on pp. 308–9. The molecular point group is T_d; since operations of the same class have the same character, the characters of the irreps spanned by the $3N=15$ displacements can be established by considering the effects of a single operation of each class. Therefore, consider the effects of E, C_3, C_2, S_4, σ_d. Use eqn (7.8.5) to identify the symmetry species spanned by the displacements. Subtract the translations (which transform as x, y, z) and the rotations (R_x, R_y, R_z). Infrared-active modes have the same symmetries as x, y, z; Raman-active modes have the same symmetries as the quadratic forms.

- *Answer.* Under E all 15 displacements are unchanged, hence $\chi(E)=15$. Under C_3 no displacements are left unchanged, hence $\chi(C_3)=0$. (There is a subtlety here: x, y, z on the central atom are mixed by C_3, so are x, y, z on the H atom lying on the C_3 axis, but $\chi(C_3)=0$ for each x, y, z basis—see eqn (7.10.2)—and so the net number of unchanged coordinates is 0.) Under S_4 the z-displacement of the central atom is reversed while all others are interchanged, and so $\chi(S_4)=-1$. Under σ_d z-displacements on C, H_3, H_4 are unchanged, so are their x-displacements, but their y-displacements change sign. The displacements on H_1, H_2 are interchanged. Hence $\chi(\sigma_d)=3+3-3=3$. The characters 15, 0, -1, 3 span $A_1+E+T_1+3T_2$. Remove translations (T_2) and rotations (T_1), leaving A_1+E+2T_2 for the vibrations. Infrared-active vibrations have T_2 symmetry. Raman-active vibrations have A_1+E+T_2 symmetry.

• *Comment.* Note that methane has no centre of inversion, and the T_2 vibrations are both infrared and Raman active.

There are several aspects of this discussion which should be noted. In the first place the symmetry selection rule does not rule out overtones. For example, since $H_2(y_i)$ is symmetrical under all the operations of C_{2v}, the $v=2$ levels of all the normal modes are symmetric, the z-component of the electric dipole moment transforms as A_1, and so the transition dipole moment from the ground state (A_1) to the $v=2$ level of any of the three modes would appear to be non-zero because $A_1 \times A_1 \times A_1 = A_1$. It must be stressed, however, that group theory *asserts when an integral must be zero but says nothing about the values of any integrals that need not be zero,* and sometimes it is found on closer inspection that such integrals happen to be zero. This is the case with overtones when there is neither electrical nor mechanical anharmonicity. Then the z-component of the electric dipole has the form

$$\mu_z = \mu_{0z} + \sum_i (\partial\mu_z/\partial Q_i)Q_i \tag{11.8.1}$$

and there are no other terms in the expansion. The transition dipole moment of the first overtone of the mode k is therefore

$$\langle 2_k|\,\mu_z\,|0_k\rangle = (\partial\mu_z/\partial Q_k)_0\langle 2_k|\,Q_k\,|0_k\rangle, \tag{11.8.2}$$

and the matrix element of Q_k vanishes if the wavefunctions are those of a harmonic oscillator (because only $\langle v\pm 1|\,Q\,|v\rangle$ integrals are non-zero, p. 299). The overtone becomes allowed if there is electrical anharmonicity, because then there are additional terms in the dependence of the dipole moment on the normal coordinate:

$$\mu_z = \mu_{0z} + \sum_i (\partial\mu_z/\partial Q_i)_0 Q_i + \tfrac{1}{2}\sum_{ij} (\partial^2\mu_z/\partial Q_i\,\partial Q_j)_0 Q_i Q_j + \ldots . \tag{11.8.3}$$

In this case the transition moment for the overtone of the k-mode is equal to eqn (11.8.2) together with an additional term proportional to $\langle 2_k|\,Q_k^2\,|0_k\rangle$, which does not vanish. Therefore, even in the absence of mechanical anharmonicity, this overtone may be visible. Note that the appearance of the second overtone requires a still higher degree of electrical anharmonicity because it can only occur if the expansion of the dipole moment contains Q^3 terms. Group theory tells us nothing about where the Taylor expansion of the dipole moment should end: in effect it supposes that all possible terms of the given symmetry are present. We require physical information to know whether or not the series ends after a couple of terms, and on that group theory has nothing to say.

All this can be illustrated by the appearance of *combination bands* in spectra, when more than one mode is excited simultaneously. The group theoretical *possibility* of the process is quite easy to see. The ground state is A_1. Consider the y-component of the electric dipole moment; it transforms as B_2. Therefore the excited state molecule must have B_2 symmetry in its vibrational wavefunctions. This can be achieved by the single excitation of the B_2 mode; but it can also be achieved by the *simultaneous* excitation of the modes B_2 and A_1 because their joint symmetry is $B_2 \times A_1 = B_2$ also. But can such transitions actually occur? To answer this we look more closely at the detailed form of the transition moment. The transition requires the non-vanishing of $\langle 1_a 1_b | \mu_y | 0_a 0_b \rangle$ (where a denotes the A_1 mode and b denotes the B_2 mode). If there is no electrical anharmonicity this matrix element is proportional to

$$\langle 1_a 1_b | \mu_y | 0_a 0_b \rangle = (\partial \mu_y / \partial Q_a)_0 \langle 1_a 1_b | Q_a | 0_a 0_b \rangle + (\partial \mu_y / \partial Q_b)_0 \langle 1_a 1_b | Q_b | 0_a 0_b \rangle$$

which vanishes in the absence of mechanical anharmonicity, because in the first term $\langle 1_b | 0_b \rangle = 0$ and in the second $\langle 1_a | 0_a \rangle = 0$. Only when there is electrical anharmonicity and we include the terms $(\partial^2 \mu_y / \partial Q_a \, \partial Q_b)_0 Q_a Q_b$ do we get a matrix element proportional to $\langle 1_a 1_b | Q_a Q_b | 0_a 0_b \rangle = \langle 1_a | Q_a | 0_a \rangle \langle 1_b | Q_b | 0_b \rangle$, which does not vanish. (We shall see later that mechanical anharmonicity also allows combination bands.)

11.9 Anharmonicities

Almost all that has been said so far has been based on the harmonic approximation to the potential energy. In real molecules the potential energy is not a simple parabolic function of the displacements (we saw that in the case of diatomic molecules, Fig. 11.17), and additional contributions to the potential energy of the form

$$V_{\text{anharmonic}} = \left(\frac{1}{3!}\right) \sum_{i,j,k} (\partial^3 V / \partial q_i \, \partial q_j \, \partial q_k)_0 q_i q_j q_k \tag{11.9.1}$$

must be taken into account. The presence of these terms removes the independence of the normal modes of vibration, for although the quadratic part of the potential energy can be diagonalized, the normal coordinates do not simultaneously diagonalize the cubic and higher parts. Therefore the hamiltonian does not separate, and the wavefunctions consequently do not factorize into independent components, one for each mode. If therefore, we continue to insist on talking in terms of the normal coordinates and the normal modes of vibration, we have to accept that they are mixed together by $V_{\text{anharmonic}}$.

Group theory provides a simplification at this point because the potential energy (whether or not it is harmonic) must be totally symmetric under any symmetry transformation of the molecule. (The hamiltonian, of which the potential energy is a part, always has the full symmetry of the point group: the energy cannot depend on how the molecule is orientated in field-free space.) Therefore the potential energy, *and each term in its expansion*, must be a basis for the totally symmetric representation of the group. It follows that the anharmonic potential mixes states of the same symmetry, because $\int \psi'^* V_{\text{an}} \psi \, \mathrm{d}\tau$ is

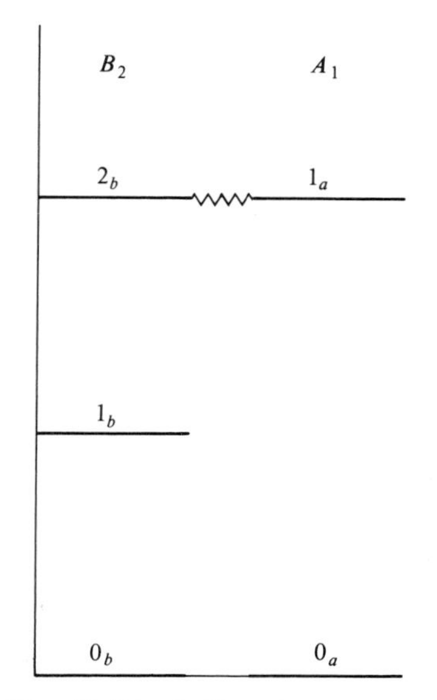

Fig. 11.24. A Fermi resonance between an overtone of B_2 and the A_1 fundamental.

only non-zero if ψ' transforms the same way as ψ when $V_{\text{anharmonic}}$ transforms as A_1.

As an example of an interaction caused by anharmonicity, consider the case where an overtone of one mode coincides with the fundamental of another, Fig. 11.24. We could then have a matrix element of the hamiltonian of the form $\langle 2_b | V_{\text{an}} | 1_a \rangle$ if the symmetries of the states are right. In the case of C_{2v} suppose the fundamental is A_1; the first overtone of either of the other modes is also A_1; therefore this matrix element may be non-zero. Whether or not it is non-zero depends on the result of calculation. When there is a term in the anharmonic potential energy of the form $(\partial^3 V/\partial Q_b\, \partial Q_b\, \partial Q_a)_0 Q_b^2 Q_a$, the matrix element is proportional to $\langle 2_b | Q_b^2 | 0_b \rangle \langle 0_a | Q_a | 1_a \rangle$, which is non-zero. Therefore, so long as the relevant third derivative of V is non-zero, the two states will mix. The special case of an interaction between a fundamental and a combination band is called a *Fermi resonance.*

The consequence of interactions of the kind just described is that the energy levels change as a result of their mixing under the influence of the perturbation (Fig. 8.1 again), and their transitions take on different intensities because the wavefunctions mix and so acquire aspects of each other's natures. This is most striking in the case of an allowed fundamental and a forbidden combination, for the latter may acquire intensity by virtue of the component of the allowed fundamental the anharmonicity mixes into it, Fig. 11.25.

11.10 Coriolis forces

Another type of perturbation that can affect the appearance of vibrational spectra is the interaction of the vibrational and rotational modes of the molecule. Such interactions are called *Coriolis interactions.*

In classical physics the Coriolis force is a force that appears to be necessary to an observer in a rotating system in order to account for the motion of particles. In particular it is the tangential component of the force, the radial component being the *centrifugal force.* The source of the tangential effective force can be appreciated by considering the paths taken by balls rolled outwards from the centre of a rotating plate. An external observer sees them roll in a straight line towards the edge. An observer stationed at the centre of the disc and rotating with it misinterprets this straight line as an arc, Fig. 11.26, and therefore imagines there to be a tangential force acting on the particle. A standard illustration of this Coriolis force is the fact that, because the earth rotates from west to east, a projectile fired towards the equator from the north pole seems to drift to the west. Both meteorology and, it is said, the motion of bath water though the plug hole depend on the Coriolis force, or so it seems to us, the rotating, earth-bound bathing observer.

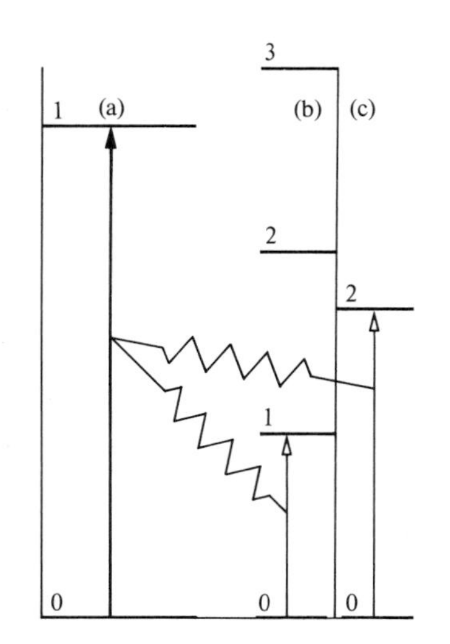

Fig. 11.25. The (b), (c) combination band borrows intensity from the allowed (a) fundamental.

Consider now the rotation of a mass on a spring, Fig. 11.27. As the mass moves out radially the rotating observer perceives it as moving in an arc, and concludes that a Coriolis force has retarded its motion. As

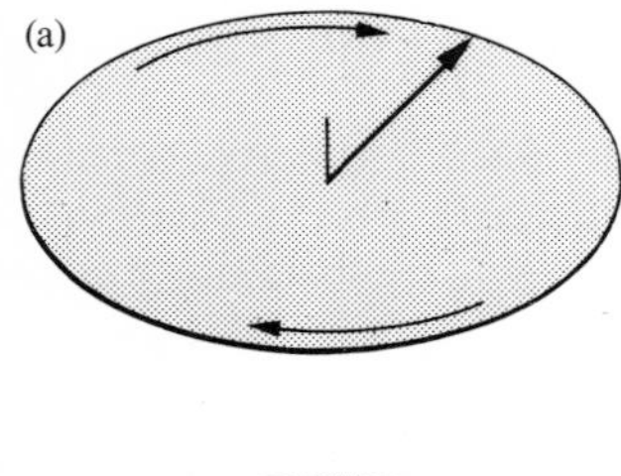

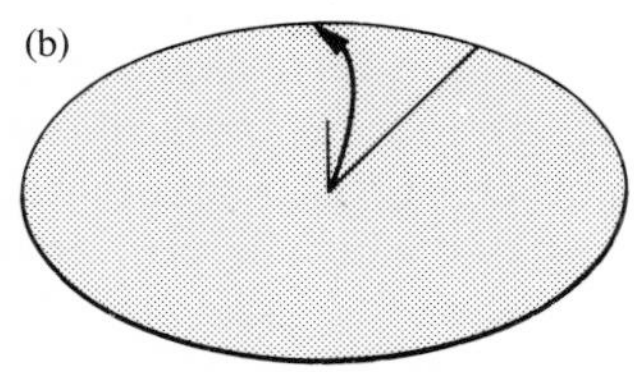

Fig. 11.26. An external observer sees (a) motion in a straight line, but an observer in the rotating frame sees (b) apparently curving motion.

the particle moves in, it appears to accelerate in the direction of travel, Fig. 11.27. Therefore, if it is vibrating the rotation of the particle is periodically accelerated and decelerated.

Consider how the Coriolis force affects a molecule when it is both rotating and also vibrating in the antisymmetric stretching mode, Fig. 11.28. As one of the bonds stretches it experiences a retarding Coriolis force; at the same time the bond that is shortening experiences an effective accelerating force. Therefore the molecule bends, Fig. 11.28(a). As the first bond shortens and the second lengthens, the accelerations change places, and the molecule bends the other way, Fig. 11.28(b). The effect of the rotation on the antisymmetric stretch is therefore to induce one of the bending modes. Quantum mechanically we say that the rotation provides a perturbation that mixes the antisymmetric stretch with one of the components of the doubly-degenerate pair of bending vibrations. As a consequence of the perturbation these two levels move apart in energy, Fig. 11.29 (which is Fig. 8.1 yet again), and so the bending mode in the plane of rotation is no longer degenerate with the bending mode perpendicular to the plane. Therefore transitions to these two levels no longer fall at the same energy, and the lines are doubled by the rotation. This effect is called *l-type doubling.*†

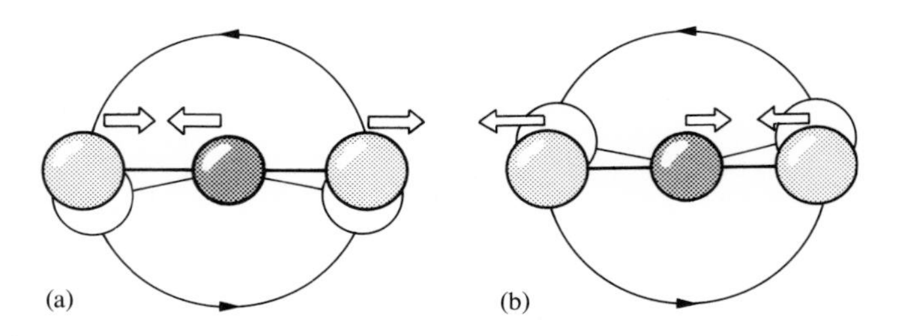

Fig. 11.28. Normal-mode coupling in a rotating molecule: (a) and (b) show different stages of the antisymmetric stretch.

Fig. 11.27. Coriolis forces on a rotating, oscillating mass.

11.11 Inversion doubling

Consider a pyramidal AB_3 molecule. If we were to plot its potential energy as it changes from pyramidal through planar to pyramidal again

† Why *l-type*? When the molecule is not rotating, the two bending modes form a doubly-degenerate pair, and we can take any linear combination of them. Two such combinations correspond to rotations of a bent molecule around an axis, Fig. 11.30. The excitation of one of these linear combinations endows the molecule with angular momentum about its axis, and when one quantum of vibration is excited the angular momentum corresponds to $\pm\hbar$. When several quanta are excited the rotation corresponds to an angular momentum $\pm l\hbar$. The degeneracy of the bending vibrations is removed by the Coriolis interaction, and so the l-type combinations can no longer be used when the transitions are doubled. *Doubling* is a general term signifying the removal of degeneracy.

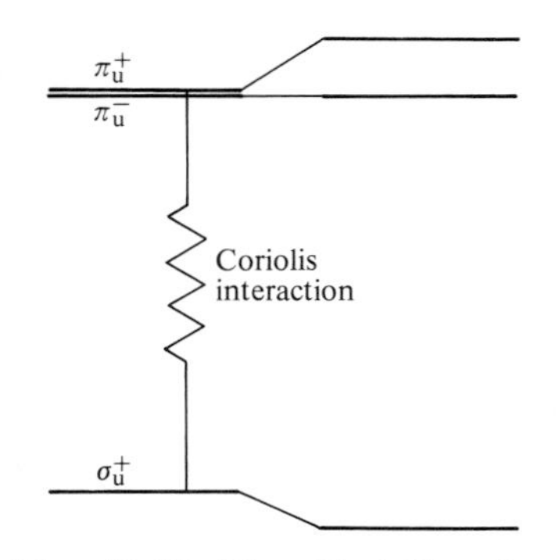

Fig. 11.29. The Coriolis effect and l-type doubling.

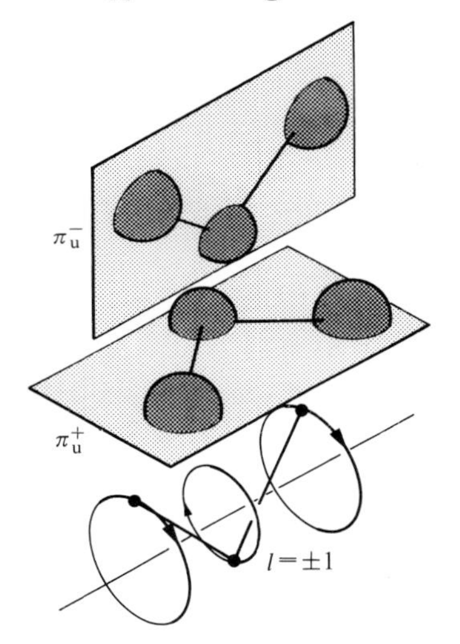

Fig. 11.30. The bending modes and one of their linear combinations.

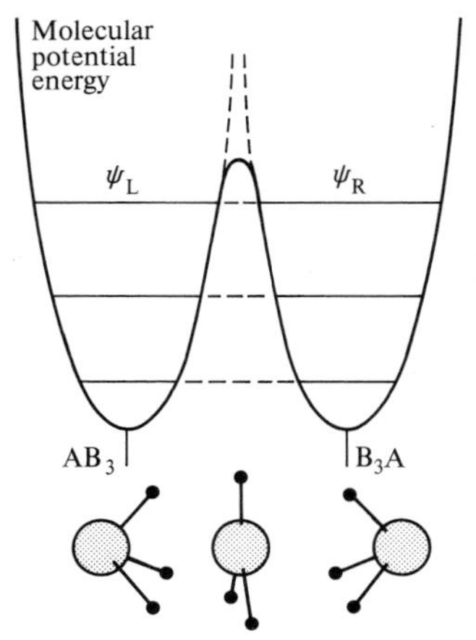

Fig. 11.31. The molecular potential energy curve for inversion.

we would obtain the molecular potential energy curve for the *inversion* $AB_3 \rightarrow B_3A$, Fig. 11.31. Either the barrier is high and the inversion is energetically very difficult (as for a well-made umbrella), or the barrier is low and the inversion is easy. In the first case the molecule vibrates around its AB_3 equilibrium conformation, and does not undergo inversion except perhaps at high excitations. The wavefunctions for these vibrations of AB_3 we denote ψ_L. If the molecule did invert, its wavefunctions would correspond to vibrations of B_3A. These we write ψ_R. The two ladders of vibrational energy levels for the two wells match, and so for a given quantum number, ψ_L and ψ_R are degenerate. When the barrier is infinite, as far as the molecule AB_3 is concerned the wavefunctions ψ_R represent states of an inaccessible other world, and it is completely oblivious of them. The interesting case, however, is when the barrier is so low that AB_3 can invert into B_3A and take on the character of ψ_R.

Suppose there is only one level on the left and one on the right, Fig. 11.32(a). The wavefunction of the (almost) harmonic oscillation on the left seeps through the barrier, and has non-zero amplitude where ψ_R is also non-zero. The hamiltonian has non-vanishing matrix elements of the form $\int \psi_R^* H \psi_L \, d\tau$, and so the two levels perturb each other. They are degenerate, and so very responsive even to small perturbations. Figure 8.1, now as Fig. 11.32(b), expresses what happens: the two wavefunctions mix to produce $\psi_L + \psi_R$ and $\psi_L - \psi_R$, and their energies move apart. Where we had a pair of degenerate levels, we now have two non-degenerate levels spreading between both wells: the level has *doubled*, the process in this case being *inversion doubling*.

In a more realistic case there are several levels in each well, but the matching pairs of degenerate states interact most strongly, and so the principal effect of the doubling can be thought of in terms of the pairwise doubling of the levels as a result of their interaction. This results in the spectrum of energy levels shown in Fig. 11.33, the differences in the extent of doubling in each case being a reflection of the increasing penetration of the barrier as its top is approached. The magnitude of the splitting depends on the state and the molecule: in the lowest state of NH_3 it corresponds to 0.793 cm^{-1}. This wavenumber corresponds to a frequency 2.4×10^{10} Hz, known as the *inversion frequency*. The source of this name can be traced back to the discussion of the time-dependent behaviour of a two-level system in Chapter 8, where we saw that if initially the system is in one state it periodically visits another degenerate state with a frequency determined by the strength of the perturbation mixing them (Fig. 8.7). In the present case, an NH_3 molecule in its vibrational ground states can be thought of as visiting the conformation H_3N with a frequency 2.4×10^{10} Hz.

The combinations $\psi_R \pm \psi_L$ are respectively even and odd under the inversion operation, and hence electric dipole transitions may take place between them. This is the basis of *maser action*, the early forerunner of lasers. The ammonia maser operates at 0.793 cm^{-1} (13 mm wavelength), in the microwave region of the spectrum.

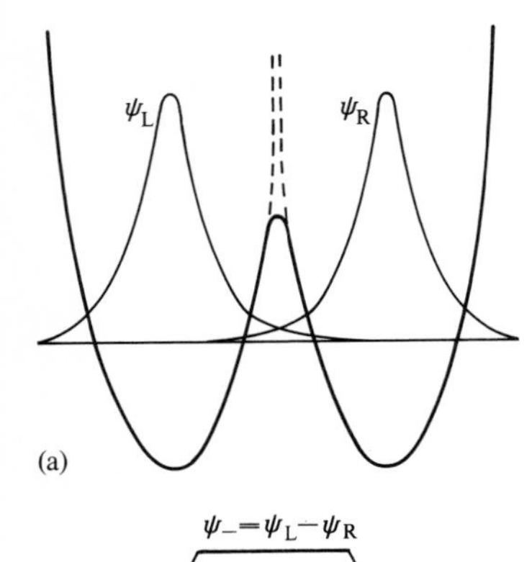

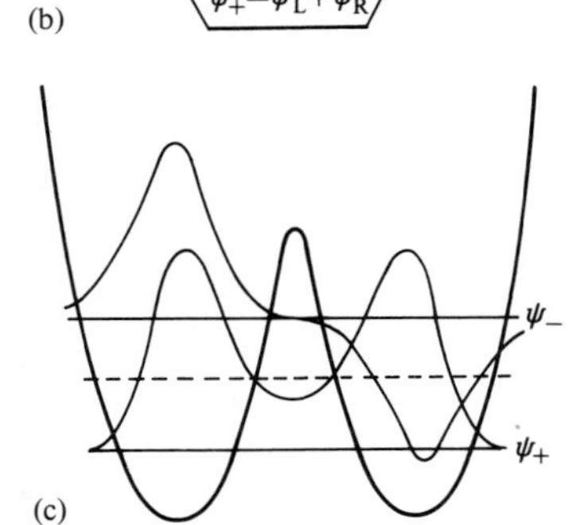

Fig. 11.32. (a) Modelling the true potential energy and wavefunctions by harmonic oscillators; (b) the energy levels arising from ψ_L, ψ_R interaction; (c) the corresponding eigenfunctions.

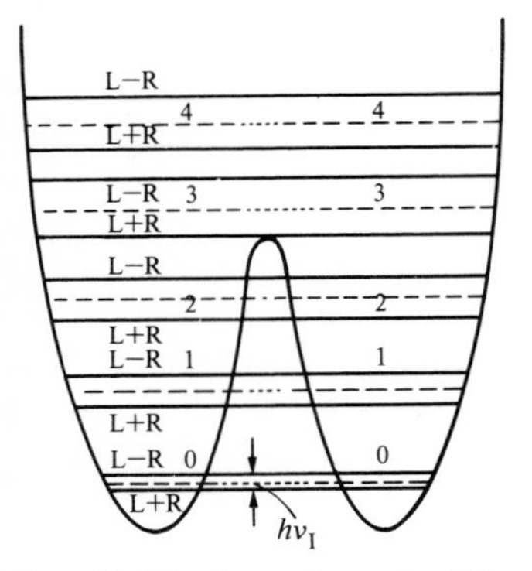

Fig. 11.33. Inversion doubling; broken lines are the original, independent harmonic oscillator energy levels.

Further reading

General spectroscopy

Physical chemistry. P. W. Atkins; Oxford University Press and W. H. Freeman, 2nd edition, 1982.
Introduction to molecular spectroscopy. G. M. Barrow; McGraw-Hill, New York, 1962.
Spectroscopy. D. H. Whiffen; Longman, London, 1972.
Spectroscopy and molecular structure. G. W. King; Holt, Rinehart, and Winston, New York, 1964.
Molecular structure and dynamics. W. H. Flygare; Prentice-Hall, Englewood Cliffs, 1978.

Microwave (rotational) spectroscopy

Microwave spectroscopy of gases. T. M. Sugden and C. N. Kenney; Van Nostrand, London, 1965.
Microwave spectroscopy. C. H. Townes and A. L. Schawlow; McGraw-Hill, New York, 1955.

Infrared (vibrational) spectroscopy

Introduction to the theory of molecular vibrations and vibrational spectroscopy. L. A. Woodward; Oxford University Press, 1972.
Infrared and Raman spectra of polyatomic molecules. G. Herzberg; Van Nostrand, Princeton, 1955.
Molecular vibrations. E. B. Wilson, J. C. Decius, and P. C. Cross; McGraw-Hill, New York, 1955.
The infrared spectra of complex molecules. L. J. Bellamy; Chapman and Hall, London, 1975.

Raman spectroscopy

Raman spectroscopy. D. A. Long; McGraw-Hill, New York, 1977.
Raman spectroscopy. H. A. Szymanski; Plenum, New York, 1967.

Molecular data

Interatomic distances. L. E. Sutton (ed.). The Chemical Society, London, 1958.
Spectra of diatomic molecules. G. Herzberg; Van Nostrand, Princeton, 1950.
Infrared characteristic group frequencies. G. Socrates; Wiley, New York, 1980.

Problems

11.1. Now that lasers are readily available there is considerable interest in the non-linear response of molecules. Show that if a time-dependent electric field $E_0 \cos \omega t$ can induce a non-linear response, that the scattered light may contain a frequency-doubled (2ω) component. *Hint.* Write $\mu(t) = \alpha E + \frac{1}{2}\beta E^2$ and consider an argument like that relating to eqn (11.1.1).

11.2. Show that the moment of inertia of a diatomic molecule formed from atoms of masses m_A and m_B and bond length R is given by $I = \mu R^2$, where $\mu = m_A m_B/(m_A + m_B)$. Calculate the moments of inertia of (a) $^1H_2$, $R =$ 75.09 pm, (b) $^2H_2$, $R = 75.09$ pm, (c) $^1H^{35}Cl$, $R = 127.5$ pm. ($M_r(^1H) = 1.0078$, $M_r(^2H) = 2.0141$, $M_r(^{35}Cl) = 34.9688$.)

11.3. The microwave spectrum of $^1H^{127}I$ consists of a series of lines separated by $12.8\ \text{cm}^{-1}$. Compute its bond length. What would be the separation of $^2H^{127}I$? ($M_r(^{127}I) = 126.9045$.)

11.4. The $J+1 \leftarrow J$ rotation transitions of $^{16}O^{12}C^{32}S$ and $^{16}O^{12}C^{34}S$ occur at the following frequencies (ν/GHz):

J:	1	2	3	4
$^{16}O^{12}C^{32}S$	24.325 92	36.488 82	48.651 64	60.814 08
$^{16}O^{12}C^{34}S$	23.732 23		47.462 40	

Find the rotational constants, the moments of inertia, and the C—S and C—O bond lengths. *Hint.* Begin by finding expressions for the moment of inertia I through $I = m_A R_A^2 + m_B R_B^2 + m_C R_C^2$ where R_X is the distance of atom X from the centre of mass. The easiest procedure is to recall the result of mechanics that the moment of inertia I′ about an axis parallel to one passing through the centre of mass, but at a distance R from it, is related to I by $I' = I + mR^2$, where m is the total mass of the body. This leads to $I = (m_A m_C/m)(R_{AB} + R_{BC})^2 + (m_B/m)(m_A R_{AB}^2 + m_C R_{BC}^2)$. Clearly, R_{AB} and R_{BC} may be found only if two values of I are known. Assume the bond lengths are the same in isotopically related molecules.

11.5. Confirm that the electric dipole selection rules are $\Delta J = \pm 1$, $\Delta M_J = 0, \pm 1$. *Hint.* Refer to Section 11.3 and evaluate the transition moment.

11.6. In PCl_3 the bond length is 204.3 pm and the ClPCl angle is 100°6′. Predict the form of (a) its microwave spectrum, (b) its rotational Raman spectrum, including the general structure of the line intensities. Ignore the effects of nuclear spin statistics. *Hint.* Establish that $I_\perp = m_B R^2(1 - \cos\theta) + (m_A m_B/m)R^2(1 + 2\cos\theta)$ for AB_3, with $m = m_A + 3m_B$, and $I_\parallel = 2m_B R^2(1 - \cos\theta)$. Suppose that the intensities are governed predominantly by the Boltzmann distribution.

11.7. In fact, the square of the electric transition dipole depends on J as $|\mu_{J+1,J}|^2 = \mu_e^2(J+1)/(2J+1)$. Predict the form of the $^1H^{35}Cl$ spectrum at 300 K (a) without taking account of this dependence, (b) taking this dependence into account. Estimate the values of J in each case corresponding to the most intense transition. *Hint.* Only relative intensities are important. Find the relative populations from the Boltzmann factor and the degeneracies. For (a) examine $(2J+1)e^{-hcBJ(J+1)/kT}$; for (b) examine $(J+1)/(2J+1)$ times this factor.

11.8. The ethyne molecule (HC≡CH) consists of two fermions (^1H) and two bosons, (^{12}C). What implications are there for the statistical weights of the levels of various J? What are the implications of replacing (a) one ^{12}C by ^{13}C, (b) both ^{12}C by ^{13}C. (The ^{13}C nucleus is a fermion, $I = \frac{1}{2}$.)

11.9. Adapt the general arguments of Appendix 6 to show that the reduced mass $\mu = m_A m_B/m$ should be used in the expression $\omega = (k/\mu)^{\frac{1}{2}}$ for the vibrational frequency of a diatomic molecule. Calculate the reduced masses of (a) $^1H_2$, (b) $^1H^{19}F$, (c) $^1H^{35}Cl$, (d) $^1H^{127}I$. The wavenumbers of the vibrations of these molecules are (a) 4400.39 cm^{-1}, (b) 4138.32 cm^{-1}, (c) 2990.95 cm^{-1}, (d) 2648.98 cm^{-1}, (e) 2308.09 cm^{-1}; calculate the force constants of the bonds. Predict the vibrational wavenumbers of the deuterium halides. ($M_r(^{19}F) = 18.9984$, $M_r(^{81}Br) = 80.9163$.)

11.10. Find the force constant of the hydrogen molecule ion on the basis of its LCAO-MO description. *Hint.* Evaluate $k = (d^2V/dR^2)_{R_e}$ where $V(R)$ is to be identified with E_+ in eqn (10.3.11a); the R-dependence of the integrals was the subject of Problem 10.2. The experimental value is 160.0 N m^{-1}.

11.11. One way of establishing the harmonic oscillator selection rules is described in the *Example* on p. 299. Another way is to use the recursion relation for the Hermite polynomials: $zH_v(z) = vH_{v-1}(z) + \frac{1}{2}H_{v+1}(z)$. Calculate the transition moment for transitions commencing in the state with quantum number v. *Hint.* The integral $\int \psi_{v'} x \psi_v \, dx$ can be evaluated very simply using the orthonormality of the oscillator functions that arise from using the recursion relation; M. Abramowitz and I. A. Stegun, *Handbook of mathematical functions*, Dover (1965), is a rich source of further relations.

11.12. The rotational constant of $^{1}H^{35}Cl$ has the value $10.4400\ cm^{-1}$ in the ground vibrational state and $10.1366\ cm^{-1}$ in the state $v=1$. Plot the wavenumbers of the P-, Q-, and R-branches against J as a representation of the structure of the 1–0 transition. (The Q-branch is not observed.)

11.13. The effect of vibrational excitation on the rotational constant can be modelled as follows. First, interpret $B=\hbar/4\pi c\mu R^2$ as the expectation value $(\hbar/4\pi c\mu)\langle 1/R^2\rangle$. Model the vibrational wavefunction by a rectangular probability amplitude, a constant from $R_e-\frac{1}{2}\delta R$ to $R_e+\frac{1}{2}\delta R$, and zero elsewhere. Evaluate $\langle 1/R^2\rangle$, and explore the approximation $\delta R^2 \ll 4R_e^2$. The magnitude of δR^2 can be estimated from $\langle (R-R_e)^2\rangle$ calculated from harmonic oscillator wavefunctions, and expressed in terms of v. Hence arrive at B in terms of v.

11.14. The three fundamental vibrations of CO_2 are observed at $1340\ cm^{-1}$, $667\ cm^{-1}$, and $2349\ cm^{-1}$, the second being the bending mode. Find the force constant of the C—O stretch.

11.15. Show that the vibrations of any non-linear AB_2 molecule span $2A_1+B_2$ in C_{2v}. Which vibrations are (a) infrared, (b) Raman active?

11.16. Establish the symmetries of the vibrations of the ethene molecule, and classify their activities.

11.17. Consider a two-dimensional harmonic oscillator with displacements in the x- and y-directions, the force constants being the same for each direction (the two bending modes of CO_2 is an example). Show that the state resulting from the excitation of the oscillator to its first excited state can be regarded as possessing one unit of angular momentum about the z-axis. *Hint*. Show that $\psi(x)\psi(y) \propto e^{i\phi}$.

12 Molecular electronic transitions

THE complexity of electronic spectra (which occur in the visible and ultraviolet regions) arises in part from the stimulation of simultaneous vibrational and rotational transitions when the electronic transition takes place. An electronic transition changes the distribution of the electrons, and the nuclei respond to the new force field by breaking into vibration. Just as in ice-skating the speed of rotation is changed by the pulling in or throwing out of the skater's arms, so the stimulation of a vibration of the molecule causes it to change its rotational state. A further complication is that in molecules there are many sources of angular momentum, and in order to make any headway it is necessary to understand how they couple together. We shall pick our way through this forest of complication by concentrating, initially at least, on diatomic molecules.

12.1 The Hund coupling cases

There are four sources of angular momentum in a diatomic molecule: the spin of the electrons, **S**, their orbital angular momenta, **L**, the rotation of the nuclear framework, **O**, and the nuclear spin, **I**.

There are interactions that couple these momenta together to varying extents. For example, the electric field due to the nuclei couples the orbital angular momenta of the electrons to the internuclear axis and only the component of momentum around that axis is well-defined. In highly excited rotational states, however, the nuclear framework is moving so fast that the electrons may be unable to follow the nuclear motion precisely, and the orbital angular momentum is *decoupled* from the axis: this is a breakdown of the Born–Oppenheimer approximation. When the spin–orbit coupling interaction is strong the spin is coupled to the orbital angular momentum, and when the latter is coupled to the molecular axis, then so too, at second hand, is the spin. If the spin–orbit coupling is very weak, a stronger coupling may be that between the spin and the magnetic moment arising from the rotation of the molecule as a whole. The nuclear spin may couple to any of the basic types of momenta or it may couple to any of their resultants. If there is an external field applied to the molecule, then that is an additional coupling option for any of the angular momenta as a result of their magnetic moments. Nuclear spin gives rise to *nuclear hyperfine effects:* we shall have quite enough on our hands without these small effects, and so henceforth we disregard their presence.

Hund attempted to impose order on the discussion of all these possibilities by focusing attention on four basic types of coupling.

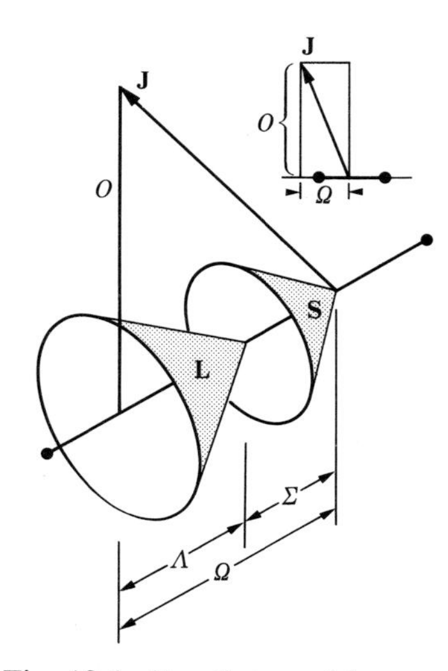

Fig. 12.1. Hund's case (a).

Hund's case (a), Fig. 12.1. The total angular momentum of the diatomic molecule is **J**. This has a component of magnitude O perpendicular to the molecular axis, which arises from the rotation of the

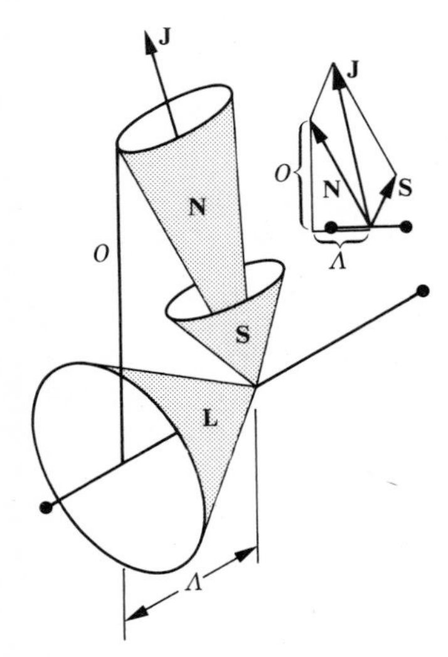

Fig. 12.2. Hund's case (b).

nuclear framework. It also has a component of magnitude $\Omega\hbar$ parallel to the axis, which is due to the electronic angular momentum. This arises as follows. The orbital angular momentum is coupled by the strong axial electrostatic field to the internuclear axis, and the spin–orbit coupling is sufficiently strong for the orbital momentum to pin the electron spin to the molecular axis too. Only the components of **L** and **S** on the axis are well defined and are ascribed the quantum numbers Λ and Σ respectively: the components are then $\Lambda\hbar$ and $\Sigma\hbar$. (Don't confuse this Σ with the Σ that occurs in the term symbol denoting $\Lambda = 0$: this Σ is the *spin* projection.) It follows that the total electronic angular momentum is directed along the axis of the molecule, and $\Omega = \Lambda + \Sigma$.

Hund's case (*b*), Fig. 12.2. When $\Lambda = 0$ the spin is no longer constrained to the internuclear axis because the spin–orbit coupling is absent. The same is true when $\Lambda \neq 0$ but the spin–orbit coupling is weak, as is the case in molecules of light atoms (e.g. the first period hydrides). In this case the orbital angular momentum may still couple to the axis with the projection $\Lambda\hbar$ but the spin now couples to the resultant of $\Lambda\hbar\hat{\mathbf{k}}$ and **O**, which we call **N**. The coupling of **S** and **N** gives **J**, the total angular momentum of the molecule, Fig. 12.2. Now the component of **J** on the axis is no longer well-defined, and so Ω is no longer a good quantum number.

Hund's case (*c*), Fig. 12.3. If the spin–orbit interaction is very strong (e.g. if the diatomic has at least one heavy atom) the spin and orbital momenta may couple strongly together to give a resultant $\mathbf{J}_e$; this resultant then couples to the axis with a component $\Omega\hbar$. The total angular momentum of the molecule, **J**, has a component $\Omega\hbar$ on the internuclear axis, and a component represented by the vector **O** perpendicular to the axis.

Hund's case (*d*), Fig. 12.4. The final case, which is quite rare in practice, arises when the coupling between the electrons and the molecular axis is so weak that they do not follow the molecule's rotation closely. The orbital angular momentum **L** about the centre of mass of the molecule couples to the molecular rotation angular momentum **O** to form the resultant **N**, which then couples with the electron spin, **S**, to form the total angular momentum **J**. This coupling is appropriate to the *Rydberg levels* of molecules, in which an electron has been excited from the valence shell orbitals into orbitals of higher principal quantum number (e.g. in H_2, into an MO formed from hydrogen 2s-orbitals). Rydberg orbitals are very diffuse, and the electron in one is so far from the nuclei of the molecule that it experiences a potential characteristic of a single point charge; hence the shape of the molecule is not transmitted to the excited electron and its rotation is barely noticed.

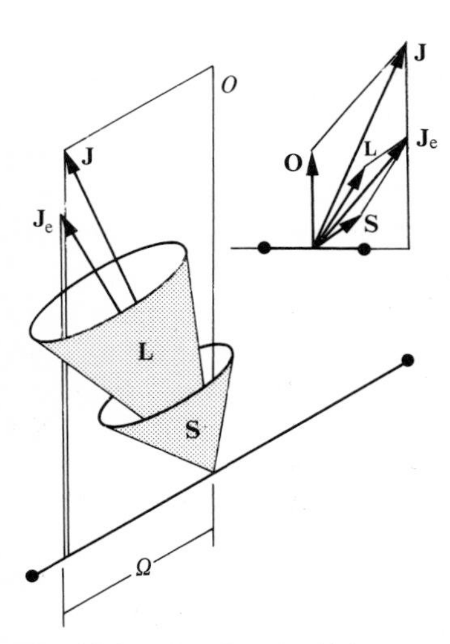

Fig. 12.3. Hund's case (c).

12.2 Decoupling and Λ-doubling

No real molecule has a structure that corresponds precisely to one of the Hund cases, although generally one scheme describes the structure

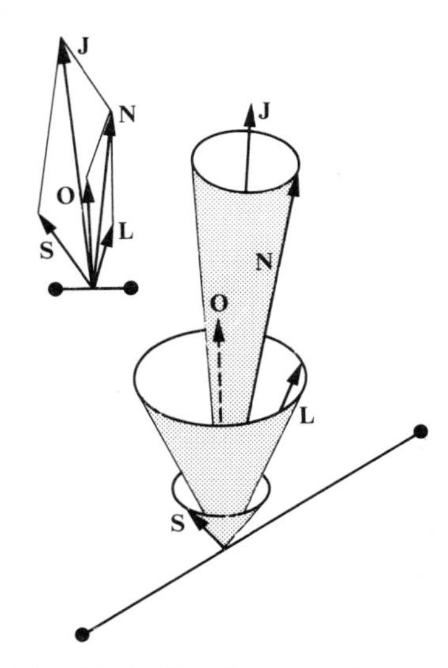

Fig. 12.4. Hund's case (d).

more closely than any other. (What this actually means is explained in Appendix 9: the hamiltonian for the molecule has off-diagonal terms in each of the representations of the molecule, but one representation normally gives rise to significantly smaller off-diagonal elements than any other.) Since no molecule is described by a pure coupling case, it follows that if we select one case, some of the characteristics of the other cases will be apparent in the spectrum of the molecule. The tendency of one scheme to be contaminated by another is called *decoupling* of the momenta. Decoupling often increases as J increases, because the electrons become increasingly unable to follow the nuclear motion: this is the phenomenon of *electron slip*. In quantum mechanical terms electron slip is expressed as the mixing of different terms. As an example, consider a $^1\Pi$ term, which in case (a) is doubly degenerate ($\Lambda = 1$ and $\Lambda = -1$ are degenerate, the orbital angular momentum differing only in the sense of the motion about the axis). That degeneracy is lost when the molecule is rotating because the $^1\Pi^+$ component of $^1\Pi$, Fig. 12.5 (which is a linear combination of the two counter-rotating states of the electron, just as p_x is a linear combination of the counter-rotating orbitals p_{+1} and p_{-1}) mixes with a nearby $^1\Sigma^+$ term. The $^1\Pi^-$ component of $^1\Pi$ remains unaffected by the rotation because it cannot mix with $^1\Sigma^+$. Therefore, since there is a perturbation mixing the two terms, they move apart in energy; consequently the degeneracy of $^1\Pi$ is removed and the energy levels 'double'. This effect is called *Λ-doubling*. It can be regarded as the contamination of case (a) by the incipient appearance of case (d).

There are two ways of imagining the mixing of terms. The first is shown in Fig. 12.6, where we see that if the electrons slip as the molecule rotates there is a tendency for the nuclei to move out of the nodal region of the $^1\Pi^+$ component, but they remain in the node in the case of $^1\Pi^-$. The shift out of the nodal region is equivalent to the addition of a little $^1\Sigma^+$ character into $^1\Pi^+$. The formal demonstration of the effect (which also leads to an estimation of its magnitude) runs as follows.

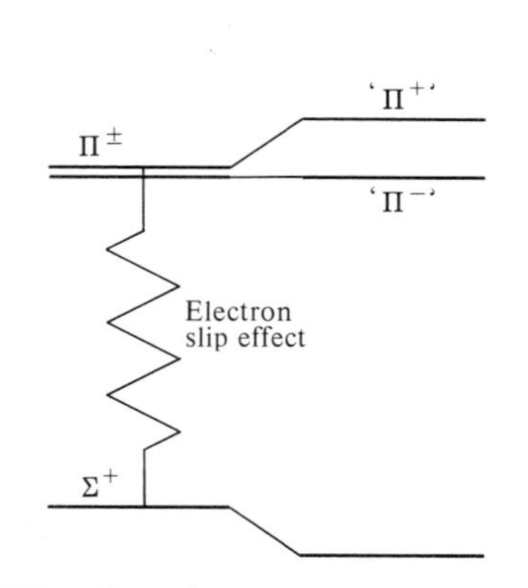

Fig. 12.5. Λ-doubling and electron slip.

The hamiltonian for the rotation of the framework of the molecule should be expressed in terms of the molecular rotation angular momentum $\mathbf{O}$. For a diatomic molecule the only components of $\mathbf{O}$ are O_x and O_y in the molecular frame, and $O_x = J_x - L_x$, $O_y = J_y - L_y$ (see the discussion of case (a)). Therefore the rotational hamiltonian is

$$H = (1/2I)(O_x^2 + O_y^2) = (1/2I)\{(J_x - L_x)^2 + (J_y - L_y)^2\}$$
$$= (1/2I)\{J^2 - J_z^2 + (L_x^2 + L_y^2) - 2(J_xL_x + J_yL_y)\}.$$

$L_x^2 + L_y^2$ is independent of the rotational state, and so we ignore its contribution. $J_xL_x + J_yL_y$ can be expressed in terms of raising and lowering operators as $\frac{1}{2}(J^+L^- + J^-L^+)$, and is plainly off-diagonal in Λ because states differing in Λ by unity are connected by the shift operators.† The off-diagonal terms may be regarded as a perturbation on the major part of the hamiltonian, and we make the division

$$H^{(0)} = (1/2I)(J^2 - J_z^2); \qquad H^{(1)} = -(1/2I)(J^+L^- + J^-L^+). \tag{12.2.1}$$

A term with rotational and orbital quantum numbers J and Λ has the eigenvalues

$$E_{J\Lambda} = hcB\{J(J+1) - \Lambda^2\}, \tag{12.2.2}$$

and states with opposite signs of Λ are degenerate. The remaining part of the hamiltonian, $H^{(1)}$, acts as a perturbation. The second-order contribution to the energy is

$$E^{(2)} = (1/2I)^2 \left\{ \frac{\langle J\Lambda|\, J^+L^- + J^-L^+ \,|J'\Lambda'\rangle\langle J'\Lambda'|\, J^+L^- + J^-L^+ \,|J\Lambda\rangle}{E_{J\Lambda} - E_{J'\Lambda'}} \right\}.$$

This expression can be used to evaluate the correction to the energy of the linear combinations $|\Pi^{\pm}\rangle = |\Lambda\rangle \pm \mathrm{i}\,|-\Lambda\rangle$ with $\Lambda = 1$, and it turns out that only the Σ_g^+, Π^+ terms are mixed in by the perturbation and that the Λ-doubling, the difference in energy between Π^+ and Π^-, has the magnitude

$$\Delta E^{(2)} = 2(hcB)^2 L(L+1)J(J+1)/(E_\Pi - E_\Sigma). \tag{12.2.3}$$

The separation of the components increases as the value of J increases, showing the increased mixing that occurs as the speed of rotation increases.

12.3 Selection rules

Now that we have some idea of the energy levels of a diatomic molecule, and the way they can be displaced by perturbations, we can move on to the prediction of the spectrum by examining the selection rules. These arise from considerations of the conservation of angular momentum during the emission or absorption and the changes of parity that accompany electric dipole transitions. They may be summarized as follows:

$$\begin{aligned}
&g \leftrightarrow u, \qquad g \not\leftrightarrow g, \qquad u \not\leftrightarrow u;\\
&\Delta J = 0, \pm 1 \text{ but } J = 0 \not\leftrightarrow J = 0 \text{ and for } \Omega = 0 \rightarrow \Omega = 0,\ \Delta J \neq 0\\
&\Delta\Lambda = 0, \pm 1; \qquad \Delta\Omega = 0, \pm 1;\\
&\Delta S = 0; \qquad \Delta\Sigma = 0 \text{ (for weak spin–orbit coupling)}\\
&\Sigma^+ \leftrightarrow \Sigma^+, \qquad \Sigma^- \leftrightarrow \Sigma^-, \qquad \Sigma^+ \not\leftrightarrow \Sigma^-.
\end{aligned} \tag{12.3.1}$$

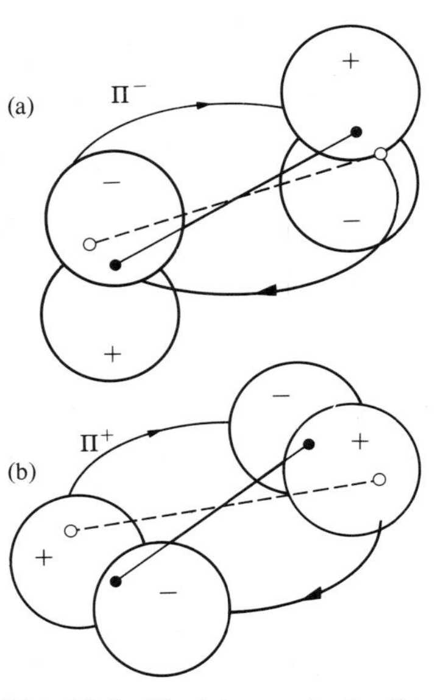

Fig. 12.6. Nuclei remain in the nodal plane in (a), but not in (b).

† It may be surprising at first sight that in the rotating frame J^+ is a *lowering* operator and J^- a *raising* operator. This reversal of their normal roles follows from the fact that we know the commutation relations in a fixed frame, and in order to establish them in a rotating frame (where we have already seen that strange things may appear to happen) we have to transform the operators to the rotating set of coordinates and work out their commutation relations. When this is done it turns out that $[J_x, J_y] = -\mathrm{i}\hbar J_z$, a change of sign relative to the static frame, conventional rules. This reversal of sign interchanges the roles of the shift operators, and leads to some engaging complexities.

The details of these rules can all be established on the basis of explicit examination of the symmetry properties of the relevant transition dipole moments.

12.4 Vibronic transitions and the Franck–Condon principle

When an electron is excited it moves into a different distribution and so exerts a different force on the nuclei. The nuclei respond by breaking into more vigorous vibration. Hence electronic transitions are accompanied by vibrational transitions, and so they are called *vibronic*, and the absorption spectrum shows many lines. The analysis of which vibrations are stimulated is based on the view that nuclei move much more slowly than electrons, and so the electron rearrangement occurs in a virtually static nuclear framework. The *Franck–Condon principle* is the approximation that *the nuclear conformation readjusts after the electronic transition, and not during it.*

How the principle is used qualitatively can be understood on the basis of Fig. 12.7 which shows the molecular potential energy curves for two electronic states of a diatomic molecule and a selection of their vibrational energy levels. The upper curve is usually displaced to the right relative to the lower curve because excitation of the electron normally sends it into a more antibonding state. The force constants also differ on account of the change in the electron distribution. We confine our attention to the *fundamental progression*, the transitions starting at the ground vibrational state of the lower electronic state. In this case the most probable conformation of the molecule when the transition begins is when the nuclei have their equilibrium separation (that corresponds to the location of the maximum of the vibrational wavefunction). At the completion of the electronic transition the nuclei are still in their original positions, and therefore the bond length of the excited state lies somewhere on the vertical line drawn on the diagram. This is the source of the expression *vertical transition*. In the upper state this conformation corresponds to a compressed state of the bond, and so the molecule responds by breaking into vibration. But which of the numerous upper vibrational states of the upper electronic state does it occupy? The brief answer is *the one which has a vibrational wavefunction with the greatest overlap with the original state's vibrational wavefunction.* The reasoning behind this remark runs as follows.

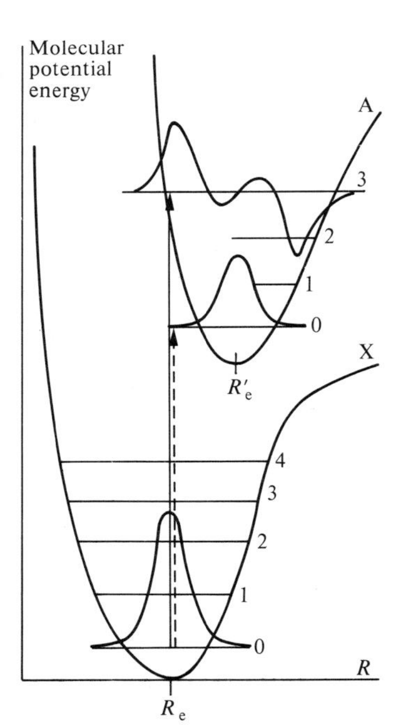

Fig. 12.7. The basis of the Franck–Condon principle.

The intensity of the transition from the joint electronic and vibrational state (i.e. the *vibronic state*) $|\varepsilon v\rangle$ to the upper vibronic state $|\varepsilon' v'\rangle$ depends on the magnitude of the transition moment $\langle \varepsilon' v' | \boldsymbol{\mu} | \varepsilon v\rangle$. In a molecule the electric dipole moment operator depends on the locations $\mathbf{r}_i$ of the electrons and the locations $\mathbf{R}_s$ and charges $Z_s e$, of the nuclei:

$$\boldsymbol{\mu} = -e\sum_i \mathbf{r}_i + e\sum_s Z_s \mathbf{R}_s = \boldsymbol{\mu}_e + \boldsymbol{\mu}_N. \tag{12.4.1}$$

The vibronic wavefunction corresponding to the vibronic state $|\varepsilon v\rangle$ is $\psi_\varepsilon(\mathbf{r}, \mathbf{R})\psi_v(\mathbf{R})$, where $\mathbf{r}$ and $\mathbf{R}$ denote the electronic and nuclear coordinates collectively. Note that the electronic wavefunction depends parametrically on the nuclear coordinates (i.e. there is a different wavefunction for each nuclear conformation). The transition moment is therefore

$$\begin{aligned}\langle\varepsilon'v'|\boldsymbol{\mu}|\varepsilon v\rangle &= \int \psi^*_{\varepsilon'}(\mathbf{r},\mathbf{R})\psi^*_{v'}(\mathbf{R})(\boldsymbol{\mu}_\mathrm{e}+\boldsymbol{\mu}_\mathrm{N})\psi_\varepsilon(\mathbf{r},\mathbf{R})\psi_v(\mathbf{R})\,\mathrm{d}\tau_\mathrm{e}\,\mathrm{d}\tau_\mathrm{N}\\ &= \int \psi^*_{v'}(\mathbf{R})\left\{\int \psi^*_{\varepsilon'}(\mathbf{r},\mathbf{R})\boldsymbol{\mu}_\mathrm{e}\psi_\varepsilon(\mathbf{r},\mathbf{R})\,\mathrm{d}\tau_\mathrm{e}\right\}\psi_v(\mathbf{R})\,\mathrm{d}\tau_\mathrm{N}\\ &\quad+\int \psi^*_{v'}(\mathbf{R})\boldsymbol{\mu}_\mathrm{N}\left\{\int \psi^*_{\varepsilon'}(\mathbf{r},\mathbf{R})\psi_\varepsilon(\mathbf{r},\mathbf{R})\,\mathrm{d}\tau_\mathrm{e}\right\}\psi_v(\mathbf{R})\,\mathrm{d}\tau_N.\end{aligned}$$

The second integral is zero because the electronic states are orthogonal whatever the value of **R**. The integral over the electronic coordinates in the first integral is the transition moment for an electronic transition when the molecule has some fixed conformation **R**. To a fair first approximation this transition moment is independent of the locations of the nuclei (so long as they are not displaced by a large amount from equilibrium), and so the integral may be approximated by a constant, $\boldsymbol{\mu}_{\varepsilon'\varepsilon}$. Therefore the overall transition dipole moment is

$$\langle\varepsilon'v'|\boldsymbol{\mu}|\varepsilon v\rangle \approx \boldsymbol{\mu}_{\varepsilon'\varepsilon}\int \psi^*_{v'}(\mathbf{R})\psi_v(\mathbf{R})\,\mathrm{d}\tau_\mathrm{N} = \boldsymbol{\mu}_{\varepsilon'\varepsilon}S_{v'v} \qquad (12.4.2)$$

where $S_{v'v}$ is the overlap integral of the two vibrational states of interest. The transition moment is therefore largest when the vibrational wavefunction of the vibrational states of the upper electronic level has the greatest overlap with the wavefunction of the initial vibrational level of the lower electronic level. Therefore, in order to assess which upper vibrational states will be the end point of the transition, we have to judge which of them most closely resemble the bell-shaped gaussian function of the initial state. Harmonic oscillator wavefunctions have their peaks close to the walls of the potential (recall Fig. 3.10) when v is large, and so the ones to which the transition occurs are those having energies close to where the vertical transition line cuts through the upper potential energy curve. Several vibrational states are usually similar enough to the initial state for them all to be stimulated, and so a *progression* of vibrations is stimulated, their relative intensities being proportional to the *Franck–Condon factors* $S^2_{v'v}$.

Example. Consider the case where two electronic states have the same force constant but in which the equilibrium bond lengths differ by ΔR. Find an expression for the intensity of the 0–0 transition as a function of ΔR.

• *Method.* The intensity is proportional to S^2_{00}; evaluate S_{00} using the harmonic oscillator wavefunctions listed in Table 3.1, one centred on $x=0$ and the other centred on $x=\Delta R$.

• *Answer.*

$$\psi_0(x) = (\alpha/\pi)^{\frac{1}{4}} e^{-\alpha x^2/2}, \qquad \psi_0'(x) = (\alpha/\pi)^{\frac{1}{4}} e^{-\alpha(x-\Delta R)^2/2}, \qquad \alpha = m\omega/\hbar.$$

$$S_{00} = (\alpha/\pi)^{\frac{1}{2}} \int_{-\infty}^{\infty} e^{-\alpha x^2/2} e^{-\alpha(x-\Delta R)^2/2}\, dx$$

$$= (\alpha/\pi)^{\frac{1}{2}} e^{-\alpha(\Delta R/2)^2} \int_{-\infty}^{\infty} e^{-\alpha(x-\Delta R/2)^2}\, dx = e^{-\alpha(\Delta R/2)^2}.$$

Therefore, the intensities are proportional to

$$S_{00}^2 = e^{-\frac{1}{2}\alpha\Delta R^2}.$$

• *Comment.* The intensity depends very strongly on ΔR. When $\Delta R = 0$ the Franck–Condon factor is unity; when $\omega = 6\times10^{14}\,\mathrm{s}^{-1}$ and $k = 500\,\mathrm{N\,m}^{-1}$, so that $\alpha = 8.4\times10^{21}\,\mathrm{m}^{-2}$, the factor has fallen to 0.01 when $\Delta R = 33$ pm. The sum of $S_{v'v}^2$ over all v' is unity (use $\sum_{v'} S_{v'v}^2 = \sum_{v'} \langle v \mid v'\rangle\langle v' \mid v\rangle = \langle v \mid v\rangle = 1$); hence the total transition intensity to all levels of the upper electronic state is independent of the shape and position of its potential energy curve.

12.5 The rotational structure of vibronic transitions

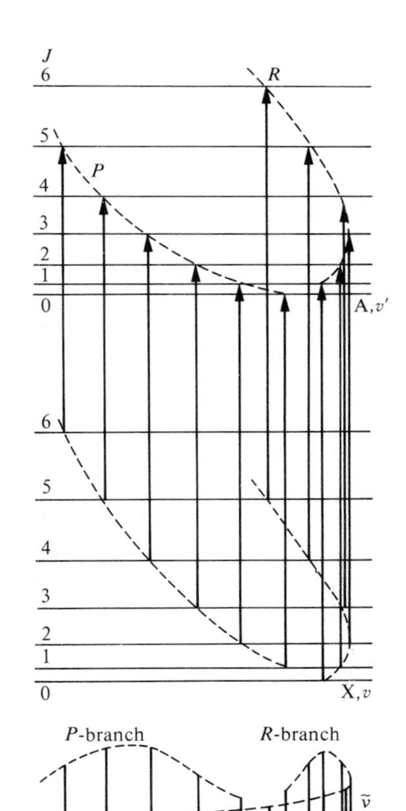

Fig. 12.8. P-, Q-, and R-branch formation (with an R-branch head) for the vibronic transition A, $v' \leftarrow X$, v.

Superimposed on vibronic transitions are rotational transitions occurring according to the selection rules set out in eqn (12.3.1). The $\Delta J = 0, \pm1$ transitions give rise to P-, Q-, and R-branches in the spectrum, and their appearances are similar, with one exception, to the branches observed in vibration–rotation spectra. The exception arises from the likelihood that the rotational constants of the upper and lower electronic states differ considerably because the bond lengths of the species are likely to be significantly different. *Head formation* (as described in Section 11.6) is therefore likely to occur, and the spectrum often has a structure like that shown in Fig. 12.8.

The presence of Λ-doubling affects the spectrum in a subtle way. In a $^1\Pi \leftarrow {^1\Sigma}$ transition, the P- and R-branches arise from the combination of the ground term with one of the components of the Π-doublet while the Q-branch results from the combination of the ground term with the other. This means that the Q-branch is slightly shifted relative to the other branches, the magnitude of the shift giving the magnitude of the Λ-doubling.

Further complications arise when one state is perturbed by another, and perturbations can be very effective in distorting the energy levels. For example, the *non-crossing rule* must be obeyed, and states of the same symmetry that approach each other in energy are strongly perturbed (Fig. 8.2 again). Another phenomenon that tends to obscure regions of the spectrum is *predissociation.* The process is illustrated in Fig. 12.9. The upper electronic state is perturbed by a dissociative state, and a molecule excited to an energy level lying near the crossing of the two terms may take on dissociative character, and fly apart. This reduces the lifetimes of molecules in the states in the region of the intersection of the curves, and by the lifetime-broadening effect the levels (and the transitions) there are diffuse.

Although it is very clear that the electronic spectra of molecules are very complex, each complexity provides information about the structure of the molecule, and the molecule can be explored in detail if the perturbations, interactions, and complications can be understood and accounted for. The principal information obtained includes bond lengths, force constants in various states, and the potential energy curves for a variety of molecular states.

12.6 The electronic spectra of polyatomic molecules

We have seen the complexity of diatomic molecule electronic spectra, and can therefore imagine the complexity of polyatomic molecule spectra. Happily there is a simplification. Many polyatomic molecules are studied in solution, and as a result of the intermolecular collisions the rotational structure of the bands is blurred. In weakly interacting solvents, such as hydrocarbons, the vibrational structure of the bands may still be present, but in interacting solvents even that may be lost. Therefore we are mainly concerned with spectra in which most of the details of vibrational and rotational structure have disappeared. This means that we need consider only the width of the absorption bands, and the position of the maximum absorption depends strongly on the structure caused by the blurred vibrational and rotational transitions. Another simplification, especially in the spectra of organic molecules, is that often the absorption in a particular region of the spectrum may be ascribed to a transition involving a particular group of atoms in the molecule. Such a group, a *chromophore*, may occur in a number of different types of molecule and gives rise to an absorption band at about the same wavenumber. This means that a simple discussion of the spectra of molecules may be based on their chromophores, and the perturbations caused by the presence of other groups.

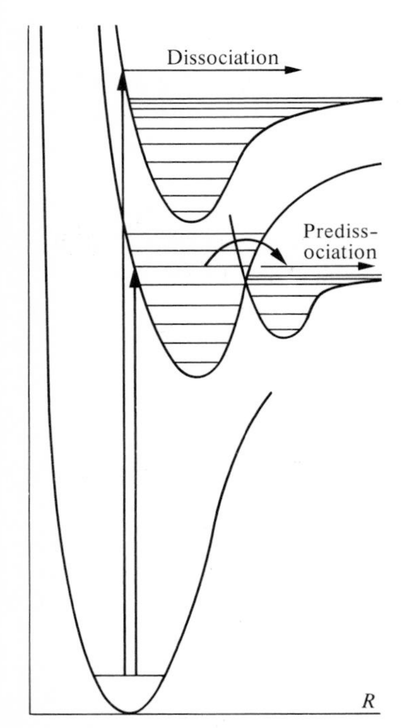

Fig. 12.9. Dissociation and predissociation on electronic excitation.

In *small molecules* we must take into account the symmetry of the whole molecule because the electronic excitation involves the entire structure. Therefore the selection rules must be expressed in terms of the molecular symmetry group. This is a simple task, for if the irreducible representations spanned by the electric dipole moment operator are known, then the selection rules may be expressed in terms of the direct product rule. As an example, consider the odd-electron, bent (C_{2v}) molecule NO_2 which has the configuration $\ldots b_2^2 a_2^2 a_1^1$ and the term symbol $^2A_1$. In C_{2v} the dipole moment operator spans $A_1 + B_1 + B_2$ (corresponding to the z-, x-, and y-components respectively). This implies that, from the ground state, transitions to the states $^2A_1$, $^2B_1$, and $^2B_2$ may be induced by light polarized with its electric vector parallel to the z-, x-, and y-directions respectively. Since excitation leading to these states involves a considerable rearrangement of the electrons in the molecule, each electronic transition is accompanied by an extensive vibrational structure. For example, the transition $\ldots b_2^2 a_2^2 b_2$, $^2B_2 \leftarrow \ldots b_2^2 a_2^2 a_1$, $^2A_1$ redistributes the electron that may be considered to be responsible for forcing the molecule into a bent configuration in its ground state. The normal convention is to write the upper term first and the lower second, then the direction of the arrow indicates emission ($\mathrm{A} \rightarrow \mathrm{X}$) or absorption ($\mathrm{A} \leftarrow \mathrm{X}$). The ground state is usually labelled X, and the excited states are labelled A, B, C, . . . , unless their symmetry label is given.

For quantitative discussion we need a measure of the intensity of a spectral line: the transition moment is the theoretical quantity, but how is it related to the measured intensity? The basic relation we require is the

Beer–Lambert law: $I_l = I_0 e^{-\alpha C l}$ (12.6.1)

relating the intensities of the light before (I_0) and after (I_l) it has passed through a length l of a sample of absorbing species at a concentration C. The derivation of this law is described in Appendix 17. The frequency-dependent quantity α is the *absorption coefficient*, and it can be measured by forming $\ln(I_l/I_0)$. The *molar absorption coefficient* (or *extinction coefficient*) $\varepsilon = \alpha/\ln 10$ is also widely encountered. The product $A = \alpha C l$ is called the *absorbance* of the sample (and sometimes the *optical density*). Therefore another form of the law is

$$\varepsilon C l = A = \lg(I_0/I_l) \tag{12.6.2}$$

which can be used to find ε and α.

The absorption coefficient and the absorbance depend on the frequency of the incident light. The *total intensity* of the transition is the integral of the absorption coefficient over the entire range of frequencies it spans. Therefore we introduce the

Integrated absorption coefficient: $\mathscr{A} = \int \alpha \, \mathrm{d}\nu = c \int \alpha \, \mathrm{d}\tilde{\nu}$. (12.6.3)

$\mathscr{A}$ is related to the transition moment: we show in Appendix 17 that

$$\mathscr{A} = (\pi \nu_{\mathrm{fi}}/3\varepsilon_0 \hbar c) L \, |\boldsymbol{\mu}_{\mathrm{fi}}|^2. \tag{12.6.4}$$

Hence we now have the connection between an observable ($\mathscr{A}$) and a calculable ($\boldsymbol{\mu}_{\mathrm{fi}}$) quantity. When we are interested in the theoretical analysis of an absorption we can estimate $\boldsymbol{\mu}_{\mathrm{fi}}$ from the measured absorbance; if we can estimate $\boldsymbol{\mu}_{\mathrm{fi}}$ on the basis of some model of the transition, then we can predict the value of $\mathscr{A}$. Sometimes it is convenient to express the transition moment in terms of the *oscillator strength*. The connection (Appendix 17) is

$$f = (4\pi m_e \nu_{\mathrm{fi}}/3e^2\hbar) \, |\mu_{\mathrm{fi}}|^2. \tag{12.6.5}$$

Intense, allowed, transitions have oscillator strengths approaching unity; forbidden transitions usually have $f \ll 1$, typically 10^{-5}–10^{-4} (Table 12.1).

Table 12.1 Oscillator strengths and extinction coefficients

	f	$\varepsilon/\text{cm}^{-1}\,\text{dm}^3\,\text{mol}^{-1}$
Electric-dipole allowed	1	10^4–10^5
Magnetic-dipole allowed	10^{-5}	10^{-2}–10
Electric-quadrupole allowed	10^{-5}	10^{-4}–10^{-1}
Spin forbidden (S–T)	10^{-5}	10^{-2}–10
Parity forbidden	10^{-1}	10^3

Further information: Photochemistry. J. G. Calvert and J. N. Pitts; Wiley, New York, 1966; *Photochemistry.* R. P. Wayne; Butterworth, London, 1970; *Introduction to applied quantum chemistry.* S. P. McGlynn, L. G. Vanquickenborne, M. Konoshita, and D. G. Carroll; Holt, Rinehart, and Winston, New York, 1972.

Example. A *charge-transfer transition* in a molecule can be modelled very crudely by the following one-dimensional process. The initial state is described by a rectangular wavefunction (see Fig. 12.10) centred on $x = 0$ and of width a. The final state is described by a wavefunction of the same size and shape but centred on $x = R$. Calculate the transition dipole moment for the transition.

- *Method.* Normalize each wavefunction to unity. Then calculate the integral $\int \psi_f x \psi_i \,d\tau$, which is trivial for wavefunctions of such simplicity.
- *Answer.* For normalization to unity, write $\psi_i = N$ for $-\frac{1}{2}a \leq x \leq \frac{1}{2}a$, and $\psi_i = 0$ elsewhere; then

$$\int \psi_i^2 \,d\tau = N^2 \int_{-\frac{1}{2}a}^{\frac{1}{2}a} dx = N^2 a; \quad \text{hence } N = 1/\sqrt{a}.$$

Likewise for ψ_f. For the transition moment there is a contribution to the integral only where both ψ_f and ψ_i are simultaneously non-zero (where they overlap, for otherwise the product $\psi_f x \psi_i$ is zero. Therefore, if $R > a$, $\mu_{fi} = 0$. If $R \leq a$,

$$\mu_{fi} = -e\int \psi_f x \psi_i \,d\tau = -e\int_{R-\frac{1}{2}a}^{\frac{1}{2}a} NxN\,dx = -eN^2\tfrac{1}{2}\{(\tfrac{1}{2}a)^2 - (R - \tfrac{1}{2}a)^2\}$$
$$= -eR(a-R)/a = -eR(1 - R/a).$$

The intensity of the transition is proportional to μ_{fi}^2.

- *Comment.* The transition moment is equivalent to a charge $q = -e(1 - R/a)$ shifting between the centres of the two distributions. The overlap integral for the pair of states is 0 for $R > a$ and $S = 1 - R/a$ for $R < a$, and so the transition moment can also be thought of as arising from the motion of a charge $-eS$ through the distance R. A long distance of charge migration may not give rise to an intense transition because the initial and final states may overlap only weakly; this is also true of real molecules.

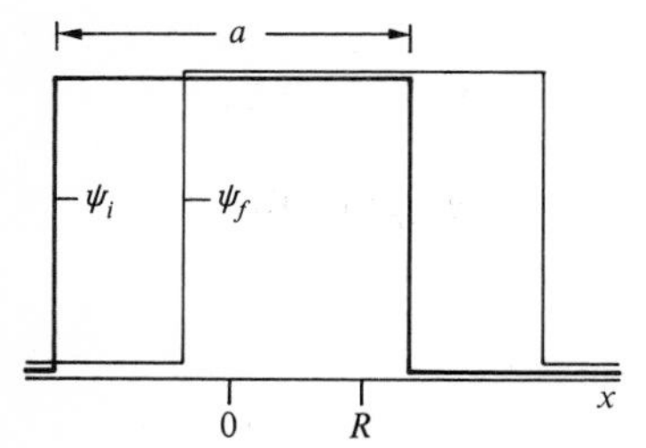

Fig. 12.10. The model for the charge-transfer transition.

12.7 Chromophores

The spectra of other molecules may often be discussed in terms of the chromophores they contain. Among the commonest are the carbonyl and nitro groups, and the ethylenic double bond. The transitions that give rise to their characteristic absorptions may be classified into $\pi^* \leftarrow n$ ('n to π-star') and $\pi^* \leftarrow \pi$ ('π to π-star'), where n represents a non-bonding orbital. An $\pi^* \leftarrow n$ transition in the carbonyl group, which occurs in the vicinity of 290 nm, involves a removal of the electron density on the oxygen to the carbon to a slight extent, because the non-bonding orbital is confined to the oxygen whereas the π^*-orbital extends over both atoms (Fig. 12.11). This explains the shift to higher transition energies that occurs when the chromophore is immersed in a polar solvent or a hydrogen-bonding solvent: the ground state favours a particular conformation of the surrounding solvent but the transition occurs in a time too short for complete reorientation of the solvent, and so while the lower state is stabilized by an interacting solvent the upper state is stabilized to only a lesser extent; consequently the transition moves to higher energies. Other properties of the chromophore may be explained in terms of the variation of the π^*-energies of the upper or the lower state brought about by substituents elsewhere in the molecule.

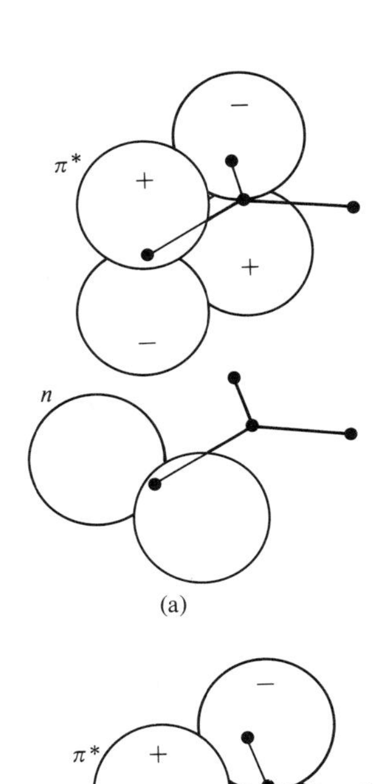

There is, however, one difficulty: the $\pi^* \leftarrow n$ transition is forbidden. This is because the transition dipole moment comes almost entirely from the transition moment on the oxygen atom, because the non-bonding orbital is confined to this atom (to a good approximation). Therefore the integral $\int \psi^*_{\pi^*}\boldsymbol{\mu}\psi_n \, d\tau$ is about equal to $c\int \psi_{p_z}\boldsymbol{\mu}\psi_{p_x} \, d\tau$ if the ψ^*-orbital is taken to be constructed as $c(\mathrm{O2p}_z)+c'(\mathrm{C2p}_z)$ from p_z-orbitals on the carbon and the oxygen. The latter integral is zero, as may easily be verified, and so the $\pi^* \leftarrow n$ transition is forbidden. Intensity may be acquired by the transition by virtue of the fact that the non-bonding orbital is not strictly localized and is not pure p_x, but a major source of intensity is the coupling of the electronic and vibrational modes of the molecule to give rise to a vibronic transition. These we shall discuss shortly.

The $\pi^* \leftarrow \pi$ transition is allowed, and the transition dipole moment is directed along the direction of the bond (Fig. 12.11). The transition reduces the strength of the bond because a bonding electron is transferred to an antibonding orbital. This reduction may be so important that the bonded groups twist about the bond direction in order to reduce the antibonding effect. Thus, in ethene itself, the CH_2 groups are perpendicular in the π^* excited state.

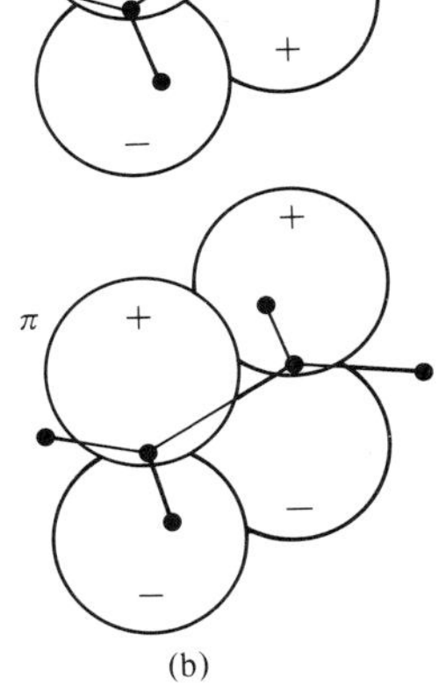

Fig. 12.11. (a) The $\pi^* \rightarrow n$ transition; (b) the $\pi^* \rightarrow \pi$ transition.

The benzene molecule is an interesting but complex example of the transitions involving the π-electrons of a molecule. There are three major bands, one at about 260 nm, the *benzenoid band*, which is weak because it is symmetry-forbidden (but we must investigate shortly why it has any intensity at all), another at 185 nm which, by virtue of its strength, appears to be symmetry-allowed, and a band at about 200 nm, which is still the matter of some dispute. The ground state of the benzene molecule is $^1A_{1g}$, when labelled according to the group D_{6h}. In this group the electric dipole moment operator spans $A_{2u}+E_{1u}$, and so the allowed transitions should be $^1E_{1u} \leftarrow {}^1A_{1g}$ and $^1A_{2u} \leftarrow {}^1A_{1g}$. The strong band has been identified as the former, because the first excited configuration is $a_{2u}^2 e_{1g}^3 e_{2u}$ which gives rise to the states of symmetry $B_{1u}+B_{2u}+E_{1u}$, and not to one of symmetry A_{1u}. This assignment may be confirmed by checking the polarization of the transition moment in a crystalline sample. The disputed band is possibly $^1B_{1u} \leftarrow {}^1A_{1g}$, and the benzenoid band has been ascribed to the transition to the other singlet state of the configuration, $^1B_{2u} \leftarrow {}^1A_{1g}$. There are also triplet terms corresponding to the three singlet terms, but the transition probability between singlets and triplets, the *intercombination transitions*, is very small in molecules built from light atoms with small spin–orbit coupling.

12.8 Vibronically allowed transitions

Our understanding of the benzene spectrum is incomplete unless we can explain why the 'forbidden' bands are partially allowed. The reason is the same as for the carbonyl chromophore.

The potential experienced by an electron in a molecule depends on the nuclear conformation. Therefore the electronic hamiltonian also depends on the conformation, and we may express the dependence in terms of the Taylor expansion of the potential with respect to distortions of the nuclear framework along normal modes. The kinetic energy operators for the electrons are independent of the nuclear coordinates, and so

$$H_e = H_e^{(0)} + \sum_i (\partial V/\partial Q_i)_0 Q_i + \ldots = H_e^{(0)} + \sum_i (\partial H_e/\partial Q_i)_0 Q_i + \ldots \tag{12.8.1}$$

The eigenfunctions ψ_ε of the operator $H_e^{(0)}$ are the electronic wavefunctions for the molecule with energy E_ε and the presence of the other terms in the hamiltonian causes a perturbation that mixes them. The function $\psi_{\varepsilon'}$ becomes

$$\psi = \psi_{\varepsilon'} + \sum_\varepsilon a_\varepsilon \psi_\varepsilon \tag{12.8.2}$$

and by first-order perturbation theory

$$a_\varepsilon = \int \psi_\varepsilon^* \left\{ \sum_i (\partial H_e/\partial Q_i) Q_i \right\} \psi_{\varepsilon'} \, d\tau_e / (E_{\varepsilon'} - E_\varepsilon). \tag{12.8.3}$$

Suppose that only the upper state of a pair is perturbed, then the transition dipole for the excitation $\varepsilon' \leftarrow \varepsilon''$ is

$$\langle \varepsilon' | \boldsymbol{\mu} | \varepsilon'' \rangle = \int \psi^* \boldsymbol{\mu} \psi_{\varepsilon''} \, d\tau_e = \int \psi_{\varepsilon'}^* \boldsymbol{\mu} \psi_{\varepsilon''} \, d\tau_e + \sum_\varepsilon a_\varepsilon^* \int \psi_\varepsilon^* \boldsymbol{\mu} \psi_{\varepsilon''} \, d\tau_e.$$

If the transition between the unperturbed levels ε' and ε'' is forbidden the first integral in this expression vanishes, and we are left with

$$\langle \varepsilon' | \boldsymbol{\mu} | \varepsilon'' \rangle = \sum_\varepsilon a_\varepsilon^* \int \psi_\varepsilon^* \boldsymbol{\mu} \psi_{\varepsilon''} \, d\tau_e. \tag{12.8.4}$$

If, therefore, the transitions $\varepsilon'' \rightarrow \varepsilon$ are allowed, and the perturbation can mix the functions ψ_ε and $\psi_{\varepsilon'}$, the $\varepsilon' \leftarrow \varepsilon''$ transition *borrows* intensity from the allowed transitions.

The next task is to see which states can be mixed. Group theory provides the answer. The hamiltonian must transform as A_1, therefore so too must each term in its expansion. Therefore so too must the second term in eqn (12.8.1) taken as the perturbation. But the normal coordinates Q_i span Γ^i, which is not necessarily A_1; therefore $(\partial H_e/\partial Q_i)_0$ must itself span the irreducible representation Γ^i if its product with Q_i is to be totally symmetric. The part of the hamiltonian in the perturbation that contains the electron coordinates is $(\partial H_e/\partial Q_i)_0$, and so the integral in eqn (12.8.3) is non-zero only if the direct product $\Gamma^{\varepsilon} \times \Gamma^i \times \Gamma^{\varepsilon''}$ contains A_1.

One final important point may be made before we proceed to an example. The presence of the factor Q_i in the perturbation implies that there is a mixing of electronic states and a mixing of vibrational states. Therefore eqn (12.8.2) should read

$$\psi = \psi^{\mathrm{el}}\psi^{\mathrm{vib}} = \psi_{\varepsilon'}\psi_{v'} + \sum_{v}\sum_{\varepsilon} a_{\varepsilon v}\psi_{\varepsilon}\psi_{v} \tag{12.8.5}$$

where

$$a_{\varepsilon v} = \sum_i \left\{ \int \psi_v^* Q_i \psi_{v'}\, \mathrm{d}\tau_{\mathrm{N}} \int \psi_\varepsilon^* (\partial H_e/\partial Q_i)_0 \psi_{\varepsilon'}\, \mathrm{d}\tau_e \right\} (E_{\varepsilon' v'} - E_{\varepsilon v})^{-1}, \tag{12.8.6}$$

and a continuation of this calculation leads to the conclusion that *a vibrational transition is excited during the electronic transition*, and so transitions are vibronic in the sense introduced in connection with diatomic molecules (p. 323). The interpretation of the breakdown of the symmetry selection rule then appears in a new light: *we should apply the rule to the vibronic states, not simply to the electronic states*, and therefore consider the *overall* symmetry of the states.

As an example, consider the forbidden $\mathrm{B}_{2u} \leftarrow A_{1g}$ band in benzene. If an E_{2g} vibration can be excited at the same time as an electronic transition occurs, then since $E_{2g} \times B_{2u} = E_{1u}$, and the transition $\mathrm{E}_{1u} \leftarrow \mathrm{A}_{1g}$ is electric-dipole allowed, the vibronic transition is allowed whereas the electronic transition $\mathrm{B}_{2u} \leftarrow \mathrm{A}_{1g}$ alone is forbidden. Therefore the $\mathrm{B}_{2u} \leftarrow \mathrm{A}_{1g}$ transition acquires intensity through its coupling to the vibrations of the molecule. The alternative viewpoint is that the E_{2g} vibration can mix the electronic state B_{2u} with the electronic E_{1u} by virtue of the appearance of A_{1g} in the direct product $E_{1u} \times E_{2g} \times B_{2u}$; therefore the transition $\mathrm{B}_{2u} \leftarrow \mathrm{A}_{1g}$ borrows intensity from the allowed $\mathrm{E}_{1u} \leftarrow \mathrm{A}_{1g}$ transition.

The same argument and interpretations may be applied to other cases, for example the carbonyl intensity stems from an out-of-plane bending vibration. The intensity of transitions in octahedral transition metal complexes also arises from vibronic effects. It is easy to see that some such mechanism is necessary because these complexes have a centre of inversion, and so a g, u-classification of the states is appropriate. But d-orbitals are *gerade*, and so only states with g symmetry may be constructed from them. As the dipole moment operator is *ungerade* no transitions between unperturbed states are allowed. This corresponds to the Laporte selection rule in atoms, which prohibits d–d transitions. If there is a coupling of the electronic states to a vibration that destroys the inversion symmetry of the molecule there may occur some mixing between orbitals that are *gerade* and *ungerade* in the undistorted molecule. That is, some p-orbitals may be mixed with d-orbitals and so provide an excitation route $d \rightarrow p \rightarrow d$. Alternatively we may view the coupling of the electronic and vibrational modes as requiring the discussion to be based on the symmetry of the vibronic states and not the electronic states alone. In this way we are led to the same conclusion that there has to be a coupling between electronic states and *ungerade* vibrations.

12.9 Singlet–triplet transitions

Another type of selection rule which is often disobeyed is the one forbidding intercombination bands, for example, *singlet–triplet transitions.* Such bands occur when in the molecule there is present a heavy atom having a large spin–orbit coupling constant. We have to demonstrate that the spin–orbit coupling hamiltonian can act as a perturbation able to mix singlets and triplets.

Consider an operator that can be written in the form

$$\Omega = \sum_i R(i) s_z(i), \tag{12.9.1}$$

where the sum is over the electrons in the molecule and R is an operator for the spatial component of the wavefunction. The effect of this operator on a singlet function for two electrons may be demonstrated quite easily as follows.

$$\begin{aligned}\Omega\, {}^1\psi &= \{R(1)s_z(1) + R(2)s_z(2)\}\phi(1)\phi(2)(\tfrac{1}{2})^{\frac{1}{2}}\{\alpha_1\beta_2 - \beta_1\alpha_2\} \\ &= (1/2\sqrt{2})[\{R(1)\phi(1)\}\phi(2)\{\alpha_1\beta_2 + \beta_1\alpha_2\} \\ &\quad - \phi(1)\{R(2)\phi(2)\}\{\alpha_1\beta_2 + \beta_1\alpha_2\}] \\ &= \frac{1}{2\sqrt{2}}\{\phi'(1)\phi(2) - \phi(1)\phi'(2)\}\{\alpha_1\beta_2 + \beta_1\alpha_2\} \propto {}^3\psi,\end{aligned}$$

where $\phi' \equiv R\phi$. We see that the effect of Ω is to mix the $M_s = 0$ state of the *triplet* with the $M_s = 0$ of the *singlet.*

The spin–orbit interaction is

$$H_{so} = \sum_i \xi_i \mathbf{l}_i \cdot \mathbf{s}_i \tag{12.9.2}$$

and has the form of the operator Ω (it also contains terms in s_x, s_y); therefore we should expect it to induce singlet–triplet mixing. In the case of two electrons $H_{so} = \xi_1 \mathbf{l}_1 \cdot \mathbf{s}_1 + \xi_2 \mathbf{l}_2 \cdot \mathbf{s}_2$, which can be written more usefully as

$$H_{so} = \tfrac{1}{2}(\xi_1\mathbf{l}_1 + \xi_2\mathbf{l}_2) \cdot (\mathbf{s}_1 + \mathbf{s}_2) + \tfrac{1}{2}(\xi_1\mathbf{l}_1 - \xi_2\mathbf{l}_2) \cdot (\mathbf{s}_1 - \mathbf{s}_2). \tag{12.9.3}$$

The operator $\mathbf{s}_1+\mathbf{s}_2$ commutes with S^2, the total spin operator, because $\mathbf{S}=\mathbf{s}_1+\mathbf{s}_2$, and all operators commute with themselves; therefore we may neglect it because it cannot mix multiplets. The operator $\mathbf{s}_1-\mathbf{s}_2$ does not commute with S^2 and this component of the perturbation is the one responsible for the mixing. The structure of the operator also lets us predict the symmetry of the orbital state mixed by the orbital part of the perturbation. The orbital angular momentum operators transform as rotations, and so the irreducible representations they span are easily determined by reference to a character table. If the symmetry species of the state of interest is known, the symmetry species of the states which can be mixed with it may be determined by the familiar direct product rule.

For example, in a C_{2v} molecule such as ClO_2^- or SO_2 the $^1A_1$ ground state contains an admixture of $^3A_2$, $^3B_1$, and $^3B_2$ (because the orbital angular momenta transform as A_2, B_1, and B_2), and so too will any $^1A_1$ excited state. A $^1B_1$ state is mixed with $^3B_2$ among others, and as $^1B_1 \leftarrow {}^1A_1$ is electric-dipole allowed, the triplet mixture in the upper ($^1B_1$) state lends intensity to $^3B_2 \leftarrow {}^1A_1$. In benzene the $^3B_{1u}$ state is mixed by the spin–orbit coupling with the states $^1E_{2u}$ and $^1B_{2u}$: vibronic coupling then provides intensity for transitions to these states from the $^1A_{1g}$ ground state, and so a route is provided for the transition $^3B_{1u} \leftarrow {}^1A_{1g}$. This transition is known in benzene itself, but its intensity is enhanced if a heavy atom substituent is present.

12.10 The decay of excitation

Most of the electronic spectra considered in this chapter have been in absorption. Polyatomic molecules are usually studied in this way because the ground state is generally the only one significantly populated at normal temperatures. Excited molecules do not remain excited, and there are a number of interesting features connected with their return to the ground state.

The most common mode of return is by *thermal decay* involving *radiationless transitions.* If the excited molecule is subjected to frequent collisions, then the colliding species may act as an acceptor for the excess energy. Collisions transfer electronic energy of one molecule into the vibrational modes of the other molecules in its environment, and this distributed energy is dispersed ever more widely as the vibrational energy is transferred by further collisions into the rotational and translational degrees of freedom of the system. In this way the excitation energy is degraded into thermal motion, and the direction of this *relaxation process* is determined by the direction of increasing entropy. The initial energy transfer process into the vibrational modes of the system is the consequence of a perturbation that mixes the two modes. Therefore we see that the efficiency of the transfer, and therefore the lifetime of the electronic state, depends upon how well the electronic energy of the excited molecule matches the vibrational energy, or one of the overtones, of the acceptor molecule. The lifetime of an excited state may be affected profoundly by choosing a solvent carefully. Water has rather high vibrational frequencies and its harmonics coincide with a range of typical electronic excitation energies. A solvent such as selenium oxychloride, in which the wavenumber of the highest fundamental is only 995 cm^{-1}, provides only a poorly matching set of acceptor levels.

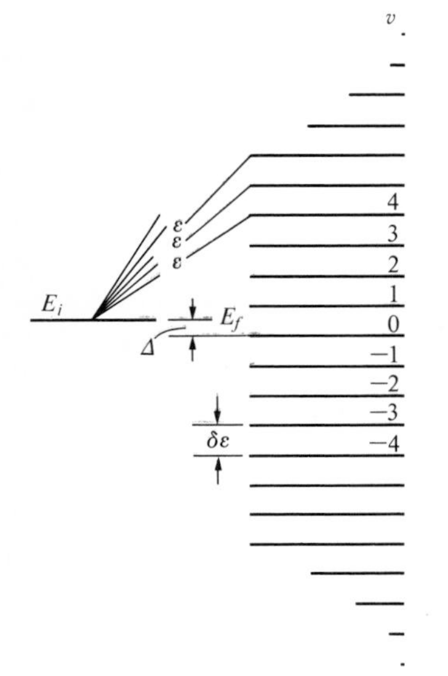

Fig. 12.12. The model for energy degradation into a reservoir.

Example. Consider the following model of a non-radiative transition. Let the initial state be i, and let there be a uniform ladder of states v of spacing $\delta\varepsilon$ in the system that acts as its thermal reservoir (Fig. 12.12). Let the matrix elements of the perturbation linking the two systems be ε for all values of v. Calculate the probability of finding the original excitation in the reservoir.

- *Method.* Using perturbation theory, the probability amplitude for the system being in the reservoir state v is $\varepsilon/(E_i - E_v)$ (from eqn (8.1.23)). The probability is the square of this, and the total probability is the sum over all states v. Use $E_v = E_f + v\,\delta\varepsilon$ with $v = 0, \pm1, \pm2, \ldots$ and write $\Delta = E_i - E_f$.
- *Answer.*

$$P = \varepsilon^2 \sum_v \{1/(E_i - E_v)^2\}$$

$$= \varepsilon^2 \sum_v \{1/(\Delta - v\,\delta\varepsilon)^2\} = (\varepsilon/\delta\varepsilon)^2 \sum_v \{1/[(\Delta/\delta\varepsilon) - v]^2\}$$

$$= (\varepsilon/\delta\varepsilon)^2 \pi^2 \operatorname{cosec}^2(\pi\Delta/\delta\varepsilon).$$

(For the sum, see M. Abramowitz and I. A. Stegan, *Handbook of mathematical functions*, Dover (1965), eqn (4.3.92).)

- *Comment.* This is a greatly simplified version of the Bixon–Jortner theory of radiationless transitions. (M. Bixon and J. Jortner, *J. Chem. Phys.*, 3284, **50** (1969)). The quantity $1/\delta\varepsilon = \rho$, the density of states in the receptor system. Therefore an alternative version is $P = (\varepsilon\pi\rho)^2 \operatorname{cosec}^2(\pi\rho\Delta)$. If E_i lies half way between the $v = 0$ and $v = 1$ receptor levels, $\Delta = \frac{1}{2}\delta\varepsilon$ and so $P = (\pi\varepsilon\rho)^2$ and the probability of the excitation appearing in the reservoir depends strongly on its density of states.

Decay processes involving the radiation of the excitation energy may occur, and give rise to the phenomena of *fluorescence* and *phosphorescence.* The phenomenological difference between them is one of lifetime: fluorescence ceases when the exciting irradiation ceases, whereas phosphorescence continues, but usually for only a short time (fractions of seconds). There is a deeper *mechanistic* distinction related to the forbidden nature of singlet–triplet transitions: fluorescence involves transitions through a sequence of singlet states, and is rapid because no intercombination transitions are involved; phosphorescence depends on a path involving a triplet state, and as the transition out of this state is slow, the lifetime of the effect is long.

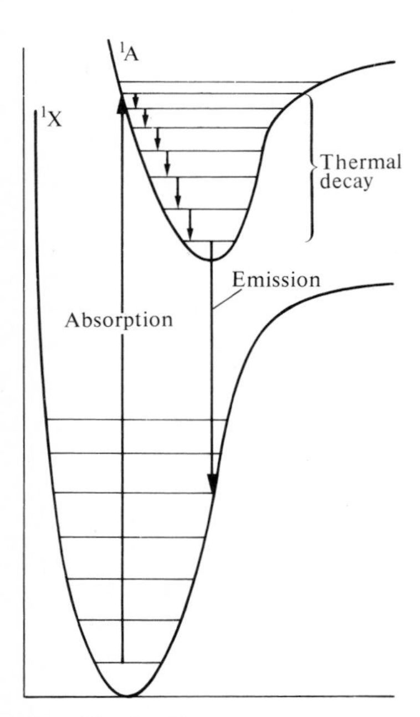

Fig. 12.13. The mechanism of fluorescence.

The transitions that give rise to *fluorescence* are shown in Fig. 12.13. The absorption is $^1A \leftarrow {}^1X$. The Franck–Condon principle governs the vibronic transitions, and so a range of vibrationally excited levels is populated. Intermolecular collisions transfer this vibrational energy into other modes of the system, but the solvent may be such that the excess electronic energy is too large to be degraded efficiently. Therefore the molecules are degraded only into the lowest vibrational level of the excited singlet. If the radiationless transition lifetime is sufficiently long, spontaneous emission of radiation occurs from this level into a range of vibrational levels of the lower singlet, again in accord with the Franck–Condon principle. This $^1A \rightarrow {}^1X$ transition will therefore appear as an emission line at a lower wavenumber than the incident radiation (the net flow of energy from the solvent or lattice into the molecules is improbable) and with a vibrational structure that is characteristic of the electronic ground state. This means that fluorescence spectra can be used to obtain information about the potential energy curve of the ground electronic state, and as the intensity depends upon the competition between the radiationless decay induced by the solvent and the spontaneous emission process, valuable information may be obtained on the role of energy transfer in solution.

The radiation step in fluorescence is a *spontaneous* emission process because the only photons present are the incident photons, and they have a different frequency from the fluorescence photons, and so do not stimulate the emission. If photons of the fluorescence frequency are present, stimulated emission can occur, and this enhances the intensity of the fluorescence by causing *resonance fluorescence.* If the population of the upper singlet can be made greater than that of the lower singlet, that is, if we raise so many molecules into the ^1A state that the population becomes equalized or even inverted† then the stimulated emission will be very strong, and a few incident photons stimulate a great deal of radiation and so amplify the incident (resonant) light. This process of **l**ight **a**mplification by the **s**timulated **e**mission of **r**adiation is the basis of *laser action.* Many of the original laser materials were solids or gases because then the population inversion is easier to obtain, but liquid lasers are now readily available because intense light sources (other lasers) are available to pump the absorption.

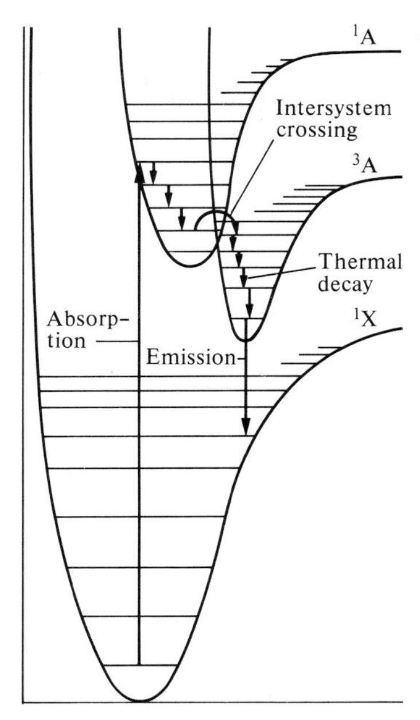

Fig. 12.14. The mechanism of phosphorescence.

The first solid state laser was ruby (Al_2O_3 with some replacement of Al^{+3} by Cr^{+3}). The ground state of the Cr^{+3} ion in the distorted octahedral field of the lattice is $^4A_{2g}$, and when this is irradiated with green (550 nm) light it fluoresces red (693 nm) owing to the absorption $^4T_{2g} \leftarrow {}^4A_{2g}$ being followed by a radiationless decay. $^4T_{2g} \rightarrow {}^2E_g$, which then emits into the ground state. If the ruby crystal is irradiated with a very intense light the population of the ^2E_g state may exceed that of the ground state, and so amplification may occur.

The transitions leading to *phosphorescence* are illustrated in Fig. 12.14. The first step is an absorption from the singlet ground state to an excited singlet, $^1A \leftarrow {}^1X$. Thermal degradation of the vibrational levels of the singlet state then occurs, and a number of courses are then

† If we were to pretend that the populations still obeyed a Boltzmann distribution, then for equality of population of states of different energy the temperature would effectively be infinite, and for a population inversion it would need to be negative. Since the inverted system possesses more energy than normal states we should be further forced to conclude that systems with $T < 0$ are 'hotter' than those with $T \geq 0$. This is, of course, only a formal temperature.

open to the system. If the degradation is fast, it may go as far as the vibrational ground state and radiationless decay or fluorescence to the ground state may follow. If the vibrational damping of the upper singlet is not too fast there may be a sufficiently strong perturbation present to cause a mixing of the singlet and triplet terms ^1A and ^3A (by Hund's multiplicity rule the triplet lies below the singlet), and instead of stepping down the vibrational ladder of ^1A the molecule might undergo an *intersystem crossing* and transfer to the triplet. The probability of this crossing depends upon the strength of the perturbation (for example, a spin–orbit interaction due to the presence of a heavy atom) and the overlap of the vibrational wavefunctions. The crossing occurs with the greatest probability at the point of intersection of the curves. If crossing does occur, thermal degradation down the triplet's vibrational levels will continue, and the molecules get trapped in the potential well of the triplet. Now there is little that the molecule can do. It cannot return to the ground state because the singlet–triplet transition is forbidden, and it cannot return to the upper singlet unless an excessively strong collision provides it with sufficient vibrational energy to approach the intersection. It is not quite true that it can do nothing, because the fact that intersystem crossing occurred indicates the presence of a perturbation that can mix the multiplicities. Therefore the singlet–triplet transition is not completely forbidden. Consequently there is a non-zero, but fairly small probability that the transition $^3A \rightarrow {}^1X$ can occur, and if it occurs before thermal degeneration of the triplet's electronic energy, an emission line at lower frequencies than the incident radiation, and with vibrational structure characteristic of the ground singlet state, will be observed. The emission may occur even after the irradiating illumination is extinguished, until the triplet level is depleted, and in some cases this may take seconds or even minutes.

12.11 The conservation of orbital symmetry

A knowledge of the way that electron distributions are reorganized is essential for understanding how molecules undergo reactions. One of the most significant illustrations in recent years of how theoretical ideas can be used to explain and systematize a wide range of observations in organic chemistry has come from the *Woodward–Hoffmann rules* and the application of the idea of the conservation of orbital symmetry. We shall see in this section the interplay of ideas stemming from molecular orbital theory, electron transition processes, and group theory, and their application to organic reaction mechanism.

We consider *pericyclic reactions*. A pericyclic reaction is a *concerted process* (i.e. one in which bond breaking and formation occur simultaneously) which takes place by the reorganization of electron pairs within a closed chain of interacting atomic orbitals. We shall concentrate on two types of this class of reaction: on *cycloaddition*, in which two or more molecules condense to form a ring and form new

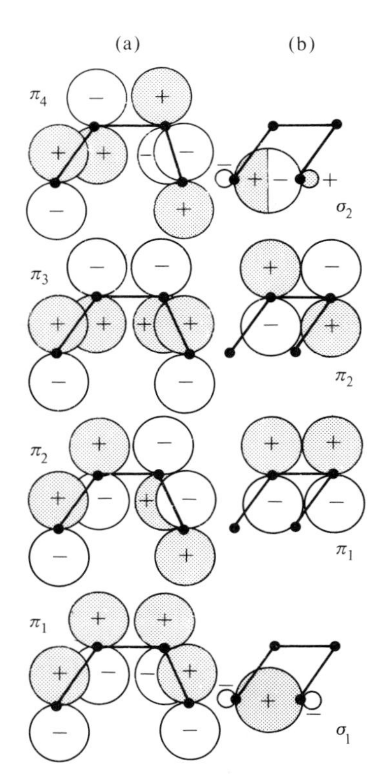

Fig. 12.15. A schematic representation of the molecular orbitals of (a) butadiene and (b) cyclobutene.

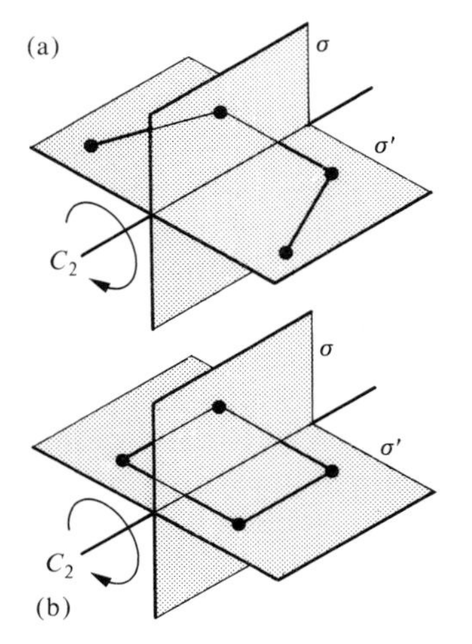

Fig. 12.16. Common symmetry elements of (a) butadiene and (b) cyclobutene.

σ-bonds at the expense of breaking old π-bonds; and on *electrocyclic reactions,* which involve ring opening or closure in a single molecule. An example of a cycloaddition reaction is the *Diels–Alder reaction* (e.g. the reaction of ethene and butadiene to form cyclohexene), and an example of an electrocyclic reaction is the ring-opening of cyclobutene to form butadiene (and vice versa). Each of these reactions has interesting features which can be explained very readily on the basis that orbital symmetry is conserved in the course of the reaction.

Consider first the electrocyclic reaction butadiene $\rightarrow$ cyclobutene, $\rightarrow$. The four butadiene π-orbitals have already been described (Fig. 10.22), and they are drawn again on the left of Fig. 12.15. As a result of the formation of the ring a π-bond turns into a σ-bond, and the orbital scheme for cyclobutene is shown on the right of Fig. 12.15 (we disregard throughout this discussion all the small changes that may also occur in the original σ-bonding framework of the molecule). Now, both butadiene and cyclobutene have symmetry elements in common. For instance, both have a C_2 axis, and both have mirror planes, Fig. 12.16. Therefore, it should be possible to trace the molecular orbitals as they change from their initial to their final form simply by keeping an eye on their symmetries with respect to the elements common to both the reactants and the products. In other words, we should be able to set up a *correlation diagram* showing how the orbitals on the left of Fig. 12.15 relate to those on the right. When that has been done we shall be in a position to describe the orbital scheme of the *transition state,* a state of the molecule intermediate between the reactant and product form and through which it must pass if the reaction is to occur.

There is, however, a crucial complication, but it is this complication that makes pericyclic reactions so interesting. We have to consider in slightly more detail how the initial molecule changes into the product. We can see that there are two pathways, Fig. 12.17. In one, the *conrotatory* path, the CH_2 groups rotate in the same sense. In the other, the *disrotatory* path, they rotate in opposite senses. Neither transition state possesses the full common symmetry of the initial and final molecules: the conrotatory path allows the C_2 axis to survive throughout the reaction, but the mirror planes are present only at the beginning and end; the disrotatory path preserves one of the mirror planes, but the C_2 axis and the second plane are present only at the beginning and end. Therefore, in order to construct the correlation diagram, we must examine the evolution of the orbitals in these two different *reduced* symmetry groups: the full symmetries of the molecule in its reactant or product forms are not relevant during the reaction process itself.

Consider the conrotatory path, the path conserving the C_2 element. The four orbitals $\pi_1, \ldots, \pi_4$ of butadiene have characters $-1, 1, -1, 1$ under C_2, Fig. 12.15. In the application of group theory to organic reaction mechanisms it is conventional to be less formal with the

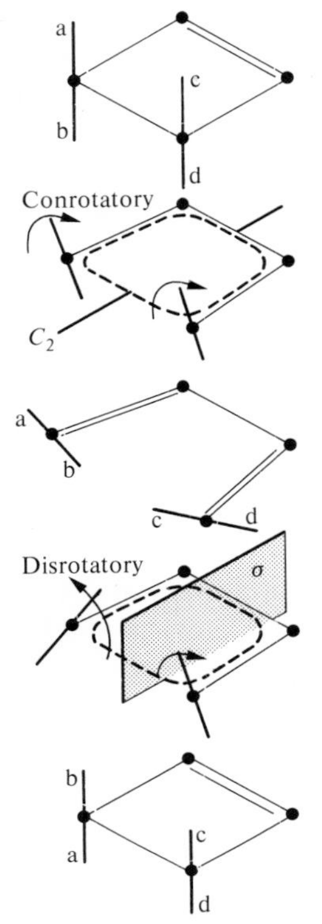

Fig. 12.17. The conrotatory and disrotatory ring closures of butadiene. (The labels a, b, c, d identify the protons; they do not necessarily indicate substituents.)

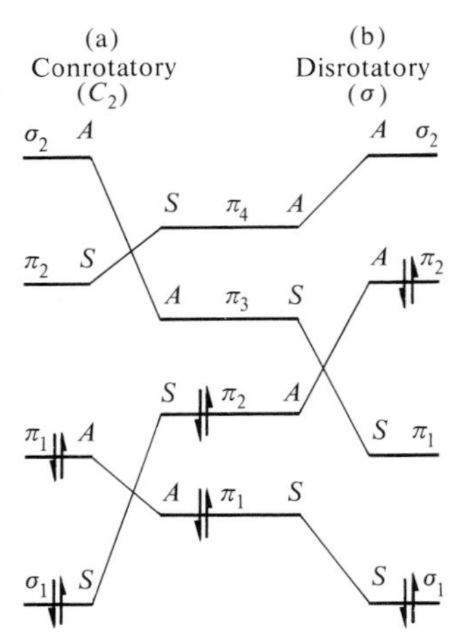

Fig. 12.18. The correlation diagram for (a) the conrotatory and (b) the disrotatory butadiene–cyclobutene interconversion.

notation, and orbitals are classified as S (for symmetric; character $+1$) or A (for antisymmetric; character -1). We adopt this notation from now on. The four orbitals of the cyclobutene molcule may also be classified under C_2, and are S, A, S, A, respectively, Fig. 12.15. The classification is summarized on the left of Fig. 12.18. The C_2 element is common to the reactant, the transition state, and the product, and so the symmetry labels S and A are applicable throughout the course of the reaction: they are 'good quantum numbers'. Therefore, we know that S orbitals of butadiene correlate with S orbitals of cyclobutene, and A correlate with A. The ambiguity about which S orbital correlates with which S (and which A with which A) is resolved by the *non-crossing rule*, which states that lines joining orbitals of the same symmetry may never cross. The basis of this rule lies in a combination of the ubiquitous Figs. 8.1 and 8.2 with the group theoretical criterion for non-zero matrix elements. Figure 8.2 shows that two levels do not cross as their unperturbed energies approach each other if the hamiltonian for the system has non-vanishing matrix elements between them; the group theoretical criterion for the non-vanishing of $\int \psi_a H \psi_b \, d\tau$ is that $\Gamma^a \times \Gamma^H \times \Gamma^b$ contains A_1. But the hamiltonian is totally symmetric (under the operations that are preserved throughout the reaction pathway, such as C_2 in the present example), and if $\Gamma^a = \Gamma^b$ (which is the case if the symmetries of the orbitals are the same with respect to the preserved symmetry elements), then the product $\Gamma^a \times \Gamma^H \times \Gamma^b$ does contain A_1. Consequently, levels of the same symmetry do not cross. It follows that the correlation diagram for the conrotatory electrocyclic reaction is that shown on the left of Fig. 12.18.

Before putting the diagram to work, consider the disrotatory path. The C_2 symmetry classification is no longer relevant because C_2 is not a preserved symmetry element during the reaction. Now we have to consider the classification under the mirror plane σ. The molecular orbitals of butadiene are S, A, S, A respectively, and those of cyclobutene are S, S, A, A. The non-crossing rule then lets us construct the correlation diagram on the right of Fig. 12.18.

Now we see the difference between the two pathways. Suppose there is insufficient energy available for the electrons to be excited out of the ground state of the reactant molecule (that is, we are considering a *thermal* reaction pathway, not a photochemical pathway). Then follow the course of the ring formation reaction. Initially the lowest two molecular orbitals of butadiene are occupied and its configuration is $\pi_1^2\pi_2^2$; under the conrotatory reaction scheme this configuration goes smoothly over into the ground state configuration $\sigma_1^2\pi_1^2$ of the product, and the energy demands of the reaction are minimal. On the other hand, under the disrotatory scheme, one of the electron pairs ends up in a high energy orbital, and the product is in the excited configuration $\sigma_1^2\pi_2^2$. There is insufficient energy available for this process to occur; therefore we conclude that in the thermal cyclization of butadiene, only the conrotatory path is taken. Conversely, since we can also interpret the reaction correlation diagram backwards, we can also conclude that in thermally induced cyclobutene ring opening, it is the

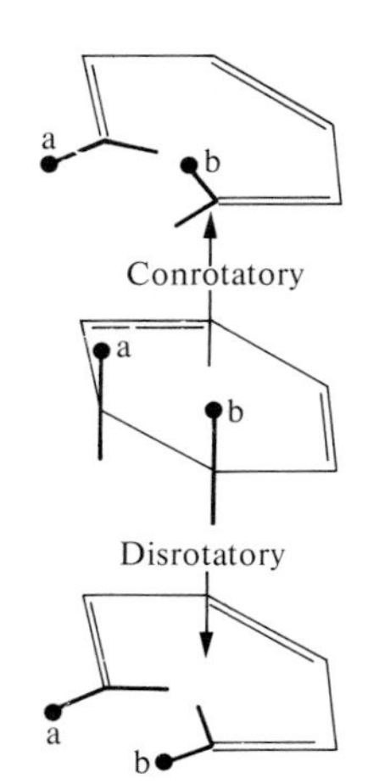

Fig. 12.19. The stereochemical consequences of different reaction paths.

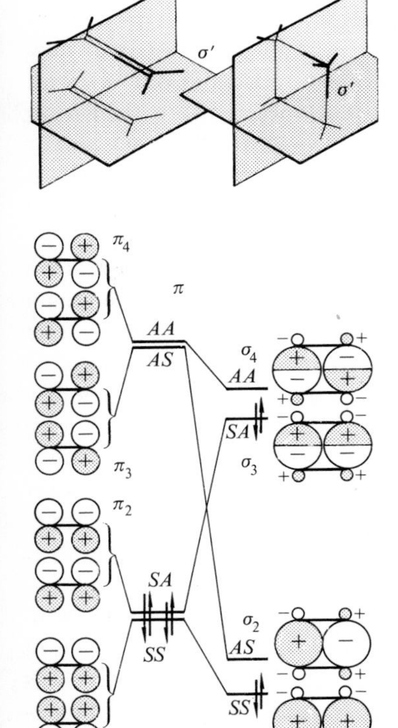

Fig. 12.20. The correlation diagram for the dimerization of ethene. The (*SA*, *SS*) pair is degenerate when the ethenes are far apart; likewise the (*AA*, *AS*) pair. The symmetry classification refers to the $\sigma\sigma'$.

conrotatory pathway that is taken (under conrotatory opening $\sigma_1^2\pi_1^2$ evolves into ground state $\pi_1^2\pi_2^2$ while under disrotatory opening $\sigma_1^2\pi_1^2$ evolves into $\pi_1^2\pi_3^2$). Notice, furthermore, how the highest occupied molecular orbital (which is referred to as the HOMO) dominates the conclusions based on the correlation diagram: the HOMO correlates strongly upwards in energy for the thermally forbidden reaction. This is a general feature, and accounts for the importance of the *frontier orbitals* (i.e. the HOMO and the LUMO, the lowest unoccupied molecular orbital) in reaction mechanisms.

There are two experimentally verifiable predictions that come from this account. In the first place we expect the activation energy for ring scission to be quite small because it can occur while the electrons remain in an overall ground state configuration. The experimental value is in fact only about 80 kJ mol^{-1}, and can be ascribed to changes in the σ-framework of the molecule and changes of hybridization, which are effects ignored in the correlation scheme. Secondly, the conrotatory path has specific stereochemical implications. Take, for instance, the analogous six π-electron reaction shown in Fig. 12.19 where the labels a and b signify substituents. The presence of substituents is normally insufficient to disturb the orbital correlation diagram significantly, and so they act simply as labels. An analysis of the appropriate correlation diagram shows that the thermally feasible reaction takes place along the *disrotatory* path, and it gives a stereochemically distinct product from the thermally forbidden reaction, which in this case is along the conrotatory path. This is confirmed experimentally. The alternation conrotatory, disrotatory, . . . for the thermally allowed reaction as the number of electrons in the π-system changes along the series 4, 6, . . . is a general prediction for electrocyclic reactions, and constitutes one of the Woodward–Hoffmann rules.

The same type of argument can be used to explain the relative ease, and the stereochemical consequences, of cycloadditions. We shall consider two contrasting examples: the thermal dimerization of ethene to cyclobutane and the Diels–Alder addition of ethene to butadiene (which occurs very readily). Why does the latter occur so much more readily than the former? The answer is again to be found in the orbital correlation diagrams: in other words, reactions, like spectral transitions, obey selection rules.

Consider the face-to-face approach of two ethene molecules. In the arrangement shown in Fig. 12.20 the two mirror planes are preserved throughout the reaction, they occur in the initial encounter, in the product, and in the transition state. They can therefore be used for a symmetry analysis of the orbitals. The bonding and antibonding orbitals of the ethenes are *A* or *S* with respect to each of the two mirror planes, and their joint classification and energies are shown on the left of Fig. 12.20. (*SA*, for instance, signifies a joint m.o. that is *S* with respect to σ and *A* with respect to σ'.) The σ-bonds that they form may also be classified as *A* or *S* with respect to each plane, and their order of energies can be judged by assessing the importance of their nodes (this can generally be done intuitively and by supposing that there is very little interaction between different σ-bonds across the cyclobutane ring). The correlation diagram is then constructed by connecting orbitals of the same symmetry but by avoiding crossings. This has been done in Fig. 12.20. Quite clearly, the HOMO of the reactants rises sharply and the dimerization process leads to a cyclobutane molecule in an excited electronic configuration, if the electron populations slide along the relevant connecting lines (technically: if the process is *adiabatic*, the word signifying a change occurring without a transition between states). Therefore we conclude that the ethene–ethene cycloaddition (and the reverse cycloreversion reaction) with the face-to-face geometry is thermally forbidden, which is in accord with observation.

Now consider the same argument for the ethene–butadiene reaction, which is the prototype of the wide class of Diels–Alder additions. We consider the face-to-face approach of the two molecules, which preserves the mirror plane shown in Fig. 12.21 throughout the course of the reaction. The orbitals of ethene and butadiene are depicted jointly on the left of the diagram and are classified with respect to the mirror plane. The left side of the diagram is simply the superposition of the butadiene and ethene orbital energy levels, with the disregarded σ-framework added throughout. In the course of the reaction two new σ-bonds are formed at the expense of two π-bonds (and one π-bond is relocated). The orbitals of the product (cyclohexene) are shown on the right of the diagram, and classified with respect to the same mirror plane. The correlation diagram can now be constructed on the basis of the non-crossing rule, and we can trace the evolution of the bonding electron pairs of the reactants as they evolve adiabatically into products. The obvious feature is that the ground state configuration correlates with the ground state configuration of the product, and, as a consequence, the product molecule is formed in its ground state. The smooth transition from reactants to products involves no excitation of electrons, and its activation energy can be expected to be sufficiently low for the process to be feasible thermally. This is in full accord with the ease with which the Diels–Alder reaction takes place: it is a thermally *allowed* reaction. Note how this $4+2$ π-electron cycloaddition is thermally allowed while the $2+2$ ethene reaction is thermally forbidden in the same face-to-face geometry. This alternation is a general feature and is another of the Woodward–Hoffmann rules.

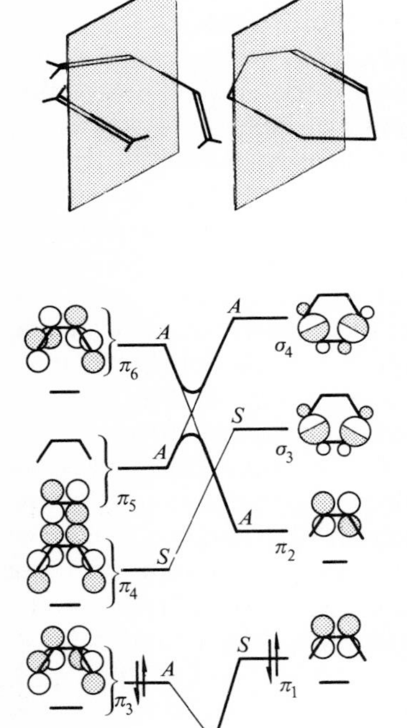

Fig. 12.21. The correlation diagram for the cycloaddition of ethene and butadiene.

Reactions are thermally allowed when there is a transfer of electron pairs from bonding orbitals in the ground state of the reactant molecules to bonding orbitals in the products. A reaction which is thermally forbidden may become photochemically allowed when electrons are excited into higher orbitals. Excitation permits reaction not only because more energy is available to overcome activation barriers to reactions but also because the consequences of orbital symmetry are different. In other words, because the initially occupied orbitals are different the same selection rules permit the exploration of different reaction channels.

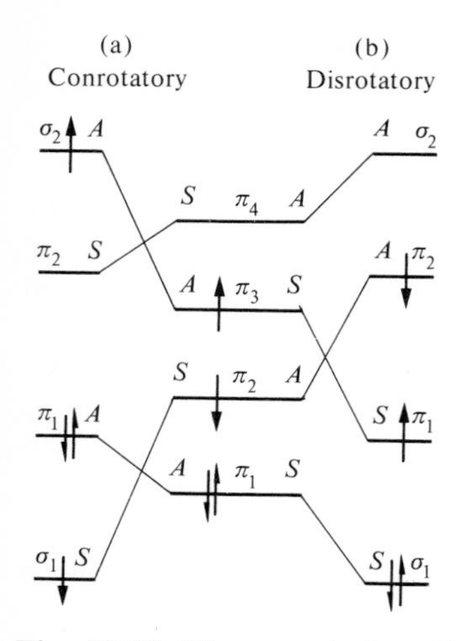

Fig. 12.22. The correlation diagram for the photochemical interconversion of butadiene and cyclobutene.

Consider once again the ring closure of butadiene. This time, however, consider a photochemical mechanism in which the absorption of a photon has led to the excitation of a single electron, Fig. 12.22. The disrotatory adiabatic correlation of the excited butadiene configuration leads to an excited cyclobutene configuration of similar energy to the starting point, while the conrotatory path involves a significant increase in energy. Hence, in contrast to the thermal electrocyclic reaction, the disrotatory path is open to the photochemically induced reaction, and the conrotatory path is closed. This reversal of the thermal prediction is another general feature of electrocyclic reactions, and is another of the Woodward–Hoffmann rules. The photochemical ring closure of butadiene is known, and it does in fact proceed by the predicted disrotatory path; nevertheless there are complications (as in most photochemical processes), and the cyclobutene that is produced is in its electronic ground state, not in an excited state as the correlation diagram suggests. We have to resolve this discrepancy.

A problem with correlation diagrams is that they focus attention on individual *orbitals*: we consider the symmetries of orbitals, and then apply the non-crossing rule to orbitals of the same symmetry. This is only an approximation. What we should deal with are the *overall states* of molecules. How this affects the conclusions can be seen as follows.

Consider the first few excited configurations of butadiene and cyclobutene. This symmetry species are obtained in the normal way, by taking the product of the symmetry species of the individual occupied orbitals, all closed shells being totally symmetric. Since the disrotatory path preserves the single mirror plane, the relevant state classification is in terms of S and A with respect to that plane. Use $S \times S = S$, $A \times A = S$, and $A \times S = A$ (which follow from the characters 1 and -1 for S and A respectively). The ground states are S (they are closed-shell species). The first excited configuration of butadiene is $\pi_1^2\pi_2\pi_3$, which is of symmetry $S \times S \times A \times S = A$. Since the two outermost electrons occupy different orbitals we may have a singlet or a triplet version of this configuration, with the triplet slightly lower in energy than the singlet. The next higher configuration is $\pi_1^2\pi_3^2$, which is S overall, and a singlet. The cyclobutene states are as set out on the diagram in Fig. 12.23. The correlation diagram in Fig. 12.18 can be used to make a first approximation to the state correlation diagram because since butadiene(π_1, π_2) correlates with cyclobutene(σ_1, π_2) it follows that butadiene($\pi_1^2\pi_2^2$, ${}^1S$) correlates with cyclobutene($\sigma_1^2\pi_2^2$, ${}^1S$). This lets us draw the correlation lines in Fig. 12.18. Now we see an important point: the *states* of the same overall symmetry cross. Like orbital crossing this is also forbidden, and so the true correlation diagram should be as depicted by the heavy lines in the diagram. The interaction of the two states of the same overall symmetry is another example of the configuration interaction (CI) encountered in Chapter 10, where we saw that it modified the description of molecules.

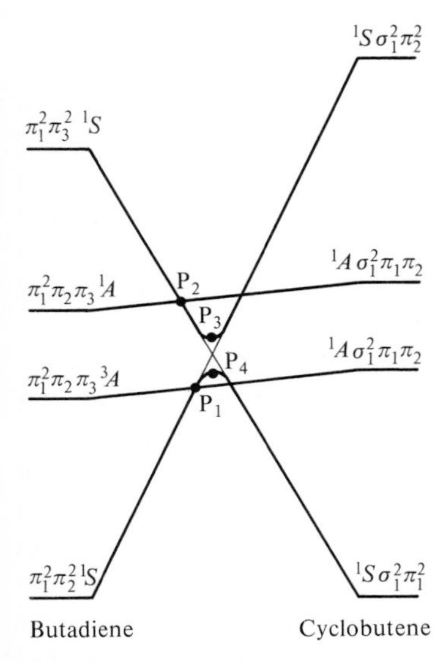

Fig. 12.23. The state correlation diagram for the butadiene–cyclobutene interconversion.

Now consider the disrotatory ring closure in terms of the overall states of the molecule. If the butadiene is in its ground state and we are considering a thermal reaction, then although in principle the ground state of cyclobutene can be reached without electronic excitation (that is, the two molecular ground states are on the same potential surface), the reaction involves a considerable activation energy, and is therefore forbidden. This modifies the earlier discussion where we concluded that it is because the disrotatory path leads to an excited state of cyclobutene, that it is forbidden. We now see that the forbidden nature of the reaction stems from the activation barrier, and that that barrier exists for two reasons: the rise in energy is a consequence of orbital correlation (and so that remains an important part of the argument), and the existence of the peak is a consequence of CI and the non-crossing of states of the same symmetry.

If the butadiene is in a triplet excited state, then disrotatory motion moves it to the point marked P_1 on the correlation diagram (Fig. 12.23). There is sufficient spin-orbit coupling to induce intersystem crossing, and so it switches to the lower ^1S curve. It cannot go forward to cyclobutene because that requires a further injection of energy to overcome the barrier at P_4; therefore it loses its energy non-radiatively, and converts back to ground state butadiene. This behaviour is observed. Now suppose that light absorption results in the formation of the first excited singlet of butadiene. Then the simple conclusion would be that it can pass over into the first excited singlet of cylcobutene, as we concluded from the simple orbital correlation diagram. The crossing at P_2, however, plays a significant role because there may be strong enough perturbations present (e.g. nuclear motions and the failure of the Born–Oppenheimer approximation) to induce an internal conversion between curves at P_2, and so the ^1S converts to ^1S when its geometry corresponds to P_2. As the reaction proceeds, the state of the molecule moves on to P_3 where it is sufficiently close to the lower surface for nuclear motions to induce a second internal conversion to the lower ^1S curve (the jumping between curves is an example of a *non-adiabatic process*). This conversion leads to the molecule being in the state represented by the point P_4. Now it needs no activation energy to go on to ground state cyclobutene (or back to ground state butadiene); hence ground state cyclobutene appears in the products of singlet excited butadiene, exactly as observed.

The same kind of analysis can account for the features of photochemically induced cycloaddition reactions. The strategy is to use the orbital correlation diagrams to construct the first approximation to the state correlation diagrams (for the lowest few configurations), then to allow for configuration interaction between states of the same overall symmetry (so that crossings are eliminated from the diagrams), and finally to recognize that all the intersections and the non-crossings are leaky on account of the presence of forgotten perturbations (such as spin–orbit coupling being able to induce intersystem crossing and the failure of the Born–Oppenheimer approximation being responsible for conversions between states of the same multiplicity).

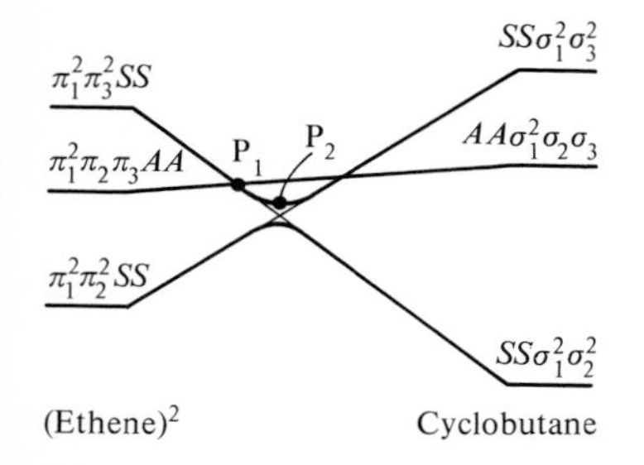

Fig. 12.24. The state correlation diagram for the dimerization of ethene.

In order to see the strategy in operation, consider the dimerization of ethene again. The orbital correlation diagram lets us construct the state correlation diagram in Fig. 12.24. The ground states of the ethene pair and the cyclobutane are each of *SS* symmetry (with respect to the two preserved mirror planes), and the forbidden nature of the thermal reaction can be ascribed to the existence of the high activation barrier. On the other hand, the first excited singlet configuration $(\pi_1^2\pi_2\pi_3)$ correlates, with little change of energy, with the first excited state $(\sigma_1^2\sigma_2\sigma_3)$ of the cyclobutane, and so a simple analysis would lead us to expect the dimerization to be photochemically allowed (which it is) and the products to be excited (which they are not). In order to explain the last point, consider conversions at intersections. The intersection of the curves at P_1 permits one internal conversion, and the close approach of the lower two configurationally interacting curves near P_2 allows a second conversion to the lower curve, the molecule then being able to slide down to either starting material or to product, each being formed in its ground state, as observed.

The face-to-face ethene dimerization is thermally forbidden but photochemically allowed. This reversal of cycloaddition behaviour is a general feature, and is yet another of the Woodward–Hoffmann rules. The Diels–Alder ethene–butadiene addition is thermally allowed. We can see that it is photochemically forbidden by reference to the state correlation diagram. On the basis of the orbital correlation diagram we can construct the diagram in Fig. 12.25. The most obvious feature is the absence of any energy barrier in the correlation of the two ground state configurations: the reaction is predicted to be thermally allowed. The first excited configuration $(\pi_1^2\pi_2^2\pi_3\pi_4)$ correlates with a highly excited configuration $(\sigma_1^2\sigma_2\pi_1^2\sigma_3)$ of the addition product, and so on simple grounds we do not expect it to occur. To some extent configuration interaction alleviates the energy requirements because there is a crossing with a configuration $(\sigma_1^2\sigma_2^2\pi_1\pi_2)$ of the same symmetry, and so the adiabatic evolution of the first excited state ends up in the first excited state of the cyclohexene. Nevertheless, this still leaves a barrier, and so the photochemical process remains forbidden, as observed.

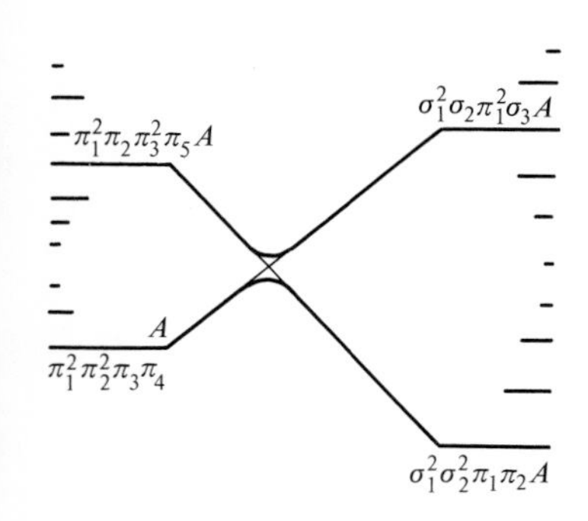

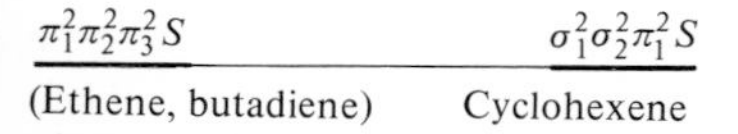

Fig. 12.25. The state correlation diagram for the cycloaddition of ethene and butadiene.

The consequences of orbital correlation diagrams, and their more sophisticated interpretation in terms of state correlations, has led to a much deeper understanding of some aspects of organic chemistry: orbital correlation is a prime example of how much theory can contribute to experimental chemistry.

Further reading

Electronic spectroscopy

The theory of the electronic spectra of organic molecules. J. N. Murrell; Chapman and Hall, London, 2nd edition, 1971.

Spectra of diatomic molecules. G. Herzberg; Van Nostrand, Princeton, 1950.

Electronic spectra of polyatomic molecules. G. Herzberg; Van Nostrand, Princeton, 1967.

The theory of transition metal ions. J. S. Griffith; Cambridge University Press, 1964.
Inorganic electronic spectroscopy. A. B. P. Lever; Elsevier, Amsterdam, 1968.
Introduction to applied quantum chemistry. S. P. McGlynn, L. G. Vanquickenborne, M. Kinoshita, and D. G. Carroll; Holt, Rinehart, and Winston, New York, 1972.

Photochemical processes

Photochemistry. J. G. Calvert and J. N. Pitts; Wiley, New York, 1966.
Photophysics of aromatic molecules. J. B. Birks; Wiley, London, 1970.
Dissociation energies. A. G. Gaydon; Chapman and Hall, London, 1952.
Non-radiative decay of ions and molecules in solids. R. Englman; North-Holland, Amsterdam, 1979.

Lasers

Lasers. B. A. Lengyel; Wiley-Interscience, New York, 1971.
Handbook of lasers. R. J. Pressley; Chemical Rubber Co., Cleveland, 1971.

Orbital symmetry

The conservation of orbital symmetry. R. B. Woodward and R. Hoffmann; Verlag Chemie and Academic Press, 1970.
Organic reactions and orbital symmetry. T. L. Gilchrist and R. C. Storr; Cambridge University Press, 2nd edition 1979.

Problems

12.1. Which of the following transitions are electric-dipole allowed: (a) ${}^2\Pi$–${}^2\Pi$, (b) ${}^1\Sigma$–${}^1\Sigma$, (c) Σ–Δ, (d) Σ^+–Σ^-, (e) Σ^+–Σ^+, (f) ${}^1\Sigma_g^+$–${}^1\Sigma_u^+$, (g) ${}^3\Sigma_g^-$–${}^3\Sigma_u^+$?

12.2. Show that in the carbonyl group the $\pi^* \leftarrow \pi$ transition is allowed, its transition moment lying along the bond.

12.3. Show that the transition ${}^1A_2 \leftarrow {}^1A_1$ is electric-dipole forbidden in H_2O, but may become allowed as a vibronic transition involving one of the molecule's vibrational modes.

12.4. Assess the polarization of the ${}^1A_2 \leftarrow {}^1A_1$ transition in H_2CO and of the $B_{2u} \leftarrow A_g$ transition in $CH_2{=}CH_2$. *Hint.* Use C_{2v} and D_{2h} respectively; consider the role of vibrational coupling.

12.5. In a diamagnetic octahedral complex of Co^{3+}, two transitions can be assigned to ${}^1T_{1g} \leftarrow {}^1A_{1g}$ and ${}^1T_{2g} \leftarrow {}^1A_{1g}$. Are these transitions forbidden? If they are forbidden, what symmetries of vibrations would provide intensity. Can the intensities be ascribed to the admixture of configurations involving p-orbitals?

12.6. Consider the molecular potential energy curves of two electronic states; let their force constants be the same, but the minima offset by a distance δR. Find an expression for the Franck–Condon factors S_{v0}^2 for $v = 0, 1, 2$ as a function of δR. What value of δR is needed for the transition intensity to $v = 1$ to dominate the other two?

12.7. Light of wavelength 256 nm passes through a 1 mm cell containing an 0.05 mol dm^{-3} solution of benzene; the light emerging has only 16 per cent its incident intensity. Calculate the absorption coefficient at that wavelength. Observations at other wavelengths showed that the absorption was roughly constant over a range of 4000 cm^{-1}. Estimate the oscillator strength and the transition moment. *Hint.* Find α from eqn (12.6.1) and approximate $\mathscr{A}$ (eqn (12.6.3)) by $\mathscr{A} \approx c\alpha\,\Delta\tilde{\nu}$ with $\Delta\tilde{\nu} \approx 4000$ cm^{-1}.

12.8. Calculate the transition dipole, the oscillator strength, and the integrated absorption coefficient for a one-dimensional harmonic oscillator consisting of two particles, one of charge e and the other of charge $-e$.

12.9. Calculate the same quantities as in Problem 12.8 for the lowest energy transition in a chain of N carbon atoms according to the free-electron molecular-orbital approximation; refer to Problem 3.14. Evaluate the expressions for the molecule carotene.

12.10. Deduce the effect of the operator H_{so} in eqn (12.9.2) on a two-electron singlet state. *Hint.* Proceed as in the discussion following eqn (12.9.1), but include s^+ and s^-.

12.11. At time $t=0$ a molecule is known to be in a singlet state. The energy separation of the singlet and triplet states is $\hbar J$. Deduce an expression for the time dependence of the probability that the system is in any of the three states of the triplet at some later time as a result of the spin–orbit interaction. Suppose that the sample consists of a large number of molecules which are excited photochemically to a singlet state over a range of time $0 \leq t_0 \leq T$ with equal probability. What is the probability that any molecule is in a triplet state at some time later than T? *Hint.* The basic equation to use is eqn (8.5.10). For the second part, average this equation over a uniform distribution of starting times in the range $0 \leq t_0 \leq T$; i.e. multiply by $\mathrm{d}t/T$ and integrate between the appropriate limits.

12.12. Which states of benzene may be mixed with $^3\mathrm{B}_{1u}$ and $^3\mathrm{B}_{2u}$ by spin–orbit coupling?

12.13. In an aromatic molecule of D_{2h} symmetry the lowest triplet term was identified as $^3\mathrm{B}_{1u}$. What is the polarization of its phosphorescence? *Hint.* Decide which singlet terms can mix with $^3\mathrm{B}_{1u}$ and assess the polarization of the light involved in the return of that state to the $^1\mathrm{A}_g$ ground state.

12.14. The Bixon–Jortner approach to radiationless transitions was sketched in a very simplified form in the *Example* on p. 334. The following is a slightly more elaborate version. Let ψ, an eigenstate of the system hamiltonian $H(\text{sys})$, be the state populated initially, and let ϕ, an eigenstate of the bath hamiltonian $H(\text{bath})$, be a state of the bath. Let $\Psi = a\psi + \sum_n b_n\phi_n$ be an eigenstate of the true hamiltonian H with energy $\mathscr{E}$. Let $\langle\psi \mid \phi_n\rangle = 0$, and $H' = H - H(\text{sys}) - H(\text{bath})$ have constant matrix elements $\langle\phi_n| H' |\psi\rangle = V$ for all n. Show that $H\Psi = \mathscr{E}\Psi$ leads to $Va + (E_n - \mathscr{E})b = 0$ and $(E - \mathscr{E})a + V\sum_n b_n = 0$. Hence find an expression for a and b_n. Let $\mathscr{E} - E_n = \gamma\varepsilon - n\varepsilon$, using $\sum_{n=-\infty}^{\infty} \{1/(\gamma - n)\} = -\pi \cot \pi\gamma$ and $\rho = 1/\varepsilon$, show that $E - \mathscr{E} - \pi\rho V^2 \cot \pi\gamma = 0$, an equation for $\mathscr{E}$. Go on to show on the basis that $a^2 + \sum_n b_n^2 = 1$, that $a^2 = V^2/\{(E - \mathscr{E})^2 + V^2 + (\pi V^2\rho)^2\}$. *Hint.* See M. Bixon and J. Jortner, *J. Chem. Phys.*, 3284, **50** (1969).

Part 4: Advanced applications

THE final part is a foretaste of the kinds of calculations quantum chemists do when they are interested in molecular properties. The material goes beyond the elementary applications we have seen so far, and beyond what is needed in undergraduate courses on quantum chemistry, but it uses no more material than was set out in Parts 1 and 2. You will see that, with a willingness to get involved in slightly more intricate calculations, molecular properties like refractive index, optical activity, magnetic suceptibilities, and magnetic resonance parameters, such as the g-value of e.s.r. and the shielding constant and the spin–spin coupling constant of n.m.r., can be calculated. As in the earlier parts, we concentrate on the processes involved rather than the sophisticated techniques that are employed in order to get results of high precision and accuracy. The material that follows is an extended exercise in the application of perturbation theory: bear in mind throughout that the electric and magnetic properties of molecules are simply their responses to stress. The electric properties can be thought of as a molecule's response to a stretching force; the magnetic properties can be thought of as their response to a twisting force.

13 The electric properties of molecules

THIS chapter explores the properties of molecules in electric fields which may be either static or oscillatory, and generated either by an external apparatus or by another molecule. These properties let us understand the magnitudes of the relative permittivities (dielectric constants) of media, their refractive indices, and their optical activities. They also let us see what controls the strengths of intermolecular forces. Throughout the chapter we draw on the techniques of perturbation theory set out in Chapter 8: *molecular properties are responses to perturbations.*

13.1 The electric polarizability

The electric *polarizability* of a molecule is a measure of its ability to respond to an electric field and to acquire an electric dipole moment. The perturbation in this case is the interaction of the electric field $\mathbf{E}$ with the electric dipole $\boldsymbol{\mu}$ formed by the charges making up the molecule:

Electric dipole interactions: $$H^{(1)} = -\boldsymbol{\mu}\cdot\mathbf{E}, \quad \boldsymbol{\mu} = \sum_i e_i\mathbf{r}_i. \qquad (13.1.1)$$

The higher moments (e.g. the electric quadrupole moment) can also interact with the electric field, but only if it is not uniform (a *quadrupole*, for instance, responds to the *field gradient*, not to the field itself). In order not to get bogged down in subscripts we suppose that the electric field is applied in the z-direction and write $\mathbf{E} = E\hat{\mathbf{k}}$, where $\hat{\mathbf{k}}$ is a unit vector in the z-direction. Then

$$H^{(1)} = -\mu_z E. \qquad (13.1.2)$$

Although we have set up perturbation theory to find expressions for the corrections to the energy, in this chapter we are interested in properties other than the energy. Therefore we have to adapt the results of perturbation theory. There are two approaches. One is to set up the operator for the property of interest and then to evaluate its expectation value using the perturbed wavefunctions. The other is to find an expression for the energy in terms of the properties of interest, and then to compare it with the expressions for the energy given by perturbation theory. Both techniques will be illustrated.

The key to the extraction of the polarizability from the energy is the Hellmann–Feynman theorem, Section 8.4. In the present case the 'parameter' P is the electric field strength, and so eqn (8.4.1) becomes

$$\mathrm{d}\mathscr{E}/\mathrm{d}E = \langle(\partial H/\partial E)\rangle. \qquad (13.1.3)$$

(There is an obvious difficulty in continuing to use E for the energy! In this chapter we use $\mathscr{E}$ for energy and E for electric field strength.) The partial derivative of the hamiltonian is simply

$$\partial H/\partial E = -\mu_z,$$

and so the dependence of the energy on the electric field strength is

$$\mathrm{d}\mathscr{E}/\mathrm{d}E = -\langle \mu_z \rangle. \tag{13.1.4}$$

The field-dependence of the energy can be expressed in two ways. One involves the Taylor series

$$\mathscr{E}(E) = \mathscr{E}_0 + (\mathrm{d}\mathscr{E}/\mathrm{d}E)_0 E + \tfrac{1}{2}(\mathrm{d}^2\mathscr{E}/\mathrm{d}E^2)_0 E^2 + \left(\frac{1}{3!}\right)(\mathrm{d}^3\mathscr{E}/\mathrm{d}E^3)_0 E^3 + \ldots \tag{13.1.5}$$

where the subscript 0 indicates that the derivatives are to be evaluated at $E = 0$. The expectation value of the z-component of the dipole moment operator when E is nonzero is therefore

$$\begin{aligned}\langle \mu_z \rangle &= -(\mathrm{d}/\mathrm{d}E)\mathscr{E}(E) \\ &= -(\mathrm{d}\mathscr{E}/\mathrm{d}E)_0 - (\mathrm{d}^2\mathscr{E}/\mathrm{d}E^2)_0 E - \tfrac{1}{2}(\mathrm{d}^3\mathscr{E}/\mathrm{d}E^3)_0 E^2 + \ldots.\end{aligned} \tag{13.1.6}$$

The magnitude of the dipole is also expressed in terms of the polarizability of the molecule. Specifically

$$\langle \mu_z \rangle = \mu_{0z} + \alpha_{zz} E + \tfrac{1}{2}\beta_{zzz} E^2 + \ldots \tag{13.1.7}$$

where μ_{0z} is the z-component of the permanent moment $\boldsymbol{\mu}_0$, α_{zz} is the *polarizability*† in the z-direction, and β_{zzz} is the *first hyperpolarizability*. It follows that we can make the identifications:

Permanent dipole moment: $\mu_{0z} = -(\mathrm{d}\mathscr{E}/\mathrm{d}E)_0$ (13.1.8a)

Polarizability: $\alpha_{zz} = -(\mathrm{d}^2\mathscr{E}/\mathrm{d}E^2)_0$ (13.1.8b)

First hyperpolarizability: $\beta_{zzz} = -(\mathrm{d}^3\mathscr{E}/\mathrm{d}E^3)_0$, (13.1.8c)

and so on. Then eqn (13.1.5) becomes

$$\mathscr{E}(E) = \mathscr{E}_0 - \mu_{0z} E - \tfrac{1}{2}\alpha_{zz} E^2 - \left(\frac{1}{3!}\right)\beta_{zzz} E^3 - \ldots. \tag{13.1.9}$$

If we can find the power expansion of the energy from perturbation

† There are two subscripts on α_{zz} because the polarizability is properly regarded as a matrix or, more loosely, as a second rank *tensor*. When a field is applied along z a dipole may be induced having components along x, y, and z, the magnitudes being $\alpha_{xz}E$, $\alpha_{yz}E$, and $\alpha_{zz}E$ respectively. Normally the diagonal element α_{zz} dominates, and so the induced moment is parallel to the inducing field. There are in general three directions which, when the field is along them, give rise to strictly parallel induced dipoles: these are the *principal axes* of the polarizability. β is a *third rank tensor*: a field along z may lead to a dipole with components $\beta_{xzz}E^2$ etc. A field with both x and y components can lead to a z-component of dipole through $\beta_{zyx}E_yE_x$, and so on

theory, it follows that we can identify the polarizability and the hyperpolarizabilities from the successive terms.

The perturbation expansion of the energy is also a power series in E. This is because $H^{(1)} \propto E$, and so successive terms are proportional to successive powers of E. The basic expression for the energy was derived in Chapter 8.

$$\mathscr{E} = \mathscr{E}^{(0)} + \langle 0|\, H^{(1)}\, |0\rangle + \sum_n{}' \left\{ \frac{\langle 0|\, H^{(1)}\, |n\rangle\langle n|\, H^{(1)}\, |0\rangle}{\mathscr{E}_0 - \mathscr{E}_n} \right\} + \dots. \tag{13.1.10}$$

Therefore, with $H^{(1)} = -\mu_z E$,

$$\mathscr{E} = \mathscr{E}^{(0)} - \langle 0|\, \mu_z\, |0\rangle E + \sum_n{}' \left\{ \frac{\langle 0|\, \mu_z\, |n\rangle\langle n|\, \mu_z\, |0\rangle}{\mathscr{E}_0 - \mathscr{E}_n} \right\} E^2 + \dots. \tag{13.1.11}$$

Compare this with eqn (13.1.9). The first-order term lets us identify the permanent dipole moment:

$$\mu_{0z} = \langle 0|\, \mu_z\, |0\rangle, \tag{13.1.12}$$

which is simply the expectation value of the dipole moment operator in the unperturbed state of the system. We knew that anyway, but it is good to check consistency. The second-order term gives us the polarizability in the z-direction:

$$\alpha_{zz} = -2\sum_n{}' \left\{ \frac{\langle 0|\, \mu_z\, |n\rangle\langle n|\, \mu_z\, |0\rangle}{\mathscr{E}_0 - \mathscr{E}_n} \right\}. \tag{13.1.13}$$

We shall write the energy difference $\mathscr{E}_n - \mathscr{E}_0$ as Δ_{n0} (a positive quantity if 0 denotes the ground state of the molecule, the only one we consider) and the matrix elements $\langle 0|\, \mu_z\, |n\rangle$ as $\mu_{z,0n}$. Then

$$\alpha_{zz} = 2\sum_n{}' \mu_{z,0n}\mu_{z,n0}/\Delta_{n0}. \tag{13.1.14}$$

Similar expressions for the x- and y-components of the polarizability are obtained when the field is applied along the x- and y-directions.

The *mean polarizability* is measured when the molecule is freely rotating in a fluid, and is defined as $\alpha = \frac{1}{3}(\alpha_{xx} + \alpha_{yy} + \alpha_{zz})$; therefore, in the present case

$$\alpha = \tfrac{2}{3}\sum_n{}' (\mu_{z,0n}\mu_{z,n0} + \mu_{y,0n}\mu_{y,n0} + \mu_{x,0n}\mu_{x,n0})/\Delta_{n0}.$$

That is,

$$\alpha = \tfrac{2}{3}\sum_n{}' \boldsymbol{\mu}_{0n} \cdot \boldsymbol{\mu}_{n0}/\Delta_{n0} = \tfrac{2}{3}\sum_n{}' |\mu_{n0}|^2/\Delta_{n0} \tag{13.1.15}$$

where we have used the usual definition of the scalar product of two vectors. When μ is expressed in C m and Δ in J the polarizability has the disagreeable units $J^{-1}\ C^2\ m^2$. It is therefore common to introduce the

Polarizability volume: $\alpha' = \alpha/4\pi\varepsilon_0$ (13.1.16)

which has the dimensions of volume (and is typically of the order of $10^{-24}\ cm^3$, similar to molecular volumes). Some expressions have a simpler form in terms of α', and we shall use it as appropriate.

We now have simple expressions for the polarizability of the molecule either when it is freely rotating (eqn (13.1.15)) or when it is stationary and the field is applied along some chosen direction (eqn (13.1.14)). In order to use them, all we have to know is everything about the excited state wavefunctions and energies. Usually (but not always) the formidable task of evaluating all the contributions to the sum is also impossible, and so it is necessary to resort to some kind of simplification. One possibility is to use the closure approximation described on p. 179; the other is to express the polarizability in terms of measureable quantities.

Example. Calculate the polarizability of a one-dimensional system of two charges e and $-e$ bound together by a spring of force constant k, the electric field being applied parallel to the x-axis.

- *Method.* The equilibrium separation is x_e and the displacement is $\xi = x - x_e$. The dipole moment operator is $\mu = ex = e\xi + ex_e$. The states of the system are harmonic oscillator states ψ_v of energy $(v+\frac{1}{2})\hbar\omega$. The matrix elements of ξ were calculated on p. 299: note that $\mu_{v'v} = 0$ unless $v' = v \pm 1$, and so there are only two terms in the sum for α_{xx}, eqn (13.1.14), hence it may be evaluated exactly. Let the initial state be ψ_v, and at the end of the calculation specialize to the ground state, $v = 0$.

- *Answer.*

$$\mu_{v'v} = e\int_{-\infty}^{\infty} \psi_{v'}(\xi)\xi\psi_v(\xi)\,d\xi$$

$$\mu_{v+1,v} = e(v+1)^{\frac{1}{2}}(\hbar/2m\omega_0)^{\frac{1}{2}}, \qquad \mu_{v-1,v} = ev^{\frac{1}{2}}(\hbar/2m\omega_0)^{\frac{1}{2}}$$

$$\alpha_{xx} = 2\sum_{v'} |\mu_{vv'}|^2/\Delta_{v'v}, \qquad \Delta_{v'v} = (v'-v)\hbar\omega_0$$

$$= (2/\hbar\omega_0)\{\mu^2_{v,v+1} - \mu^2_{v,v-1}\} = 2(e^2\hbar/2m\omega_0)\{(v+1)-v\}/\hbar\omega_0 = e^2/k.$$

- *Comment.* The polarizability is independent of the state. Notice that it depends on the force constant but not on the masses of the particles: this is because the polarizability depends on the degree of stretching of the system but not on the rate at which the stretching occurs (i.e. it is a static, not a dynamical, property). The dynamic polarizability, the response to an oscillating field, would depend on the masses (see below, p. 360). This calculation models the *atomic* or *distortion contribution* to the polarizability of a molecule, the contribution to the polarizability of a distortion of the molecular geometry.

First, consider what happens under closure. As decided in Chapter 8 we assume that all the excitation energies may be replaced by a single magnitude Δ (which might be taken to be of the same order of magnitude as the ionization energy, or $\Delta \approx \eta I$ with $\eta \approx 1$); then using eqn (8.1.26).

$$\alpha \approx \frac{2}{3\Delta} \sum_n{}' \boldsymbol{\mu}_{0n} \cdot \boldsymbol{\mu}_{n0} = (2/3\Delta)\{\langle 0|\, \mu^2 \,|0\rangle - \langle 0|\, \mu \,|0\rangle^2\}.$$

On writing $\delta\mu^2 = \langle \mu^2 \rangle - \langle \mu \rangle^2$ we find

$$\alpha \approx (2/3\Delta)\delta\mu^2. \qquad (13.1.17)$$

We shall refer to $\delta\mu$, the root mean square deviation of the dipole moment from its mean value, as the *fluctuation* of the dipole. Even in a molecule with no permanent dipole moment ($\mu_0 = 0$) the expectation value of the square of the dipole moment operator is non-zero, and so every molecule has a non-vanishing dipole fluctuation. We can think of this fluctuation as being a true fluctuation, when the electric dipole flickers in magnitude and direction but with a time average which is either zero (in non-polar molecules) or non-zero (in polar molecules). The polarizability of the molecule is related to the magnitude of these fluctuations, and molecules with large dipole fluctuations are predicted to have high polarizabilities. This is consistent with the view that a molecule is easily distorted if the electrons are not under the tight control of their nuclei. The connection between the response of a system (in this case measured by the polarizability) and the magnitude of the fluctuations in the unperturbed system (in this case in the charge distribution) is a special case of the *fluctuation–dissipation theorem.* A reference is given in *Further reading* (p. 374).

The last paragraph suggests that the polarizability of molecules is expected to increase as their radius and number of electrons increases, since both aspects can be expected to lead to a bigger fluctuation in the charge distribution. Consider first the role of size in the case of a one-electron atom. Since the dipole moment operator is then $\boldsymbol{\mu} = -e\mathbf{r}$, in the closure approximation the polarizability is

$$\alpha \approx (2/3\Delta)e^2\langle r^2 \rangle. \qquad (13.1.18)$$

$\langle r^2 \rangle$ is the mean square radius of the electron's orbital, and so it is clear that as the mean radius increases the polarizability increases. This is consistent with the central nucleus progressively losing control over the electron as the orbital expands. In a many-electron atom we can expect each electron to contribute a similar term, and therefore the polarizability to be proportional to $N_e\langle r^2 \rangle$ where now $\langle r^2 \rangle$ is the mean square radius of all the orbitals occupied in the atom and N_e is the total number of electrons. Since $\langle r^2 \rangle \approx R_a^2$, where R_a is the radius of the atom

$$\alpha \approx (2/3\Delta)e^2 N_e R_a^2. \qquad (13.1.19)$$

Therefore, as atom size increases, either as a result of an increase in the number of electrons or by the expansion of its radius, so the polarizability increases.

The polarizability depends on the square of the matrix elements of the electric dipole moment. But we have already met this quantity in the discussion of spectral intensities. Therefore we should expect to be able to express the polarizabilities in terms of intensities. The most direct procedure is to use the definition of the oscillator strength of the electric dipole transitions $n \leftarrow 0$. According to the definition in Chapter 12,

$$f_{n0} = (4\pi m_e/3e^2\hbar)\nu_{n0}\,|\mu_{n0}|^2. \qquad (13.1.20)$$

Therefore

$$\alpha = (\hbar^2 e^2/m_e)\sum_n{}' f_{n0}/\Delta_{n0}^2. \qquad (13.1.21)$$

In the first place this expression lets us estimate the polarizability of a molecule from spectroscopic data, because the oscillator strengths can be assessed from the absorbances of the spectral lines ($f \propto \mathcal{A}$, combine eqns (12.6.4) and (12.6.5)) and the Δ_{n0} transition energies from their locations ($\Delta_{n0} = hc\tilde{\nu}_{n0}$). In the second place it lets us conclude that if a molecule has intense (large f), low energy (Δ small) transitions in its spectrum then it is highly polarizable. Hence, intensely coloured molecules should be strongly polarizable. In contrast, molecules that absorb only weakly and at high energies (e.g. the colourless hydrocarbons, which absorb only in the ultraviolet, and then only weakly) are expected to be only weakly polarizable. This is confirmed in practice.

The exact expression in eqn (13.1.21) can be developed by making the approximation that all the excitation energies are equal. Then

$$\alpha \approx (\hbar^2 e^2/m_e\Delta^2)\sum_n{}' f_{n0}.$$

The sum over the oscillator strengths is a standard result known as the

Kuhn–Thomas sum rule: $\sum_n f_{n0} = N_e$. (13.1.22)

It is proved in Appendix 18. Therefore

$$\alpha \approx (\hbar^2 e^2/m_e\Delta^2)N_e. \qquad (13.1.23)$$

This shows that α increases as the number of electrons increases. α also depends sharply on the mean excitation energy of the molecule, and increases as Δ decreases. Since the mean excitation energy of molecules can be expected to decrease as the number of electrons increases, the ratio N_e/Δ^2 can be expected to increase more strongly than N_e itself. This is the basic reason why molecules composed of heavy atoms are so strongly polarizable.

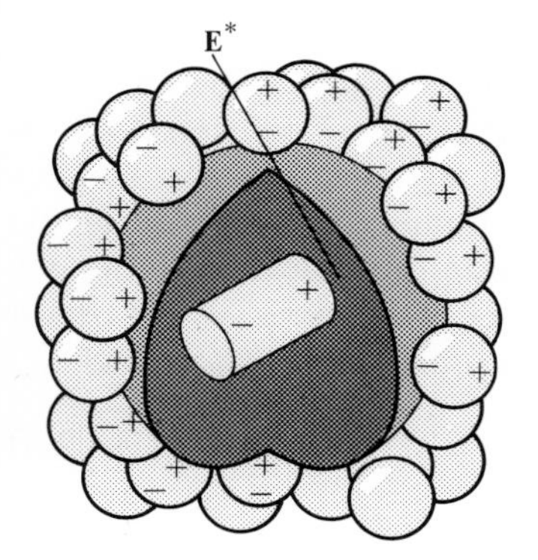

Fig. 13.1. The polarization of the surroundings contributes to the total local electric field.

13.2 Bulk electrical properties

Once we have established an expression for the polarizability we can use it to discuss some of the interesting bulk properties of non-conducting, *dielectric*, media. The immediate problem that confronts us is that the electric field experienced by a molecule in the interior of a sample is not equal to the electric field applied at its surface. The polarization of the molecules in the bulk causes them to give rise to their own fields, Fig. 13.1, and so each molecule experiences the total local field, E^*, and the dipole induced in it depends on its value. We have to relate the local electric field to the electrical properties of the dielectric.

We need the concept of the *permittivity* of a medium. In a vacuum an electric charge q gives rise to a Coulomb potential $q/4\pi\varepsilon_0 r$, where ε_0 is the ***vacuum permittivity*** ($\varepsilon_0 = 8.854\times10^{-12}\,\mathrm{J^{-1}\,C^2\,m^{-1}}$). When the same charge is in a medium it gives rise to the potential $q/4\pi\varepsilon r$, where ε is the medium's permittivity. The ratio $\varepsilon_r = \varepsilon/\varepsilon_0$ is called the *relative permittivity* (or the *dielectric constant*). In practice the relative permittivity is measured as the ratio of the capacitances of a capacitor with and without the dielectric between the plates.

Now consider the electric field between two plates of area A, each carrying a charge density σ (so that the charge on each is σA) and of opposite signs on opposite plates. A result from elementary electrostatics is that the electric field strength between the plates is σ/ε_0 when the dielectric is absent, but $E = \sigma/\varepsilon$ when it is present. Instead of thinking of the reduction of the electric field as arising from the modification due to the permittivity, we can think of the surface charge as polarizing the medium and inducing on it a surface charge density P, Fig. 13.2. This induced charge density is called the *polarization* of the medium. Then the electric field between the plates is written $E = (\sigma - P)/\varepsilon_0$. Since the two electric fields are the same but merely expressed differently, we can equate them and find an expression for P: $(\sigma - P)/\varepsilon_0 = \sigma/\varepsilon = \sigma/\varepsilon_r\varepsilon_0$ implies

$$P = (\varepsilon_r - 1)\sigma/\varepsilon_r = (\varepsilon_r - 1)\varepsilon_0 E.$$

The *electric susceptibility*, χ_e, of the medium is defined through

Polarization: $P = \chi_e \varepsilon_0 E$; (13.2.1)

therefore,

Electric susceptibility: $\chi_e = \varepsilon_r - 1$. (13.2.2)

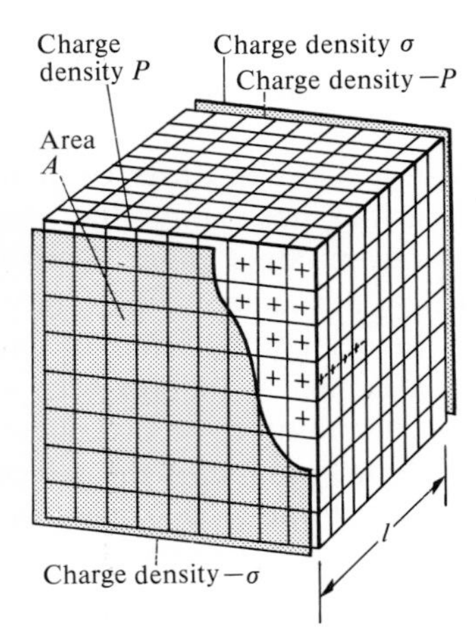

Fig. 13.2. The polarization of a medium and the mean dipole density.

The next stage involves relating the polarization to the polarizability of the medium. As well as being the induced surface charge density, $\boldsymbol{P}$ is also the *dipole moment density* of the medium (the mean dipole moment per unit volume). The relation can be established by referring to Fig. 13.2 which shows that the medium can be regarded as the charges PA and $-PA$ separated by the distance l; in other words, the medium acquires a dipole moment PAl. The volume of the sample is Al; and so the dipole moment density is $PAl/Al = P$, as we set out to show. In molecular terms the dipole moment density is the number density of molecules in the sample, $\mathcal{N}$, multiplied by the mean dipole moment of each molecule. If we concentrate on molecules without a permanent dipole moment this is $\langle\mu\rangle = \alpha E^*$ where α is the mean molecular polarizability and E^* is the total local field. Therefore

$$P = \alpha\mathcal{N}E^*. \tag{13.2.3}$$

What, though, is E^*? The *Lorentz local field* is an approximate relation between E^* and the applied field E obtained by assuming that the medium is a continuous dielectric:

Lorentz local field: $E^* = E + P/3\varepsilon_0$. (13.2.4)

This expression can be combined with the last and E^* eliminated:

$$P = \left\{\frac{3\alpha\mathcal{N}}{3\varepsilon_0 - \alpha\mathcal{N}}\right\}\varepsilon_0 E. \tag{13.2.5}$$

Comparison of this with eqn (13.2.1) lets us identify the susceptibility as

$$\chi_e = \frac{3\alpha\mathcal{N}}{3\varepsilon_0 - \alpha\mathcal{N}} = \frac{\alpha\mathcal{N}/\varepsilon_0}{1 - \alpha\mathcal{N}/3\varepsilon_0}. \tag{13.2.6}$$

It immediately follows that

Relative permittivity: $\varepsilon_r = 1 + \chi_e = \dfrac{1 + (2\alpha\mathcal{N}/3\varepsilon_0)}{1 - (\alpha\mathcal{N}/3\varepsilon_0)}$, (13.2.7)

and so now we have a route from the polarizability to the relative permittivity of the medium.

So far we have considered only non-polar species. In order to deal with polar molecules we have to allow for the possibility that the polarization of the medium, the mean dipole moment density, may have a contribution from the polar molecules' permanent dipole moments. Although it is a fluid, on account of the orientating effect of the applied field the sample may have a predominance of dipoles with a particular orientation, and therefore their contribution to the mean dipole moment density may be non-zero. The magnitude of the effect can be calculated from the Boltzmann distribution, because the favoured orientations are the ones with lower energy. The energy of a permanent dipole $\boldsymbol{\mu}_0$ in a local field E^* in the z-direction is $-\mu_{0z}E^*$ where μ_{0z} is the component of $\boldsymbol{\mu}_0$ in the field direction. We can express this as $\mathscr{E}(\theta) = -\mu_0 E^* \cos\theta$ in the usual spherical polar coordinates. At a temperature T the proportion of molecules with the orientation θ in the range θ to $\theta + \mathrm{d}\theta$ is given by the Boltzmann distribution as

$$\frac{\mathrm{d}N(\theta)}{N} = \frac{\mathrm{e}^{-\mathscr{E}(\theta)/kT} \sin\theta\, \mathrm{d}\theta}{\int_0^{\pi} \mathrm{e}^{-\mathscr{E}(\theta)/kT} \sin\theta\, \mathrm{d}\theta} = \frac{\mathrm{e}^{\mu_0 E^* \cos\theta/kT} \sin\theta\, \mathrm{d}\theta}{\int_0^{\pi} \mathrm{e}^{\mu_0 E^* \cos\theta/kT} \sin\theta\, \mathrm{d}\theta}.$$

The denominator can be evaluated readily (we write $x = \mu_0 E^*/kT$):

$$\int_0^{\pi} \mathrm{e}^{x\cos\theta} \sin\theta\, \mathrm{d}\theta = \int_{-1}^{1} \mathrm{e}^{x\cos\theta}\, \mathrm{d}\cos\theta = (\mathrm{e}^{x} - \mathrm{e}^{-x})/x.$$

Then the Boltzmann distribution is

$$\mathrm{d}N(\theta)/N = x\mathrm{e}^{x\cos\theta} \sin\theta\, \mathrm{d}\theta/(\mathrm{e}^{x} - \mathrm{e}^{-x}). \tag{13.2.8}$$

The average dipole density is simply the average of $\mu_{0z} = \mu_0 \cos\theta$ weighted by the Boltzmann factor and divided by the volume of the sample:

$$P = \frac{(N/V)\mu_0 \int_0^{\pi} \cos\theta\, \mathrm{e}^{x\cos\theta} \sin\theta\, \mathrm{d}\theta}{(1/x)(\mathrm{e}^{x} - \mathrm{e}^{-x})} = \mu_0 \mathscr{N} \mathscr{L}(x) \tag{13.2.9}$$

where $\mathscr{L}(x)$ is the

Langevin function: $$\mathscr{L}(x) = \frac{\mathrm{e}^{x} + \mathrm{e}^{-x}}{\mathrm{e}^{x} - \mathrm{e}^{-x}} - \frac{1}{x}. \tag{13.2.10}$$

When $x \ll 1$ (as it is at all normal temperatures and field strengths) this reduces to $\mathscr{L}(x) \approx \frac{1}{3}x = \mu_0 E^*/3kT$. Therefore the permanent dipoles contribute

$$P \approx \mu_0^2 E^* \mathscr{N}/3kT \tag{13.2.11}$$

to the polarization. Equation (13.2.3) should therefore be changed to

$$P = \{\alpha + \mu_0^2/3kT\}\mathscr{N}E^* \tag{13.2.12}$$

when the molecules are polar. It follows that for a polar medium:

Electric susceptibility: $$\chi_{\mathrm{e}} = \frac{(\alpha + \mu_0^2/3kT)\mathscr{N}/\varepsilon_0}{1 - (\alpha + \mu_0^2/3kT)\mathscr{N}/3\varepsilon_0} \tag{13.2.13}$$

Relative permittivity: $$\varepsilon_{\mathrm{r}} = \frac{1 + 2(\alpha + \mu_0^2/3kT)\mathscr{N}/3\varepsilon_0}{1 - (\alpha + \mu_0^2/3kT)\mathscr{N}/3\varepsilon_0}. \tag{13.2.14}$$

The practical form of this expression is obtained by expressing the number density in terms of the mass density ρ:

$$\mathscr{N} = N/V = L(M/M_{\mathrm{m}})/V = L\rho/M_{\mathrm{m}}, \tag{13.2.15}$$

M_m being the molar mass and L Avogadro's constant. Then with the polarizability volume $\alpha' = \alpha/4\pi\varepsilon_0$,

$$\varepsilon_r = \frac{1+2a}{1-a}, \qquad a = (4\pi L/3)(\rho/M_m)\{\alpha' + \mu_0^2/12\pi\varepsilon_0 kT\}. \quad (13.2.16)$$

The dependence of the permittivity of a medium on the characteristics of its molecules can now be discussed in the same way as before, because we know how they determine α. We expect a medium to have a high permittivity if α is large or, if it is polar, if the permanent dipole moments are large. Hence media composed of brightly coloured molecules, or of molecules formed from heavy atoms, should be expected to have high relative permittivities (but notice that ρ and M_m work in opposite directions).

13.3 Refractive index

It follows from Maxwell's equations (Appendix 19) that the speed of light in a medium is inversely proportional to the square root of the permittivity. Therefore the *refractive index*, $n_r = c/(\text{speed in medium})$, can be expressed in terms of the relative permittivity, $n_r = \varepsilon_r^{\frac{1}{2}}$. We have an expression for ε_r in terms of the molecular properties α and μ_0, and so in principle we can calculate n_r. There is one simplification and one complication.

The simplification is that the permanent dipole is too sluggish to respond to the high frequency alternation of the direction of the electric field vector in a light ray (a molecule needs about 10^{-12} s to rotate, but $\mathbf{E}$ reverses in about 10^{-15} s). Therefore we can ignore the contribution of μ_0 to the permittivity and use eqn (13.2.7) for ε_r. Then

$$n_r = \varepsilon_r^{\frac{1}{2}} = \left\{\frac{1+(2\alpha\mathcal{N}/3\varepsilon_0)}{1-(\alpha\mathcal{N}/3\varepsilon_0)}\right\}^{\frac{1}{2}} \approx 1 + \alpha\mathcal{N}/2\varepsilon_0 \quad (13.3.1)$$

where we have assumed that the polarizability terms are small enough to use $1/(1-x) \approx 1+x$ and $(1+x)^{\frac{1}{2}} \approx 1+\frac{1}{2}x$. On inserting the relations $\mathcal{N} = L\rho/M_m$ and $\alpha' = \alpha/4\pi\varepsilon_0$ we obtain

$$n_r \approx 1 + 2\pi L\alpha'\rho/M_m \quad (13.3.2)$$

which shows that the refractive index increases with the polarizability volume and density of the medium.

The complication (apart from the restriction of the validity of these results as a result of the use of the Lorentz local field) is that the refractive index is a property related to an *oscillating* electric field. Therefore we cannot use eqn (13.1.15) for α directly, but must recalculate it on the basis of time-dependent perturbation theory.

The strategy is to calculate the expectation value of $\boldsymbol{\mu}$ using the perturbed functions (this was the alternative approach mentioned in the introduction). The perturbation is now

$$H^{(1)}(t) = -\boldsymbol{\mu}\cdot\mathbf{E}(t) = -2\mu_z E\cos\omega t \tag{13.3.3}$$

if the field $2E\cos\omega t$ is in the z-direction and oscillating with a frequency ω. The expectation value of the dipole moment is

$$\langle\boldsymbol{\mu}\rangle = \int \Psi^*(t)\boldsymbol{\mu}\Psi(t)\,\mathrm{d}\tau \tag{13.3.4}$$

and the time-dependent wavefunction is given by eqn (8.6.2) as

$$\Psi(t) = \psi_0 \mathrm{e}^{-\mathrm{i}E_0 t/\hbar} + \sum_n{}' a_n(t)\psi_n \mathrm{e}^{-\mathrm{i}E_n t/\hbar}, \tag{13.3.5}$$

where, as usual, the prime signifies the omission of the term with $n = 0$. Since we are looking for the field-induced contribution to the dipole moment we need to evaluate $\langle\boldsymbol{\mu}\rangle$ to first-order in E:

$$\begin{aligned}\langle\boldsymbol{\mu}\rangle &= \int \psi_0^*\boldsymbol{\mu}\psi_0\,\mathrm{d}\tau + \sum_n{}'\left\{\int \psi_0^*\boldsymbol{\mu}\psi_n\,\mathrm{d}\tau a_n(t)\mathrm{e}^{-\mathrm{i}\omega_{n0}t}\right.\\ &\qquad \left. + \int \psi_n^*\boldsymbol{\mu}\psi_0\,\mathrm{d}\tau a_n^*(t)\mathrm{e}^{\mathrm{i}\omega_{n0}t}\right\}\\ &= \boldsymbol{\mu}_0 + \sum_n{}'\left\{\boldsymbol{\mu}_{0n}a_n(t)\mathrm{e}^{-\mathrm{i}\omega_{n0}t} + \boldsymbol{\mu}_{n0}a_n^*(t)\mathrm{e}^{\mathrm{i}\omega_{n0}t}\right\}.\end{aligned} \tag{13.3.6}$$

where $\hbar\omega_{n0} = E_n - E_0$. One problem with this approach is that when a field is suddenly applied it may result in the generation of transient oscillation in the charge density of the molecules, and this confuses the analysis. Therefore we ensure that all transients have died away by starting to switch on the oscillating field long ago and allowing it to rise to full strength slowly. This is achieved in the same way as we switched on the static perturbation in Chapter 8, by modifying the perturbation to

$$H^{(1)}(t) = -2\mu_z E(1-\mathrm{e}^{-t/\tau})\cos\omega t = -\mu_z E(1-\mathrm{e}^{-t/\tau})(\mathrm{e}^{\mathrm{i}\omega t}+\mathrm{e}^{-\mathrm{i}\omega t})$$

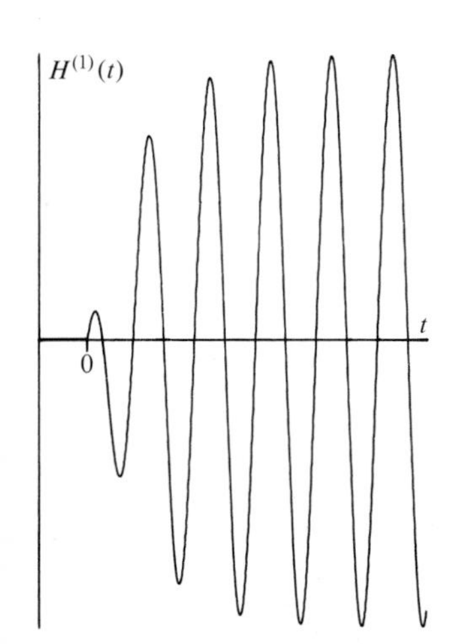

Fig. 13.3. The exponentially switched oscillating perturbation.

as illustrated in Fig. 13.3. Since we are interested in times very long compared with the switching time τ, eqn (8.5.5) for the coefficients $a(t)$ in eqn (13.3.6) gives (for $t \gg \tau$)

$$\begin{aligned}a_n(t) &= (1/\mathrm{i}\hbar)\int_0^t H_{n0}^{(1)}(t)\mathrm{e}^{\mathrm{i}\omega_{n0}t}\,\mathrm{d}t\\ &= (\mu_{z,n0}E/\hbar)\left\{\left(\frac{\mathrm{e}^{\mathrm{i}(\omega+\omega_{n0})t}}{\omega+\omega_{n0}}\right) - \left(\frac{\mathrm{e}^{-\mathrm{i}(\omega-\omega_{n0})t}}{\omega-\omega_{n0}}\right)\right\}\end{aligned}$$

and consequently

$$\langle\mu_z\rangle = \mu_{ez} + (2/\hbar)\sum_n\left\{\frac{\omega_{n0}\mu_{z,0n}\mu_{z,n0}}{\omega_{n0}^2-\omega^2}\right\}2E\cos\omega t. \tag{13.3.7}$$

where we have used the hermiticity of $\boldsymbol{\mu}$ ($\boldsymbol{\mu}_{0n}^* = \boldsymbol{\mu}_{n0}$). If this expression is compared with

$$\langle\mu_z\rangle = \mu_{0z} + \alpha_{zz}(\omega)E(t) + \ldots$$

we arrive at the z-component of the

$$\textit{Dynamic polarizability:}\ \alpha_{zz}(\omega)=(2/\hbar)\sum_n\left\{\frac{\omega_{n0}\,|\mu_{z,0n}|^2}{\omega_{n0}^2-\omega^2}\right\}. \quad (13.3.8)$$

The mean polarizability is the average $\frac{1}{3}(\alpha_{xx}+\alpha_{yy}+\alpha_{zz})$, and so

$$\textit{Mean polarizability:}\ \alpha(\omega)=(2/3\hbar)\sum_n\left\{\frac{\omega_{n0}\,|\mu_{0n}|^2}{\omega_{n0}^2-\omega^2}\right\} \quad (13.3.9)$$

where $|\mu_{0n}|^2=\boldsymbol{\mu}_{0n}\cdot\boldsymbol{\mu}_{n0}$. Notice how this expression reduces to the static polarizability α when $\omega\to 0$. Furthermore, when ω is so high that $\omega^2\gg\omega_{n0}^2$,

$$\begin{aligned}\alpha(\omega)&\approx(2/3\hbar)\sum_n(-\omega_{n0}/\omega^2)\,|\mu_{0n}|^2\\&\approx-(e^2/m_e\omega^2)\sum_n f_{n0}=-(e^2/m_e\omega^2)N_e.\end{aligned} \quad (13.3.10)$$

(We have used eqns (13.1.20) and (13.1.22).) When $\omega\to\infty$ the polarizability vanishes because even the electrons cannot contribute to the induced moment if the field changes direction too quickly for them to follow. We shall return to the interpretation of the negative polarizability (signifying that $\langle\mu_z\rangle$ is in the opposite direction to the field) in a moment.

Example. Calculate the polarizability of the same system as in the last *Example* (p. 352) but now exposed to an electric field oscillating at the frequency ω.

- *Method.* Use eqn (13.3.8) with change of notation $z\to x$. All the matrix elements needed are listed in the first *Example*.
- *Answer.*

$$\begin{aligned}\alpha_{xx}(\omega)&=(2/\hbar)\sum_{v'}{}'\left\{\frac{\omega_{v'v}\mu_{v'v}^2}{\omega_{v'v}^2-\omega^2}\right\},\qquad \omega_{v'v}=(v'-v)\omega_0\\&=(2/\hbar)\left\{\frac{\omega_0\mu_{v+1,v}^2}{\omega_0^2-\omega^2}-\frac{\omega_0\mu_{v-1,v}^2}{\omega_0^2-\omega^2}\right\}\\&=\left\{\frac{(2/\hbar)\omega_0}{\omega_0^2-\omega^2}\right\}(\mu_{v+1,v}^2-\mu_{v-1,v}^2)\\&=e^2/m(\omega_0^2-\omega^2)=e^2/(k-m\omega^2).\end{aligned}$$

- *Comment.* Once again the calculation is exact; it reduces to that in the former *Example* when $\omega=0$. In this case the polarizability depends on the masses of the particles because it measures how rapidly they respond to the oscillating perturbation. If the masses are infinite the polarizability is zero except when $\omega=0$. Notice that the atomic contribution to the dynamic polarizability is negligible at optical frequencies, this being an example of the case $\omega\gg\omega_0$.

We can now use eqn (13.3.9) in eqn (3.3.2) to find the frequency dependence (the *dispersion*) of the refractive index:

$$n_r(\omega) \approx 1 + (2\pi L\rho/M_m)\alpha'(\omega)$$
$$\approx 1 + (L\rho/3\hbar\varepsilon_0 M_m)\sum_n \left\{\frac{\omega_{n0}\,|\mu_{0n}|^2}{\omega_{n0}^2 - \omega^2}\right\} \qquad (13.3.11)$$

When the correction $(2\pi L\rho/M_m)\alpha'(\omega)$ is not small enough for the approximations involved in eqn (13.3.1) to be used we can use instead

$$n_r^2(\omega) = \frac{1 + 2\alpha(\omega)\mathcal{N}/3\varepsilon_0}{1 - \alpha(\omega)\mathcal{N}/3\varepsilon_0} \qquad (13.3.12)$$

which is a version of the

Lorenz–Lorentz formula: $$\frac{n_r^2 - 1}{n_r^2 + 2} = \tfrac{1}{3}\alpha(\omega)\mathcal{N}/\varepsilon_0 = (4\pi/3)\alpha'(\omega)L\rho/M_m. \qquad (13.3.13)$$

This is normally expressed in terms of the

Molar refractivity: $$R_m = \left(\frac{M_m}{\rho}\right)\left\{\frac{n_r^2 - 1}{n_r^2 + 2}\right\} \qquad (13.3.14)$$

when it takes the form

$$R_m = (4\pi/3)L\alpha'(\omega). \qquad (13.3.15)$$

Note that R_m has the dimensions of a molar volume.

The advantage of concentrating on the molar refractivity is that it eliminates the molar mass and density dependence of the refractive index itself, and hence focuses attention on the molecular property, the polarizability volume α'. This property is more likely to be additive than the refractive index itself, in the sense that the refractivity of a molecule may be expressed as the sum of the refractivities of its component atoms or groups. This is confirmed to some extent, and tables of atomic refractivities have been compiled. The molar refractivity of the molecule is obtained as their sum, and then the refractive index is obtained by the appropriate manipulation of eqn (13.3.14).

Now consider the dispersion characteristics of the refractive index. We shall suppose that the density is always small enough for eqn (13.3.11) to be applicable. Suppose ω is so close to one of the electronic transition frequencies that its contribution dominates the frequency dependence (the terms in the sum go as $1/(\omega_{n0}^2 - \omega^2)$); then

$$n_r \approx 1 + A\left\{\frac{\omega_{n0}\,|\omega_{0n}|^2}{\omega_{n0}^2 - \omega^2}\right\}; \qquad A = (L\rho/3\hbar\varepsilon_0 M_m)\,|\mu_{n0}|^2. \qquad (13.3.16)$$

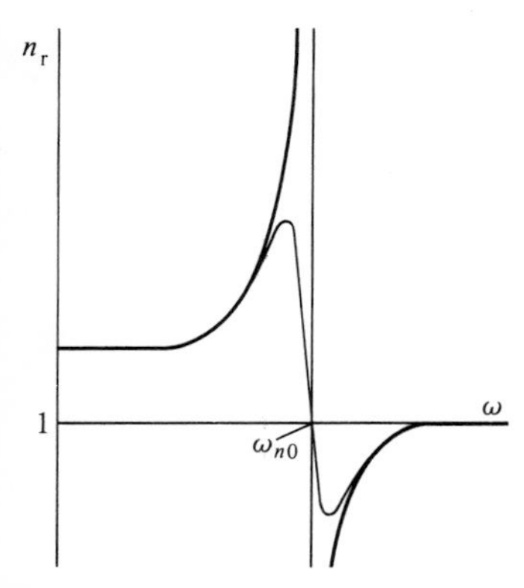

Fig. 13.4. The dispersion of the refractive index.

This is sketched in Fig. 13.4. When $\omega < \omega_{n0}$, $n_r > 1$ and increases as ω increases. This behaviour is a reflection of the effective degeneracy brought about by an oscillating perturbation, as described in Chapter 8. The increase of n_r with frequency means that blue light is refracted more than red, and hence we see the underlying reason why a prism disperses white light (and the source of the name 'dispersion' for the frequency dependence of a property). At *resonance*, when $\omega = \omega_{n0}$, something dramatic appears to happen, but in fact the perturbation calculation is invalid in this region and the true behaviour is like that shown in Fig. 13.4 as the feint line.

13.4 Dispersion forces

There are many interactions between molecules. In this section we consider the one that dominates the interactions between uncharged non-polar species when they are not actually in contact. This is the *dispersion interaction.* It arises from the coupling of instantaneous fluctuations in the charge distribution of neighbouring molecules. In particular, there may be a fluctuation in the charge distribution on one molecule that gives it a dipole briefly; that dipole may induce another in a neighbouring molecule, and the two dipoles will interact favourably. We have seen already that the polarizability is a measure both of the charge fluctuation in a molecule, and of the response of a molecule to an applied field; therefore both the initial transient dipole and the response of the neighbouring molecule can be expected to be related to their polarizabilities, and hence that the energy of interaction should be proportional to $\alpha_A\alpha_B$. This is the relation we establish here.

We use perturbation theory to calculate the lowering of the energy when two spherical molecules or atoms are brought to a separation R. The perturbation hamiltonian is the interaction between two dipole moment operators centred on the two atoms, Fig. 13.5:

$$H^{(1)} = (1/4\pi\varepsilon_0 R^3)\{\boldsymbol{\mu}_A \cdot \boldsymbol{\mu}_B - 3\boldsymbol{\mu}_A \cdot \hat{\mathbf{R}}\hat{\mathbf{R}} \cdot \boldsymbol{\mu}_B\} \tag{13.4.1}$$

where $\hat{\mathbf{R}}$ is the unit vector $\mathbf{R}/R$. (This is the generalized version of the simpler expression $\mu_A\mu_B(1-3\cos^2\theta)/4\pi\varepsilon_0 R^3$ appropriate when the dipoles are parallel.) The unperturbed hamiltonian is the sum of the electronic hamiltonians of the two atoms:

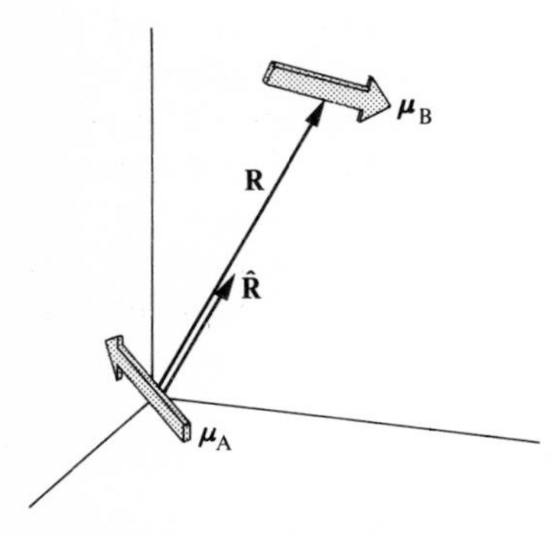

Fig. 13.5. The interaction between point dipoles.

$$H^{(0)} = H_A + H_B. \tag{13.4.2}$$

The wavefunction of the pair is $\psi_{n_A}\psi_{n_B}$, corresponding to the joint state $|n_A n_B\rangle$, and the corresponding energy levels of the joint system are the eigenvalues in the equation

$$(H_A + H_B)\,|n_A n_B\rangle = H_A\,|n_A n_B\rangle + H_B\,|n_A n_B\rangle = (\mathscr{E}_{n_A} + \mathscr{E}_{n_B})\,|n_A n_B\rangle. \tag{13.4.3}$$

We write these joint energies as $\mathscr{E}_{n_A n_B}$. The ground state is $n_A = 0$, $n_B = 0$, which we write simply $|0\rangle$, its energy being $\mathscr{E}_{00}$.

The first-order interaction energy is the expectation value of $H^{(1)}$:

$$\begin{aligned}\mathscr{E}^{(1)} &= \langle 0| H^{(1)} |0\rangle \\ &= (1/4\pi\varepsilon_0 R^3)\langle 0| \boldsymbol{\mu}_A \cdot \boldsymbol{\mu}_B - 3\boldsymbol{\mu}_A \cdot \hat{\mathbf{R}}\hat{\mathbf{R}} \cdot \boldsymbol{\mu}_B |0\rangle \\ &= (1/4\pi\varepsilon_0 R^3)\{\boldsymbol{\mu}_{0A} \cdot \boldsymbol{\mu}_{0B} - \boldsymbol{\mu}_{0A} \cdot \hat{\mathbf{R}}\hat{\mathbf{R}} \cdot \boldsymbol{\mu}_{0B}\}\end{aligned} \quad (13.4.4)$$

where $\boldsymbol{\mu}_{0A} = \langle 0_A| \boldsymbol{\mu}_A |0_A\rangle$, the permanent dipole moment of A, and likewise for B. But in an atom the permanent dipole moment is zero; therefore there is no first-order contribution to the interaction energy. (This term survives in the interaction between polar species.)

As the first-order contribution is zero, we must investigate the second-order energy. Physically this means that we allow for the distortion of the wavefunction of each molecule as a result of the presence of the other. The second-order energy is

$$\mathscr{E}^{(2)} = \sum_{n_A, n_B}{}' \left\{\frac{\langle 0| H^{(1)} |n_A n_B\rangle\langle n_A n_B| H^{(1)} |0\rangle}{\mathscr{E}_{00} - \mathscr{E}_{n_A n_B}}\right\}. \quad (13.4.5)$$

We write $\mathscr{E}_{n_A n_B} - \mathscr{E}_{00} = \Delta_{n_A 0} + \Delta_{n_B 0}$, the sum of the excitation energies of the two atoms.

The perturbation hamiltonian has quite a complicated form. It is easier to handle if it is expressed in terms of a system of coordinates in which the interatomic separation defines the z-axis, Fig. 13.6. Then

$$H^{(1)} = (1/4\pi\varepsilon_0 R^3)\{\mu_{Ax}\mu_{Bx} + \mu_{Ay}\mu_{By} - 2\mu_{Az}\mu_{Bz}\}. \quad (13.4.6)$$

Even so, the second-order energy takes the formidable form

$$\begin{aligned}\mathscr{E}^{(2)} = -(1/4\pi\varepsilon_0 R^3)^2 \sum_{n_A, n_B}{}' &\langle 0| \mu_{Ax}\mu_{Bx} + \mu_{Ay}\mu_{By} - 2\mu_{Az}\mu_{Bz} |n_A n_B\rangle \\ &\times\langle n_A n_B| \mu_{Ax}\mu_{Bx} + \mu_{Ay}\mu_{By} - 2\mu_{Az}\mu_{Bz} |0\rangle/(\Delta_{n_A 0} + \Delta_{n_B 0}).\end{aligned} \quad (13.4.7)$$

Happily, most of the nine terms vanish. Consider a term such as $\langle 0| \mu_{Ax}\mu_{Bx} |n_A n_B\rangle\langle n_A n_B| \mu_{Ay}\mu_{By} |0\rangle$. This includes the factor $\langle 0_A| \mu_{Ax} |n_A\rangle\langle n_A| \mu_{Ay} |0_A\rangle$, which is zero in an atom (or a spherical molecule). The quickest way of establishing this is to note that we are free to choose an alternative coordinate system on A with the y-axis pointing in the opposite direction but with the x-axis remaining the same. The product then changes sign. But its contribution to the energy cannot depend on our choice of an axis system; therefore it must be zero. The same argument applies to all the cross terms in eqn (13.4.7), and so the only terms that survive are the three terms of the form $(x \ldots x)(x \ldots x)$, etc. In the case of atoms (and spherical tops) all three surviving terms are equal. This is because in a spherical system

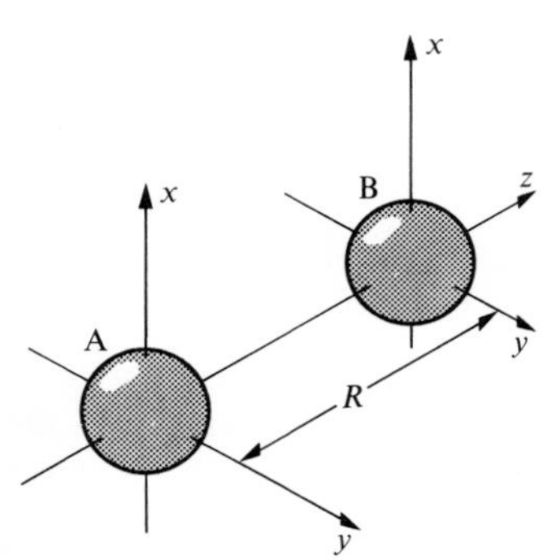

Fig. 13.6. The coordinate system for the dispersion energy calculation.

$$\begin{aligned}\langle 0_A| \mu_{Ax} |n_A\rangle\langle n_A| \mu_{Ax} |0_A\rangle &= \langle 0_A| \mu_{Ay} |n_A\rangle\langle n_A| \mu_{Ay} |0_A\rangle \\ &= \langle 0_A| \mu_{Az} |n_A\rangle\langle n_A| \mu_{Az} |0_A\rangle\end{aligned}$$

and so

$$\langle 0_A| \mu_{Ax} |n_A\rangle\langle n_A| \mu_{Ax} |0_A\rangle = \tfrac{1}{3}\langle 0_A| \boldsymbol{\mu}_A |n_A\rangle \cdot \langle n_A| \boldsymbol{\mu}_A |0_A\rangle,$$

and likewise for the other two components, and for the three components labelled B. Therefore the entire expression reduces to a single term:

$$\mathscr{E}^{(2)} = -\tfrac{2}{3}(1/4\pi\varepsilon_0 R^3)^2 \sum_{n_A,n_B}{}' \left\{\frac{(\boldsymbol{\mu}_{A,0n_A}\cdot\boldsymbol{\mu}_{A,n_A0})(\boldsymbol{\mu}_{B,0n_B}\cdot\boldsymbol{\mu}_{B,n_B0})}{\Delta_{n_A0}+\Delta_{n_B0}}\right\}. \tag{13.4.8}$$

This confirms that there is a non-zero interaction energy which is attractive ($\mathscr{E}^{(2)}$ is negative) and depends on the separation of the atoms as $1/R^6$.

We can obtain an approximate, revealing, and useful form of the expression by making use of the closure approximation (other interesting forms can also be obtained in terms of the oscillator strengths of the transitions on the two atoms). Therefore we write $\Delta_{n_A0}\approx\Delta_A$ and likewise for B, and obtain

$$\begin{aligned}\mathscr{E}^{(2)} &\approx -\tfrac{2}{3}(1/4\pi\varepsilon_0 R^3)^2\left\{\frac{1}{\Delta_A+\Delta_B}\right\}\sum_{n_A,n_B}{}'\{(\boldsymbol{\mu}_{A,0n_A}\cdot\boldsymbol{\mu}_{A,n_A0})(\boldsymbol{\mu}_{B,0n_B}\cdot\boldsymbol{\mu}_{B,n_B0})\}\\ &\approx -(1/24\pi^2\varepsilon_0^2)\left(\frac{1}{R^6}\right)\left(\frac{1}{\Delta_A+\Delta_B}\right)\langle\mu_A^2\rangle\langle\mu_B^2\rangle\end{aligned} \tag{13.4.9}$$

where $\langle\mu_A^2\rangle=\langle 0_A|\,\mu_A^2\,|0_A\rangle$, and likewise for B. This expression can be taken further by using the relation between the mean square dipole moment and the polarizability of a molecule, eqn (13.1.17) with $\delta\mu^2=\langle\mu^2\rangle$. On inserting $\langle\mu_A^2\rangle\approx 3\alpha_A\Delta_A/2$ we arrive at

$$\mathscr{E}^{(2)} \approx -(3/32\pi^2\varepsilon_0^2)\left\{\frac{\Delta_A\Delta_B}{\Delta_A+\Delta_B}\right\}\left\{\frac{\alpha_A\alpha_B}{R^6}\right\} = -\frac{3}{2}\left\{\frac{\Delta_A\Delta_B}{\Delta_A+\Delta_B}\right\}\left\{\frac{\alpha_A'\alpha_B'}{R^6}\right\}. \tag{13.4.10}$$

A rough indication of the magnitudes of the mean excitation energies is provided by the ionization energies of the atoms, and if we write $\Delta_A\approx\eta I_A$, where η is of the order of unity, we arrive at the

London formula: $$\mathscr{E}^{(2)} \approx -\tfrac{3}{2}\eta\left\{\frac{I_A I_B}{I_A+I_B}\right\}\left\{\frac{\alpha_A'\alpha_B'}{R^6}\right\}. \tag{13.4.11}$$

Example. Calculate the dispersion energy between two electrons harmonically oscillating in three dimensions and separated by a distance R, and express the answer in terms of the polarizabilities.

• *Method.* Use eqn (13.4.8). For the matrix elements use the values in the preceding two *Examples* but distinguish the systems by difference force constants k_A, k_B and frequencies ω_A, ω_B. On account of the selection rules, there are only four terms in the sum over v_A, v_B. Take one system to be in the state v_A and the other in the state v_B, then specialize to $v_A=v_B=0$ at the end of the calculation. Use the *Example* on p. 352 to express the interaction energy in terms of the polarizabilities through $\alpha_A=e^2/k_A$ and $\alpha_B=e^2/k_B$.

• *Answer.* The sum has the following four non-zero terms:

$$\mathscr{E} = -\tfrac{2}{3}(1/4\pi\varepsilon_0 R^3)^2\left\{\frac{|\mu^{A}_{v,v+1}|^2\,|\mu^{B}_{v,v+1}|^2}{\hbar(\omega_A+\omega_B)} + \frac{|\mu^{A}_{v,v+1}|^2\,|\mu^{B}_{v,v-1}|^2}{\hbar(\omega_A-\omega_B)}\right.$$
$$\left. + \frac{|\mu^{A}_{v,v-1}|^2\,|\mu^{B}_{v,v+1}|^2}{\hbar(-\omega_A+\omega_B)} + \frac{|\mu^{A}_{v,v-1}|^2\,|\mu^{B}_{v,v-1}|^2}{\hbar(-\omega_A-\omega_B)}\right\}$$
$$= -\tfrac{2}{3}(1/4\pi\varepsilon_0 R^3)^2(3\hbar e^2/2m_A\omega_A)(3\hbar e^2/2m_B\omega_B)$$
$$\times\left\{\frac{(v_A+1)(v_B+1)-v_Av_B}{\hbar(\omega_A+\omega_B)} + \frac{(v_A+1)v_B - v_A(v_B+1)}{\hbar(\omega_A-\omega_B)}\right\}$$

because

$$|\mu_{v,v+1}|^2 = |\mu_{x;v,v+1}|^2 + |\mu_{y;v,v+1}|^2 + |\mu_{z;v,v+1}|^2$$
$$= 3e^2(\hbar/2m\omega_0)(v+1), \text{ etc.}$$

Therefore

$$\mathscr{E} = -(\tfrac{3}{2})(1/4\pi\varepsilon_0 R^3)^2(e^4\hbar/m_A m_B\omega_A\omega_B)\left\{\frac{\omega_A(1+2v_B)-\omega_B(1+2v_A)}{\omega_A^2-\omega_B^2}\right\}$$
$$= -(\tfrac{3}{2})(1/4\pi\varepsilon_0 R^3)^2 e^4\hbar/m_A m_B\omega_A\omega_B(\omega_A+\omega_B) \quad \text{when} \quad v_A = v_B = 0.$$

Since $e^2 = \alpha_A k_A$, $e^2 = \alpha_B k_B$, and $\alpha' = \alpha/4\pi\varepsilon_0$, we have

$$\mathscr{E} = -\tfrac{3}{2}\hbar\left\{\frac{\omega_A\omega_B}{\omega_A+\omega_B}\right\}\left\{\frac{\alpha'_A\alpha'_B}{R^6}\right\}.$$

• *Comment.* Notice how this exact calculation mirrors the form of the London formula but with the ionization energies replaced by $\hbar\omega_A$ and $\hbar\omega_B$, in this case the only excitation energies allowed (so that the closure 'approximation' is exact).

The London formula, while only approximate, reveals the essential character of the dispersion energy and may be used to make rough estimates of its magnitude. We see that *the greater the polarizabilities of the atoms, the stronger their interaction* by the dispersion mechanism. We have already seen how the polarizability is related to structure, and the remarks made earlier in the chapter may be extended to the interactions between atoms and molecules. Thus we expect intensely coloured, highly polarizable, large, many-electron molecules to have strong dispersion interactions. One consequence of this is the high volatility of the colourless (and only slightly polarizable) low molecular mass hydrocarbons.

13.5 Retardation effects

Not only is the London formula an approximation, even the starting point of the calculation, the hamiltonian in eqn (13.4.1), is also only an approximation. This is because the actual interaction of two atoms is by way of the electromagnetic field. Consider how a fluctuation in the charge distribution on A is transmitted to B. When the charge distribution shakes it generates an electromagnetic wave which is transmitted towards B at the speed of light. The charge distribution on B shakes in

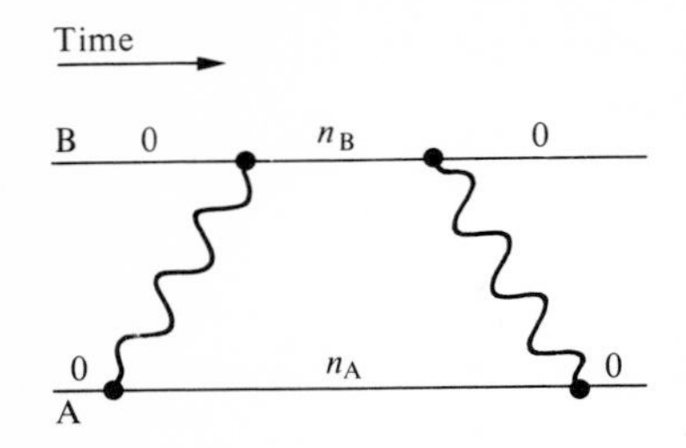

Fig. 13.7. A Feynman diagram depicting a contribution to the dispersion interaction.

response to the incoming signal, and generates a wave which returns to A. The true interaction is therefore the exchange of photons between the atoms, Fig. 13.7. Now, it takes a time R/c for the signal from A to arrive at B, and the response takes the same time to return. The fluctuations on A occur with a frequency of the order of $\Delta/\hbar$, and if the time interval for the round trip, $2R/c$, is longer than this, the dipole on A will have fluctuated to a new position. Therefore the interaction is weakened. Only when the atoms are so close that $2R/c$ is very much shorter than $\Delta/\hbar$ is there an effectively instantaneous interaction, and the dipolar interaction have the full strength as represented by the hamiltonian in eqn (13.4.1). When $2R/c \gg \Delta/\hbar$ (when R exceeds about 10 nm) the weakening effect of the *retardation* of the interaction is so great that the $1/R^6$ form of the interaction changes to the more rapidly decaying $1/R^7$ and

$$\mathscr{E}^{(2)} \approx -(23\hbar c/4\pi)\alpha'_A\alpha'_B/R^7, \qquad (2R\hbar/c\Delta \gg 1). \tag{13.5.1}$$

When $2R\hbar/c\Delta \approx 1$ the expressions are much more complicated because the $1/R^6$ term is then in the middle of evolving into the long-distance $1/R^7$ form. Retardation effects are important in colloids and macromolecules.

13.6 Optical activity

In an experiment to measure optical activity a plane polarized beam of light (that is, one in which the electric vector is oscillating in a plane) is passed through the active medium and the ray emerges with its plane of polarization rotated. This behaviour can be accounted for in terms of the medium having different refractive indices for left- and right-circularly polarized radiation, and hence is an aspect of *optical birefringence*, in this case, *circular birefringence.*

Consider the diagram shown in Fig. 13.8 which shows how a plane polarized ray may be decomposed into a superposition of two counter-rotating circularly polarized components $\mathbf{E}^+$ and $\mathbf{E}^-$. The expressions for the components in terms of the time and the location along the propagation direction (z) are

$$\mathbf{E}^- = E\hat{\mathbf{i}}\cos\phi_- - E\hat{\mathbf{j}}\sin\phi_-; \quad \mathbf{E}^+ = E\hat{\mathbf{i}}\cos\phi_+ + E\hat{\mathbf{j}}\sin\phi_+, \tag{13.6.1}$$

with $\hat{\mathbf{i}}$ and $\hat{\mathbf{j}}$ unit vectors perpendicular to the propagation direction, and

$$\phi_\pm = \omega t - 2\pi z/\lambda_\pm, \qquad 1/\lambda_\pm = \nu/v_\pm = n_\pm \nu/c \tag{13.6.2}$$

because the speed of light in the medium is c/n, and we have allowed for the possibility that the refractive indices are different for left ($\mathbf{E}^+$) and right ($\mathbf{E}^-$) circularly polarized light. Since $2\pi\nu = \omega$,

$$\phi_\pm = \omega t - n_\pm\omega z/c = \phi \mp (\omega z/2c)\Delta n \begin{cases} \phi = \omega t - n\omega z/c \\ n = \tfrac{1}{2}(n_+ + n_-) \\ \Delta n = n_+ - n_- = n_L - n_R. \end{cases} \tag{13.6.3}$$

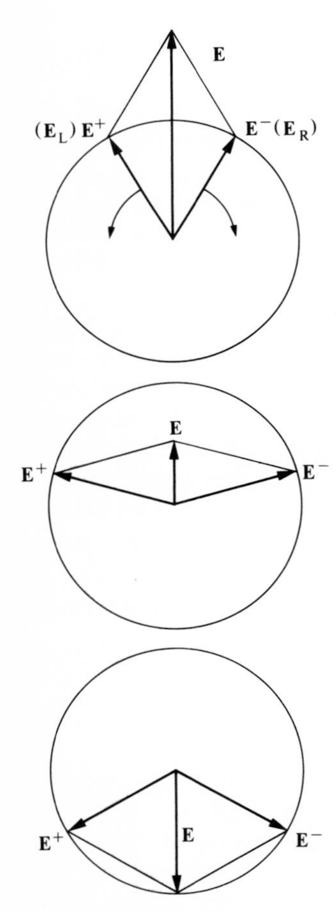

Fig. 13.8. The composition of counter-rotating components into a linearly polarized field.

When the medium is absent or not circularly birefringent, $\Delta n = 0$ and $\mathbf{E}^{\pm} = E\hat{\mathbf{i}} \cos \phi \pm E\hat{\mathbf{j}} \sin \phi$. The superposition $\mathbf{E}^{+} + \mathbf{E}^{-}$ gives a ray with the electric vector

$$\mathbf{E} = \mathbf{E}^{+} + \mathbf{E}^{-} = 2E\hat{\mathbf{i}} \cos \phi, \qquad (13.6.4)$$

which oscillates in the plane defined by the direction of propagation and the polarization vector $\hat{\mathbf{i}}$. When the beam enters a birefringent medium, one of the components propagates faster than the other because their speeds are $v_{\pm} = c/(n \pm \frac{1}{2}\Delta n)$. On account of this difference the phases of the two components are no longer the same at the same instant and location in the medium, and the superposition is a ray with the electric vector

$$\begin{aligned}\mathbf{E} = \mathbf{E}^{+} + \mathbf{E}^{-} &= E\{(\cos \phi_{+} + \cos \phi_{-})\hat{\mathbf{i}} + (\sin \phi_{+} - \sin \phi_{-})\hat{\mathbf{j}}\} \\ &= 2E \cos \phi\{\hat{\mathbf{i}} \cos(z\omega\Delta n/2c) - \hat{\mathbf{j}} \sin(z\omega\Delta n/2c)\}. \qquad (13.6.5)\end{aligned}$$

This is still a plane polarized ray, but its plane of polarization is inclined at an angle $\theta = z\omega\Delta n/2c$ to the original direction, Fig. 13.9. The rotation angle for a path length l through the medium is therefore $\theta = l\omega\Delta n/2c$. Hence we have confirmed that there is an angle of rotation when the medium is circularly birefringent ($\Delta n \neq 0$). Moreover, if $n_{\mathrm{L}} > n_{\mathrm{R}}$ (so that $\Delta n > 0$), $\theta > 0$ corresponding to *dextrorotation* (+). When $n_{\mathrm{L}} < n_{\mathrm{R}}$ (so that $\Delta n < 0$) we have *laevorotation* (−).

When there is also absorption as the ray passes through the medium the left and right circularly polarized components may be absorbed differentially (i.e. the system may be *circularly dichroic*). Then the simple addition that led to eqn (13.6.5) is no longer valid and the decay of intensity of one component relative to the other results in the superposition becoming *elliptically polarized*, the tip of the electric vector then tracing out an ellipse. In the extreme case of one component being totally absorbed, the emergent light is purely circularly polarized in the opposite sense. Whenever there is circular birefringence there is always an accompanying circular dichroism, but it is normally significant only when the incident radiation is close to an absorption frequency, and so the small degree of ellipticity can normally be ignored.

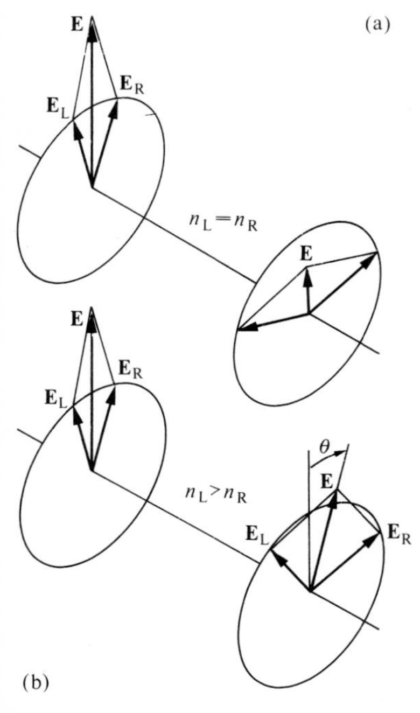

Fig. 13.9. Propagation of light through (a) an optically inactive medium, (b) an optically active ($n_{\mathrm{L}} > n_{\mathrm{R}}$) medium.

The next job is to find an explanation of the sample's circular birefringence. Since we have seen that the refractive index of a sample is related to the polarizability of its molecules, we should expect birefringence when the molecules respond differently to left and right circularly polarized light. But why should this be so?

The fundamental reason lies in the spatial extension of the molecule and the variation of the electromagnetic field over it. This is because the left and right circularly polarized components of the light have different spatial characteristics, and under some circumstances (which we later investigate) it can be expected that they interact with the medium to different extents, but only if the molecules can distinguish the components by sampling them over a range. According to Maxwell's equations (Appendix 19), the spatial variation of the electric field of a light wave is proportional to the time variation of the magnetic field ($\partial\mathbf{E}/\partial x \propto \partial\mathbf{B}/\partial t$). Therefore, the part of the electric dipole (or the electric polarization $\mathbf{P}$) induced by the spatial variation of the electric field ($\mathbf{P} \propto \partial\mathbf{E}/\partial x$, etc.) can be expressed in terms of the time variation of the magnetic field ($\mathbf{P} \propto \partial\mathbf{B}/\partial t$). Therefore the total polarization of the medium should be written

$$\mathbf{P} = \mathcal{N}\alpha\mathbf{E} - \mathcal{N}\beta\dot{\mathbf{B}} \tag{13.6.6}$$

when the spatial variation is taken into account, β being some coefficient (not the hyperpolarizability). That the polarization does in fact depend on $\dot{\mathbf{B}}$ will be shown more formally below.

We can see that we are on the right track. In the first place the magnetic field of an electromagnetic wave is perpendicular to the electric field. Therefore, while the $\alpha\mathbf{E}$ component of this expression induces an electric dipole in the molecule in the same plane as $\mathbf{E}$ itself (and therefore generates a wave in the same plane), the $\beta\dot{\mathbf{B}}$ component induces an electric dipole *perpendicular* to the plane of $\mathbf{E}$, and that generates an oscillating electric field in the same plane. The resultant of these two components lies in a plane rotated from the initial orientation, and so the plane of polarization is rotated. This is confirmed when we solve the Maxwell equations for a medium with a polarization given by eqn (13.6.6). The actual calculation is described in Appendix 19, where we see that whereas in the absence of the β term the refractive index turns out to be given by $n_r = (1 + \alpha\mathcal{N}/\varepsilon_0)^{\frac{1}{2}} \approx 1 + \alpha\mathcal{N}/2\varepsilon_0$; when β is non-zero the medium is birefringent, and the refractive indexes of the circularly polarized components are

$$n_{\pm} \approx 1 + \alpha\mathcal{N}/2\varepsilon_0 \pm \omega\beta\mathcal{N}/c\varepsilon_0 \qquad (n_+ = n_L;\ n_- = n_R). \tag{13.6.7}$$

It follows that the difference of refractive indexes is $\Delta n = 2\omega\beta\mathcal{N}/c\varepsilon_0$, and that the angle of rotation (in radians) after passage through a medium of length l is

$$\theta = \omega^2\beta\mathcal{N}l/c^2\varepsilon_0 = \beta\mu_0\omega^2 l\mathcal{N} \tag{13.6.8}$$

(where we have used $\varepsilon_0\mu_0 = 1/c^2$).

The calculation of the angle of rotation now boils down to calculating β. That is, we must calculate the polarization of the medium in response to the changing magnetic field of the light ray. The strategy involves adapting the calculation of n_r, which was based on $H^{(1)} = -\boldsymbol{\mu} \cdot \mathbf{E}(t)$, to the case where

$$H^{(1)}(t) = -\boldsymbol{\mu} \cdot \mathbf{E}(t) - \mathbf{m} \cdot \mathbf{B}(t), \tag{13.6.9}$$

$\mathbf{m}$ being the magnetic dipole moment operator for the molecule. For all cases of interest to us this is equal to $\gamma_e\mathbf{l}$, where $\mathbf{l}$ is the orbital angular momentum operator and γ_e the magnetogyric ratio ($-e/2m_e$). The precise form of the perturbation depends on which component of circular polarization we are considering, and so we write

$$H^{\pm}(t)=-\boldsymbol{\mu}\cdot\mathbf{E}^{\pm}(t)-\mathbf{m}\cdot\mathbf{B}^{\pm}(t), \tag{13.6.10}$$

with

$$\mathbf{E}^{\pm}(t)=2E(\hat{\mathbf{i}}\cos\omega t\pm\hat{\mathbf{j}}\sin\omega t);\ \mathbf{B}^{\pm}(t)=2B(\pm\hat{\mathbf{i}}\sin\omega t-\hat{\mathbf{j}}\cos\omega t) \tag{13.6.11}$$

(the magnetic field vector is in step with the electric vector, but perpendicular to it).

Therefore the hamiltonians are

$$\begin{aligned}H^{\pm}(t)=&-E\{(\mathrm{e}^{\mathrm{i}\omega t}+\mathrm{e}^{-\mathrm{i}\omega t})\mu_x\mp\mathrm{i}(\mathrm{e}^{\mathrm{i}\omega t}-\mathrm{e}^{-\mathrm{i}\omega t})\mu_y\}\\&-B\{-(\mathrm{e}^{\mathrm{i}\omega t}+\mathrm{e}^{-\mathrm{i}\omega t})m_y\mp\mathrm{i}(\mathrm{e}^{\mathrm{i}\omega t}-\mathrm{e}^{-\mathrm{i}\omega t})m_x\}.\end{aligned} \tag{13.6.12}$$

This entire expression should be multiplied by the factor $(1-\mathrm{e}^{-t/\tau})$ used to switch on the electric field slowly, as explained before. Then the coefficients to first-order in the perturbation are given by the usual perturbation expression:

$$a_n^{\pm}(t)=(1/\mathrm{i}\hbar)\int_0^t H_{n0}^{\pm}(t)\mathrm{e}^{\mathrm{i}\omega_{n0}t}\,\mathrm{d}t. \tag{13.6.13}$$

With this form of the perturbed wavefunction we calculate the induced dipole moment, just as we did for n_r:

$$\begin{aligned}\langle\boldsymbol{\mu}^{\pm}\rangle&=\boldsymbol{\mu}_0+\sum_n{}'\{\boldsymbol{\mu}_{0n}a_n^{\pm}(t)\mathrm{e}^{-\mathrm{i}\omega_{n0}t}+\boldsymbol{\mu}_{n0}a_n^{\pm *}(t)\mathrm{e}^{\mathrm{i}\omega_{n0}t}\}\\&=(4/\hbar)\mathrm{re}\sum_n{}'\left\{\boldsymbol{\mu}_{0n}(E\mu_{x,n0}-Bm_{y,0n})\left(\frac{\omega_{n0}\cos\omega t-\mathrm{i}\omega\sin\omega t}{\omega_{n0}^2-\omega^2}\right)\right.\\&\qquad\left.\mp\mathrm{i}\boldsymbol{\mu}_{0n}(E\mu_{y,n0}+Bm_{x,n0})\left(\frac{\mathrm{i}\omega_{n0}\sin\omega t-\omega\cos\omega t}{\omega_{n0}^2-\omega^2}\right)\right\}.\end{aligned}$$

All the unperturbed wavefunctions may be taken to be real; therefore all $\boldsymbol{\mu}_{0n}$ are real while all $\mathbf{m}_{0n}$ are imaginary (because $\boldsymbol{\mu}$ is real and $\mathbf{m}=\gamma_e\mathbf{l}$ is imaginary). The real part of the expression is therefore

$$\begin{aligned}\langle\boldsymbol{\mu}^{\pm}\rangle&=(4/\hbar)\mathrm{re}\sum_n{}'\left\{\frac{E\omega_{n0}}{\omega_{n0}^2-\omega^2}\right\}\boldsymbol{\mu}_{0n}(\mu_{x,n0}\cos\omega t\pm\mu_{y,n0}\sin\omega t)\\&\quad-(4/\hbar)\mathrm{im}\sum_n{}'\left\{\frac{B\omega}{\omega_{n0}^2-\omega^2}\right\}\boldsymbol{\mu}_{0n}(m_{y,n0}\sin\omega t\pm m_{x,n0}\cos\omega t)\\&=(4/\hbar)\mathrm{re}\sum_n{}'\left\{\frac{E\omega_{n0}}{\omega_{n0}^2-\omega^2}\right\}\boldsymbol{\mu}_{0n}\boldsymbol{\mu}_{n0}\cdot(\hat{\mathbf{i}}\cos\omega t\pm\hat{\mathbf{j}}\sin\omega t)\\&\quad-(4\hbar)\mathrm{im}\sum_n{}'\left\{\frac{B\omega}{\omega_{n0}^2-\omega^2}\right\}\boldsymbol{\mu}_{0n}\mathbf{m}_{n0}\cdot(\hat{\mathbf{j}}\sin\omega t\pm\hat{\mathbf{i}}\cos\omega t)\\&=(2/\hbar)\mathrm{re}\sum_n{}'\left\{\frac{\omega_{n0}}{\omega_{n0}^2-\omega^2}\right\}\boldsymbol{\mu}_{0n}\boldsymbol{\mu}_{n0}\cdot\mathbf{E}^{\pm}(t)\\&\quad-(2/\hbar)\mathrm{im}\sum_n{}'\left\{\frac{1}{\omega_{n0}^2-\omega^2}\right\}\boldsymbol{\mu}_{0n}\mathbf{m}_{n0}\cdot\dot{\mathbf{B}}^{\pm}(t).\end{aligned} \tag{13.6.14}$$

On comparing this expression with eqn (13.6.6), we find

$$\boldsymbol{\beta} = (2/\hbar)\mathrm{im}\sum_n{}' \left\{\frac{\boldsymbol{\mu}_{0n}\mathbf{m}_{n0}}{\omega_{n0}^2 - \omega^2}\right\}. \qquad (13.6.15)$$

We can readily pick out the β_{xx}, β_{yy}, and β_{zz} components of β, and therefore arrive at the expression for the rotational average in solution, $\beta = \frac{1}{3}(\beta_{xx} + \beta_{yy} + \beta_{zz})$:

$$\beta = \tfrac{1}{3}(2/\hbar)\mathrm{im}\sum_n{}' \left\{\frac{\boldsymbol{\mu}_{0n} \cdot \mathbf{m}_{n0}}{\omega_{n0}^2 - \omega^2}\right\}. \qquad (13.6.16)$$

It follows that the angle of rotation is given by the

$$\textit{Rosenfeld equation:}\ \theta = \mathcal{N}l(2\mu_0/3\hbar) \sum_n{}' \left\{\frac{\omega^2 R_{n0}}{\omega_{n0}^2 - \omega^2}\right\} \qquad (13.6.17)$$

R_{n0} is the

$$\textit{Rotational strength:}\ R_{n0} = \mathrm{im}(\boldsymbol{\mu}_{0n} \cdot \mathbf{m}_{n0}) \qquad (13.6.18)$$

of the $n \leftarrow 0$ transition.

13.7 Rotational strengths

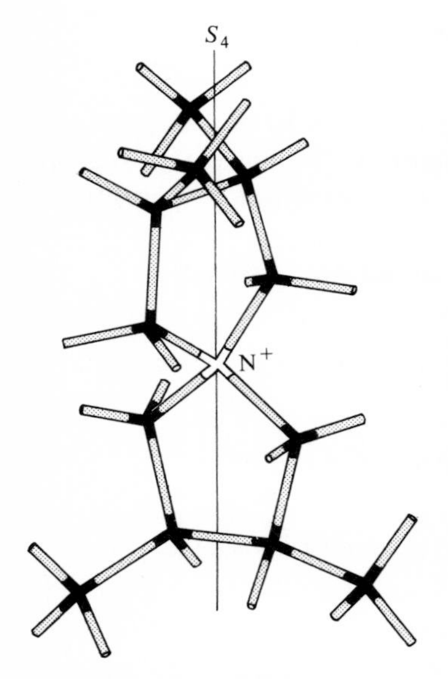

Fig. 13.10. An optically inactive molecule that lacks a centre of inversion and a mirror plane.

The optical activity of molecules may be discussed largely in terms of the rotational strengths of transitions (which play a role similar to oscillator strengths of absorptions). The first property is that *the rotational strengths of all the transitions in a molecule are zero if the molecule has a plane of symmetry or a centre of inversion.*† The symmetry argument behind this requirement depends on $\boldsymbol{\mu}$ being a polar vector while $\mathbf{m}$ is an axial vector. When the molecule is inverted $\boldsymbol{\mu}_{0n} \cdot \mathbf{m}_{n0}$ changes to $-\boldsymbol{\mu}_{0n} \cdot \mathbf{m}_{n0}$ because $\boldsymbol{\mu} \to -\boldsymbol{\mu}$ while $\mathbf{m} \propto \mathbf{r} \wedge \mathbf{p} \to (-\mathbf{r}) \wedge (-\mathbf{p}) \propto \mathbf{m}$, and the wavefunctions occur an even number of times. Hence the rotational strength must be zero (because if $R_{n0} = -R_{n0}$, $R_{n0} = 0$). A similar argument eliminates the rotational strength if there is a mirror plane.

The second property is that if a pair of molecules are each other's mirror image, then under a reflection the wavefunctions of one are taken into the wavefunctions of the other. But under a reflection R_{n0} changes sign. Therefore *enantiomeric pairs* (molecules related by a reflection) *have equal but opposite optical rotations.*

† The precise requirement, of which these are special cases, is that the molecule should contain no improper rotation axis. Some molecules may have no centre of inversion or mirror plane and yet be optically inactive; closer inspection shows that they have an improper rotation axis. An example is given in Fig. 13.10.

A third property comes from the following *sum rule*:

$$\sum_n R_{n0} = \text{im}\sum_n \boldsymbol{\mu}_{0n} \cdot \mathbf{m}_{n0} = \text{im}\sum_n \langle 0|\,\boldsymbol{\mu}\,|n\rangle \cdot \langle n|\,\mathbf{m}\,|0\rangle$$
$$= \text{im}\,\langle 0|\,\boldsymbol{\mu}\cdot\mathbf{m}\,|0\rangle = 0 \qquad (13.7.1)$$

because $\boldsymbol{\mu}\cdot\mathbf{m} \propto \mathbf{r}\cdot\mathbf{r}\wedge\mathbf{p} = \mathbf{r}\wedge\mathbf{r}\cdot\mathbf{p} = 0$ from the properties of vectors and the fact that a vector product of a vector with itself is zero. The sum rule has the important consequence that the angle of optical rotation tends to zero at both high and low frequencies. At very high frequencies, when $\omega^2 \gg \omega_{n0}^2$ the rotation angle is

$$\theta \approx \mathcal{N}l(2\mu_0/3\hbar)\sum_n{}' \omega^2 R_{n0}/(-\omega^2)$$
$$\approx -\mathcal{N}l(2\mu_0/3\hbar)\sum_n{}' R_{n0}$$
$$\approx -\mathcal{N}l(2\mu_0/3\hbar)\left\{\sum_n R_{n0} - R_{00}\right\} = 0 \qquad (13.7.2)$$

because the first sum (over all n) disappears by the sum rule and R_{00} is zero because it is the imaginary part of the expectation value of the product of two expectation values, and expectation values of hermitian operators are real. At the other extreme of frequency, when $\omega^2 \ll \omega_{n0}^2$, we have

$$\theta \approx \mathcal{N}l(2\mu_0/3\hbar)\sum_n{}' \omega^2 R_{n0}/\omega_{n0}^2 \approx 0 \qquad (13.7.3)$$

because of the ω^2 factor in the numerator.

The variation of the angle of rotation with the frequency of the incident light is called *optical rotatory dispersion* (ORD). A typical shape of an ORD curve is shown in Fig. 13.11. The rotation is close to zero at frequencies well away from absorption bands, but may become very large close to a transition frequency, where the denominator $\omega_{n0}^2 - \omega^2$ approaches zero. The rotation does not actually rise to infinity as this perturbation result suggests: in the region of an absorption band perturbation theory fails, and special techniques have to be used instead (just as in the case of the dispersion of the refractive index).

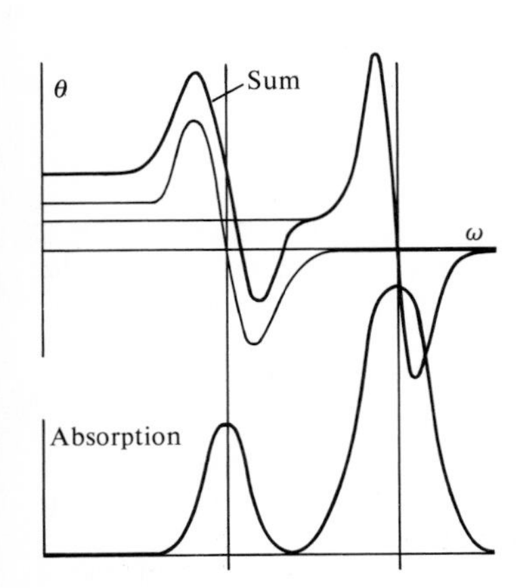

Fig. 13.11. Optical rotatory dispersion in the vicinity of two absorption bands.

When the incident frequency lies close to one transition frequency of the molecule, its contribution to the optical rotation dominates, and then the angle of rotation is given by

$$\theta \approx \mathcal{N}l\,(2\omega^2\mu_0/3\hbar)R_{k0}/(\omega_{k0}^2 - \omega^2) \qquad (13.7.4)$$

when $k \leftarrow 0$ is the dominating transition. The area under the dispersion curve in this region can then be used to estimate the value of the rotational strength R_{k0} (just as oscillator strengths are obtained from absorbances, the integrals of absorption bands). Much theoretical work has gone into the calculation of the rotational strengths of transitions, and the type of calculation depends on whether the molecule as a whole contributes to the transition or whether it is localized on a particular chromophore. The carbonyl group has received much attention, and we shall consider it briefly in order to illustrate the basic ideas and difficulties.

We saw in Chapter 12 that the transition in the region of 290 nm in carbonyl compounds can be ascribed to the $\pi^* \leftarrow n$ transition of the carbonyl group itself. The non-bonding orbital n is almost pure p_y-orbital on oxygen. The transition is electric-dipole forbidden, but it is magnetic-dipole allowed because p_y is rotated into p_x by a rotation around the C—O axis, and this is brought about by $m_z = \gamma_e l_z$. The motion of the electron during the transition can be thought of as lying in a circle around the C—O bond, Fig. 13.12(a). Since it is electric-dipole forbidden, this transition has no rotational strength because $\boldsymbol{\mu}_{0k} \cdot \mathbf{m}_{k0}$ is zero. However, we should take into account the environment of the carbonyl group which may distort its orbitals (and also have other effects, such as spreading its orbitals over other atoms, and so allowing them to explore the asymmetry of the molecule as a whole). We therefore have to investigate how the distortion of the chromophore can lead the π^*–n transition to acquire some electric dipole character parallel to the axis, so that the scalar product $\boldsymbol{\mu}_{0k} \cdot \mathbf{m}_{k0}$ no longer vanishes. This will happen if the transition becomes *helical*, the electron spiralling around the axis instead of moving in a circle alone, Fig. 13.12(b).

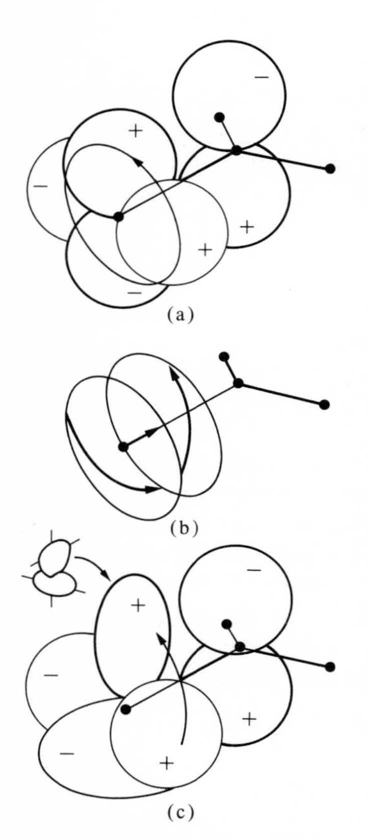

Fig. 13.12. (a) A magnetic-dipole allowed $\pi^* \leftarrow n$ transition; (b) a helical transition has both magnetic dipole and electric dipole components; (c) the $\pi^* \leftarrow n$ transition becomes helical with the admixture of d-orbitals.

One way of achieving a helical transition is for the π^*-orbital to have some d_{yz}-character. The distortion this introduces is sketched in Fig. 13.12(c), which suggests that the π^*-orbtial is formed between a p-orbital on C and a $p_x d_{yz}$-hybrid on O, the latter being a twisted distribution of charge density. If the π^*-orbital contains some d_{yz}-character, and is of the form $c_p p_x + c_d d_{yz}$ on the oxygen, the electric dipole moment operator has a non-vanishing matrix element $\langle \pi^* | \mu_z | n \rangle = c_d \langle d_{yz} | \mu_z | p_y \rangle$. Since the magnetic moment operator has the non-vanishing component $\langle \pi^* | m_z | n \rangle = c_p \langle p_x | m_z | p_y \rangle$, the $\pi^* - n$ transition has a non-vanishing rotational strength $R_{n0} \propto c_p c_d$. Hence the distorted carbonyl group is optically active. We also note that the effect of the d-orbital admixture is to remove its plane of symmetry, which is consistent with the earlier discussion about the criteria for optical activity.

Example. Consider in more detail the model of a helical transition described above. Let there be a single centre to which the electron is confined, and in the ground state it occupies a pure $2p_y$-orbital while the upper orbital is the pd-hybrid $\psi = p_x \cos \eta + d_{yz} \sin \eta$ (where $0 \leq \eta \leq \pi/2$); p_x, p_y, d_{yz} being Slater orbitals. Evaluate the rotational strength of the transition as a function of the parameter η.

• *Method.* The rotational strength, eqn (13.6.18), has contributions only from the z-components of the transitions. For the matrix element of $m_z = \gamma_e l_z$, use $l_z = (\hbar/i)(\partial/\partial\phi)$ and recognize that $p_x \propto \cos \phi$ and $p_y \propto \sin \phi$. For the matrix elements of $\mu_z = -ez$ write $z = r \cos \theta$ and use the form of the STOs specified on p. 234, their angular components being the spherical harmonics of Table 4.1. Use $n = 2$ for the p-orbitals (i.e. 2p-orbitals) and $n = 3$ for the d-orbital (i.e. a 3d-orbital).

• *Answer.* Since $l_z p_y \propto (\hbar/i)(\partial/\partial\phi)\sin\phi = (\hbar/i)\cos\phi$, we have $l_z p_y = -i\hbar p_x$, and so

$$\langle\psi|\, m_z\, |0\rangle = \gamma_e\langle p_x|\, l_z\, |p_y\rangle\cos\eta = -i\hbar\gamma_e\langle p_x \mid p_x\rangle\cos\eta$$
$$= i\mu_B\cos\eta \ (\text{since } \mu_B = -\gamma_e\hbar).$$

For the electric dipole moment we need the explicit form of the orbitals:

$$p_y = (3/4\pi)^{\frac{1}{2}}\sin\theta\sin\phi(2^5\zeta_p^5/4!)^{\frac{1}{2}}re^{-\zeta_p r}, \qquad \zeta_p = Z_p^*/n_p a_0$$
$$d_{yz} = \tfrac{1}{2}(15/4\pi)^{\frac{1}{2}}\sin 2\theta\sin\phi(2^7\zeta_d^7/6!)^{\frac{1}{2}}r^2e^{-\zeta_d r}, \qquad \zeta_d = Z_d^*/n_d a_0.$$
$$\langle 0|\, \mu_z\, |\psi\rangle = -e\langle p_y|\, z\, |d_{yz}\rangle\sin\eta = -e\{a_0 2^6(6\zeta_p^5\zeta_d^7)^{\frac{1}{2}}/(\zeta_p+\zeta_d)^7\}\sin\eta.$$

Therefore, from eqn (13.6.18) and with $\sin\eta\cos\eta = \frac{1}{2}\sin 2\eta$,

$$R = -ea_0\mu_B\{2^5(6\zeta_p^5\zeta_d^7)^{\frac{1}{2}}/(\zeta_p+\zeta_d)^7\}\sin 2\eta$$
$$= -(6.16\times10^{-51}\ \text{C}^2\,\text{m}^3\,\text{s}^{-1})A\sin 2\eta, \qquad A = (\zeta_p^5\zeta_d^7)^{\frac{1}{2}}/(\zeta_p+\zeta_d)^7 a_0.$$

• *Comment.* R is greatest when $\eta = \pi/4$ and $5\pi/4$. The angle of rotation can be evaluated from eqn (13.7.4) by writing $\mathcal{N} = CL$ (where C is the concentration), for then

$$\theta \approx -3.0\times10^3(C/\text{mol dm}^{-3})(l/\text{dm})\{\omega^2/(\omega_{k0}^2-\omega^2)\}A\sin 2\eta.$$

For an oxygen atom, $\zeta_p = 2.25/a_0$ and $\zeta_d = 0.33/a_0$ (p. 235), and so $A = 2.11\times10^{-4}$. This corresponds to $R \approx (1.30\times10^{-54}\ \text{C}^2\,\text{m}^3\,\text{s}^{-1})\sin 2\eta$ and to $\theta \approx -36°\ (C/\text{mol dm}^{-3})(l/\text{dm})\{\omega^2/(\omega_{k0}^2-\omega^2)\}\sin 2\eta$. Therefore, for $C \approx 1$ mol dm^{-3}, $l \approx 1$ dm, $\eta \approx \pi/4$, and $\omega^2/(\omega_{k0}^2-\omega^2) \approx 0.6$ (corresponding to $\lambda_{k0} \approx 300$ nm, $\lambda \approx 500$ nm), we have $\theta \approx -20°$.

The principal difficulty with the calculation is the estimation of the extent of distortion induced in a chromophore by the asymmetry of the remainder of the molecule. This delicate problem can be explored via *Further reading*, p. 374.

13.8 Induced birefringence

Even though molecules that are not dissymmetric are not naturally optically active, activity may be imposed upon them by the application of electric and magnetic fields. For example, in the Faraday effect a magnetic field is applied along the direction of propagation of a plane polarized beam of light, and the sample rotates the plane of polarization to an extent proportional to the field strength:

$$\textit{Faraday effect}: \ \theta = VH. \qquad (13.8.1)$$

The constant of proportionality is called the *Verdet constant*, and is the measure of the extent to which the field distorts the structure of the molecule or affects the rotational averaging procedure by orientating the molecules. All molecules rotate the beam of light in a Faraday experiment, because only electric dipole interactions with the electromagnetic field of the light beam are involved, and so molecular dissymmetry is not necessary (the static field imposes dissymmetry).

When the magnetic field is applied perpendicular to the propagation direction and at 45° to the plane of polarization, the beam acquires an elliptical polarization, but it is not rotated. This magnetically imposed circular dichroism is the *Cotton–Mouton effect*, and arises through distortion and orientation of the molecules in the sample. The electrical analogue of the Cotton–Mouton effect is the *Kerr effect*, in which ellipticity is induced by an externally applied electric field at right angles to the propagation direction. Instead of measuring the ellipticity itself, the absorption of the beam when the field is applied perpendicular and parallel to the polarization direction is monitored. The Kerr effect can also be used in a very fast light switch, for the absorption characteristics of a sample (e.g. nitrobenzene) can be changed very rapidly. Such switches are used to obtain very intense pulses from lasers.

Now that lasers have made available very intense beams of light it is possible to use the electric fields of the light itself, and in this way to induce the *optical Kerr effect*, for example, in which ellipticity is acquired by a weak measuring beam as a result of the presence of a parallel intense beam. Other effects also become apparent when very intense beams are employed. For example, the optical rotation angle in a naturally active molecule becomes intensity dependent. The *inverse Faraday effect*, in which a solution is magnetized by passing a circularly polarized beam through it has been detected. These effects are called collectively *non-linear optics*.

All optical birefringence properties may be interpreted in a unified way by viewing them as a manifestation of a scattering process in which the projectiles are photons which scatter from targets, molecules, with a probability of a change occurring in their polarization direction. All these phenomena can therefore be discussed in terms of the powerful language of *scattering theory*.

Further reading

Dipole moments and polarizabilities

Physical chemistry. P. W. Atkins; Oxford University Press and W. H. Freeman, 2nd edition, 1982.

The theory of the electric and magnetic properties of molecules. D. W. Davies; Wiley, New York, 1967.

The theory of electric and magnetic susceptibilities. J. H. Van Vleck; Oxford University Press, 1932.

Theory of electric polarization (2 volumes). C. J. F. Böttcher and P. Bordewijk; Elsevier, Amsterdam, 1978.

Electromagnetic fields. R. K. Wangsness; Wiley, New York, 1979.

Electrodynamics of continuous media. L. D. Landau and E. M. Lifshiftz, Pergamon, Oxford, 1960.

Intermolecular forces

Molecular forces. B. Chu; Interscience, New York, 1967.

Molecular theory of gases and liquids. J. O. Hirschfelder, C. F. Curtiss, and R. B. Bird; Wiley, New York, 1964.

Intermolecular forces. G. C. Maitland, M. Rigby, E. B. Smith, and W. A. Wakeham; Oxford University Press, 1981.

Optical activity

The theory of optical activity. D. J. Caldwell and H. Eyring; Wiley-Interscience, New York, 1971.

The molecular basis of optical activity. E. Charney; Wiley, New York, 1979.

Problems

13.1. The polarizability volume of tetrachloromethane is $10.5 \times 10^{-24}\,\text{cm}^3$. Calculate (a) the magnitude of the dipole moment induced by a $10\,\text{kV m}^{-1}$ electric field, (b) the change in energy per unit amount of substance (e.g. per mole of CCl_4).

13.2. Model an atom by an electron in a one-dimensional box of length L (there is assumed to be an 'invisible' positive charge at the centre of the box which provides the positive end of the dipole but does not affect the wavefunctions). Calculate the polarizability of the system parallel to its length. *Hint.* Use eqn (13.1.14); the wavefunctions are given in eqn (3.3.5). The procedure and results of Problem 8.5 can be used.

13.3. Repeat the calculation, but use the closure approximation. Explore the validity of using the value of Δ calculated in Problem 8.9.

13.4. Evaluate the polarizability and polarizability volume of a hydrogen atom; for simplicity, confine the perturbation sum to the 2p-orbitals. Explore the consequences of making the closure approximation, and compare the calculations with the approximate expression eqn (13.1.23).

13.5. Devise a variational calculation of the polarizability of the hydrogen atom. *Hint.* A simple procedure would be to take as a trial function the linear combination $c_{1s}\psi_{1s} + c_{2p_z}\psi_{2p_z}$ (the basis could be lengthened in a more sophisticated treatment) or alternatively $(1+az)\psi_{1s}$, with a the variation parameter. The hamiltonian is $H = H_0 + ezE$ in each case. Find the best energy and identify α_{zz}.

13.6. Establish a perturbation theory expression for the components of the first hyperpolarizability of a molecule. *Hint.* Refer to eqn (13.1.8c). You need to return to Chapter 8 and to derive an expression for the third-order correction to the energy:

$$E^{(3)} = \sum_{mn}{}' \{H^{(1)}_{0m}H^{(1)}_{mn}H^{(1)}_{n0}/(E_m - E_0)(E_n - E_0)\} - H^{(1)}_{00}\sum_{n}{}' \{H^{(1)}_{0n}H^{(1)}_{n0}/(E_n - E_0)^2\}.$$

13.7. Show group theoretically that in a tetrahedral molecule (a) the mean hyperpolarizability is zero, (b) the only non-zero components are β_{xyz} and the permutations of its indices. *Hint.* The mean is defined as $(3/5)(\beta_{xxz} + \beta_{yyz} + \beta_{zzz})$; and so (b) implies (a). For (b) consider the symmetry characteristics of $\mathscr{E} = -(1/3!)\sum_{abc}\beta_{abc}E_aE_bE_c$, the generalization of eqn (13.1.9).

13.8. Evaluate the first hyperpolarizability of a one-dimensional system of two charges e and $-e$ bound together by a spring of force constant k, the electric field being applied parallel to the x-axis. *Hint.* Use the matrix elements set out in the *Example* on p. 299: the result can be obtained by inspection.

13.9 Prove the sum rule $\sum_f x_{mf}x_{fn}\omega_{fn} = (\hbar/m)\delta_{mn} + \frac{1}{2}\omega_{mn}(x^2)_{mn}$. *Hint.* Consider the matrix elements of the commutator $[H, x^2]$.

13.10. Use the closure expressions to estimate the contribution to the polarizability of a carbon atom of one of its 2p-electrons when the field is applied (a) parallel, (b) perpendicular to the axis. Assess the contributions of the 1s-electrons and the 2s-electrons, and estimate the total mean polarizability by adding all the contributions. *Hint.* Use Slater atomic orbitals and eqn

(13.1.18). The 2s,2p energy separation is about 75 000 cm^{-1}; the first ionization energy corresponds to 11.264 eV. The energies of the 1s-electrons can be estimated by regarding them as hydrogen-like. Take C: $1s^2 2s2p_x 2p_y 2p_z$.

13.11. The oscillator strength of a transition at about 160 nm in ethene is about 0.3. Estimate the mean polarizability volume of the molecule. (The experimental value is $4.22\times10^{-24}\ \text{cm}^3$.)

13.12. Deduce an expression for the refractive index of a gas of free electrons. *Hint.* Take the limit of eqn (13.3.11) when $\omega^2 \gg \omega_{n0}^2$ and refer to eqn (13.3.10); this leads to the *Thomson formula* for the refractive index.

13.13. A region of interstellar space contained a diffuse gas of hydrogen atoms at a density 10^5 atoms m^{-3}. What is the refractive index for visible (590 nm) light in the region?

13.14. Some molar refractivities (in $\text{cm}^3\ \text{mol}^{-1}$) of groups at 589 nm are as follows: C—H, 1.65; C—C, 1.20; C=C, 2.79; C—O, 1.41; C=O, 3.34; O—H, 1.85. Estimate the refractive index of (a) acetic acid (ethanoic acid), (b) ethane, (c) ethene, (d) water. Take the gases at 1 atm and 298 K; the density of acetic acid is 1.046 g cm^{-3}.

13.15. Consider two particles, each in a one-dimensional box, with the centres of the boxes separated by a distance R. Each system may be regarded as a model of an atom in the same sense as in Problem 13.2. Calculate the dispersion energy when the boxes are (a) in line, (b) broadside on. *Hint.* Base the calculation on eqn (13.4.5), noting that the dipole moment operators have only one component in a one-dimensional system. Much of the calculational work has been done in Problem 13.2.

13.16. Investigate the usefulness of the closure approximation in the calculation of the dispersion energy of the system described in Problem 13.15. What values of Δ_A and Δ_B should be used?

13.17. Estimate the dispersion energy between two hydrogen atoms.

13.18. Devise a variational calculation of the dispersion interaction between two hydrogen atoms. Start by using the trial functions suggested in Problem 13.5, but note that the dipolar hamiltonian also introduces distortions perpendicular to the line of centres of the atoms; ignore this initially, but include it in an improved trial function. The hamiltonian to use in the evaluation of the Rayleigh ratio (or the secular determinant, depending on the trial function) is $H_a + H_b + H^{(1)}$, $H^{(1)}$ being given in eqn (13.4.1).

13.19. Evaluate the rotational strength of a transition of an electron from a $2p_x$-orbital to a $2p_z$,$3d_{xy}$-hybrid orbital. Assume the orbitals are on a carbon atom. Estimate the optical rotation angle for 590 nm light. *Hint.* Follow the *Example* on p. 372, with changes of detail.

13.20. An electron is bound to a nucleus and undergoes harmonic vibrations in three dimensions, the frequencies being ω_X, ω_Y, and ω_Z. It is subjected to a perturbation of the form $H^{(1)} = Axyz$. Calculate the rotational strength and the optical rotation angle of first order in A. *Hint.* Base the answer on eqn (13.6.18), evaluating the matrix elements using the first-order perturbed wavefunctions, eqn (8.1.23). Express all the operators (including the angular momentum operators, via eqn (6.1.2)) in terms of annihilation and creation operators, and use the matrix elements established on p. 299.

14 The magnetic properties of molecules

THE difference between electric and magnetic perturbations is that while the former stretch a molecule, the latter twist it (this will be demonstrated explicitly in due course). The effect of a twisting perturbation is to induce electronic currents that circulate through the nuclear framework of the molecule. These currents give rise to their own magnetic fields. One effect is to modify the magnetic flux density in the material: if the induced field augments the applied field the flux density is increased and the substance is *paramagnetic*. Conversely (and more commonly) if the flux density is reduced on account of an opposing effect of the induced currents, the substance is *diamagnetic*. The distinction gives rise to different signs of the *magnetic susceptibility*, as we shall see. If there are unpaired electrons present, their spin magnetic moments may interact with the induced currents and give rise to the *g-value* in e.s.r. Any magnetic nuclei present may interact with the current, and hence we account for the *shielding constant* of n.m.r. A nuclear spin can itself give rise to a magnetic field which can also generate electronic current, and this current can interact with another magnetic nucleus to give the *fine structure* characteristic of *spin–spin coupling* in n.m.r. Two unpaired electrons can also be coupled together by the currents they induce as a result of their magnetic moments: this is a contribution to *spin–spin coupling* in a triplet state e.s.r. The picture of a circulating electronic current induced by the twisting effect of a magnetic perturbation is a basis for unifying all these properties, Fig. 14.1, and will be expressed quantitatively later in the chapter.

(a)

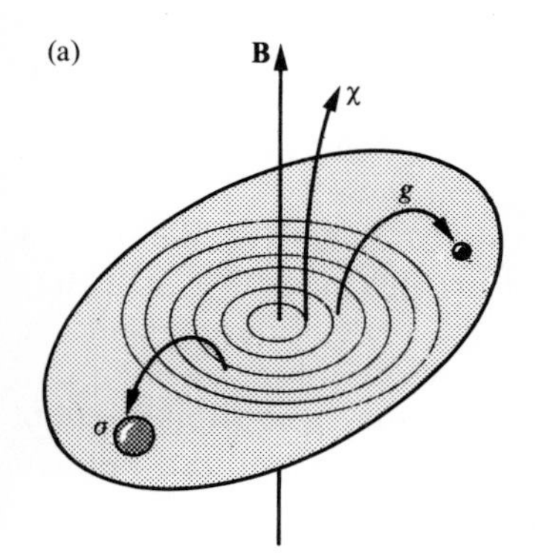

(b)

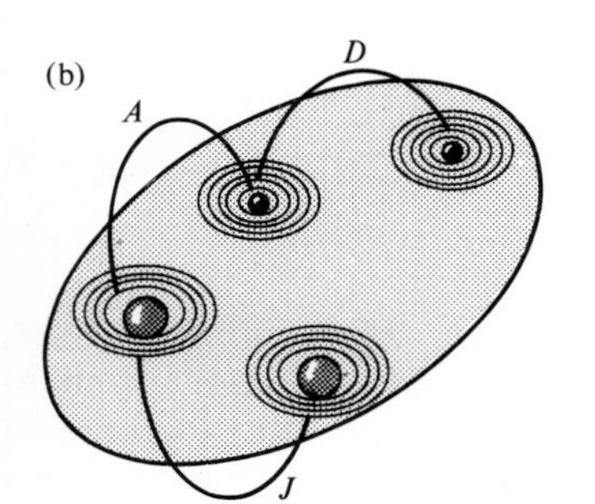

Fig. 14.1. Schematic representation of the roles of magnetic currents: (a) their contribution to susceptibility (χ), g-value, and shielding constant (σ); (b) their contribution to electron spin–spin interaction (D), hyperfine interactions (A), and nucleus spin–spin coupling (J).

14.1 The magnetic susceptibility

The electric properties we discussed in Chapter 13 have analogies in magnetism. A molecule may possess a *permanent magnetic dipole moment* $\mathbf{m}_0$. It may also acquire a magnetic moment in the presence of an applied magnetic field. The analogue of the electric polarizability is the *magnetizability*, ξ. A bulk sample subjected to a magnetic field of strength H (usually expressed in $\mathrm{A\,m^{-1}}$) acquires a *magnetization*, M, just as a dielectric acquires a polarization, and we write

Magnetization: $$\mathbf{M} = \kappa \mathbf{H}, \qquad (14.1.1)$$

where κ is the *magnetic susceptibility*. The quantity κ is referred to more precisely as the *volume magnetic susceptibility*; it is more commonly converted into the

Mass magnetic susceptibility:

$$\chi = \kappa/(\rho/\mathrm{kg\,m^{-3}}) = 10^{-3}\kappa(\rho/\mathrm{g\,cm^{-3}}) \qquad (14.1.2a)$$

by factoring out the density of the sample. We shall write $\rho^{\ominus} \equiv 1\ \mathrm{kg\ m^{-3}}$, and express this as

$$\chi = \kappa\rho^{\ominus}/\rho. \tag{14.1.2b}$$

Both κ and χ are dimensionless, and it is important to establish which quantity is being reported in tables of data (χ is the one normally listed). The *magnetic flux density* (or *induction*) within a medium is proportional to the magnetic field strength:

Flux density: $\mathbf{B} = \mu\mathbf{H}$ (14.1.3)

where μ is the *permeability* of the medium. This is normally expressed in terms of the *vacuum permeability* μ_0 ($\mu_0 = 4\pi \times 10^{-7}\ \mathrm{N\ A^{-2}}$; μ_0 and ε_0 are related by $\mu_0\varepsilon_0 = 1/c^2$) and the *relative permeability* $\mu_r = \mu/\mu_0$, the analogue of the relative permittivity of a dielectric. Since the flux density can be regarded as arising from the field $\mathbf{H}$ through $\mu_0\mathbf{H}$ together with a correction arising from the magnetization that field has induced, we write

$$\mathbf{B} = \mu_0(\mathbf{H} + \mathbf{M}) = \mu_0(1 + \kappa)\mathbf{H}. \tag{14.1.4}$$

Therefore $\mu_r = 1 + \kappa$ (the analogue of eqn (13.2.2), the relation between the electric susceptibility and the permittivity).

The susceptibility may be either positive or negative. When it is positive the magnetization opposes $\mathbf{H}$ and the flux density is less than in a vacuum. Materials for which $\chi < 0$ are called *diamagnetic.* In contrast, in a paramagnetic material the susceptibility is positive and the flux density is greater than in a vacuum: the magnetization the field induces has given rise to 'lines of force' that augment the applied field.†

The magnetization of the medium is its *magnetic dipole moment density* (recall the analogous interpretation of the polarization as the electric dipole moment density):

$$\mathbf{M} = \mathcal{N}\langle\mathbf{m}\rangle, \tag{14.1.5}$$

where $\langle\mathbf{m}\rangle$ is the average value of the molecular magnetic moment and $\mathcal{N}$ the number density.

There are two contributions to the average dipole moment. First, there is the contribution from the permanent magnetic dipole moment of the molecules. This depends on the orientating effect of the applied field as expressed through the Boltzmann distribution. The energy of a magnetic dipole in a magnetic field $\mathbf{B}$ is $-\mathbf{m}\cdot\mathbf{B}$, which reduces to $-m_zB$ if the field lies in the z-direction. The Boltzmann-weighted average of m_z for a fluid sample at a temperature T is therefore

$$\langle m_z\rangle = m_0\mathcal{L}(x), \qquad x = m_0B/kT, \tag{14.1.6}$$

† The name comes from the behaviour of a long cylindrical sample which, if supported in the field of a magnet of small cross-section, tends to lie across (*dia*: across) the field so as to minimize its energy. If its energy decreases with field, as in the case of a paramagnetic sample, the same sample would tend to lie parallel to the field (*para*: beside, along).

where we have been able to use eqn (13.2.9) directly ($\mathscr{L}$ is the Langevin function). The magnetization of the sample is therefore

$$\mathbf{M} = \mathbf{m}_0 \mathcal{N} \mathscr{L}(x) \approx \mathcal{N} m_0^2 \mathbf{B}/3kT \tag{14.1.7}$$

because normally $x \ll 1$. By combining $\mathbf{B} = \mu_0(\mathbf{H}+\mathbf{M})$ and $\mathbf{M} = \kappa\mathbf{H}$ we find

$$\mathbf{M} = (1/\mu_0)\left(\frac{\kappa}{1+\kappa}\right)\mathbf{B} \approx (\kappa/\mu_0)\mathbf{B} \quad \text{when} \quad \kappa \ll 1.$$

Therefore the permanent moment contributes

$$\kappa = \mu_0 m_0^2 \mathcal{N}/3kT \tag{14.1.8}$$

to the susceptibility. This is positive, and so the permanent moments contribute to the paramagnetic susceptibility.

The last expression depends on the density of the medium through $\mathcal{N} = L\rho/M_\text{m}$, and hence its temperature dependence is quite complicated. The explicit $1/T$ dependence can be isolated by dividing by ρ, and so we arrive at the reason for introducing the mass susceptibility. In the present case

$$\chi = (L\rho/M_\text{m})(\mu_0 m_0^2/3kT)(\rho^\ominus/\rho) = L\mu_0 m_0^2 \rho^\ominus/3kTM_\text{m}.$$

That is, we have the

Curie law: $\chi = C/T, \qquad C = \mu_0 \rho^\ominus m_0^2 L/3kM_\text{m}.$ (14.1.9)

All that remains is to calculate the magnitude of the permanent magnetic moment. This is easiest when there is no orbital magnetic moment, in which case the magnetic moment arises from the spin of the electrons, and we have *spin-only paramagnetism*. Then $m_\text{s}^2 = g_\text{e}^2\gamma_\text{e}^2 S^2$, and the expectation value of S^2 is $S(S+1)\hbar^2$. Therefore the spin-only mass susceptibility is

$$\begin{aligned}\chi &= \{g_\text{e}^2\gamma_\text{e}^2\hbar^2\mu_0\rho^\ominus L/3kM_\text{m}T\}S(S+1)\\ &= \{g_\text{e}^2\mu_\text{B}^2\mu_0\rho^\ominus L/3kM_\text{m}\}S(S+1)/T\\ &= 6.300\times10^{-3}\{S(S+1)/M_\text{r}(T/\text{K})\},\end{aligned} \tag{14.1.10}$$

where M_r is the relative molecular mass of the species. Some idea of the magnitudes of magnetic susceptibilities can therefore be obtained by taking typical values for S, M_r, and T. In the case $S = \frac{1}{2}$, $M_\text{r} \approx 100$, and $T \approx 300$ K, we have $\chi \approx 2\times10^{-7}$. The corresponding value of the volume susceptibility if $\rho \approx 1$ g cm^{-3} would be 1000 times greater, and so $\kappa \approx 2\times10^{-4}$.

When should we expect the spin-only formula to be applicable? The orbital angular momentum of a molecule is *quenched* when its electrons are described by real wavefunctions. This can be proved by making use of the hermiticity of the orbital angular momentum operator. Suppose the molecule is in some state $|0\rangle$ described by real wavefunctions, then by hermiticity $\langle 0|\mathbf{l}|0\rangle = \langle 0|\mathbf{l}|0\rangle^*$. Since the wavefunctions are real, this is equal to $\langle 0|\mathbf{l}^*|0\rangle$. But since $\mathbf{l}$ is an imaginary operator $(\mathbf{l}=(\hbar/\mathrm{i})\mathbf{r}\wedge\boldsymbol{\nabla})$, $\mathbf{l}^*=-\mathbf{l}$. Therefore, when the wavefunctions are real, $\langle 0|\mathbf{l}|0\rangle=-\langle 0|\mathbf{l}|0\rangle$ and so the expectation value of all the components of l must be zero. Since the wavefunctions of electrons in orbitally non-degenerate states may always be chosen to be real, it follows that orbitally non-degenerate system show only their spin-only paramagnetism.

When we turn to the case of molecules in *orbitally non-degenerate singlet states* there is neither spin nor orbital magnetic moment. The sole contribution to the magnetic dipole density, and therefore the magnetization of the sample, comes from the magnetic moments induced by the applied field. Since now $\langle\mathbf{m}\rangle=\xi\mathbf{H}$ and $\mathbf{M}=\mathcal{N}\langle\mathbf{m}\rangle$ we can identify the induced contribution to the susceptibility from $\mathbf{M}=\kappa\mathbf{H}$ as

$$\kappa=\mathcal{N}\xi=(\rho L/M_{\mathrm{m}})\xi \tag{14.1.11a}$$

$$\chi=\kappa\rho^{\ominus}/\rho=(\rho^{\ominus}L/M_{\mathrm{m}})\xi. \tag{14.1.11b}$$

Therefore, in order to make progress, we must be able to calculate the magnetizability ξ.

The principal complication in the calculation of magnetic properties is the fact that the hamiltonian for the interaction between matter and a magnetic field is not simply $-\mathbf{m}\cdot\mathbf{B}$, as might be expected by analogy with the electric interaction, because there is an additional second-order contribution. Since magnetizabilities (like polarizabilities) are obtained from the second-order correction to the energy, this additional term is just as important as $H^{(1)}$ in its contribution, and we cannot dismiss it as a 'second-order irrelevance'. The first task is to see how the second-order term arises; the second is to use it.

14.2 The vector potential

Just as the electric field $\mathbf{E}$ can be expressed as a derivative of a *scalar potential*, so the magnetic field can be expressed as a derivative of the *vector potential* (*potentials* store information about the electromagnetic field as a function of position: this information is 'potential' in the sense that it can be released by evaluating its derivatives: fields are derived from potentials). Just as the Schrödinger equation is expressed in terms of a scalar potential when an electric field (e.g. that of the nucleus) is present, so we can expect it to be expressed in another type of potential when a magnetic field is present. This turns out to be the vector potential.

The idea of a *scalar function* should be familiar: a scalar function associates a single number with each point in space. For example, $y=x^2$ defines a scalar function having the values 1 at $x=1$, 4 at $x=2$ and so on. The temperature of a body can be expressed as a scalar function. The Coulomb potential, the scalar potential for the electric field, is another example of a scalar function. A *vector function*, on the other hand, associates three numbers with each point. We can think of these three numbers as the components of a vector, and therefore

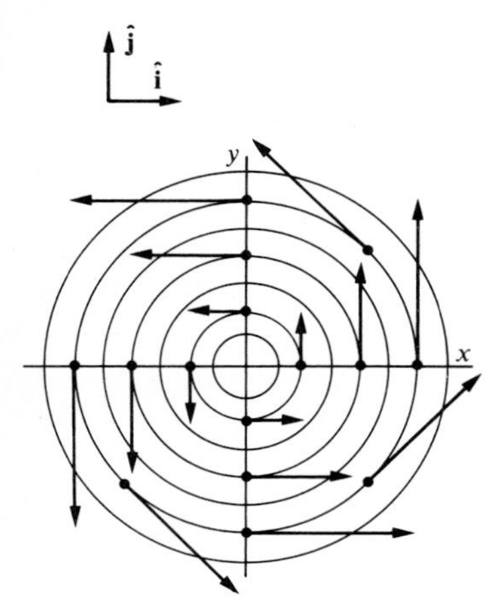

Fig. 14.2. The vector function $\mathbf{V}=-y\hat{\mathbf{i}}+x\hat{\mathbf{j}}$, with zero divergence.

think of the vector function as associating a vector of some magnitude and orientation with each point of space. The electric field in a plane polarized light ray is an example of a vector function.

Vector functions are more difficult to represent diagrammatically than scalar functions because we have to denote their direction as well as their magnitude at each point. Take, for example, the vector function

$$\mathbf{V}=-y\hat{\mathbf{i}}+x\hat{\mathbf{j}} \tag{14.2.1}$$

where $\hat{\mathbf{i}}$ and $\hat{\mathbf{j}}$ are unit vectors in the x, y-plane. This function is drawn in Fig. 14.2. It can be constructed by concentrating first on the values it takes along the line at $y=0$; then $\mathbf{V}=x\hat{\mathbf{j}}$, and so its magnitude increases with $|x|$ and it points in the direction of $\hat{\mathbf{j}}$ for $x>0$ and along $-\hat{\mathbf{j}}$ for $x<0$. These values are denoted by the arrows sprouting from the x-axis. Then take $x=0$, when $\mathbf{V}=-y\hat{\mathbf{i}}$. The magnitude varies as $|y|$, and the direction is along $-\hat{\mathbf{i}}$ for $y>0$ but along $\hat{\mathbf{i}}$ for $y<0$. The same technique may be used to represent the values at all points in the plane, and the overall function can be expressed as a series of contours carrying directional arrows. The function $\mathbf{V}$ obviously represents a circulation of some kind about the z-axis perpendicular to the plane (about the direction $\hat{\mathbf{k}}$). In contrast the vector function

$$\mathbf{V}'=x\hat{\mathbf{i}}+y\hat{\mathbf{j}}, \tag{14.2.2}$$

which is drawn in Fig. 14.3, suggests a radial flow away from some central point.

We need to define the derivatives of vector functions. There are two of importance for us. Suppose we write the vector function as

$$\mathbf{F}=f_x\hat{\mathbf{i}}+f_y\hat{\mathbf{j}}+f_z\hat{\mathbf{k}}$$

where f_x, f_y, f_z may be functions of x, y, and z (so that in $\mathbf{V}$, $f_x=-y$, $f_y=x$, and $f_z=0$), then the divergence of the function is defined as

$$\textit{Divergence}:\ \nabla\cdot\mathbf{F}=(\partial f_x/\partial x)+(\partial f_y/\partial y)+(\partial f_z/\partial z). \tag{14.2.3}$$

The reason for this name can be seen by evaluating the divergences of $\mathbf{V}$ and $\mathbf{V}'$: we find $\nabla\cdot\mathbf{V}=0$ but $\nabla\cdot\mathbf{V}'=2$. This reflects their appearance in the diagrams: $\mathbf{V}$ does not diverge, it circulates around the centre point; $\mathbf{V}'$ does diverge. The other derivative is the curl, defined as

$$\textit{Curl}:\ \nabla\wedge\mathbf{F}=\begin{vmatrix}\hat{\mathbf{i}} & \hat{\mathbf{j}} & \hat{\mathbf{k}}\\ (\partial/\partial x) & (\partial/\partial y) & (\partial/\partial z)\\ f_x & f_y & f_z\end{vmatrix}. \tag{14.2.4}$$

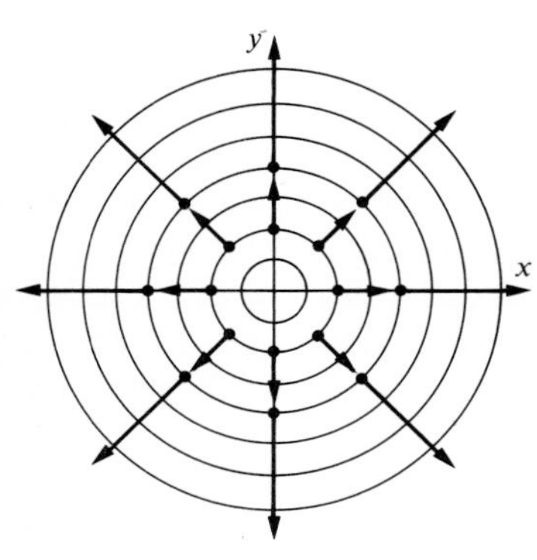

Fig. 14.3. The vector function $\mathbf{V}'=x\hat{\mathbf{i}}+y\hat{\mathbf{j}}$, with zero curl.

For $\mathbf{V}$ we have

$$\nabla\wedge\mathbf{V}=\begin{vmatrix}\hat{\mathbf{i}} & \hat{\mathbf{j}} & \hat{\mathbf{k}}\\ (\partial/\partial x) & (\partial/\partial y) & (\partial/\partial z)\\ -y & x & 0\end{vmatrix}=2\hat{\mathbf{k}}. \tag{14.2.5}$$

For $\mathbf{V}'$, $\nabla \wedge \mathbf{V}' = 0$. Once again the name 'curl' fits the appearance of the vector function: $\mathbf{V}$ curls round $\hat{\mathbf{k}}$; $\mathbf{V}'$ diverges without curling. Note that the divergence of a vector function is a scalar while the curl is another vector. In the case of the curl, the vector denotes the sense of rotation according to the right-hand screw rule.

Now we are in a position to introduce the vector potential. The vector potential corresponding to a magnetic field $\mathbf{B}$ is defined such that

$$\mathbf{B} = \nabla \wedge \mathbf{A}. \tag{14.2.6}$$

For example, suppose we are given the vector potential

$$\mathbf{A} = \tfrac{1}{2}B\mathbf{V} = \tfrac{1}{2}B(-y\hat{\mathbf{i}} + x\hat{\mathbf{j}}), \tag{14.2.7}$$

then

$$\mathbf{B} = \nabla \wedge \mathbf{A} = \tfrac{1}{2}B\nabla \wedge \mathbf{V} = B\hat{\mathbf{k}}. \tag{14.2.8}$$

In other words, *the vector potential $\tfrac{1}{2}B\mathbf{V}$ describes a uniform magnetic field of magnitude B pointing in the direction of $\hat{\mathbf{k}}$*. It is quite easy to generalize this important result and to deduce that

$$\mathbf{A} = \tfrac{1}{2}\mathbf{B} \wedge \mathbf{r} \tag{14.2.9}$$

corresponds to a uniform field $\mathbf{B}$. (Some of the vector relations needed for evaluating derivatives of vector functions are given in Appendix 20.) Therefore, we can always set up the vector potential for a given uniform field $\mathbf{B}$ simply by forming $\tfrac{1}{2}\mathbf{B} \wedge \mathbf{r}$.

There are two points. One is that not all magnetic fields are uniform; then $\mathbf{A}$ takes on a more complicated form. We shall see an example of this later. The other is that the choice of a vector potential corresponding to a given field is not unique. For example, since $\nabla \wedge \mathbf{V}' = 0$ the same field $B\hat{\mathbf{k}}$ is represented by the vector potential

$$\mathbf{A} = \tfrac{1}{2}B\mathbf{V} + \lambda\mathbf{V}' \tag{14.2.10}$$

for all values of the constant λ, Fig. 14.4. Different choices of λ correspond to different choices of *gauge*, and changing λ corresponds to a *gauge transformation*. In passing we note that it is always possible to choose a value of λ such that $\nabla \cdot \mathbf{A} = 0$ (in the present case that corresponds to $\lambda = 0$), and such a choice is called the *Coulomb gauge*. Everything that follows will be in the Coulomb gauge, as may be checked explicitly at any stage. There are advantages that come from working in other gauges, especially in relativity theory. When a different gauge is chosen for each point in space we arrive at a formalism that gives rise to the *gauge theories* now being used to describe the properties of elementary particles.

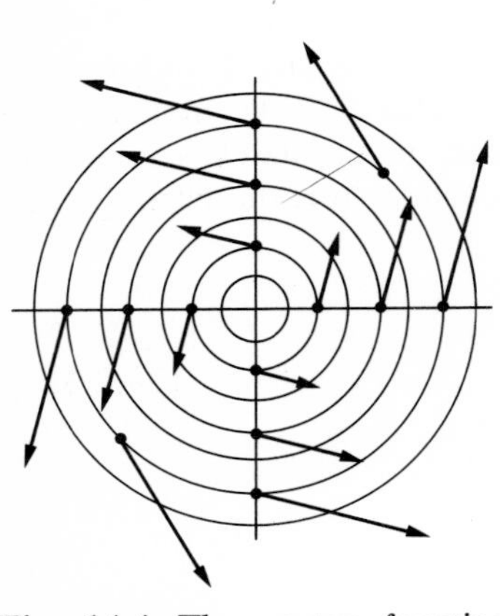

Fig. 14.4. The vector function $\mathbf{V} + \lambda\mathbf{V}'$, with non-zero divergence and curl.

Having established the description of the magnetic field in terms of the vector potential, it is time to do something with it. The point of introducing $\mathbf{A}$ is that the hamiltonian in the presence of a magnetic field can be obtained very simply. We shall adopt the following rule: its basis is explained in Appendix 21. *In the presence of a magnetic field described by a vector potential* $\mathbf{A}$, *the hamiltonian is obtained by replacing the linear momentum* $\mathbf{p}$ *wherever it occurs by* $\mathbf{p}+e\mathbf{A}$.

Consider the unperturbed hamiltonian for an electron with potential energy $V(\mathbf{r})$:

$$H^{(0)}=(1/2m_e)p^2+V.$$

In the presence of a magnetic field, p^2 is replaced by

$$(\mathbf{p}+e\mathbf{A})^2=(\mathbf{p}+e\mathbf{A})\cdot(\mathbf{p}+e\mathbf{A})=p^2+e(\mathbf{p}\cdot\mathbf{A}+\mathbf{A}\cdot\mathbf{p})+e^2A^2.$$

Some care is required with the term $\mathbf{p}\cdot\mathbf{A}$ because $\mathbf{p}$ is a differential operator and operates both on $\mathbf{A}$ and the unwritten wavefunction. However, we can write

$$\mathbf{p}\cdot\mathbf{A}\psi=(\hbar/\mathrm{i})(\nabla\cdot\mathbf{A}\psi)=(\hbar/\mathrm{i})\{(\nabla\cdot\mathbf{A})\psi+\mathbf{A}\cdot\nabla\psi\}=\mathbf{A}\cdot\mathbf{p}\psi \qquad (14.2.11)$$

because in the Coulomb gauge $\nabla\cdot\mathbf{A}=0$. It follows that the hamiltonian in the presence of the magnetic field is

$$H=(1/2m_e)p^2+V+(e/m_e)\mathbf{A}\cdot\mathbf{p}+(e^2/2m_e)A^2. \qquad (14.2.12)$$

This differs from the original hamiltonian not only by the presence of the first-order term but also by the presence of the second-order term. Therefore we write

$$H=H^{(0)}+H^{(1)}+H^{(2)}; \qquad H^{(1)}=(e/m_e)\mathbf{A}\cdot\mathbf{p}, \qquad H^{(2)}=(e^2/2m_e)A^2. \qquad (14.2.13)$$

The first-order contribution can be written in a more familiar form by considering the case of a uniform magnetic field. Then

$$H^{(1)}=(e/m_e)\tfrac{1}{2}\mathbf{B}\wedge\mathbf{r}\cdot\mathbf{p}=(e/2m_e)\mathbf{B}\cdot\mathbf{r}\wedge\mathbf{p}=(e/2m_e)\mathbf{B}\cdot\mathbf{l}$$

(since $\mathbf{a}\wedge\mathbf{b}\cdot\mathbf{c}=\mathbf{a}\cdot\mathbf{b}\wedge\mathbf{c}$.). Therefore, since $\gamma_e=-e/2m_e$,

$$H^{(1)}=-\gamma_e\mathbf{B}\cdot\mathbf{l}=-\mathbf{m}\cdot\mathbf{B}, \qquad \mathbf{m}=\gamma_e\mathbf{l}. \qquad (14.2.14)$$

Note that the spin does not appear in this expression: for that we would need to work from the Dirac equation.

The second-order contribution to the hamiltonian can also be expressed very simply when the field is uniform. Suppose it lies in the z-direction; then $\mathbf{A}=\tfrac{1}{2}B\mathbf{V}$, and so

$$A^2=\tfrac{1}{4}B^2(-y\hat{\mathbf{i}}+x\hat{\mathbf{j}})\cdot(-y\hat{\mathbf{i}}+x\hat{\mathbf{j}})=\tfrac{1}{4}B^2(x^2+y^2). \qquad (14.2.15)$$

Therefore,

$$H^{(2)}=(e^2/8m_e)B^2(x^2+y^2). \qquad (14.2.16)$$

The second-order contribution to the energy of the system has to be calculated using the perturbation expression

$$\mathscr{E}^{(2)}=\langle 0|\,H^{(2)}\,|0\rangle+\sum_n{}'\left\{\frac{\langle 0|\,H^{(1)}\,|n\rangle\langle n|\,H^{(1)}\,|0\rangle}{\mathscr{E}_0-\mathscr{E}_n}\right\}$$

on account of the presence of the second-order hamiltonian (which is absent in the electric case). In the case of a field $B\hat{\mathbf{k}}$,

$$\mathscr{E}^{(2)}=(e^2/8m_e)\langle 0|\,x^2+y^2\,|0\rangle B^2+(e/2m_e)^2\sum_n{}'\left\{\frac{\langle 0|\,l_z\,|n\rangle\langle n|\,l_z\,|0\rangle}{\mathscr{E}_0-\mathscr{E}_n}\right\}B^2,$$

which can be expressed more neatly if we write $\langle 0|\,x^2+y^2\,|0\rangle=\langle x^2+y^2\rangle$, $l_{z,0n}=\langle 0|\,l_z\,|n\rangle$, and $\Delta_{n0}=\mathscr{E}_n-\mathscr{E}_0$:

$$\mathscr{E}^{(2)}=\left\{(e^2/8m_e)\langle x^2+y^2\rangle-(e/2m_e)^2\sum_n{}'(l_{z,0n}l_{z,n0}/\Delta_{n0})\right\}B^2. \qquad (14.2.17)$$

Notice that the first term is positive, and increases the energy of the system as the field is increased; the second term is negative and decreases the energy.

Now consider how $\mathscr{E}^{(2)}$ may be expressed in terms of the magnetizability ξ_{zz}. The latter is the coefficient of proportionality between the magnetic field strength and the induced dipole:

$$\langle m_z\rangle=m_{0z}+\xi_{zz}H+\ldots\approx m_{0z}+\xi_{zz}B/\mu+\ldots \qquad (14.2.18)$$

(we have used eqn (14.1.3) to bring in B in place of H). The energy of a magnetic dipole in a magnetic field is $-m_zB$, where B is the flux density; but we cannot simply write $\mathscr{E}^{(2)}=-\langle m_z\rangle B$ because $\langle m_z\rangle$ changes as the field is increased from zero. We proceed as follows. At some value of the field the magnetic moment has the value $\langle m_z\rangle_B$. An infinitesimal increase $\mathrm{d}B$ in field changes the energy by $\mathrm{d}\mathscr{E}^{(2)}=-\langle m_z\rangle_B\,\mathrm{d}B$. Therefore the total change of energy caused by increasing the field from zero (when $\mathscr{E}^{(2)}=0$) to B is

$$\int_0^{\mathscr{E}^{(2)}}\mathrm{d}\mathscr{E}^{(2)}=-\int_0^B\langle m_z\rangle_B\,\mathrm{d}B.$$

Therefore

$$\begin{aligned}\mathscr{E}^{(2)}&=-\int_0^B\{m_{0z}+\xi_{zz}B/\mu+\ldots\}\,\mathrm{d}B\\&=-m_{0z}B-\tfrac{1}{2}\xi_{zz}B^2/\mu_0+\ldots\end{aligned} \qquad (14.2.19)$$

where we have supposed that μ is so close to its vacuum value that we can write $\mu=\mu_0$.

It follows that we can identify ξ_{zz} by comparing the two expressions for $\mathscr{E}^{(2)}$. This gives

$$\xi_{zz}=-(e^2\mu_0/4m_e)\langle x^2+y^2\rangle+(e^2\mu_0/2m_e^2)\sum_n{}'(l_{z,0n}l_{z,n0}/\Delta_{n0}). \qquad (14.2.20)$$

The *mean magnetizability* of a freely rotating molecule is the average value of its magnetizabilities when the field is applied in turn along the x, y, and z-directions, $\xi = \frac{1}{3}(\xi_{xx} + \xi_{yy} + \xi_{zz})$. Therefore

$$\begin{aligned}\xi &= -(e^2\mu_0/4m_e)\tfrac{1}{3}\langle(x^2+y^2)+(y^2+z^2)+(x^2+z^2)\rangle \\ &\quad + \tfrac{1}{3}(e^2\mu_0/2m_e^2)\sum_n{}' (l_{z,0n}l_{z,n0} + l_{y,0n}l_{y,n0} + l_{x,0n}l_{x,n0})/\Delta_{n0} \\ &= -(e^2\mu_0/6m_e)\langle r^2\rangle + (e^2\mu_0/6m_e^2)\sum_n{}' (\mathbf{l}_{0n}\cdot\mathbf{l}_{n0}/\Delta_{n0}). \qquad (14.2.21)\end{aligned}$$

The *mean volume magnetic susceptibility* is therefore

$$\kappa = -(e^2\mu_0\mathcal{N}/6m_e)\langle r^2\rangle + (e^2\mu_0\mathcal{N}/6m_e^2)\sum_n{}' (\mathbf{l}_{0n}\cdot\mathbf{l}_{n0}/\Delta_{n0}), \qquad (14.2.22)$$

and the *mean mass magnetic susceptibility* is

$$\chi = -(e^2\mu_0 L\rho^{\ominus}/6m_eM_m)\langle r^2\rangle + (e^2\mu_0 L\rho^{\ominus}/6m_e^2M_m)\sum_n{}' (\mathbf{l}_{0n}\cdot\mathbf{l}_{n0}/\Delta_{n0}). \qquad (14.2.23)$$

The expressions for the susceptibility apparently (we shall say why 'apparently' shortly) fall into two contributions, one positive and the other negative. Therefore we write

$$\kappa = \kappa^d + \kappa^p \quad \text{or} \quad \chi = \chi^d + \chi^p \qquad (14.2.24)$$

and identify the *diamagnetic susceptibility* as

$$\begin{aligned}\kappa^d &= -(e^2\mu_0/6m_e)\mathcal{N}\langle r^2\rangle; \\ \chi^d &= -(e^2\mu_0 L\rho^{\ominus}/6m_eM_m)\langle r^2\rangle \qquad (14.2.25)\end{aligned}$$

and the *paramagnetic susceptibility* as

$$\begin{aligned}\kappa^p &= (e^2\mu_0/6m_e)\mathcal{N}\sum_n{}' (\mathbf{l}_{0n}\cdot\mathbf{l}_{n0}/\Delta_{n0}); \\ \chi^p &= (e^2\mu_0 L\rho^{\ominus}/6m_e^2M_m)\sum_n (\mathbf{l}_{0n}\cdot\mathbf{l}_{n0}/\Delta_{n0}). \qquad (14.2.26)\end{aligned}$$

Notice that this paramagnetic susceptibility has nothing to do with the electron spin and, unlike spin paramagnetism, is independent of the temperature (in its χ form); hence it is known as *temperature independent paramagnetism* (TIP). An alternative name is the *high-frequency paramagnetism* (since it involves excited states); the diamagnetic contribution is called the *Langevin term.*

Example. Consider a model system in which one electron occupies a $2p_x$-orbital and where the $2p_y$-orbital lies an energy Δ above it. Calculate the magnetic susceptibility in the z-direction.

- *Method.* Use $\chi_{zz} = L\rho^{\ominus}\xi_{zz}$ with ξ_{zz} given by eqn (14.2.20). For the evaluation of $\langle x^2+y^2\rangle$ use STOs (p. 234): for $2p_y$ use the $2p_x$-STO in the last *Example* (p.

372) but with $\sin\phi$ replaced by $\cos\phi$. There is only one non-zero term in the high-frequency contribution. Use $l_z = (\hbar/\mathrm{i})(\partial/\partial\phi)$ and note that $\mathrm{p}_x \propto \cos\phi$ while $\mathrm{p}_y \propto \sin\phi$.

- *Answer.* Since $l_z \cos\phi = \mathrm{i}\hbar \sin\phi$, $l_z \mathrm{p}_x = \mathrm{i}\hbar \mathrm{p}_y$. Therefore

$$\sum_n{}' \{l_{z,0n} l_{z,n0}/\Delta_{n0}\} = |\langle \mathrm{p}_y|\, l_z\, |\mathrm{p}_x\rangle|^2/\Delta = \hbar^2/\Delta$$

$$\langle x^2+y^2\rangle = \int \psi_{\mathrm{p}_x}^2 (x^2+y^2)\,\mathrm{d}\tau = \int \psi_{\mathrm{p}_x}^2 r^2 \sin^2\theta\,\mathrm{d}\tau$$

$$= (3/4\pi)(2^5\zeta^5/4!)\int_0^{2\pi}\cos^2\phi\,\mathrm{d}\phi \int_0^{\pi}\sin^4\theta\sin\theta\,\mathrm{d}\theta\int_0^{\infty} r^6 \mathrm{e}^{-2\zeta r}\,\mathrm{d}r$$

$$= 6a_0^2(n/Z^*)^2.$$

Consequently

$$\xi_{zz} = -6a_0^2(e^2\mu_0/4m_\mathrm{e})(n/Z^*)^2 + (e^2\mu_0\hbar^2/2m_\mathrm{e}^2\Delta)$$

$$\chi_{zz} = -\tfrac{3}{2}(e^2a_0^2\mu_0\rho^{\ominus}L/m_\mathrm{e})\{(n/Z^*)^2 - (\hbar^2/3m_\mathrm{e}a_0^2)/\Delta\}.$$

- *Comment.* A convenient practical form of this expression is

$$\chi_{zz} = -8.96\times10^{-11}\{(n/Z^*)^2 - 9.07/(\Delta/\mathrm{eV})\},$$

which becomes paramagnetic when Δ drops below about $(3Z^*/n)^2$ eV.

The observed magnetic susceptibility of a sample depends on the competition between the diamagnetic and paramagnetic contributions. In *atoms* the paramagnetic contribution is zero. This is because we are free to choose the z-direction as the axis of quantization of the z-component of angular momentum, hence $|0\rangle$ and $|n\rangle$ are eigenstates of l_z, and so l_z is diagonal; but $n=0$ is excluded by the prime on the summation defining the paramagnetic susceptibility, and so $\chi^\mathrm{p}=0$. The *total magnetic susceptibility of a sample of atoms* is therefore

$$\kappa = -(e^2\mu_0/6m_\mathrm{e})\mathcal{N}\langle r^2\rangle; \tag{14.2.27a}$$

$$\chi = -(e^2\mu_0 L\rho^{\ominus}/6m_\mathrm{e}M_\mathrm{m})\langle r^2\rangle = -3.554\,1\times10^{-6}(R/\mathrm{nm})^2/M_r \tag{14.2.27b}$$

with $R^2 = \langle r^2\rangle$, so long as it has no unpaired electrons. For a typical atom with $M_\mathrm{r}\approx 20$ and a radius of 0.15 nm we find $\chi\approx -4.0\times10^{-9}$. If an unpaired spin is present, as in the case of the hydrogen atom, its paramagnetic contribution ($\chi\approx1.6\times10^{-5}$ at 300 K) overwhelms the orbital diamagnetism ($\chi\approx1.0\times10^{-8}$ for the 1s-orbital). In the absence of spin, however, there is only the diamagnetic component, and all atoms have non-zero (but small) diamagnetic polarizabilities.

In the case of molecules, the axis of quantization of the orbital angular momentum is no longer the direction of the applied field, except fortuitously. Now the susceptibility is a sum of the Langevin and the high-frequency terms.† In most molecules the former dominates, and most molecules without unpaired electrons are diamagnetic,

† Classical mechanics cannot account for the susceptibilities of molecules. This is the content of a theorem courteously referred to by van Vleck as 'Miss van Leeuwen's theorem' which demonstrates that χ^p and χ^d exactly cancel each other in a classical mechanical calculation. This is a late but interesting illustration of the inadequacy of classical physics.

their susceptibilities varying as $\langle r^2 \rangle$ but inversely as M_m. Only when there are low-lying excited states may the high-frequency contribution dominate the Langevin term and the molecule be weakly paramagnetic. This can be demonstrated from eqn (14.2.23) by making the closure approximation and setting $\langle 0|\, l^2\, |0\rangle = l(l+1)\hbar^2$:

$$\begin{aligned}\chi &\approx (e^2\mu_0 L\rho^{\ominus}/6m_e M_m)\{-\langle r^2\rangle + l(l+1)\hbar^2/m_e\Delta\}\\ &\approx 3.554\times 10^{-6}(1/M_r)\{-(R/\text{nm})^2 + 0.076\,2l(l+1)/(\Delta/\text{eV})\}.\end{aligned} \tag{14.2.28}$$

Therefore, for $R \approx 0.3$ nm and $l \approx 1$, the second term dominates the first only when Δ is less than about 1.7 eV ($= 13\,500\ \text{cm}^{-1}$), which requires very low lying energy levels.

The evaluation of the Langevin contribution is much easier than the high-frequency contribution because it depends on knowing only the ground-state wavefunction and then evaluating $\langle r^2 \rangle$. Since the wavefunctions of atoms are only slightly changed on bond formation, there is some hope that the diamagnetic susceptibility of a molecule can be expressed as the sum of the susceptibilities of its atoms (the mass susceptibility then being appropriate, because density differences have been eliminated). This additivity is expressed in *Pascal's rules* for ascribing individual susceptibilities to the components of a molecule. The predicted and experimental values are agreeably close, except in the case of conjugated hydrocarbons. The failure then has been ascribed to an effect connected with the delocalization of the π-electrons, and is referred to as the *delocalization susceptibility* or the *ring current effect.* The latter name is based on the view that the conjugated system provides a pathway through which the electrons can circulate and hence the molecule can support significantly larger electronic currents than in analogous unconjugated systems. Furthermore, the symmetry of π-systems leads one to suspect that the magnetic susceptibility of a planar conjugated system should be appreciably anisotropic. The calculation of the effect is very complicated even in benzene because note has to be taken of the delocalization susceptibility, the anisotropies of the Langevin and the high-frequency susceptibilities of the p-electrons, and the anisotropy of the susceptibility of the σ-orbitals. Sources for further information will be found in *Further reading*, p. 414.

One of the pitfalls in interpreting magnetic susceptibilities in terms of paramagnetic and diamagnetic susceptibilities is as follows. The susceptibility of a molecule is given by eqn (14.2.23) as a whole, but, we have discussed its two components separately. In fact, the only proper way of regarding the expression is as a single unit, and it should not be discussed as if it consisted of two separable parts. The reason for this is quite subtle, but is connected with the arbitrary nature of vector potentials. In fact, the magnitudes of the two components of eqn (14.2.23) change when different gauges are chosen for the vector potential, and it is even possible to choose a gauge where the high-frequency contribution disappears completely. Of course the other term then changes its magnitude, because the overall susceptibility, being a physical observable, remains the same. The point of the remark, though, is to emphasize that *the division of a susceptibility into diamagnetic and paramagnetic contributions is entirely arbitrary.* To some extent this is not so in the case of atoms, because the choice of a gauge is dictated by whether or not there is a natural choice of origin for a coordinate system. There the nucleus acts as the origin, but in molecules there is no such natural origin. Consequently, although it may be helpful to think in terms of the susceptibility being the sum of two physically interpretable contributions, we have to be extremely guarded in the significance we allow them.

14.3 The current density

A better discussion of the difference (in so far as it is significant) between the diamagnetic and paramagnetic susceptibilities can be obtained by considering the nature of the electronic currents induced in a molecule by a magnetic field. In order to establish a quantitative theory we need a measure of the *current density* at any point. A suitable expression can be constructed as follows.

Current is charge × velocity; velocity is momentum/mass; therefore current has the form $-e\mathbf{p}/m_e$ for an electron of charge $-e$ and mass m_e. Current density is obtained by weighting this expression with the probability density that the electron is at the point of interest. This points to defining the current density as $-(e/m_e)\psi^*\mathbf{p}\psi$. But we want current density to be a real physical observable whereas $\mathbf{p}$ is imaginary. Therefore we define

$$\mathbf{j}_0 = -\tfrac{1}{2}(e/m_e)(\psi^*\mathbf{p}\psi + \psi\mathbf{p}^*\psi^*) \tag{14.3.1}$$

because the sum of a function and its complex conjugate is real. This is the form when there is no magnetic field present (as signified by the subscript zero). When the field is present we replace $\mathbf{p}$ by $\mathbf{p}+e\mathbf{A}$. Therefore in the presence of a magnetic field described by a real vector potential $\mathbf{A}$,

Current density: $$\mathbf{j} = -(e/2m_e)(\psi^*\mathbf{p}\psi - \psi\mathbf{p}\psi^*) - (e^2/m_e)A^2\psi^*\psi \tag{14.3.2}$$

since $\mathbf{p}^* = -\mathbf{p}$.† This definition can be used to map the currents present in a molecule when there is a magnetic field present, and shows in a

† A more rigorous way of constructing the current density is to look for an expression such that, if Ψ satisfies the time-dependent Schrödinger equation, the charge density $\rho = -e\,|\Psi|^2$ satisfies the *continuity equation* $\partial\rho/\partial t = -\nabla\cdot\mathbf{j}$. This equation is well known in fluid dynamics and expresses the flow of an incompressible fluid of density ρ. If the equation is integrated over some volume element, it expresses the fact that the rate of depletion of fluid within the region is equal to the rate at which the fluid flows through the walls of the element.

striking fashion the differences between molecules with diamagnetic and paramagnetic susceptibilities.

Consider first the current density in a molecule where the single electron of interest is described by a *real wavefunction and there is no magnetic field present.* Then, since $\mathbf{A}=0$,

$$\mathbf{j}_0=-(e/2m_e)(\psi\mathbf{p}\psi-\psi\mathbf{p}\psi)=0. \qquad (14.3.3)$$

There is zero current density at every point in the molecule. It will be recalled that we have already seen that a molecule in a non-degenerate orbital state is described by a real wavefunction and that it has zero angular momentum. This current-density picture is an alternative way of visualizing what that means.

Example. Calculate the current density for (a) an electron in a one-dimensional square well, (b) a stream of electrons each with linear momentum $k\hbar$ along $+x$, the particle density in the beam being $\mathcal{N}$ (electrons per unit length).

- *Method.* In the absence of a magnetic field $A=0$, and so use eqn (14.3.1). For the particle in a box use $\psi=N\sin(n\pi x/l)$, $N=(2/l)^{\frac{1}{2}}$. For the free electrons use $\psi=Ne^{ikx}$ with $N=1/l^{\frac{1}{2}}$ and $l\to\infty$. Use $p=(\hbar/i)(d/dx)$.
- *Answer.* (a) Since $p\sin(n\pi x/l)=(n\pi/l)(\hbar/i)\cos(n\pi x/l)$,

$$j_x=-(e/2m_e)N^2(n\pi/l)(\hbar/i)\{\sin(n\pi x/l)\cos(n\pi x/l)-\sin(n\pi x/l)\cos(n\pi x/l)\}=0.$$

(b) Since $pe^{ikx}=\hbar ke^{ikx}$ we have

$$\begin{aligned}j_x&=-(e/2m_e)N^2\{e^{-ikx}pe^{ikx}-e^{ikx}pe^{-ikx}\}\\&=-(e/2m_e)(1/l)\{\hbar ke^{-ikx}e^{ikx}+\hbar ke^{ikx}e^{-ikx}\}=-(e\hbar k/m_el).\end{aligned}$$

Let the total number of electrons present be N_e, then the total current density

$$j_x=-(e\hbar k/m_e)(N_e/l)=-(e\hbar k/m_e)\mathcal{N}$$

and $\mathcal{N}=N_e/l$ is a constant when $l\to\infty$ and $N_e\to\infty$ simultaneously.

- *Comment.* The vanishing of the current density in (a) reflects the fact that the particle in a box is not in an eigenstate of the linear momentum, and emphasizes the general result that real wavefunctions correpond to the absence of motion. The *mass current density* is opposite in sign to the charge current density (for electrons; for positively charged particles they lie in the same direction).

If the electron is in an *orbitally degenerate state,* its wavefunction is not necessarily real. For example, suppose the electron occupies a π-orbital in a molecule with cylindrical symmetry (e.g. a diatomic molecule, or an ion at the centre of a complex with three-fold symmetry, or higher, Fig. 14.5). Then if it has a definite component of orbital angular momentum about the z-axis, its wavefunction is of the form $f(r,\theta)e^{\pm i\phi}$ for $\Lambda=\pm1$. We can calculate the current density as follows using $\mathbf{p}=(\hbar/i)\nabla$. For the $\Lambda=-1$ state:

$$\begin{aligned}\mathbf{j}_0&=-(e/2m_e)(\hbar/i)\{fe^{i\phi}\nabla fe^{-i\phi}-fe^{-i\phi}\nabla fe^{i\phi}\}\\&=(ie\hbar/2m_e)\{fe^{i\phi}(\nabla f)e^{-i\phi}-ife^{i\phi}(\nabla\phi)fe^{-i\phi}-fe^{-i\phi}(\nabla f)e^{i\phi}\\&\quad-ife^{i\phi}(\nabla\phi)fe^{-i\phi}\}\\&=(e\hbar/m_e)f^2\nabla\phi.\end{aligned}$$

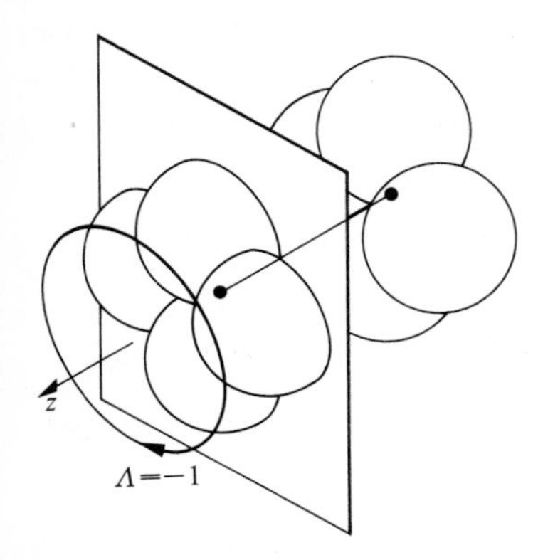

Fig. 14.5. The electronic angular momentum of a Π-molecule ($\Lambda=-1$).

The gradient of ϕ is evaluated most easily by noting that $\phi = \arctan(x/y)$; for then

$$\nabla\phi = (\partial\phi/\partial x)\hat{\mathbf{i}} + (\partial\phi/\partial y)\hat{\mathbf{j}} + (\partial\phi/\partial z)\hat{\mathbf{k}}$$
$$= (-y\hat{\mathbf{i}} + x\hat{\mathbf{j}})/(x^2 + y^2) = \mathbf{V}/(x^2 + y^2), \qquad (14.3.4)$$

where $\mathbf{V}$ is the swirling function of Fig. 14.2. Therefore the current density takes the form

$$\mathbf{j}_0 = (e\hbar/m_e)\{f^2/(x^2 + y^2)\}\mathbf{V}, \qquad (14.3.5)$$

which is proportional to $\mathbf{V}$ but has a different dependence on the distance from the origin, Fig. 14.6, because $\mathbf{V}$ is weighted by the probability density of the electron. The flow lines of the current density are obvious. Notice that they flow clockwise seen from below (from a point on $-z$), opposite in sense to the angular momentum: the difference in sign reflects the negative charge of the electron, for charge and mass flow in opposite directions.

Now we turn to the remaining case of interest: when the molecule is *orbitally non-degenerate but is exposed to a magnetic field.* Because the vector potential is now non-zero *and* because the wavefunctions are distorted by the applied field (and are no longer real), the current density is non-zero. We can obtain an expression for the current density correct to first-order in the strength of the applied field.

In the presence of a field the wavefunctions are perturbed from ψ_0 to $\psi_0 + \psi^{(1)}$, where

$$\psi^{(1)} = \sum_n{}' a_n\psi_n, \qquad a_n = -H^{(1)}_{n0}/\Delta_{n0} \qquad (14.3.6)$$

as deduced in Chapter 8, eqn (8.1.22). The perturbation is proportional to l_z, which is imaginary, and so the perturbed wavefunction is complex, which is what we need for non-zero current density. (This is another example of how the nature of the perturbation is impressed on the original states of the molecule.) In order to calculate the first-order correction to the current density we need the distortion of the wavefunction only to first-order in the perturbation, and need not trouble about the role of $H^{(2)}$. Similarly, since $\mathbf{A}$ is first-order in the field, the expression $\mathbf{A}\psi^*\psi$ in the current density is first-order when we replace ψ by ψ_0. Therefore, the first-order current density is

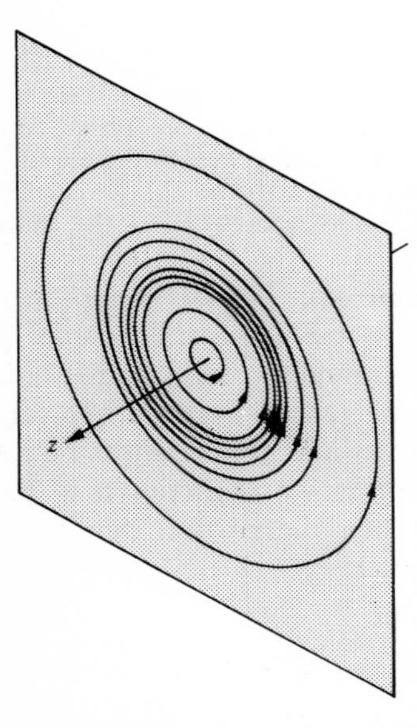

Fig. 14.6. The current density in the plane drawn in Fig. 14.5.

$$\mathbf{j} = -(e/2m_e)\{(\psi_0 + \psi^{(1)})^*\mathbf{p}(\psi_0 + \psi^{(1)}) - (\psi_0 + \psi^{(1)})\mathbf{p}(\psi_0 + \psi^{(1)})^*\} - (e^2/m_e)\mathbf{A}\psi_0^2$$
$$= -(e/2m_e)\{\psi_0\mathbf{p}\psi^{(1)} + \psi^{(1)*}\mathbf{p}\psi_0 - \psi_0\mathbf{p}\psi^{(1)*} - \psi^{(1)}\mathbf{p}\psi_0\} - (e^2/m_e)\mathbf{A}\psi_0^2$$

because ψ_0 is real. Since all the ψ_n in the sum expressing $\psi^{(1)}$ are real, we obtain

$$\mathbf{j} = -(ie\hbar/2m_e)\sum_n{}' (a_n - a_n^*)(\psi_n\nabla\psi_0 - \psi_0\nabla\psi_n) - (e^2/m_e)\mathbf{A}\psi_0^2. \qquad (14.3.7)$$

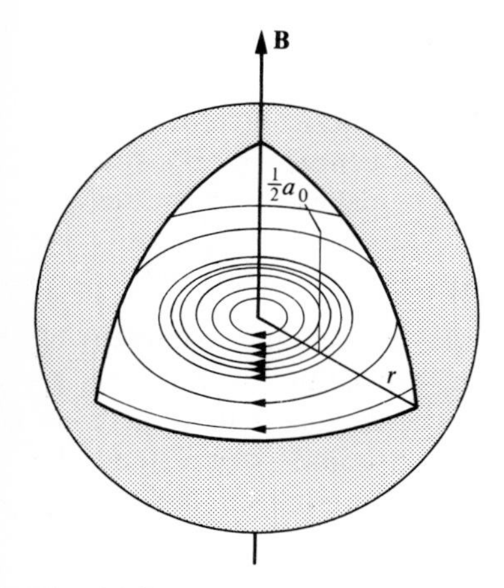

Fig. 14.7. The current density in the equatorial plane of a hydrogen atom in a magnetic field.

The natural apparent division of this expression is into the part that depends only on the ground-state wavefunction, the

Diamagnetic current density: $\mathbf{j}^{\mathrm{d}} = -(e^2/m_\mathrm{e})\mathbf{A}\psi_0^2$, (14.3.8)

and a part that depends on the involvement of the excited-state wavefunctions, the

Paramagnetic current density:

$$\mathbf{j}^{\mathrm{p}} = -(\mathrm{i}e\hbar/2m_\mathrm{e})\sum_n{}' (a_n - a_n^*)(\psi_n\nabla\psi_0 - \psi_0\nabla\psi_n). \qquad (14.3.9)$$

We stress again, that while this may seem a natural division, it is in fact only natural for the choice of gauge used for the vector potential: just as the magnetic susceptibility should be thought of as a whole and not divided into its arbitrary components, so too only the total current density has physical significance and its division into two components, while convenient, is arbitrary.

14.4 The diamagnetic current density

When the applied magnetic field lies in the z-direction the vector potential takes the form $\mathbf{A} = \frac{1}{2}B\mathbf{V}$, and so

$$\mathbf{j}^{\mathrm{d}} = -(e^2B/2m_\mathrm{e})\psi_0^2\mathbf{V}. \qquad (14.4.1)$$

Although $\mathbf{j}^{\mathrm{d}}$ has the characteristic swirling form of $\mathbf{V}$, it is swirling in the opposite direction (the − sign) and its shape is modified by the presence of the factor ψ_0^2. If ψ_0 is the 1s-orbital of atomic hydrogen, then $\psi_0 = (1/\pi a_0^3)^{\frac{1}{2}}\mathrm{e}^{-r/a_0}$ and

$$\mathbf{j}^{\mathrm{d}} = -(e^2B/2\pi m_\mathrm{e}a_0^3)(-y\hat{\mathbf{i}} + x\hat{\mathbf{j}})\mathrm{e}^{-2r/a_0}, \qquad (14.4.2)$$

This is sketched in Fig. 14.7. Note that the current density depends on the applied field, and that its magnitude is greatest in the equatorial plane of the atom, and then at a radius $\frac{1}{2}a_0$. On this circle even in a field as small as 10^{-4} T (1 gauss), the current density reaches the enormous value of 8.01×10^{7} A m^{-2}. (Drama subsides when this figure is expressed in atomic dimensions: taking a typical area element as 1 pm^2, it corresponds to 8.01×10^{-14} A pm^{-2}, or 0.5 electrons pm^{-2} μs^{-1}.)

When the magnetic field is applied to a p_x-orbital the shapes of the contour lines are more complicated. Suppose we take $\psi_0 = Nr\sin\theta\cos\phi\,\mathrm{e}^{-br}$ for a typical 2p-orbital, then the diamagnetic current density is

$$\mathbf{j}^{\mathrm{d}} = -(e^2B/2m_\mathrm{e})(-y\hat{\mathbf{i}} + x\hat{\mathbf{j}})N^2r^2\cos^2\phi\sin^2\theta\,\mathrm{e}^{-2br}. \qquad (14.4.3)$$

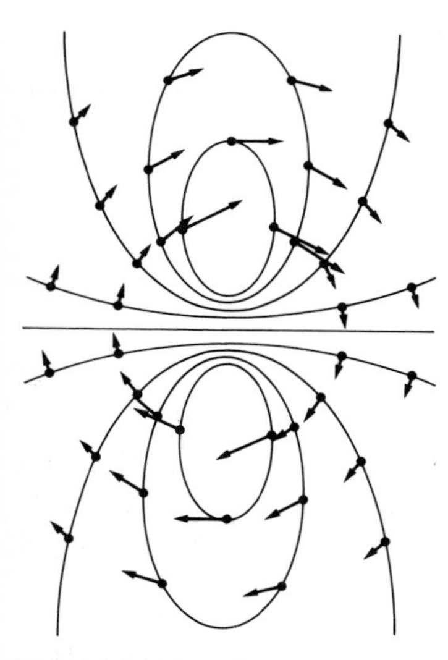

Fig. 14.8. The diamagnetic current density in a p-orbital (with the field perpendicular to the plane).

The direction at any point is still determined by $-y\hat{\mathbf{i}} + x\hat{\mathbf{j}}$, but the radial dependence is much more complicated. Its form in the equatorial plane ($\sin\theta = 1$) is sketched in Fig. 14.8. The point to note is that the current is confined to each lobe of the orbital and there is no flow across the node.

The central feature characteristic of the diamagnetic current is that it is a circulating distortion confined to the zone occupied by the orbital, and it vanishes where the orbital amplitude vanishes (at its nodes).

14.5 The paramagnetic current density

We can discover the principal features of the paramagnetic current by concentrating on a simple model system consisting of the two p-orbitals with their degeneracy removed by a crystal field, Fig. 14.9. The magnetic field is applied along the z-axis, and so we need the matrix elements of l_z in order to evaluate the mixing coefficients a_n. Since $l_z = (\hbar/\mathrm{i})(\partial/\partial\phi)$,

$$l_z\psi_{\mathrm{p}_x} = (\hbar/\mathrm{i})(\partial/\partial\phi)\sin\theta\cos\phi f(r) = -(\hbar/\mathrm{i})\sin\theta\sin\phi f(r) = -(\hbar/\mathrm{i})\psi_{\mathrm{p}_y}.$$

Therefore

$$\langle \mathrm{p}_y|\, l_z\, |\mathrm{p}_x\rangle = \mathrm{i}\hbar; \qquad \langle \mathrm{p}_x|\, l_z\, |\mathrm{p}_y\rangle = \langle \mathrm{p}_y|\, l_z\, |\mathrm{p}_x\rangle^* = -\mathrm{i}\hbar. \tag{14.5.1}$$

The coefficient for the admixture of the p_y-orbital into the p_x under the influence of the applied magnetic field is therefore

$$a_{\mathrm{p}_y} = \gamma_e l_{z,n0} B/\Delta_{n0} = -\mathrm{i}\mu_\mathrm{B} B/\Delta. \tag{14.5.2}$$

The paramagnetic current density then consists of a single term ($n = \mathrm{p}_y$):

$$\begin{aligned}\mathbf{j}^\mathrm{p} &= -(\mathrm{i}e\hbar/2m_e)(a_{\mathrm{p}_y} - a^*_{\mathrm{p}_y})\{\mathrm{p}_y\nabla\mathrm{p}_x - \mathrm{p}_x\nabla\mathrm{p}_y\} \\ &= -(e\hbar/m_e)(\mu_\mathrm{B}B/\Delta)\{\mathrm{p}_y\nabla\mathrm{p}_x - \mathrm{p}_x\nabla\mathrm{p}_y\}.\end{aligned}$$

The remaining work is to evaluate the gradients:

$$\begin{aligned}\mathrm{p}_y\nabla\mathrm{p}_x - \mathrm{p}_x\nabla\mathrm{p}_y &= \sin\theta\sin\phi f\nabla\sin\theta\cos\phi f - \sin\theta\cos\phi f\nabla\sin\theta\sin\phi f \\ &= \sin^2\theta f^2\{\sin\phi\nabla\cos\phi - \cos\phi\nabla\sin\phi\} \\ &= f^2\sin^2\theta\{-\sin^2\phi\nabla\phi - \cos^2\phi\nabla\phi\} = -f^2\sin^2\theta\nabla\phi.\end{aligned}$$

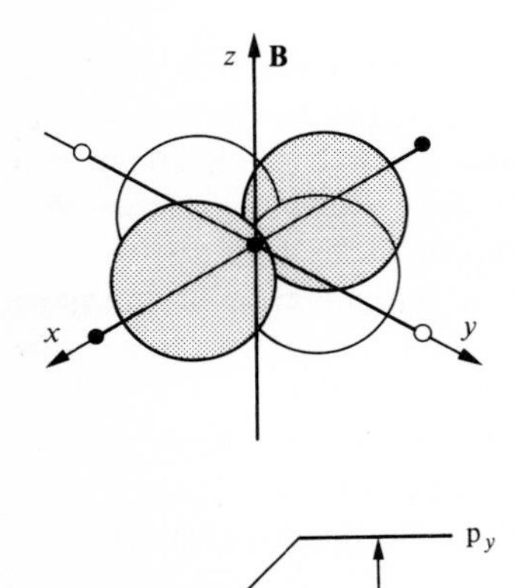

Fig. 14.9. The model for the paramagnetic susceptibility calculation.

Therefore the current density is

$$\begin{aligned}\mathbf{j}^\mathrm{p} &= (e\hbar/m_e)(\mu_\mathrm{B}B/\Delta)f^2\sin^2\theta\nabla\phi \\ &= (e\hbar/m_e)(\mu_\mathrm{B}B/\Delta)\left\{\frac{f^2\sin^2\theta}{x^2+y^2}\right\}\mathbf{V},\end{aligned} \tag{14.5.3}$$

and the ubiquitous swirling function $\mathbf{V}$ is back on stage again. This expression is the same as that for the current density in the degenerate p-orbital case eqn (14.3.5), apart from the presence of the factor $\mu_\mathrm{B}B/\Delta$; therefore

$$\mathbf{j}^\mathrm{p} = (\mu_\mathrm{B}B/\Delta)\mathbf{j}_0. \tag{14.5.4}$$

We can now construct a picture of the induced paramagnetic current: its form is exactly the same as the current that exists when the orbitals are degenerate and the molecule is in a state of well-defined orbital angular momentum, the difference being in its magnitude. The factor $\mu_B B/\Delta$ represents the degree of success of the perturbation (which has a strength $\mu_B B$) in overcoming the energy separation (Δ) which tends to lock the electron into its original location. At a field of 1 T (10 kG) $\mu_B B = 9.27 \times 10^{-24}$ J, corresponding to a wavenumber 0.47 cm^{-1}. Hence the ratio is $0.47/(\tilde{\nu}/\text{cm}^{-1})$, which is a small quantity for most typical energy separations. That is why the paramagnetic susceptibility is usually dominated by the diamagnetic except when energy separations are small. Note that the paramagnetic and diamagnetic currents flow in opposite directions around the direction of the applied field. This accounts for the difference between the diamagnetic and paramagnetic susceptibilities: the diamagnetic current acts as a source of field that opposes the applied field, while the paramagnetic current acts to augment the applied field. The contributions to the magnetizations of the sample in the two cases are opposite.

14.6 The relation of the susceptibility to the current

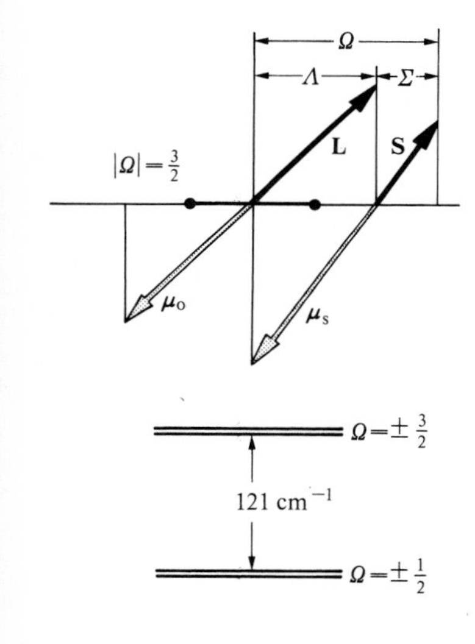

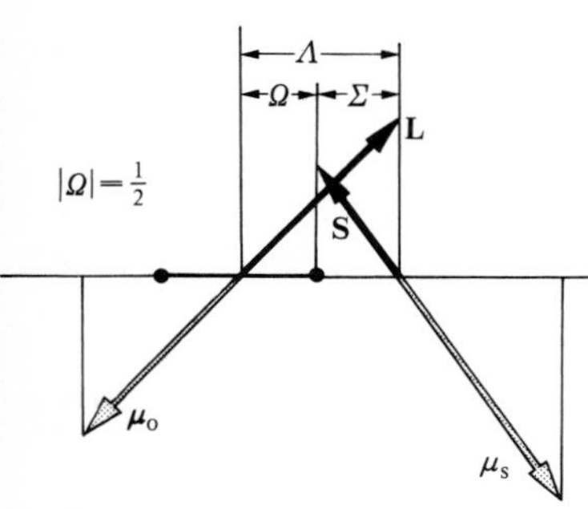

Fig. 14.10. The levels of NO, $\pi^{*}\,{}^2\Pi$ and (in $|\Omega| = \frac{1}{2}$) the cancellation of the components of magnetic moments along the axis.

The discussion can be summarized as follows. In an *orbitally degenerate state* in a definite state of orbital angular momentum, there is an electronic current even in the absence of an applied magnetic field. This accounts for the *permanent orbital magnetic moment* of such molecules. An example is NO, ... π^{*1}, ${}^2\Pi$, which has the two levels ${}^2\Pi_{\frac{1}{2}}$ and ${}^2\Pi_{\frac{3}{2}}$ (the level depending on $\Omega = \Lambda + \Sigma$ in case (a), Fig. 14.10). The orbital angular momentum about the axis gives rise to a magnetic moment $-\mu_B \Lambda$ along the axis. There is also a spin magnetic moment, and its projection on the axis is $-g_e \mu_B \Sigma$ on account of the anomalous g-value of the spin. The total magnetic moment of the molecule is therefore $-(\mu_B \Lambda + g_e \mu_B \Sigma) \approx -(\Lambda + 2\Sigma)\mu_B$, directed parallel to the axis. In the ${}^2\Pi_{\frac{1}{2}}$ level of the term when $\Lambda = 1$ and $\Sigma = -\frac{1}{2}$ (or -1 and $+\frac{1}{2}$), the magnetic moment is $-\{1 + 2(-\frac{1}{2})\}\mu_B = 0$: hence we have the remarkable conclusion that the *although the molecule has a single unpaired electron, it has no permanent magnetic moment*: the spin moment is almost exactly cancelled by the permanent orbital magnetic moment arising from the unquenched orbital current around the axis. In the other level ${}^2\Pi_{\frac{3}{2}}$ the moments add to each other, and the total magnetic moment is $-\{1 + 2(\frac{1}{2})\}\mu_B$. This level lies quite close to the lower ${}^2\Pi_{\frac{1}{2}}$ level (the separation, due to the weak spin–orbit coupling, is 121 cm^{-1}), and so the magnetic moment, and therefore the magnetic susceptibility of a sample of nitrogen(II) oxide, shows a pronounced and complicated temperature dependence which tends towards zero at low temperatures.

In a molecule in an *orbitally non-degenerate state* and with *no unpaired electrons*, an applied magnetic field can stir up a paramagnetic current if it can induce mixing of orbitals to provide a route for the circulation of the electrons around the axis defined by the direction of the applied field. The importance of this process depends on the closeness (in energy) of the orbitals that can be mixed into the ground state. This depends in turn on the orientation of the molecule in the field. For example, if the field lies perpendicular to a p_x-orbital, it may induce its mixing with a p_y-orbital; but if it lies parallel to a p_x-orbital, the perturbation matrix element depends on $l_x p_x = 0$, and so no current is induced. Likewise, s-orbitals do not contribute to the paramagnetic current (when we are thinking of the orbitals on a central ion in a complex) because $l_q s = 0$ for all $q = x, y, z$. Only rarely does the paramagnetic current exceed the diamagnetic, and the molecule acquire a paramagnetic moment even though it has no unpaired electron spins. Confirmation that the paramagnetism then observed is of the high-frequency variety comes from the observation of the temperature independence of the magnetic susceptibility. Examples of species showing TIP include the tetraoxomanganate(VII) ion, a few rare-earth compounds, and some cobalt ammine complexes.

The most common case is the possession of a diamagnetic susceptibility by orbitally non-degenerate molecules. The source of the diamagnetism is the diamagnetic current induced within its orbitals by the applied magnetic field, and the opposing magnetic field which that current generates.

14.7 Chemical shifts and shielding constants

Modern interest in magnetic currents and magnetic susceptibilities centres on the shielding and coupling constants of n.m.r. and their role in the interpretation of spectra. The aim of this section is to see how the fields generated by induced currents are transmitted to nuclei embedded in the molecule.

First, a few remarks on n.m.r. Many nuclei possess spin with a magnitude determined by the spin quantum number I, which for different nuclei ranges from 0 to about 6. For a proton $I = \frac{1}{2}$; for ^{14}N, $I = 1$. With the nuclear spin there is associated a magnetic moment, and we write

$$\mathbf{m} = \gamma_N \mathbf{I} \qquad (14.7.1)$$

where γ_N is the *magnetogyric ratio* of the nucleus in question. It is often expressed in terms of the *nuclear magneton* $\mu_N = e\hbar/2m_p$ ($\mu_N = 5.050\,824 \times 10^{-27}\ \mathrm{J\,T^{-1}}$; m_p is the mass of the proton) and the nuclear g-value, then

$$\mathbf{m} = g_N(\mu_N/\hbar)\mathbf{I}. \qquad (14.7.2)$$

Some typical values of g_N are listed in Table 14.1: the actual values depend on details of the structures of nuclei. The product $g_N I$ is called the *magnetic moment* of the nucleus, although this is a loose use of the term. Notice that the neutron, although uncharged, has a magnetic moment, and that the magnetic moment of a complex nucleus is not

Table 14.1 Nuclear spin properties

Isotope (* radio-active)	Natural abundance (per cent)	Spin I	Magnetic moment μ/μ_N	g-value g_N
^1n*	—	$\frac{1}{2}$	−1.9130	−3.8260
^1H	99.9844	$\frac{1}{2}$	2.792 85	5.5857
^2H	0.0156	1	0.857 45	0.857 45
^3H*	—	$\frac{1}{2}$	−2.127 65	−4.2553
^{13}C	1.108	$\frac{1}{2}$	0.7023	1.4046
^{14}N	99.635	1	0.403 56	0.403 56
^{17}O	0.037	$\frac{5}{2}$	−1.893	−0.757 20
^{19}F	100	$\frac{1}{2}$	2.628 35	5.2567
^{35}Cl	75.4	$\frac{3}{2}$	0.8218	0.5479
^{37}Cl	24.6	$\frac{3}{2}$	0.6841	0.4561

The nuclear magneton has the value $\mu_N = e\hbar/2m_p = 5.050\,82 \times 10^{-27}\,\mathrm{J\,T^{-1}}$; μ is the magnetic moment of the spin state with the largest value of m_I: $\mu = g_N\mu_N I$.
Further information: G. W. C. Kaye and T. H. Laby, *Tables of physical and chemical constants*, Longman; *American Institute of Physics Handbook*, McGraw-Hill, New York.

the sum of the magnetic moments of its constituents (and may be negative). The first point emphasizes the naivety of interpreting the magnetic moment of a particle with spin in terms of the classical idea of a rotating charge; the second underlines the complexity of nuclear structure. Even the proton has a structure so complicated that it has not yet been possible to account for the value $g = 5.586$. As a simple rule of thumb, note that nuclei with even mass and charge numbers have zero spin, which probably indicates some kind of spin pairing in the ground states of nuclei.

In a magnetic field the energy of a nucleus depends on its orientation. The hamiltonian for the interaction is $-\mathbf{m}\cdot\mathbf{B}$, and so when $\mathbf{B}$ lies in the z-direction

$$H^{(1)} = -\gamma_N \mathbf{I}\cdot\mathbf{B} = -\gamma_N B I_z. \tag{14.7.3}$$

The energy of a nucleus in the state $|I, m_I\rangle$ is therefore the eigenvalue in

$$H^{(1)}|I, m_I\rangle = -\gamma_N B I_z\,|I, m_I\rangle = -\gamma_N \hbar B m_I\,|I, m_I\rangle, \tag{14.7.4}$$

or

$$E(m_I) = -\gamma_N \hbar B m_I = -g_N \mu_N B m_I, \qquad m_I = I, I-1, \ldots, -I. \tag{14.7.5}$$

When a nucleus is bathed in electromagnetic radiation with its magnetic vector perpendicular to the direction of the applied field, it is stimulated to undergo transitions between states in accord with the selection rule $\Delta m_I = \pm 1$. The transition rate is greatest when the field and the nucleus are in resonance (Chapter 8), and so an intense absorption is observed when the frequency of the applied field satisfies the

$$\textit{Resonance condition:}\ \hbar\omega = E(m_I - 1) - E(m_I) = g_N \mu_N B. \quad (14.7.6)$$

In an actual experiment the frequency of the radiation is kept constant (in early spectrometers at about 60 MHz but now at a few hundred MHz) and the absorption is monitored as the field is swept. A field of 1.4 T is needed for proton resonance at 60 MHz, and this is readily achievable with a laboratory electromagnet. It is observed that, at a given radiofrequency, nuclei of the same species but in different parts of a molecule or in different molecules, resonate at different applied fields. The most famous example of this *chemical shift* is in the proton resonance spectrum of ethanol, where three regions of absorption are observed with intensities in the ratio 1:2:3. These regions are ascribed to the protons in OH, CH_2, and CH_3 respectively.

The explanation lies in the currents the external field induces in the electrons of the molecule, and the local magnetic fields they in turn generate. The total local field, B_{loc}, differs from the applied field, and to a good approximation the difference is proportional to the applied field. We therefore write

$$B_{loc} = B - \sigma B, \quad (14.7.7)$$

where σ is the *shielding constant* for the nuclei under consideration. The resonance condition should be expressed in terms of the local field, for that is what is experienced by the nuclei buried in the molecule. Therefore

$$\hbar\omega = g_N \mu_N B_{loc} = g_N \mu_N (1 - \sigma) B, \quad (14.7.8)$$

and different nuclei come into resonance at different values of the applied field. The magnitude of σ depends on the chemical environment of the nucleus, and for protons it generally has a value close to 10^{-5} (which is normally quoted as '10 p.p.m.', 10 parts per million). Our task is to evaluate σ.

Consider a molecule containing a single magnetic nucleus and any number of non-magnetic nuclei. There are two magnetic fields present. One is the uniform applied field described by the vector potential $\mathbf{A}_{ex} = \frac{1}{2}\mathbf{B} \wedge \mathbf{r}$. The other is the magnetic field arising from the nuclear magnetic dipole; it is described by a vector potential $\mathbf{A}_{nuc}$. Our first job is to establish its form.

The classical expression for the energy of interaction of a dipole $\mathbf{m}_1$ with a dipole $\mathbf{m}_2$ separated by a vector $\mathbf{r}$ is

$$\mathscr{E} = (\mu_0/4\pi r^3)\{\mathbf{m}_1 \cdot \mathbf{m}_2 - 3\mathbf{m}_1 \cdot \hat{\mathbf{r}}\hat{\mathbf{r}} \cdot \mathbf{m}_2\} \quad (14.7.9)$$

where $\hat{\mathbf{r}} = \mathbf{r}/r$. This is illustrated in Fig. 14.11(a), and is a more general form of the $(1 - 3\cos^2\theta)$ dependence characteristic of the interaction of

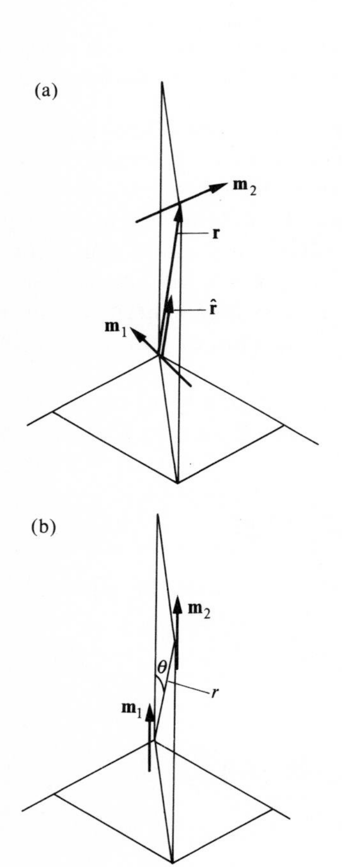

Fig. 14.11. (a) The general point dipole interaction; (b) the special case when the dipoles are parallel.

two parallel dipoles (to which it reduces, Fig. 14.11(b)). The energy of interaction of a dipole $\mathbf{m}_1$ with a magnetic field is $-\mathbf{m}_1 \cdot \mathbf{B}$, and so the dipolar field arising from $\mathbf{m}_2$ is

$$\mathbf{B}(\mathbf{m}_2) = -(\mu_0/4\pi r^3)\{\mathbf{m}_2 - 3\hat{\mathbf{r}}\hat{\mathbf{r}} \cdot \mathbf{m}_2\}, \tag{14.7.10}$$

which is a non-uniform vector field with a complicated directional and distance dependence. It is sketched in Fig. 14.12. We have to find $\mathbf{A}_{\text{nuc}}$ such that its curl, $\nabla \wedge \mathbf{A}_{\text{nuc}}$, is equal to the field $\mathbf{B}(\mathbf{m}_2)$. It is fairly easy to confirm (Appendix 22) that

$$\mathbf{A}_{\text{nuc}} = (\mu_0/4\pi r^3)\mathbf{m}_2 \wedge \mathbf{r}. \tag{14.7.11}$$

Since the dipole $\mathbf{m}_2$ is to be identified as the nuclear magnetic moment $\gamma_{\text{N}}\mathbf{I}$, it follows that

$$\mathbf{A}_{\text{nuc}} = (\gamma_{\text{N}}\mu_0/4\pi r^3)\mathbf{I} \wedge \mathbf{r}. \tag{14.7.12}$$

The hamiltonian for the molecule in an applied magnetic field is obtained from the unperturbed hamiltonian by replacing $\mathbf{p}$ wherever it occurs by $\mathbf{p} + e\mathbf{A}$, where now $\mathbf{A} = \mathbf{A}_{\text{nuc}} + \mathbf{A}_{\text{ex}}$. It proves sensible to proceed in two stages: first to consider the molecule with no applied field, and then to switch on the field. Therefore we begin by replacing $\mathbf{p}$ by $\mathbf{p} + e\mathbf{A}_{\text{nuc}}$:

$$H = (1/2m_{\text{e}})(\mathbf{p} + e\mathbf{A}_{\text{nuc}})^2 + V = H^{(0)} + H^{(1)} + H^{(2)},$$

$$H^{(1)} = (1/2m_{\text{e}})(\mathbf{p} \cdot \mathbf{A}_{\text{nuc}} + \mathbf{A}_{\text{nuc}} \cdot \mathbf{p}); \qquad H^{(2)} = (e^2/2m_{\text{e}})A^2_{\text{nuc}}.$$

We disregard contributions to the energy that are quadratic in the nuclear magnetic moment, and therefore ignore $H^{(2)}$. The nuclear vector potential is divergenceless ($\nabla \cdot \mathbf{A}_{\text{nuc}} = 0$) and so, using the same argument as before, $\mathbf{p} \cdot \mathbf{A}_{\text{nuc}} = \mathbf{A}_{\text{nuc}} \cdot \mathbf{p}$. Hence the first-order hamiltonian is

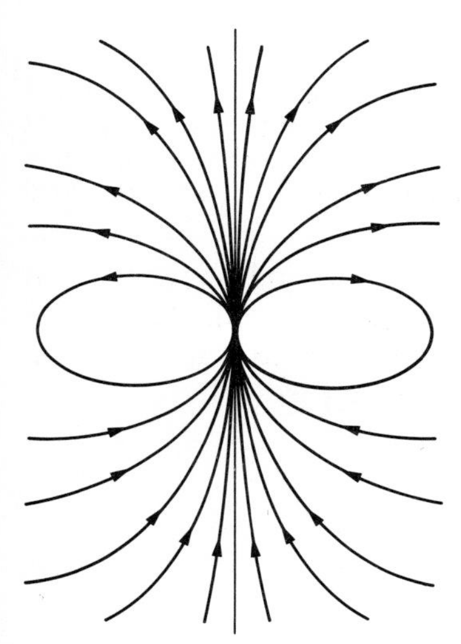
Fig. 14.12. The magnetic field from a point dipole (N towards top of page, S towards bottom).

$$H^{(1)} = (e/m_{\text{e}})\mathbf{A}_{\text{nuc}} \cdot \mathbf{p}. \tag{14.7.13}$$

The first-order energy correction due to this term in the hamiltonian is

$$\begin{aligned}\mathscr{E}^{(1)} &= \int \psi^* H^{(1)} \psi \, \mathrm{d}\tau = (e/m_{\text{e}}) \int \psi^* \mathbf{A}_{\text{nuc}} \cdot \mathbf{p}\psi \, \mathrm{d}\tau \\ &= (e/2m_{\text{e}})\left\{\int \mathbf{A}_{\text{nuc}} \cdot \psi^* \mathbf{p}\psi \, \mathrm{d}\tau + \int \mathbf{A}_{\text{nuc}} \cdot \psi^* \mathbf{p}\psi \, \mathrm{d}\tau\right\}.\end{aligned}$$

Since $\mathbf{p}$ is hermitian and $\mathbf{A}_{\text{nuc}}$ is real the second integral can be manipulated as follows:

$$\begin{aligned}\int \psi^* \mathbf{A}_{\text{nuc}} \cdot \mathbf{p}\psi \, \mathrm{d}\tau &= \int (\mathbf{p}^* \cdot \mathbf{A}_{\text{nuc}}\psi^*)\psi \, \mathrm{d}\tau \ (\mathbf{p} \text{ hermitian}) \\ &= \int (\mathbf{p}^* \cdot \mathbf{A}_{\text{nuc}})\psi^*\psi \, \mathrm{d}\tau + \int \mathbf{A}_{\text{nuc}} \cdot (\mathbf{p}^*\psi^*)\psi \, \mathrm{d}\tau \\ &= \int \mathbf{A}_{\text{nuc}} \cdot \psi \mathbf{p}^* \psi^* \, \mathrm{d}\tau \ (\mathbf{A}_{\text{nuc}} \text{ divergenceless}).\end{aligned}$$

Therefore

$$\mathscr{E}^{(1)} = (e/2m_e)\int \mathbf{A}_{\text{nuc}}\cdot(\psi^*\mathbf{p}\psi + \psi\mathbf{p}^*\psi^*)\,d\tau$$

and the current density operator has arrived on the scene; therefore

$$\mathscr{E}^{(1)} = -\int \mathbf{A}_{\text{nuc}}\cdot\mathbf{j}\,d\tau. \tag{14.7.14}$$

Example. A beam of electrons of number density $\mathcal{N}$, each electron having a momentum $k\hbar$ in the z-direction, moves in a straight line and makes a distance of closest approach b (i.e. b is the *impact parameter*) to a neutron. Calculate the energy of the interaction of the neutron magnetic moment and the magnetic field generated by the electron's motion. Find an expression for the magnetic field produced by the beam.

- *Method.* Use eqn (14.7.14) to calculate the energy; the neutron's vector potential is given by eqn (14.7.12); the electron current density was calculated in the preceding *Example*, p. 389, and is a constant. Use the set of coordinates depicted in Fig. 14.13, and note that only the x-component of $\mathbf{r}\wedge\hat{\mathbf{k}}$ is nonzero.
- *Answer.* Since $\mathbf{j} = -(e\hbar k/m_e)(N_e/l)\hat{\mathbf{k}}$ with $l\to\infty$ and $N_e\to\infty$ such that $N_e/l = \mathcal{N}$, a constant, $\mathbf{A}_{\text{nuc}} = (\gamma_N\mu_0/4\pi r^3)\mathbf{I}\wedge\mathbf{r}$, and $\hat{\mathbf{k}}\wedge\mathbf{r} = \hat{\mathbf{i}}r\sin\theta$,

$$\begin{aligned}
\mathscr{E}^{(1)} &= (e\hbar k/m_e)(N_e/l)(\gamma_N\mu_0/4\pi)\mathbf{I}\wedge\int\{(\mathbf{r}\cdot\hat{\mathbf{k}})/r^3\}\,d\tau\\
&= -(e\hbar k\gamma_N\mu_0 I_x N_e/4\pi m_e l)\int(1/r^2)\sin\theta\,d\tau, \qquad d\tau = dz\\
&= -(e\hbar k\gamma_N\mu_0 I_x N_e/4\pi m_e l)b\int_{-\frac{1}{2}l}^{\frac{1}{2}l}(b^2+z^2)^{-\frac{3}{2}}\,dz, \qquad \sin\theta = b/r\\
&= -(e\hbar k\gamma_N\mu_0 I_x N_e/4\pi m_e l)b\{l/b^2(b^2+\tfrac{1}{4}l^2)^{\frac{1}{2}}\} \quad\text{and}\quad l\to\infty\\
&= -(e\hbar k\gamma_N\mu_0 I_x/2\pi m_e b)\mathcal{N}.
\end{aligned}$$

The energy of interaction can be written $\mathscr{E}^{(1)} = -\gamma_N\mathbf{I}\cdot\mathbf{B}$, and so $\mathbf{B} = B\mathbf{i}$ with

$$B = (e\hbar k\mu_0/2\pi m_e b)\mathcal{N}.$$

- *Comment.* We specified a neutron so that the electron's path is not distorted by electrostatic forces. Since $k\hbar = \sqrt{(2m_e eV)}$, when $b\approx 100$ pm (1 Å) and $V\approx 10$ keV, $B\approx 9.5\times10^{-9}\,\mathcal{N}$ T m; hence if $\mathcal{N}\approx 10^5\ \text{m}^{-1}$ (electrons in the beam having an average separation of 0.01 mm), $B\approx 9.5\times10^{-4}T \mathrel{\hat{=}} 9.5$ G. This field points in the x-direction.

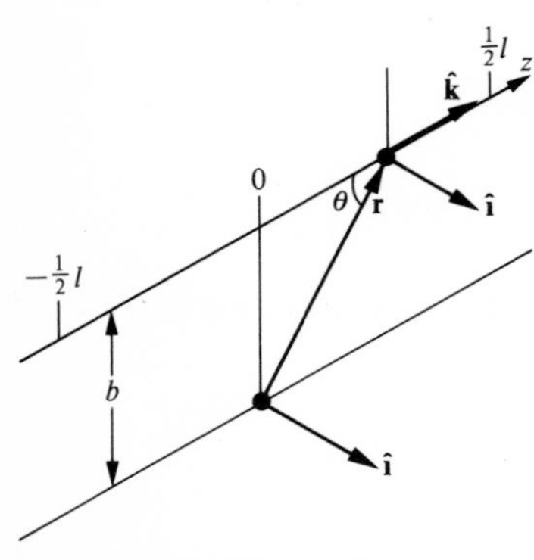

Fig. 14.13. The coordinate system for the interaction energy calculation.

Now turn on the external magnetic field. The only quantity that changes is the current density operator, for it now takes the form given in eqn (14.3.2) with $\mathbf{A}_{\text{ex}}$ standing in for $\mathbf{A}$. In order to emphasize that only the external vector potential is involved in $\mathbf{j}$ we write

$$\mathbf{j}_{\text{ex}} = -(e/2m_e)\{\psi^*\mathbf{p}\psi - \psi\mathbf{p}\psi^*\} - (e^2/2m_e)\mathbf{A}_{\text{ex}}\psi^*\psi$$

and the first-order correction to the energy is simply

$$\mathscr{E}^{(1)} = -\int \mathbf{A}_{\text{nuc}}\cdot\mathbf{j}_{\text{ex}}\,d\tau, \tag{14.7.15}$$

which represents a coupling between the nucleus and the currents induced by the applied field. A more explicit form can be obtained by introducing the dipolar vector potential and using the vector identity $\mathbf{a}\cdot\mathbf{b}\wedge\mathbf{c}=\mathbf{a}\wedge\mathbf{b}\cdot\mathbf{c}$:

$$\mathscr{E}^{(1)} = -(\mu_0/4\pi)\gamma_N\mathbf{I}\wedge\int(1/r^3)\mathbf{r}\cdot\mathbf{j}_{ex}\,d\tau$$
$$= -(\gamma_N\mu_0/4\pi)\mathbf{I}\cdot\int\left\{\frac{\mathbf{r}\wedge\mathbf{j}_{ex}}{r^3}\right\}d\tau.$$

Since the energy of a magnetic dipole in a magnetic field is $-\mathbf{m}\cdot\mathbf{B}$, $\mathscr{E}^{(1)}$ can be interpreted as the interaction of the nuclear dipole moment $\gamma_N\mathbf{I}$ with the magnetic field

$$\mathbf{B}' = (\mu_0/4\pi)\int(\mathbf{r}\wedge\mathbf{j}_{ex}/r^3)\,d\tau. \tag{14.7.16}$$

This is the additional field previously referred to as the chemical shift, $-\sigma\mathbf{B}$. Therefore, by expressing the right-hand side in terms of the applied field, the coefficient of the term linear in $\mathbf{B}$ can be identified as the shielding constant $-\sigma$. (Terms proportional to higher powers of $\mathbf{B}$ also occur on the right, but they are very small, and the shielding constant is independent of the field to a very good approxiamtion.)

The current density can be decomposed (arbitrarily) into diamagnetic and paramagnetic contributions. This suggests that we make the same decomposition for the shielding constant, and write

$$\sigma = \sigma^p + \sigma^d,$$
$$\sigma^p\mathbf{B} = -(\mu_0/4\pi)\int\{\mathbf{r}\wedge\mathbf{j}^p_{ex}/r^3\}\,d\tau; \qquad \sigma^d\mathbf{B} = -(\mu_0/4\pi)\int\{\mathbf{r}\wedge\mathbf{j}^d_{ex}/r^3\}\,d\tau. \tag{14.7.17}$$

We have seen that the two components of the current travel in opposite directions (i.e. that $\mathbf{j}^p$ and $\mathbf{j}^d$ have opposite signs); the two components of the shielding constant therefore have different signs, and so the shift of the resonance may be to high or to low field.

Consider the diamagnetic contribution. The diamagnetic current is given in eqn (14.3.8). For a field in the z-direction,

$$\mathbf{r}\wedge\mathbf{j}^d_{ex} = -(e^2/m_e)\mathbf{r}\wedge\mathbf{A}_{ex}\psi_0^2$$
$$= -(e^2B/2m_e)\psi_0^2(x\hat{\mathbf{i}}+y\hat{\mathbf{j}}+z\hat{\mathbf{k}})\wedge(-y\hat{\mathbf{i}}+x\hat{\mathbf{j}})$$
$$= -(e^2B/2m_e)\psi_0^2\{-xz\hat{\mathbf{i}}+yz\hat{\mathbf{j}}+(x^2+y^2)\hat{\mathbf{k}}\}.$$

This shows that the induced field experienced by the nucleus has components in all three directions. We are interested in the z-component of the local field (the coefficient of $\hat{\mathbf{k}}$); then use of eqn (14.7.17) gives

$$\sigma^d_{zz}B = -(e^2\mu_0B/8\pi m_e)\int\{(x^2+y^2)/r^3\}\psi_0^2\,d\tau,$$

which identifies the diamagnetic shielding constant as

$$\sigma^d_{zz} = (e^2\mu_0/8\pi m_e)\int\{(x^2+y^2)/r^3\}\psi_0^2\,d\tau. \tag{14.7.18}$$

This is the shielding constant in the z-direction. If, as is usually the case in high-resolution n.m.r., the sample is fluid, the observable is the mean shielding constant $\sigma = \frac{1}{3}(\sigma_{xx} + \sigma_{yy} + \sigma_{zz})$. Since $(x^2+y^2)+(y^2+z^2)+(z^2+x^2) = 2r^2$, we arrive at the

Lamb formula:

$$\sigma^{\mathrm{d}} = (e^2\mu_0/12\pi m_{\mathrm{e}}) \int (1/r)\psi_0^2\,\mathrm{d}\tau = (e^2\mu_0/12\pi m_{\mathrm{e}})\langle 1/r\rangle. \qquad (14.7.19)$$

The magnitude of the diamagnetic contribution to the shielding constant therefore depends on the average distance of the electrons from the nucleus in question. This is an easy quantity to estimate (especially for atoms), and so σ^{d} can be calculated quite readily. In the case of the ground state of the hydrogen atom, for instance, $\langle 1/r\rangle = 1/a_0$, and so $\sigma^{\mathrm{d}} = \frac{1}{3}\alpha^2 \approx 1.8\times 10^{-5}$ (α being the fine structure constant, p. 217). We can think of this shift as arising when the nucleus experiences the currents set up within atomic orbitals, of the type illustrated in Fig. 14.7.

Example. An electron is confined to a two-dimensional disc-like region of radius R. Model its distribution by ψ = constant. A magnetic dipole lies at the centre of the disc (and does not perturb the wavefunction). Calculate the diamagnetic contribution to the shielding constant for the dipole when the field is perpendicular to the disc; ignore electron spin.

- *Method.* Use eqn (14.7.19). The wavefunction should be normalized; the volume element in two dimensions is $r\,\mathrm{d}r\,\mathrm{d}\phi$, $0 \le r \le R$, $0 \le \phi \le 2\pi$.

- *Answer.* $\psi = 1/R\sqrt{\pi}$;

$$\sigma^{\mathrm{d}} = (e^2\mu_0/12\pi m_{\mathrm{e}})\langle 1/r\rangle$$
$$= (e^2\mu_0/12\pi^2 m_{\mathrm{e}} R^2)\int_0^{2\pi}\mathrm{d}\phi \int_0^R (1/r) r\,\mathrm{d}r = e^2\mu_0/6\pi m_{\mathrm{e}} R.$$

- *Comment.* The result can be expressed in the form σ^{d}/p.p.m. = 1.879/(R/nm). Therefore, a disc of the order of magnitude of an atom ($R \approx 0.1$ nm) gives $\sigma^{\mathrm{d}} \approx 19$ p.p.m., which is of the order of magnitude of the shifts observed. Note that σ^{d} decreases as R increases because the currents induced by the field are increasingly far from the dipole The paramagnetic contribution to σ (see below) is identically zero because $l_z\psi = 0$. The true wavefunctions are Bessel functions: see the *Example* on p. 59.

The paramagnetic shielding constant arises from an interaction of the nucleus with the field generated by currents of the form shown in Fig. 14.6, and therefore depends on the ability of the applied field to mix excited states into the ground state of the molecule. It follows from the earlier discussion of the paramagnetic current that in free atoms and ions there is no paramagnetic contribution to the chemical shift (because in such species there is no paramagnetic current). In molecules the paramagnetic contribution is not necessarily zero, and in many cases it is dominant.

The strategy for the calculation of the chemical shift arising from the paramagnetic current is to use eqn (14.3.9) in eqn (14.7.17) for the current density, and then to extract the term linear in the applied field and pointing along the z-direction. The coefficient of B is then identified as $-\sigma_{zz}$. We can use eqn (14.3.9) for the first-order current:

$$-\sigma^{\mathrm{p}}\mathbf{B}=(e\mu_0/8\pi m_{\mathrm{e}})\sum_n{}'(a_n-a_n^*)\int\{\mathbf{r}\wedge(\psi_n\mathbf{p}\psi_0-\psi_0\mathbf{p}\psi_n)/r^3\}\,\mathrm{d}\tau.$$

Now notice that $\mathbf{r}\wedge\mathbf{p}=\mathbf{l}$, the orbital angular momentum operator; then

$$\begin{aligned}\sigma^{\mathrm{p}}\mathbf{B}&=-(e\mu_0/8\pi m_{\mathrm{e}})\sum_n{}'(a_n-a_n^*)\int\{\psi_n(1/r^3)\mathbf{l}\psi_0-\psi_0(1/r^3)\mathbf{l}\psi_n\}\,\mathrm{d}\tau\\&=-(e\mu_0/8\pi m_{\mathrm{e}})\sum_n{}'(a_n-a_n^*)\{\langle n|\,(1/r^3)\mathbf{l}\,|0\rangle-\langle 0|\,(1/r^3)\mathbf{l}\,|n\rangle\}.\end{aligned}$$

Since $\mathbf{l}$ is hermitian and imaginary ($\mathbf{l}^*=-\mathbf{l}$),

$$\langle 0|\,(1/r^3)\mathbf{l}\,|n\rangle=\langle n|\,\mathbf{l}(1/r^3)\,|0\rangle^*=-\langle n|\,\mathbf{l}(1/r^3)\,|0\rangle=-\langle n|\,(1/r^3)\mathbf{l}\,|0\rangle.$$

(The last step follows from $(\mathbf{l}r^{-3})=0$; the easiest way of seeing this is to remember that $\mathbf{l}$ is the generator of infinitesimal rotations (p. 164), but $1/r^3$ is invariant under rotation. A lengthier way involves evaluating $(\mathbf{l}r^{-3})$ explicitly.) Therefore

$$\sigma^{\mathrm{p}}\mathbf{B}=-(e\mu_0/4\pi m_{\mathrm{e}})\sum_n{}'(a_n-a_n^*)\langle n|\,(1/r^3)\mathbf{l}\,|0\rangle.$$

Now we have to tackle the mixing coefficients $a_n=-H_{n0}/\Delta_{n0}$ where $H^{(1)}$ is the perturbation due to the applied field. Since the field lies in the z-direction, $H^{(1)}=-\gamma_{\mathrm{e}}Bl_z$. Therefore

$$a_n=\gamma_{\mathrm{e}}Bl_{z,n0}/\Delta_{n0};\qquad a_n^*=\gamma_{\mathrm{e}}Bl_{z,n0}^*/\Delta_{n0}=-\gamma_{\mathrm{e}}Bl_{z,n0}/\Delta_{n0}$$

where in the final equality we have used $\mathbf{l}^*=-\mathbf{l}$ and the reality of the basis functions ψ_n and ψ_0. It follows that

$$a_n-a_n^*=2\gamma_{\mathrm{e}}Bl_{z,n0}=2\gamma_{\mathrm{e}}Bl_{z,0n}^*=-2\gamma_{\mathrm{e}}Bl_{z,0n}$$

(hermiticity for $l_{z,0n}^*=l_{z,n0}$; imaginary $\mathbf{l}$ for $l_{z,0n}^*=-l_{z,0n}$).

Now we can knot everything together. We require the z-component of the induced field, and so

$$\sigma_{zz}^{\mathrm{p}}B=(e\gamma_{\mathrm{e}}\mu_0B/2\pi m_{\mathrm{e}})\sum_n{}'(l_{z,0n}/\Delta_{0n})\langle n|\,(1/r^3)l_z\,|0\rangle,$$

hence, with $\gamma_{\mathrm{e}}=-e/2m_{\mathrm{e}}$,

$$\sigma_{zz}^{\mathrm{p}}=-(e^2\mu_0/4\pi m_{\mathrm{e}}^2)\sum_n{}'\{l_{z,0n}(r^{-3}l_z)_{n0}/\Delta_{n0}\}.\qquad(14.7.20)$$

The *mean paramagnetic shielding constant* is therefore

$$\sigma^{\mathrm{p}}=-(e^2\mu_0/12\pi m_{\mathrm{e}}^2)\sum_n{}'\{\mathbf{l}_{0n}\cdot(r^{-3}\mathbf{l})_{n0}/\Delta_{n0}\}.\qquad(14.7.21)$$

This is the expression we have been seeking.

The sign of σ^{p} is negative, which reflects an increase of flux density at the nucleus ($B_{\mathrm{loc}} > B$) and therefore a shift of resonance to low fields. A simple interpretation of the form of the expression is that $\{(\mu_{\mathrm{B}}B)/\Delta\}\mathbf{l}_{0n}$ represents the driving effect of the applied field stirring up angular momentum (and so inducing the current pattern illustrated in Fig. 14.6), and the other factor, $\langle n|\,(1/r^3)\mathbf{l}\,|0\rangle$ represents the transmission of this current by a dipolar interaction to a nucleus at a distance r. The relative magnitude of the paramagnetic and diamagnetic contributions is

$$|\sigma^{\mathrm{p}}/\sigma^{\mathrm{d}}| = (1/m_{\mathrm{e}}\langle 1/r\rangle)\sum_n{}' \{\mathbf{l}_{0n}\cdot(r^{-3}\mathbf{l})_{n0}/\Delta_{n0}\}.$$

The matrix elements of $\mathbf{l}$ are each of order $\hbar$, and so, if $|\mathbf{l}_{n0}| = \eta\hbar$, $\eta \approx 1$,

$$|\sigma^{\mathrm{p}}/\sigma^{\mathrm{d}}| = \eta^2(\hbar^2/m_{\mathrm{e}})\langle 1/r^3\rangle/\Delta\langle 1/r\rangle. \tag{14.7.22}$$

If we write $\langle 1/r\rangle = 1/R$, then very crudely $\langle 1/r^3\rangle \approx 1/R^3$, and so

$$|\sigma^{\mathrm{p}}/\sigma^{\mathrm{d}}| \approx (\hbar^2/2m_{\mathrm{e}}\Delta)/R^2 \approx 300/(R/\mathrm{nm})^2(\Delta/\mathrm{cm}^{-1}).$$

Taking $\Delta \approx 30\,000\ \mathrm{cm}^{-1}$ and $R \approx 0.05$ nm, we have $|\sigma^{\mathrm{p}}/\sigma^{\mathrm{d}}| \approx 4$ which suggests that paramagnetic shifts are of greater importance than diamagnetic (so long as there are low-lying excited states available). This is because the $1/r^3$ term magnifies the effects of the currents when they lie close to the nucleus, but there is no such magnification effect for an external observer measuring magnetic susceptibility.

14.8 Shielding by a neighbouring group

When the nucleus of interest is some way removed from the region of greatest current density, such as a proton near a benzene ring, the shielding constant can be expressed in terms of the anisotropy of the magnetic susceptibility of the neighbouring group. This result is established as follows.

Consider some group in a molecule which has a magnetizability ξ. When it is exposed to a magnetic field it acquires an induced dipole moment $\langle\mathbf{m}\rangle = \xi\mathbf{H}$. This can be expressed in terms of the flux density through $H \approx B/\mu_0$ and in terms of the magnetic susceptibility through $\xi = M_{\mathrm{m}}\chi/L\rho^{\ominus}$, eqn. (14.1.11b). Hence $\langle\mathbf{m}\rangle = (M_{\mathrm{m}}\chi/L\mu_0\rho^{\ominus})\mathbf{B}$. If the magnetizability (and susceptibility) is anisotropic, the magnitude of the induced dipole depends on the orientation of the molecule with respect to the applied field. Let there be a nucleus of interest at a distance R from this magnetizable group, Fig. 14.14. If the distance is large, the group appears to be a point magnetic dipole, and so it gives rise to a local magnetic field of the form

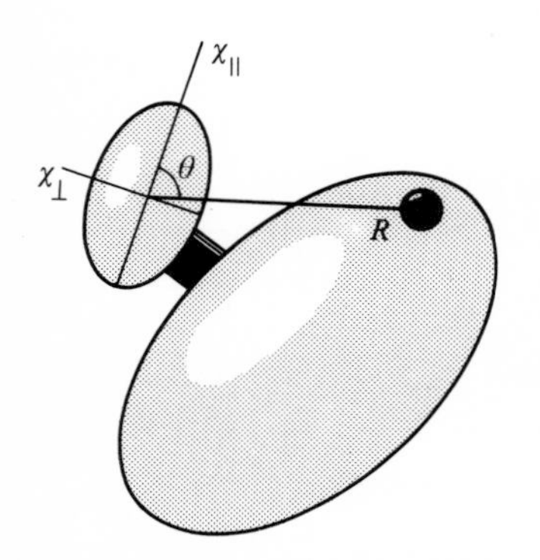

Fig. 14.14. Schematic representation of a magnetizable group and its location relative to a resonant nucleus.

$$\mathbf{B}(\mathbf{m}) = -(\mu_0/4\pi R^3)\{\mathbf{m} - 3\mathbf{m}\cdot\hat{\mathbf{R}}\hat{\mathbf{R}}\}, \qquad \mathbf{m} = \xi\cdot\mathbf{H} \approx \xi\cdot\mathbf{B}/\mu_0, \tag{14.8.1}$$

which in some orientations decreases the total field and in others increases it. If the induced magnetic moment is the same for all orientations, the average magnetic field at the nucleus is zero because the dipolar interaction averages to zero. But if the magnetic moment changes in magnitude as the molecule rotates, this averaging procedure is upset and the local field does not average to zero over a full rotation of the molecule. If the group is axially symmetric so that it has a magnetizability $\xi_{\parallel}$ when the field is along its axis and $\xi_{\perp}$ when it is perpendicular, and the vector joining the centre of the shielding group to the proton nucleus makes an angle θ to the symmetry axis, Fig. 14.14, it is not too hard to show that the magnitude of the average shielding constant by this mechanism is

$$\begin{aligned}\sigma &= (\xi_{\parallel}-\xi_{\perp})(1-3\cos^2\theta)/12\pi R^3.\\ &= (M_{\mathrm{m}}/12\pi\rho^{\ominus}L)(\chi_{\parallel}-\chi_{\perp})(1-3\cos^2\theta)/R^3.\end{aligned} \tag{14.8.2}$$

In a compound like benzene, $\theta = 90°$, and so the shielding due to the susceptibility of the ring is $(M_{\mathrm{m}}/12\pi L\rho^{\ominus})(\chi_{\parallel}-\chi_{\perp})/R^3$, although it is probably true that the protons are too close to the ring for the point-dipole approximation to be reliable. In benzene $\chi_{\parallel}-\chi_{\perp}$ is positive; consequently, although the susceptibilities are both diamagnetic, the shielding constant is paramagnetic.

A summary of these results is as follows. The chemical shift arises from the interaction of the nucleus with the currents induced by the applied field. These currents may be either diamagnetic or paramagnetic. When the nucleus of interest inhabits the group where the currents are flowing, a diamagnetic current gives a diamagnetic shift and a paramagnetic current gives a paramagnetic shift. When the nucleus of interest inhabits a neighbouring group, the diamagnetic susceptibility of the group leads to a dipolar field that does not average to zero if the susceptibility is anisotropic. Whether this field is shielding or anti-shielding (and the shielding constant correspondingly positive or negative) depends on the angular disposition of the nucleus relative to the group (because $1-3\cos^2\theta$ may be positive in some molecules but negative in others).

14.9 The *g*-value

Now we turn to electron spin resonance. In e.s.r. the g-value serves a role analogous to the shielding constant in n.m.r. Both measure the magnitude of currents induced in the electrons of a molecule by an applied field and the extent of their coupling to a spin magnetic moment.

The basic principles of the e.s.r. technique are as follows. An electron has a magnetic moment $\mathbf{m} = g_{\mathrm{e}}\gamma_{\mathrm{e}}\mathbf{s}$, and in a magnetic field along the z-direction its energies are restricted to the eigenvalues of the operator $-\mathbf{m}\cdot\mathbf{B} = -g_{\mathrm{e}}\gamma_{\mathrm{e}}Bs_z$, which are $-g_{\mathrm{e}}\gamma_{\mathrm{e}}B\hbar m_s$ with $m_s = \pm\frac{1}{2}$. Since $\gamma_{\mathrm{e}} = -\mu_{\mathrm{B}}/\hbar$, these eigenvalues are $g_{\mathrm{e}}m_s\mu_{\mathrm{B}}B$. The system is bathed in radiation of frequency ω and the magnetic field is swept until the

$$\textit{Resonance condition:}\ \hbar\omega = E(\tfrac{1}{2}) - E(-\tfrac{1}{2}) = g_e\mu_B B \tag{14.9.1}$$

is reached, when a strong, resonant absorption is observed. Typical frequencies are around 9.4 GHz ($\omega = 5.9\times10^{10}\,\mathrm{s}^{-1}$) and resonance occurs in the region of 0.3 T (3 kG). Hence the technique is a form of *microwave spectroscopy* (9.4 GHz corresponds to 3 cm wavelength radiation).

The electron responsible for the resonant absorption is normally part of a molecule, and so it experiences not simply the applied field but the sum of the applied and the induced fields. The resonance condition is modified to

$$\hbar\omega = g_e\mu_B(1-\sigma)B, \tag{14.9.2}$$

and the quantity $g_e(1-\sigma)$ is set equal to the parameter g, the *g-factor* of the molecule.

While it is possible to proceed in the same way as in the calculation of the shielding constant, it is instructive to take a different route and to introduce the *spin hamiltonian*, a concept widely used in e.s.r. The idea behind the spin hamiltonian is that while the true hamiltonian for an electron involves a lot of different operators, it may be possible to express it as an effective hamiltonian in which the effect of all operators other than the spin have been absorbed into various parameters. For instance, the true hamiltonian for an electron spin contains a term representing the effect of the magnetic field on the electron spin moment, $-g_e\gamma_e\mathbf{s}\cdot\mathbf{B}$, another for the spin–orbit coupling $(hc\zeta/\hbar^2)\mathbf{l}\cdot\mathbf{s}$, and another for the interaction of the applied field with the orbital angular momentum $-\gamma_e\mathbf{l}\cdot\mathbf{B}$. The quantity $hc\zeta/\hbar^2$ is awkward to carry around (ζ is a wavenumber, $hc\zeta$ an energy), and so we write $\lambda = hc\zeta/\hbar^2$. The true hamiltonian therefore contains the terms

$$H^{(1)} = -g_e\gamma_e\mathbf{s}\cdot\mathbf{B} + \lambda\mathbf{l}\cdot\mathbf{s} - \gamma_e\mathbf{l}\cdot\mathbf{B}. \tag{14.9.3}$$

The spin hamiltonian absorbs the effects of the orbital angular momentum operators into a single parameter, and takes the form

$$\textit{Spin hamiltonian:}\ H^{(\mathrm{spin})} = -g\gamma_e\mathbf{s}\cdot\mathbf{B} \tag{14.9.4}$$

where the only eigenvalues that remain to be found are those of the spin operator. If the field lies in the z-direction, $H^{(\mathrm{spin})} = -g\gamma_e Bs_z$, and the resonance condition then becomes

$$\hbar\omega = g\mu_B B. \tag{14.9.5}$$

Hence the parameter g can be determined. The eigenstates between which the resonance transitions occur should properly be regarded as the eigenstates of the complete hamiltonian $H^{(1)}$; but we are playing a trick, and expressing them as the eigenstates of the effective operator $H^{(\mathrm{spin})}$.

How can we collect the effects of all the terms of the true hamiltonian into a single parameter, the g-value? The idea is to use $H^{(1)}$ in a perturbation series for the true energies of the system, and then to find the form of the spin hamiltonian that matches the expansion term by term. Since we shall work only to second-order in $H^{(1)}$, this matching turns out to be surprisingly easy.

Let the unperturbed eigenstates of $H^{(0)}$ be $|n\rangle$ with $|0\rangle$ the ground state (the state of interest). The first-order correction to the energy is the expectation value of $H^{(1)}$ within the orbitally non-degenerate (real) state $|0\rangle$:

$$\mathscr{E}^{(1)} = -g_e\gamma_e\langle 0|\, s_z\, |0\rangle B - \gamma_e\langle 0|\, l_z\, |0\rangle B + \lambda\langle 0|\, \mathbf{l}\cdot\mathbf{s}\, |0\rangle.$$

The second two terms are proportional to expectation values of the components of the orbital angular momentum; but in a real, orbitally non-degenerate state such expectation values are zero (p. 379). Therefore the first-order energy is

$$\mathscr{E}^{(1)} = -g_e\gamma_e B\hbar m_s. \tag{14.9.6}$$

Clearly, we can achieve the same result by using the spin hamiltonian $H_1^{(\text{spin})} = -g_e\gamma_e B s_z$ and operating on the spin states alone. The idea of the spin hamiltonian is beginning to emerge.

Now consider the energy correction to second-order in the perturbation. The basic expression is

$$\mathscr{E}^{(2)} = \sum_n{}' \left\{\frac{\langle 0|\, H^{(1)}\, |n\rangle\langle n|\, H^{(1)}\, |0\rangle}{\mathscr{E}_0 - \mathscr{E}_n}\right\}$$

and when the perturbation hamiltonian is inserted there will be nine terms. However, we are looking for a contribution that can be expressed like eqn (14.9.4), and therefore we are interested only in terms that are bilinear in the spin and external field ($\mathbf{B}\ldots\mathbf{s}$). Only the cross-terms arising from the orbital and the spin–orbit interactions, which have the structure $\mathbf{B}\cdot\mathbf{l}\ldots\mathbf{l}\cdot\mathbf{s}$, have the appropriate form. Therefore only two contributions to the expression for $\mathscr{E}^{(2)}$ are relevant to the present problem. They are

$$\mathscr{E}^{(2)} = \sum_n{}' \left\{\frac{\lambda(-\gamma_e B)(\langle 0|\, l_z\, |n\rangle\langle n|\, \mathbf{l}\cdot\mathbf{s}\, |0\rangle + \langle 0|\, \mathbf{l}\cdot\mathbf{s}\, |n\rangle\langle n|\, l_z\, |0\rangle)}{\mathscr{E}_0 - \mathscr{E}_n}\right\}.$$

Furthermore, in this simple introduction, we are interested only in an effective hamiltonian containing the operator s_z for the spin (in more advanced work the g-factor turns out to be a matrix: we ignore that complication). Therefore, with $\mathscr{E}_n - \mathscr{E}_0 = \Delta_{n0}$,

$$\begin{aligned}\mathscr{E}^{(2)} &= +\gamma_e B\lambda \sum_n{}' \{l_{z,0n}l_{z,n0}m_s\hbar + m_s\hbar l_{z,0n}l_{z,n0}\}/\Delta_{n0}\\ &= 2\gamma_e B\lambda \sum_n{}' \{l_{z,n0}l_{z,n0}/\Delta_{n0}\}m_s\hbar.\end{aligned}$$

Exactly the same contribution to the energy is obtained if we use an operator

$$H_2^{(\text{spin})} = 2\gamma_e B\lambda \sum_n{}' (l_{z,0n}l_{z,n0}/\Delta_{n0})s_z \tag{14.9.7}$$

which acts on only the spin states. This is the second-order contribution to the spin hamiltonian.

That is as far as we need go. It follows that the total spin hamiltonian is the effective operator

$$\begin{aligned} H^{(\text{spin})} &= H_1^{(\text{spin})} + H_2^{(\text{spin})} + \dots \\ &= -g_e\gamma_e Bs_z + 2\gamma_e B\lambda \sum_n{}' (l_{z,0n}l_{z,n0}/\Delta_{n0})s_z + \dots \\ &= -g_{zz}\gamma_e Bs_z, \end{aligned} \tag{14.9.8}$$

with

$$g_{zz} = g_e - 2\lambda \sum_n{}' (l_{z,0n}l_{z,n0}/\Delta_{n0}). \tag{14.9.9}$$

The quantity of interest in a fluid solution is the mean g-value, $g = \frac{1}{3}(g_{xx} + g_{yy} + g_{zz})$, and so

$$g = g_e + \delta g; \qquad \delta g = -\tfrac{2}{3}\lambda \sum_n{}' (\mathbf{l}_{0n} \cdot \mathbf{l}_{n0}/\Delta_{n0}). \tag{14.9.10}$$

As an example of the application of these expressions, consider the model system depicted in Fig. 14.15 where the single unpaired electron occupies the p_x-orbital, and there is a totally unoccupied p_y-orbital at an energy Δ above it. When the field is along the z-axis the g-value of the electron is

$$g_{zz} = g_e - 2\lambda\langle p_x| l_z |p_y\rangle\langle p_y| l_z |p_x\rangle/\Delta.$$

Since $\langle p_y| l_z |p_x\rangle = i\hbar$ and $\lambda = hc\zeta/\hbar^2$,

$$g_{zz} = g_e - 2hc\zeta(-i)i/\Delta = g_e - 2hc\zeta/\Delta = g_e - 2\zeta/\tilde{\nu}. \tag{14.9.11}$$

($\tilde{\nu}$ is the wavenumber corresponding to Δ: $\tilde{\nu} = \Delta/hc$.) This expression shows that $g < g_e$ in this case, and therefore that the resonance (at fixed microwave frequency) lies at higher values of the magnetic field than for a free electron: the electron is shielded by its orbital field.

When the field is aligned parallel to the x-axis the g-value depends on matrix elements of the form $\langle n| l_x |p_x\rangle$, which all vanish because $l_x |p_x\rangle = 0$. Therefore in this direction $g_{xx} = g_e$.

The extent of the deviation of the g-value from g_e is quite easy to understand on the basis of the perturbation expressions. The term $\mu_B B/\Delta$ in the second-order energy expression reflects the extent to which the applied field can mix in excited states and so provide a pathway through the molecule for electrons to acquire orbital angular momentum. This momentum is then transmitted to the electron spin as an effective field *via* the spin–orbit coupling ζ. As the energy separation decreases the field can stir up currents more readily, and so g

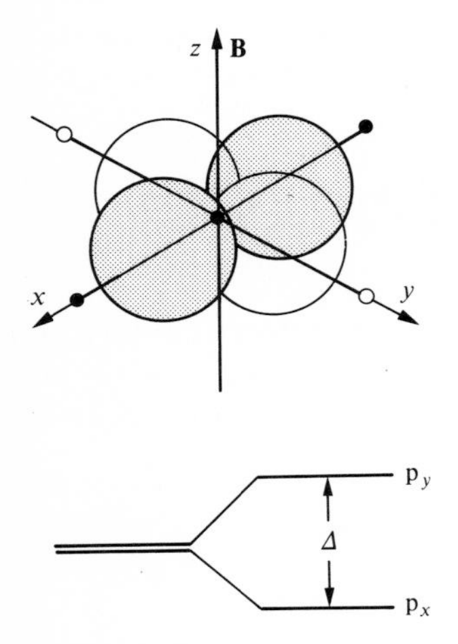

Fig. 14.15. The model for the g-value calculation.

deviates more strongly from g_e. The same is true as the spin–orbit coupling constant increases, because when it is large even a small induced angular momentum may give rise to a strong field at the electron spin.

Applications of ideas like these to real molecules will be found in the books cited in *Further reading*, p. 414. As an example, consider the nitrogen (IV) oxide molecule NO_2. Its unpaired electron occupies an a_1-orbital, an sp-hybrid along the external bisector of the ONO angle, Fig. 14.16. When the field is parallel to this bisector, the g-shift is expected to be zero (and $g \approx g_e = 2.0023$) because it corresponds to the g_{xx} calculation of the model system in Fig. 14.15. The measured value is in fact 2.002. When the field is parallel to the O . . . O direction, the arrangement resembles the calculation of g_{zz} for the model system, and so we expect the g-value to be $g_e - 2hc\zeta/\Delta$, where Δ is the separation of the a_1- and b_1-orbitals (strictly the A_1 and B_1 states), the latter being a molecular orbital constructed from p-orbitals perpendicular to the molecule plane. We therefore expect a g-value of less than the free-spin value, and the measured value is 1.995, as expected.

14.10 Spin–spin coupling

There are three types of spin–spin coupling of interest in molecules:

(1) *Electron–electron coupling*, which gives rise to the *fine-structure* in e.s.r. when there is more than one unpaired electron present in the molecule (in triplet molecules).
(2) *Electron–nucleus coupling*, which gives rise to the *hyperfine structure* of spectra. It is observed in atomic and molecular electronic spectra, but its major importance lies in e.s.r., where it accounts for the complicated line structure of the spectra. In n.m.r. the presence of an unpaired electron may give rise to the *Knight shift* of the spectrum.
(3) *Nucleus–nucleus coupling*, which gives rise to the important *fine structure* of n.m.r. spectra.

We shall deal cursorily with the first of these, and then build up towards a discussion of the spin–spin coupling that gives rise to n.m.r. fine structure.

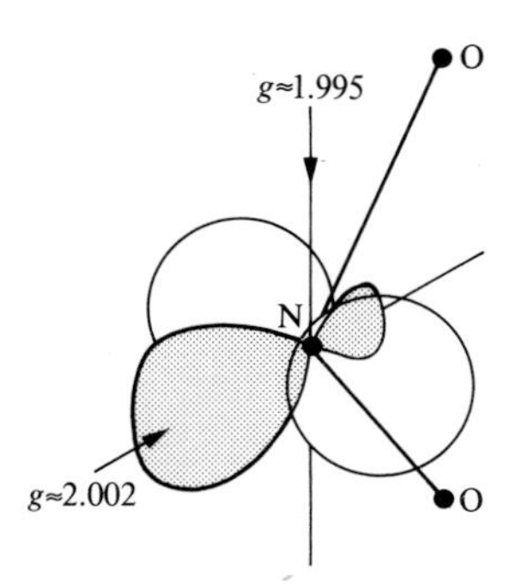

Fig. 14.16. g-values in NO_2.

One mechanism of interaction between electron spins is the direct dipole–dipole interaction of their magnetic moments. The hamiltonian for the interaction has the same form as the energy expression given in eqn (14.7.9). If the electron spins are parallel and aligned along the z-direction,

$$H = (\mu_0 g_e^2 \gamma_e^2 / 4\pi R^3)(1 - 3\cos^2\theta) s_{1z} s_{2z}. \qquad (14.10.1)$$

The energy of this interaction is therefore

$$\mathscr{E} = (\mu_0 g_e^2 \mu_B^2 / 4\pi R^3)(1 - 3\cos^2\theta) m_{s1} m_{s2}. \qquad (14.10.2)$$

In a freely rotating molecule in fluid solution only the average value of this expression is observed; but the average of $1 - 3\cos^2\theta$ over a sphere is zero; hence there is no net dipole–dipole interaction energy in a freely rotating molecule. The interaction does not average to zero in a solid (unless the molecule is itself spherical), and so investigating the energy of interaction (using triplet-state e.s.r.) is a way of exploring the distribution of the two electrons.

Another mechanism of interaction between electron spins has the same directional dependence as the dipolar interaction. Each of the two electrons interacts with its orbital angular momentum through terms of the form $\zeta_i \mathbf{s}_i \cdot \mathbf{l}_i$. When these perturbations are used in second-order perturbation theory they give rise to a second-order contribution that can be modelled by a term in the spin-hamiltonian that is bilinear $(\mathbf{s}_1 \ldots \mathbf{s}_2)$ in the electron spin operators. This term has the form $\mathbf{s}_1 \cdot \mathbf{s}_2 - 3\mathbf{s}_1 \cdot \hat{\mathbf{r}}\hat{\mathbf{r}} \cdot \mathbf{s}_2$, exactly as for the dipolar interaction (but not with the latter's simple $1/r^3$ dependence). It can be thought of as expressing the energy of interaction of the electron spins that are communicating via their orbital angular momenta: a spin stirs its own orbital momentum, which is experienced by the other electron, which transmits it to its spin *via* the spin–orbital coupling. The direct dipole mechanism dominates this alternative route in most inorganic radicals.

14.11 Hyperfine interactions

The term 'hyperfine interaction' denotes any interaction between electrons and nuclei other than the Coulombic interaction. It may be either electric (e.g. the interaction of an electric quadrupole moment of a nucleus with the gradient of the electric field arising from an anisotropic electron distribution) or magnetic. We concentrate on the latter.

There are two types of magnetic interaction between an electron spin and a nuclear spin. One is a direct dipolar interaction between the two magnetic moments. The magnetic field from a nuclear magnetic dipole $\mathbf{m}_I = \gamma_N \mathbf{I}$ is given by the expression we first saw as eqn (14.7.10):

$$\mathbf{B}(\mathbf{I}) = -(\gamma_N \mu_0 / 4\pi r^3)(\mathbf{I} - 3\hat{\mathbf{r}}\hat{\mathbf{r}} \cdot \mathbf{I}). \qquad (14.11.1)$$

Since the energy of interaction of this field with an electron spin magnetic moment $\mathbf{m}_s = g_e \gamma_e \mathbf{s}$ is $-\mathbf{m}_s \cdot \mathbf{B}(\mathbf{I})$, the interaction hamiltonian is

$$H_{hf} = -\mathbf{m}_s \cdot \mathbf{B}(\mathbf{I}) = (\mu_0 g_e \gamma_e \gamma_N / 4\pi r^3)(\mathbf{s} \cdot \mathbf{I} - 3\mathbf{s} \cdot \hat{\mathbf{r}}\hat{\mathbf{r}} \cdot \mathbf{I}). \qquad (14.11.2)$$

When the electron and nuclear spins are strongly aligned by an applied field so that only their z-components are of interest, this reduces to

$$H_{hf} = (\mu_0 g_e \gamma_e \gamma_N / 4\pi r^3)(1 - 3\cos^2\theta) I_z s_z, \qquad (14.11.3)$$

and its eigenvalues are

$$\begin{aligned} \mathscr{E}_{hf} &= (\mu_0 g_e \gamma_e \gamma_N / 4\pi r^3)(1 - 3\cos^2\theta) m_I m_s \hbar^2 \\ &= -(g_e g_N \mu_B \mu_N \mu_0 / 4\pi r^3)(1 - 3\cos^2\theta) m_I m_s. \end{aligned}$$

If the electron occupies an orbital ψ on the nucleus of interest, the average energy of interaction is

$$\mathscr{E}_{\text{hf}} = -\left\{(g_e g_N \mu_B \mu_N \mu_0/4\pi)\int \psi^2(1/r^3)(1-3\cos^2\theta)\,d\tau\right\} m_s m_I. \tag{14.11.4}$$

If that orbital is an s-orbital, the angular integration can be done immediately, with the result that the integral over $1-3\cos^2\theta$ vanishes. Hence *an electron in an s-orbital has no net dipole–dipole interaction with its nucleus.* If the unpaired electron occupies a p-orbital on the nucleus in question, the angular dependence of the wavefunction upsets the angular integration and the dipolar interaction does not average to zero. For instance, if the electron occupies a p_z-orbital, we have $\psi_{p_z} = f(r)\cos\theta$ and

$$\begin{aligned}\int \psi^2(1/r^3)(1-3\cos^2\theta)\,d\tau &= \int_0^{2\pi} d\phi \int_0^{\infty} (1/r^3) f^2(r) r^2\,dr \\ &\quad \times \int_0^{\pi} \cos^2\theta(1-3\cos^2\theta)\sin\theta\,d\theta \\ &= -(8\pi/15)\int_0^{\infty} (f^2/r)\,dr,\end{aligned} \tag{14.11.5}$$

and the radial integral can be evaluated by substituting the appropriate form of the radial part of the orbital (e.g. hydrogen-like, Slater, or SCF-HF). Typical interaction energies calculated in this way correspond to the electron experiencing a nuclear hyperfine field of around 10 mT (100 gauss) or so.

The second hyperfine mechanism is the *Fermi contact interaction.* It is only an approximation that the magnetic field from a nucleus is characteristic of a point magnetic dipole. If we imagine the nucleus to have non-zero radius, then the magnetic field it generates is characteristic of a loop of current, Fig. 14.17. At large distances this field is indistinguishable from a field generated by a point dipole. No electron, except an s-electron, ever approaches the nucleus sufficiently closely for the difference to be significant, but since s-electrons have non-zero probability density of being at the nucleus, for them the deviation from a point dipole field must be taken into account. Furthermore, the field close to the loop does not average to zero over a spherical distribution, Fig. 14.17, and so this contribution to the total magnetic interaction does not vanish.

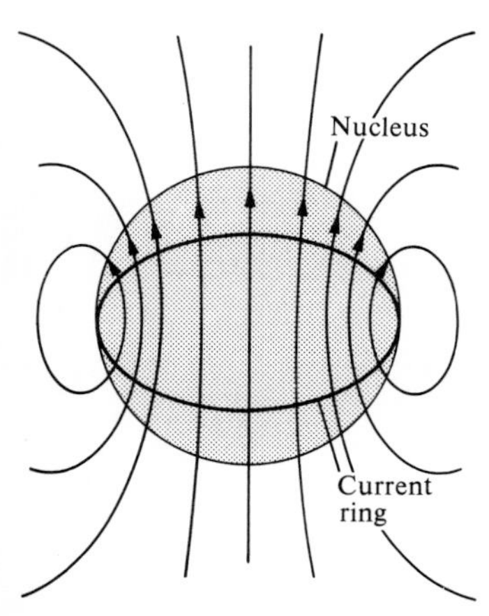

Fig. 14.17. An electron close to a magnetic nucleus discerns that its magnetic field is not point-dipolar.

This simple picture may be expressed quantitatively in terms of the Dirac equation. It then turns out that the energy of interaction between an electron and a nucleus by the contact mechanism is expressed by the following term in the true hamiltonian:

$$H_{\text{hf(contact)}} = -(2\mu_0/3) g_e \gamma_e \gamma_N \delta(\mathbf{r}_N)\mathbf{s}\cdot\mathbf{I}, \tag{14.11.6}$$

where $\delta(\mathbf{r}_N)$ is the δ-function. A δ-function has the property that

$$\int f(x)\,\delta(x)\,dx = f(0), \tag{14.11.7}$$

and so it picks out the value of the function at $x = 0$. Therefore, when we evaluate the first-order correction to the energy we find

$$\begin{aligned}\mathscr{E}^{(1)} &= -(2\mu_0/3)g_e\gamma_e\gamma_N \int \psi^*\,\delta(\mathbf{r}_N)\psi\mathbf{s}\cdot\mathbf{I}\,d\tau \\ &= -(2\mu_0/3)g_e\gamma_e\gamma_N\,|\psi(0)|^2\,\hbar^2 m_s m_I.\end{aligned} \tag{14.11.8}$$

The same first-order energy is obtained by selecting a spin hamiltonian of the form

$$H^{(\text{spin})}_{\text{hf(contact)}} = -(2\mu_0/3)g_e\gamma_e\gamma_N\,|\psi(0)|^2\,\mathbf{s}\cdot\mathbf{I} \tag{14.11.9}$$

where $|\psi(0)|^2$ is the probability density of finding the electron at the nucleus. (If the existence of spin is accepted, then it is unnecessary to turn to the Dirac equation. A careful extraction of the magnetic field corresponding to the nuclear vector potential delivers a contact term. This is illustrated in Appendix 22.) In the case of a hydrogen atom, $\psi(0)^2 = (1/\pi a_0^3)$ for the 1s-orbital, and so the energy contribution from the contact interaction is reproduced by the hamiltonian

$$H^{(\text{spin})}_{\text{hf(contact)}} = -(2\mu_0/3)g_e\gamma_e\gamma_N(1/\pi a_0^3)I_z s_z \tag{14.11.10}$$

if only the z-components of the spins are relevant (when the applied field is strong). The eigenvalues are therefore

$$\mathscr{E} = (2g_e g_N\mu_B\mu_N\mu_0/3\pi a_0^3)m_I m_s; \qquad \mathscr{E}/h = (1422.8\text{ MHz})m_I m_s. \tag{14.11.11}$$

This corresponds to the 1s-electron experiencing a nuclear hyperfine field of about 0.5 mT (500 gauss).

The effect of hyperfine interactions on an e.s.r. spectrum can be calculated on the basis of first-order perturbation theory, although when its effect is strong this may be insufficient. The overall form of the spin hamiltonian is

$$H^{(\text{spin})} = g(\mu_B/\hbar)Bs_z + A(\theta)I_z s_z + CI_z s_z \tag{14.11.12a}$$

where

$$A(\theta) = -(g_e g_N\mu_B\mu_N\mu_0/4\pi\hbar^2)\left\langle\left\{\frac{1-3\cos^2\theta}{r^3}\right\}\right\rangle \tag{14.11.12b}$$

$$C = (2g_e g_N\mu_B\mu_N\mu_0/3\hbar^2)\psi(0)^2. \tag{14.11.12c}$$

The eigenstates of this hamiltonian are $|m_s m_I\rangle$, and the corresponding eigenvalues are

$$E = g\mu_B Bm_s + A(\theta)m_I m_s + Cm_I m_s = \{g\mu_B B + [A(\theta) + C]m_I\}m_s. \tag{14.11.13}$$

The energy separation is the difference of such terms for $m_s = \pm\frac{1}{2}$, and so the resonance condition is

$$\hbar\omega = g\mu_B B + \{A(\theta) + C\}m_I, \quad \text{or} \quad g\mu_B B = \hbar\omega - \{A(\theta) + C\}m_I. \tag{14.11.14}$$

Therefore, for a given applied frequency, there will be $2I+1$ field values ($m_I = I, I-1, \ldots, -I$) which satisfy this condition, and therefore there will be that number of equally spaced lines in the spectrum. The spacing of the lines, $C + A(\theta)$, depends on the orientation of the radical, and in a fluid (where the average value of $A(\theta)$ is zero) the splitting is simply C. From the measurement of A and C it is possible to map the distribution of the unpaired electron through the molecule.

Example. Use Slater atomic orbitals to estimate the magnitude of the dipolar interaction between a nitrogen nucleus and an electron in a $2p_z$-orbital with the applied field (a) parallel, (b) perpendicular to the orbital's axis.

- *Method.* The STOs are specified on p. 234. For valence shell nitrogen orbitals $Z^* = 3.90$. Evaluate eqn (14.11.12b) with $g(^{14}N) = 0.403\,56$. Ensure that θ is defined appropriately. When the field lies parallel to the orbital's axis, the θ in $(1 - 3\cos^2\theta)$ is the same as the θ in $p_z \propto \cos\theta$. When the field lies perpendicular to the axis, continue to use $(1 - 3\cos^2\theta)$ for the interaction but regard the p-orbital as a p_x-orbital using $p_x \propto \sin\theta\cos\phi$.
- *Answer.* (a) $\psi = (Z^{*5}/32\pi a_0^5)^{\frac{1}{2}} r \cos\theta e^{-Z^*r/2a_0}$.

$$\langle(1/r^3)(1 - 3\cos^2\theta)\rangle = (Z^{*5}/32\pi a_0^5)\int_0^{2\pi} d\phi \times \int_0^{\pi} \cos^2\theta(1 - 3\cos^2\theta)\sin\theta\, d\theta \int_0^{\infty} re^{-Z^*r/a_0}\, dr$$
$$= -Z^{*3}/30a_0^3.$$

(b) $\psi = (Z^{*5}/32\pi a_0^5)^{\frac{1}{2}} r \sin\theta\cos\phi e^{-Z^*r/2a_0}$.

$$\langle(1/r^3)(1 - 3\cos^2\theta)\rangle = +Z^{*3}/60a_0^3.$$

Therefore (a) $\hbar^2 A = (g_e g_N \mu_B \mu_N \mu_0/120\pi a_0^3)Z^{*3} = (2.11 \times 10^{-27}\ \text{J})g_N Z^{*3}$.

(b) $\hbar^2 A = -(g_e g_N \mu_B \mu_N \mu_0/240\pi a_0^3)Z^{*3} = -(1.05 \times 10^{-27}\ \text{J})g_N Z^{*3}$.

The interaction constants are normally expressed as frequencies; on dividing by h we have

(a) $\hbar^2 A/h = (3.18\ \text{MHz})g_N Z^{*3}$; (b) $\hbar^2 A/h = (-1.59\ \text{MHz})g_N Z^{*3}$.

Since $g_N = 0.403\,56$ and $Z^* = 3.90$,

(a) $\hbar^2 A/h = 76.2$ MHz; (b) $\hbar^2 A/h = -38.1$ MHz.

- *Comment.* The values obtained from SCF orbitals are 134 MHz and -67 MHz respectively. Slater orbitals are not very good close to the nucleus (which is where $1/r^3$ is important).

14.12 Nuclear spin–spin coupling

There are several interactions in molecules that may give rise to interactions between nuclear magnetic moments. One possibility is the direct dipole–dipole interaction of the kind discussed for electrons. This interaction is important in solids, but in liquids it averages to zero as a result of the molecular rotation. In *liquid crystals* molecules may rotate freely only around a single axis, and so the dipolar interaction does not average completely to zero. The location and separation of magnetic nuclei can then be inferred from the remnant of dipolar interaction that survives to affect the spectrum.

There are also indirect coupling mechanisms involving the transmission of spin information through the electrons. One such mechanism is illustrated in Fig. 14.18. The first step is a hyperfine interaction between one nucleus and an electron: the interaction has the effect of favouring one orientation of the electron spin rather than the other. The other electron in the bond must have the opposite spin. It bathes the other nucleus. The latter responds to the spin orientation of the electron through another hyperfine interaction. Consequently one of its orientations is favoured more than the other. As a result there is an energy difference between the different relative spin orientations of the nuclei, and a term of the form $J\mathbf{I}_1 \cdot \mathbf{I}_2$ can be expected in the spin hamiltonian. ($\mathbf{I}_1 \cdot \mathbf{I}_2$ can be thought of temporarily as a classical scalar product: when $\mathbf{I}_1$ and $\mathbf{I}_2$ are parallel, $\mathbf{I}_1 \cdot \mathbf{I}_2$ is positive; when they are antiparallel, it is negative; therefore terms like $\mathbf{I}_1 \cdot \mathbf{I}_2$ in hamiltonians reproduce the energy dependence of the relative orientations of angular momenta. We have already seen this in the case of $\mathbf{l} \cdot \mathbf{s}$ for spin–orbit coupling and $\mathbf{s} \cdot \mathbf{I}$ for hyperfine interactions.)

The explicit calculation runs as follows. The hyperfine mechanism must first be specified. In many cases the contact interaction is the more important, and so we shall confine attention to it. (The dipolar interaction can make a contribution, even though the molecule may be rotating, because two such interactions, one with each nucleus, give rise to the square term $(1-3\cos^2\theta)^2$, which does not average to zero.) The contact interactions with the two nuclei take the form

$$H^{(1)} = -(2\mu_0/3)g_e\gamma_e\left\{\gamma_A \sum_i \mathbf{I}_A \cdot \mathbf{s}_i\delta(\mathbf{r}_{iA}) + \gamma_B \sum_i \mathbf{I}_B \cdot \mathbf{s}_i\delta(\mathbf{r}_{iB})\right\} \tag{14.12.1}$$

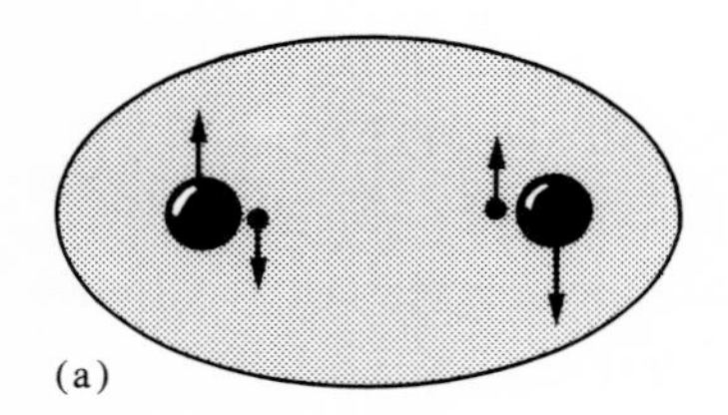
(a)

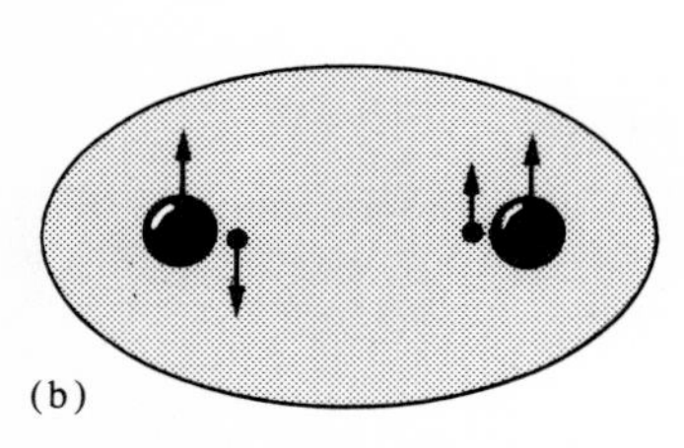
(b)

Fig. 14.18. (a) Low energy, (b) higher energy relative nuclear spin orientations.

where the sum over i is the sum over all the electrons (each one in the molecule may have a hyperfine interaction with each of the nuclei A and B). When this operator is integrated over the wavefunctions of the molecule, the δ-functions pick out the value of $|\psi(0)|^2$ at each nucleus for each electron, and so we get the familiar form of the spin hamiltonian for contact interactions with the two nuclei. For simplicity of notation we write

$$\mathbf{A} = \sum_i \mathbf{s}_i\,\delta(\mathbf{r}_{iA}), \qquad \mathbf{B} = \sum_i \mathbf{s}_i\,\delta(\mathbf{r}_{iB});$$

then the hamiltonian becomes

$$H^{(1)} = -(2\mu_0/3)g_e\gamma_e\{\gamma_A\mathbf{I}_A \cdot \mathbf{A} + \gamma_B\mathbf{I}_B \cdot \mathbf{B}\}. \tag{14.12.2}$$

When this first-order perturbation hamiltonian is used to calculate the second-order correction to the energy, it gives rise to four terms of the form $\mathbf{I} \ldots \mathbf{I}$. We are interested in the contribution to the spin hamiltonian for the spin–spin coupling between I_A and I_B, and therefore seek a term of the form $J\mathbf{I}_A \cdot \mathbf{I}_B$. Therefore we retain only the two terms $\mathbf{I}_A \ldots \mathbf{I}_B$ and $\mathbf{I}_B \ldots \mathbf{I}_A$. The second-order energy then takes the form

$$H^{(\text{spin})} = -2(2\mu_0/3)^2 g_e^2 \gamma_e^2 \gamma_A \gamma_B \sum_n{}' \left\{ \frac{\mathbf{I}_A \cdot \langle 0|\,\mathbf{A}\,|n\rangle\langle n|\,\mathbf{B}\,|0\rangle \cdot \mathbf{I}_B}{\mathscr{E}_n - \mathscr{E}_0} \right\}.$$

We make the usual replacement $\mathscr{E}_n - \mathscr{E}_0 = \Delta_{n0}$. The spherical average of this expression is the relevant quantity for a molecule freely rotating in fluid solution. The average value of $\mathbf{I}_A \cdot \mathbf{AB} \cdot \mathbf{I}_B$ is $\frac{1}{3}\mathbf{I}_A \cdot \mathbf{I}_B \mathbf{A} \cdot \mathbf{B}$, and so

$$H^{(\text{spin})} = J\mathbf{I}_A \cdot \mathbf{I}_B,$$

$$J = -\tfrac{2}{3}(2\mu_0/3)^2 g_e^2 \gamma_e^2 \gamma_A \gamma_B \sum_n{}' \{\langle 0|\,\mathbf{A}\,|n\rangle \cdot \langle n|\,\mathbf{B}\,|0\rangle/\Delta_{n0}\}. \tag{14.12.3}$$

This is the basic expression for the calculation of the spin–spin coupling constant J, but it is obviously much too complicated to talk about in this form (although in a computation of J this is the expression that would have to be manipulated). The major difficulty lies in the effects of the operators $\mathbf{A}$ and $\mathbf{B}$. These have the form of the operators Ω discussed on p. 332, and which were shown there to have the effect of mixing in *triplet* excited states. Therefore the excitation energy Δ_{n0} is a triplet excitation energy, which adds another complication. We can proceed with a discussion of the factors influencing the properties of J only by making quite drastic approximations, and so as not merely to come to a stop, we introduce the closure approximation. Then

$$J \approx -\tfrac{2}{3}(2\mu_0/3)^2 g_e^2 \gamma_e^2 \gamma_A \gamma_B \langle 0|\,\mathbf{A} \cdot \mathbf{B}\,|0\rangle/\Delta_T \tag{14.12.4}$$

where Δ_T is the mean triplet excitation energy.

Consider a two-electron bond between A and B. Then

$$\begin{aligned}\mathbf{A} \cdot \mathbf{B} &= \sum_{i,j} \mathbf{s}_i \cdot \mathbf{s}_j \delta(\mathbf{r}_{Ai})\delta(\mathbf{r}_{Bj}) \\ &= s_1^2 \delta(\mathbf{r}_{A1})\delta(\mathbf{r}_{B1}) + s_2^2 \delta(\mathbf{r}_{A2})\delta(\mathbf{r}_{B2}) \\ &\quad + \mathbf{s}_1 \cdot \mathbf{s}_2 \{\delta(\mathbf{r}_{A1})\delta(\mathbf{r}_{B2}) + \delta(\mathbf{r}_{A2})\delta(\mathbf{r}_{B1})\}.\end{aligned}$$

The first two terms give zero when integrated over the wavefunction because a single electron cannot simultaneously be at two different nuclei. The action of $\mathbf{s}_1 \cdot \mathbf{s}_2$ in the remaining terms is found by writing it as

$$\mathbf{s}_1 \cdot \mathbf{s}_2 = \tfrac{1}{2}(\mathbf{s}_1 + \mathbf{s}_2)^2 - \tfrac{1}{2}s_1^2 - \tfrac{1}{2}s_2^2 = \tfrac{1}{2}(S^2 - s_1^2 - s_2^2). \tag{14.12.5}$$

When this operates on the ground-state singlet term of the molecule,

S^2 vanishes, and s_1^2 and s_2^2 each give $\frac{1}{2}(\frac{1}{2}+1)\hbar^2=\frac{3}{4}\hbar^2$. Therefore

$$\mathbf{s}_1\cdot\mathbf{s}_2\,|0\rangle=-\tfrac{3}{4}\hbar^2\,|0\rangle. \qquad (14.12.6)$$

The expression for the coupling constant has therefore been reduced to

$$J\approx\tfrac{1}{2}(2\mu_0/3)^2g_e^2\gamma_e^2\gamma_A\gamma_B\hbar^2\langle 0|\,\delta(\mathbf{r}_{A1})\delta(\mathbf{r}_{B2})+\delta(\mathbf{r}_{B1})\delta(\mathbf{r}_{A2})\,|0\rangle/\Delta_T.$$

Now suppose that both electrons occupy the MO $\psi=c_A\psi_A+c_B\psi_B$, so that the overall space function for the ground state is

$$\psi_0(1,2)=\{c_A\psi_A(1)+c_B\psi_B(1)\}\{c_A\psi_A(2)+c_B\psi_B(2)\}.$$

When this is inserted and the trivial δ-function integrations carried out (but with all two-centre terms neglected) we find

$$J\approx(2\mu_0g_e\mu_B/3)^2\gamma_A\gamma_B\,|\psi_A(0)|^2\,|\psi_B(0)|^2c_A^2c_B^2/\Delta_T. \qquad (14.12.7)$$

where $\psi_A(0)$ is the amplitude of ψ_A evaluated at the nucleus A, and likewise for B. Since only s-orbitals have non-vanishing amplitude at their nucleus the orbitals ψ_A and ψ_B must each contain s-character (e.g. be sp^n-hybrids or pure s-orbitals).

There are other mechanisms of nuclear spin–spin coupling that do not involve the electron spins. For example, the nucleus can induce orbital currents which couple with the other nucleus. Such mechanisms are discussed in the books referred to below. At this point, though, it seems to me appropriate to pause.

Further reading

Background electromagnetic theory

Introduction to electromagnetic fields and waves. P. Lorrain and D. R. Corson; W. H. Freeman, San Francisco, 2nd edition, 1970.

Electricity and magnetism. B. I. Bleaney and B. Bleaney; Oxford University Press, 3rd edition, 1976.

Electromagnetic fields. R. K. Wangsness; Wiley, New York, 1979.

Magnetic susceptibilities

Physical chemistry. P. W. Atkins; Oxford University Press and W. H. Freeman, San Francisco, 2nd edition, 1982.

Introduction to magnetochemistry. A. Earnshaw; Academic Press, New York, 1968.

The theory of the electric and magnetic properties of molecules. D. W. Davies; Wiley, New York, 1967.

The theory of electric and magnetic susceptibilities. J. H. Van Vleck, Oxford University Press, 1932.

Magnetic resonance parameters

Introduction to magnetic resonance. A. Carrington and A. D. McLachlan; Harper and Row, New York, 1967.

Principles of magnetic resonance. C. P. Slichter; Springer, Berlin, 2nd edition, 1978.

Theoretical foundations of electron spin resonance. J. E. Harriman, Academic Press, New York, 1978.

Problems

14.1. Calculate the spin contribution to the mass magnetic susceptibility of hydrogen atoms at 298 K.

14.2. Consider a molecule in which there is an excited state at an energy Δ above the non-degenerate ground state. Show that the angular momentum is no longer completely quenched when a magntic field is present. *Hint.* Review the argument on p. 379, and consider how it is modified when the ground state is perturbed.

14.3. Calculate the average values of S_z^2, S_xS_z, and S_z^4 for a state with spin quantum number S and with all M_s states equally occupied. *Hint.* Use $\sum_{r=1}^{n} r^2 = \frac{1}{6}n(n+1)(2n+1)$. For the sum over higher powers, see M. Abramowitz and I. A. Stegun, *Handbook of mathematical functions*, Dover (1965), Chapter 23, especially Section 23.1.4.

14.4. The average value of S_z^2 can also be evaluated very simply by noting that in the absence of fields $\langle S_x^2\rangle = \langle S_y^2\rangle = \langle S_z^2\rangle$. Find the average value of S_z^2 in this way.

14.5. Sketch the form of the vector function $\mathbf{V} = x\hat{\mathbf{k}} - z\hat{\mathbf{i}}$ and calculate its divergence and curl.

14.6. Confirm that the vector potential $\mathbf{A} = \frac{1}{2}\mathbf{B} \wedge \mathbf{r}$ describes a uniform magnetic field $\mathbf{B}$, and show that it has zero divergence. Find expressions for vector potentials corresponding to a uniform magnetic field (a) parallel to the x-axis, (b) along the direction of the unit vector (1, 1, 1). Find an expression for A^2 for the vector potential $\mathbf{A}$, and evaluate it for the two special cases.

14.7. Take a vector potential of the form in eqn (14.2.10) and find expressions for the hamiltonian in the presence of the corresponding magnetic field but for general values of the gauge transformation parameter λ. Is it possible to choose a value of λ such that $H^{(2)}$ is absent?

14.8. Calculate the contribution to the mass magnetic susceptibility of (a) a 1s-electron, (b) a 2s-electron, taking Slater orbitals. Specialize to (i) the hydrogen atom, (ii) the carbon atom.

14.9. Estimate the contribution to the diamagnetic susceptibility of a 2p-electron when the field is (a) parallel, (b) perpendicular to the axis. Use Slater orbitals, and then specialize to the carbon atom. What is the mean value?

14.10. Confirm the remark made in the footnote on p. 388, that the current density and the charge density satisfy the continuity equation, $\partial\rho/\partial t = -\nabla \cdot \mathbf{j}$.

14.11. An electron occupies one of a doubly degenerate pair of d-orbitals, and its orbital angular momentum corresponds to $\Lambda = +2$. Compute an expression for the current density and plot it for a 3d Slater atomic orbital on carbon (take $Z^* \approx 1$).

14.12. Plot contour diagrams of the type shown in Fig. 14.7 for planes parallel to the equatorial plane of the hydrogen atom at heights 0, a_0, and $2a_0$ above the nucleus.

14.13. Calculate the form of the diamagnetic and paramagnetic contributions to the current density induced by a magnetic field in the z-direction when the electron occupies (a) a $3d_{xy}$-orbital, (b) a $3d_{x^2-y^2}$-orbital. Suppose that all the degeneracies have been removed by a crystal field. Sketch the form of the current in the equatorial plane. *Hint.* For the diamagnetic contribution, follow Section 14.4, and for the paramagnetic, follow Section 14.5.

14.14. Sketch the form of the diamagnetic and paramagnetic current densities for an electron in (a) a 2s-orbital, (b) a $3p_z$-orbital.

14.15. Consider a nitrogen oxide molecule (NO) in which the unpaired electron occupies a $2p\pi^*$-orbital formed from a linear combination of the nitrogen and oxygen 2p-orbitals. For simplicity, take the molecular orbital to be $(\psi_N - \psi_O)/\sqrt{2}$; we have ignored the overlap integral. Consider a plane containing both nuclei. Plot contours of the magnitude of the diamagnetic current density taking the p-orbitals to be Slater atomic orbitals: note that this is a broadside view of the current density.

14.16. Suppose that the NO molecule treated in Problem 14.15 is trapped in a matrix that removes the degeneracy of the π^*-orbitals and separates them by 1 eV. What magnetic flux density is needed to restore 10 per cent of the original current density?

14.17. Find an expression for the energy of interaction of the current density computed in Problem 14.15 with the magnetic moment of the nitrogen nucleus. What magnetic flux density does the current give rise to? *Hint.* Use eqn (14.7.14); $g(^{14}N) = 0.403\,56$ and $I(^{14}N) = 1$.

14.18. Calculate the diamagnetic contribution to the mean shielding constant of an electron in (a) a 2s-orbital, (b) a 2p-orbital. Take Slater orbitals, and then specialize to an electron of a carbon atom.

14.19. Calculate the magnitude of the paramagnetic contribution to the mean shielding constant for the same species as in Problem 14.18. Assume that the field mixes in an orbital lying about 5 eV above the orbital of interest.

14.20. The ground state of the NO_2 molecules is $^2A_1$, and that of the ClO_2 molecule is $^2B_1$. What states contribute to the deviations of the g-value of the radicals from g_e? *Hint.* The perturbation transforms as a rotation; both molecules are C_{2v}.

14.21. Long ago, in Problem 10.10, the structure of H_2O was investigated. Take the same molecular orbitals for the radical H_2O^+ and estimate its g-values.

14.22. In tetrahedral complexes of Ti^{3+} (configuration d^1), a tetragonal distortion removes the degeneracy of the d-orbitals almost completely. The lowest orbital is d_{z^2}, and the d_{xz}- and d_{yz}-orbitals, which remain degenerate, are at an energy Δ above it. Find an expression for the g-values when the field is applied along the x-, y-, and z-axes of the complex, and estimate their values. Take $\Delta \triangleq 10^4\ \text{cm}^{-1}$ and $\zeta = 154\ \text{cm}^{-1}$.

14.23. Show that the energy of dipolar interaction of two electron spin magnetic moments may be expressed as $\mathbf{S} \cdot \mathbf{D} \cdot \mathbf{S}$, where $\mathbf{S} = \mathbf{s}_1 + \mathbf{s}_2$. *Hint.* The energy is proportional to $\mathbf{s}_1 \cdot \mathbf{s}_2 - 3\mathbf{s}_1 \cdot \hat{\mathbf{r}}\hat{\mathbf{r}} \cdot \mathbf{s}_2$. Expand this in terms of its cartesian components and employ relations such as $s_{1x}^2 = \frac{1}{4}\hbar^2$, $S_x^2 = 2s_{1x}s_{2x} + \frac{1}{2}\hbar^2$, etc.

14.24. A Slater 2s-orbital has a node at the nucleus. Adopt the orthogonalization procedure mentioned in the *Example* on p. 236, which also removes the node, and find a relation for the Fermi contact interaction first for general Z^*, and then for ^{14}N.

14.25. Find an expression for the dipolar hyperfine interaction constant for an electron in a Slater $3d_{z^2}$-orbital when the field is (a) parallel, (b) perpendicular to the axis.

14.26. Estimate the spin–spin coupling constant for the molecule $^1H^2H$. *Hint.* Use eqn (14.12.7) with a simple LCAO-MO. Take $\Delta_T \triangleq 10$ eV. Express J as a frequency. The experimental value is 40 Hz.

14.27. Write the n.m.r. spin hamiltonian for a molecule containing two protons, one in an environment with shielding constant σ_A and the other with shielding constant σ_B. Let them be coupled through a constant J. Evaluate the matrix elements of the hamiltonian for the states $|m_{IA}m_{IB}\rangle$, and construct and solve the 4×4 secular determinant for the eigenvalues and eigenstates. Determine the allowed magnetic dipole transitions (they correspond to matrix elements of $I_{Ax}+I_{Bx}$) and find their relative intensities. Draw a diagram of the spectrum expected when (a) $J=0$, (b) $J \ll (\sigma_A-\sigma_B)\omega_0$, (c) $J \approx (\sigma_A-\sigma_B)\omega_0$, (d) $\sigma_A=\sigma_B$. *Hint.* The spin hamiltonian will have the form $B(1-\sigma_A)I_{Az}+B(1-\sigma_B)I_{Bz}+J\mathbf{I}_A\cdot\mathbf{I}_B$, with the appropriate constants. Construct the matrix of the hamiltonian and evaluate its eigenvalues and eigenvectors. Intensities are proportional to the squares of the matrix elements of $I_{Ax}+I_{Bx}$.

Appendices

Appendix 1 The density of states

A WAVE equation in three dimensions is the differential equation

$$(\partial^2 f/\partial x^2)+(\partial^2 f/\partial y^2)+(\partial^2 f/\partial z^2)-(1/c^2)(\partial^2 f/\partial t^2)=0, \tag{A1.1}$$

where c is the speed of propagation. We consider waves trapped in a cubic box of side L and volume $V=L^3$: the shape and size can be shown to be unimportant. On trying a solution of the form $f=A\sin(n_1\pi x/L)\sin(n_2\pi y/L)\sin(n_3\pi z/L)$, with A a function only of time and the n_i integers, the equation becomes

$$(n_1^2+n_2^2+n_3^2)(\pi^2/L^2)A-(1/c^2)(\mathrm{d}^2A/\mathrm{d}t^2)=0.$$

This corresponds to the equation for a wave of frequency ω, (i.e. it has a solution of the form $A\propto\cos\omega t$) where

$$\omega^2=(c^2\pi^2/L^2)(n_1^2+n_2^2+n_3^2). \tag{A1.2}$$

The total number of modes with frequencies less than or equal to ω is the number of ways the integers n_1, n_2, and n_3 can be chosen to satisfy the inequality

$$n_1^2+n_2^2+n_3^2\leq\omega^2L^2/c^2\pi^2. \tag{A1.3}$$

If we imagine a three-dimensional space with the axes labelled n_1, n_2, and n_3, Fig. A1.1, then the inequality defines the interior of a sphere of radius $n=\omega L/c\pi$, or rather the octant of the sphere corresponding to the positive values of the integers (the octant rather than the whole is the basis of Jeans's correction of Rayleigh's original calculation). The number of ways of choosing the integers to fill this octant is equal to its volume because each triple of integers (n_1, n_2, n_3) defines a unit cube. Its volume is $\frac{1}{8}(\frac{4}{3}\pi n^3)=\frac{1}{6}(\omega^3/c^3\pi^2)V$. The number density of oscillators, the number per unit volume, is therefore $\mathcal{N}=\frac{1}{6}(\omega^3/c^3\pi^2)$. The number density of oscillators with frequencies in the range ω to $\omega+\mathrm{d}\omega$ is the difference of the number densities at the frequencies $\omega+\mathrm{d}\omega$ and ω:

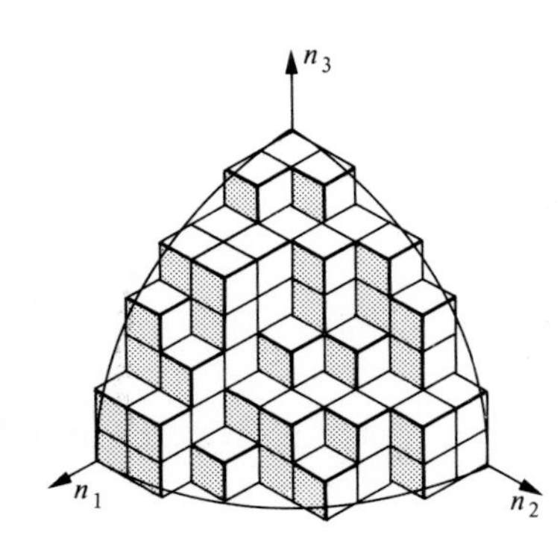

Fig. A1.1. The volume of the octant and the number of states.

$$\begin{aligned}\mathrm{d}\mathcal{N}&=\mathcal{N}(\omega+\mathrm{d}\omega)-\mathcal{N}(\omega)=\tfrac{1}{6}(1/c^3\pi^2)\{(\omega+\mathrm{d}\omega)^3-\omega^3\}\\&=\tfrac{1}{6}(1/c^3\pi^2)\{\omega^3+3\omega^2\,\mathrm{d}\omega+\ldots-\omega^3\}=\tfrac{1}{2}(\omega^2/c^3\pi^2)\,\mathrm{d}\omega.\end{aligned} \tag{A1.4}$$

(Only first-order infinitesimals are retained.) Since $\omega=2\pi\nu$, it follows that $\mathrm{d}\omega=2\pi\,\mathrm{d}\nu$, and so

$$\mathrm{d}\mathcal{N}=(4\pi\nu^2/c^3)\,\mathrm{d}\nu. \tag{A.15}$$

An electromagnetic wave may also have either of two polarizations, and so the total number density is twice the value just quoted. Hence we find

$$\mathrm{d}\mathcal{N} = (8\pi\nu^2/c^3)\,\mathrm{d}\nu, \tag{A1.6}$$

the result used in the text.

Appendix 2 Action

Hamilton's principle asserts that *the path taken by a particle is the one that involves the least action.* The *action*, S, can be expressed as an integral of another function, the *lagrangian*, L, which depends on the position (x) and the speed ($\dot{x}$) of the particle at each point along its path:

$$\textit{Action}: \; S = \int_{t_1}^{t_2} L(x, \dot{x})\,\mathrm{d}t. \tag{A2.1}$$

For the systems of interest to us at this stage, L is the difference of the kinetic and potential energies of the particle:

$$\textit{Lagrangian}: \; L(x, \dot{x}) = T - V = \tfrac{1}{2}m\dot{x}^2 - V(x). \tag{A2.2}$$

We have to show that the path corresponding to minimum action is the one that the particle would take if it followed Newton's equation of motion at each point.

Suppose that x and $\dot{x}$ are varied a little at each point of the particle's path (except at its fixed end points). The lagrangian changes by δL, and so the action, its integral, changes by

$$\delta S = \int_{t_1}^{t_2} \delta L(x, \dot{x})\,\mathrm{d}t. \tag{A2.3}$$

Since L depends on x and $\dot{x}$ a change in these quantities leads to a change in L given by

$$\delta L = (\partial L/\partial x)\,\delta x + (\partial L/\partial \dot{x})\,\delta \dot{x}.$$

Therefore, on integrating by parts,

$$\begin{aligned}\delta S &= \int_{t_1}^{t_2} (\partial L/\partial x)\,\delta x\,\mathrm{d}t + \int_{t_1}^{t_2} (\partial L/\partial \dot{x})(\mathrm{d}\,\delta x/\mathrm{d}t)\,\mathrm{d}t \\ &= \int_{t_1}^{t_2} (\partial L/\partial x)\,\delta x\,\mathrm{d}t + \left\{ (\partial L/\partial \dot{x})\,\delta x \Big|_{t_1}^{t_2} - \int_{t_1}^{t_2} [(\mathrm{d}/\mathrm{d}t)(\partial L/\partial \dot{x})]\,\delta x\,\mathrm{d}t \right\}.\end{aligned}$$

Since the end points of the path are fixed, the middle term is zero (δx is zero at the end points). Therefore the change in S is

$$\delta S = \int_{t_1}^{t_2} \{(\partial L/\partial x) - (\mathrm{d}/\mathrm{d}t)(\partial L/\partial \dot{x})\}\,\delta x\,\mathrm{d}t.$$

According to Hamilton's principle, the action is a minimum for the actual path; hence any small variation of the path corresponds to $\delta S = 0$, the usual condition for the extremum (maximum or minimum) of a function. In this case $\delta S = 0$ is obtained for small but otherwise arbitrary variations δx only if the integrand vanishes. Thus we arrive at the

Euler–Lagrange equations of motion: $(\partial L/\partial x) - (\mathrm{d}/\mathrm{d}t)(\partial L/\partial \dot{x}) = 0$ (A2.4)

which have to be satisfied everywhere along the true path.

When we test this equation on the form of the lagrangian given in eqn (A2.2) we have

$$(\partial L/\partial x) = -\mathrm{d}V/\mathrm{d}x = F; \qquad (\mathrm{d}/\mathrm{d}t)(\partial L/\partial \dot{x}) = (\mathrm{d}/\mathrm{d}t)m\dot{x} = m\ddot{x};$$

hence $F - m\ddot{x} = 0$, which is Newton's equation. In other words, Hamilton's principle is equivalent to Newton's equation.

Consider now the total derivative of the action with respect to the time t_2: this tells us how the action varies when the location of the end point changes:

$$\mathrm{d}S/\mathrm{d}t_2 = (\partial S/\partial t_2) + (\partial S/\partial x_2)(\partial x_2/\partial t_2). \tag{A2.5}$$

From eqn (A2.1)

$$\mathrm{d}S/\mathrm{d}t_2 = L, \text{ evaluated at the end point,}$$

and from eqn (A2.4)

$$(\partial S/\partial x_2) = \int_{t_1}^{t_2} (\partial L/\partial x_2)\,\mathrm{d}t = \int_{t_1}^{t_2} (\mathrm{d}/\mathrm{d}t)(\partial L/\partial \dot{x}_2)\,\mathrm{d}t = (\partial L/\partial \dot{x}_2),$$

also evaluated at the same end point. Since $(\partial L/\partial \dot{x}) = m\dot{x}$, we arrive at

$$L = (\partial S/\partial t_2) + m\dot{x}^2 = (\partial S/\partial t_2) + 2T.$$

Then as $T - V = L$,

$$T + V = -(\partial S/\partial t_2).$$

The sum $T + V$ is the total energy of the system, E; and so

$$E = -(\partial S/\partial t_2), \tag{A2.6}$$

as used in the text.

Appendix 3 The motion of wavepackets

The time-dependent form of the wavefunction of a particle of mass m in a state of linear momentum $p = k\hbar$ is given in eqn (3.1.6) as

$$\Psi_k(x, t) = A\mathrm{e}^{\mathrm{i}kx - \mathrm{i}E(k)t/\hbar}, \qquad E(k) = k^2\hbar^2/2m. \tag{A3.1}$$

Such a particle may be regarded as having a *phase velocity* $v_p = p/m = k\hbar/m = h/\lambda m$. The wavepacket of an imprecisely prepared system is the superposition mentioned in eqn (3.1.7):

$$\Psi(x, t) = \int g(k)\Psi_k(x, t)\,\mathrm{d}k. \qquad \text{(A3.2)}$$

We suppose that the *shape function* $g(k)$ peaks sharply around k_0 and falls to zero for values of $|k - k_0|$ significantly larger than Γ. For example, $g(k)$ might be the gaussian function

$$g(k) = N\mathrm{e}^{-(k-k_0)^2/2\Gamma^2}, \qquad N = 1/(2\pi)^{\frac{1}{2}}\Gamma, \qquad \text{(A3.3)}$$

N being a normalization constant. Because g is peaked sharply around k_0 the only values of $E(k)$ that contribute significantly to the integral are those near $E(k_0)$. Therefore we may expand $E(k)$ as a Taylor series, and discard all except the first few terms:

$$\begin{aligned} E(k) &= E(k_0) + (k - k_0)(\mathrm{d}E/\mathrm{d}k)_{k_0} + \tfrac{1}{2}(k - k_0)^2(\mathrm{d}^2E/\mathrm{d}k^2)_{k_0} + \ldots \\ &= E(k_0) + (k - k_0)v_g\hbar + \tfrac{1}{2}(k - k_0)^2 w_g\hbar + \ldots \end{aligned}$$

where $v_g = (\mathrm{d}E/\mathrm{d}k)_{k_0}/\hbar$ is called the *group velocity* (for reasons that will shortly become clear), and $w_g = (\mathrm{d}^2E/\mathrm{d}k^2)_{k_0}/\hbar$. In the present case $v_g = k_0\hbar/m = p_0/m$ and $w_g = \hbar/m$. The wavepacket therefore has the form

$$\Psi(x, t) = AN\int \{\mathrm{e}^{-(k-k_0)^2/2\Gamma^2 + \mathrm{i}kx - \mathrm{i}E(k_0)t/\hbar - \mathrm{i}(k-k_0)v_g t - \frac{1}{2}\mathrm{i}(k-k_0)^2 w_g t - \ldots}\}\,\mathrm{d}k.$$

If we wait only for short times we can neglect the term $\frac{1}{2}(k - k_0)^2 w_g t$ relative to $\frac{1}{2}(k - k_0)^2/\Gamma^2$ (the precise condition for the validity of this is that $\Gamma^2 w_g t \ll 1$). Then

$$\Psi(x, t) \approx AN\mathrm{e}^{\mathrm{i}k_0x - \mathrm{i}E(k_0)t/\hbar}\int \mathrm{e}^{-(k-k_0)^2/2\Gamma^2 + \mathrm{i}(k-k_0)(x - v_g t)}\,\mathrm{d}k. \qquad \text{(A3.4)}$$

At $t = 0$ the packet has the form

$$\Psi(x, 0) = AN\mathrm{e}^{\mathrm{i}k_0x}\int \mathrm{e}^{-(k-k_0)^2/2\Gamma^2 + \mathrm{i}(k-k_0)x}\,\mathrm{d}k = A\mathrm{e}^{\mathrm{i}k_0x}G(x) \qquad \text{(A3.5)}$$

where

$$G(x) = N\int \mathrm{e}^{-(k-k_0)^2/2\Gamma^2 + \mathrm{i}(k-k_0)x}\,\mathrm{d}k = \mathrm{e}^{-x^2\Gamma^2/2}. \qquad \text{(A3.6)}$$

The probability density of finding the particle at x at $t = 0$ is therefore

$$|\Psi(x, 0)|^2 = A^2\,|G(x)|^2 = A^2\mathrm{e}^{-x^2\Gamma^2}, \qquad \text{(A3.7)}$$

which is a gaussian function centred on $x = 0$ with a width δx of the order of $1/\Gamma$.

At a later time the wavepacket has the form given in eqn (A3.4):

$$\Psi(x, t) = A\mathrm{e}^{\mathrm{i}k_0x - \mathrm{i}E(k_0)t/\hbar}G(x - v_g t) \qquad \text{(A3.8)}$$

and its probability density is

$$|\Psi(x,t)|^2 = A^2\,|G(x-v_{g}t)|^2 = A^2\mathrm{e}^{-(x-v_{g}t)^2\Gamma^2}. \tag{A3.9}$$

This is the same function as in eqn (A3.7) but centred on $x = v_g t$. That is, *the packet has moved without change of shape from $x = 0$ to $x = v_g t$, and is therefore moving uniformly with the group velocity v_g*. Note that $v_g = p_0/m$, which is just the classical velocity of the particle.

The conclusion is valid so long as $\Gamma^2 w_g t \ll 1$, or $\hbar\Gamma^2 t/m \ll 1$. When sufficient time has elapsed for this condition to be invalidated we may no longer neglect terms in w_g. These additional terms result in the spreading of the packet. For example, $G(x)$ becomes

$$G(x) = N\int \mathrm{e}^{-(k-k_0)^2/2\Gamma^2+\mathrm{i}(k-k_0)(x-v_g t)-\frac{1}{2}\mathrm{i}(k-k_0)^2 w_g t}\,\mathrm{d}k \tag{A3.10}$$

so that

$$|G(x)|^2 = (1/\gamma)\mathrm{e}^{-(x-v_g t)^2\Gamma^2/\gamma^2}, \qquad \gamma^2 = 1 + w_g^2 t^2 \Gamma^4 \tag{A3.11}$$

which is the same gaussian function of x as in eqn (A3.9) but with $1/\Gamma^2$ replaced by γ^2/Γ^2; and so *the width of the packet increases as time passes*, and if its initial uncertainty of location is $\delta x_0 = 1/\Gamma$, at a time t

$$\delta x = \gamma/\Gamma = (1 + w_g^2 t^2\Gamma^4)^{\frac{1}{2}}/\Gamma = (1 + w_g^2 t^2/\delta x_0^4)^{\frac{1}{2}}\delta x_0. \tag{A3.12}$$

Since $w_g = \hbar/m$,

$$\delta x = (1 + t^2\hbar^2/m^2\delta x_0^4)^{\frac{1}{2}}\delta x_0 \tag{A3.13}$$

and so the time for the uncertainty in location to spread from δx_0 to δx is

$$t = (m/\hbar)\delta x_0(\delta x^2 - \delta x_0^2)^{\frac{1}{2}}. \tag{A3.14}$$

If $\delta x \gg \delta x_0$ this simplifies to

$$t \approx (m/\hbar)\delta x_0 \delta x, \tag{A3.15}$$

This means that the location of even a static particle spreads with time, but the effect is negligible for most macroscopic objects. For instance, if $m \approx 1$ g and $\delta x_0 \approx 10^{-9}$ m, then the uncertainty in the position approaches 10^{-6} m only when t reaches about 10^{16} s, which is about 10 million years. On the other hand, for an electron located to within 1 pm, an uncertainty of position of about 0.1 nm (the size of an atom) is reached after only 10^{-18} s. That is another reason why it is not possible to speak of trajectories of microscopic particles.

Appendix 4 The harmonic oscillator

The Schrödinger equation is given in eqn (3.5.3). Make the substitutions $\lambda = 2E/\hbar\omega$ and $y = (m\omega/\hbar)^{\frac{1}{2}}x$, with $\omega = (k/m)^{\frac{1}{2}}$. It becomes

$$(\mathrm{d}^2/\mathrm{d}y^2 - y^2)\psi = -\lambda\psi. \tag{A4.1}$$

The left-hand side may be factorized by noting that

$$(\mathrm{d}/\mathrm{d}y + y)(\mathrm{d}/\mathrm{d}y - y)\psi = (\mathrm{d}^2/\mathrm{d}y^2 - y^2 - 1)\psi$$

and

$$(\mathrm{d}/\mathrm{d}y - y)(\mathrm{d}y + y)\psi = (\mathrm{d}^2/\mathrm{d}y^2 - y^2 + 1)\psi.$$

For simplicity we introduce abbreviations for $(\mathrm{d}/\mathrm{d}y \pm y)$:

$$a \equiv \mathrm{d}/\mathrm{d}y + y, \qquad a^+ \equiv \mathrm{d}/\mathrm{d}y - y \tag{A4.2}$$

when the last pair of equations become

$$(\mathrm{d}^2/\mathrm{d}y^2 - y^2)\psi = (aa^+ + 1)\psi,$$
$$(\mathrm{d}^2/\mathrm{d}y^2 - y^2)\psi = (a^+a - 1)\psi.$$

Therefore the Schrödinger equation may be written in either of the following forms:

$$aa^+\psi^{(\lambda)} = -(\lambda + 1)\psi^{(\lambda)}, \tag{A4.3a}$$
$$a^+a\psi^{(\lambda)} = -(\lambda - 1)\psi^{(\lambda)}, \tag{A4.3b}$$

where $\psi^{(\lambda)}$ signifies the wavefunction corresponding to the energy λ. It follows that

$$(a^+a - aa^+)\psi^{(\lambda)} = 2\psi^{(\lambda)}. \tag{A4.4}$$

This is true for any wavefunction, and so

$$a^+a - aa^+ = 2. \tag{A4.5}$$

Now take eqn (A4.3b) and operate on both sides with a:

$$aa^+a\psi^{(\lambda)} = -(\lambda - 1)a\psi^{(\lambda)}.$$

Using eqn (A4.5) in the form $aa^+ = a^+a - 2$ this turns into

$$a^+aa\psi^{(\lambda)} = -(\lambda - 1)a\psi^{(\lambda)} + 2a\psi^{(\lambda)};$$

that is,

$$a^+aa\psi^{(\lambda)} = -(\lambda - 3)a\psi^{(\lambda)}. \tag{A4.6}$$

The Schrödinger equation for the function corresponding to the energy $\lambda - 2$ is

$$a^+a\psi^{(\lambda-2)} = -(\lambda - 3)\psi^{(\lambda-2)}.$$

Hence $a\psi^{(\lambda)}$ must be proportional to $\psi^{(\lambda-2)}$. Therefore the function corresponding to the energy $\lambda - 2$ can be generated by the application of the operator $a = (\mathrm{d}/\mathrm{d}y + y)$ to the function corresponding to the energy λ. The process may be repeated, and in a similar way the functions corresponding to the energies $\lambda - 4$, $\lambda - 6, \ldots$ generated. This cannot be repeated indefinitely, because the energy of a harmonic oscillator cannot be negative (it is the sum of a mean kinetic energy and a mean potential energy $\frac{1}{2}kx^2$, both of which are positive); therefore there must be a minimum value of λ. Since there is no function corresponding to an energy lower than $\lambda_{\min}$, when a is applied to $\psi^{(\lambda_{\min})}$ we must get zero; therefore $a\psi^{(\lambda_{\min})} = 0$. If both sides of this equation are multiplied by a^+ we obtain $a^+a\psi^{(\lambda_{\min})} = 0$. Then using eqn (A4.3b)

$$0 = a^+a\psi^{(\lambda_{\min})} = -(\lambda_{\min} - 1)\psi^{(\lambda_{\min})};$$

consequently $\lambda_{\min} = 1$. As we have already demonstrated that λ may take integral values differing by 2, we conclude that $\lambda = 1, 3, 5, \ldots$; that is, $\lambda = 2v + 1$, where $v = 0, 1, 2, \ldots$ Since $E = \lambda(\hbar\omega/2)$, we conclude that

$$E = (2v + 1)(\hbar\omega/2) = (v + \tfrac{1}{2})\hbar\omega. \tag{A4.7}$$

Finding the wavefunctions is now simple, for it is an easy matter to show that, just as the effect of multiplying $\psi^{(\lambda)}$ by a was to generate the function corresponding to the next lower energy (which is why a is called an *annihilation operator*), when a^+ is applied, the next higher function is generated (hence a^+ is a *creation operator*). In order to prove this, take eqn (A4.3a) and multiply from the left by a^+ and proceed as in the discussion above. Instead of labelling ψ with $\lambda = 1, 3, 5, \ldots$ we shall now label it with v and write them ψ_v, $v = 0, 1, 2, \ldots$.

By applying a^+ in succession v times we generate ψ_v from the ground state ψ_0:

$$\psi_v = N_v(a^+)^v\psi_0 = N_v\{(\mathrm{d}/\mathrm{d}y - y)(\mathrm{d}/\mathrm{d}y - y)\ldots(\mathrm{d}/\mathrm{d}y - y)\}_{(v\ \text{times})}\psi_0 \tag{A4.8}$$

where N_v is a normalization constant. We therefore need to find ψ_0 because the others can be generated from it. In order to find ψ_0 recall that it satisfies $a\psi_0 = 0$; therefore the equation we must solve is simply

$$(\mathrm{d}/\mathrm{d}y + y)\psi_0 = 0, \quad \text{or} \quad (\mathrm{d}\psi_0/\psi_0) = -y\,\mathrm{d}y.$$

The solution is the gaussian function:

$$\psi_0 = N_0 e^{-y^2/2}, \tag{A4.9}$$

where N_0 is a constant to be determined by normalization. If a^+ is applied successively we obtain

$$\psi_1 = N_1 2y e^{-y^2/2}, \qquad \psi_2 = N_2(4y^2 - 2)e^{-y^2/2},$$

and so on. The polynomials in y, $(1, 2y, 4y^2 - 2, \ldots)$ are the Hermite polynomials $H_0, H_1, H_2, \ldots$ Therefore the wave functions are

$$\psi_v = N_v H_v(y)e^{-y^2/2}; \qquad y = (m\omega/\hbar)^{\frac{1}{2}}x. \tag{A4.10}$$

Appendix 5 The virial theorem

In classical mechanics the proof of the virial theorem is based on the disappearance of the time average of the time derivative of the product

$\mathbf{p} \cdot \mathbf{r}$, where $\mathbf{p}$ is the linear momentum and $\mathbf{r}$ the position. In quantum mechanics we use the same type of proof but consider the time derivative of the expectation value of the operator $\mathbf{p} \cdot \mathbf{r}$. From eqn (5.5.1)

$$(\mathrm{d}/\mathrm{d}t)\langle \mathbf{p} \cdot \mathbf{r} \rangle = (\mathrm{i}/\hbar)\langle [H, \mathbf{p} \cdot \mathbf{r}] \rangle. \tag{A5.1}$$

For simplicity we limit the discussion to a one-dimensional system when $\mathbf{p} \cdot \mathbf{r} = px$, and use the relations

$$[H, px] = [H, p]x - p[x, H],$$
$$[H, p] = [T + V, (\hbar/\mathrm{i})\,\mathrm{d}/\mathrm{d}x] = \mathrm{i}\hbar\,\mathrm{d}V/\mathrm{d}x,$$
$$[H, x] = [T + V, x] = -2(\hbar^2/2m)\,\mathrm{d}/\mathrm{d}x = -(\hbar\mathrm{i}/m)p.$$

Therefore, since $p^2/2m = T$, we have

$$(\mathrm{d}/\mathrm{d}t)\langle px \rangle = 2\langle T \rangle - \langle x\,\mathrm{d}V/\mathrm{d}x \rangle. \tag{A5.2}$$

The time average is

$$(1/\tau)\int_0^\tau (\mathrm{d}/\mathrm{d}t)\langle px \rangle\,\mathrm{d}t = (1/\tau)\int_0^\tau \{2\langle T \rangle - \langle x\,\mathrm{d}V/\mathrm{d}x \rangle\}\,\mathrm{d}t.$$

Therefore

$$(1/\tau)\langle px \rangle \Big|_{t=0}^{t=\tau} = 2\langle T \rangle - \langle x\,\mathrm{d}V/\mathrm{d}x \rangle. \tag{A5.3}$$

The left-hand side is zero, for if the motion is periodic we may choose τ to be the period, and if the motion is not periodic we may choose τ to be infinite, and so the quotient disappears because for a bounded system $\langle px \rangle_{t=\infty} - \langle px \rangle_{t=0}$ is finite and $\tau \to \infty$. Therefore

$$2\langle T \rangle = \langle x\,\mathrm{d}V/\mathrm{d}x \rangle. \tag{A5.4}$$

But the force on the particle in the system is $-\mathrm{d}V/\mathrm{d}x$, and so

$$2\langle T \rangle = -\langle xF \rangle, \tag{A5.5}$$

and in three dimensions this is the

Virial theorem: $2\langle T \rangle = -\langle \mathbf{r} \cdot \mathbf{F} \rangle$. (A5.6)

If the potential energy has the form $V = ax^s$ then eqn (A5.4) gives

$$\langle T \rangle = \tfrac{1}{2}s\langle V \rangle, \tag{A5.7}$$

the result used in the text. The theorem may be extended to operators other than $\mathbf{p} \cdot \mathbf{r}$: different choices give rise to the *hypervirial theorems*.

Appendix 6 Reduced mass

We seek to show that the motion of two particles may be separated into the motion of the centre of mass and their relative motion.

Let the masses be m_1 and m_2, their locations $\mathbf{r}_1$ and $\mathbf{r}_2$, and the total mass $m = m_1 + m_2$. Their separation is

$$\mathbf{r} = \mathbf{r}_1 - \mathbf{r}_2 \tag{A6.1}$$

and the location of the centre of mass is

$$\mathbf{R} = (1/m)(m_1\mathbf{r}_1 + m_2\mathbf{r}_2). \tag{A6.2}$$

The hamiltonian for the system, given that the potential energy depends only on their separation, is

$$H = -(\hbar^2/2m_1)\nabla_1^2 - (\hbar^2/2m_2)\nabla_2^2 + V(\mathbf{r}_1 - \mathbf{r}_2). \tag{A6.3}$$

We want to show that this can be transformed into

$$H = -(\hbar^2/2m)\nabla_{\mathbf{R}}^2 - (\hbar^2/2\mu)\nabla_{\mathbf{r}}^2 + V(\mathbf{r}) \tag{A6.4}$$

in what should be an obvious notation. If this is so, then it is easy to see that the wavefunction factorizes into $\Psi(\mathbf{R})\psi(\mathbf{r})$ and the motions are separable (using the separation of variables argument in Section 3.4).

The transformation of the potential energy is trivial. Consider the derivative $\partial/\partial x_1$ expressed in terms of the new coordinates. Since from (A6.1–2)

$$x = x_1 - x_2, \qquad X = (m_1/m)x_1 + (m_2/m)x_2$$

it follows that

$$\partial/\partial x_1 = (\partial X/\partial x_1)(\partial/\partial X) + (\partial x/\partial x_1)(\partial/\partial x) = (m_1/m)(\partial/\partial X) + (\partial/\partial x)$$
$$\partial/\partial x_2 = (m_2/m)(\partial/\partial X) - (\partial/\partial x).$$

Therefore the x-component of the sum of the two laplacians is

$$\frac{1}{m_1}\left(\frac{\partial^2}{\partial x_1^2}\right) + \frac{1}{m_2}\left(\frac{\partial^2}{\partial x_2^2}\right) = \frac{1}{m_1}\left(\frac{m_1}{m}\frac{\partial}{\partial X} + \frac{\partial}{\partial x}\right)^2 + \frac{1}{m_2}\left(\frac{m_2}{m}\frac{\partial}{\partial X} - \frac{\partial}{\partial x}\right)^2$$
$$= \frac{1}{m}\left(\frac{\partial^2}{\partial X^2}\right) + \left(\frac{1}{m_1} + \frac{1}{m_2}\right)\left(\frac{\partial^2}{\partial x^2}\right).$$

The y- and z-component can be dealt with in the same way, and so

$$(1/m_1)\nabla_1^2 + (1/m_2)\nabla_2^2 = (1/m)\nabla_{\mathbf{R}}^2 + (1/\mu)\nabla_{\mathbf{r}}^2 \tag{A6.5}$$

with

$$1/\mu = 1/m_1 + 1/m_2. \tag{A6.6}$$

This establishes the equivalence of eqns (A6.3) and (A6.4) and the separability of the motion.

Appendix 7 The factorization method

The factorization method is a way of solving a wide variety of differential equations that occur in quantum mechanics. We see it in operation in the cases of the harmonic oscillator (Appendix 4) and for angular momentum (Chapter 6). In this appendix we sketch the general scheme and illustrate it in the case of the hydrogen atom. Full details can be found in H. S. Green, *Matrix mechanics,* P. Noordhoff (1965) and in L. Infeld and T. E. Hull, *Rev. Mod. Phys.*, **23**, 21 (1951).

Suppose the hamiltonian H can be expressed in the form

$$H = H_1 = a_1^+ a_1 + \varepsilon^{(1)} \tag{A7.1}$$

where $\varepsilon^{(1)}$ is a number. Define H_2 (and in general H_j) through

$$H_2 = a_1 a_1^+ + \varepsilon^{(1)} \qquad (H_{j+1} = a_j a_j^+ + \varepsilon^{(j)} \text{ in general}) \tag{A7.2}$$

and express it in the form

$$H_2 = a_2^+ a_2 + \varepsilon^{(2)} \qquad (H_{j+1} = a_{j+1}^+ a_{j+1} + \varepsilon^{(j+1)} \text{ in general}). \tag{A7.3}$$

Then the $\varepsilon^{(j)}$ are the eigenvalues of the system. This is demonstrated as follows. Let $|\varepsilon\rangle$ be an eigenstate of H with eigenvalue ε. Consider $a_1|\varepsilon\rangle$ and its conjugate $(a_1|\varepsilon\rangle)^+ = \langle\varepsilon| a_1^+$; the quantity $\langle\varepsilon| a_1^+ a_1 |\varepsilon\rangle$ is non-negative, and so

$$\langle\varepsilon| a_1^+ a_1 |\varepsilon\rangle = \langle\varepsilon| H_1 - \varepsilon^{(1)} |\varepsilon\rangle = \varepsilon - \varepsilon^{(1)} \geq 0. \tag{A7.4}$$

Now consider $\langle\varepsilon| a_1^+ a_2^+ a_2 a_1 |\varepsilon\rangle$, which is also non-negative. Since $a_2^+ a_2 = H_2 - \varepsilon^{(2)}$, it is equal to $\langle\varepsilon| a_1^+(H_2 - \varepsilon^{(2)}) a_1 |\varepsilon\rangle$. But from eqn (A7.2) and (A7.3),

$$H_2 a_1 = (a_1 a_1^+ + \varepsilon^{(1)}) a_1 = a_1(a_1^+ a_1 + \varepsilon^{(1)}) = a_1 H_1.$$

Therefore

$$\begin{aligned}\langle\varepsilon| a_1^+ a_2^+ a_2 a_1 |\varepsilon\rangle &= \langle\varepsilon| a_1^+ a_1 (H_1 - \varepsilon^{(2)}) |\varepsilon\rangle = (\varepsilon - \varepsilon^{(2)})\langle\varepsilon| a_1^+ a_1 |\varepsilon\rangle \\ &= (\varepsilon - \varepsilon^{(2)})(\varepsilon - \varepsilon^{(1)}) \geq 0.\end{aligned}$$

This generalizes into

$$(\varepsilon - \varepsilon^{(n)})(\varepsilon - \varepsilon^{(n-1)}) \ldots (\varepsilon - \varepsilon^{(2)})(\varepsilon - \varepsilon^{(1)}) \geqslant 0. \tag{A7.5}$$

When the eigenvalue spectrum is unbounded (so that there is no upper bound to the values of $\varepsilon^{(j)}$) the only way of satisfying this expression is for ε to be equal to one of the $\varepsilon^{(j)}$. Hence the $\varepsilon^{(j)}$ are the eigenvalues of the system. That being the case, and we have $\varepsilon = \varepsilon^{(1)}$, eqn (A7.4) is satisfied if $a_1|\varepsilon\rangle = 0$. Hence we need solve this equation (a first-order differential equation) for the eigenfunction ψ_1. The other eigenfunctions are then found by solving $a_j|\varepsilon^{(j-1)}\rangle = 0$ in turn.

The technique can be illustrated as follows. The radial wavefunction P satisfies eqn (4.3.4):

$$\mathrm{d}^2P/\mathrm{d}r^2 + (2\mu/\hbar^2)\{(e^2/4\pi\varepsilon_0 r) - l(l+1)\hbar^2/2\mu r^2\}P = -(2\mu E/\hbar^2)P.$$

We write $p_r = (\hbar/\mathrm{i})\mathrm{d}/\mathrm{d}r$, $\varepsilon = 2\mu E$, $c = \mu e^2/4\pi\varepsilon_0$; then

$$\{p_r^2 + l(l+1)\hbar^2/r^2 - 2c/r\}P = \varepsilon P. \tag{A7.6}$$

The operator on the left takes the form in eqn (A7.1) when

$$a_1 = p_r + \mathrm{i}\{(l+1)\hbar/r - c/(l+1)\hbar\} \tag{A7.7a}$$

$$a_1^+ = p_r - \mathrm{i}\{(l+1)\hbar/r - c/(l+1)\hbar\} \tag{A7.7b}$$

$$\varepsilon^{(1)} = -c^2/(l+1)\hbar^2. \tag{A7.7c}$$

Since then

$$a_1 a_1^+ + \varepsilon^{(1)} = p_r^2 - 2c/r + (l+1)(l+2)\hbar^2/r^2,$$

if we introduce

$$a_2 = p_r + \mathrm{i}\{(l+2)\hbar/r - c/(l+2)\hbar\}$$

$$a_2^+ = p_r - \mathrm{i}\{(l+2)\hbar/r - c/(l+2)\hbar\}$$

this expression is equal to $a_2^+ a_2 + \varepsilon^{(2)}$ if $\varepsilon^{(2)} = -c^2/(l+2)\hbar^2$. Therefore, in general, if

$$a_j = p_r + \mathrm{i}\{(l+j)\hbar/r - c/(l+j)\hbar\} \tag{A7.8a}$$

$$a_j^+ = p_r - \mathrm{i}\{(l+j)\hbar/r - c/(l+j)\hbar\} \tag{A7.8b}$$

we require

$$\varepsilon^{(j)} = -c^2/(l+j)\hbar^2, \qquad j = 1, 2, \ldots \tag{A7.9}$$

These are the eigenvalues we seek. They are normally expressed in terms of $n = 1, 2, \ldots$, which entails $l = 0, 1, \ldots n-1$, and written

$$E_n = \varepsilon/2\mu = -(\mu e^4/32\pi^2\varepsilon_0^2\hbar^2)(1/n^2) \tag{A7.10}$$

in accord with eqn (4.3.7). The eigenfunctions are then obtained by solving $a_j\,|\varepsilon^{(j-1)}\rangle = 0$, or

$$[p_r + \mathrm{i}\{(l+j)\hbar/r - c/(l+j)\hbar\}]\psi_{j-1} = 0, \tag{A7.11}$$

a first-order differential equation.

Appendix 8 Matrices

A *matrix* is an array of numbers. Matrices may be combined together (by addition of multiplication) according to generalizations of the rules for ordinary numbers.

Consider a square matrix $\mathbf{M}$ of n^2 numbers arranged in n rows and n columns. These n^2 numbers are the *elements* of the matrix and may be specified by stating the row, r, and the column, c, at which they may be found. The matrix elements are therefore denoted M_{rc} with $r = 1, 2, \ldots, n$ and $c = 1, 2, \ldots, n$. Thus, in the matrix

$$\mathbf{M} = \begin{pmatrix} 1 & 2 \\ 3 & 4 \end{pmatrix},$$

the elements are $M_{11} = 1$, $M_{12} = 2$, $M_{21} = 3$, and $M_{22} = 4$. (We shall deal almost exclusively with square matrices although rectangular matrices of nm elements are also encountered.) Two matrices $\mathbf{M}$ and $\mathbf{N}$

may be added to give the sum $\mathbf{S}=\mathbf{M}+\mathbf{N}$ using the rule that

Sum: $S_{rc}=M_{rc}+N_{rc}$. (A8.1)

Thus if

$$\mathbf{N}=\begin{pmatrix}5 & 6\\ 7 & 8\end{pmatrix},$$

$$\mathbf{S}=\begin{pmatrix}1 & 2\\ 3 & 4\end{pmatrix}+\begin{pmatrix}5 & 6\\ 7 & 8\end{pmatrix}=\begin{pmatrix}1+5 & 2+6\\ 3+7 & 4+8\end{pmatrix}=\begin{pmatrix}6 & 8\\ 10 & 12\end{pmatrix}.$$

Two matrices may be multiplied to give the product $\mathbf{P}=\mathbf{MN}$ according to the rule

Product: $P_{rc}=\sum_n M_{rn}N_{nc}$. (A8.2)

Therefore, with the above matrices we obtain

$$P_{11}=M_{11}N_{11}+M_{12}N_{21}=1.5+2.7=19,$$
$$P_{12}=M_{11}N_{12}+M_{12}N_{22}=1.6+2.8=22,$$

and so on, whence

$$\mathbf{P}=\begin{pmatrix}19 & 22\\ 43 & 50\end{pmatrix}.$$

Notice, however, that the matrix $\mathbf{R}=\mathbf{NM}$ is not equal to the matrix $\mathbf{P}=\mathbf{MN}$ (two matrices are *equal* only if $P_{rc}=R_{rc}$ for all elements):

$$\mathbf{R}=\begin{pmatrix}23 & 34\\ 31 & 46\end{pmatrix}\neq\mathbf{P}.$$

Therefore in this example $\mathbf{MN}\neq\mathbf{NM}$, and so matrix multiplication is not necessarily *commutative*. (Examples could be chosen for which $\mathbf{MN}=\mathbf{NM}$; the multiplication would then be commutative and the matrices would be said to *commute*.)

The *determinant* of a matrix is written $|\mathbf{M}|$ and for the matrix in the first example,

$$|\mathbf{M}|=\begin{vmatrix}1 & 2\\ 3 & 4\end{vmatrix}=1.4-2.3=-2.$$

If $\mathbf{P}=\mathbf{MN}$ then $|\mathbf{P}|=|\mathbf{M}|\,|\mathbf{N}|$.

A *diagonal matrix* is one in which the only non-zero elements lie on the *major diagonal*; for example

$$\mathbf{M}=\begin{pmatrix}1 & 0 & 0\\ 0 & 2 & 0\\ 0 & 0 & 1\end{pmatrix}$$

is diagonal. This condition may be written succintly as

Diagonal matrix: $M_{rc} = m_r\delta_{rc}$, (A8.3)

where δ_{rc} is the *Kronecker delta*, which is equal to 1 when $r = c$ and is zero otherwise. In the above example $m_1 = 1$, $m_2 = 2$, $m_3 = 1$.

The *unit matrix*, **1**, is a special case of a diagonal matrix, for all elements are zero except the diagonal elements, which are all unity:

Unit matrix (**1**): $1_{rc} = \delta_{rc}$. (A8.4)

The *inverse* of a matrix **M** is written $\mathbf{M}^{-1}$ and is defined such that

Inverse ($\mathbf{M}^{-1}$): $\mathbf{M}\mathbf{M}^{-1} = \mathbf{M}^{-1}\mathbf{M} = \mathbf{1}$. (A8.5)

In order to construct the inverse of a given matrix **M** proceed as follows:

(a) Form the determinant $|\mathbf{M}|$; for example $|\mathbf{M}| = -2$.

(b) Form the transpose $\tilde{\mathbf{M}}$; for example $\tilde{\mathbf{M}} = \begin{pmatrix} 1 & 3 \\ 2 & 4 \end{pmatrix}$.

(c) Form $\tilde{\mathbf{M}}'$ where $\tilde{M}'_{rc}$ is the cofactor (signed minor, the determinant formed from the matrix with row r and column c struck out); for example

$$\tilde{\mathbf{M}}' = \begin{pmatrix} 4 & -2 \\ -3 & 1 \end{pmatrix}.$$

(d) Construct $\mathbf{M}^{-1} = \tilde{\mathbf{M}}'/|\mathbf{M}|$; for example,

$$\mathbf{M}^{-1} = (1/-2)\begin{pmatrix} 4 & -2 \\ -3 & 1 \end{pmatrix} = \begin{pmatrix} -2 & 1 \\ \frac{3}{2} & -\frac{1}{2} \end{pmatrix}.$$

The *transpose* of a matrix is written $\tilde{\mathbf{M}}$ and is defined by

Transpose ($\tilde{\mathbf{M}}$): $\tilde{M}_{ab} = M_{ba}$ (A8.6)

where in each case the first index labels the row and the second the column. Thus, if

$$\mathbf{M} = \begin{pmatrix} a & c + \mathrm{i}d \\ e + if & b \end{pmatrix}, \quad \tilde{\mathbf{M}} = \begin{pmatrix} a & e + if \\ c + \mathrm{i}d & b \end{pmatrix}.$$

The *complex conjugate* of a matrix is written $\mathbf{M}^*$ and defined by

Complex conjugate ($\mathbf{M}^*$): $M^*_{rc} = (M_{rc})^*$ (A8.7)

so that with the matrix above,

$$\mathbf{M}^* = \begin{pmatrix} a & c - \mathrm{i}d \\ e - if & b \end{pmatrix}$$

if a, b, c, d, d, e, and f are supposed to be real numbers.

The *adjoint* of a matrix is written $\mathbf{M}^+$ and is equal to $\tilde{\mathbf{M}}^*$:

$$\textit{Adjoint } (\mathbf{M}^+)\text{: } M^+_{ab} = (\tilde{M}^*)_{ab} = (M^*)_{ba} = M^*_{ba}. \quad \text{(A8.8)}$$

Therefore, for example,

$$\mathbf{M}^+ = \begin{pmatrix} a & e-\mathrm{i}f \\ c-\mathrm{i}d & b \end{pmatrix}.$$

A matrix is *unitary* if $\mathbf{M}^{-1} = \mathbf{M}^+$:

$$\textit{Unitary matrix } (\mathbf{M}^{-1} = \mathbf{M}^+)\text{: } (M^{-1})_{ab} = M^*_{ba}. \quad \text{(A8.9)}$$

A matrix is *hermitian* or *self-adjoint* if $\mathbf{M}^+ = \mathbf{M}$:

$$\textit{Hermitian matrix } (\mathbf{M}^+ = \mathbf{M})\text{: } M^+_{ab} = M^*_{ba} = M_{ab}. \quad \text{(A8.10)}$$

In the example, $\mathbf{M}$ is hermitian if $c = e$, $d = -f$.

A set of n *simultaneous equations*

$$\left.\begin{array}{l} a_{11}x_1 + a_{12}x_2 + \ldots a_{1n}x_n = b_1 \\ a_{21}x_1 + a_{22}x_2 + \ldots a_{2n}x_n = b_2 \\ \ldots \quad \ldots \quad \ldots \quad \ldots \quad \ldots \\ a_{n1}x_1 + a_{n2}x_2 + \ldots a_{nn}x_n = b_n \end{array}\right\} \quad \text{(A8.11)}$$

may be written in matrix notation if we introduce the *column vectors* $\mathbf{x}$ and $\mathbf{b}$:

$$\textit{Column vectors: } \mathbf{x} = \begin{bmatrix} x_1 \\ x_2 \\ \vdots \\ x_n \end{bmatrix}, \qquad \mathbf{b} = \begin{bmatrix} b_1 \\ b_2 \\ \vdots \\ b_n \end{bmatrix} \quad \text{(A8.12)}$$

(so that $\mathbf{x}$ and $\mathbf{b}$ are $n \times 1$ matrices), for then if

$$\mathbf{A} = \begin{pmatrix} a_{11} & a_{12} \ldots a_{1n} \\ a_{21} & a_{22} \ldots a_{2n} \\ \vdots & \vdots \quad\quad \vdots \\ a_{n1} & a_{n2} \quad\quad a_{nn} \end{pmatrix} \quad \text{(A8.13)}$$

the n equations may be written in the succinct form

$$\mathbf{A}\mathbf{x} = \mathbf{b}. \quad \text{(A8.14)}$$

The formal solution is obtained by multiplying both sides by $\mathbf{A}^{-1}$:

$$\mathbf{x} = \mathbf{A}^{-1}\mathbf{b}. \quad \text{(A8.15)}$$

and so the equations are solved once the inverse of the matrix of the coefficients is known.

Suppose that $\mathbf{b} = \lambda\mathbf{x}$, then we have an

Eigenvalue equation: $\mathbf{Ax} = \lambda\mathbf{x}$, (A8.16)

where λ is the *eigenvalue* and the $\mathbf{x}$ is the *eigenvector.* In general there are n eigenvalues which satisfy the n simultaneous equations

$$(\mathbf{A} - \lambda\mathbf{1})\mathbf{x} = 0, \tag{A8.17}$$

and so there will be n corresponding eigenvectors: we denote the eigenvalues λ_i and the corresponding eigenvectors $\mathbf{x}^{(i)}$. Equation (A8.17) defines the set of n simultaneous equations

$$\left.\begin{array}{l} (a_{11} - \lambda)x_1 + a_{12}x_2 + \ldots a_{1n}x_n = 0 \\ a_{21}x_1 + (a_{22} - \lambda)x_2 + \ldots a_{2n}x_n = 0 \\ \ldots \quad \ldots \quad \ldots \quad \ldots \\ a_{n1}x_1 + a_{n2}x_2 + \ldots (a_{nn} - \lambda)x_n = 0 \end{array}\right\} \tag{A8.18}$$

and these have a non-trivial solution only if the determinant of the coefficients disappears: but this determinant is just $|\mathbf{A} - \lambda\mathbf{1}|$, and so the n eigenvalues may be found from the solution of the

Secular equation: $|\mathbf{A} - \lambda\mathbf{1}| = 0.$ (A8.19)

The n eigenvalues this yields may be used to find the n eigenvectors. These eigenvectors (which are $n \times 1$ matrices) may be used to form an $n \times n$ matrix $\mathbf{X}$. Thus, since

$$\mathbf{x}^{(1)} = \begin{bmatrix} x_1^{(1)} \\ x_2^{(1)} \\ \vdots \\ x_n^{(1)} \end{bmatrix}, \qquad \mathbf{x}^{(2)} = \begin{bmatrix} x_1^{(2)} \\ x_2^{(2)} \\ \vdots \\ x_n^{(2)} \end{bmatrix}, \text{ etc.,}$$

we may form

$$\mathbf{X} = (\mathbf{x}^{(1)}, \mathbf{x}^{(2)}, \ldots, \mathbf{x}^{(n)}) = \begin{pmatrix} x_1^{(1)} & x_1^{(2)} & \ldots & x_1^{(n)} \\ x_2^{(1)} & x_2^{(2)} & \ldots & x_2^{(n)} \\ \vdots & \ldots & \ldots & \end{pmatrix} \tag{A8.20}$$

so that $X_{ab} = x_a^{(b)}$. If further we write $\Lambda_{ab} = \lambda_a\delta_{ab}$, so that $\mathbf{\Lambda}$ is a diagonal matrix with the elements $\lambda_1, \lambda_2, \ldots, \lambda_n$ along the diagonal, all the eigenvalue equations $\mathbf{Ax}^{(i)} = \lambda_i\mathbf{x}^{(i)}$ may be combined into the single equation

$$\mathbf{AX} = \mathbf{X\Lambda} \tag{A8.21}$$

because this is equal to

$$\sum_n A_{rn}X_{nc} = \sum_n X_{rn}\Lambda_{nc}$$

or

$$\sum_n A_{rn}x_n^{(c)} = \sum_n x_r^{(n)}\lambda_n\delta_{nc} = \lambda_c x_r^{(c)}$$

or

$$\mathbf{A}\mathbf{x}^{(c)} = \lambda_c \mathbf{x}^{(c)},$$

as required.

Therefore, if we form $\mathbf{X}^{-1}$ from $\mathbf{X}$, we have a

Similarity transformation: $\boldsymbol{\Lambda} = \mathbf{X}^{-1}\mathbf{A}\mathbf{X}$. (A8.22)

which makes $\mathbf{A}$ diagonal (because $\boldsymbol{\Lambda}$ is diagonal). Therefore *if the matrix* $\mathbf{X}$ *which causes the matrix* $\mathbf{X}^{-1}\mathbf{A}\mathbf{X}$ *to be diagonal can be found, the problem of solving the eigenvalue equation is solved,* for the diagonal matrix $\boldsymbol{\Lambda}$ that is produced has the eigenvalues as its only non-zero elements, and the matrix $\mathbf{X}$ used to bring about the transformation has the corresponding eigenvectors as its columns. *The eigenvalue equation may therefore be solved by diagonalizing the matrix.* There are analytical methods for diagonalizing matrices, but they are normally dealt with using a library program on a computer.

Appendix 9 Vector coupling coefficients

The following three tables of vector coupling coefficients give an idea of their form: we also present a short example of their application. For more extensive tables see, for example, Heine or Rotenberg *et al.*, referred to in the bibliography of Chapter 6.

1. $j_1 = j_2 = \frac{1}{2}$. $|jm_j\rangle$

m_{j1}	m_{j2}	$\|1,1\rangle$	$\|1,0\rangle$	$\|0,0\rangle$	$\|1,-1\rangle$
$\frac{1}{2}$	$\frac{1}{2}$	1			
$\frac{1}{2}$	$-\frac{1}{2}$		$\sqrt{\frac{1}{2}}$	$\sqrt{\frac{1}{2}}$	
$-\frac{1}{2}$	$\frac{1}{2}$		$\sqrt{\frac{1}{2}}$	$-\sqrt{\frac{1}{2}}$	
$-\frac{1}{2}$	$-\frac{1}{2}$				1

2. $j_1 = 1$, $j_2 = \frac{1}{2}$. $|jm_j\rangle$

m_{j1}	m_{j2}	$\|\frac{3}{2},\frac{3}{2}\rangle$	$\|\frac{3}{2},\frac{1}{2}\rangle$	$\|\frac{1}{2},\frac{1}{2}\rangle$	$\|\frac{3}{2},-\frac{1}{2}\rangle$	$\|\frac{1}{2},-\frac{1}{2}\rangle$	$\|\frac{3}{2},-\frac{3}{2}\rangle$
1	$\frac{1}{2}$	1					
1	$-\frac{1}{2}$		$\sqrt{\frac{1}{3}}$	$\sqrt{\frac{2}{3}}$			
0	$\frac{1}{2}$		$\sqrt{\frac{2}{3}}$	$-\sqrt{\frac{1}{3}}$			
0	$-\frac{1}{2}$				$\sqrt{\frac{2}{3}}$	$\sqrt{\frac{1}{3}}$	
-1	$\frac{1}{2}$				$\sqrt{\frac{1}{3}}$	$-\sqrt{\frac{2}{3}}$	
-1	$-\frac{1}{2}$						1

3. $j_1 = 1,\ j_2 = 1.$ $|jm_j\rangle$

m_{j1}	m_{j2}	$\|2,2\rangle$	$\|2,1\rangle$	$\|1,1\rangle$	$\|2,0\rangle$	$\|1,0\rangle$	$\|0,0\rangle$	$\|2,-1\rangle$	$\|1,-1\rangle$	$\|2,-2\rangle$
1	1	1								
1	0		$\sqrt{\frac{1}{2}}$	$\sqrt{\frac{1}{2}}$						
0	1		$\sqrt{\frac{1}{2}}$	$-\sqrt{\frac{1}{2}}$						
1	−1				$\sqrt{\frac{1}{6}}$	$\sqrt{\frac{1}{2}}$	$\sqrt{\frac{1}{3}}$			
0	0				$\sqrt{\frac{2}{3}}$	0	$-\sqrt{\frac{1}{3}}$			
−1	1				$\sqrt{\frac{1}{6}}$	$-\sqrt{\frac{1}{2}}$	$\sqrt{\frac{1}{3}}$			
−1	0							$\sqrt{\frac{1}{2}}$	$\sqrt{\frac{1}{2}}$	
0	−1							$\sqrt{\frac{1}{2}}$	$-\sqrt{\frac{1}{2}}$	
−1	−1									1

The example is the determination of the energy of an atom in a magnetic field B in the z-direction, its single p-electron having a spin–orbit coupling constant ζ. The hamiltonian is

$$H = (\mu_B/\hbar)(l_z + 2s_z)B + (hc\zeta/\hbar^2)\mathbf{l}\cdot\mathbf{s}. \tag{A9.1}$$

The possible energies of the atom are the eigenvalues of the matrix of H (p. 99), which we may set up in either the coupled or the uncoupled representation. For the calculation in the uncoupled representation it is convenient to write H in the form

$$H = (\mu_B B/\hbar)(l_z + 2s_z) + (hc\zeta/\hbar^2)\{l_z s_z + \tfrac{1}{2}(l^+s^- + l^-s^+)\}. \tag{A9.2}$$

When all the matrix elements are calculated using the results in eqn (6.3.18–20) we obtain

	$\|1,\frac{1}{2}\rangle$	$\|0,0\rangle$	$\|-1,\frac{1}{2}\rangle$	$\|1,-\frac{1}{2}\rangle$	$\|0,-\frac{1}{2}\rangle$	$\|-1,-\frac{1}{2}\rangle$
$\langle 1,\frac{1}{2}\|$	$2\mu_B B + \frac{1}{2}\zeta hc$	0	0	0	0	0
$\langle 0,\frac{1}{2}\|$	0	$\mu_B B$	0	$\sqrt{2}\zeta hc$	0	0
$\langle -1,\frac{1}{2}\|$	0	0	$-\frac{1}{2}\zeta hc$	0	$\sqrt{2}\zeta hc$	0
$\langle 1,-\frac{1}{2}\|$	0	$\sqrt{2}\zeta hc$	0	$-\frac{1}{2}\zeta hc$	0	0
$\langle 0,-\frac{1}{2}\|$	0	0	$\sqrt{2}\zeta hc$	0	$-\mu_B B$	0
$\langle -1,-\frac{1}{2}\|$	0	0	0	0	0	$-2\mu_B B + \frac{1}{2}\zeta hc$

in the notation $|m_l, m_s\rangle$.

For the coupled representation calculations it is sensible to write H in the form

$$\begin{aligned} H &= (\mu_B B/\hbar)(l_z + s_z) + (\mu_B B/\hbar)s_z + \tfrac{1}{2}(hc\zeta/\hbar^2)\{(\mathbf{l}+\mathbf{s})^2 - l^2 - s^2\} \\ &= (\mu_B B/\hbar)j_z + (\mu_B B/\hbar)s_z + \tfrac{1}{2}(hc\zeta/\hbar^2)(j^2 - j^2 - s^2). \end{aligned} \tag{A9.3}$$

In the coupled representation the states are eigenstates of j^2, j_z, l^2, and s^2, and so most of the matrix elements of this hamiltonian may be evaluated very simply. The difficulty is associated with the effect of s_z, for the states in the coupled representation are not eigenstates of this operator. The effect of s_z may be determined by expanding the coupled states in terms of the uncoupled states, using the vector coupling coefficients. As an example, consider the particular matrix element $\langle\frac{1}{2},\frac{1}{2}|\, s_z\,|\frac{3}{2},\frac{1}{2}\rangle$. From the table of coefficients we write

$$|\tfrac{3}{2},\tfrac{1}{2}\rangle=(\tfrac{1}{3})^{\frac{1}{2}}\,|1,-\tfrac{1}{2}\rangle+(\tfrac{2}{3})^{\frac{1}{2}}\,|0,\tfrac{1}{2}\rangle,$$

where the right-hand side uses the notation $|m_l, m_s\rangle$, and the left-hand side the notation $|j, m_j\rangle$. The effect on this of the operator s_z is as follows:

$$s_z\,|\tfrac{3}{2},\tfrac{1}{2}\rangle=(\tfrac{1}{3})^{\frac{1}{2}}\{-\tfrac{1}{2}\hbar\,|1,-\tfrac{1}{2}\rangle+\tfrac{1}{2}\hbar 2^{\frac{1}{2}}\,|0,\tfrac{1}{2}\rangle\}.$$

The state $|\frac{1}{2},\frac{1}{2}\rangle$ in the coupled representation is

$$|\tfrac{1}{2},\tfrac{1}{2}\rangle=(\tfrac{2}{3})^{\frac{1}{2}}\,|1,-\tfrac{1}{2}\rangle-(\tfrac{1}{3})^{\frac{1}{2}}\,|0,\tfrac{1}{2}\rangle,$$

and so the required matrix element, using the orthonormality of the states in the uncoupled representation, is

$$\langle\tfrac{1}{2},\tfrac{1}{2}|\, s_z\,|\tfrac{3}{2},\tfrac{1}{2}\rangle=-\tfrac{1}{3}\sqrt{2}\,\mu_B B.$$

In this way the entire matrix may be constructed, and we obtain

	$\lvert\frac{3}{2},\frac{3}{2}\rangle$	$\lvert\frac{3}{2},\frac{1}{2}\rangle$	$\lvert\frac{3}{2},-\frac{1}{2}\rangle$	$\lvert\frac{3}{2},-\frac{3}{2}\rangle$	$\lvert\frac{1}{2},\frac{1}{2}\rangle$	$\lvert\frac{1}{2},-\frac{1}{2}\rangle$
$\langle\frac{3}{2},\frac{3}{2}\rvert$	$2\mu_B B+\frac{1}{2}\zeta hc$	0	0	0	0	0
$\langle\frac{3}{2},\frac{1}{2}\rvert$	0	$\frac{2}{3}\mu_B B+\frac{1}{2}\zeta hc$	0	0	$-\frac{1}{3}\sqrt{2}\mu_B B$	0
$\langle\frac{3}{2},-\frac{1}{2}\rvert$	0	0	$-\frac{2}{3}\mu_B B+\frac{1}{2}\zeta hc$	0	0	$-\frac{1}{3}\sqrt{2}\mu_B B$
$\langle\frac{3}{2},-\frac{3}{2}\rvert$	0	0	0	$-2\mu_B B+\frac{1}{2}\zeta hc$	0	0
$\langle\frac{1}{2},\frac{1}{2}\rvert$	0	$-\frac{1}{3}\sqrt{2}\mu_B B$	0	0	$\frac{1}{2}\mu_B B-\zeta hc$	0
$\langle\frac{1}{2},-\frac{1}{2}\rvert$	0	0	$-\frac{1}{3}\sqrt{2}\mu_B B$	0	0	$-\frac{1}{2}\mu_B B-\zeta hc$

The point of the calculation now becomes clear. In order to determine the energy levels of the electron we must diagonalize the matrix. But if the externally applied field is very weak (so that $\mu_B B \ll hc\zeta$) then the matrix of H has much smaller off-diagonal elements in the coupled representation than in the uncoupled representation: only B appears in these elements in the former whereas ζ occurs in them in the latter. Conversely, if the field is strong (so that $\mu_B B \gg hc\zeta$) then the uncoupled representation has the smaller off-diagonal elements and is more nearly diagonal. Therefore, for practical convenience, it is better to set up the matrix in a representation that reflects the physics of the problem, because then it is much easier to diagonalize. When the spin–orbit coupling is strong the coupled representation should be used; when the field is very strong the uncoupled representation is better. The representation that most nearly diagonalizes the hamiltonian is the closest to the 'true' description of the system, and so we conclude that the coupled representation, with vectors adopting precise orientations to each other, is better when there is spin–orbit coupling. The uncoupled representation, when the vectors do not make precise angles to each other but where their projection on the z-axis (the field direction) is well-defined, is better when the external field is present and dominates the spin–orbit coupling. In the absence of spin–orbit coupling and external fields, both representations are equally good.

Appendix 10 Character tables

The operations. The operations of a group (of order h) are designated as follows:

E: the identity;
C_n: a rotation by $2\pi/n$;
σ_v: a reflection in a plane containing the major symmetry axis;
σ_h: a reflection in a plane perpendicular to the major symmetry axis;
σ_d: a reflection in a plane containing the major symmetry axis and bisecting the angle between two-fold axes perpendicular to the symmetry axis;
S_n: a rotation by $2\pi/n$ followed by a reflection in a plane perpendicular to the axis of rotation.

The irreducible representations. The general designation of the symmetry species of an irreducible representation is Γ_i, with $i = 1, 2, \ldots$. A common system of nomenclature employs the letters A, B, E, and T as follows:

A: a one-dimensional irrep in which the character of the major rotation is $+1$. If several occur in the same group they are distinguished by subscripts, e.g. $A_1, A_2, \ldots$;
B: a one-dimensional irrep, but one that has odd parity (changes sign) under the principal rotation. Distinguishing suffices may also be used, e.g. $B_1, B_2, \ldots$;
E: a two-dimensional irrep (so that the character of the identity is 2);
T: a three-dimensional irrep;
$\Sigma, \Pi, \Delta, \ldots$ are used in the axial rotation group ($C_{\infty v}$) and all except Σ are two dimensional;
$S, P, D, \ldots$ are used in the full rotation group R_3 and have the dimensionalities $1, 3, 5, \ldots$.

A prime, e.g. A' or E', denotes that a reflection has even parity; a double prime, e.g. A'' or E'', denotes a reflection with odd parity;

A subscript g or u denotes that the character of the inversion is even or odd respectively;

A superscript + or −, e.g. Σ^+, denotes a reflection σ_v with even or odd parity.

The groups. The nomenclature of the groups is based on the following *Schoenflies scheme*:

C_n: contains only a simple n-fold axis;
C_{nv}: contains an n-fold axis and n vertical mirror planes;
C_{nh}: contains an n-fold axis and a horizontal mirror plane;
D_n: contains n two-fold axes perpendicular to the major C_n axis;
D_{nh}: contains D_n and a horizontal mirror plane;

D_{nd}: D_n and n σ_d mirror planes;
S_n: contains an S_n axis;
T: the group of the regular tetrahedron (E, C_2, C_3);
T_d: T and mirror planes (T, σ_d, S_4);
T_h: T and a centre of inversion;
O: the group of a regular octahedron or cube;
O_h: O and a centre of inversion;
$C_{\infty v}$: the infinite axial rotation group;
$D_{\infty h}$: σ_v and C_2 anywhere within the σ_h plane.

The *Hermann–Mauguin symbols* (e.g., $2mm$) are given for crystallographic groups.

The tables.

C_{2v}, $2mm$	E	C_2	σ_v	σ_v'	$h=4$	
A_1	1	1	1	1	z, z^2, x^2, y^2	
A_2	1	1	−1	−1	xy	R_z
B_1	1	−1	1	−1	x, xz	R_y
B_2	1	−1	−1	1	y, yz	R_x

C_{3v}, $3m$	E	$2C_3$	$3\sigma_v$	$h=6$	
A_1	1	1	1	z, z^2, x^2+y^2	
A_2	1	1	−1		R_z
E	2	−1	0	$(x, y), (xy, x^2-y^2)(xz, yz)$	(R_x, R_y)

C_{4v}, $4mm$	E	C_2	$2C_4$	$2\sigma_v$	$2\sigma_d$	$h=8$	
A_1	1	1	1	1	1	z, z^2, x^2+y^2	
A_2	1	1	1	−1	−1		R_z
B_1	1	1	−1	1	−1	x^2-y^2	
B_2	1	1	−1	−1	1	xy	
E	2	−2	0	0	0	$(x, y), (xz, yz)$	(R_x, R_y)

C_{5v}	E	$2C_5$	$2C_5^2$	$5\sigma_v$	$h=10, \alpha=72°$	
A_1	1	1	1	1	z, z^2, x^2+y^2	
A_2	1	1	1	−1		R_z
E_1	2	$2\cos\alpha$	$2\cos 2\alpha$	0	$(x, y), (xz, yz)$	(R_x, R_y)
E_2	2	$2\cos 2\alpha$	$2\cos\alpha$	0	(xy, x^2-y^2)	

$C_{6v}, 6mm$	E	C_2	$2C_3$	$2C_6$	$3\sigma_d$	$3\sigma_v$	$h = 12$	
A_1	1	1	1	1	1	1	z, z^2, x^2+y^2	
A_2	1	1	1	1	−1	−1		R_z
B_1	1	−1	1	−1	−1	1		
B_2	1	−1	1	−1	1	−1		
E_1	2	−2	−1	1	0	0	$(x, y), (xz, yz)$	(R_x, R_y)
E_2	2	2	−1	−1	0	0	(xy, x^2-y^2)	

$C_{\infty v}$	E	C_2	$2C_\phi$	σ_v	$h = \infty$	
$A_1(\Sigma^+)$	1	1	1	1	z, z^2, x^2+y^2	
$A_2(\Sigma^-)$	1	1	1	−1		R_z
$E_1(\Pi)$	2	−2	$2\cos\phi$	0	$(x, y), (xz, yz)$	(R_x, R_y)
$E_2(\Delta)$	2	2	$2\cos 2\phi$	0	(xy, x^2-y^2)	
...	.	.	.	.	...	

$D_2, 222$	E	C_2^z	C_2^y	C_2^x	$h = 4$	
A	1	1	1	1	x^2, y^2, z^2	
B_1	1	1	−1	−1	z, xy	R_z
B_2	1	−1	1	−1	y, xz	R_y
B_3	1	−1	−1	1	x, yz	R_x

$D_3, 32$	E	$2C_3$	$2C_2'$	$h = 6$	
A_1	1	1	1	z^2, x^2+y^2	
A_2	1	1	−1	z	R_z
E	2	−1	0	$(x, y), (xz, yz)(xy, x^2-y^2)$	(R_x, R_y)

$D_4, 422$	E	C_2	$2C_4$	$2C_2'$	$2C_2''$	$h = 8$	
A_1	1	1	1	1	1	z^2, x^2+y^2	
A_2	1	1	1	−1	−1	z	R_z
B_1	1	1	−1	1	−1	x^2-y^2	
B_2	1	1	−1	−1	1	xy	
E	2	−2	0	0	0	$(x, y), (xz, yz)$	(R_x, R_y)

$D_{3h}, \bar{6}m2$	E	σ_h	$2C_3$	$2S_3$	$3C_2'$	$3\sigma_v$	$h=12$	
A_1'	1	1	1	1	1	1	z^2, x^2+y^2	
A_2'	1	1	1	1	−1	−1		R_z
A_1''	1	−1	1	−1	1	−1		
A_2''	1	−1	1	−1	−1	1	z	
E'	2	2	−1	−1	0	0	$(x, y), (xy, x^2-y^2)$	
E''	2	−2	−1	1	0	0	(xz, yz)	(R_x, R_y)

$D_{\infty h}$	E	$2C_\phi$	C_2'	i	$2iC_\phi$	iC_2'	$h=\infty$	
$A_{1g}(\Sigma_g^+)$	1	1	1	1	1	1	z^2, x^2+y^2	
$A_{1u}(\Sigma_u^+)$	1	1	1	−1	−1	−1		
$A_{2g}(\Sigma_g^-)$	1	1	−1	1	1	−1		R_z
$A_{2u}(\Sigma_u^-)$	1	1	−1	−1	1	1	z	
$E_{1g}(\Pi_g)$	2	$2\cos\phi$	0	0	$-2\cos\phi$	0	(xz, yz)	(R_x, R_y)
$E_{1u}(\Pi_u)$	2	$2\cos\phi$	0	0	$2\cos\phi$	0	(x, y)	
$E_{2g}(\Delta_g)$	2	$2\cos 2\phi$	0	0	$2\cos 2\phi$	0	(xy, x^2-y^2)	
$E_{2u}(\Delta_u)$	2	$2\cos 2\phi$	0	0	$-2\cos 2\phi$	0		

$T_d, \bar{4}3m$	E	$8C_3$	$3C_2$	$6\sigma_d$	$6S_4$	$h=24$	
A_1	1	1	1	1	1	$x^2+y^2+z^2$	
A_2	1	1	1	−1	−1		
E	2	−1	2	0	0	$(3z^2-r^2, x^2-y^2)$	
T_1	3	0	−1	−1	1		
T_2	3	0	−1	1	−1	(x, y, z)	(R_x, R_y, R_z) (xy, xz, yz)

$O, 432$	E	$8C_3$	$3C_2$	$6C_2'$	$6C_4$	$h=24$	
A_1	1	1	1	1	1	$x^2+y^2+z^2$	
A_2	1	1	1	−1	−1		
E	2	−1	2	0	0	$(x^2-y^2, 3z^2-r^2)$	
T_1	3	0	−1	−1	1	(x, y, z)	(R_x, R_y, R_z)
T_2	3	0	−1	1	−1	(xy, yz, zx)	

Appendix 11 Direct products

In general $g \times g = g$, $g \times u = u$, $u \times u = g$;

$$\Gamma' \times \Gamma' = \Gamma', \qquad \Gamma' \times \Gamma'' = \Gamma'', \qquad \Gamma'' \times \Gamma'' = \Gamma'$$

For C_2, C_{2v}, C_{2h}; C_3, C_{3v}, C_{3h}; D_3, D_{3h}, D_{3d}; C_6, C_{6v}, C_{6h}, D_6, S_6

	A_1	A_2	B_1	B_2	E_1	E_2
A_1	A_1	A_2	B_1	B_2	E_1	E_2
A_2		A_1	B_2	B_1	E_1	E_2
B_1			A_1	A_2	E_2	E_1
B_2				A_1	E_2	E_1
E_1					$A_1+[A_2]+E_2$	$B_1+B_2+E_1$
E_2						$A_1+[A_2]+E_2$

For T, T_h, T_d; O, O_h:

	A_1	A_2	E	T_1	T_2
A_1	A_1	A_2	E	T_1	T_2
A_2		A_1	E	T_2	T_1
E			$A_1+[A_2]+E$	T_1+T_2	T_1+T_2
T_1				$A_1+E+[T_1]+T_2$	$A_2+E+T_1+T_2$
T_2					$A_1+E+[T_1]+T_2$

For $C_{\infty v}$, $D_{\infty h}$:

	Σ^+	Σ^-	Π	Δ	...
Σ^+	Σ^+	Σ^-	Π	Δ	...
Σ^-		Σ^+	Π	Δ	...
Π			$\Sigma^++[\Sigma^-]+\Delta$	$\Pi+\Phi$	...
Δ				$\Sigma^++[\Sigma^-]+\Gamma$	...
$\vdots$					...

In each case $[\Gamma]$ denotes the representation spanned by the *antisymmetrized product* of a degenerate irrep with itself.

Appendix 12 The orthogonality of the basis functions

1. Theorem. *Two functions are orthogonal if they are basis functions for different irreducible representations of a group, or if they are members of a basis of a particular irreducible representation but are in different positions in the row.*

Proof. Let the set of functions $f_i^{(l)}$ with $i = 1, 2, \ldots d_l$ be the basis of irreducible representation of symmetry species $\Gamma^{(l)}$ and the set $f_{i'}^{(l')}$ with

$i' = 1, 2, \ldots d_{l'}$ be a basis for $\Gamma^{(l')}$. This means that for all the operations R of the group

$$Rf_i^{(l)} = \sum_j f_j^{(l)} D_{ji}^{(l)}(R); \qquad Rf_{i'}^{(l')} = \sum_{j'} f_{j'}^{(l')} D_{j'i'}^{(l')}(R).$$

A scalar product, or an integral, must be independent of any symmetry transformation. Therefore

$$\langle f_i^{(l)} \mid f_{i'}^{(l')} \rangle = R\langle f_i^{(l)} \mid f_{i'}^{(l')} \rangle = \langle Rf_i^{(l)} \mid Rf_{i'}^{(l')} \rangle$$

for all operations R. Therefore, since there are h elements in the group and each one leaves the integral unchanged,

$$\begin{aligned}\langle f_i^{(l)} \mid f_{i'}^{(l')} \rangle &= (1/h)\sum_R \langle Rf_i^{(l)} \mid Rf_{i'}^{(l')} \rangle \\ &= (1/h)\sum_R \sum_j \sum_{j'} D_{ji}^{(l)}(R)^* D_{j'i'}^{(l')}(R) \langle f_j^{(l)} \mid f_{j'}^{(l')} \rangle.\end{aligned}$$

The great orthogonality theorem, eqn (7.7.1), may be used to write this as

$$\begin{aligned}\langle f_i^{(l)} \mid f_{i'}^{(l')} \rangle &= \sum_j \sum_{j'} (1/d_l)\delta_{ll'}\, \delta_{jj'}\, \delta_{ii'} \langle f_j^{(l)} \mid f_{j'}^{(l')} \rangle \\ &= \delta_{ll'}\, \delta_{ii'} (1/d_l) \sum_j \langle f_j^{(l)} \mid f_j^{(l)} \rangle. \end{aligned} \qquad \text{(A12.1)}$$

Therefore $\langle f_i^{(l)} \mid f_{i'}^{(l')} \rangle \propto \delta_{ll'}\, \delta_{ii'}$, which completes the proof of the theorem.

2. Theorem. *The scalar product or the integral* $\langle f_i^{(l)} \mid f_i^{(l)} \rangle$ *is independent of* i.

Proof. From the preceding theorem we have

$$\langle f_i^{(l)} \mid f_i^{(l)} \rangle = (1/d_l) \sum_j \langle f_j^{(l)} \mid f_j^{(l)} \rangle$$

and the sum on the right-hand side is independent of i.

Appendix 13 Electric dipole transitions

Consider a molecule exposed to light with its electric vector lying in the z-direction and oscillating at a frequency $\omega = 2\pi\nu$. The perturbation is

$$H^{(1)}(t) = -\mu_z E(t), \qquad E(t) = 2E \cos \omega t. \qquad \text{(A13.1)}$$

The transition rate is given by eqn (8.7.7):

$$W_{\mathrm{f}\leftarrow\mathrm{i}} = (1/\hbar)^2 \, |H_{\mathrm{if}}^{(1)}|^2 \, \hat{\rho}(\nu_{\mathrm{fi}}) = (E^2/\hbar^2) \, |\mu_{z,\mathrm{fi}}|^2 \, \hat{\rho}(\nu_{\mathrm{fi}}). \qquad \text{(A13.2)}$$

In a fluid sample the z-direction corresponds to all possible directions in the molecules, and so we should replace $|\mu_{z,\mathrm{fi}}|^2$ by its mean value $\frac{1}{3}|\mu_{\mathrm{fi}}|^2$. (In a solid where the crystal lattice freezes the molecular orientations, $\mu_{z,\mathrm{fi}}$ corresponds to a single component of the transition dipole in the molecular axis system.) The energy contained in a classical electromagnetic field is

$$\mathscr{E}=\frac{1}{2}\int(\varepsilon_0\langle E(t)^2\rangle+\mu_0\langle H(t)^2\rangle)\,\mathrm{d}\tau \tag{A13.3}$$

where $\langle E(t)^2\rangle$ is the time average of the squared electric field strength. In the present case, since the period is $2\pi/\omega$,

$$\langle E(t)^2\rangle=4E^2\left\{\int_0^{2\pi/\omega}\cos^2\omega t\,\mathrm{d}t\right\}\Big/\int_0^{2\pi/\omega}\mathrm{d}t=2E^2$$

and from electromagnetic theory $\mu_0\langle H(t)^2\rangle=\varepsilon_0\langle E(t)^2\rangle$ (use the Maxwell equations in Appendix 19). Therefore, for a uniform field in a region of volume V,

$$\mathscr{E}=2\varepsilon_0E^2V,\quad\text{or}\quad E^2=\mathscr{E}/2\varepsilon_0V. \tag{A13.4}$$

Consequently

$$W_{\mathrm{f}\leftarrow\mathrm{i}}=\tfrac{1}{3}\,|\mu_{\mathrm{fi}}|^2(1/2\varepsilon_0\hbar^2)\mathscr{E}\hat{\rho}(\nu_{\mathrm{fi}})/V.$$

$\mathscr{E}\hat{\rho}(\nu_{\mathrm{fi}})/V$ is the product of the energy of a mode times the number of modes per unit volume in the range ν to $\nu+\mathrm{d}\nu$; hence it is the *energy density of states* at the frequency ν, $\rho(\nu)$. Therefore

$$W_{\mathrm{f}\leftarrow\mathrm{i}}=(1/6\varepsilon_0\hbar^2)\,|\mu_{\mathrm{fi}}|^2\rho(\nu). \tag{A13.5}$$

and we can identify the coefficient of stimulated absorption as

$$B_{\mathrm{fi}}=(1/6\varepsilon_0\hbar^2)\,|\mu_{\mathrm{fi}}|^2. \tag{A13.6}$$

Therefore, from eqn (8.8.7), the coefficient of spontaneous emission is

$$A_{\mathrm{fi}}=8\pi h(\nu_{\mathrm{fi}}/c)^3(1/6\varepsilon_0\hbar^2)\,|\mu_{\mathrm{fi}}|^2=(8\pi^2/3)(\nu_{\mathrm{fi}}^3/\varepsilon_0\hbar c^3)\,|\mu_{\mathrm{fi}}|^2. \tag{A13.7}$$

Appendix 14 Molecular integrals

In the case of the *molecular orbital description of* H_2^+, the energy is given by the expression quoted in eqn (10.3.11) with

$$j'/j_0=\int a^2(1)(1/r_{1b})\,\mathrm{d}\tau_1=(1/R)\{1-(1+s)\mathrm{e}^{-2s}\} \tag{A14.1}$$

$$k'/j_0=\int a(1)b(1)(1/r_{1b})\,\mathrm{d}\tau_1=(1/a_0)\{1+s\}\mathrm{e}^{-s} \tag{A14.2}$$

$$S=\int a(1)b(1)\,\mathrm{d}\tau_1=\{1+s+\tfrac{1}{3}s^2\}\mathrm{e}^{-s} \tag{A14.3}$$

where $j_0=e^2/4\pi\varepsilon_0$ and $s=R/a_0$. We have taken 1s-orbitals on each atom. For techniques of integration see J. C. Slater, *Quantum theory of molecules and solids*, Vol. 1, McGraw-Hill (1963), especially p. 50, and S. P. McGlynn, L. G. Vanquickenborne, M. Kinoshita, and D. G.

Carroll, *Introduction to applied quantum chemistry*, Holt, Rinehart, and Winston (1972).

In the case of the *molecular orbital description of* H_2, the energy is obtained by evaluating the expectation value of the hamiltonian given in eqn (10.2.1) using the wavefunction given in eqn (10.3.12). This gives

$$E = 2E_{1s} + (e^2/4\pi\varepsilon_0 R) - 2\left(\frac{j' + k'}{1+S}\right) + \left(\frac{j + 2k + m + 4l}{2(1+S)^2}\right) \tag{A14.4}$$

where j', k', and S are given above and

$$j/j_0 = \int a^2(1)(1/r_{12})b^2(2)\,\mathrm{d}\tau_1\,\mathrm{d}\tau_2$$
$$= 1/R - (1/2a_0)\{(2/s) + 11/4 + \tfrac{3}{2}s + \tfrac{1}{3}s^2\}e^{-2s} \tag{A14.5}$$

$$k/j_0 = \int a(1)b(1)(1/r_{12})a(2)b(2)\,\mathrm{d}\tau_1\,\mathrm{d}\tau_2$$
$$= (1/5a_0)(A - B), \text{ see below.} \tag{A14.6}$$

$$l/j_0 = \int a^2(1)(1/r_{12})a(2)b(2)\,\mathrm{d}\tau_1\,\mathrm{d}\tau_2$$
$$= (1/2a_0)\{(2s + \tfrac{1}{4} + 5/8s)e^{-s} - (\tfrac{1}{4} + 5/8s)e^{-3s}\} \tag{A14.7}$$

$$m/j_0 = \int a^2(1)(1/r_{12})a^2(2)\,\mathrm{d}\tau_1\,\mathrm{d}\tau_2 = 5/8a_0 \tag{A14.8}$$

$$V = \int a^2(1)(1/r_{1a})\,\mathrm{d}\tau_1 = 1/a_0 \tag{A14.9}$$

with

$$A = (6/s)\{S^2(\gamma + \ln s) - S'^2E_1(4s) + 2SS'E_1(2s)\} \tag{A14.10a}$$

$$B = \{-25/8 + 23s/4 + 3s^2 + \tfrac{1}{3}s^3\}e^{-2s}; \quad S'(s) = S(-s) \tag{A14.10b}$$

where γ is Euler's constant ($\gamma = 0.577\,22\ldots$) and $E_1(x)$ is the exponential integral, a function listed in M. Abramowitz and I. A. Stegun, *Handbook of mathematical functions*, Dover (1965), p. 228. These conclusions give some idea of the complexity of integrals that will arise when a different basis set is taken or when the molecule is polyatomic.

In the *valence bond description of* H_2 the energies are given by

$$E = 2E_{1s} + (e^2/4\pi\varepsilon_0 R) - \left(\frac{J \pm K}{1 \pm S^2}\right) \tag{A14.11}$$

with

$$J = j - 2j' \tag{A14.12a}$$

$$K = k - 2k'S. \tag{A14.12b}$$

All integrals are as listed above.

Appendix 15 Classification of molecular terms

The *terms* of a molecule arising from a given configuration are labelled with the symmetry species of the irreducible representations they span. This is an extension of the method for classifying atomic terms where the spherical harmonics span irreps of the full rotation group and are labelled S, P, D, ... In molecules we use the irreps of the relevant point group and, if several electrons are present, determine the permissible terms by decomposing the direct product (use Appendix 11). A roman letter (e.g. $\mathrm{A_1}$) denotes the term of symmetry species A_1. In the spectroscopic literature the notation $X, A, B, C, \ldots$ denotes the lowest (X) and successive terms of the same multiplicity. In order to distinguish these from the symmetry species labels they are sometimes written $\tilde{X}, \tilde{A}, \tilde{B}, \tilde{C}, \ldots$. The labels $\tilde{a}, \tilde{b}, \tilde{c}, \ldots$ are used similarly to denote successive terms of multiplicity different from $\tilde{X}$. Closed shells transform according to the totally symmetric representation of the group (A_1) and so may generally be neglected. The value of $2S+1$ is denoted by a left superscript.

Consider a molecule belonging to the group C_{2v} with the configuration $\ldots a_1^2b_1^2b_2^2$. This is a closed-shell species and so gives rise to the term $\mathrm{A_1}$. The same conclusion would have been arrived at if we had evaluated the direct product, because $B_1\times B_1 = B_2\times B_2 = A_1\times A_1 = A_1$, and so the configuration spans $\ldots (A_1\times A_1)\times(B_1\times B_1)\times(B_2\times B_2)$, or $\ldots A_1\times A_1\times A_1 = A_1$. It follows that if there is a single electron outside a closed shell the term is labelled with the irreducible representation spanned by the orbital that the extra electron occupies. Thus we could have the following terms: $\ldots a_1^2b_1^2b_2^2a_1\ {}^2\mathrm{A_1}, \ldots a_1^2b_1^2b_2^2b_2\ {}^2\mathrm{B_2}$, and so on. For two electrons outside a closed shell take the direct product. As an example, since $B_1\times B_2 = A_2$ in the group C_{2v} the configuration $\ldots a_1^2b_1^2b_2^2b_1b_2$ spans ${}^1\mathrm{A_2}$ or ${}^3\mathrm{A_2}$.

This simple scheme may be extended to the other point groups. Thus, in the group O_h a configuration e_g^2 gives rise to the terms $E_g\times E_g = A_{1g} + A_{2g} + E_g$. The multiplicity of the terms must accord with the Pauli principle. This means that if e_g^2 is a triplet (spin symmetric, space *antisymmetric*) we have to take the *antisymmetrized direct product* (which is denoted in Appendix 11 by $[\Gamma]$). Since $E_g\times E_g = A_{1g}+[A_{2g}]+E_g$, we know that A_{2g} is a triplet, hence ${}^3\mathrm{A_{2g}}$, while the others, being space symmetric, must be singlets (${}^1\mathrm{A_{1g}}, {}^1\mathrm{E_g}$).

For diatomic molecules use $C_{\infty v}$ (heteronuclear), $D_{\infty h}$ (homonuclear). Then closed shells are Σ_g^+ and $\ldots \pi_g^1$ is ${}^2\Pi_g$. The configuration $\ldots \pi_u^2$ gives rise to $\Sigma_g^+ + [\Sigma_g^-] + \Delta_g$ (Appendix 11) and hence to the terms ${}^1\Sigma_g^+$, ${}^3\Sigma_g^-$, and ${}^1\Delta_g$.

Appendix 16 Normal modes: an example

Consider a linear triatomic molecule BAB in which the mass of A is m_A and the mass of B is m_B. For simplicity we confine attention to displacement along the axis of the molecule, and the displacement of the atoms B, A, and B are written ξ_1, ξ_2, and ξ_3 respectively. Since the relative displacements of the atoms A and B in the pairs B–A and A–B are $\xi_1-\xi_2$ and $\xi_3-\xi_2$ respectively, and each bond has the same force constant k, the potential energy is

$$V=\tfrac{1}{2}k(\xi_1-\xi_2)^2+\tfrac{1}{2}k(\xi_3-\xi_2)^2. \qquad \text{(A16.1)}$$

The force constant matrix, eqn (11.7.2), is

$$k_{ij}=(\partial^2 V/\partial\xi_i\,\partial\xi_j)_0.$$

Therefore

$$\mathbf{k}=\begin{pmatrix} k & -k & 0\\ -k & 2k & -k\\ 0 & -k & k\end{pmatrix}. \qquad \text{(A16.2)}$$

We shall work with the mass-weighted coordinates q_i,

$$q_i=m_i^{\frac{1}{2}}\xi_i. \qquad \text{(A16.3)}$$

The force constant matrix is now $\mathbf{K}$, where (eqn (11.7.4))

$$K_{ij}=(\partial^2 V/\partial q_i\,\partial q_j)_0=(m_i m_j)^{-\frac{1}{2}}k_{ij}.$$

Therefore

$$\mathbf{K}=\begin{pmatrix} k/m_{\mathrm{B}} & -k/(m_{\mathrm{A}}m_{\mathrm{B}})^{\frac{1}{2}} & 0\\ -k/(m_{\mathrm{A}}m_{\mathrm{B}})^{\frac{1}{2}} & 2k/m_{\mathrm{A}} & -k/(m_{\mathrm{A}}m_{\mathrm{B}})^{\frac{1}{2}}\\ 0 & -k/(m_{\mathrm{A}}m_{\mathrm{B}})^{\frac{1}{2}} & k/m_{\mathrm{B}}\end{pmatrix}. \qquad \text{(A16.4)}$$

We seek a linear combination of the coordinates that makes this matrix diagonal: this involves finding the eigenvalues (see Appendix 8) by solving the secular determinant:

$$0=|\mathbf{K}-\kappa\mathbf{1}|=\begin{vmatrix} k/m_{\mathrm{B}}-\kappa & -k/(m_{\mathrm{A}}m_{\mathrm{B}})^{\frac{1}{2}} & 0\\ -k/(m_{\mathrm{A}}m_{\mathrm{B}})^{\frac{1}{2}} & 2k/m_{\mathrm{A}}-\kappa & -k/(m_{\mathrm{A}}m_{\mathrm{B}})^{\frac{1}{2}}\\ 0 & -k/(m_{\mathrm{A}}m_{\mathrm{B}})^{\frac{1}{2}} & k/m_{\mathrm{B}}-\kappa\end{vmatrix}.$$

The roots of this cubic equation for κ are

$$\kappa_1=0, \qquad \kappa_2=k/m_{\mathrm{B}}, \qquad \kappa_3=k/\mu; \qquad \mu=m_{\mathrm{A}}m_{\mathrm{B}}/(2m_{\mathrm{B}}+m_{\mathrm{A}}).$$

μ is a reduced mass. The effective force constants κ_l depend on the masses, and so the vibrational frequencies of the corresponding normal modes also depend on the masses of the atoms involved. The mode with zero force constant (no restoring force) corresponds to the translation of the entire molecule parallel to its axis.

The eigenvectors Q_l of $\mathbf{K}$ are the combinations

$$Q_l=\sum_l c_{il}q_i$$

and are found by solving the simultaneous equations

$$\sum_j (K_{ij}-\kappa_l\delta_{ij})c_{jl}=0$$

with $l = 1, 2, 3$ in turn. As the simplest example, consider the mode Q_1 which corresponds to $\kappa_1 = 0$. The equations for c_{i1} reduce to

$$\sum_j K_{ij} c_{j1} = 0$$

or

$$K_{11}c_{11} + K_{12}c_{21} + K_{13}c_{31} = 0 \qquad (i = 1),$$
$$K_{21}c_{11} + K_{22}c_{21} + K_{23}c_{31} = 0 \qquad (i = 2),$$
$$K_{31}c_{11} + K_{32}c_{21} + K_{33}c_{31} = 0 \qquad (i = 3).$$

The coefficients K_{ij} are given in eqn (A16.4), and so

$$c_{11} = c_{21}(m_A/m_B)^{\frac{1}{2}}, \qquad c_{31} = c_{23}(m_A/m_B)^{\frac{1}{2}}.$$

We also require

$$c_{11}^2 + c_{21}^2 + c_{31}^2 = 1,$$

and so

$$c_{11} = c_{31} = m_B/(2m_B + m_A)^{\frac{1}{2}}, \qquad c_{21} = m_A/(2m_B + m_A)^{\frac{1}{2}}. \qquad \text{(A16.5)}$$

If the total mass of the molecule is written m it follows that

$$Q_1 = (1/m^{\frac{1}{2}})(m_B^{\frac{1}{2}}q_1 + m_A^{\frac{1}{2}}q_2 + m_B^{\frac{1}{2}}q_3)$$
$$= (1/m^{\frac{1}{2}})(m_B\xi_1 + m_A\xi_2 + m_B\xi_3) \qquad \text{(A16.6)}$$

The modes corresponding to κ_2 and κ_3 (Q_2 and Q_3) are found in a similar way:

$$Q_2 = (\tfrac{1}{2})^{\frac{1}{2}}(q_1 - q_3); \qquad \text{(A16.7)}$$
$$Q_3 = (2m)^{-\frac{1}{2}}\{m_A^{\frac{1}{2}}q_2 - 2m_B^{\frac{1}{2}}q_2 + m_A^{\frac{1}{2}}q_3\}. \qquad \text{(A16.8)}$$

The former is a symmetrical mode (the atoms B move in opposite directions) and involves no motion of the central atom; the second involves the central atom moving against the outer pair, which move in phase. See Fig. 11.21.

It may be verified that the kinetic energy may be expressed in the form $\sum_i \frac{1}{2}\dot{Q}_i^2$, and so both energy distributions are diagonal, as required.

Appendix 17 Oscillator strength

In this appendix we establish the relation between the *absorbance* of a band ($\mathscr{A}$) and the *transition dipole moment* (μ_{fi}). First we need the relation between *energy density* ($\mathscr{U}$), *flux* (Φ), and *intensity* (I).

Consider a plane of area A at x with light approaching perpendicularly from the left. All the photons within a distance $c\Delta t$ (and therefore in a volume $Ac\Delta t$) pass through the plane in the interval Δt. If the energy density in the field is $\mathscr{U}$, the total electromagnetic energy passing through the plane in that interval is $\mathscr{U}Ac\Delta t$. The *energy flux density* (Φ) is the energy per unit time per unit area, and so $\Phi = \mathscr{U}$

$Ac\Delta t/A\Delta t = c\mathcal{U}$. The energy density in the range ν to $\nu + \mathrm{d}\nu$ is $\mathrm{d}\mathcal{U} = \rho(\nu)\,\mathrm{d}\nu$, and so the flux density in the same range is $\mathrm{d}\Phi = c\rho(\nu)\,\mathrm{d}\nu$; write $\mathrm{d}\Phi = I(\nu)\,\mathrm{d}\nu$, where $I(\nu)$ is the *intensity*; then $I = c\rho$.

Now consider the processes that occur within the slab of thickness $\mathrm{d}l$ and which reduce the energy density of the field and therefore lower the intensity of the beam. Let the number of molecules per unit volume able to absorb light of frequency ν in the range ν to $\nu + \mathrm{d}\nu$ be $n(\nu)\,\mathrm{d}\nu$ (so that the total number density of absorbers is $\int n(\nu)\,\mathrm{d}\nu = \mathcal{N}$). The rate at which any one molecule absorbs a photon is $W = B\rho$, and as each photon has an energy $h\nu$ the rate of change of energy density per unit volume for frequencies in the range ν to $\nu + \mathrm{d}\nu$ is

$$(\mathrm{d}/\mathrm{dt})\,\mathrm{d}\mathcal{U} = -h\nu W n(\nu)\,\mathrm{d}\nu = -n(\nu)h\nu B\rho\,\mathrm{d}\nu,$$

or

$$(\mathrm{d}/\mathrm{d}t)\rho = -n(\nu)h\nu B\rho. \tag{A17.1}$$

The energy entering the slab from the left during the interval $\mathrm{d}t$ is $\Phi(x)A\,\mathrm{d}t$, the energy leaving on the right (at $x + \mathrm{d}l$) is $\Phi(x + \mathrm{d}l)A\,\mathrm{d}t$; by the conservation of energy the difference is equal to the change of energy in the slab:

$$\mathrm{d}\mathcal{U}A\,\mathrm{d}l = \Phi(x + \mathrm{d}l)A\,\mathrm{d}t - \Phi(x)A\,\mathrm{d}t$$

or

$$\mathrm{d}\mathcal{U}/\mathrm{d}t = \{\Phi(x + \mathrm{d}l) - \Phi(x)\}/\mathrm{d}l = \mathrm{d}\Phi/\mathrm{d}l.$$

$\mathcal{U}$ and Φ are the total energy density and flux density; but the relation is true for each frequency component in the beam, and so for frequencies in the range ν to $\nu + \mathrm{d}\nu$

$$(\mathrm{d}/\mathrm{d}t)(\rho(\nu)\,\mathrm{d}\nu) = (\mathrm{d}/\mathrm{d}l)(I(\nu)\,\mathrm{d}\nu), \quad \text{or} \quad \mathrm{d}\rho/\mathrm{d}t = \mathrm{d}I/\mathrm{d}l. \tag{A17.2}$$

Therefore eqn (A17.1) becomes

$$\mathrm{d}I = -n(\nu)h\nu B\rho\,\mathrm{d}l = -n(\nu)(h\nu/c)BI\,\mathrm{d}l. \tag{A17.3}$$

The reduction of intensity when a beam passes through a solution of length $\mathrm{d}l$ where the absorbers are at a concentration C is proportional to $\mathrm{d}l$, C, and the incident intensity I, the coefficient of proportionality being the absorption coefficient (p. 327):

$$\mathrm{d}I = -\alpha(\nu)CI\,\mathrm{d}l. \tag{A17.4}$$

Comparison of the two expressions leads to

$$\alpha(\nu)C = (Bh\nu/c)n(\nu), \quad \text{or} \quad \alpha(\nu)/\nu = (Bh/c)n(\nu)/C.$$

Multiplication of both sides by $\mathrm{d}\nu$ and integration over all the frequencies of the band leads to $\mathcal{N}/C$ on the right. But $\mathcal{N} = CL$, where L is Avogadro's constant; hence

$$\int \{\alpha(\nu)/\nu\}\,\mathrm{d}\nu = BhL/c. \tag{A17.5}$$

For typical bands the frequency is virtually constant over the range for which $\alpha(\nu)$ is non-zero, and so we set $\nu \approx \nu_{fi}$ on the left, and recognize $\int \alpha(\nu)\,d\nu = \mathcal{A}$, the integrated absorption coefficient of the band. Hence

$$\mathcal{A} = (h\nu_{fi}/c)LB. \tag{A17.6}$$

The Einstein coefficient for electric dipole transitions is calculated in Appendix 13 and shown to be $(1/6\varepsilon_0\hbar^2)\,|\mu_{fi}|^2$; therefore

$$\mathcal{A} = (\pi\nu_{fi}/3\varepsilon_0\hbar c)L\,|\mu_{fi}|^2, \tag{A17.7}$$

which is a direct link between a measureable quantity $\mathcal{A}$ and a calculable quantity μ_{fi}.

It turns out to be useful to introduce the dimensionless *oscillator strength* of a transition:

Oscillator strength:
$$f = (4m_e c\varepsilon_0/Le^2)\mathcal{A} = 6.257\times10^{-19}\times(\mathcal{A}/\text{m}^2\,\text{mol}^{-1}\,\text{s}). \tag{A17.8}$$

Then, for an electric dipole transition,

$$\begin{aligned} f &= (4m_e c\varepsilon_0/Le^2)\times(\pi\nu_{fi}/3\varepsilon_0 c\hbar)L\,|\mu_{fi}|^2 \\ &= (4\pi m_e\nu_{fi}/3e^2\hbar)\,|\mu_{fi}|^2. \end{aligned} \tag{A17.9}$$

In the case of a one-dimensional simple harmonic oscillator, $f = \frac{1}{3}$, and for an electron harmonically oscillating in three dimensions (which was an early model of the atom), $f = 1$. The observed oscillator strength is therefore the ratio of the strength of the transition to the strength of an ideally oscillating electron. In practice $f \approx 1$ for allowed electric dipole transitions, and $f \ll 1$ if the transition is forbidden (there is almost always some perturbation to break a selection rule). For spin-forbidden transitions f may be as low as 10^{-6}. Some typical values are listed in Table 12.1.

Appendix 18 Sum rules

In this appendix we establish the

Kuhn–Thomas sum rule:
$$\sum_n f_{n0} = N_e. \tag{A18.1}$$

As a first step we derive the relation between the matrix elements of the position and momentum operators known as the

Velocity-dipole relation:
$$\mathbf{p}_{mn} = -\mathrm{i}(m_e\omega_{mn}/e)\boldsymbol{\mu}_{mn}. \tag{A18.2}$$

Consider the one-dimensional form of this expression for a one-electron system in which $\mu_x = -ex$:

$$p_{x,mn} = \mathrm{i}m_e\omega_{mn}x_{mn}. \tag{A18.3}$$

The proof depends on the evaluation of the commutator of the hamiltonian and the position operator:

$$\langle m|\,[H, x]\,|n\rangle = \langle m|\,Hx\,|n\rangle - \langle m|\,xH\,|n\rangle$$
$$= (E_m - E_n)\langle m|\,x\,|n\rangle = \hbar\omega_{mn}x_{mn}. \quad \text{(A18.4)}$$

We have to show that the left-hand side of this expression is proportional to $p_{x,mn}$. Since

$$H = -(\hbar^2/2m_e)(d^2/dx^2) + V(x),$$

and V commutes with x, we have

$$[H, x] = -(\hbar^2/2m_e)[d^2/dx^2, x] = -(\hbar^2/2m_e)\{(d^2/dx^2)x - x(d^2/dx^2)\}$$
$$= -(\hbar^2/2m_e)\{2(d/dx) + x(d^2/dx^2) - x(d^2/dx^2)\}$$
$$= -(\hbar^2/m_e)\,d/dx = (\hbar/im_e)p_x.$$

The matrix element $\langle m|\,[H, x]\,|n\rangle$ is therefore equal both to $(\hbar/im_e)p_{x,mn}$ and to $\hbar\omega_{mn}x_{mn}$, which are therefore equal. This confirms eqn (A18.3).

Now express *one* of the transition dipole moments in f_{n0} in terms of the linear momentum using the velocity dipole relation:

$$f_{n0} = (4\pi m_e \nu_{n0}/3e^2\hbar)\,|\mu_{fi}|^2$$
$$= \tfrac{1}{2}(2m_e/3e^2\hbar)\omega_{n0}\{\boldsymbol{\mu}_{e,0n}\cdot\boldsymbol{\mu}_{e,n0} + \boldsymbol{\mu}_{e,0n}\cdot\boldsymbol{\mu}_{e,n0}\}$$
$$= \tfrac{1}{2}(2m_e/3e^2\hbar)(e/m_e)i\{\boldsymbol{\mu}_{e,0n}\cdot\mathbf{p}_{n0} - \mathbf{p}_{0n}\cdot\boldsymbol{\mu}_{e,n0}\}.$$

The sum over n produces the commutator

$$\sum_n f_{n0} = (-i/3\hbar)\langle 0|\,\mathbf{r}\cdot\mathbf{p} - \mathbf{p}\cdot\mathbf{r}\,|0\rangle.$$

But for each term in the scalar product we have $xp_x - p_x x = i\hbar$, etc., and so

$$\sum_n f_{n0} = 1. \quad \text{(A18.5)}$$

This is for a single electron. Therefore, if the system consists of N_e electrons, each one gives the same contribution, and so overall the sum of the oscillator strengths of all possible electronic transitions in the molecule is equal to N_e, as we set out to prove.

Appendix 19 The Maxwell equations

The Maxwell equations describe the properties of electric and magnetic fields in various media. They involve six quantities:

E: electric field strength ($V\,m^{-1}$);
D: electric displacement ($C\,m^{-2}$);
ρ: charge density ($C\,m^{-3}$);
H: magnetic field strength ($A\,m^{-1}$);
B: magnetic flux density (T or $Wb\,m^{-2}$);
J: current density ($A\,m^{-2}$).

The electric displacement and the magnetic flux density are related to the magnetic and electric field strengths by the polarization, **P**, and the magnetization, **M**:

$$\mathbf{D} = \varepsilon_0 \mathbf{E} + \mathbf{P}, \qquad \mathbf{B} = \mu_0 \mathbf{H} + \mu_0 \mathbf{M}. \tag{A19.1}$$

Then we have the

Maxwell equations:

$$\text{(a) } \nabla \cdot \mathbf{D} = \rho; \qquad \text{(b) } \nabla \cdot \mathbf{B} = 0,$$
$$\text{(c) } \nabla \wedge \mathbf{E} = -\partial \mathbf{B}/\partial t; \qquad \text{(d) } \nabla \wedge \mathbf{H} = \mathbf{J} + \partial \mathbf{D}/\partial t. \tag{A19.2}$$

We may describe **B** and **E** in terms of two potentials, ϕ and **A**. The divergence of a curl is identically zero, and so eqn (A19.2b) is satisfied if

$$\mathbf{B} = \nabla \wedge \mathbf{A}, \tag{A19.3}$$

A is called the *vector potential*. It then follows from eqn (A19.2c) that

$$\nabla \wedge (\mathbf{E} + \partial \mathbf{A}/\partial t) = 0,$$

and so

$$\mathbf{E} = -\partial \mathbf{A}/\partial t + \mathbf{f},$$

where **f** is a vector with zero curl. But the curl of a gradient of a scalar function is identically zero, and so $\mathbf{f} = -\nabla \phi$. Therefore

$$\mathbf{E} = -\partial \mathbf{A}/\partial t - \nabla \phi. \tag{A19.4}$$

ϕ is the *scalar potential*. When **A** is independent of time, the electric field is the negative gradient of the scalar potential, $\mathbf{E} = -\nabla \phi$.

The Maxwell equations take on a special importance in free space (which has $\mathbf{P} = 0$, $\mathbf{M} = 0$, $\rho = 0$, and $\mathbf{J} = 0$) for then $\mathbf{D} = \varepsilon_0 \mathbf{E}$ and $\mathbf{B} = \mu_0 \mathbf{H}$, and the equations are

$$\text{(a) } \nabla \cdot \mathbf{E} = 0, \qquad \text{(b) } \nabla \cdot \mathbf{H} = 0,$$
$$\text{(c) } \nabla \wedge \mathbf{E} = -\mu_0\, \partial \mathbf{H}/\partial t, \qquad \text{(d) } \nabla \wedge \mathbf{H} = \varepsilon_0\, \partial \mathbf{E}/\partial t. \tag{A19.5}$$

If the curl of eqn (A19.5c) is taken we obtain

$$\nabla \wedge (\nabla \wedge \mathbf{E}) + (\partial/\partial t) \nabla \wedge \mathbf{B} = 0,$$

and since $\nabla \wedge \nabla \wedge \mathbf{E} = \nabla(\nabla \cdot \mathbf{E}) - \nabla^2 \mathbf{E}$ (Appendix 20),

$$\nabla(\nabla \cdot \mathbf{E}) - \nabla^2 \mathbf{E} + (\partial/\partial t) \nabla \wedge \mathbf{B} = 0.$$

The first term in this equation disappears by virtue of eqn (A19.5a), and the third term is

$$(\partial/\partial t) \nabla \wedge \mathbf{B} = \varepsilon_0 \mu_0\, \partial^2 \mathbf{E}/\partial t^2$$

by eqn (A19.5d). Therefore in free space the electric field strength satisfies the equation

$$\nabla^2\mathbf{E} - \varepsilon_0\mu_0\ddot{\mathbf{E}} = 0, \tag{A19.6}$$

which is the equation for a wave propagating with a velocity $c = 1/(\varepsilon_0\mu_0)^{\frac{1}{2}}$.

For light propagating in a medium we must allow for the polarization. Suppose

$$\mathbf{P} = \mathcal{N}\boldsymbol{\alpha}\cdot\mathbf{E} - \mathcal{N}\boldsymbol{\beta}\cdot\dot{\mathbf{B}}. \tag{A19.7}$$

(In optically inactive media the β-term is absent and the calculation is simpler; nevertheless, we need the more general case, the simpler being obtained at any stage by setting $\boldsymbol{\beta} = 0$.) From eqns (A19.2d), eqn (A19.1), and $\mathbf{B} \approx \mu_0\mathbf{H}$,

$$\nabla\wedge\mathbf{B} \approx \mu_0\dot{\mathbf{D}} = \mu_0\varepsilon_0\dot{\mathbf{E}} + \mu_0\mathcal{N}\boldsymbol{\alpha}\cdot\dot{\mathbf{E}} - \mu_0\mathcal{N}\boldsymbol{\beta}\cdot\ddot{\mathbf{E}}.$$

Taking the curl of both sides and using $\nabla\wedge\nabla\wedge\mathbf{B} = \nabla(\nabla\cdot\mathbf{B}) - \nabla^2\mathbf{B} = -\nabla^2\mathbf{B}$ (by eqn (A19.2b)), taking $\boldsymbol{\beta}$ to be a diagonal matrix so that $\boldsymbol{\beta}\cdot\ddot{\mathbf{E}} = \beta\ddot{\mathbf{E}}$, then by eqn (A19.2c),

$$\nabla^2\mathbf{B} = \mu_0\varepsilon_0(1 + \alpha\mathcal{N}/\varepsilon_0)\ddot{\mathbf{B}} - \mu_0\beta\mathcal{N}\nabla\wedge\ddot{\mathbf{B}}. \tag{A19.8}$$

If $\beta = 0$ we see that the magnetic field propagates at a speed $v = [1/\mu_0\varepsilon_0(1 + \alpha\mathcal{N}/\varepsilon_0)]^{\frac{1}{2}}$; hence the refractive index is

$$n_r = c/v = (1 + \alpha\mathcal{N}/\varepsilon_0)^{\frac{1}{2}} \approx 1 + \alpha\mathcal{N}/2\varepsilon_0 \tag{A19.9}$$

as in eqn (13.3.1).

When $\beta \neq 0$ we expect birefringence ($n_+ \neq n_-$). The circularly polarized electric vector has the form $\mathbf{E}^{\pm} \propto (\hat{\mathbf{i}} \mp \mathrm{i}\hat{\mathbf{j}})\mathrm{e}^{\mathrm{i}\phi_{\pm}}$, $\phi_{\pm} = \omega t - \mathbf{k}_{\pm}\cdot\mathbf{r}$ (we must allow for the wavevector to depend on the sense of polarization because $k = 2\pi/\lambda$ and λ depends on refractive index through $\lambda = v/\nu = c/n\nu$). It proves convenient to work with the magnetic component of the electromagnetic field, and because $\nabla\wedge\mathbf{E}^{\pm} \propto \partial\mathbf{H}^{\pm}/\partial t$ it follows that

$$\mathbf{B}^{\pm} = B(\hat{\mathbf{j}} \pm \mathrm{i}\hat{\mathbf{i}})\mathrm{e}^{\mathrm{i}\phi_{\pm}} \tag{A19.10}$$

has the appropriate polarization characteristics. (We use $\mathbf{B} = \mu_0\mathbf{H}$ throughout because the magnetization is negligible.) Then since

$$\nabla\wedge\mathbf{B}^{\pm} = \mathrm{i}k_{\pm}B(\hat{\mathbf{i}} \mp \mathrm{i}\hat{\mathbf{j}})\mathrm{e}^{\mathrm{i}\phi_{\pm}}$$

(because $\hat{\mathbf{i}}$, $\hat{\mathbf{j}}$, $\hat{\mathbf{k}}$ form a right-handed set of unit vectors), $\nabla^2\mathbf{B}^{\pm} = -k_{\pm}^2\mathbf{B}^{\pm}$, and $\ddot{\mathbf{B}}^{\pm} = -\omega^2\mathbf{B}^{\pm}$, eqn (A19.8) becomes

$$k_{\pm}^2 = \mu_0\varepsilon_0\omega^2(1 + \alpha\mathcal{N}/\varepsilon_0) \pm \mu_0\beta\mathcal{N}k_{\pm}$$

for the coefficients of both $\hat{\mathbf{i}}$ and $\hat{\mathbf{j}}$. (This shows that $\hat{\mathbf{i}} \pm \mathrm{i}\hat{\mathbf{j}}$ are 'normal modes' for the propagation of radiation through a birefringent medium.) Since $\mu_0\varepsilon_0 = 1/c^2$ and $k_{\pm} = 2\pi\nu n_{\pm}/c$ (from the remark above), it follows that

$$n_{\pm}^2 = 1 + \alpha\mathcal{N}/\varepsilon_0 \pm \omega\beta\mathcal{N}n_{\pm}/c\varepsilon_0. \tag{A19.11}$$

This is a quadratic equation for $n_{\pm}$. The solution to first-order in β is

$$n_{\pm} \approx 1 + \alpha\mathcal{N}/2\varepsilon_0 \pm \omega\beta\mathcal{N}/2c\varepsilon_0, \tag{A19.12}$$

which is the expression used in the text (eqn (13.6.7)).

Appendix 20 Vector properties

1. *Vector multiplication*

The products of two vectors $\mathbf{F}$ and $\mathbf{G}$, where

$\mathbf{F} = F_x\hat{\mathbf{i}} + F_y\hat{\mathbf{j}} + F_z\hat{\mathbf{k}}$ and $\mathbf{G} = G_x\hat{\mathbf{i}} + G_y\hat{\mathbf{j}} + G_z\hat{\mathbf{k}}$

are defined as

Scalar product: $\mathbf{F}\cdot\mathbf{G} = F_xG_x + F_yG_y + F_zG_z.$ (A20.1)

Vector product: $\mathbf{F}\wedge\mathbf{G} = \begin{vmatrix} \hat{\mathbf{i}} & \hat{\mathbf{j}} & \hat{\mathbf{k}} \\ F_x & F_y & F_z \\ G_x & G_y & G_z \end{vmatrix}.$ (A20.2)

$\mathbf{F}\cdot\mathbf{G}$ is a scalar, $\mathbf{F}\wedge\mathbf{G}$ is a vector.

The following relations are useful:

(a) $\mathbf{F}\wedge\mathbf{G} = -\mathbf{G}\wedge\mathbf{F}$

(b) $\mathbf{F}\cdot(\mathbf{G}\wedge\mathbf{H}) = \mathbf{G}\cdot(\mathbf{H}\wedge\mathbf{F}) = \mathbf{H}\cdot(\mathbf{F}\wedge\mathbf{G}) = (\mathbf{F}\wedge\mathbf{G})\cdot\mathbf{H}$

(c) $\mathbf{F}\wedge(\mathbf{G}\wedge\mathbf{H}) = \mathbf{G}(\mathbf{F}\cdot\mathbf{H}) - \mathbf{H}(\mathbf{F}\cdot\mathbf{G})$

(d) $(\mathbf{F}\wedge\mathbf{G})\cdot(\mathbf{H}\wedge\mathbf{I}) = (\mathbf{F}\cdot\mathbf{H})(\mathbf{G}\cdot\mathbf{I}) - (\mathbf{F}\cdot\mathbf{I})(\mathbf{G}\cdot\mathbf{H}).$ (A20.3)

2. *Vector differentiation*

The differentiation of a scalar is denoted ∇ or grad, and is defined by

Gradient: $\operatorname{grad} U = \nabla U = (\partial U/\partial x)\hat{\mathbf{i}} + (\partial U/\partial y)\hat{\mathbf{j}} + (\partial U/\partial z)\hat{\mathbf{k}}$ (A20.4)

∇U is a vector.

The differentiations of vectors are defined as

Divergence: $\operatorname{div}\mathbf{F} = \nabla\cdot\mathbf{F} = (\partial F_x/\partial x) + (\partial F_y/\partial y) + (\partial F_z/\partial z)$ (A20.5)

Curl: $\operatorname{curl}\mathbf{F} = \nabla\wedge\mathbf{F} = \begin{vmatrix} \hat{\mathbf{i}} & \hat{\mathbf{j}} & \hat{\mathbf{k}} \\ \partial/\partial x & \partial/\partial y & \partial/\partial z \\ F_x & F_y & F_z \end{vmatrix}.$ (A20.6)

$\nabla\cdot\mathbf{F}$ is a scalar and $\nabla\wedge\mathbf{F}$ is a vector.

The following relations are useful (U, V are scalar functions; $\mathbf{F}$, $\mathbf{G}$

are vectors):

$$
\begin{aligned}
&\text{(a)} \quad \nabla(UV) = U\nabla V + V\nabla U,\\
&\text{(b)} \quad \nabla\cdot(U\mathbf{F}) = U\nabla\cdot\mathbf{F} + \mathbf{F}\cdot\nabla U,\\
&\text{(c)} \quad \nabla\wedge(U\mathbf{F}) = U(\nabla\wedge\mathbf{F}) + (\nabla U)\wedge\mathbf{F},\\
&\text{(d)} \quad \nabla\cdot(\mathbf{F}\wedge\mathbf{G}) = \mathbf{G}\cdot(\nabla\wedge\mathbf{F}) - \mathbf{F}\cdot(\nabla\wedge\mathbf{G}),\\
&\text{(e)} \quad \nabla\wedge(\nabla\wedge\mathbf{F}) = \nabla(\nabla\cdot\mathbf{F}) - \nabla^2\mathbf{F},\\
&\text{(f)} \quad \nabla\wedge(\mathbf{F}\wedge\mathbf{G}) = \mathbf{F}(\nabla\cdot\mathbf{G}) - (\nabla\cdot\mathbf{F})\mathbf{G} + (\mathbf{G}\cdot\nabla)\mathbf{F} - (\mathbf{F}\cdot\nabla)\mathbf{G},\\
&\text{(g)} \quad \nabla(\mathbf{F}\wedge\mathbf{G}) = (\mathbf{F}\cdot\nabla)\mathbf{G} + (\mathbf{G}\cdot\nabla)\mathbf{F} + \mathbf{F}\wedge(\nabla\wedge\mathbf{G}) + \mathbf{G}\wedge(\nabla\wedge\mathbf{F}),\\
&\text{(h)} \quad \nabla^2 U = \nabla\cdot\nabla U,\\
&\text{(i)} \quad \nabla\wedge(\nabla U) = 0. \qquad \text{(A20.7)}
\end{aligned}
$$

Appendix 21 The canonical momentum

'Canon' means rule. The following is a rule for finding the momentum for any system of particles. As an example we construct the expression for the linear momentum in the presence of a magnetic field.

Step 1. Choose a lagrangian L such that the Euler–Lagrange equations (eqn (A2.4)) correspond to the known equations of motion.

The equation of motion of an electron in the presence of electric and magnetic fields is given by the

Lorentz force law: $m_e\ddot{\mathbf{r}} = \mathbf{F} = -e(\mathbf{E} + \dot{\mathbf{r}}\wedge\mathbf{B}).$ (A21.1)

This is reproduced by the Euler–Lagrange equations if we take

Lagrangian: $L = \frac{1}{2}m_e\dot{r}^2 + e\phi - e\dot{\mathbf{r}}\cdot\mathbf{A}$ (A21.2)

because

$$\partial L/\partial x = e(\partial\phi/\partial x) - e(\partial/\partial x)(\dot{\mathbf{r}}\cdot\mathbf{A})$$

and so in three dimensions

$$\nabla L = e\nabla\phi - e\nabla(\mathbf{r}\cdot\mathbf{A}) = e\nabla\phi - e\dot{\mathbf{r}}\cdot\nabla\mathbf{A} - e\dot{\mathbf{r}}\wedge(\nabla\wedge\mathbf{A})$$

(using eqn (A20.7g));

$$
\begin{aligned}
(\mathrm{d}/\mathrm{d}t)(\partial L/\partial\dot{x}) &= (\mathrm{d}/\mathrm{d}t)(m_e\dot{x} - eA_x) = m_e\ddot{x} - e(\mathrm{d}A_x/\mathrm{d}t)\\
&= m_e\ddot{x} - e(\partial A_x/\partial t) - e\{(\partial x/\partial t)(\partial A_x/\partial x) + \ldots\};
\end{aligned}
$$

and so in three dimensions

$$(\mathrm{d}/\mathrm{d}t)(\partial L/\partial\dot{\mathbf{r}}) = m_e\ddot{\mathbf{r}} - e\dot{\mathbf{A}} - e(\dot{\mathbf{r}}\cdot\nabla)\mathbf{A}.$$

Then the Euler–Lagrange equations are

$$e\nabla\phi - e\dot{\mathbf{r}}\cdot\nabla\mathbf{A} - e\dot{\mathbf{r}}\wedge(\nabla\wedge\mathbf{A}) = m_e\ddot{\mathbf{r}} - e\dot{\mathbf{A}} - e(\dot{\mathbf{r}}\cdot\nabla)\mathbf{A},$$

which reduces to the Lorentz law using eqns (A19.3) and (A19.4). Hence the lagrangian in eqn (A21.2) is acceptable.

Step 2. Form the

$$\textit{Canonical momentum:}\ p_q = \partial L/\partial \dot{q}. \tag{A21.3}$$

Then, with the lagrangian in eqn (A21.2) we have

$$p_x = m_e\dot{x} - eA_x$$

and so in three dimensions

$$\mathbf{p} = m_e\dot{\mathbf{r}} - e\mathbf{A}. \tag{A21.4}$$

Step 3. Form the

$$\textit{Hamiltonian:}\ H = \mathbf{p}\cdot\dot{\mathbf{r}} - L \tag{A21.5}$$

and express it in terms of $\mathbf{p}, \mathbf{r}$ as variables.

In the present example, since $\dot{\mathbf{r}} = (\mathbf{p} + e\mathbf{A})/m_e$,

$$\begin{aligned} H &= \mathbf{p}\cdot(\mathbf{p}+e\mathbf{A})/m_e - (1/2m_e)(\mathbf{p}+e\mathbf{A})^2 - e\phi + (e/m_e)(\mathbf{p}+e\mathbf{A})\cdot\mathbf{A} \\ &= (\mathbf{p}+e\mathbf{A})^2/2m_e - e\phi \end{aligned} \tag{A21.6}$$

which could be obtained from the hamiltonian $p^2/2m + e\phi$ by replacing $\mathbf{p}$ everywhere by $\mathbf{p} + e\mathbf{A}$, the rule used in the text.

Appendix 22 The dipolar potential

In this appendix we deduce the form of the magnetic field corresponding to the vector potential

$$\mathbf{A} = a\mathbf{m}\wedge\mathbf{r}/r^3, \qquad a = \mu_0/4\pi \tag{A22.1}$$

used in Section 14.7 to represent the effect of a nucleus (eqn (14.7.11)).

First, note that as $\nabla(1/r) = -\mathbf{r}/r^3$,

$$\mathbf{A} = -a\mathbf{m}\wedge\nabla(1/r).$$

From eqn (A20.7f) with $\mathbf{F} = \mathbf{m}$, $\mathbf{G} = (\nabla r^{-1})$,

$$\begin{aligned} \nabla\wedge\mathbf{A} &= -a\nabla\wedge(\mathbf{m}\wedge\nabla\mathbf{r}^{-1}) \\ &= -a\{\mathbf{m}(\nabla\cdot\nabla r^{-1}) - (\nabla\cdot\mathbf{m})(\nabla r^{-1}) + (\nabla r^{-1})(\nabla\mathbf{m}) - (\mathbf{m}\cdot\nabla)(\nabla r^{-1})\} \\ &= -a\{\mathbf{m}(\nabla^2 r^{-1}) - (\mathbf{m}\cdot\nabla)(\nabla r^{-1})\} \end{aligned} \tag{A22.2}$$

because $\mathbf{m}$ is a constant and $\nabla\cdot\nabla = \nabla^2$. The second term may be written

$$\begin{aligned} (\mathbf{m}\cdot\nabla)(\nabla r^{-1}) &= -\mathbf{m}\cdot\nabla(\mathbf{r}/r^3) \\ &= -\{m_x(\partial/\partial x) + m_y(\partial/\partial y) + m_z(\partial/\partial z)\}(1/r^3)(x\hat{\mathbf{i}} + y\hat{\mathbf{j}} + z\hat{\mathbf{k}}) \\ &= -(\mathbf{m}/r^3) - \mathbf{r}\mathbf{m}\cdot(\nabla r^{-3}) = -(\mathbf{m}/r^3) + 3\mathbf{r}(\mathbf{m}\cdot\mathbf{r})/r^5 \\ &= -(1/r^3)\{\mathbf{m} - 3\mathbf{m}\cdot\hat{\mathbf{r}}\hat{\mathbf{r}}\}, \end{aligned}$$

and so this part of the vector potential accounts for the contribution

$$\mathbf{B}_{\text{dipolar}} = -(a/r^3)\{\mathbf{m} - 3\mathbf{m}\cdot\hat{\mathbf{r}}\hat{\mathbf{r}}\} = -(\mu_0/4\pi r^3)\{\mathbf{m} - 3\mathbf{m}\cdot\hat{\mathbf{r}}\hat{\mathbf{r}}\} \qquad \text{(A22.3)}$$

to the field, as in eqn (14.7.11).

When the system is spherically symmetrical the first term in eqn (A22.2) does not necessarily vanish when it is averaged over the appropriate wavefunctions. Furthermore, the spherical average of the second term produces

$$\langle(\mathbf{m}\cdot\nabla)(\nabla r^{-1})\rangle = \tfrac{1}{3}\mathbf{m}(\nabla^2 r^{-1}).$$

Therefore, in this case,

$$\nabla\wedge\mathbf{A} = -\tfrac{2}{3}a\mathbf{m}\nabla^2(1/r).$$

A standard property is

$$\nabla^2(1/r) = -4\pi\delta(\mathbf{r}) \qquad \text{(A22.4)}$$

where $\delta(\mathbf{r})$ is the Dirac δ-function (p. 410). Therefore this term contributes

$$\mathbf{B}_{\text{contact}} = (8\pi/3)a\mathbf{m}\delta(\mathbf{r}) = (2\mu_0/3)\mathbf{m}\delta(\mathbf{r}), \qquad \text{(A22.5)}$$

as in eqn (14.11.6).

List of worked *Examples*

Index

T signifies a Table

Useful formulas

Boltzmann distribution: $p_i = (1/q)\mathrm{e}^{-\varepsilon_i/kT}, \quad q = \sum_j \mathrm{e}^{-\varepsilon_j/kT}$

Planck distribution: $\mathrm{d}\mathcal{U} = \rho(\nu)\,\mathrm{d}\nu = \rho(\lambda)\,\mathrm{d}\lambda$

$\rho(\nu) = (8\pi h\nu^3/c^3)/(\mathrm{e}^{h\nu/kT} - 1); \quad \rho(\lambda) = (8\pi hc/\lambda^5)/(\mathrm{e}^{hc/\lambda kT} - 1)$

de Broglie relation: $p = h/\lambda$

Schrödinger equation: $H\Psi = \mathrm{i}\hbar(\partial\Psi/\partial t), \quad H = -(\hbar^2/2m)\nabla^2 + V$

$\Psi(x, t) = \psi(x)\mathrm{e}^{-\mathrm{i}Et/\hbar}$ for $H\psi = E\psi$

Wavefunctions and eigenvalues

Free particle: $\psi_k = (1/L)^{\frac{1}{2}}\mathrm{e}^{\mathrm{i}kx}, L \to \infty; p = k\hbar$

Particle in a box: $\psi_n = (2/L)^{\frac{1}{2}}\sin(n\pi x/L); \; E_n = n^2(h^2/8mL^2); \; n = 1, 2, \ldots$

Particle on a ring: $\psi_{m_l} = (1/2\pi)^{\frac{1}{2}}\mathrm{e}^{\mathrm{i}m_l\phi}; \; l_z = m_l\hbar; \; m_l = 0, \pm 1, \pm 2, \ldots$

Particle on a sphere: $\psi = Y_{lm_l}(\theta, \phi); \quad l = 0, 1, 2, \ldots; \; m_l = 0, \pm 1, \ldots \pm l$

Harmonic oscillator: $\psi_v = N_v H_v(y)\mathrm{e}^{-\frac{1}{2}y^2}, \; N_v = (1/2^v v!\,\pi^{\frac{1}{2}})^{\frac{1}{2}}, \; y = (m\omega/\hbar)^{\frac{1}{2}}x$

$E_v = (v + \tfrac{1}{2})\hbar\omega, \; \omega = (k/m)^{\frac{1}{2}}, \; v = 0, 1, 2, \ldots$

Hydrogen-like atom: $\psi = R_{nl}(r)Y_{lm_l}(\theta, \phi); \; n = 1, 2, \ldots; \; l = 0, 1, \ldots (n-1); \; m_l = 0, \pm 1, \ldots \pm l$

$E_{nlm_l} = -Z^2(\mu e^4/32\pi^2\varepsilon_0^2\hbar^2)(1/n^2)$; n^2-fold degenerate

Rotating molecule: $E_{JKM_J} = hcBJ(J+1) + hc(A - B)K^2; \; J = 0, 1, 2, \ldots; \; K = 0$(linear);

$K = 0, \pm 1, \pm 2, \ldots \pm J; \; M_J = 0, \pm 1, \ldots \pm J$

$A = \hbar/4\pi cI_{\parallel}, \; B = \hbar/4\pi cI_{\perp}$

Operators

Hermiticity: $\langle m|\,\Omega\,|n\rangle = \langle n|\,\Omega\,|m\rangle^*$; Completeness: $\sum_n |n\rangle\langle n| = 1$

Commutation relation: $[x, p] = \mathrm{i}\hbar$; Uncertainty relation: $\delta A\,\delta B \geq \tfrac{1}{2}|\langle C\rangle|, \; \mathrm{i}C = [A, B]$

Angular momentum

Commutation rules: $[j_x, j_y] = \mathrm{i}\hbar j_z, \; [j^2, j_q] = 0$

Magnitude: $j^2\,|j, m_j\rangle = j(j+1)\hbar^2\,|j, m_j\rangle, \quad j = 0, \tfrac{1}{2}, 1, \ldots$

z-component: $j_z\,|j, m_j\rangle = m_j\hbar\,|j, m_j\rangle, \quad m_j = j, j-1, j-2, \ldots, -j$ ($2j+1$ values)

Matrix elements: $\langle j, m_j \pm 1|\, j^{\pm}\,|j, m_j\rangle = \{j(j+1) - m_j(m_j \pm 1)\}^{\frac{1}{2}}\hbar; \quad j^{\pm} = j_x \pm \mathrm{i}j_y$

Clebsch–Gordan series: $j = j_1 + j_2, j_1 + j_2 - 1, \ldots |j_1 - j_2|$

Perturbation theory: $H = H^{(0)} + H^{(1)} + H^{(2)}$

Two-level system: $\psi_+ = \psi_1 \cos\beta + \psi_2 \sin\beta$; $\psi_- = -\psi_1 \sin\beta + \psi_2 \cos\beta$

$E_\pm = \frac{1}{2}(E_1 + E_2) \pm \frac{1}{2}\{(E_1 - E_2)^2 + 4\varepsilon^2\}^{\frac{1}{2}}$, $\tan 2\beta = 2\varepsilon/(E_1 - E_2)$

Many-level system: $\psi_0^{(1)} = \sum_n a_n \psi_n$, $a_n = H_{n0}^{(1)}/(E_0 - E_n)$

$E_0^{(1)} = H_{00}^{(1)}$; $E_0^{(2)} = H_{00}^{(2)} + \sum'_n H_{0n}^{(1)} H_{n0}^{(1)}/(E_0 - E_n)$

Time-dependent two-level system for $P_1(0) = 1$ and $H_{21}^{(1)} = \hbar V$ for $t \geq 0$:

$P_2(t) = (2V/\Omega)^2 \sin^2 \frac{1}{2}\Omega t$; $\Omega^2 = \omega_{21}^2 + 4V^2$

Time-dependent many-level system for $a_i(0) = 1$: $\psi = \sum_f a_f(t)\Psi_f(t)$,

$$a_f(t) = (1/i\hbar)\int_0^t H_{fi}^{(1)}(t)e^{i\omega_{fi}t}\,dt$$

Perturbation hamiltonians

Electric dipole: $-\boldsymbol{\mu}\cdot\mathbf{E}$; $\boldsymbol{\mu} = \sum_i e_i \mathbf{r}_i$;

Magnetic dipole: $-\mathbf{m}\cdot\mathbf{B}$; $\mathbf{m} = \gamma_e \mathbf{l} + g_e \gamma_e \mathbf{s}$, $\gamma_e = -e/2m_e$

Dipole–dipole: $(K/r^3)\{\mathbf{d}_A \cdot \mathbf{d}_B - 3\mathbf{d}_A \cdot \hat{\mathbf{r}}\hat{\mathbf{r}} \cdot \mathbf{d}_B\}$;

$K = 1/4\pi\varepsilon_0$ for $\mathbf{d} = \boldsymbol{\mu}$; $K = \mu_0/4\pi$ for $\mathbf{d} = \mathbf{m}$

Second-order magnetic: $(e^2/2m_e)A^2 = (e^2/8m_e)B^2(x^2 + y^2)$ for $\mathbf{B} = B\hat{\mathbf{k}}$

Current density: $\mathbf{j} = -(e/2m_e)(\psi^*\mathbf{p}\psi - \psi\mathbf{p}\psi^*) - (e^2/2m_e)\mathbf{A}\psi^*\psi$

Group theory

Character: $\chi(R) = \mathrm{tr}\,\mathbf{D}(R)$; Order: h; Dimension: d_l

Similarity: $\mathbf{D}'(R) = \mathbf{c}^{-1}\mathbf{D}(R)\mathbf{c}$ for $\mathbf{f}' = \mathbf{fc}$

Conjugacy: $R' = S^{-1}RS$; implying $\chi(R') = \chi(R)$, R', R same class

LOT: $\sum_c g(c)\chi^{(l)*}(c)\chi^{(l')}(c) = h\delta_{ll'}$

Reduction: $a_l = (1/h)\sum_c g(c)\chi^{(l)}(c)^*\chi(c)$

Projection operator: $p^{(l)} = (d_l/h)\sum_R \chi^{(l)*}(R)R$; $p^{(l)}f_j = \sum_i f_i^{(l)}$

Character of rotation in R_3: $\chi(C_\alpha) = \sin(l + \frac{1}{2})\alpha/\sin\frac{1}{2}\alpha$

Integrals over symmetric range: $\int f^{(l)*}f^{(l')}\,d\tau = 0$ unless $\Gamma^{(l)} \times \Gamma^{(l')}$ includes A_1

Spectroscopy

Beer–Lambert law: $I_l = I_0 e^{-\alpha Cl}$; $\varepsilon = \alpha/\ln 10$; $\mathscr{A} = \int \alpha\,d\nu$

Integrated absorption coefficient: $\mathscr{A} = (\pi\nu_{fi}/3\varepsilon_0\hbar c)L\,|\mu_{fi}|^2$; $\boldsymbol{\mu}_{fi} = \int \psi_f^*\boldsymbol{\mu}\psi_i\,d\tau$

Oscillator strength: $f_{fi} = (4\pi m_e/3e^2\hbar)\nu_{fi}\,|\mu_{fi}|^2$

Einstein coefficients: $A_{fi}/B_{fi} = 8\pi h(\nu_{fi}/c)^3$; $B_{fi} = (1/6\varepsilon_0\hbar^2)\,|\mu_{fi}|^2$